Preface

This volume contains the papers presented at the Ninth International Conference on Automated Deduction (CADE-9) held on May 23–26 at Argonne National Laboratory, Argonne, Illinois. The conference commemorates the twenty-fifth anniversary of the discovery of the resolution principle, which took place during the summer of 1963. Alan Robinson, Larry Wos, and Bill Miller, all of whom were at Argonne during that summer, presented invited talks at the conference.

The CADE conferences are a forum for reporting on research on all aspects of automated deduction, including theorem proving, logic programming, unification, deductive databases, term rewriting, ATP for non-standard logics, and program verification. Preceding conferences have been held at

- Argonne National Laboratory, USA, 1974 (IEEE Transactions on Computers Vol. C-25, no. 8)
- Oberwolfach, West Germany, 1976
- Massachusetts Institute of Technology, USA, 1977 (MIT, Cambridge, MA)
- University of Texas, Austin TX, USA (University of Texas, Austin, TX)
- Les Arcs, France, 1980 (Springer-Verlag Lecture Notes in Computer Science, vol. 87)
- Courant Institute, New York, USA, 1982 (Springer-Verlag Lecture Notes in Computer Science, vol. 138)
- Napa, California, USA, 1984 (Springer-Verlag Lecture Notes in Computer Science, vol. 170)
- Oxford University, Oxford, UK, 1986 (Springer-Verlag Lecture Notes in Computer Science, vol. 230)

For this ninth conference, 109 papers were submitted. All papers were refereed by at least two referees, and the program committee accepted the 52 that appear here. Also included in this volume are abstracts of 21 implementations of automated deduction systems.

We would like to thank first the authors of the papers, without whom there would be no conference. Next we thank the program committee, who handled more submissions than expected, with both industry and grace. Special thanks are due to W. W. Bledsoe for hosting the Program Committee meeting in Austin, and to Suezette Branton, who handled the local arrangements for that meeting. We also thank Argonne's Conference Services for their work on local arrangements for the conference, and our secretary Teri Huml for miscellaneous items of assistance too numerous to mention.

Ewing Lusk and Ross Overbeek
Conference Co-Chairmen

Lecture Notes in Computer Science

Edited by G. Goos and J. Hartmanis

310

Springer-Verlag

Berlin Heidelberg New York London Paris Tokyo

Editors

Ewing Lusk
Ross Overbeek
Mathematics and Computer Science Division
Argonne National Laboratory
9700 South Cass Avenue, Argonne, IL 60439, USA

CR Subject Classification (1987): I.2.3

ISBN 3-540-19343-X Springer-Verlag Berlin Heidelberg New York
ISBN 0-387-19343-X Springer-Verlag New York Berlin Heidelberg

© Springer-Verlag Berlin Heidelberg 1988
Printed in Germany

Printing and binding: Druckhaus Beltz, Hemsbach/Bergstr.
2145/3140-543210

Table of Contents

Session 19

Session 20

System Abstracts

FIRST-ORDER THEOREM PROVING USING
CONDITIONAL REWRITE RULES

Hantao Zhang[†]
Department of Computer Science
Rensselaer Polytechnic Institute
Troy, NY 12180, U.S.A.

Deepak Kapur[†]
Department of Computer Science
State University of New York at Albany
Albany NY 12222, U.S.A.

Abstract. A method based on superposition on maximal literals in clauses and conditional rewriting is discussed for automatically proving theorems in first-order predicate calculus with equality. First-order formulae (clauses) are represented as conditional rewrite rules which turn out to be an efficient representation. The use of conditional rewriting for reducing search space is discussed. The method has been implemented in *RRL*, a *Rewrite Rule Laboratory*, a theorem proving environment based on rewriting techniques. It has been tried on a number of examples with considerable success. Its performance on bench-mark examples, including Schubert's Steamroller problem, SAM's lemma, and examples from set theory, compared favorably with the performance of other methods reported in the literature.

1. Introduction

A method using the conditional rewriting paradigm is discussed for automatically proving theorems in first-order predicate calculus with equality. A first-order formula in the clausal form is represented as a (conditional) rewrite rule. The method involves two key inference rules - (i) *superposition* of two conditional rules, an operation crucial for completion-procedure based approaches to automated reasoning, and (ii) *reduction* or *simplification*, another key operation used in most symbolic computation and automated reasoning systems. We first illustrate the method using the following contrived example.

A set S of unsatisfiable clauses are firstly transformed into a set R of (conditional) rewrite rules.

$$
S: \begin{array}{ll}
(c1). & P(f(x),a) \mid \neg Q(x) \\
(c2). & \neg P(f(e),a) \mid e = c \\
(c3). & \neg P(f(a),x) \\
(c4). & f(a) = f(c) \\
(c5). & Q(e)
\end{array}
\qquad
R: \begin{array}{ll}
(r1). & P(f(x),a) \rightarrow true \text{ if } Q(x) \\
(r2). & P(f(e),a) \rightarrow false \text{ if } \neg(e = c) \\
(r3). & P(f(a),x) \rightarrow false \\
(r4). & f(a) \rightarrow f(c) \\
(r5). & Q(e) \rightarrow true
\end{array}
$$

It is easy to see that the rule $(r3)$ can be reduced by the rule $(r4)$ into

$$(r3'). \quad P(f(c),x) \rightarrow false;$$

further, the rule $(r3)$ can be removed altogether.

† Partially supported by the National Science Foundation Grant no. CCR-8408461.

The rule $(r1)$ also reduces the left-hand side of the rule $(r2)$ to *true*, since the condition of $(r1)$ can be established by the rule $(r5)$. This results in the new clause $(e = c)$, which is transformed into the rule

$(r2')$. $e \to c$.

This rule in turn can reduce the rule $(r5)$ into

$(r5')$. $Q(c) \to true$,

and again the rules $(r2)$ and $(r5)$ can be removed.

Now a superposition between $(r1)$ and $(r3')$, which is equivalent to a resolution on the literal $P(f(x),a)$, produces the clause $\neg Q(c)$, which can be further simplified to an empty clause by $(r5')$. That is, a contradiction, *true* = *false*, is derived. As the reader would have noticed, the method needs only one unification for the above example.

The key steps of the method as illustrated by the above example are:

(a) Every clause is transformed into a conditional rewrite rule, which can be viewed as just another presentation of a clause. For simplicity, we assume throughout this paper that existential quantifiers from a formula are already removed using Skolemization and the result is converted into the clausal form, even though our method does not depend upon the input being in the clausal form. A unit clause makes an unconditional rule while a non-unit clause makes a conditional rule. An ordering is imposed on literals and terms, and the head of a rewrite rule comes from a maximal literal in the corresponding clause.

(b) Reduction can be performed easily with rewrite rules. Once a rule gets reduced, it can be thrown away which could result in a considerable saving in the search space. As will be shown later in the paper, conditional rewriting as defined here is more powerful than subsumption and demodulation. For instance, the left-hand side of the rule $(r2)$ can be reduced only by conditional rewriting.

(c) Rules are superposed. A superposition between two rules, a classical concept introduced by [Knuth&Bendix 70] and extended to conditional rewrite rules by [Brand et al 78], subsumes both resolution and paramodulation on the maximal literals of the corresponding two clauses. In this sense, resolution and paramodulation can be treated uniformly. Further, rewriting concepts which serve as powerful mechanisms for equational reasoning can be easily incorporated in the method.

We refer to the superposition used in this method as *clausal superposition* for the emphasis that the rules are made from clauses and the result of a superposition is a clause. Clausal superposition is a general operation including "order-resolution" and "oriented-paramodulation" given in [Hsiang&Rusinowitch 86], in which a proof system consisting of these two rules of inference has been proved to be refutationally complete. In other words, clausal superposition is a single complete inference rule for first-order theorem proving with equality, when viewing each clause as a rewrite rule.

The method has been implemented in *RRL, Rewrite Rule Laboratory*, a theorem proving environment based on rewrite rules [Kapur et al 86]. It has been tried on a number of examples and results are extremely encouraging. Later, we discuss the performance of the method on a number of examples including Schubert's Steamroller problem, SAM's lemma, and examples from set theory.

1.1. History and Background

This method is mainly based on a well-found ordering on atoms and function terms appearing in each formulae (later in the paper, *term* is used for both atom and function term). The idea to order terms in a clause can be found in [Kowalski&Hayes 69], [Boyer 71], [Reiter 71], [Lankford 75], [Lankford&Ballantyne 79], [Fribourg 85], and more recently in [Hsiang&Rusinowitch 86]. Boyer's locking method assigned arbitrary weights to literals. Lankford and Ballantyne extended locking from resolution to paramodulation. Reiter investigated the linear strategy when one of the two literals in a binary resolution is maximal in the clause containing it. Fribourg combined the locking resolution with a strong restriction of paramodulation in which paramodulation is performed only on the greater argument of the equality predicate. Hsiang and Rusinowitch defined "order resolution" as a binary resolution where the two literals involved in the unification are maximal, and refined Fribourg's restricted paramodulation as "oriented-paramodulation" that is performed on the greater argument in an equality atom only if this equality atom is maximal in the clause. Using transfinite semantic trees, they proved that "order-resolution" and "oriented-paramodulation" are complete inference rules for first-order predicate calculus with equality (without the need of the functional reflexive axioms).

The clausal superposition inference rule introduced in this paper is different from the clausal superposition defined by [Fribourg 85] in that a superposition in our case is performed between maximal literals of the two clauses, while Fribourg's superposition is performed on the rightmost literals of the two clauses. Moreover, we do not need functional reflexive axioms, while Fribourg's method needs them. Our clausal superposition is theoretically equivalent to "order-resolution" and "oriented-paramodulation" together. However, as clauses are transformed into rewrite rules in our method, rewriting techniques which have been found useful for equational theories can be easily integrated. More importantly, the inference rule of reduction proposed in this paper is more powerful than "demodulation" and "subsumption" together, as we use non-unit clauses as well as unit clauses to perform rewriting.

Rewriting methods for theorem proving in first-order predicate calculus have been studied by [Hsiang 82], [Kapur&Narendran 84], [Paul 85] and [Hsiang 87]. Hsiang's as well as Kapur and Narendran's approaches treat logical formulas as sums (using exclusive or) of products, namely, polynomials over a boolean ring. A rewrite rule can be made from such a polynomial which is used to simplify other polynomials. A completion procedure is used to deduce a contradiction in which the superposition among rules plays the role of resolution.

Our method does not rely on the data structure of polynomials; instead conditional rules are made from clauses. Unit clauses make unconditional rules (those rules are also made in the systems of [Hsiang 82], [Kapur&Narendran 84], etc.). Non-unit clauses make conditional rewrite rules, with the maximal literal of the clause as the head of the rule. Once those rules are made, the reduction process looks much like the one used in Boyer and Moore's prover [Boyer&Moore 79].

For first-order predicate calculus with equality, the refutational completeness proofs for inference systems based on resolution and paramodulation can be found in [Peterson 83] and [Hsiang&Rusinowitch 86], where no functional reflexive axioms are needed. The proof techniques in these papers relied on the use of semantic trees, that is, an atom may be interpreted as *true* and *false* respectively to construct a fork. Peterson's proof required the use of a reduction ordering which is order-isomorphic to ω on ground terms. Hsiang and Rusinowitch relaxed this condition so that trans-finite semantic trees are not excluded.

Bachmair et al [1986] developed a technique called *proof ordering* for establishing completeness of different versions of completion procedures. The basic idea of this technique is to establish a criterion under which every application of each inference rule produces a *smaller* proof. Bachmair and Dershowitz [1987] used this technique for rewrite-based methods for first-order predicate calculus without the equality predicate.

Our proof techniques for showing refutational completeness are syntactic as well as semantic. Like classical methods, we first prove the ground case, then use a lemma to lift the completeness proof to the general case. Unlike the conventional methods, we do not use semantic trees for the ground case. Rather than showing that each inference rule produces a "simpler" proof as Bachmair et al's method does, we prove that if the input is inconsistent, then the empty clause must be produced by a procedure similar to the Knuth-Bendix completion procedure. As in [Peterson 83] and [Hsiang 87], this set of inference rules does not need the functional reflexive axioms for handling equality. Because of lack of space, the details of these novel proof techniques are not given in this paper; an interested reader may consult a full version of the paper.

2. Definitions and Notation

We will assume the standard definitions of a *function term*, an *atomic formula (atom)*, and a *literal*. Function terms and atoms are referred as *terms*. We will also assume the standard definitions of *ground term*, *subterm*, *positions* in a term, *equation*, *substitution* and the *most general unifier (mgu)*. The subterm of t at a position u is denoted by t/u. If s is another term, then $t[u \leftarrow s]$ is the term obtained when t/u is replaced by s. A substitution is called *ground* if for $\sigma(x) \neq x$, $\sigma(x)$ is a ground term. A substitution σ is usually extended to a mapping from terms to terms by natural injection.

In forming literals from atoms, the symbol $\neg$ will be used to represent negation. It is also used to represent the operator that forms a literal of the opposite *polarity*, e.g. if P is the literal $\neg A$ then $\neg P$ denotes A. The two constant atoms *true* and *false* will be used explicitly with conventional meaning, i.e., $\neg$ *true* = *false* and $\neg$ *false* = *true*.

A clause is a disjunction of literals and will be written as $P_1 | P_2 | \cdots | P_n$, where the disjunction connective "|" is assumed to be associative, commutative and idempotent; so the ordering of the literals in a disjunction is immaterial and duplicate literals in a disjunction are always removed. The atom *false* is the identity element for "|" and (*true* $|F$) is equivalent to *true* for any formula F. A clause containing *true* as a literal is said to be *trivial*. The empty clause will be denoted by *false*. A clause consisting of a single literal (except *false*) will be called a *unit clause* and no distinction will be made between a unit clause and the single literal of which it is composed.

Since we shall frequently write the negation of literals in a clause, for the convenience of the presentation, we will represent a clause in its negated form, that is, a clause $P_1 | P_2 | \cdots | P_n$ will be written as **not**{ $\neg P_1, \neg P_2, \cdots, \neg P_n$ } and referred as *negated clause*. Hence from now on, **not**{ $Q_1, Q_2, \cdots, Q_n$ } denotes the clause $\neg Q_1 | \neg Q_2 | \cdots | \neg Q_n$. A substitution σ can also be extended in a natural way to a mapping from clauses to clauses. We shall use $\sigma(c)$ to denote **not**{ $\sigma(P_1), \sigma(P_2), \cdots, \sigma(P_n)$ } for a clause $c =$ **not**{ $P_1, P_2, \cdots, P_n$ }.

We use "=" as the equality symbol both in the meta-language and in the language of first-order predicate calculus with equality. Equality atoms will usually be enclosed in parentheses to avoid confusion. A literal $(t \neq s)$ denotes $\neg (t = s)$. Conventional equality axioms are assumed and only $(x = x)$ and $(x = y) | (y \neq x)$ are built into our proof system. An atom of the form $(x = x)$ is always reduced to *true*; $(s = t)$ and $(t = s)$ are considered as the same formula. A formula $(P = true)$ is equivalent to P, $(P = false)$ to $\neg P$. By this notation, (*true* = *false*) is the same as *false*.

We will assume the standard definitions of an *interpretation*, and an interpretation *satisfying* or *falsifying* a formula or a set of formulas, as well as a formula or a set of formulas being *satisfiable* or *unsatisfiable*. An interpretation is *consistent* if no atom under this interpretation has more than one value.

2.1. Reduction Ordering

A relation R is said to be *stable under substitution* if for any terms t_1 and t_2 and any substitution σ, $t_1 \, R \, t_2$ implies $\sigma(t_1) \, R \, \sigma(t_2)$. A *reduction ordering* is a transitive monotonic relation on terms, which is *terminating* (or *Noetherian*) and stable under substitution. Almost all well-behaved orderings in the term rewriting literature (see [Dershowitz 87] for a survey), such as the original Knuth-Bendix ordering, the recursive path ordering, the path of subterm ordering, the recursive decomposition ordering and the term path ordering, can be used in our method. Henceforth, when a precedence relation on operators is given, the symbol "=" is assumed to be less than every operator except *true* and *false*, and the recursive path ordering is assumed to be used to extend the precedence relation to terms.

Throughout this paper, we assume that a total reduction ordering > exists on all ground terms. The extension of > to variable terms cannot be and does not need to be total. For any ordering > on terms, we can use the extension of > to sets of terms, $>^s$, to compare two clauses.

3. Superposition

3.1. Making Rewrite Rules

Given a reduction ordering and a clause $c = \mathbf{not}\{\ P_1, P_2, \cdots, P_n\ \}$, a literal P_i (or a subterm t of P_i) in c is said to be maximal if no literal in c is greater than P_i (or t). Clauses are transformed into conditional rewrite rules as follows:

Rule Making:

$$(M) \quad \frac{\mathbf{not}\{\ P_1, P_2, \cdots, P_i, \cdots, P_n\ \}}{lhs \rightarrow rhs \ \mathbf{if}\ \{\ P_1, P_2, \cdots, P_{i-1}, P_{i+1}, \cdots, P_n\ \}} \quad \text{if } P_i \text{ is maximal.}$$

where (a) $lhs = A$, $rhs = false$ if $P_i = A$, an atom;

(b) if $P_i = \neg\ (t = s)$, then

    (i) if $t > s$, then $lhs = t$, $rhs = s$;

    (ii) if $s > t$, then $lhs = s$, $rhs = t$;

    (iii) if s and t are non-comparable, then two rules are generated, one in which $lhs = s$, $rhs = t$, and another in which $lhs = t$, $rhs = s$;

(c) $lhs = B$, $rhs = true$, if $P_i = \neg B$, where B is a non-equality atom.

The literal $lhs \rightarrow rhs$ will be referred as the *head* of the rewrite rule. The condition of a rewrite rule will sometimes be abbreviated as **c**. If $\mathbf{c} = \{\ P_1, ..., P_n\ \}$ and $\mathbf{c}' = \{\ Q_1, ..., Q_m\ \}$, then $\{\ \mathbf{c}, \mathbf{c}'\ \}$ denotes $\{\ P_1, ..., P_n, Q_1, ..., Q_m\ \}$.

A rule made by the case (a) or (c) is called a *boolean rule*. A rule made by the case (b) is called a *domain rule* since the type of the left-hand side of the rule is not boolean. Unconditional rewrite rules are made from unit clauses while conditional rules are made from non-unit clauses. Note that conditional rewrite rules used by Fribourg [Fribourg 85b] are also made from clauses. The difference from ours is that the head of a rule in Fribourg's method is made from positive literals while the condition is made from negative literals. His clausal superposition performed on these rules looks much like a resolution between two clauses and further, these rules were not used.

Example 3.1 Given the following two ground clauses

$$\text{(c1)} \quad \textbf{not}\{ \ (f(a) \neq a), \ (g(a) = a), \ \neg p(a) \ \} \ ,$$
$$\text{(c2)} \quad \textbf{not}\{ \ p(a), \ (f(a) = g(a)) \ \} \ ,$$

different rules can be made from the same clause under different orderings. A few of examples are illustrated in Fig. 3.1.

Ordering	Rules
$p > f > g > a$	(r1) $p(a) \rightarrow true$ **if** $\{ \ (f(a) \neq a), \ (g(a) = a) \ \}$, (r2) $p(a) \rightarrow false$ **if** $\{ \ (f(a) = g(a)) \ \}$.
$f > g > p > a$	(r1') $f(a) \rightarrow a$ **if** $\{ \ (g(a) = a), \ \neg p(a) \ \}$ (r2') $(f(a) = g(a)) \rightarrow false$ **if** $\{ \ p(a) \ \}$
$g > f > p > a$	(r1'') $(g(a) = a) \rightarrow false$ **if** $\{ \ (f(a) \neq a), \ \neg p(a) \ \}$ (r2'') $(g(a) = f(a)) \rightarrow false$ **if** $\{ \ p(a) \ \}$

Fig. 3.1.

Note also that if a clause has more than one maximal literal, then more than one rule must be made from the same clause. For example, to express the statement that each object is either a or b, the (negated) clause is

$$\textbf{not}\{ \ x \neq a, \ x \neq b \ \} \ .$$

If $b > a$, then two rules will be made from the above clause, that is,

$$\text{(r1)} \quad x \rightarrow b \ \textbf{if} \ \{ \ x \neq a \ \}, \qquad \text{(r2)} \quad b \rightarrow x \ \textbf{if} \ \{ \ x \neq a \ \}.$$

The variables in each rule are assumed to be different from other rules, even when those rules are made from the same clause.

3.2. Inference Rules

Clausal Superposition:

$$(S) \quad \frac{\begin{array}{l} lhs_1 \rightarrow rhs_1 \ \textbf{if} \ \mathbf{c_1} \\ lhs_2 \rightarrow rhs_2 \ \textbf{if} \ \mathbf{c_2} \end{array}}{\sigma(\textbf{not}\{ \ \neg \ (lhs_1[u \leftarrow rhs_2] = rhs_1), \ \mathbf{c_1}, \ \mathbf{c_2} \ \})} \quad \textbf{if} \ \begin{cases} u \text{ is a non-variable position of } lhs_1 \\ \sigma \text{ is an mgu of } lhs_1/u \text{ and } lhs_2 \end{cases}$$

The result of a clausal superposition is thus a clause. We call this clause a *critical clause* since it is conventionally called *conditional critical pair*, (consider $lhs_1[u \leftarrow rhs_2]$ and rhs_1 as a pair of terms) which is a natural extension of Knuth and Bendix's definition to conditional rewrite systems. We say that the rule $lhs_2 \rightarrow rhs_2$ if $\mathbf{c_2}$ superposed into the rule $lhs_1 \rightarrow rhs_1$ if $\mathbf{c_1}$. This extension appears first, to the best of our knowledge, in [Brand&al 78]. See [Remy 82], [Zhang 84], and [Kaplan 84] for the refinements of this concept. Fribourg [1985] also used the name "clausal superposition" which subsumes resolution and paramodulation in the locking method.

Example 3.2 Let us consider the rewrite rules in Fig 3.1. A clausal superposition between $(r1)$ and $(r2)$ produces the critical clause

$$\mathbf{not}\{\ \neg\ (true = false),\ (f(a) \neq a),\ (g(a) = a),\ (f(a) = g(a))\ \}$$

or $\qquad \mathbf{not}\{\ (f(a) \neq a),\ (g(a) = a),\ (f(a) = g(a))\ \}$.

This superposition is equivalent to a resolution, more precisely, to an oriented resolution [Hsiang&Rusinowitch 86], since $p(a)$ is the maximal atom in that clause. A superposition between $(r1')$ and $(r2')$ produces the critical clause

$$\mathbf{not}\{\ \neg\ ((a = g(a)) = false),\ p(a),\ (g(a) = a),\ \neg p(a)\ \}$$

which is trivial after reformulation. This superposition is equivalent to a paramodulation. No clausal superposition can be done between the rules $(r1'')$ and $(r2'')$.

If both rules in a clausal superposition are boolean rules, then a superposition is equivalent to a binary resolution. If one of them is a domain rule, then a superposition is the same as a paramodulation. The soundness of superposition thus follows from the soundness of resolution and paramodulation.

Theorem 3.3 (soundness of clausal superposition) If $r3$ is a rule made from a critical clause of $r1$ and $r2$, then for any interpretation Π, $\Pi(r1) = true$ and $\Pi(r2) = true$ implies $\Pi(r3) = true$.

In the resolution approach, *factoring* is a necessary rule of inference to guarantee the completeness. This is true for our method also.

Factoring:

$$(F) \quad \frac{\mathbf{not}\{\ P_1, P_2, .., P_n\ \}}{\sigma(\mathbf{not}\{\ P_1, P_2, .., P_n\ \})} \qquad \text{if}\ \ P_i \neq P_j\ \text{and}\ \sigma(P_i) = \sigma(Pj)$$

This inference rule is performed at the clause level before a conditional rewrite rule is made from a clause. This inference rule can be certainly defined in terms of rewrite rules rather than clauses. For simplicity, we leave it as is. In addition, we have:

$$(F') \quad \frac{\mathbf{not}\{\ P_1, P_2, .., P_n\ \}}{\sigma(\mathbf{not}\{\ P_1, P_2, .., P_n\ \})} \qquad \text{if}\ \ P_i\ \text{is}\ t_1 = t_2\ \text{and}\ \sigma\ \text{is a mgu of}\ t_1 \text{and} t_2.$$

The result of performing (F') is the same as superposing the rule $(x = x) \rightarrow true$ into the literal P_i. (F') is needed since $(x = x)$ is always reduced to *true* by the system.

3.3. Refutational Completeness

Before giving the main procedure for establishing a refutational proof of a formula, let us introduce two subroutines.

Make−rule (S, R): takes a set S of clauses as input and outputs a set of rules, which are not in R and are made from each non-trivial clause in S.

Superposition (R): returns all critical clauses between any pair of rules in R.

Procedure **PROC**(S):
    1. $R := Make−rule (S, \varnothing)$;
    2. If $false \in R$, stop with "S is unsatisfiable";
    3. New $:= Make−rule (Superposition (R), R)$;
    4. If New $= \varnothing$, stop with "S is satisfiable";
    5. $R = R \cup New$;
    6. Go to 2.

PROC looks like the Knuth-Bendix completion procedure [Knuth&Bendix 70], except that no reduction is performed here.

Theorem 3.4 If **PROC** stops at Step 2, then the input S is inconsistent.

For refutational completeness, it needs to be shown that whenever the procedure **PROC** stops after finitely many steps with the output "S is satisfiable" or loops forever, then S is satisfiable. The proof can be factored as follows: (i) prove the above statement considering S as a set of ground clauses, and (ii) lift the ground case to the general case.

Given an unsatisfiable set S of non-ground clauses, by Herbrand's Theorem, there is an unsatisfiable finite set GI of ground instances of the clauses in S. If **PROC** produces a proof on GI, and if it can be shown that all ground instances appearing in deriving a contradiction from GI can be represented as instances of general clauses derived from S, then the clausal superposition inference is refutationally complete for the general case, too.

Given a set S of ground clauses, and a total ordering on ground terms, we have an important lemma:

Lemma 3.5 If two rewrite rules $r1$ and $r2$ are ground and a rule $r3$ is made from a critical clause of $r1$ and $r2$, then either $r1 >^s r3$ or $r2 >^s r1$.

Applying this lemma to all rewrite rules R produced by **PROC** on S, we may systematically construct an interpretation from R in which S is true. An interpretation is constructed starting from the minimal rule in R (under the ordering $>^s$) and making the head of this rule to be true. Next we choose the second minimal rule of R, and evaluate this rule in the interpretation constructed so far. If it is true under the interpretation, we do nothing; otherwise, we make the rule to be true, and so on. No matter whether R is infinite or finite, an interpretation can be constructed in this way. We state the main result and omit the construction; the details can be found in the full version of the paper.

Theorem 3.6 If S is ground and **PROC** stops at Step 4 or loops forever, then S is satisfiable.

Theorems 3.4 and 3.6 guarantee that **PROC** can be used as a semi-decision procedure on ground atoms. Our next task is to prove that clausal superposition inference is refutationally complete without the use of functionally reflexive axioms and superposition at variable positions. We first show below that the lifting from ground case to the general case works if clausal superposition is performed at a non-variable position. Subsequently, we show that in other cases, if a proof of a contradiction from ground instances cannot be lifted to the general case, then it can be successively transformed into another proof of a contradiction from ground instances which can be lifted to the general case.

Lemma 3.7 (Superposition Lifting at Non-variable Positions) If $r1'$ and $r2'$ are instances of $r1$ and $r2$, respectively, and if c' is a critical clause produced by superposing $r2'$ into a position p of $r1'$ where p is a non-variable position of $r1$, then there exists a critical clause c by superposing $r2$ into p of $r1$ such that c' is an instance of c.

This lemma can be viewed as a combination of *Resolution Lifting Lemma* [Robinson 65] and *Paramodulation Lifting Lemma* [Wos&Robinson 70]. Similar proofs can be found in [Robinson&Wos 70], [Chang&Lee 73] and [Peterson 83].

Now let us concentrate on the case that p is not a non-variable position of $r1$. In [Robinson&Wos 70], paramodulation into variables and reflexive functions are not excluded. For any term t and any ground substitution σ, by paramodulating reflexive functions into the variables of t as long as necessary, we can always get a new term t' such that $t' = \sigma(t)$.

In [Peterson 83] as well as in [Hsiang&Rusinowitch 86], paramodulation into variables and functionally reflexive functions are excluded. They show that if we *properly* choose the ground instances of an unsatisfiable set of general clauses, then we can get a proof of a contradiction on these ground instances; moreover, each new clause obtained in the proof can be obtained either by a resolution, or by a paramodulation into a non-variable position of the original general clause. The criterion of [Peterson 83], as well as of [Hsiang&Rusinowitch 86], to choose *proper* instances of the input clauses depends on the notion of *reducibility* of an atom *under an interpretation*.

We have another lifting proof which is based on a well-founded ordering on sets of ground instances of the input. For an inconsistent set S of clauses, there exists a finite inconsistent set GI of ground instances of S. We show that if a ground critical clause generated from *GI* cannot be lifted to a critical clause generated from *S* using Lemma 3.7, we can always choose another inconsistent set of ground instances smaller than *GI*. Since the ordering is well-founded, we can finally find a set *GI'* of ground instances, which is inconsistent and each critical clause produced from *GI'* can be always lifted to a general critical clause. As we do not have enough space to go into details, we give only the main result without proof.

Theorem 3.8 (Superposition Lifting): For any inconsistent set *S* of clauses, there exists an inconsistent set *GI* of ground instances of *S* and there exists a superposition proof *P* on *GI*, such that every critical clause in *P* can be lifted (by Lemma 3.7) to a critical clause generated from *S*.

With this theorem, we can say that clausal superposition (with factoring) constitutes a refutationally complete set of inference rules even without functionally (and predicate) reflexive axioms and without superposition into variables.

4. Reduction

In this section we discuss how reduction can be incorporated into these inference rules since reduction is a very powerful rule of inference widely used in almost every automatic reasoning systems. Reduction is a simplification process in which bigger terms in an expression are repeatedly substituted by smaller equivalent ones. Because of reduction, the search space can be reduced tremendously in automatic reasoning. We can say that the success of the Knuth-Bendix completion procedure for equational theories basicly relies on reduction. Concepts used closest to reduction in resolution-based inference systems are subsumption and demodulation; see also [Lankford 75; Lankford and Ballantyne, 1979] for integrating rewriting concepts in a resolution-based inference system.

4.1. Contextual Rewriting

A *congruence* is an equivalence relation on $T(F,X)$ satisfying *superterm property*. That is, $\equiv$ is a congruence relation iff
1. (reflexivity) $t \equiv t$,
2. (symmetry) $t_1 \equiv t_2$ implies $t_2 \equiv t_1$,
3. (transitivity) $t_1 \equiv t_2$ and $t_2 \equiv t_3$ implies $t_1 \equiv t_3$, and
4. (superterm property) $m \equiv n$ implies $f(\cdots,m,\cdots) \equiv f(\cdots,n,\cdots)$.
Let $M = \{ <t_1,t_2> \mid t_1 \equiv t_2 \}$. Then M is said to be the set of equations *satisfying* $\equiv$ and, reciprocally, $\equiv$ is the *relation induced* by M.

The *constant congruence* $\cong_E$ generated by a set E of equations is the minimal congruence relation defined by the above four rules plus the following:
5. (basis case) For any equation $m = n$ of E, $m \cong_E n$;

The *equational congruence* $=_E$ generated by a set E of equations is the minimal congruence relation defined by the above five rules plus the following:

    6. (substitution) If $m =_E n$, then for any substitution σ, $\sigma(m) =_E \sigma(n)$.

Remark: In the constant congruence $\cong_E$, a variable is treated as a constant, that is, no substitutions for variables are allowed. This relation is rarely discussed in the publication because of its limited use. It is obvious that $\cong_E \subseteq =_E$.

Definition 4.1 (contextual rewriting $\to^C$) Given a set C of equations and a rewriting system R, a rewriting under the context C is recursively defined as follows: a term t is reduced to t' by R under the context C, noted as $t \to^C t'$, if there exists a term t_1, a subterm s of t_1, a rewrite rule $lhs \to rhs$ **if** $\{ P_1, P_2, \cdots, P_k \}$ in R and a substitution σ such that

    (a) $t \cong_C t_1$,   (c) for each $P_i, 1 \le i \le k$, $\sigma(P_i) \to^C true$,

    (b) $\sigma(lhs) = s$,   (d) $t' = t_1[s / \sigma(rhs)]$.

Example 4.2 Let R contain the definition of $+$, $*$ (*multiplication*), gcd on natural numbers. Let R also contain a rule

    (r1) $gcd(u*v, u) \to u$

which states that the gcd of any number and its multiplication is the number itself. Let the context C be $\{ z = (x*y) \}$ and the term be $gcd(z, x)$. Replacing z by $x*y$ in $gcd(z, x)$, we have $gcd(z, x) \cong_C gcd(x*y, x)$. Then applying the rule (r1) on $gcd(x*y, x)$, we get x. That is, $gcd(z, x)$ is rewritten to x by (r1) under the context $\{ z = (x*y) \}$.

By definition, the contextual rewriting is equivalent to the composition of the relation $\cong_C$ and the classical conditional rewriting $\to$, namely, $\to^C \equiv \cong_C . \to$. Conditional rewriting was studied in different forms by [Brand&al 78], [Lankford 79], [Remy 82], [Drosten 83], [Kaplan 84] and [Zhang 84]. The contextual rewriting proposed in [Zhang&Remy 85] is different from the one given in this paper by that the context given in [Zhang&Remy 85] is a boolean expression and was not used for simplification; while the context in this paper is a set of equalities and is used for reduction under the relation *constant congruence*.

4.2. Termination of Rewriting

Like other kinds of rewriting relations, the reduction process may go into loops without stopping. For example, using the rule

    (r1) $less(x, y) \to true$ **if** $\{ less(x, z), less(z, y) \}$,

the transitivity rule of a partial order *less*, on the term $less(A, B)$, the process may never terminate. Since when $less(x, y)$ matches $less(A, B)$, we want to establish the premises of the rule (r1), i.e., $less(A, z)$ and $less(z, B)$, to be reduced to $true$. Again the rule (r1) can apply and the process keeps going on.

Definition 4.3 (depth of contextual rewriting) If $t \to^C t'$ by a rewrite rule *lhs* $\to$ *rhs* if $\{ P_1, P_2, \cdots, P_k \}$ with the substitution σ, under the context C, the *depth* of this rewriting is

0 if the condition $\{ P_1, ..., P_n \}$ is empty;

$1 + \max\{ depth(P_i \to^C true) \}$ otherwise.

By definition, the depth of a rewrite is the depth of recursive applications of contextual rewriting over conditions to establish them to be true. For example, if we have

$x \leq x \to true$

$x \leq s(y) \to true$ **if** $x \leq y$

The depth of $0 \leq 0 \to^C true$ is 0. The depth of $0 \leq s(0) \to^C true$ is 1 and the depth of $0 \leq s^i(0) \to^C true$ is i. In the *less* example above, the depth of a rewriting can be infinity.

In [Kaplan 84] and [Zhang 84], a sufficient condition is imposed on each rule: there exists a reduction ordering $>$ such that for each rule *lhs* $\to$ *rhs* if $\{ P_1, P_2, \cdots, P_k \}$ of R, (i) *lhs* $>$ *rhs* and (ii) *lhs* $> P_i$ for $1 \leq i \leq k$. It is the condition (ii) that prevents the depth of a rewriting from being infinity. A rewrite rule satisfying the conditions (i) and (ii) is said to be *terminating*. Under this condition, the contextual rewriting will always terminate.

Unfortunately some conditional rewrite rules are inherently non-terminating even when they are ground and there is a total ordering on terms. For instance, the rule

$a \to b$ **if** $a \neq c$,

where $a > b > c$, is not terminating. Terminating condition also excludes the rules like (r1) above that have variables in the condition but not in the head.

There are two choices to handle this problem: one is to use only terminating rules for reduction; the other is to limit the depth of each rewriting to a reasonable number and to use the rule satisfying (i) above. Both criteria work well to guarantee the termination and the completeness when it is integrated in **PROC**. For pragmatic reasons, we prefer the second criterion.

4.3. Simplification of Clauses

The (negated) clauses presented above are of the form

not$\{ P_1, P_2, \cdots, P_k \}$.

In order to explain how to apply contextual rewriting, we prefer to transform each clause by writing each literal as an equality. This can be done as follows: If P_i is a negated atom, i.e., $P_i = \neg A$, then write P_i as $A = false$; If P_i is already an equality, do nothing; Otherwise write it as $A = true$ where $P_i = A$. Now each (negated) clause can be written as

not$\{ m_1 = n_1, \cdots, m_k = n_k \}$

and we refer it as *equality clause*.

To simplify an equality $m_i = n_i$ in an equality clause, we use the rest of equalities as the context, then apply the contextual rewriting on m_i and n_i. This is illustrated by the following examples.

Example 4.4 (con't of Example 4.2) The clause to be simplified is
$$c = (z \neq x^*y \mid gcd(z,x) = x).$$
The negated form is
$$c' = \mathbf{not}\{ z = x^*y, gcd(z,x) \neq x \},$$
and the equality clausal form is
$$e = \mathbf{not}\{ z = x^*y, (gcd(z,x)=x) = false \}.$$

To simplify $gcd(z,x) = x$, we make $\{ z = x^*y \}$ as the context. As in Example 4.2, $gcd(z,x)$ is rewritten to x. That is, e is simplified to a trivial clause.

Example 4.5 Let $R = \{ (r1)\ hasteeth(x) \to true\ \mathbf{if}\ istiger(x),$
$$(r2)\ isanimal(x) \to true\ \mathbf{if}\ istiger(x) \}$$
Given a clause
$$c = (\neg istiger(y) \mid \neg isanimal(y) \mid \neg hasteeth(y)),$$
the negated clause is
$$c' = \mathbf{not}\{ istiger(y), isanimal(y), hasteeth(y) \},$$
and the equality clause is
$$e = \mathbf{not}\{ istiger(y) = true, isanimal(y) = true, hasteeth(y) = true \},$$

To simplify $hasteeth(y) = true$, let the context be
$$C = \{ istiger(y) = true, isanimal(y) = true \}$$
and we match the left-hand side of (r1) onto $hasteeth(y)$ to obtain the substitution $\sigma = (x \leftarrow y)$. The condition of (r1) is $istiger(y)$, which can be simplified to $true$ by the context, i.e, by constant congruence. So (r1) is applicable and $hasteeth(y)$ is reduced to $true$. Similarly, $isanimal(y)$ can also be reduced to $true$ by (r2). After removing $true = true$, we have a simplified clause e' of e, i.e.,
$$e' = \mathbf{not}\{ istiger(y) \}$$

It is not surprising that any consequence of reduction can be obtained by superposition. The significance of reduction is that if a clause gets simplified, then the original clause can be thrown away without violating the completeness of the system.

Theorem 4.6 (Soundness of rewriting $\to^C$) Suppose a clause c is rewritten (once) to c' by R. If there exists an interpretation I such that each rewrite rule of R is interpreted to $true$, then $I(c) = I(c')$.

Theorem 4.7 Given a set S of clauses and $c \in S$. If c is rewritten to c' by the rules made from other clauses in S, then S is unsatisfiable iff $S - \{ c \} \cup \{ c' \}$ is unsatisfiable.

With the above results, we can show that the integration of contextual rewriting into the procedure **PROC** does not affect the refutational completeness of **PROC**: the empty clause (*true* = *false*) can never be reduced; if **PROC** stops with "*S* is satisfiable" or loops forever, we can construct an interpretation using the same construction as in the proof of Theorem 3.6 (omitted in this paper), such that the output *R* is true under the interpretation. Since every deleted clause has the same truth value as the corresponding reduced clause (*true* in this case), the input *S* is satisfiable.

The idea of using the rest of the literals in a clause as the context to simplify one literal was first described in [Boyer&Moore 79]. In the reduction process of their theorem prover, each statement is first transformed into clausal form, a literal is simplified under the assumption that the rest of the literals are *false* (we assume the literals to be *true* in a negated clause). Our contribution is that we gave a formal description of the idea under the form of contextual rewriting and emphasized that the constant congruence should be used for the equalities in the context, not the congruence relation generated by the equalities. More importantly, we show that this rewriting can apply to the first-order theorem proving when viewing each clause as a (conditional) rewrite rule.

5. Experimental Results

We have implemented the method in *RRL*, a *Rewrite Rule Laboratory*, a theorem prover based on rewriting techniques being developed at General Electric Corporate Research and Development and Rensselaer Polytechnic Institute. We implemented **PROC** (Sec. 3.3) and integrated the reduction procedure (Def. 4.1) into it. Reduction is done on each clause before *Make−Rule* is invoked; whenever a new rule is made, this rule is used to simplify other rules. Many optimizations can be made in the implementation. For example, if two rules have the same right-hand side, there is no need to superpose one of them at the top position of the left-hand side of the other, because the resulting critical clause is trivial. It is for the same reason that there is no need to apply the factoring rule (F') on literals with positive polarity. Heuristics developed in *RRL* can be used to select rules for superposition, if these heuristics are fair strategy for a completion procedure.

The method has been tried on a number of examples including the seventy-five problems given in [Pelletier 86] for testing automatic theorem provers. Preliminary results are very encouraging as our method compares favorably with resolution method, natural deduction method, as well as Kapur and Narendran's method. We are preparing a list of results for all 75 problems given in [Pelletier 86].

These 75 problems include SAM's Lemma on lattice theory. We got a proof of length 13, using functions instead of predicates for the definitions of *min* and *max*. AC-unification and AC-matching algorithms are used as *min* and *max* have commutative and associative properties. That is, **PROC** acts as the AC-completion procedure as described in [Lankford&Ballantyne, 77] and [Peterson&Stickel 81]. One hundred twenty formulae were derived and all of them were retained. The best time is 298 seconds on a Symbolics LM 3600. See *Appendix* **A** for the proof found by *RRL*.

In the proof, the deletion of the input clause

#9 $max(x,z) = x \mid \neg (min(x,y) = z)$

may illustrate the power of contextual rewriting. The rewrite rules involved are:

[6] $min(x,x) \rightarrow x$,

[7] $min(x,max(x,y)) \rightarrow x$,

[8] $max(x,min(y,z)) \rightarrow min(y,max(x,z))$ **if** $min(x,y) = x$.

The literal $max(x,z) = x$ was first reformulated as $max(x,min(x,y)) = x$ by the context { $z = min(x,y)$ }, which is made from the other literal. Next, we can apply Rule [8] on $max(x,min(x,y))$, since the condition of [8] can be established by Rule [6]. The result of this rewriting is $min(x,max(x,y))$, which can be further reduced to x by Rule [7]. In short, the first literal of the clause #9 is reduced to $x = x$, or *true*. #9 becomes trivial and can be thrown away safely. Note that neither demodulation nor subsumption can remove this clause.

Schubert's Steamroller challenge problem is also included in [Pelletier 86]. Using our method, we got a proof of length 15 (not including reductions), 133 formulas are derived and 116 of them are retained, and successful unifications are 206. It takes one minute on a Symbolics LM3600 which compares favorably with results reported in Stickel's survey article on the performance of various theorem provers on the untyped version of this problem. The proof found by *RRL* is given in *Appendix* **C**.

We have also been able to automatically prove many lemmas in Boyer et al's article on finite axiomatization of set theory. For instance, the first lemma in that article can be automatically proved using this method in 4 seconds on Symbolics LM3600. Its proof is given in *Appendix* **B**.

Preliminary experiments suggest that our method needs less unifications but this saving may be at the cost of making rules from clauses and performing rewritings. We will not elaborate on this issue, since we could not compare different methods by a fair criterion, such as the effort involved and the machine used for implementation, the user's assistance in a proof, etc.

Buchberger gave in [Buchberger 85] a synopsis of the critical-pair/completion approach for solving algorithmic problems in first-order predicate calculus, polynomial ideal theory and rewrite systems. He showed that the *critical-pair computation* (or *superposition*) is a common key idea in the resolution principle (by formal distortion), in the Groebner basis generation, as well as in the Knuth-Bendix completion procedure. In fact, one basic step in the resolution algorithm can be viewed as a superposition and a resolvent can be viewed as a critical pair. Buchberger wrote that "it is exciting to see that the research activities that started twenty years ago from very different roots, finally meet and merge.'' We are glad to see that our result provides one more evidence to Buchberger's opinion. Many techniques and methods developed in resolution approach and rewrite rule approach can be borrowed from each other and enhance further study of both approaches.

Acknowledgement: We would like to thank David Musser and David Plaisted for helpful comments on an earlier draft of this paper.

REFERENCES

[Bachmair&al 86] Bachmair, L., Dershowitz, N., and Hsiang, J., "Orderings for equational proofs," *Proc. IEEE Symp. Logic in Computer Science*, Cambridge, MA, 336-357, 1986.

[Bachmair&Dershowitz 87] Bachmair, L., and Dershowitz, N., "Inference rules for rewrite-based first-order theorem proving," *Proc. IEEE Symp. Logic in Computer Science*, Ithaca, New York, 1987.

[Boyer 71] Boyer, R.S., *Locking: A restriction of resolution*. Ph.D. Thesis, University of Texas, Austin, 1971.

[Boyer&Moore 79] Boyer, R.S. and Moore, J.S., *A computational logic*. Academic Press, New York, 1979.

[Brand&al 78] Brand, D., Darringer, J.A., and Joyner, W.H., *Completeness of conditional reductions*. Research Report RC 7404 IBM, 1978.

[Buchberger 85] Buchberger, B., "History and basic feature of the critical-pair/completion procedure," *Rewriting Techniques and Applications*, Lect. Notes in Comp. Sci., Vol. 202, Springer-Verlag, Berlin, 301-324, 1985.

[Chang&Lee 73] Chang, C-L. and Lee, R.C., *Symbolic logic and mechanical theorem proving*. Academic Press, New York, 1973.

[Dershowitz 87] Dershowitz, N., "Termination of rewriting," J. of Symbolic Computation Vol. 3, 1987, 69-116.

[Drosten 83] Drosten K. *Toward executable specifications using conditional axioms*. Report 83-01, t.U. Braunschweig, 1983.

[Fribourg 85] Fribourg, L., "A superposition oriented theorem prover" *Theoretical Computer Science*, Vol. 35, 129-164, 1985.

[Fribourg 85b] Fribourg, L., "SLOG: A logic programming language interpreter based on clausal superposition and rewriting," In *Proc. 1985 Intl. Symposium on Logic Programming*, pp. 172-184, Boston, 1985.

[Hsiang 82] Hsiang, J., *Topics in automated theorem proving and program generation*. Ph.D. Thesis, UIUCDCS-R-82-1113, Univ. of Illinois, Urbana-Champaigne, 1982.

[Hsiang 87] Hsiang, J., "Rewrite method for theorem proving in first order theory with equality," *J. Symbolics Computation*, Vol. 3, 133-151, 1987.

[Hsiang&Rusinowitch 86] Hsiang, J., and Rusinowitch, M., "A new method for establishing refutational completeness in theorem proving," *Proc. 8th Conf. on Automated Deduction*, LNCS No. 230, Springer-Verlag, 141-152, 1986.

[Huet&Oppen 80] Huet, G. and Oppen, D., "Equations and rewrite rules: a survey," in: *Formal languages: Perspectives and open problems*. (R. Book, ed.), Academic Press, New York, 1980.

[Kaplan 84] Kaplan S., *Fair conditional term rewriting systems: unification, termination and confluence*. LRI, Orsay, 1984.

[Kapur&Narendran 85] Kapur, D. and Narendran, P., "An equational approach to theorem proving in first-order predicate calculus," *9th IJCAI*, Los Angeles, CA, August 1985.

[Kapur et al 86] Kapur, D., Sivakumar, G., and Zhang, H., "RRL: A Rewrite Rule Laboratory," *Proc. 8th*

Intl Conf. on Automated Deduction (CADE-8), Oxford, U.K., LNCS 230, Springer-Verlag, 1986.

[Knuth&Bendix 70] Knuth, D., and Bendix, P., "Simple Word Problems in Universal Algebras," in: *Computational problems in abstract algebra*. (Leech, ed.) Pergamon Press, 263-297, 1970.

[Kowalski&Hayes 69] Kowalski, R., and Hayes, P., "Semantic trees in automatic theorem proving," in *Machine Intelligence*, Vol. 5, (Meltzer, B. and Michie, D., eds.) American Elsevier, 1969.

[Lankford 75] Lankford, D.S., *Canonical inference*. Report ATP-32, Dept. of Mathematics and Computer Sciences, Univ. of Texas, Austin, Texas, 1975.

[Lankford&Ballantyne 77] Lankford, D.S., and Ballantyne, A.M., *Decision procedures for simple equational theories with commutative-associative axioms: Complete sets of commutative-associative reductions*. Memo ATP-39, Department of Mathematics and Computer Sciences, University of Texas, Austin, TX, August 1977.

[Lankford&Ballantyne 79] Lankford, D.S., and Ballantyne, A.M., "The refutational completeness of blocked permutative narrowing and resolution," *4th Conf. on Automated Deduction*, Austin, Texas, 1979.

[Lankford 79] Lankford, D.S., *Some new approaches to the theory and applications of conditional term rewriting systems*. Report MTP-6, Math. Dept., Lousiana Tech. University, 1979.

[Pelletier 86] Pelletier F.J., "Seventy-five problems for testing automatic theorem provers," *J. of Automated reasoning* Vol. 2, 191-216, 1986.

[Pletat&al 82] Pletat, U., Engels, G., and Ehrich, H.D., "Operational semantics of algebraic specifications with conditional equations," *7th, CAAP '82*, LNCS, Springer-Verlag, 1982.

[Reiter 71] Reiter, R., "Two results on ordering fro resolution with merging and linear format," *JACM*, 630-646, 1971.

[Remy 82] Remy, J.L., *Etudes des systemes reecriture conditionnelles et applications aux types abstraits algebriques*. These d'etat, Universite Nancy I, 1982.

[Peterson 83] Peterson, G.E., "A technique for establishing completeness results in theorem proving with equality," *SIAM J. Computing* Vol. 12, No. 1. 82-99, 1983.

[Peterson&Stickel 81] Peterson, G.L., and Stickel, M.E., "Complete sets of reductions for some equational theories," *JACM* Vol. 28 No. 2, 233-264, 1981.

[Stickel 85] Stickel, M.E., "Automated deduction by theory resolution," *J. of Automated Reasoning* Vol. 1, 333-355, 1985.

[Walther 84] Walther, C., "A mechanical solution of Schubert's steamroller by many-sorted resolution," *Proc. of the AAAI-84 National Conf. on Artificial Intelligence*, Austin, Texas, 330-334, 1984.

[Wos&Robinson 70] Wos, L.R., and Robinson, G., "Paramodulation and set of support," *Proc. IRIA Symposium on Automatic Demonstration*, Versailles, France, 1968, Springer-Verlag, Berlin, 276-310, 1970.

[Zhang 84] Zhang, H., *Reveur 4: Etude et Mise en Oeuvre de la Reecriture Conditionelle*. Thesis of "Doctorat de 3me cycle", Universite de Nancy I, France, 1984.

[Zhang&Remy 85] Zhang, H., and Remy, J.L., "Contextual rewriting," *Proc. of rewriting techniques and application*, Dijon, France, 1985.

Appendix A.
Output of SAM's Lemma from RRL

Input #3, produced: [3] MAX(X, 0) ---> X
Input #6, produced: [6] MIN(X, X) ---> X
Input #7, produced: [7] MIN(X, MAX(X, Y)) ---> X
Input #8, produced: [8] MAX(X, MIN(Y, Z)) ---> MIN(Y, MAX(X, Z)) if { (MIN(X, Y) = X) }
Input #10, produced: [10] MAX(A, B) ---> D1
Input #11, produced: [11] MIN(A, B) ---> D2
Input #12, produced: [12] MIN(C1, D1) ---> 0

Input #13, produced: [13] MIN(C2, D2) ---> 0
Input #14, produced: [14] MIN(B, C2) ---> E
Input #15, produced: [15] MIN(A, C2) ---> F
Input #16, produced: [16] MAX(C1, E) ---> G
Input #17, produced: [17] MAX(C1, F) ---> H
Input #18, produced: [18] (MIN(G, H) = C1) ---> FALSE
Rule [10] superposed with Rule [7], produced: [19] MIN(A, D1) ---> A
Rule [10] superposed with Rule [7], produced: [20] MIN(B, D1) ---> B
Rule [12] superposed with Rule [8], reduced by Rule [3], produced:
    [25] MIN(D1, MAX(X, C1)) ---> X if { (MIN(X, D1) = X) }
Rule [6] superposed with Rule [14], reduced by Rule [14], produced: [28] MIN(C2, E) ---> E
Rule [11] superposed with Rule [14], reduced by Rule [13], produced: [31] MIN(A, E) ---> 0
Rule [16] superposed with Rule [7], produced: [32] MIN(C1, G) ---> C1
Rule [15] superposed with Rule [19], reduced by Rule [15], produced: [37] MIN(D1, F) ---> F
Rule [14] superposed with Rule [20], reduced by Rule [14], produced: [40] MIN(D1, E) ---> E
Rule [16] superposed with Rule [25], reduced by Rule [40], produced: [47] MIN(D1, G) ---> E
Rule [15] superposed with Rule [28], reduced by Rule [31], produced: [51] MIN(E, F) ---> 0
Rule [37] superposed with Rule [47], reduced by Rule [51], produced: [77] MIN(F, G) ---> 0
Rule [77] superposed with Rule [8], reduced by Rule [3], produced:
    [104] MIN(G, MAX(X, F)) ---> X if { (MIN(X, G) = X) }
Rule [17] superposed with Rule [104], reduced by Rules [32], and [18], produced: [121] TRUE ---> FALSE

Length of the proof (unifications)	*= 13*	*Number of rules generated*	*= 121*
Number of rules retained	*= 120*	*Number of critical pairs*	*= 807*
Time used	*= 298.1 sec*		

Appendix B.
Output of a Problem in Set Theory from RRL

Input #1, produced: [1] MEM(X, SET2(X, Y)) ---> TRUE if { M(X) }
Input #2, produced: [2] MEM(U, SET2(X, Y)) ---> FALSE if { (X = U) <=> FALSE, (Y = U) <=> FALSE }
Input #3, produced: [3] M(SET2(X, Y)) ---> TRUE
Input #4, produced: [4] SET2(X, X) ---> SET1(X)
Input #5, produced: [5] SET2(SET1(X), SET2(X, Y)) ---> PAIR(X, Y)
Input #6, produced: [6] M(AX) ---> TRUE
Input #8, produced: [8] (AX = AU) ---> FALSE
Input #9, produced: [9] PAIR(AX, AY) ---> PAIR(AU, AV)
Rule [4] superposed with Rule [1], produced: [10] MEM(Y, SET1(Y)) ---> TRUE if { M(Y) }
Rule [4] superposed with Rule [2], produced: [11] MEM(U, SET1(Y)) ---> FALSE if { (Y = U) <=> FALSE }
Rule [3] superposed with Rule [4], produced: [12] M(SET1(X)) ---> TRUE
Rule [5] superposed with Rule [2], produced:
    [15] MEM(U, PAIR(X, Y)) ---> FALSE if { (SET2(X, Y) = U) <=> FALSE, (SET1(X) = U) <=> FALSE }
Rule [9] superposed with Rule [15], produced:
    [16] MEM(U, PAIR(AU, AV)) ---> FALSE if { (SET1(AX) = U) <=> FALSE, (SET2(AX, AY) = U) <=> FALSE }
Rule [5] superposed with Rule [1], reduced by Rule [12], produced: [17] MEM(SET1(X), PAIR(X, Y)) ---> TRUE
Rule [16] superposed with Rule [17], produced:
    [19] SET1(AX) ---> SET1(AU) if { (SET2(AX, AY) = SET1(AU)) <=> FALSE }
Rule [19] superposed with Rule [10], reduced by Rules [6], [11], and [8], produced: [21] SET2(AX, AY) ---> SET1(AU)
Rule [21] superposed with Rule [1], reduced by Rules [6], [11], and [8], produced: [22] TRUE ---> FALSE

Length of the proof (unifications)	*= 9*	*Number of rules generated*	*= 22*
Number of rules retained	*= 22*	*Number of critical pairs*	*= 31*
Time used	*= 4.13 sec*		

Appendix C.
Output of Steamroller from RRL

Input #1, produced: [1] ISWOLF(AWOLF) ---> TRUE
Input #2, produced: [2] ISFOX(FOX) ---> TRUE
Input #3, produced: [3] ISBIRD(BIRD) ---> TRUE
Input #4, produced: [4] ISSNAIL(SNAIL) ---> TRUE
Input #5, produced: [5] ISGRAIN(GRAIN) ---> TRUE
Input #6, produced: [6] ISPLANT(SPLANT) ---> TRUE
Input #8, produced: [8] ISANIMAL(X) ---> TRUE if { ISWOLF(X) }
Input #9, produced: [9] ISANIMAL(X) ---> TRUE if { ISFOX(X) }
Input #10, produced: [10] ISANIMAL(X) ---> TRUE if { ISBIRD(X) }
Input #11, produced: [11] ISANIMAL(X) ---> TRUE if { ISSNAIL(X) }
Input #13, produced: [13] ISPLANT(X) ---> TRUE if { ISGRAIN(X) }
Input #14, produced: [14] SMALLER(X, Y) ---> TRUE if { ISSNAIL(X), ISBIRD(Y) }
Input #15, produced: [15] SMALLER(X, Y) ---> TRUE if { ISBIRD(X), ISFOX(Y) }
Input #16, produced: [16] SMALLER(X, Y) ---> TRUE if { ISFOX(X), ISWOLF(Y) }
Input #17, produced: [17] EATS(X, Y) ---> FALSE if { ISWOLF(X), ISFOX(Y) }
Input #18, produced: [18] EATS(X, Y) ---> FALSE if { ISWOLF(X), ISGRAIN(Y) }
Input #19, produced: [19] EATS(X, Y) ---> FALSE if { ISBIRD(X), ISSNAIL(Y) }
Input #23, produced: [23] EATS(Y, SPLANT) ---> TRUE if { ISSNAIL(Y) }
Input #25, produced: [24] SMALLER(Z, X) ---> FALSE if
        { ISANIMAL(X), ISPLANT(Y), ISANIMAL(Z), ISPLANT(W),
        EATS(Z, W), EATS(X, Y) <=> FALSE, EATS(X, Z) <=> FALSE }
Input #26, produced: [25] EATS(X, Y) ---> FALSE if { EATS(Y, Z), ISANIMAL(X), ISANIMAL(Y), ISGRAIN(Z) }
Rule [24] superposed with Rule [16], reduced by Rules [17], [9], [8], produced:
    [28] EATS(X, W) ---> FALSE if
        { ISWOLF(Y), ISFOX(X), EATS(Y, Y1) <=> FALSE, ISPLANT(W), ISPLANT(Y1) }
Rule [24] superposed with Rule [15], reduced by Rules [17], [9], [8], [10], [9], produced:
    [29] EATS(X, W) ---> FALSE if
        { ISFOX(Y), ISBIRD(X), EATS(Y, X) <=> FALSE,
        EATS(Y, Y1) <=> FALSE, ISPLANT(W), ISPLANT(Y1) }
Rule [24] superposed with Rule [14], reduced by Rules [10], [9], [19], [11], [10], produced:
    [30] EATS(X, W) ---> FALSE if
        { ISBIRD(Y), ISSNAIL(X), EATS(Y, Y1) <=> FALSE, ISPLANT(W), ISPLANT(Y1) }
Rule [23] superposed with Rule [30], reduced by Rules [4], [6], produced:
    [31] EATS(Y, Y1) ---> TRUE if { ISPLANT(Y1), ISBIRD(Y) }
Rule [29] superposed with Rule [31], reduced by Rule [6], produced:
    [32] EATS(Y, Y1) ---> TRUE if { ISPLANT(Y2), EATS(Y, Y2) <=> FALSE, ISBIRD(Y1), ISFOX(Y) }
Rule [25] superposed with Rule [32], reduced by Rules [3], [5], [31], [13], [10], [9], produced:
    [33] EATS(Y, Y1) ---> TRUE if { ISFOX(Y), ISPLANT(Y1) }
Rule [28] superposed with Rule [33], reduced by Rules [6], [2], produced:
    [34] EATS(Y, Y1) ---> TRUE if { ISPLANT(Y1), ISWOLF(Y) }
Rule [18] superposed with Rule [34], reduced by Rules [5], [13], produced: [35] ISWOLF(Y) ---> FALSE
Rule [35] deleted Rule [1], produced: [36] TRUE ---> FALSE

Length of the proof (unifications)	= 9	*Number of rules generated*	= 36
Number of rules retained	= 35	*Number of critical pairs*	= 68
Time used	= 12.1 sec		

Elements of Z-Module Reasoning

Tie Cheng Wang

Kestrel Institute
1801 Page Mill Road
Palo Alto, CA 94304

Abstract

Z-module reasoning is a new approach to automate the equality-oriented reasoning required to attack problems from ring theory (associative and nonassociative). This approach obtains most of its efficiency by employing two distinct types of reasoning for completing a proof. One type of reasoning is carried out by a vector generator which uses inference rules based on paramodulation. However, the vector generator does not attempt to prove a theorem by itself; rather, it merely produces a set of identity vectors from the hypotheses of the theorem to be proved. The other type of reasoning is carried out by an integer Gaussian elimination program and a proof-procedure. These programs are responsible for finally proving the theorem by "calculating" the truth of the theorem in terms of the Z-module generated by the set of identity vectors produced by the vector generator. This paper presents algorithms needed for designing a Z-module reasoning system and discusses issues concerning the soundness and completeness of these algorithms. The paper will show how to use the Z-module reasoning to prove a general equality theorem whose hypothesis does not contain a set of axioms of ring theory.

1   Introduction

We begin a description of Z-module reasoning with the initial ideas concerning to prove a basic theorem from ring theory. By a basic ring theorem, we mean a statement that, given a set H of defining identities for a variety of rings, asserts that an identity holds in every ring of that variety. Examples of basic ring theorems include the x-cube ring theorem: if $xxx = x$ for all x in an associative ring, then the ring is commutative; a Moufang theorem in alternative rings: if $x(yy) = xyy$ and $x(xy) = xxy$ for all x and y in a ring, then $x(yzy) = xyzy$ for all x, y, and z in the ring, etc. With equality predicate, a basic ring theorem can be symbolized to be a well formed formula, whose hypothesis consists of a set A of basic axioms of (nonassociative) ring theory and a set

H of identity equations, whose conclusion is an identity equation. The righthand side of each identity equation is zero, and the lefthand side of the equation is a polynomial on a set $V[X]$, where $V[X]$ is the entire set of words on a set X of variables. Each word of $V[X]$ is a term containing no function symbols other that $\times$.

Given a basic ring theorem, $A \cup H \to C$, we shall prove it by deriving a contradiction to an assumption that the theorem is false. Thus, we assume that there exists a ring R satisfying A and H, but for which the conclusion C is false. Let $g_0 \neq 0$ be the negation of C, we shall call g_0 a goal vector of the given theorem. Without losing generality, let C be $\forall x \forall y (xy - yx = 0)$, then $g_0 = ab - ba$. This means that there exist two (constant) elements a and b in R such that $ab - ba \neq 0$. Let X_0 be the set $\{a, b\}$, then $ab - ba$ must be contained in the subring R_0 of R that is generated by X_0. If we can prove $ab - ba = 0$ in R_0, then a contradiction is obtained.

From ring theory, we know that there exists a ring F_0 freely generated by X_0 such that R_0 is a homomorphic image of F_0, i.e,

$$R_0 = \Phi F_0,$$

where the restriction of Φ on X_0 is an identity mapping (so that we can simply write Φt as t for every $t \in F_0$). Let K be the set $\{v \in F_0 | \Phi v = 0\}$, then K is called the kernel of Φ. One approach to prove $g_0 = 0$ in R_0 is by proving that g_0 is an element of the kernel K of Φ. The Z-module reasoning is based on this approach.

The kernel K is a two-sided ideal of F_0 that is determined by the given set H of defining identities. Stated more precisely, K is an ideal of F_0 generated by a set T_p,

$$T_p = \{f(p_1, ..., p_n) \mid f(x_1, ..., x_n) = 0 \in H, p_i \in F_0 \text{ for } 1 \leq i \leq n\}.$$

If T_p is $\emptyset$, then we shall have $K = \{0\}$. That is, R_0 is isomorphic to F_0. Hence g_0 holds in R_0 if and only if $g_0 = 0$ modulo the set of basic axioms of a ring. If T_p is not empty, then we can construct K recursively from T_p in the following way: K contains every element of T_p; for every x, y in F_0 and every k in K and every z in Z, K contains all sums of elements xk, ky, xky and zk.

We note that each of F_0 and R_0 has a structure of a Z-module generated by $V[X_0]$. A Z-module M is said to be generated by a set S if only if each element of M can be written as a polynomial on S — a linear combination (with integer coefficients only) of the elements of S. A Z-module can be described by means of a set of generators and a set of defining relations. Without being specified precisely, [S] will be used by us to denote the Z-module generated by the set S with the empty set of defining relations.

We shall specify the ideal K of F_0, generated by T_p, by means of a Z-module generated by a subset of elements of F_0. Let W be a word in $V[X_0]$. Denote the term obtained by replacing one occurrence of a subterm s of W by t' with $W[s]|_s^{t'}$. We construct the following set G_p, which is called the set of Z-module generators determined by H on $V[X_0]$.

$$G_p = \{W[s]|_s^{t'} \mid W[s] \in V[X_0], t' \in T_p\}.$$

The advantage of studying the kernel K in terms of a set of Z-module generators is that, once a set of Z-module generators is given, it is usually much easier to determine whether or not an element belongs to the Z-module generated by the set. However, the set G_p appears to be too complicated and too large to be generated. In particular, we note that a term $p_i, 1 \leq i \leq n$, contained in a polynomial $f(p_1, ..., p_n)$ of T_p may require the use of a very long polynomial (i.e, the polynomial containing a large number of words).

We prefer such a set of Z-module generators for the kernel K that contains only those elements of F_0 that can be written with short polynomials. We shall call an expression obtained from a polynomial $f(x_1, ..., x_n)$ a simple instance of that polynomial if it is obtained by replacing each variable x_i of the polynomial with a word w_i on $V[X_0]$. Let $W[s]$ be any word W that contains a subword s, and let t' be a simple instance of a polynomial t on $V[X_0]$. Then the result $W[s]|_s^{t'}$ is called a simple Z-module obtained from t. Let

$$G_w = \{W[s]|_s^{f(w_1, ..., w_n)} \mid W[s] \in V[X_0], f(x_1, ..., x_n) = 0 \in H, w_i \in V[X_0]\}.$$

G_w is called the set of simple Z-module generators determined by H on $V[X_0]$. $W[s]|_s^{f(w_1, ..., w_n)}$ is called a simple Z-module generator obtained from $f(x_1, ..., x_n) = 0$. We note that a simple Z-module generator obtained from an identity equation $t = 0$ is a short polynomial in the sense that the number of monomials contained in the generator is the same as that of the polynomial t in $t = 0$. In comparison with the set G_p, the structure of G_w is much simpler. The fundamental approach of Z-module reasoning is to deduce a subset of simple generators for the Z-submodule of the kernel K that is related to the goal vector g_0.

However, since the set G_w is a subset of G_p, the Z-module $[G_w]$ generated by G_w may be only a proper subset of $[G_p]$. In that case, one will not be able to prove $g_0 \in K$ in terms of the set G_w if g_0 is not contained in $[G_w]$, even though $g_0 \in [G_p]$ holds. For example, if H is $\{x(yy) - xyy = 0\}$ and $X_0 = \{a, b, c\}$, then K contains an element $(a((b + c)(b + c)) - a(b + c)(b + c)$, whose polynomial is $a(bc) + a(cb) - abc - acb$ on $V[\{a, b, c\}]$. But the polynomial $a(bc) + a(cb) - abc - acb$ is not a simple Z-module generator obtained from $x(yy) - xyy = 0$. Thus, if g_0 is $a(bc) + a(cb) - abc - acb$, then $g_0 \in K$ cannot be proved in terms of G_w.

A set H of identity equations is said to be *Z-module complete* if the kernel K determined by H on any set V of words is a Z-module generated by a set of simple Z-module generators determined by H on V. Generally speaking, the set H of special hypotheses contained in the original statement of a ring theorem may not be Z-module complete. However, our experiments have shown that, for many practical theorems from ring theory, such as the example we just given, we can deduce from the set H of special hypotheses of the theorem a set L of additional identities such that $H \cup L$ is Z-module complete. A useful algorithm for such a deduction is called linearization.

Of course, in order to determine if g_0 is contained in the kernel K, we usually do not need to create explicitly the whole set of Z-module generators for the entire ideal K. Actually, if g_0 is indeed contained in K, it should be contained in a Z-submodule of

K and this Z-submodule must be finite. The vector generator to be presented later is designed for deducing the set of simple Z-module generators for such a Z-submodule.

Once a set G of Z-module generators is created, the next task of the Z-module reasoning method is to check if g_0 is contained in the Z-module $[G]$. This can be done in two steps. First, we deduce a basis for $[G]$ by using an integer Gaussian elimination algorithm. Then the existence of g_0 in the Z-module $[G]$ can be determined by a proof-procedure in terms of this basis.

To summarize our approach, the Z-module reasoning method proves a basic ring theorem in four steps. The first step is to deduce a Z-module complete set from the set of special hypotheses of the given theorem. The second step is to deduce a set of simple Z-module generators related to the goal vector derived from the denial of the theorem. The third step is to deduce a basis for the Z-module generated by these generators. The final step is to determine the existence of the goal vector in terms of the resulting basis. In the following sections, we shall present the algorithms for each of the main components of the Z-module reasoning method, and discuss issues related to the completeness and soundness of these algorithms.

Although the Z-module reasoning method is introduced in this section by the consideration of proving basic ring theorems, this method is useful for proving many other theorems from ring theory. We shall describe in a later section of this paper how the Z-module reasoning method can also be used for proving those theorems from equality theory whose hypotheses do not contain a set of axioms of ring theory.

2   A Linearization Algorithm

It is well known that the existence of some identities in a ring implies the existence of some additional identities in the ring. A method given by mathematicians for deducing these additional identities is called linearization [12]. The linearization algorithm we now describe is designed to enable the user of a Z-module reasoning system to prepare input data for the vector generator of the system. Given a set H of identity equations, to obtain LINEARIZATION(H):

(1) Pick an equation $t = 0$ from H and a variable x that is a subword of a word of a monomial of t.

(2) Substitute $u + v$ for x, where u and v are variables that do not occur in $t = 0$, and then expand the resulting equation by using the distributive laws of ring theory.

(3) Remove those summands from the result whose sum is a variant of the lefthand side of an equation in H.

(4) If the final result is not $0 = 0$, then add it to H. Repeat, starting with step 1, with another variable different from x of $t = 0$ or with another equality of H

that has not been considered.

Example 2.1 Let H be the set consisting of $\{x(yy) - xyy = 0\}$. We deduce an additional equation from H with the given linearization algorithm by the following steps:

By replacing y with $u+v$ in $x(yy) - xyy$, we obtain $x((u+v)(u+v)) - x(u+v)(u+v)$. By distributing the result, we obtain $x(uu) + x(uv) + x(vu) + x(vv) - xuu - xuv - xvu - xvv = 0$. By removing the identities $x(uu) - xuu$ and $x(vv) - xvv$, we obtain an additional equation $x(uv) + x(vu) - xuv - xvu = 0$.

The linearization algorithm will terminate at this step since no other additional equations can be deduced.

For illustrating the usage of the linearization algorithm, we consider the set H given in example 2.1. Let G_p be the entire set of Z-module generators determined by H on $V[X_0]$, where $X_0 = \{a, b, c\}$. Let G_w be the set of simple Z-module generators determined by H on $V[X_0]$. As we have shown in the preceding section, $[G_w] \neq [G_p]$.

Denote $x(yy) - xyy$ by e_1. Denote the linearization of e_1, $x(yz) + x(zy) - xyz - xzy$, by e_2. Let $H' = \{e1, e2\}$. It is easy to verify that the set of Z-module generators determined by H' on $V[X_0]$ is the same as the set G_p. Let G'_w be the set of simple Z-module generators determined by H' on $V[X_0]$. We can prove $[G'_w] = [G_p]$ by showing that each element of G_p can be rewritten, by using the distributive laws of ring theory, into a sum of one or more simple Z-module generators obtained from H'.

For example, consider an element $W[s]\big|_s^{t_1(t_2 t_2) - t_1 t_2 t_2}$ of G_p. If t_2 is (or can be rewritten to be) a polynomial $a_2 w_2 + t_3$, where w_2 is in $V[X_0]$ and a_2 is in Z, we can rewrite the entire expression with the following steps:

$$W[s]\big|_s^{t_1(t_2 t_2) - t_1 t_2 t_2}$$
$$= W[s]\big|_s^{t_1((a_2 w_2 + t_3)(a_2 w_2 + t_3)) - t_1(a_2 w_2 + t_3)(a_2 w_2 + t_3)}$$
$$= W[s]\big|_s^{t_1(a_2 w_2 a_2 w_2) + t_1(a_2 w_2 t_3) + t_1(t_3 a_2 w_2) + t_1(t_3 t_3) - t_1 a_2 w_2 a_2 w_2 - t_1 a_2 w_2 t_3 - t_1 t_3 a_2 w_2 - t_1 t_3 t_3}$$
$$= a_2 a_2 W[s]\big|_s^{t_1(w_2 w_2) - t_1 w_2 w_2}$$
$$\quad + a_2 W[s]\big|_s^{t_1(w_2 t_3) + t_1(t_3 w_2) - t_1 w_2 t_3 - t_1 t_3 w_2}$$
$$\quad + W[s]\big|_s^{t_1(t_3 t_3) - t_1 t_3 t_3}$$

The final expression is a sum of three Z-module generators obtained from e_1, e_2, and e_1 of H', respectively. If t_1 and t_3, both are words, then the expression is a sum of three simple Z-module generators obtained from H'. Otherwise, we can repeatedly rewrite each of these non-simple Z-module generators until all summands become a simple Z-module generator from H'.

For solving the problems from associative ring theory, the associative law may be used as a demodulator for rewriting each word contained in a linearization of an equation to be left associative, thus reducing the number of additional equalities that can be deduced. Some examples of incorporating demodulation into linearization have been given in [9].

3 Vector Generation

The vector generator to be presented in this section is an important component of a Z-module reasoning system. This component is responsible for deducing a set of simple Z-module generators related to a proof of the given theorem. In this section, we shall present an algorithm that can be used for designing a vector generator, and discuss the completeness of this algorithm.

The basic objects maintained by the vector generator are words and vectors. The phrase "word" was used in the preceding section to refer a term containing no function symbols other than $\times$. We now define a "word" in general. A term t is called a *word* if t is a term whose top-level function symbol is not $+, -$, or $\times$, or if t is a term $U \times V$ (written as UV) such that U and V are words. For example, $a(p(a + b, c, d))$ is a word, but $a(a + b)cd$ is not. A *monomial* is a product of an integer (with default value 1) and a word. A *polynomial* is a sum of monomials. A term is called *a word of a polynomial* if the term is the word (not a subword) of a monomial of the polynomial. In our Z-module reasoning system, all words maintained by the vector generator are indexed by an indexing mechanism. A polynomial is called a *vector* if all monomials of it are ordered on the indices of their words in ascending order.

3.1 A Lemma on Related-words and Related-vectors

Definition (*related-words* and *related-generators*). Let g_0 be a polynomial and G be a set of polynomials. We define recursively a set $RW(g_0, G)$ of those words that we shall call the words of G related to g_0. If g_0 is 0, the $RW(g_0, G) = \emptyset$; otherwise the first word of g_0 belongs to $RW(g_0, G)$, and if $w \in RW(g_0, G)$, then for each t of G that contains a word w, every word of t belongs to $RW(g_0, G)$; t is called an element of G *immediately related* to w. Each element of G that contains a word in $RW(g_0, G)$ is called an element related to g_0. The set consisting of those elements in G that are related to g_0 is denoted by $RG(g_0, G)$.

Lemma 3.1 If g_0 is a polynomial and G is a set of polynomials, then $g_0 \in [G]$ iff $g_0 \in [RG(g_0, G)]$.

Because of space limit, we shall not present proofs for most of the lemmas and theorems given in this paper. The readers may prove them by themselves or request the proofs that are included in the original draft of this paper from the author.

By Lemma 3.1, in order to determine if a vector g is contained in a Z-module generated by a set S, one needs only to determine if the vector is contained in a subset of S. This subset can be obtained by collecting all elements of S immediately related to the first word of g, then by recursively collecting all elements of S immediately related to the words that have been collected in the preceding steps.

By the definition of related-vectors, each simple Z-module generator immediately related to a word must contain the word itself. On the basis of this fact and the property of being a simple Z-module generator, we can design an inference rule to deduce efficiently all of those Z-module generators from the set of special hypotheses of the given theorem that are immediately related to a given word. This inference rule is called identity paramodulation (ide-paramodulation). A special version of ide-paramodulation has been defined in [9]. The definition given in this paper is more general.

3.2 Identity Paramodulation

Definition *ide-paramodulation*: Let $L = R$ be an equation, $W[X]$ be a word, and X be a subterm of W. If θ is a unifier of L and X, then $(W[L] - W[R])\theta$ is called an ide-paramodulant of the word W and the equation $L = R$ on X. We shall also say that $(W[L] - W[R])\theta$ is obtained from the word W and the equation $L = R$ by ide-paramodulating $L = R$ into the subterm X of W.

Note the difference between ide-paramodulation and standard paramodulation. The expression considered by standard paramodulation is a clause, and the result is a new clause. Whereas the expression considered by ide-paramodulation must be a word, and the result is a term. For example, by paramodulating the lefthand side of the equality $a + b = 0$ into the subterm $a + b$ of the clause $a + b + c \neq e$, we obtain the paramodulant $0 + c \neq e$. However, with the same equation and the term $a + b + c$, there will be no corresponding ide-paramodulant that can be produced because, by definition, $a + b + c$ is not a word.

3.3 Input Data Preparation of the Vector Generator

The input data of the vector generator mainly consists of a set of input equations and a goal vector. Each input equation is obtained from a special hypothesis of the given theorem, or obtained from an additional equation obtained by linearizing these special hypotheses. Let $t_1 \neq t_2$ be the denial of the conclusion, then the vector obtained from the term $t_1 - t_2$ is the goal vector of the theorem.

The remaining input data consists of a set of demodulators and a set of reverse-demodulators. Demodulators will be used by a normalization subroutine of the vector generator for rewriting the words, deduced from applying ide-paramodulation, into canonical forms, resulting in a reduction of the number of words maintained by the vector generator. Reverse-demodulators are used by the vector generator to retrieve the set of words that were, or could have been written into, a canonical form by usng the set of given demodulators. Reverse-demodulators are used to remedy the possible incompleteness that might result from the use of demodulation. Given a set R of reverse-demodulators, we denote the output of reverse-demodulation applied to a word w_0 by $Reverse\_demo(w_0)$. Then $Reverse\_demo(w_0)$ contains the word w_0, and, for any element w of $Reverse\_demo(w_0)$, that can be rewritten into w' by a reverse-demodulator, $Reverse\_demo(w_0)$ contains w'. We shall write a reverse-demodulator in the form $L \Leftarrow R$ to mean rewrite R to L.

Although the ide-paramodulation is defined in general, the current version of Z-module reasoning requires that each ide-paramodulant contain no variables. Thus, the input data of the vector generator is required to satisfy the following.

(1) The righthand side of each input equation does not contain variables that do not occur in the lefthand side of the equation.

(2) The lefthand side of each input equation is not 0.

(3) The set of demodulators must have a finite termination property, and so must the set of reverse-demodulators.

(4) The goal vector contains no variables.

To prove a theorem with a Z-module reasoning system, the user is required to prepare a set of input data for the vector generator. Various alternatives for preparing input data for an advanced use of Z-module reasoning have been described in [10]. For a naive use of the system, besides deducing some additional equations by means of the linearization algorithm, the user needs only to reformulate the input hypotheses to satisfy the preceding given requirements. The reformulation can be accomplished with the following operations:

(OP1) From an identity equation t=0 and a word w contained in t, obtain an equation $w = w - t$ or obtain an equation $w = w + t$.

(OP2) From $L = R$, obtain an equation $L = R$ and/or an input equation $R = L$.

(OP3) Replace each equation whose righthand side contains a variable that does not occur in the lefthand side by some instances of that equation. Each instance is obtained from the equation by replacing each variable that has no occurrence in the lefthand side of the equation with a ground term.

In a later section, we shall show that one can use a Z-module reasoning system to prove some equality theorems that contain no axioms of ring theory. However, in that case, linearization and the operation OP1 shall not be used in preparing the set of input data for the vector generator, because they implicitly use the ring axioms. More discussion about this subject will be given in a later section.

Since, when the vector generator deduces an ide-paramodulant from a word and an input equation by ide-paramodulation, only the lefthand side of an equation will be considered for unification with a subterm of a word, one may need to obtain more than one input equation for the vector generator from a given equation (OP2). For example, if there exists a hypothesis $xxx = x$, then the user may input both $xxx = x$ and $x = xxx$ to allow ide-paramodulation to consider both sides of the hypothesis equation.

3.4 The Vector Generator Procedure

With the preceding discussion as background, we can outline the vector generator procedure by follows:

With a given a set of input equations and a goal vector, the vector generator deduces a set of vectors by initially focusing on the words contained in the goal vector. From each word on which it focuses, it produces all legal (in terms of some given restriction to be described) vectors by applying ide-paramodulation to that word and each input equation, and then normalizes the result. During the process, new words are extracted for producing more vectors in later steps. Each vector deduced by the vector generator is called an *identity vector.*

Since an identity vector is an ordered polynomial whose monomials are ordered according to the ordering of their words, the indexing mechanism of the vector generator, which gives an ordering for the set of words contained in the goal vector and the vectors generated later, is important. By presenting a part of this indexing mechanism, we can describe an algorithm for designing the vector generator in the following way:

Let T be the set of input equations of the vector generator. Let *index_table* be an array whose smallest index is 1. At the beginning of a vector generation process, the *index_table* is initialized to be null for each index of the table. A variable *focus_index* is initialized to be 1. Then each word contained in the goal vector is stored in the first null entry of the *word_table*. Starting from *focus_index* = 1,

(1) Let $W = word\_table[focus\_index]$;
    If W is null, then exist;
(2) For each W_r in $Reverse\_demo(W)$,
    for each equation $L = R$ in T ,
        for each ide-paramodulant Q of W_r and $L = R$,
            let $V = normalize(Q)$,
            If V is a new vector, then store V;
(3) Increase *focus_index* by 1;
    Go to step (1).

The subroutine $normalize(Q)$ transforms an ide-paramodulant Q into a vector in the following way:

(1) Expand Q to be a polynomial with term-rewrite rules, $x(y + z) \rightarrow xy + xz$, $(y + z)x \rightarrow yx + zx, x(\alpha y) \rightarrow \alpha(xy)$, where α is an integer and x, y, z are terms.

(2) Rewrite each word into a canonical form by using the given set of demodulators.

(3) Collect the monomials with identical words into a single monomial and delete those monomials whose coefficients become 0.

(4) For each word contained in the resulting polynomial, find the entry in the *word_table* in which the word is stored. If found, replace the word by the index of that entry, else store the word into the first null entry of the table, and then replace the word by the index of that entry.

(5) construct a vector by reordering the remaining monomials (now each monomial becomes a product of a coefficient and an index) on the indices contained in it in ascending order.

To guarantee that the vector generator terminates, the user can set a complexity limit for the acceptable words. A word generated by the vector generator whose complexity exceeds the given value will be discarded, and so will a vector that contains a discarded word. However, for some problems, such as Example 3.1 to be given next, such a restriction is not needed; the special feature of the input data guarantees the termination of the vector generator.

3.5 An Example of Vector Generation

Example 3.1 If a (nonassociative) ring satisfies $x(yy) = xyy$ for all x, y, then $2x(yzy) = 2xyzy$ for all x, y, z.

Table 3.1 The input data for proving example 3.1.

1. $x(yy) \Rightarrow xyy$	:demodulator
2. $x(yy) \Leftarrow xyy$	:reverse-demodulator
3. $x(yz) = xyz + xzy - x(zy)$	:input equation
4. $xyz = x(yz) + x(zy) - xzy$	:input equation
Goal vector: $2(a(bcb)) - 2(abcb)$	:a, b, c are constants.

As was shown in Example 2.1, by linearizing $x(yy) - xyy = 0$, we can obtain an additional equation $x(yz) + x(zy) - xyz - xzy = 0$. From this equation, we can obtain two different input equations for the vector generator: Equation 3 and Equation 4 of Table 3.1. Expressions 1 and 2 of Table 3.1 are both obtained from the hypothesis $x(yy) = xyy$, they are used as demodulator and reverse-demodulator, respectively. The following table gives the complete set of words and identity vectors produced by the vector generator. The superscript attached to each word in the table is the index of the word assigned by the indexing mechanism of the Z-module reasoning system. In the actual implementation, only these indices are included in these vectors.

Table 3.2 The set of identity vectors for proving example 3.1

$$1(a(bcb))^1 - 1(a(bc)b)^3 - 1(ab(bc))^4 + 1(a(b(bc)))^5$$
$$1(a(bcb))^1 - 1(a(b(bc)))^5 - 1(a(b(cb)))^6 + 1(a(bbc))^7$$
$$1(abcb)^2 - 1(ab(bc))^4 - 1(ab(cb))^8 + 1(abbc)^9$$
$$1(abcb)^2 - 1(a(bc)b)^3 - 1(a(cb)b)^{10} + 1(acbb)^{11}$$
$$1(a(b(cb)))^6 - 1(ab(cb))^8 - 1(a(cb)b)^{10} + 1(a(cbb))^{12}$$
$$1(a(bbc))^7 - 1(abbc)^9 - 1(acbb)^{11} + 1(a(cbb))^{12}$$

3.6 Completeness of the Vector generator

As was mentioned earlier, the fundamental approach of the Z-module reasoning method is to prove that the goal vector derived from the denial of the theorem is contained in the kernel of a homomorphism related to the denial of the theorem. To do this, the vector generator is used to deduce a set of simple Z-module generators for a Z-submodule of the kernel that is related to the goal vector.

The completeness of the vector generator is concerned with the following problem: given a set H of special hypotheses of a theorem and a goal vector, if the goal vector is contained in the Z-module generated by the entire set of simple Z-module generators determined by H, does the vector generator always deduce a Z-module (i.e. a set of generators for it) that is large enough to contain the goal vector.

Since the concept of simple Z-module generator was introduced for the basic ring theorems, the completeness theorem given here will be applied only to a special class of problems. However, we believe that, with a more general definition of simple Z-module generator, a general version of the completeness theorem can be proved.

Theorem 3.1 Let H be a set of identity equations of a basic ring theorem, and g be the goal vector derived from the denial of the theorem. if g is contained in the Z-module $[G_w]$ generated by a set G_w of simple Z-module generators determined by H on a set $V[X_0]$, where X_0 is a set consisting of all leaf subwords (constants) of g, then there exists a set T of input equations that can be obtained from H with operations OP1, OP2 and/or OP3, given in Section 3.4, such that, with the set T of input data and the goal vector g, the vector generator will deduce a Z-module that contains g.

With Lemma 3.1, this theorem can be proved without essential difficulty. Roughly speaking, since the vector generator is constructed on the basis of lemma 3.1, the key point of an induction proof of the theorem is to show that for any word w_0 of $RW(g, G'_w)$, where G'_w is a finite subset of G_w, such that $g \in [G_{w'}]$ holds, the set of vectors deduced by the vector generator with w_0 and T contains the set of those elements of $RG(g, G'_w)$ that are immediately related to w_0. This point can be proved according to the definition of ide-paramodulation and the given algorithm of vector generator.

4 Integer Gaussian Elimination (IGE)

Once a set S of identity vectors has been produced by the vector generator, the remaining objective of a Z-module reasoning system is to determine whether or not the goal vector derived from the given theorem belongs to the Z-module $[S]$ generated by S, that is, to determine whether or not the goal vector is a linear combination (with integer coefficients only) of a subset of the identity vectors of S. This process is called vector calculation.

The vector calculation is accomplished with two procedures. One is called an *integer Gaussian elimination program* (IGE), the other is called a *proof-procedure*. The IGE program is used to deduce a triangle basis for the Z-module generated by the set of given vectors. The proof-procedure is used to determine if a vector is contained in the Z-module generated by a triangle basis.

Definition (*triangle basis*). A set T of elements of a Z-module M is said to be a triangle basis of M if M can be generated by T and if no two vectors of T have identical leading words. The *leading word* (leading coefficient) of a vector is the word (coefficient) of the first non-zero monomial of that vector.

We have given a pseudo Gaussian elimination algorithm in [9] for deducing a triangle basis (it was called pseudo basis) for a Z-module generated by a set of vectors. Although this algorithm has been used successfully in proving various theorems, it is incomplete. In this section, we will present an integer Gaussian elimination algorithm (IGE) and a completeness theorem about this algorithm. First, we give the following lemma.

Lemma 4.1 Let V_1 and V_2 be two vectors with an identical leading w_1, such that

$$V_1 = a_1 w_1 + \Sigma_{i=2}^n a_i w_i,$$
$$V_2 = b_1 w_1 + \Sigma_{i=2}^n b_i w_i.$$

Let g_0 be the gcd (greatest common divisor) of a_1 and b_1, and let s_0 and t_0 be any integers satisfying $g_0 = s_0 a_1 + t_0 b_1$. Let p_0 be the gcd of s_0 and t_0. Then there exist vectors V_3 and V_4,

$$V_3 = g_0 w_1 + \Sigma_{i=2}^n (s_0 a_i + t_0 b_i) w_i,$$
$$V_4 = \Sigma_{i=2}^n (p_0 d a_i - p_0 c b_i) w_i,$$

where $c = a_1/g_0, d = b_1/g_0$, such that V_3 and V_4 are each an integer linear combinations of V_1 and V_2, and V_1 and V_2 are each integer linear combinations of V_3 and V_4.

According to this lemma, if a set S of vectors contains two vectors both having a leading word w, then we can obtain a set S' from S by replacing these vectors with two new vectors such that $[S]$ is identical to $[S']$ and only one of these new vectors has a leading word w; the other one is either 0, so it can be discarded, or has a leading word whose index is greater than the index of w. In light of this property, we can design an integer Gaussian elimination algorithm that is shown in the following:

Let S be a set of vectors, $\{w_1, w_2, ..., w_n\}$ be the list formed from the entire set of words occurring in S such that

$$index(w_i) < index(w_{i+1}), \text{ for } 1 \le i < n.$$

INTEGER GAUSSIAN ELIMINATION(S)
(0) $i = 0$;
(1) If $i > n$ /* there are total of n words in S */
    then exit with S
    else $i := i + 1$ and $w := w_i$;
(2) Find two vectors V_1 and V_2 with leading words identical to w;
    If no such two vectors found,
        then go to step (1)
        else $S := S - \{V_1, V_2\}$;

(3) Form vectors V_3 and V_4 by the rules described in lemma 4.1;
     If V_4 is zero,
          then $S := S \cup \{V_3\}$,
          else $S := S \cup \{V_3, V_4\}$;
     Go to step (1).

In deducing V_3 and V_4 from V_1 and V_2 by the rules described in lemma 4.1, a subroutine "fullgcd" will be used to calculate the gcd g_0 together with the two integers s_0 and t_0 for the two given integers a_1 and b_1 such that the relation $s_0 a_1 + t_0 b_1 = g_0$ holds. It is well known that there are usually more than one pair of integers s an t such that $sx + ty$ is the gcd of two integers x and y. For example, if $x = 1$ and $y = 1$, then every pair of integers (s, t) that satisfies $s = 1 - t$ will be suffice. We require that, if $a = 1$ and $b = 1$, 'fullgcd" will return values with $s = 0$ and $t = 1$. A "fullgcd" algorithm is given by the following, which calculates the pair of s and t along with calculating the gcd of x and y.

```
FULLGCD(x, y)
  u := 0; v := 1; s := 1; t := 0; k := 0;   /* the initial values */
loop
  If y = 0 then exit with s, t, x;   /* x will be the gcd of the input x and y */
  stemp := u; ttemp := v; xtemp := y;
  k := x/y;
  y := x mod y;
  u := s - k × u;
  v := t - k × v;
  s := stemp; t := ttemp; x := xtemp;
  go to loop.
```

Theorem 4.1 The given integer Gaussian elimination algorithm is complete. That is, given a finite set S of vectors, this algorithm will deduce in finitely many steps a triangle basis for the Z-module $[S]$ generated by S.

With Lemma 4.1, the theorem is easy to prove by induction on the number of words contained in S.

For an efficient implementation of the IGE algorithm, we shall transform the set of input vectors into a matrix. The following example shows how a matrix is formed, and a triangle basis is deduced by an IGE program.

Example 4.1 Let us consider the deduction of a triangle basis for the Z-module generated by the set of identity vectors shown in Table 3.2. The set of vectors forms a matrix: each vector corresponds to a row of the matrix, each word corresponds to a column of the matrix. The column number of a word is just the index of the word, which has been shown as its superscript of the word in Table 3.2. The input and output matrices of the IGE program for this example are shown in the following figures:

$$
\begin{pmatrix}
1 & & -1 & -1 & 1 & & & & & \\
1 & & & & -1 & -1 & 1 & & & \\
& 1 & & -1 & & & & -1 & 1 & \\
& 1 & -1 & & & & & & -1 & 1 \\
& & & & 1 & & -1 & & -1 & & 1 \\
& & & & 1 & & & -1 & & -1 & 1
\end{pmatrix}
$$

$$\Longrightarrow$$

$$
\begin{pmatrix}
1 & & -1 & -1 & 1 & & & & & \\
& 1 & & -1 & & & -1 & 1 & & \\
& & 1 & 1 & -2 & -1 & 1 & & & \\
& & 2 & -2 & -1 & 1 & 1 & -1 & -1 & 1 \\
& & & & 1 & & -1 & & -1 & & 1 \\
& & & & 1 & & & -1 & & -1 & 1
\end{pmatrix}
$$

Table 4.1 The triangle basis deduced for proving example 3.1.

$$1(a(bcb))^1 - 1(a(bc)b)^3 - 1(ab(bc))^4 + 1(a(b(bc)))^5$$
$$1(abcb)^2 - 1(ab(bc))^4 - 1(ab(cb))^8 + 1(abbc)^9$$
$$1(a(bc)b)^3 + 1(ab(bc))^4 - 2(a(b(bc)))^5 - 1(a(b(cb)))^6 + 1(a(bbc))^7$$
$$2(ab(bc))^4 - 2(a(b(bc)))^5 - 1(a(b(cb)))^6 + 1(a(bbc))^7 + 1(ab(cb))^8$$
$$\qquad -1(abbc)^9 - 1(a(cb)b)^{10} + 1(acbb)^{11}$$
$$1(a(b(cb)))^6 - 1(ab(cb))^8 - 1(a(cb)b)^{10} + 1(a(cbb))^{12}$$
$$1(a(bbc))^7 - 1(abbc)^9 - 1(acbb)^{11} + 1(a(cbb))^{12}$$

It is worth to pointing out that, although the design and use of an IGE algorithm or a PGE algorithm [9] is important to the efficiency of the Z-module reasoning, the design and use of the vector generator and the linearization algorithm as well as the indexing mechanism of the vector generator which transforms a symbolic computation problem into an integer matrix calculation problem is essential in our approach. In light of the efficiency of integer matrix calculation, our system allows to produce efficiently more than 10,000 vectors (up to 65,535 words) in a Sun 3/75 for solving those practical problems [9,10] that may be too complex to be handled by those systems that depend solely on symbolic computation.

5 The Proof Procedure of Z-module Reasoning

Lemma 5.1 Let S be a set of non-zero vectors, $\{v_1, ..., v_m\}$, and g be a non-zero vector with a leading monomial $z_g w_g$. If S is a triangle basis for the Z-module $[S]$, then the necessary condition for $g \in [S]$ is that there exists a vector $v_{g'}$ in S such that the leading word of $v_{g'}$ is the word w_g and the leading coefficient z_g of g is properly divisible by the leading coefficient of $v_{g'}$.

On the basic of this lemma, the proof-procedure is easy to be designed. Given a set S of triangle basis of $[S]$ and a vector g,

PROOF-PROCEDURE(S,g):
(1) If $g = 0$ then exit with "proved"
     else let zw be the leading monomial of g;
(2) If there is a vector v in S whose leading word is w
     then let a be the leading word of v
     else exit with "fail";
(3) If there is an integer c, such that $z = ca$
     then $g = g - cv$
     else exit with "fail";
(4) go to step (1)

Theorem 5.1 Let S be a finite set of non-zero vectors, $\{v_1, ..., v_m\}$, and g be a vector. If S is a triangle basis for a Z-module $[S]$, then proof-procedure(S,g) exits with "proved" if only if g belongs to $[S]$.

Example 5.1 For proving the theorem given in example 3.1, we obtain by the vector generator a set S of identity vectors shown in Table 3.1. A triangle basis B for the Z-module $[S]$ deduced by the IGE program has been shown in Table 4.1. Recall that the goal vector of the theorem is $2(a(bcb)) - 2(abcb)$. Denote it be g_0. Then g_0 is proved to be contained in $[S]$ in terms of the triangle basis B by the following steps:

$$g_0 = 2(a(bcb))^1 - 2(abcb)^2$$
$$g_1 = -2(abcb)^2 + 2(a(bc)b)^3 + 2(ab(bc))^4 - 2(a(b(bc)))^5$$
$$g_2 = 2(a(bc)b)^3 - 2(a(b(bc)))^5 - 2(ab(cb))^8 + 2(abbc)^9$$
$$g_3 = -2(ab(bc))^4 + 2(a(b(bc)))^5 + 2(a(b(cb)))^6 - 2(a(bbc))^7$$
$$\quad\quad -2(ab(cb))^8 + 2(abbc)^9$$
$$g_4 = 1(a(b(cb)))^6 - 1(a(bbc))^7 - 1(ab(cb))^8 + 1(abbc)^9 - 1(a(cb)b)^{10} + 1(acbb)^{11}$$
$$g_5 = -1(a(bbc))^7 + 1(abbc)^9 + 1(acbb)^{11} - 1(a(cbb))^{12}$$
$$g_6 = 0$$

6 The Application to General Equality Theory

A preliminary version of a Z-module reasoning system has been designed and used in proving many theorems from ring theory. Some experimental results have been given in [9] and [10]. These experiments show that the Z-module reasoning method is quite efficient in proving a class of theorems from ring theory. The theorems proved by our Z-module reasoning system, that may be difficult for other automated theorem provers, include (1) if $xxx = x$ for all x in an associative ring, then the ring is commutative; (2) if $xxxx = x$ for all x in an associative ring, then the ring is commutative; (3) $(x,x,y)^2 x(x,x,y)^2 = 0$ in right alternative rings (open conjecture); (4) a number of interesting theorems from nonassociative rings, including the Moufang identity mentioned in the beginning of this paper. However, the version of Z-module reasoning described in [9] and [10] is not applicable to those equality problems whose hypotheses contain no axioms of ring theory.

With the new definition of ide-paramodulation given in this paper, the Z-module reasoning method described in this paper can also be used for proving some theorems from general equality theory. We classify the equality theorems into two distinct classes. One is the class of ring equality theorems, the other is a class of non-ring equality theorems. The hypothesis of a ring equality theorem contains a set of basic axioms of ring theory. Whereas, the hypothesis of a non-ring equality theorem contains no basic axioms of ring theory.

The approach for using Z-module reasoning to solve the class of non-ring equality theorems is based on the fact that a non-ring equality theorem can be transformed into a ring equality theorem, which will be called pseudo-ring theorem, by simply adding a set of basic axioms of ring theory to the hypothesis of the non-ring theorem. If a non-ring theorem is true, then the pseudo-ring theorem obtained from it must also be true and so is provable at least theoretically by the Z-module reasoning. The problem is that, if a pseudo-ring theorem has been proved to be true by the Z-module reasoning, does it mean that the non-ring theorem from which the pseudo-ring theorem is obtained must be true? This is the soundness problem of the Z-module reasoning. In the next section, we shall prove that the Z-module reasoning is sound for both classes of equality theorems.

Actually, in proving a non-ring theorem with a Z-module reasoning system, the user does not need to add the basic axioms of rings into the system since they are already built into Z-module reasoning. As was mentioned before, when one prepares the set of input data for the vector generator for proving a ring theorem, one needs only to consider the set of special hypotheses of the given theorem. Now, if the theorem to be proved is a non-ring theorem, then the whole set of hypotheses of the theorem will be the set of special hypotheses to be used for preparing the input data for the vector generator. In addition, one needs to rename the function symbols $+, -, \times$ and 0 of a non-ring theorem to be some symbols different from $+, -, \times$ and 0, since these symbols are reserved for specifying a ring theorem. Only the operations OP2 and OP3 given in Section 3 can be used for preparing input data for the vector generator in proving a non-ring theorem.

Example 6.1 This example concerns to prove an expression is a fixed point in a theory that contains the combinators B and W.

Let T be a theory consisting of two combinators b and w, where b satisfies $\forall x \forall y \forall z (bxyz = x(yz)))$ and w satisfies $\forall x \forall y (wxy = xyy)$. Prove that $b(ww)(bw(bbb))$ is a fixed point combinator of T, i.e.,

$$\forall x(x(b(ww)(bw(bbb))x) = b(ww)(bw(bbb))x).$$

This theorem has been proved by Lusk and Overbeek with ITP [2]. More deep results related to this problem has been produced by Wos and McCune [3]. It is noted, since $wxy \equiv xyy$ when x and y are both replaced by w, there is no complete sets of reductions for the set of equalities given in this example. We use this example to explain how to use Z-module reasoning to prove a non-ring theorem. (However, we wish to point out that it is much easier to prove that a given expression is a fixed point combinator than it is to find such an expression. Indeed, it would be very difficult to use the Z-module system described in this paper for searching for such expressions. For the subject of searching for fixed point combinators, Wos and McCune [11] give an elegant solution.)

The special hypotheses of the theorem consist of the definitions of the combinators b and w. Denoting the result (xy) of applying the combinator x to y by $x * y$, we can produce the following input data for the vector generator:

Table 6.1 The set of input data for proving example 6.1.

$b * x * y * z \Rightarrow x * (y * z)$:demodulator
$w * x * y = x * y * y$
$x * y * y = w * x * y$
Goal Vector:
$a * (b * (w * w) * (b * w * (b * b * b)) * a) - b * (w * w) * (b * w * (b * b * b)) * a$
Complexity Limit: less than 16 symbols in a word (without counting '*').

The following table is taken from the print-out of our Z-module reasoning system. The set of equations together with their proof coefficients (the number contained in '[ ]') ahead of the equations were found by an additional procedure of the system. To check the proof, the user may refer to the index table listed below the set of equations. The lefthand side of each equation is an identity vector deduced by the vector generator from one of the input equations given in Table 6.1 and the given demodulator. To check that the goal vector is 0, one needs only to multiply each equation by its proof-coefficient, and then sum the lefthand sides and the righthand sides of the results, respectively. Multiplying the final result by -1, one obtains an identity equation whose lefthand side is identical to the given goal vector, whose righthand side is 0. In the next section, we shall discuss in detail why this computer proof is indeed a proof of the original theorem.

Table 6.2 The set of input data for proving example 6.1

```
Proved the theorem +1*(1)-1*(2)=0 with the following 7 equations:
[-1]  +1*(1)-1*(3)=0
[+1]  +1*(2)-1*(5)=0
[-1]  +1*(3)-1*(7)=0
[+1]  +1*(5)-1*(10)=0
[-1]  +1*(7)-1*(13)=0
[+1]  +1*(10)-1*(19)=0
[-1]  +1*(13)-1*(19)=0
Total words 64, Total equations 121, Basis size 64,
Vector Generation 9.0s, IGE and Proof 0.6s
```

```
Index_table:
 1: a*(b*(w*w)*(b*w*(b*b*b))*a
 2: b*(w*w)*(b*w*(b*b*b))*a
 3: a*(w*w*(w*(b*(b*a))))
 5: w*w*(w*(b*(b*a)))
 7: a*(w*(w*(b*(b*a)))*(w*(b*(b*a))))
10: w*(w*(b*(b*a)))*(w*(b*(b*a)))
13: a*(w*(b*(b*a))*(w*(b*(b*a)))*(w*(b*(b*a))))
19: w*(b*(b*a))*(w*(b*(b*a)))*(w*(b*(b*a)))
```

The reader may note that, in this example, the hypothesis $b*x*y*z = x*(y*z)$ is used only as a demodulator, but no corresponding reverse-demodulator of it is included in the input data of the vector generator. We shall point out that such a strategy is not complete, that is, there exist counterexamples in which an absence of the corresponding reverse-demodulators may cause proof failure. The Z-module reasoning proof for the x-cube ring theorem is such an example. A discussion on this subject has been included in [9].

The similar problem exists for the input equations of the vector generator. For example, in proving by Z-module reasoning the typical testing theorem, that a group is commutative if $xx = e$ for all x in that group, if $xx = e$ is included in the input data of the vector generator, but $e = xx$ is not, then there will be no proof of the theorem that can be obtained by the Z-module reasoning. The problem is that certain Z-module generator that are needed for a proof may only be produced from those words that contain e by replacing e with some instances of xx. Note that, since the Z-module reasoning require that each variable contained in an input equation must have an occurrence in the lefthand side of the equation, $e = xx$ is not allowed to be included in the input data. In this case, one can construct, with the operator OP3 stated in Section 3.3, some instances of such equations, and use these instances as a part of the input data of the vector generator. For this testing problem, with a complexity limit 6, our Z-module reasoning system proved the theorem when we included the following set of instances of $e = xx$: $\{e = a*a, e = b*b, e = a*b*a*b, e = b*a*b*a\}$. Of course, such sort of instances can be generated by the help of a computer program.

7   The Soundness of Z-module Reasoning

We have presented an algorithm for each component of a Z-module reasoning system and discussed the application of Z-module reasoning for proving theorems from two distinct classes of equality theorems. In this section, we shall discuss issues related to the soundness of the system. The first theorem states that Z-module reasoning is sound for the class of ring theorems.

Theorem 7.1 Let S be a set of clauses that contains a set A of basic axioms of ring theory, a set H of equalities and an inequality $t_1 \neq t_2$. Let T be a set of equations obtained from H with OP1, OP2, and/or OP3. If a proof is obtained by the Z-module reasoning with a set of input data consisting of T and a goal vector $t_1 - t_2$, then S must be E-unsatisfiable.

The theorem can be proved by using theorems 4.1 and 5.1, and by noticing that each indentity vector generated by the vector generator, or by the IGE program, must be an identity derivable from S.

Theorem 7.2 Let S be a set of clauses that contains a set H of equalities and an inequality $t_1 \neq t_2$, but S contains no function symbols $+, -, \times$ and 0. Let T be a set of input equations obtained from H by using operations OP2 and/or OP3. If a proof is obtained by the Z-module reasoning with a set of input data consisting of T and a goal vector $t_1 - t_2$, then S must be E-unsatisfiable.

Proof (outline). We shall call an equation a simple equality if both sides of the equation are words. Since only OP2 and OP3 are used in obtaining T, each element of T must be a simple equality that can be deduced from S. We shall call a vector a simple vector if it has the form $w_1 - w_2$ or $-w_1 + w_2$ such that both w_1 and w_2 are words. Since S contains no function symbols $+, -$ and $\times$, according to the definition of ide-paramodulation and the definition of the procedure normalization, each vector deduced from a word and a simple equation must be a simple vector. In addition, corresponding to each simple vector $w_1 - w_2$, or $-w_1 - w_2$ deduced by the vector generator, there must exist a simple equality $w_1 = w_2$ such that $E \cup H \rightarrow w_1 = w_2$ (E is the set of equality axioms). The similar conclusions can be made for each vector contained in the triangle basis B deduced by the IGE program from a set of such simple vectors. Actually, each simple vector $w_i - w_j$ of B corresponds to a term-rewrite rule $w_i \Rightarrow w_j$ that can be deduced from the set of input hypotheses. Thus one can show that each iteration of proof-procedure corresponds to a term-rewriting process that rewrites the leading word of the input goal vector or a residual goal vector of it into a word whose index is greater than the index of the leading word. Therefore, the success of the proof-procedure means that the lefthand side and the righthand side of the given inequality $t_1 \neq t_2$ can be rewritten into an identical word by a set of term-rewrite rules derivable from $E \cup H$. ∎

8 Summary

We presented an algorithm for each of the main components of a Z-module reasoning system and discussed issues concerning the completeness of these algorithms. We also described the application of Z-module reasoning to proving theorems from general equality theory, including the so called non-ring equality theorems. We proved that the Z-module reasoning is sound both for the class of ring equality theorems and for the class of non-ring theorems. The completeness of the entire Z-module reasoning is not addressed in this paper.

I wish to thank my colleagues from ANL-MCS (Mathematics and Computer Science Division, Argonne National Laboratory) for many fruitful discussions and, in particular, Larry Wos for his advices and valuable suggestions in this research. Most of this work was done when the author was visiting ANL-MCS, and the work was supported in part by the Applied Mathematical Sciences subprogram of the office of Energy Research, U.S. Department of Energy, under contract W-31-109-Eng-38.

References

[1] W. W. Bledsoe, "Non-resolution theorem proving," Artificial Intelligence (9), pp. 1-35, 1977.

[2] E. Lusk and R. Overbeek, "The automated reasoning system ITP", ANL-84-27, Argonne National Laboratory (April, 1984).

[3] W. McCune and L. Wos, "A case study in automated theorem proving: searching for sages in combinatory logic," Journal of Automated Reasoning, 3(1) pp. 91-107 (1987).

[4] E. Kleinfeld, "Right alternative rings," Proc. Amer. Math. Soc., Vol. 4, pp. 939-944 (1953).

[5] R. L. Stevens, "Some experiments in nonassociative ring theory with an automated theorem prover," Journal of Automated Reasoning, 3(2) PP. 211-223 (1987).

[6] A. Thedy, "Right alternative rings", Journal of Algebra, Vol. 1, pp. 1-43 (1975).

[7] M. E. Stickel, " A Unification Algorithm for associative commutative functions," J. ACM, Vol. 28, No. 3, pp 423-4343.

[8] T. C. Wang and W. W. Bledsoe, "Hierarchical Deduction", Journal of Automated Reasoning, Vol. 3, No. 1, pp. 35-71 (1987).

[9] T. C. Wang, "Case studies of Z-module reasoning: proving benchmark theorems from ring theory," Journal of Automated Reasoning, Vol. 3, No. 4, pp. 437-451 (1987).

[10] T. C. Wang and R. Stevens, "Solving open problems in right alternative rings with Z-module reasoning," Submitted for publication.

[11] L. Wos , and W. W. McCune,, "Searching for Fixed Point Combinators by Using Automated Theorem Proving: A Preliminary Report," ANL-88-10 (1988).

[12] K. A. Zhevlakov, et al., "Rings that are nearly associative," Academic Press, New York, 1982.

Learning and Applying Generalised Solutions using Higher Order Resolution *

M. R. Donat
Microsoft
16011 NE 36th Way
Redmond WA 98073, USA.

L. A. Wallen[†]
Dept. of Computer Sciences
University of Texas at Austin
Austin, TX 78712, USA.

Abstract

The performance of problem solvers and theorem provers can be improved by means of mechanisms that enable the application of old solutions to new problems. One such (learning) mechanism consists of generalising an old solution to obtain a specification of the most general problem that it solves. The generalised solution can then be applied in the solution of new problems wherever instances of the general problem it solves can be identified.

We present a method based on higher order unification and resolution for the generalisation of solutions and the flexible application of such generalisations in the solution of new problems. Our use of higher order unification renders the generalisations useful in the solution of both sub- and superproblems of the original problem. The flexibility thus gained is controlled by means of filter expressions that restrict the unifiers considered.

In this way we show how the bulk of the problem solving, generalisation and application tasks of a (learning) problem solving system can be performed by the one algorithm. This work generalises ad hoc techniques developed in the field of Explanation Based Learning and presents the results in a formal setting.

Key words and phrases. Resolution, higher order unification, Explanation Based Learning, generalisation.

---

*This work was supported in part by SERC/Alvey grants GR/D/44874 and GR/D/44270 whilst both authors were with the Dept. of Artificial Intelligence, University of Edinburgh, Scotland.

†Lincoln Wallen is supported by a grant from the British Petroleum Venture Research Unit.

1   Introduction

The performance of problem solvers and theorem provers can be improved by mechanisms that enable the application of old solutions to new problems. One such learning mechanism consists of generalising an old solution to obtain a specification of the most general problem that it solves. The generalised solution can then be applied in the solution of new problems wherever instances of the general problem can be identified.

Such techniques have been investigated in the field of Explanation Based Learning. (See for example [7]). Typically the methods developed require one algorithm to generalise solutions, one to determine the applicability of the generalisation in the context of a new problem, and one to actually apply the generalisation. These algorithms are distinct again from the basic problem solving algorithms of the system. Moreover, they are frequently ad hoc and their properties left implicit in the definitions. The work of Kedar-Cabelli and McCarty [4] is a notable exception.

In this paper we present a generalisation method based on higher order unification and resolution. The problem solving operators of the domain are represented as higher order Horn clauses in the style of Paulson [8], and Miller and Nadathur [5]. Solving a problem then corresponds to the construction of a higher order resolution proof of an atomic sentence that represents the problem. Within this framework we develop a generalisation mechanism based on higher order unification. The result of generalising a given solution (proof) is simply another problem solving operator (Horn clause) for use on future problems.

We demonstrate how higher order unification may be utilised to identify those problems in which previously generalised solutions are of use. The new problem is simply unified with the conclusion of the Horn clause representing the generalisation. The flexibility of higher order unification (as opposed to first order) ensures that if the new problem contains an instance of the old problem as a subproblem, the generalisation will be applied appropriately. Moreover, and perhaps surprisingly, the unification operation also renders the generalisation applicable if the new problem is an instance of a subproblem of the old problem. That is, if on the way to solving the old problem, a subproblem similar in structure to the new problem was encountered and solved, unifying the generalisation with the new problem will effect its solution. The details of the old solution that are irrelevant to the new problem are conveniently ignored.

This flexibility comes at a price. Used in the way outlined above, the unification operation typically produces multiple unifiers. To control the generation of unifiers the notion of a *filter expression* is introduced which rejects certain types of unifier based on their structure.

Consequently, we show that the bulk of the problem solving, generalisation and application tasks of an adaptable problem solving system may be performed by the one algorithm: higher order unification. This work both generalises the ad hoc techniques proposed for Explanation Based Learning, and sets the results in a formal framework.

The methods have been implemented in a version of Paulson's Isabelle system [8] — though the λ-Prolog system [5] would have sufficed equally well — and applied to the domain of symbolic integration [6]: a standard domain in the Explanation Based Learning field.

The structure of this paper is as follows. In section 2 we present our method of representing problem solving operators by means of higher order Horn clauses. We use Mitchell's symbolic integration domain for concreteness. We also discuss the unification-based generalisation mechanism. In section 3 we investigate the use of higher order unification in the application of generalisations to new problems. Examples are presented followed by a more detailed discussion of the application process. In section 4 we note some of the combinatorial problems inherent in our use of unification and develop the notion of a "filter expression" to control the unifiers produced. The implementation of the generalisation algorithm and filter expressions is briefly described in section 5. A more detailed description can be found in [2]. In section 6 we discuss the limitations of our approach and in section 7 discuss related work. Finally, in section 8 we discuss an extension of our method.

2   Proof and Generalisation

2.1   The Formalism

The basic framework that we use is due to Paulson [8] and is implemented in his Isabelle system. For a more detailed study of Isabelle see [8] and [1]. Our extension is the generalisation construction algorithm and the filter expression modifications.

The basic element is the rule. A *rule* consists of a list of premises and a conclusion and is typically illustrated with its premise formulae above a line and its conclusion formula below it (fig 1). Variables are denoted by capital letters, small letters represent constant terms.

Proving a theorem is the task of deriving the rule corresponding to the theorem from a trivial rule, using the other rules of the logic. (A *trivial rule* is a rule in which the single premise is identical to the conclusion.) A new rule, R_2, is derived by *composing* two existing rules, Q and R_1, in the following manner:

- The conclusion of Q is unified with a selected premise p of R_1 producing unifier U. If the formulae do not unify, composition is not possible via this premise.

$$\frac{F(B) \qquad A = B}{F(A)} \qquad \frac{\text{constant}(A) \qquad \text{not}(A = -1)}{\int X^A d(X) = \frac{X^{A+1}}{A+1}}$$

substitution rule   power rule

$$\frac{}{\cos^2 A = 1 - \sin^2 A} \qquad \frac{\text{constant}(C)}{\int C * A \, d(X) = C * \int A \, d(X)}$$

cos identity   scalar extraction

Figure 1: Inference rules

- The new rule R_2 is constructed from R_1 by replacing the selected premise p by the premises of Q.

- Lastly, the substitutions of U are applied to all variables in R_2.

Rule Q is said to be *applied to* R_1. This composition of rules is equivalent to Horn clause resolution. The premises and conclusion of a rule correspond to the antecedents and consequent of a clause respectively.

In this paper the notation:

$$r_p \Rightarrow \begin{array}{c} R_1 \\ R_2 \end{array} \qquad \text{or} \qquad R_1 \overset{r_p}{\Rightarrow} R_2$$

is used to show the derivation of rule R_2 from R_1 by applying rule r to premise p of R_1. The first notation is used in Figure 2, the second is used later in the paper.

Rule applications are illustrated in the proof shown in Figure 2.

2.2 Symbolic Integration

The symbolic integration logic used in the examples is motivated by the set of operators used in LEX2 [6] and consists of a number of mathematical identities and a substitution rule for rewriting an expression on the basis of these identities. A solution to a problem is a sequence of rewrite operations that reduce an integral to a *simple expression*: one which does not contain any integrals.

The "theorems" of the domain are simple expressions or rules whose premises are simple expressions and whose conclusion contains an integral. In this way proving a

$$\text{phase 1} \begin{cases} & \dfrac{\int 3*x^2 dx}{\int 3*x^2 dx} \\\\ \text{substitution}_{(1)} \Rightarrow & \dfrac{B \quad \int 3*x^2 dx = B}{\int 3*x^2 dx} \\\\ \begin{array}{l}\text{scalar} \\ \text{extraction}\end{array}{}_{(2)} \Rightarrow & \dfrac{3 * \int x^2 dx \quad \text{constant}(3)}{\int 3*x^2 dx} \end{cases}$$

$$\text{phase 2} \begin{cases} \text{substitution}_{(1)} \Rightarrow & \dfrac{3*B \quad \int x^2 dx = B \quad \text{constant}(3)}{\int 3*x^2 dx} \\\\ \begin{array}{l}\text{power} \\ \text{rule}\end{array}{}_{(2)} \Rightarrow & \dfrac{3 * \frac{x^3}{3} \quad \text{constant}(2) \quad \text{not}(2 = -1) \quad \text{constant}(3)}{\int 3*x^2 dx} \end{cases}$$

Figure 2: A proof of the theorem $\int 3*x^2 dx$

theorem corresponds to solving the integral in the conclusion. In the terminology of Explanation Based Learning [7], simple expressions are considered to be *operational*: the proof — and hence the generalisations obtained from them — need not proceed beyond this level of detail. Constants of integration are omitted for simplicity.

Our particular formulation of the integration domain may appear a little non-standard. Though we have access to a higher order language we have not represented an integral as a binding operator. We have chosen a first order representation so as to remain as close as possible to the original papers [6].

The solution of $\int x^2 dx$ requires the use of the power rule operator. This operator is encoded as the rule shown in Figure 1 and is based on the identity $\int X^A d(X) = \frac{X^{A+1}}{A+1}$ with the conditions that A is a constant other than -1.

In this way each operator is described as a rewrite operation with conditions. Each rewrite rule addresses the appropriate subexpression of the current problem through the use of the substitution rule (fig 1).

The proof of Figure 2 is performed in two rewrite phases. The first, scalar extraction, moves the constant outside the integration sign. Once this has been done, the integral can be solved using the power rule. The scalar extraction and power rules can only be used after the appropriate subexpressions, $\int 3*x^2 dx$ and $\int x^2 dx$, have been isolated by applying the substitution rule first.

The structure of a proof can be seen more clearly using a tree representation

$$3 * \frac{x^3}{3} \quad \frac{\dfrac{\text{constant}(2) \quad \text{not}(2 = -1)}{\int x^2 dx = \frac{x^3}{3}}}{3 * \int x^2 dx} \quad \frac{\text{constant}(3)}{\int 3*x^2 dx = 3 * \int x^2 dx}$$
$$\int 3*x^2 dx$$

Figure 3: Tree representation of the proof in Figure 2

$$F(A) \quad \text{unified with} \quad \int b*c \; da$$

F	=	$\lambda Y.Y$	$\lambda Y. \int Y da$	$\lambda Y. \int b*c \; dY$
A	=	$\int b*c \; da$	$b*c$	a
		(a)	(b)	(c)

F	=	$\lambda Y. \int b*c \; da$	$\lambda Y. \int Y *c \; da$	$\lambda Y. \int b*Y da$
A	=	A	b	c
		(d)	(e)	(f)

Figure 4: Multiple unifiers

such as the one shown in Figure 3. Each horizontal line separates the conclusion and premises of a rule.

During the proof of a particular theorem it is useful to think of the conclusion of the rule as the goal and its premises as the current subgoals. From this point of view each rule in the derivation of Figure 2 can be thought of as a separate proof state.

One of the complications of higher order unification as compared with its first order counterpart is that typically the set of most general unifiers is not a singleton. When F(A) is unified with the formula $\int b*c \; da$, an independent unifier is produced for each of the possible function substitutions for F (see fig 4). Note that it is possible for F to ignore its parameter (fig 4d). In this paper, such a substitution is referred to as a *trivial substitution*.

$$\cfrac{\text{constant}(A_4) \quad \text{not}(A_4 = -1)}{\int X_4^{A_4} d(X_4) = \frac{X_4^{A_4+1}}{A_4+1}}$$

$$\|$$

$$\cfrac{F_3(B_3) \qquad\qquad A_3 = B_3 \qquad\qquad \cfrac{\text{constant}(C_2)}{\int C_2 * A_2 \ d(X_2) = C_2 * \int A_2 \ d(X_2)}}{\cfrac{\cfrac{F_3(A_3)}{\|}}{F_1(B_1)} \qquad\qquad\qquad\qquad \cfrac{}{A_1 = B_1}}$$

$$\cfrac{}{F_1(A_1)}$$

$$\|$$

$$\frac{\int 3*x^2 dx}{\int 3*x^2 dx}$$

(a)

$$G_5 = \cfrac{F_3(\frac{X_4^{A_4+1}}{A_4+1}) \quad \text{constant}(A_4) \quad \text{not}(A_4 = -1) \quad \text{constant}(C_2)}{F_1(\int C_2 * A_2 \ d(X))}$$

$$\text{restriction:} F_1(C_2 * \int A_2 d(X_2)) = F_3(\int X_4^{A_4} d(X_4))$$

(b)

Figure 5: Generalisation of the derivation of Figure 2

2.3 Generalisation

In Figure 2 the trivial rule representing the target problem is transformed into a rule representing a solution of the problem (or theorem) by a particular sequence of rule applications. The generalisation task is to construct a single general rule that corresponds to this sequence. The properties of the rule composition operation (resolution and unification) makes this task quite simple.

The composition of the rules from Figure 2 can be represented as the exploded tree shown in Figure 5a. We wish to determine a *generalisation* G that can be applied to produce the same results (fig 5b).

Due to the associativity of unification, the order in which compositions are performed is irrelevant. The desired generalisation can be obtained by composing the rules as shown in Figure 5 with the exception that the trivial rule is not included. This effectively excludes the detail of the particular "training example" from the generalisation. An application of G_5 will produce the same results as applying each of the rules of the original training example in the same order.

The subscripts in G_5 indicate variables standardised apart. The restriction is a constraint on later unification and, in the implementation, is generated by Isabelle as a method of delaying the evaluation of the sequence of resolvents that corresponds to the rule of G_5. This delayed evaluation mechanism is used to represent possibly infinite sets of unifiers. G_5 therefore represents a set of generalisations. Multiple generalisations are discussed in more detail in [2].

The equivalence of a rule and a proof state provides a simple and natural method for constructing generalisations from a given solution. Thus, the process of constructing an explanation based generalisation (a generalised solution) can be performed purely by unification. This has also been demonstrated by Kedar-Cabelli & McCarty's PROLOG-EBG program for first order Horn clause logics [4].

3   Applying Generalisations

In this section we demonstrate the flexibility that higher order unification provides during the application of a generalisation such as G_5. Since a generalisation naturally takes the form of a rule, it can be applied as such. PROLOG-EBG could be modified easily to be able to apply its first order generalisations to new problems. However, higher order unification allows a generalisation to be used in a much more flexible way.

When a generalisation is applied, only the pertinent rewrite phases are performed. If a rewrite encoded in the generalisation does not apply to the current subgoal, it is discarded automatically. Also, these rewrite phases can be applied to any pertinent subexpression. Targeting subexpressions and omitting different portions of the generalisation automatically is something that a first order system like PROLOG-EBG cannot do for free.

The examples in Figure 6 illustrate the various contexts in which a generalisation can be applied.   In Figure 6a, the generalisation G_5 is used in the proof of $\int 2{*}x^3 dx$. This type of application can be performed by PROLOG-EBG in much the same way.

The use of G_5 in the problem of Figure 6b is outside the scope of the PROLOG-EBG program. The function variables of the higher order generalisation allow G_5 to be applied to problems (*superproblems*) that contain variations of the original integral as a subexpression (fig 6b).

In section 2 we touched on the concept of a trivial substitution. The trivial substitutions of the variables F_1 and F_3 in G_5 correspond to omissions of rewriting phases. By ignoring the parameter, a trivial substitution isolates the expression from being rewritten. Because of this phenomenon, G_5 can be applied to problems in which only a portion of the proof structure encoded in the generalisation is

(a) $$\frac{\int 2*x^3 dx}{\int 2*x^3 dx} \overset{G_5}{\Rightarrow} \frac{2*\frac{x^4}{4} \quad \text{constant(3)} \quad \text{not}(3 = -1) \quad \text{constant(2)}}{\int 2*x^3 dx}$$

(b) $$\frac{(\int 3*x^2 dx) + 1}{(\int 3*x^2 dx) + 1} \overset{G_5}{\Rightarrow} \frac{3*\frac{x^3}{3} + 1 \quad \text{constant(2)} \quad \text{not}(2 = -1) \quad \text{constant(3)}}{(\int 3*x^2 dx) + 1}$$

(c) $$\frac{\int x^2 dx}{\int x^2 dx} \overset{G_5}{\Rightarrow} \frac{\frac{x^3}{3} \quad \text{constant(2)} \quad \text{not}(2 = -1) \quad \text{constant(C)}}{\int x^2 dx}$$

(d) $$\frac{\int 3 * \sin x \, dx}{\int 3 * \sin x \, dx} \overset{G_5}{\Rightarrow} \frac{3 * \int \sin x \, dx \quad \text{constant(A)} \quad \text{not}(A = -1) \quad \text{constant(3)}}{\int 3 * \sin x \, dx}$$

(e) $$\frac{\int \sin x \, dx}{\int \sin x \, dx} \overset{G_5}{\Rightarrow} \frac{\int \sin x \, dx \quad \text{constant(A)} \quad \text{not}(A = -1) \quad \text{constant(C)}}{\int \sin x \, dx}$$

Figure 6: Applying the generalisation G_5

pertinent (*subproblems*). Two such applications are illustrated in Figures 6c and 6d.

This ability to select the appropriate subexpression, or to isolate the whole expression from being rewritten, provides flexibility in the application of generalisations. Applying a generalisation is similar to going through a list of rewrite rules, applying those that are applicable at that point. But it should be noted that each application of a generalisation is a single resolution operation involving a single unification. No other algorithms are involved. This allows the bulk of the proof, generalisation and application processes of an adaptive problem solving system to be incorporated into the one algorithm.

The contribution of higher order unification to application is most significant because previous systems that exhibited similar application capabilities required special representations for the generalisation and an additional algorithm for extracting the applicable components (*eg.*, STRIPS [3]). Higher order unification supports a more compact representation of generalisations and eliminates the need for a separate application algorithm.

4 Filter Expressions

From the examples examined in the previous section it is evident that generalisation in a higher order system provides a great deal of flexibility. Unfortunately, there are problems associated with controlling this. In this section we describe some of these problems and introduce the concept of a filter expression to solve them.

4.1 Some problems

The first problem is that of *useless subgoals*. When the application of a generalisation proceeds by omitting certain proof steps, the subgoals based on these proof steps are referred to as *useless subgoals*. Subgoals such as constant(C) (fig 6c), and constant(A) and not(A = −1) (fig 6d), are useless subgoals. These subgoals typically contain several uninstantiated variables. Although they cause no problem in actually completing the proof, they are a source of inefficiency.

The second problem is also a question of efficiency, but much more serious. Because unification with a generalisation can utilise a subsequence of the original proof, it can also utilise the empty sequence! In other words, any generalisation similar to that shown in Figure 5 will unify with any term (fig 6e).

Such use of a generalisation will be combinatorially explosive because the generalisation always applies. This is the problem of *uncontrolled unification*. An application of a generalisation in this case is referred to as a *useless application*.

A theorem prover with no mechanism to control this behaviour would be grossly inefficient.

4.2 A solution

The uncontrolled unification problem is a result of trivial substitutions of function variables. While these trivial substitutions give higher order generalisations their flexibility, they need to be regulated to prevent combinatorial explosion.

Since an instance of uncontrolled unification corresponds to an omission of all the proof steps in the training example, it would be beneficial for unification, in this circumstance, to fail rather than succeed and contribute to the explosion of the search space. Also, in some applications it may be desirable to constrain which proof steps of the training example should not be omitted. For example, although G_5 applies to $\int 3 * \sin x \, dx$ (fig 6d), the result of this application is not as useful as we might expect from a successful unification. It may be desirable for the application of G_5 to succeed only if the power rule is pertinent as this step reduces the number of integrals in the expression.

The desired effects can be achieved by attaching *filter expressions* to the generalisations. Such an expression provides greater control over rule composition by describing which trivial function substitutions are allowed in an acceptable unification of premise and conclusion. The description is given in the form of a Boolean expression (involving conjunction, disjunction and negation) with variables as atomic propositions. The truth value of the expression is used to filter the undesirable unifiers from the sequence produced by the unification algorithm. A literal in the expression is false if there is a trivial function substitution for its corresponding variable, otherwise it is true. A unifier is removed from the sequence if the filter expression is false in the context of that unifier. Unification fails if the resulting sequence is empty.

In the case of generalisation G_5, Figure 5, uncontrolled unification can be suppressed using the filter $(F_3$ or $F_1)$. Uncontrolled unification occurs when both of the variables F_3 and F_1 have trivial substitutions. In these circumstances the filter expression is false and the unifier is removed as desired.

To see the effects of the filter $(F_3$ or $F_1)$ on a typical application of G_5, consider the results of applying this generalisation to the problem:

$$\frac{1 + \int 2 * x^3 dx}{1 + \int 2 * x^3 dx}$$

The sequence of resolvents produced with no filter is:

(a)
$$\frac{1 + 2 * \frac{x^4}{4} \quad \text{constant}(3) \quad \text{not}(3 = -1) \quad \text{constant}(2)}{1 + \int 2*x^3 dx}$$

(b)
$$\frac{1 + 2 * \int x^3 dx \quad \text{constant}(A) \quad \text{not}(A = -1) \quad \text{constant}(2)}{1 + \int 2*x^3 dx}$$

(c)
$$\frac{1 + \int 2*x^3 dx \quad \text{constant}(A) \quad \text{not}(A = -1) \quad \text{constant}(C)}{1 + \int 2*x^3 dx}$$

In the sequence above, (a) is a solution of the superproblem, (b) introduces useless subgoals and (c) is an example of uncontrolled unification, a useless application. The filter (F_3 or F_1) removes only resolvent (c), eliminating the possibility of uncontrolled unification. The filter (F_3 and F_1) removes resolvents (b) and (c), eliminating the useless subgoals case. A more appropriate method of addressing the useless subgoal problem is proposed in section 8.

The filter (F_3) forces the application to fail when the power rule is not pertinent. This filter is true only when the substitution for F_3 is not a trivial substitution, i.e. whenever the power rule is used. With this filter G_5 unifies with subgoals (a) and (b) but not with (c) below.

$$\int 3*x^2 dx \quad \int x^2 dx \quad \int \sin x \, dx$$

$$\text{(a)} \qquad \text{(b)} \qquad \text{(c)}$$

By examining the rules used to construct a generalisation, an appropriate filter expression can be determined, given constraints on which rewrite steps are mandatory for a successful application. In this way a generalisation can be fine tuned to decrease the explosive behaviour of higher order generalisations.

5 Implementation

5.1 The Generalisation Algorithm

The input for a generalisation procedure is a training example. A training example is a starting rule, or head, and a sequence of rule compositions. Each rule records how it was derived. This description is referred to as the *derivation sequence*. The original Isabelle system did not support the representation of derivation sequences.

In section 2 we described the basic method of constructing a generalisation. The actual implementation of the generalisation algorithm processes a list of composition operations rather than a tree, simply because Isabelle does composition operations sequentially. Given a rule, the head of the derivation sequence is ignored. Instead of redoing the entire derivation, the first rule applied to this head is used as the head for the derivation of the generalisation. Once the first rule of the derivation has been chosen, the derivation sequence is used to apply all of the rule compositions of the training example to construct its generalisation. As a side effect of each rule containing its derivation sequence, each generalisation contains its own derivation sequence. If necessary, this information can be used to examine the details encoded in the generalisation.

5.2   Filter Expressions

To implement filter expressions, the sequence of unifiers produced by the higher order unification algorithm is passed through a filtering procedure that evaluates the truth value of the appropriate filter expression in the context of each unifier. If the filter is true, the unifier is allowed to pass. Otherwise, the unifier is removed from the sequence. An empty sequence indicates a failure to unify.

There are three operators for constructing filter expressions; AND, OR and NOT. AND and OR take a list of filter expressions as input and return the truth value of the appropriate conjunct or disjunct, respectively. NOT negates the value of the given filter expression. For example, filters (F_3 or F_1) and (F_3 and F_1) would be represented as OR($[F_1, F_3]$) and AND($[F_1, F_3]$) respectively. The power rule filter for generalisation G_5 is represented as (F_3). NOT is included to allow the specification of more complex filters such as those requiring implication. By definition, AND($[]$) is true, OR($[]$) is false.

6   Discussion

6.1   Undirected Unification

Although filters provide some control during the application of higher order generalisations, they are unable to alleviate the problem of *undirected unification*. While uncontrolled unification refers to heavy use of trivial substitutions, undirected unification refers to non-trivial substitutions that produce rewrites in random subexpressions but do not necessarily correspond to proof steps. A solution to this problem is proposed in section 8.

As an example, consider the derivation in Figure 7. Its generalisation will be referred to as G_7. The rewrite rules used to construct G_5 are quite specific in terms of the context in which they can apply. Because of this specificity the number of unifiers is quite small. In contrast, however, the trivial exponent rule used in Figure 7 can unify with any subexpression, thus increasing the number of unifiers

$$\frac{\text{constant}(A_4) \quad \text{not}(A_4 = -1)}{\int X_4^{A_4} d(X_4) = \frac{X_4^{A_4+1}}{A_4+1}}$$

$$\|$$

$$\frac{F_3(B_3) \quad A_3 = B_3}{F_3(A_3)} \qquad A_2 = (A_2)^1$$

$$\| \qquad\qquad \|$$

$$\frac{F_1(B_1)}{F_1(A_1)} \qquad A_1 = B_1$$

$$\|$$

$$\frac{\int x \, dx}{\int x \, dx}$$

Figure 7: Derivation using the trivial exponent rule

$$\frac{\int x \, dx}{\int x \, dx} \quad \overset{G_7}{\Rightarrow} \quad \left\{ \begin{array}{c} \dfrac{(\int x \, dx)^1 \quad \text{constant}(A_8) \quad \text{not}(A_8 = -1)}{\int x \, dx} \\[2em] \dfrac{\frac{x^2}{2} \quad \text{constant}(2) \quad \text{not}(2 = -1)}{\int x \, dx} \\[2em] \dfrac{\int x d(x^1) \quad \text{constant}(A_8) \quad \text{not}(A_8 = -1)}{\int x \, dx} \end{array} \right.$$

Figure 8: Applications of G_7

in each rewrite operation. Figure 8 shows a portion of the sequence of resolvents when applying G_7 to $\int x \, dx$.

As well as being inconvenient, the more serious aspect of the problem arises during the useless application of a generalisation suffering from this problem. In this case, if the generalisation has a filter, determining that no appropriate unifier exists requires the examination of each unifier in the sequence and may take a great deal of computation.

6.2   Reordering Rewritings

It is important to note that the sequence of rewritings encoded in a generalisation cannot be reordered by higher order unification. This is demonstrated in Figure 9.

A generalisation of the solution to problem (a) (fig 9) is not applicable to problem (c), and vice versa. This is because the order of operations is important in each problem. The subexpressions of problem (b) can be solved in either order. Hence, problem (b) is simple enough to be solved by the generalisations of either (a) or (c).

6.3   Other Domains

The success of the generalisations presented in this paper is due to the particular form of the substitution rule (fig 1) and its interaction with the trivial substitution phenomenon of higher order unification. This underlies the flexibility of sub- and superproblem application. The domain of symbolic integration lends itself to this flexibility because of the recursive nature of subexpressions.

The work presented in this paper is based on the assumption that the domain can be formalised as a system of rewrite rules. We consider this a weak assumption. The application of these ideas to domains not of this form has not yet been explored.

7   Related Work

7.1   The EBG Technique

An *EBG* procedure is one that constructs a definition of a "goal concept" in the terms of a predetermined language [7] b generalising an example of that concept. The particular language used is specified via the "operationality" criterion. Such a definition is therefore called an *operational definition*.

A generalisation in our sense is also an operational definition of a goal concept. The conclusion of the generalisation is the concept being defined. The premises correspond to the terms defining the concept. Both PROLOG-EBG and our generalisation system are *EBG* procedures.

7.2   PROLOG-EBG

Kedar-Cabelli and McCarty [4] present a Prolog meta-interpreter, PROLOG-EBG, that constructs generalisations from training examples. Indeed, whilst we have used Paulson's Isabelle as an implementation environment, a more natural setting

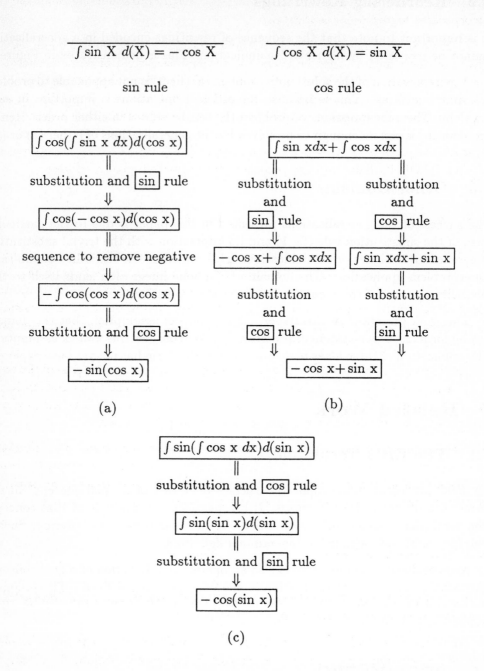

Figure 9: Unification cannot reorder a proof sequence

is perhaps Miller and Nadathur's λ-Prolog [5]: a logic programming language that incorporates higher order unification.

Our work extends that of Kedar-Cabelli and McCarty in several ways. Firstly, since we use a higher order language for the representation of domain operators, we can represent domains naturally that involve certain inherently higher order constructs such as quantifiers (and integrals for that matter). Secondly we show how a higher order language can be used to represent generalisations in such a way as to permit their flexible application via higher order unification to new problems. Thirdly, the proof in [2] provides a sound theoretical justification for the correctness of both PROLOG-EBG and the generalisation algorithm presented here.

7.3   STRIPS

STRIPS [3] was a component of a robot system that learnt how to push boxes from room to room. STRIPS comprised three main components: a problem solver to generate a solution, a generaliser to make a generalisation or *macro operator* (MACROP) of the solution, and a MACROP editor that applied the generalisation to new problems. The editor extracted portions of the macro operations that were pertinent to the problem solver's current goal.

A STRIPS macro operator typically represented over-generalisations of the training example (such as showing that it was possible to have the robot end up in two different rooms at the same time!). In our formalism, the generalisations produced are as valid as the rules in the logic from which they were derived.

Because of the representation of the macro operators, an additional algorithm was required in STRIPS — the MACROP editor — to apply an operator to new problems. Such an algorithm is not necessary here because the editing tasks are performed automatically by higher order unification.

8   Future Work

In this section we suggest an extension of our current techniques to provide a solution to the problem of useless subgoals.

Applying a generalisation using higher order unification results in the application of a particular subsequence of operators in the generalisation. To obtain the subsequence itself requires the use of an extraction procedure. This extraction can be achieved by analysing the instantiated proof tree corresponding to the application of a generalisation to a subgoal. We call this process *minimal proof extraction*. This is an important process in applications that need to examine which particular subsequence of proof steps an application of a generalisation corresponds to: e.g. a

$$F_3\left(\frac{X_4^{A_4+1}}{A_4+1}\right) \quad \frac{\text{constant}(A_4) \quad \text{not}(A_4=-1)}{\int X_4^{A_4} d(X_4)) = \frac{X_4^{A_4+1}}{A_4+1}}$$

$$\frac{F_3(\int X_4^{A_4} d(X_4))}{\| \quad}$$

$$\text{constant}(C_2)$$
$$\int C_2 * A_2 \ d(X)$$
$$= C_2 * \int A_2 \ d(X_2)$$

$$\frac{F_1(C_2 * \int A_2 \ d(X_2))}{F_1(\int C_2 * A_2 \ d(X))}$$

Figure 10: Tree representation of G_5

$$\frac{x^3}{3} \quad \frac{\text{constant}(2) \quad \text{not}(2=-1)}{\int x^2 dx = \frac{x^3}{3}}$$

$$\frac{\text{constant}(C_2)}{\int C_2 * A_2 \ d(X) = C_2 * \int A_2 \ d(X_2)}$$

$$\frac{\int x^2 dx}{\int x^2 dx}$$

Figure 11: Instantiation of G_5 applied to $\int x^2 dx$

system that controls the movements of a robot.

Figure 10 shows the tree representation for the generalisation G_5. Consider the instantiation of the proof tree corresponding to the application of the generalisation to $\int x^2 dx$ (fig 11). The substitution for F_1 in the main conclusion is trivial. Consequently, the first premise of the base rule, $\int x^2 dx$, is identical to the conclusion. Also, the variables corresponding to the scalar extraction portion of the proof remain uninstantiated.

The proof in Figure 11 can be reduced by comparing the syntactic structure of the conclusion with each of the rule's premises. The objective of a proof is to eliminate premises. If one of the premises of a rule is identical to the rule's conclusion then a proof that excludes this rule must be at least as good as the full proof. If one of the premises of a rule is syntactically identical to its conclusion, then the rule can be removed. To remove the rule from the proof, the proof tree above the identical premise can be applied to the premise originally unified with the conclusion of the removed rule. This is illustrated in Figure 12.

By removing rules the minimal proof corresponding to instantiation of the generalisation can be efficiently constructed. This is the extracted minimal proof. This procedure is also a solution to the useless subgoal problem as any useless subgoals are removed along with the rule that generates them (fig 12).

$$\cfrac{\frac{x^3}{3} \quad \cfrac{constant(2) \quad not(2 = -1)}{\int x^2 dx = \frac{x^3}{3}}}{\int x^2 dx}$$

Figure 12: Condensed proof of Figure 11

Since each rule contains a record of its derivation, minimal proof extraction can be applied recursively to generalisations that contain other generalisations in their derivation sequence.

9   Conclusions

We have presented a framework for the construction of generalisations and their application to new problems based on higher order unification and resolution. This has been implemented as a component of Paulson's Isabelle system. The advantages of using higher order generalisations and the problems they introduce were analysed. We have presented or proposed solutions to all of these problems.

We have demonstrated that our use of higher order unification allows generalisations to be applied to sub- and superproblems. This corresponds to the automatic discovery of the appropriate portion of the generalisation that is applicable. A generalisation method based on higher order unification allows much more information from the training example to be utilised in other problems than is possible in a first order system.

We have shown that higher order unification provides a single algorithm for the tasks of proof, generalisation and application and can aid in the construction of a concise, reliable system. Earlier systems such as STRIPS required three separate algorithms for these tasks and could not give the same guarantees of reliability.

The implementation of filter expressions increases control over the application of generalisations. They eliminate the problem of uncontrolled unification. Minimal proof extraction can be used to remove useless subgoals. This proposed extension provides an ability to recognise a smallest subproof corresponding to the application of a generalisation.

Our work extends generalisation beyond PROLOG-EBG to higher order Horn clause logics and explores the technology required for applying these generalisations. We use unification to provide a theoretical foundation for the use of generalisations.

Acknowledgements

We would like to thank Alan Bundy, Roberto Desimone and Larry Paulson for advice during the research. The Edinburgh University Mathematical Reasoning Group and the two referees provided helpful comments on earlier versions of this paper.

References

[1] deGroote, P. *How I spent my time in Cambridge with Isabelle*, Technical Report RR 87-1, Université Catholique de Louvain, January 1987.

[2] Donat, M.R. *Construction and Application of Generalisations Using Higher Order Unification*. Master's thesis, Department of Artificial Intelligence, University of Edinburgh, 1987.

[3] Fikes, R.E., Hart, P.E., and Nilsson, N.J. Learning and executing generalized robot plans. *Artificial Intelligence*, 3:251–288, 1972.

[4] Kedar-Cabelli, S. and McCarty, L.T. Explanation-based generalization as resolution theorem proving. In P. Langley, editor, *Proceedings of the 4th International Machine Learning Workshop*, pages 383–389, Morgan Kaufmann, 1987.

[5] Miller, D.A. and Nadathur, G. "Higher-order logic programming," Proceedings of the 3rd Int. Logic Programming Conference, London, June 1986, 448-462.

[6] Mitchell, T.M., Utgoff, P. E., and Banerji, R. Learning by experimentation: acquiring and modifying problem-solving heuristics. In *Machine Learning*, pages 163–190, Tioga Press, 1983.

[7] Mitchell, T.M., Keller, R.M., and Kedar-Cabelli, S.T. Explanation-based generalization: a unifying view. *Machine Learning*, 1(1):47–80, 1986.

[8] Paulson, L. Natural deduction as higher order resolution. *Journal of Logic Programming*, 3:237–258, 1986.

Specifying Theorem Provers in a Higher-Order Logic Programming Language

Amy Felty and Dale Miller
Department of Computer and Information Science
University of Pennsylvania
Philadelphia, PA   19104–6389   USA

Abstract

Since logic programming systems directly implement search and unification and since these operations are essential for the implementation of most theorem provers, logic programming languages should make ideal implementation languages for theorem provers. We shall argue that this is indeed the case if the logic programming language is extended in several ways. We present an extended logic programming language where first-order terms are replaced with simply-typed λ-terms, higher-order unification replaces first-order unification, and implication and universal quantification are allowed in queries and the bodies of clauses. This language naturally specifies inference rules for various proof systems. The primitive search operations required to search for proofs generally have very simple implementations using the logical connectives of this extended logic programming language. Higher-order unification, which provides sophisticated pattern matching on formulas and proofs, can be used to determine when and at what instance an inference rule can be employed in the search for a proof. Tactics and tacticals, which provide a framework for high-level control over search, can also be directly implemented in this extended language. The theorem provers presented in this paper have been implemented in the higher-order logic programming language λProlog.

1   Introduction

Logic programming languages have many characteristics that indicate that they should serve as good implementation languages for theorem provers. First, at the foundation of computation in logic programming is search, and search is also fundamental to theorem proving. The process of discovering a proof involves traversing an often very large and complex search space in some controlled manner. Second, unification, which is required in most theorem provers, is immediately and elegantly accessible in most logic programming systems. Third, if in fact theorem provers can be written directly in logic programming, the simple and declarative reading of such logic programs should help in understanding formal properties, such as completeness and soundness, of the resulting implementation. This potential advantage is very important for theorem prover implementations not only because such formal properties are important considerations but also because such implementations are often very complex and hard to understand.

Traditional logic programming languages such as Prolog [SS86] are not sufficient for providing natural implementations of several aspects of theorem provers. One deficiency, as argued in [MN87b], is that first-order terms are quite inadequate for a clean representation of formulas. For instance, first-order terms provide no mechanism for representing variable abstraction required for quantification in first-order formulas. Of course, quantification could be specially encoded. For example, in Prolog, we can represent abstractions in formulas by representing bound variables as either constants or logical variables. The formula $\forall x \exists y\, P(x,y)$ could be written as the first-order term `forall(x,exists(y,p(x,y)))`. This kind of encoding is very unnatural and spoils the elegance which logic programming often offers. We shall mention other similar shortcomings of conventional Prolog systems.

In this paper, we use a higher-order logic programming language based on *higher-order hereditary Harrop formulas* [MNS87]. This language replaces first-order terms with simply typed λ-terms. The abstractions built into λ-terms can be used to naturally represent quantification. Our extended language also permits queries and the bodies of clauses to be both implications and universally quantified. We shall show how such queries are, in fact, necessary for implementing various kinds of theorem provers. The programs that we present in this paper have been tested using λProlog which is a partial implementation of higher-order hereditary Harrop formulas [MN87a]. Various aspects of this language have been discussed in [MN86a, MN86b, MN87b, Nad86].

Our main claim in this paper is that such a language is a very suitable environment for implementing theorem provers. We will show that search and unification accommodate the tasks involved in theorem proving very naturally. Most of our theorem provers will have a clean declarative reading which provides them with implementation independent semantics and makes establishing their formal properties more tractable.

In the next section, we will briefly present higher-order hereditary Harrop formulas. In Section 3, we discuss how to directly specify various inference rules as such formulas. In Sections 4 and 5 we discuss the implementation of tactic style theorem provers which allow greater control in searching for proofs and provide means for user participation in the theorem proving process. In Section 4 we show how to implement high-level tacticals, and in Section 5 we illustrate the tactic specification of inference rules for a particular proof system. Finally, in Section 6, we discuss related work.

2   Extended Logic Programs

Higher-order hereditary Harrop formulas extend positive Horn clause in essentially two ways. The first extension permits richer logical expressions in both queries (goals) and the bodies of program clauses. In particular, this extension provides for implications, disjunctions, and universally and existentially quantified formulas, as well as conjunction. The addition of disjunctions and existential quantifiers into the bodies of clauses does not depart much from the usual presentation of Horn clauses since such extended clauses are classically equivalent to Horn clauses. The addition of implications and universal quantifiers, however, makes a significant departure. The second extension to Horn clauses makes this language *higher-order* in the sense that it is possible to quantify over predicate and function symbols. For a complete realization of this kind of extension, several other features must be added. In order to instantiate predicate and function variables with terms, first-order terms are replaced by more expressive simply typed λ-terms. The application of λ-terms is handled by λ-conversion, while the unification of λ-terms is handled by higher-order unification.

There are four major components to our extended logic programming language: types, λ-terms, definite clauses, and goal formulas. Types and terms are essentially those of the simple theory of types [Chu40]. We assume that a certain set of non-functional (primitive) types is provided. This set must contain the type symbol o which will denote the type of logic programming propositions: other primitive types are supplied by the programmer. Function types of all orders are also permitted: if α and β are types then so is $\alpha \rightarrow \beta$. The arrow type constructor associates to the right: read $\alpha \rightarrow \beta \rightarrow \gamma$ as $\alpha \rightarrow (\beta \rightarrow \gamma)$. A function symbol whose target type is o will also be considered a predicate symbol. No abstractions or quantifications are permitted in types (hence the adjective "simple").

Simply typed λ-terms are built in the usual fashion. Through the writing of programs, the programmer specifies (explicitly or implicitly) typed constants and variables. λ-terms can then be built up using constants, variables, applications, and abstractions in the usual way. Logical connectives and quantifiers are introduced into these λ-terms by introducing suitable constants; in particular, the constants $\wedge, \vee, \supset$ are all assumed to have type $o \rightarrow o \rightarrow o$, and the constants Π and Σ are given type $(\alpha \rightarrow o) \rightarrow o$ for each type replacing the "type variable" α. (Negation is not used in this programming language.) The expressions $\Pi\lambda x \ A$ and $\Sigma\lambda x \ A$ are abbreviated to be $\forall x \ A$ and $\exists x \ A$, respectively. A λ-term which is of type o is called a *proposition*.

Equality between λ-terms is taken to mean $\beta\eta$-convertible. We shall assume that the reader is familiar with the basic facts about λ-conversion. It suffices to say here that this equality check is decidable, although if the terms being compared are large, this check can be very expensive: β-reduction can greatly increase the size of λ-terms.

A proposition in λ-normal form whose head is not a logical constant will be called an *atomic* formula. In this section, A denotes a syntactic variable for atomic formulas.

We now define two new classes of propositions, called *goal formulas* and *definite clauses*. Let G be a syntactic variable for goal formulas and let D be a syntactic variable for definite clauses. These two classes are defined by the following mutual recursion.

$$G := A \mid G_1 \vee G_2 \mid G_1 \wedge G_2 \mid D \supset G \mid \exists v \ G \mid \forall c \ G$$

$$D := A \mid G \supset A \mid \forall v \ D$$

Here, the universal quantifier $\forall c \ G$ in goal formulas must be over a constant c in G and the existential quantification $\exists v \ G$ must be over a variable v in G. Similarly, to form the formula $\forall v \ D$, v is a variable in D. There is one final restriction: if an atomic formula is a definite clause, it must have a constant as its head. The heads of atomic goal formulas may be either variable or constant. A *logic program* or just simply a *program* is a finite set, generally written as $\mathcal{P}$, of closed definite formulas.

Several abstract properties of this logic programming language are presented in [MNS87]. It is stated there that although substitutions for predicates in the underlying impredicative and unramified logic can be very complex, in this setting, substitutions for predicates can be determined through unifications. Furthermore, substitution terms can be restricted to be those λ-terms whose embedded logical connectives (if any) satisfy the constraints placed on goal formulas above.

Another property known for this logic programming language is that a sound and complete (with respect to intuitionistic logic) *non-deterministic* interpreter can be implemented by employing the following six *search operations*. Here, the interpreter is attempting to determine if the goal formula G follows from the program $\mathcal{P}$. The substitution instances used by GENERIC and BACKCHAIN are those described above.

AND   If G is $G_1 \wedge G_2$ then try to show that both G_1 and G_2 follow from $\mathcal{P}$.

OR   If G is $G_1 \vee G_2$ then try to show that either G_1 or G_2 follows from $\mathcal{P}$.

AUGMENT   If G is $D \supset G'$ then add D to the current program and try to show G'.

GENERIC   If G is $\forall x\, G'$ then pick a new parameter c and try to show $[x/c]G'$.

INSTANCE   If G is $\exists x\, G'$ then pick some closed λ-term t and try to show $[x/t]G'$.

BACKCHAIN   If G is atomic, we must now consider the current program. If there is a universal instance of a definite clause which is equal to G then we are done. If there is a definite clause with a universal instance of the form $G' \supset G$ then try to show G' from $\mathcal{P}$. If neither case holds then G does not follow from $\mathcal{P}$.

In order to implement such an interpreter, it is important to make choices which are left unspecified in the high-level description above. There are, of course, many ways to make such choices. We will assume for the purposes of this paper that choices similar to those routinely used in Prolog are employed. In particular, we are following the conventions established in the λProlog system which implements most of the language we are describing (see [MN86a] and [MN87a]).

The order in which conjuncts and disjuncts are attempted and the order for backchaining over definite clauses is determined exactly as in conventional Prolog systems: conjuncts and disjuncts are attempted in the order they are presented. Definite clauses are backchained over in the order they are listed in $\mathcal{P}$ using a depth-first search paradigm to handle failures.

The non-determinism in the INSTANCE operation is extreme. Generally when an existential goal is attempted, there is very little information available as to what closed λ-term should be inserted. Instead, the Prolog implementation technique of instantiating the existential quantifier with a logical (free) variable which is later "filled in" using unification is employed. A similar use of logical variables is made in implementing BACKCHAIN: universal instances are made using new logical variables.

The addition of logical variables in our setting, however, forces the following extensions to conventional Prolog implementations. First, higher-order unification becomes necessary since these logical variables can occur inside λ-terms. Also the equality of terms is not a simple syntactic check but a more complex check of $\beta\eta$-conversion. Since higher-order unification is not in general decidable and since most general unifiers do not necessarily exist when unifiers do exist, unification can contribute to the search aspects of the full interpreter. λProlog addresses this by implementing a depth-first version of the unification search procedure described in [Hue75]. For more information on how logic programming behaves with such a unification procedure, see [MN86a, Nad86]. The higher-order unification problems we shall encounter in this paper are all rather simple: it is easy to see, for example, that all such problems are decidable.

The presence of logical variables in an implementation also requires that GENERIC be implemented slightly differently than is described above. In particular, if the goal $\forall x\, G'$

contains logical variables, the new parameter c must not appear in the terms eventually instantiated for the logical variables which appear in G' or in the current program. Without this check, logical variables would not be a sound implementation technique.

Since much of this paper is concerned with how to implement theorem provers using the class of hereditary Harrop formulas presented above, we shall need to present many such formulas. We will make such presentations by using the syntax adopted by the λProlog system, which itself borrows from conventional Prolog systems.

Variables are represented by tokens with an upper case initial letter and constants are represented by tokens with a lower case initial letter. Function application is represented by juxtaposing two terms of suitable types. Application associates to the left, except when a constant is declared to be infix and then normal infix conventions are adopted. λ-abstraction is represented using backslash as an infix symbol: a term of the form $\lambda x\ T$ is written as `X\T`. Terms are most accurately thought of as being representatives of $\beta\eta$-conversion equivalence classes of terms. For example, the terms `X\(f X)`, `Y\(f Y)`, `(F\Y\(F Y) f)`, and `f` all represent the same class of terms.

The symbols `,` and `;` represent $\wedge$ and $\vee$ respectively, and `,` binds tighter than `;`. The symbol `:-` denotes "implied-by" while `=>` denotes the converse "implies." The first symbol is used to write the top-level connective of definite clauses: the clause $G \supset A$ is written `A :- G`. Implications in goals and the bodies of clauses are written using `=>`. Free variables in a definite clause are assumed to be universally quantified, while free variables in a goal are assumed to be existentially quantified. Universal and existential quantification within goals and definite clauses are written using the constants `pi` and `sigma` in conjunction with a λ-abstraction.

Below is an example of a (first-order) program using this syntax.

```
sterile Y :- pi X\ (bug X => (in X Y => dead X)).
dead X :- heated Y, in X Y, bug X.
heated j.
```

The goal (`sterile j`), for example, follows from these clauses.

In order to base a logic programming language on the simply typed λ-calculus, all constants and variables must be assigned a type. In λProlog, types are assigned either explicitly by user declarations or by automatically inferring them from their use in programs. Explicit typings are made by adding to program clauses declarations such as:

```
type    sterile  jar -> o.
type    in       insect -> jar -> o.
```

where `jar` and `insect` are primitive types. Notice that from this declaration, the types of the variables and other constants in the example clauses above can easily be inferred.

λProlog permits a degree of polymorphism by allowing type declarations to contain type variables (written as capital letters). For example, `pi` is given the polymorphic typing `(A -> o) -> o`. It is also convenient to be able to build new "primitive" types from other types. This is done using type constructors. In this paper, we will need to have only one such type constructor, `list`. For example, (`list jar`) would be the type of lists all of whose entries are of type `jar`. Lists are represented by the following standard construction: `[]` represents an empty list of polymorphic type (`list A`), and if `X` is of type `A` and `L` is of type (`list A`) then `[X|L]` represents a list of type (`list A`) whose first element is `X` and whose tail is `L`. Complex expressions such as `[X|[Y|[]]]` are abbreviated as simply `[X,Y]`.

3 Specifying Inference Rules

In this section we briefly outline how to use definite clauses to specify inference rules in two kinds of proof systems. Although we only consider theorem provers for first-order logic, the techniques described in this section are much more general (see Section 6).

Since we wish to implement a logic within a logic, we will find it convenient to refer to the logic programming language as the *metalogic* and the logic being specified as the *object logic*. An object logic's syntax can be represented in much the same way as the metalogic's syntax is represented. To represent a first-order logic, we introduce two primitive types: `bool` for the object-level boolean and `i` for first-order individuals. Given these types, we introduce the following typed constants.

```
type   and      bool -> bool -> bool.
type   or       bool -> bool -> bool.
type   imp      bool -> bool -> bool.
type   neg      bool -> bool.
type   forall   (i -> bool) -> bool.
type   exists   (i -> bool) -> bool.
```

It is easy to identify closed λ-terms of type `bool` as first-order formulas and of type `i` as first-order terms. For example, the λ-term

```
(forall X\ (exists Y\ ((p X Y) imp (q (f X Y)))))
```

represents the first-order formula $\forall x \exists y (P(x,y) \supset Q(f(x,y)))$.

We shall outline how to specify both sequential and natural deduction inference rules for a first-order logic. To define the sequential proof system we introduce a new infix constant `-->` of type (`list bool`) `-> bool -> sequent`; that is, a sequent contains a list of formulas as its antecedent and a single formula as its succedent (much as in the LJ sequent system in [Gen35]). We also want to retain proofs as they are built, so we shall introduce another primitive type `proof_object` which will be the type of proofs.

The basic relation between a sequent and its proofs will be represented as a binary relation on the metalevel by the constant `proof` of type `sequent -> proof_object -> o`. The inference rules of sequential calculus can be considered as simple declarative facts about the `proof` relation. As we shall see, all these declarative facts are expressible as definite clauses.

Consider the $\wedge$-R inference rule which introduces a conjunction on the right side of the sequent.

$$\frac{\Gamma \longrightarrow A \qquad \Gamma \longrightarrow B}{\Gamma \longrightarrow A \wedge B} \wedge\text{-R}$$

The declarative reading of this inference rule is captured by the following definite clause.

```
proof (Gamma --> (A and B)) (and_r P1 P2) :- proof (Gamma --> A) P1,
                                             proof (Gamma --> B) P2.
```

This clause may be read as: if P1 is a proof of (Gamma --> A) and P2 is a proof of (Gamma --> B), then (and_r P1 P2) is a proof of (Gamma --> (A and B)). The rule can also be viewed as defining the constant and_r: it is a function from two proofs (the premises of the ∧-R rule) to a new proof (its conclusion). Its logic program type is proof_object -> proof_object -> proof_object.

Operationally, this rule could be employed to establish a proof-goal: using the BACK-CHAIN search command, first unify the sequent and proof in the head of this clause with the sequent and proof in the query. If there is a match, use the AND search operation to verify the two new proof-goals in the body of this clause. The unification here is essentially first-order.

Next we consider the two inference rules for proving disjunctions.

$$\frac{\Gamma \longrightarrow A}{\Gamma \longrightarrow A \vee B} \text{ V-R} \qquad\qquad\qquad \frac{\Gamma \longrightarrow B}{\Gamma \longrightarrow A \vee B} \text{ V-R}$$

These rules have a very natural rendering as the following definite clause.

```
proof (Gamma --> (A or B)) (or_r P) :- proof (Gamma --> A) P;
                                       proof (Gamma --> B) P.
```

Declaratively, this clause specifies the meaning of a proof of a disjunction. For (or_r P) to be a proof of (Gamma --> (A or B)), P must be a proof of either (Gamma --> A) or (Gamma --> B). Operationally, this clause would cause an OR search operation to be used to determine which of the proof-goals in the body should succeed.

Introductions of logical constants into the antecedent of a sequent can be achieved similarly. The main difference here is that the antecedent is a list instead of a single formula. Consider the following implication introduction rule.

$$\frac{\Gamma \longrightarrow A \qquad B,\Gamma \longrightarrow C}{A \supset B,\Gamma \longrightarrow C} \supset \text{-L}$$

This could be specified as the following definite clause.

```
proof ([(A imp B) | Gamma] --> C) (imp_l P1 P2) :- proof (Gamma --> A) P1,
                                                   proof ([B|Gamma] --> C) P2.
```

All propositional rules for Gentzen sequential systems can be very naturally understood as combining a first-order unification step with possibly an AND or an OR search operation. The structural rules of contraction, thinning, and interchange could be specified by simply manipulating lists of formulas. For example, the following clauses specify these three structural rules.

```
proof ([C|Gamma] -> A) (contract P) :- proof ([C,C|Gamma] -> A) P.
proof ([C|Gamma] -> A) (thin P) :- proof (Gamma -> A) P.
proof (Gamma1 -> A) (interchange P) :- append S1 [B,C|S2] Gamma1,
                                        append S1 [C,B|S2] Gamma2,
                                        proof (Gamma2 -> A) P.
```

Here, append is the standard Prolog procedure for appending lists.

We now look at specifying quantifier introduction rules. Here, the operational reading of definite clauses will use the INSTANCE and GENERIC search operations and higher-order unification. Consider the following ∃-R inference rule:

$$\frac{\Gamma \longrightarrow [x/t]B}{\Gamma \longrightarrow \exists x\ B}\ \exists\text{-R}$$

which can be written as the following definite clause.

```
proof (Gamma --> (exists A)) (exists_r P) :- sigma T\ (proof (Gamma --> (A T)) P).
```

The existential formula of the conclusion of this rule is written (exists A) where the logical variable A has functional type i -> bool. Since A is an abstraction over individuals, (A T) represents the formula that is obtained by substituting T for the bound variable in A. Declaratively, this clause reads: if there exists a term T (of type i) such that P is a proof of (Gamma --> (A T)), then (exists_r P) is a proof of (Gamma --> (exists A)). Operationally, we rely on higher-order unification to instantiate the logical variable A. The existential instance (A T) is obtained via the interpreter's operations of λ-application and normalization. Of course, the implementation of INSTANCE will choose a logical variable with which to instantiate T. By making T a logical variable, we do not need to commit to a specific term for the substitution. It will later be assigned a value through unification if there is a value which results in a proof.

The following consideration of the ∀-R inference figure raises a slight challenge to our specification of inference rules.

$$\frac{\Gamma \longrightarrow [x/y]B}{\Gamma \longrightarrow \forall x\ B}\ \forall\text{-R}$$

There is the additional proviso that y is not free in Γ or $\forall x\ B$. Although our programming language does not contain a check for "not free in" it is still possible to specify this inference rule. This proviso is handled by using a universal quantifier at the metalevel.

```
proof (Gamma --> (forall A)) (forall_r P) :- pi Y\ (proof (Gamma --> (A Y)) (P Y)).
```

Again A has functional type. In this case, so does P, and the type of forall_r is (i -> proof_object) -> proof_object. Declaratively, this clause reads: if we have a function P that maps arbitrary terms Y to proofs (P Y) of the sequent (Gamma --> (A Y)), then (forall_r P) is a proof of (Gamma --> (forall A)). In order to capture the proviso on y it is necessary to introduce a λ-abstraction over type i into proof objects.

Operationally, the GENERIC search operation is used to insert a new parameter of type i into the sequent. Since that parameter will not be permitted to appear in Gamma, A, or P, the proviso will be satisfied.

The following simple definite clause specifies initial sequents, that is, a sequent whose antecedent contains one formula which is also its succedent.

```
proof ([A] --> A) (initial A).
```

Here the constant initial is of type bool -> proof_object. It represents one way in which formulas get placed inside a proof.

We next briefly consider specifying inference rules in a natural deduction setting (see [Gen35] or [Pra65]). Here, the basic proof relation is between proofs and formulas (instead of sequents). Hence, for these examples, we assume that proof is of the type bool -> proof_object -> o. Several of the introduction rules for this system resemble rules that apply to succedents in the sequential system just considered. Those that correspond to the example inference rules given in the previous section are as follows:

$$\frac{A \qquad B}{A \wedge B} \wedge\text{-I} \qquad\qquad \frac{A}{A \vee B} \vee\text{-I} \qquad\qquad \frac{B}{A \vee B} \vee\text{-I}$$

$$\frac{[x/t]B}{\exists x\ B} \exists\text{-I} \qquad\qquad\qquad \frac{[x/y]B}{\forall x\ B} \forall\text{-I}$$

The ∀-I rule also has the proviso that y cannot appear in $\forall x\ B$ or in any assumptions on which that formula depends. These inference rules can be specified naturally as the following definite clauses.

```
proof (A and B) (and_i P1 P2) :- proof A P1, proof B P2.
proof (A or B) (or_i P) :- proof A P; proof B P.
proof (exists A) (exists_i P) :- sigma T\ (proof (A T) P).
proof (forall A) (forall_i P) :- pi Y\ (proof (A Y) (P Y)).
```

In natural deduction, unlike sequential systems, we have the additional task of specifying the operation of discharging assumptions. Consider the following implication introduction rule.

$$\frac{\begin{array}{c}(A)\\ B\end{array}}{A \supset B} \supset\text{-I}$$

This rule can very naturally be specified using the definite clause:

```
proof (A imp B) (imp_i P) :- pi PA\ ((proof A PA) => (proof B (P PA))).
```

This clause represents the fact that if P is a "proof function" which maps an arbitrary proof of A, say PA, to a proof of B, namely (P PA), then (imp_i P) is a proof of (A imp B). Here, the proof of an implication is represented by a function from proofs to proofs. The constant imp_i has the type (proof_object -> proof_object) -> proof_object. Notice that while sequential proofs only contain abstractions of type i, natural deduction proofs contain abstractions of both types i and proof_object.

Operationally, the AUGMENT search operation plays a role in representing the discharge of assumptions. In this case, to solve the subgoal (pi PA\ ((proof A PA) => (proof B (P PA)))), the GENERIC operation is used to choose a new object, say pa, to play the role of a proof of the formula A. The AUGMENT goal is used to add this assumption about A and pa, that is (proof A pa), to the current set of program clauses. This clause is then available to use in the search for a proof of B. The proof of B will most likely contain instances of the proof of A (the term pa). The function P is then the result of abstracting out of this proof of B this generic proof object.

There are several aspects of both sequential and natural deduction proof systems which the above discussion does not cover. We elaborate a bit further on two such aspects: the representation of proof objects and controlling search in the resulting specification.

The proof objects built in the previous examples serve only to show how proofs might be built and to illustrate the differences between the two styles of proof systems. Proof terms could contain more information. For example, it might be desirable to have two or-introduction rules in both proof systems. The two proof building constants, say or_r1 and or_r2 for the sequential system, would indicate whether the left or right disjunct had been proved. Similarly, in the introduction of existential quantifiers, it might be sensible to store inside the proof the actual substitution term used. In that case, the exists_r would be given the type i -> proof_object -> proof_object.

Depending on the later use made of proofs, it might be desirable for proof objects to contain less information than we have specified. For example, it might be desirable for a single sequential proof to be a proof of many different sequents, that is, the proof terms should be polymorphic. In that case, it might be desirable for the initial proof term to not store a formula within the proof. Instead, initial could have the simpler type proof_object. Of course, proof objects do not need to be built at all. The predicate proof could be replaced with similar predicates, say provable or true, of type sequent -> o or bool -> o.

Another aspect of these theorem provers not yet considered is control. Assume that we have a complete set of definite clauses which specify the inference rules needed to implement some sequent proof system. These definite clauses could be used, for example, to do both proof checking and theorem proving. To do proof checking, assume that we are given a sequent (Gamma --> A) and a proof term P. If the goal (proof (Gamma --> A) P) succeeds, then P is a valid proof of (Gamma --> A). In this example, the proof in the initial proof-goal is closed, and this causes all subsequent proofs in proof-goals to also be closed. Since the top-level constant of a proof term completely determines the unique definite clause which can be used in backchaining, there is little problem of controlling proof checking. Even the simple minded depth-first discipline of λProlog will work.

Such definite clauses could also be used to do theorem proving. Here, we start with a proof object which is just a logical variable which we wish to have instantiated. Since the proof object is a variable, multiple definite clauses could be applied to any one sequent. In particular, the structural rules could always be applied: as written above, they are much too non-deterministic to be useful in this setting. The same control problem is true for elimination rules in the natural deduction setting (see Section 5). It is possible to not implement thinning directly if the definition of initial sequents is extended to be sequents whose succedent is contained in its antecedent. Interchange does not need to be implemented directly if all rules introducing logical constants into the antecedent can operate on any formula in the antecedent instead of just the first formula. Contraction, of course, cannot be so simply removed. Implementing contraction is the great challenge to automating theorem provers. It is possible to build systems of definite clauses which can provide complete theorem provers under depth-first processing of backtracking. Generally, the proof system must be modified somewhat and careful controlling of contraction must be observed. For example, we have implemented a variant of Gentzen's LK sequent system [Gen35] so that it is a complete theorem prover for first-order classical logic [Fel87].

In the next two sections, we describe a different approach to specifying a set of inference rules so that it is easier to be explicit about controlling search.

4   A Logic Programming Implementation of Tacticals

Tactic style theorem provers were first built in the early LCF systems and have been adopted as a central mechanism in such notable theorem proving systems as Edinburgh LCF [GMW79], Nuprl [Con86], and Isabelle [Pau87]. Primitive tactics generally implement inference rules while compound tactics are built from these using a compact but powerful set of *tacticals*. Tacticals provide the basic control over search. Tactics and tacticals have proved valuable for several reasons. They promote modular design and provide flexibility in controlling the search for proofs. They also allow for blending automatic and interactive theorem proving techniques in one environment. This environment can also be grown incrementally.

We shall argue in this section and the next that logic programming provides a very suitable environment for implementing both tactics and tacticals. Generally tactics and tacticals have been implemented in the functional programming language ML. Here we shall show how they can be implemented in λProlog. This implementation is very natural and extends the usual meaning of tacticals by permitting them to have access to logical variables and all six search operations. A comparison between the ML and λProlog implementations is contained in Section 6.

In this section, we assume that tactics are primitive, and show how to implement the higher-level tacticals. In the next section, we show how to implement individual tactics for the proof systems considered in the previous section. Our presentation in these two sections is cursory: for more details, the reader is referred to [Fel87].

We introduce the primitive type `goalexp` to denote *goal expressions*. In the next section, we will define primitive goals which encode such propositions as "this sequent has this proof" or "this formula is provable." For this section, we wish to think more abstractly of goals: they simply denote judgments which could succeed or fail. We define the following goal constructors used to build compound forms of such judgments.

```
type  truegoal     goalexp.
type  andgoal      goalexp -> goalexp -> goalexp.
type  orgoal       goalexp -> goalexp -> goalexp.
type  allgoal      (A -> goalexp) -> goalexp.
type  existsgoal   (A -> goalexp) -> goalexp.
type  impgoal      A -> goalexp -> goalexp.
```

Here, truegoal represents the trivially satisfied goal, andgoal corresponds to the AND search operation, orgoal to OR, allgoal to GENERIC, existsgoal to INSTANCE, and impgoal to AUGMENT. Notice that allgoal, existsgoal, and impgoal are polymorphic. We will use allgoal and existsgoal in the next section with A instantiated to types i and proof_object.

The meaning of a tactic will be a *relation* between two goals: that is, its type is goalexp -> goalexp -> o. Abstractly, if a tactic denotes the relation R, then $R(g_1, g_2)$ means that to satisfy goal g_1, it is sufficient to satisfy goal g_2. Primitive tactics are implemented directly in the underlying λProlog language in a fashion similar to that used in the preceding section; they are also assumed to work only for primitive goals. Compound tactics and the application of tactics to compound goals are implemented completely by the program clauses below.

The first program we describe, called maptac, applies tactics to compound goals. It takes a tactic as an argument and applies it to the input goal in a manner consistent with the meaning of the goal structure. For example, on an andgoal structure, maptac will apply the tactic to each subgoal separately, forming a new andgoal structure to combine the results. The type of maptac is (goalexp -> goalexp -> o) -> goalexp -> goalexp -> o, that is, the metalevel predicate maptac takes as its first argument a metalevel predicate which represents a tactic.

```
maptac Tac truegoal truegoal.

maptac Tac (andgoal InGoal1 InGoal2) (andgoal OutGoal1 OutGoal2) :-
  maptac Tac InGoal1 OutGoal1, maptac Tac InGoal2 OutGoal2.

maptac Tac (orgoal InGoal1 InGoal2) OutGoal :-
  maptac Tac InGoal1 OutGoal; maptac Tac InGoal2 OutGoal.

maptac Tac (allgoal InGoal) (allgoal OutGoal) :-
  pi T\ (maptac Tac (InGoal T) (OutGoal T)).

maptac Tac (existsgoal InGoal) OutGoal :-
  sigma T\ (maptac Tac (InGoal T) OutGoal).

maptac Tac (impgoal A InGoal) (impgoal A OutGoal) :-
  (memo A) => (maptac Tac InGoal OutGoal).

maptac Tac InGoal OutGoal :- Tac InGoal OutGoal.
```

The last clause above is used once the goal is reduced to a primitive form. Note that an auxiliary predicate memo (of polymorphic type A -> o) was introduced in the clause implementing `impgoal`. This allows the introduction of new clauses into the program. The type and content of these clauses will be specific to a particular theorem prover. In the next section we will illustrate how it is used to add assumptions that are discharged during the construction of natural deduction proofs.

The following definite clauses implement several of the familiar tacticals found in many tactic style theorem provers.

```
then Tac1 Tac2 InGoal OutGoal :- Tac1 InGoal MidGoal, maptac Tac2 MidGoal OutGoal.

orelse Tac1 Tac2 InGoal OutGoal :- Tac1 InGoal OutGoal; Tac2 InGoal OutGoal.

idtac Goal Goal.

repeat Tac InGoal OutGoal :- orelse (then Tac (repeat Tac)) idtac InGoal OutGoal.

try Tac InGoal OutGoal :- orelse Tac idtac InGoal OutGoal.

complete Tac InGoal truegoal :- Tac InGoal OutGoal, goalreduce OutGoal truegoal.
```

The **then** tactical performs the composition of tactics. `Tac1` is applied to the input goal, and then `Tac2` is applied to the resulting goal. `maptac` is used in the second case since the application of `Tac1` may result in an output goal (`MidGoal`) with compound structure. This tactical plays a fundamental role in combining the results of step-by-step proof construction. The substitutions resulting from applying these separate tactics get combined correctly since `MidGoal` provides the necessary sharing of logical variables between these two calls to tactics. The **orelse** tactical simply uses the OR search operation so that `Tac1` is attempted, and if it fails (in the sense that the logic programming interpreter cannot satisfy the corresponding metalevel goal), then `Tac2` is tried. The third tactical, `idtac`, returns the input goal unchanged. This tactical is useful in constructing compound tactic expressions such as the one found in the **repeat** tactical. **repeat** is recursively defined using the three tacticals, **then**, **orelse**, and **idtac**. It repeatedly applies a tactic until it is no longer applicable. The **try** tactical prevents failure of the given tactic by using `idtac` when `Tac` fails. It might be used, for example, in the second argument of an application of the **then** tactical. It prevents failure when the first argument tactic succeeds and the second does not. Finally the **complete** tactical tries to completely solve the given goal. It will fail if there is a non-trivial goal remaining after `Tac` is applied. It requires an auxiliary procedure `goalreduce` of type `goalexp -> goalexp -> o` which simplifies compound goal expressions by removing occurrences of `truegoal` from them. The code for `goalreduce` is as follows:

```
goalreduce (andgoal truegoal Goal) OutGoal :- goalreduce Goal OutGoal.
goalreduce (andgoal Goal truegoal) OutGoal :- goalreduce Goal OutGoal.
goalreduce (allgoal T\ truegoal) truegoal.
goalreduce (impgoal A truegoal) truegoal.
goalreduce Goal Goal.
```

Although the **complete** tactical is the only one that requires the use of the `goalreduce` procedure, it is also possible and probably desirable to modify the other tacticals so that they use it to similarly reduce their output goal structures whenever possible.

Notice that the notion of success and failure of the interpreter in Section 2 carries over to the success and failure of a tactic to solve a goal. The failure of the interpreter to succeed on a goal of the form (Tac InGoal OutGoal) indicates the failure of Tac to make progress toward solving InGoal.

The definite clauses listed above provide a complete implementation of tacticals. We now illustrate how tactics for a theorem prover can be implemented.

5   Inference Rules as Tactics

In this section we illustrate how to specify the inference rules of Gentzen's NK natural deduction system [Gen35]. Each inference rule will be specified as a primitive tactic. The basic goal needed to be established in this system is that a certain formula has a certain proof. This is encoded by the constant proofgoal which is of type bool -> proof_object -> goalexp. Goals of the form (proofgoal A P) will be called *atomic* goals of the natural deduction theorem prover, in contrast to *compound* goals built from using the goal constructors from the last section.

The implementation of inference rules is done in a fashion similar to the techniques in Section 3, except the clauses are made into named facts. For example, ∧-I is specified as:

```
and_i_tac (proofgoal (A and B) (and_i P1 P2))
          (andgoal (proofgoal A P1) (proofgoal B P2)).
```

This tactic can be applied whenever the formula in the input goal is a conjunction. This clause has essentially the same meaning as the definite clause for the ∧-I rule of Section 3, except that this clause is not automatically BACKCHAINed on by the interpreter. Put another way, the inference rule is represented here declaratively instead of procedurally. The procedural representation is at the mercy of the depth-first logic programming interpreter. In this other form, however, tacticals can specify their own forms of control.

In order to handle the hypotheses in this proof system, we introduce the new primitive type assump and the additional metalevel predicate hyp of type bool -> proof_object -> assump. This new symbol will be used in conjunction with impgoal and AUGMENT to represent assumptions. For example, ⊃-I can be implemented using the clause:

```
imp_i_tac (proofgoal (A imp B) (imp_i P))
          (allgoal PA\ (impgoal (hyp A PA) (proofgoal B (P PA)))).
```

The AUGMENT goal will then add a clause of the form (memo (hyp A pa)) to the program (where pa is a new constant generated by the GENERIC goal to replace PA).

Elimination rules are used to do forward reasoning from assumptions stored as memo facts. For example, the following clause specifies the ∧-E inference rule.

```
and_e_tac (proofgoal C PC)
          (impgoal (hyp A (and_e1 P))
             (impgoal (hyp B (and_e2 P)) (proofgoal C PC))) :-
   memo (hyp (A and B) P).
```

This clause works by moving the goal (proofgoal C PC) into the expanded context containing the hypotheses (hyp A (and_e1 P)) and (hyp B (and_e2 P)) if the hypothesis (hyp (A and B) P) already exists.

The remaining inference rules for NK are given below. We use the constant **perp** of type **bool** to represent the formula $\bot$.

```
or_i_tac (proofgoal (A or B) (or_i P))
        (orgoal (proofgoal A P) (proofgoal B P)).

forall_i_tac (proofgoal (forall A) (forall_i P))
            (allgoal T\ (proofgoal (A T) (P T))).

exists_i_tac (proofgoal (exists A) (exists_i P))
            (existsgoal T\ (proofgoal (A T) P)).

neg_i_tac (proofgoal (neg A) (neg_i P))
          (allgoal PA\ (impgoal (hyp A PA) (proofgoal perp (P PA)))).

or_e_tac (proofgoal C (or_e P P1 P2))
        (andgoal (allgoal PA\ (impgoal (hyp A PA) (proofgoal C (P1 PA))))
                 (allgoal PB\ (impgoal (hyp B PB) (proofgoal C (P2 PB))))) :-
  memo (hyp (A or B) P).

imp_e_tac (proofgoal B (imp_e P PA)) (proofgoal A PA) :-
  memo (hyp (A imp B) P).

forall_e_tac (proofgoal B PB)
            (existsgoal T\ (impgoal (hyp (A T) (forall_e P)) (proofgoal B PB))) :-
  memo (hyp (forall A) P).

exists_e_tac (proofgoal B (exists_e P PB))
            (allgoal T\ (allgoal PA\ (impgoal (hyp (A T) PA)
                                              (proofgoal B (PB T PA))))) :-
  memo (hyp (exists A) P).

neg_e_tac (proofgoal perp (neg_e P PA)) (proofgoal A PA) :-
  memo (hyp (neg A) P).

perp_tac (proofgoal A (contra PA))
        (allgoal P\ (impgoal (hyp (neg A) P) (proofgoal perp (PA P)))).

close_tac (proofgoal A P) truegoal :- memo (hyp A P).
```

Given these primitive tactics, we can then write compound tactics using the tacticals of the last section. For example, consider the following query (metalevel goal):

```
repeat (orelse or_i_tac and_i_tac) (proofgoal ((r or (p imp q)) and s) P) OutGoal.
```

This goal would succeed by instantiating P to the open proof term (and_i (or_i P1) P2) and instantiating OutGoal to the open goal expression:

```
(andgoal (orgoal (proofgoal r P1) (proofgoal (p imp q) P1)) (proofgoal s P2)).
```

This latter goal could then get processed by other tactics. Such processing would then instantiate the logical variables P1 and P2 which would then further fill in the original proof variable P in the query above.

Providing a means for accommodating user interaction is one of the strengths of tactic theorem provers. One way to provide an interface to the user in this paradigm is by writing tactics that request input. The following is a very simple tactic which asks the user for direction.

```
query (proofgoal A P) OutGoal :- write A, write "Enter tactic:", read Tac,
                                 Tac (proofgoal A P) OutGoal.
```

Here we have a tactic that, for any atomic input goal, will present the formula to be proved to the user, query the user for a tactic to apply to the input goal, then apply the input tactic. As in Prolog, (`write A`) prints A to the screen and will always succeed while (`read A`) prompts the user for input and will succeed if A unifies with the input. In this case (`read Tac`) will accept any term of type `goalexp -> goalexp -> o`.

Using this tactic, the following tactic, named `interactive`, represents a proof editor for natural deduction for which the user must supply all steps of the proof.

```
interactive InGoal OutGoal :- repeat query InGoal OutGoal.
```

These additions to the tactic prover will still not be sufficient, in general, for interactive theorem proving in a natural deduction setting. For example, if there is more than one conjunction among the discharged assumptions, the ∧-I rule will be applicable in more than one way. The user needs the capability to specify which formula to apply the tactic to. One way to solve this problem is to extend the program with tactics that request input from the user. Such tactics could be written as:

```
and_e_query (proofgoal C PC)
            (impgoal (hyp A (and_e1 P))
                (impgoal (hyp B (and_e2 P)) (proofgoal C PC))) :-
  memo (hyp (A and B) P), write "Eliminate this conjunction?",
  write (A and B), read "yes".
```

The tactic would enumerate conjunctive hypotheses until the user types in the word "yes."

6 Related Work

The development of the theory of higher-order hereditary Harrop formulas was motivated by a desire to develop a clean semantics for a programming language which embraced many more aspects of logic than first-order Horn clauses embrace. Other applications of this language that have been explored using the λProlog implementation lie in the areas of program transformations [MN87b] and computational linguistics [MN86b]. In this paper we have emphasized particular aspects that are useful for the specific task of implementing theorem provers.

The UT prover is well-known for its implementation of a natural deduction style proof system [Ble77, Ble83]. Since some aspects of its implementation have been designed to handle quantifiers and substitutions in a principled fashion, it is interesting to compare it to the systems described in this paper. In the UT prover, the IMPLY procedure is based on a set of rules for a "Gentzen type" system for first-order logic. In this procedure, formulas keep their basic propositional structure although their quantifiers are removed. In the AND-SPLIT rule of this prover (which corresponds to the ∧-R rule in a sequent system), the first conjunctive subgoal returns a substitution which must then be applied to the second subgoal before it is attempted. In logic programming, such composition of substitutions obtained from separate subgoals is handled automatically via shared logical variables. An issue that arises as a result of the AND-SPLIT rule is the occurrence of "conflicting bindings" due to the need to instantiate a quantified formula more than once. The UT prover uses *generalized substitutions* [TB79] to handle such multiple instances. A substitution is the final result returned when a complete proof is found. In contrast, in our logic programming language, quantification in formulas is represented by λ-abstraction. As a result, we do not need to remove quantifiers before attempting a proof. Instead we implement the inference rules for quantified formulas as illustrated in Section 3. For example, in a sequent system for classical logic, the inference rules for a universally quantified hypothesis or an existentially quantified conclusion are two rules which must allow multiple instantiations. This is easily accomplished by introducing a new logical variable for each instantiation. As the result of a successful proof we obtain a proof term rather than a substitution. As mentioned in Section 3, the construction of such proof terms may be defined to include the substitution information if desired. In fact such proof terms could be simplified to only return substitution information. Such simplified proof terms would be very similar to the generalized substitutions of the UT prover.

Other theorem provers that are based on tactics and tacticals include LCF [GMW79], Nuprl [Con86], and Isabelle [Pau86]. The programming language ML is the metalanguage used in all of these systems. ML is a functional language with several features that are useful for the design of theorem provers. It contains a secure typing scheme and is higher-order, allowing complex programs to be composed easily. There are several differences in the implementations of tactics in ML and in λProlog. First, tactics in ML are functions that take a goal as input and return a pair consisting of a list of subgoals and a validation. In contrast, tactics in λProlog are relational, which is very natural when the relation being modeled is "is a proof of." The fact that input and output distinctions can be blurred makes it possible, as described in Section 3, for tactics to be used in both a theorem proving and proof checking context. The functional aspects of ML do not permit this dual use of tactics. The ML notion of validations is replaced in our system by (potentially much larger and more complex) proof objects.

Second, it is worth noting the differences between the ML and λProlog implementations of the **then** tactical. The λProlog implementation of **then** reveals its very simple nature: **then** is very similar to the natural join of two relations. In ML, the **then** tactical applies the first tactic to the input goal and then maps the application of the second tactic over the list of intermediate subgoals. The full list of subgoals must be built as well as the compound validation function from the results. These tasks can be quite complicated, requiring some auxiliary list processing functions. In λProlog, the analogue of a list of subgoals is a nested **andgoal** structure. These are processed by the **andgoal** clause of **maptac**. The behavior of **then** (in conjunction with **maptac**) in λProlog is also a bit richer in two ways. First of all, `maptac` is richer than the usual notion of a mapping function in that, in addition to nested `andgoal` structures, it handles all of the other goal structures corresponding to the λProlog search operations. Secondly, in the ML version of `then`, if the second tactic fails after a successful call to the first tactic, the full tactic still fails. In contrast, in λProlog, if the first tactic succeeds and the second fails, the logic programming interpreter will backtrack and try to find a new way to successfully apply the first tactic, exhausting all possibilities before completely failing. Alternatively, we could use the cut (`!`) as in Prolog which prevents backtracking beyond a specified point and thus restrict the meaning of `then` to match its ML counterpart.

A third difference with ML is that for every constructor used to build a logic, explicit discriminators and destructors must also be introduced. In logic programming, however, the purpose served by these explicit functions is achieved within unification. The difference here is particularly striking if we look at the different representations of quantified formulas. A universal formula is constructed in ML by calling a `mk_forall` function which takes a variable and a formula and returns a universally quantified formula. `is_forall` and `dest_forall` are the corresponding discriminator and destructor to test for universal formulas and to obtain its components, respectively. Manipulating quantified formulas requires that the binding be separated from its body. In logic programming, we identify a term as a universal quantification if it can be unified with the term (`forall A`). However, since terms in λProlog represent $\beta\eta$-equivalence classes of λ-terms, the programmer does not have access to bound variable names. Although such a restriction may appear to limit access to the structure of λ-terms, sophisticated analysis of λ-terms is still possible to perform using higher-order unification. In addition, there are certain advantages to such a restriction. For example, in the case of applying substitutions, all the renaming of bound variables is handled by the metalanguage, freeing the programmer from such concerns.

The Isabelle theorem prover [Pau87] uses a fragment of higher-order logic with implication and universal quantification which is used to specify inference rules. That fragment is essentially a subset of the higher-order hereditary Harrop formulas. Hence, it seems very likely that Isabelle could be rather directly implemented inside λProlog. Although such an implementation might achieve the same functionality as is currently available in Isabelle, it is not likely to be nearly as efficient. This is due partly to the fact that a λProlog implementation implements a general purpose programming language.

Although our example theorem provers have been for first-order logic, we have also considered the logic of higher-order hereditary Harrop formulas as a specification language for a wide variety of logics. In this respect, we share a common goal with the Edinburgh Logical Framework (LF) project [HHP87]. LF was developed for the purpose of capturing the uniformities of a large class of logics so that it can be used as the basis for implementing proof systems. The two approaches are actually similar in ways that go beyond simply sharing common goals. First, the LF notions of *hypothetical* and *schematic* judgments can be implemented with the GENERIC and AUGMENT search operations in the logic programming setting. The LF hypothetical judgment takes the form $J_1 \vdash J_2$ and represents the assertion that J_2 follows from J_1. Objects of this type are functions mapping proofs of J_1 to proofs of J_2. Such a judgment can be implemented by having the GENERIC search operation introduce a new proof object, and then using AUGMENT to assume the fact that this new object is a proof of J_1. A proof of J_2 would then be the intended function applied to this new object. A schematic judgment in LF is of the form $\bigwedge_{x:A} J(x)$. It is a statement about arbitrary objects x of type A and is proved (inhabited) by a function mapping such objects to proofs of $J(x)$. This is implemented by using the GENERIC and AUGMENT search operations to first introduce an arbitrary constant and then assume it to be of LF type A.

Second, and more specifically, we have developed an algorithm that systematically translates all of the example LF signatures in [HHP87] and [AHM87] to logic programs [Fel87]. In the logic programming setting, in addition to being natural specifications, the resulting definite clauses also represent non-deterministic theorem provers. LF signatures could also be translated to sets of tactics to be used in a tactic theorem prover. The formal properties of this translation have yet to be established.

Acknowledgements The authors would like to thank Robert Constable, Elsa Gunter, Robert Harper, and Frank Pfenning for valuable comments and discussions. We are also grateful to the reviewers of an earlier draft of this paper for their comments and corrections. The first author is supported by US Army Research Office grant ARO-DAA29-84-9-0027, and the second author by NSF grant CCR-87-05596 and DARPA N000-14-85-K-0018.

References

[AHM87] Arnon Avron, Furio A. Honsell, and Ian A. Mason. *Using Typed Lambda Calculus to Implement Formal Systems on a Machine.* Technical Report ECS-LFCS-87-31, Laboratory for the Foundations of Computer Science, University of Edinburgh, June 1987.

[Ble77] W. W. Bledsoe. Non-resolution theorem proving. *Artificial Intelligence*, 9:1–35, 1977.

[Ble83] W. W. Bledsoe. *The UT Prover.* Technical Report ATP-17B, University of Texas at Austin, April 1983.

[Con86] R. L. Constable et al. *Implementing Mathematics with the Nuprl Proof Development System.* Prentice-Hall, 1986.

[Chu40] Alonzo Church. A formulation of the simple theory of types. *Journal of Symbolic Logic*, 5:56–68, 1940.

[Fel87] Amy Felty. Implementing theorem provers in logic programming. November 1987. Dissertation Proposal, University of Pennsylvania.

[Gen35] Gerhard Gentzen. Investigations into logical deductions, 1935. In M. E. Szabo, editor, *The Collected Papers of Gerhard Gentzen*, pages 68–131, North-Holland Publishing Co., Amsterdam, 1969.

[GMW79] Michael J. Gordon, Arthur J. Milner, and Christopher P. Wadsworth. *Edinburgh LCF: A Mechanised Logic of Computation.* Volume 78 of *Lecture Notes in Computer Science*, Springer-Verlag, 1979.

[HHP87]   Robert Harper, Furio Honsell, and Gordon Plotkin. A framework for defining logics. In *Symposium on Logic in Computer Science*, pages 194–204, Ithaca, NY, June 1987.

[Hue75]   G. P. Huet. A unification algorithm for typed λ-calculus. *Theoretical Computer Science*, 1:27–57, 1975.

[MN86a]   Dale Miller and Gopalan Nadathur. Higher-order logic programming. In *Proceedings of the Third International Logic Programming Conference*, pages 448–462, London, June 1986.

[MN86b]   Dale Miller and Gopalan Nadathur. Some uses of higher-order logic in computational linguistics. In *Proceedings of the 24th Annual Meeting of the Association for Computational Linguistics*, pages 247–255, 1986.

[MN87a]   Dale Miller and Gopalan Nadathur. λProlog Version 2.6. August 1987. Distribution in C-Prolog code.

[MN87b]   Dale Miller and Gopalan Nadathur. A logic programming approach to manipulating formulas and programs. In *IEEE Symposium on Logic Programming*, San Francisco, September 1987.

[MNS87]   Dale Miller, Gopalan Nadathur, and Andre Scedrov. Hereditary harrop formulas and uniform proof systems. In *Symposium on Logic in Computer Science*, pages 98–105, Ithaca, NY, June 1987.

[Nad86]   Gopalan Nadathur. *A Higher-Order Logic as the Basis for Logic Programming*. PhD thesis, University of Pennsylvania, December 1986.

[Pau86]   Lawrence C. Paulson. Natural deduction as higher-order resolution. *Journal of Logic Programming*, 3:237–258, 1986.

[Pau87]   Lawrence C. Paulson. *The Representation of Logics in Higher-Order Logic*. Draft, University of Cambridge, July 1987.

[Pra65]   Dag Prawitz. *Natural Deduction*. Almqvist & Wiksell, Uppsala, 1965.

[SS86]   L. Sterling and E. Shapiro. *The Art of Prolog: Advanced Programming Techniques*. MIT Press, Cambridge MA, 1986.

[TB79]   Mabry Tyson and W. W. Bledsoe. Conflicting bindings and generalized substitutions. In *4th International Conference on Automated Deduction*, pages 14–18, Springer-Verlag, February 1979.

QUERY PROCESSING IN QUANTITATIVE LOGIC PROGRAMMING

V.S.Subrahmanian

vs@logiclab.cis.syr.edu

School of Computer & Information Science

313 Link Hall

Syracuse University

Syracuse, N.Y. 13244.

Abstract

In [12] the notion of a quantitative logic program has been introduced, and its declarative semantics explored. The operational semantics given in [12] is extended significantly in this paper - in particular, the notion of correct answer substitution is introduced and soundness and completeness results obtained. In addition, the completeness results for the and-or tree searching technique given in [12] is strengthened to be applicable to quantitative logic programs that are not well covered, thus removing one restriction in the completeness theorem obtained in [12]. In addition, the soundness and completeness results for SLDq-resolution in [12] are strengthened to apply to any nice QLP. Moreover, all these soundness and completeness results are applicable to existential queries unlike the results of [12,13] and [14] which are applicable to ground queries only. It was shown in [12] that the greatest supported model of a class of QLPs is semi-computable. In this paper, we give an explicit procedure to compute (partially) the greatest supported model, and obtain soundness and completeness results. This has applications in reasoning about beliefs.

1. Introduction

A quantitative logic program (QLP) was defined in [12] as a set of (universally closed) sentences of the form

$$A_0 : \mu_0 \Leftarrow A_1 : \mu_1 \& \dots \& A_k : \mu_k$$

where each A_i is an atom, and $\mu_i \in [0,1]$. Several different kinds of queries were (informally) introduced in [12] - we make these intuitions more formal here.

(1) A *ground query* (or g-query, for short) is a formula of the form $\Leftarrow A : \mu$ where $A \in B_Q$ and $\mu \in [0,1]$ and B_Q is the Herbrand base of the QLP Q.

(2) Similarly, if A is an atom (possibly non-ground), then $\Leftarrow A : \mu$ is an *existential query* (or ex-query for short).

(3) If $A \in B_Q$, then $\Leftarrow A :?$ is a *confidence-request query* (or cr-query, for short). Intuitively, given a ground atom A, the cr-query $\Leftarrow A :?$ asks: "What is the truth value assigned A by the least model of Q ?".

Different classes of QLPs (e.g. nice QLPs, decent QLPs, well-covered QLPs) were defined in [12], and some soundness and completeness results were obtained (for ground queries) with respect to these classes of programs. In this paper, we:

(1) Introduce the notion of answer substitution, and give an operational semantics that is both sound and complete w.r.t. computation of answer substitutions for existential queries for the class of function-free QLPs. In particular, we strengthen the completeness results of [12] by dropping the restriction that QLPs be well-covered. In addition, we prove that SLDq-resolution [12] computes *correct* answer substitutions for ex-queries. We also obtain a completeness result. This was not done in [12].

(2) In [12], it was shown that if Q is a nice, decent QLP, then the greatest supported model of Q is semi-computable (r.e.). However, no algorithm to partially compute the greatest supported model was given there. In this paper, we explicitly develop such an algorithm, and obtain soundness and completeness results. We have already shown in [12] that the semi-computability of the greatest supported model of Q allows us to reason about beliefs. To our knowledge, no such algorithm to reason about (even a limited notion of) consistency exists in the literature.

As this paper is primarily an extension of [12], we expect the reader to be familiar with the model theoretic aspects of quantitative logic programming described there. In addition, the algorithm given there for computing ground queries with respect to well-covered QLPs is similar to the approach of Van Emden [14], and we consequently see no reason to repeat that procedure here. We will, however, give some of the other main definitions and theorems of that paper in the next section. Section 3 is concerned with describing processing of existential queries. Section 4 described how the greatest supported model can be computed.

2. QLPs - An Overview

<u>DEF 1</u>: The set T of truth values is $[0,1] \cup \{\top\}$. The partial ordering $\preceq$ on T is defined as shown in the following Hasse diagram (Fig.1):

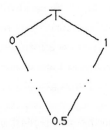

As usual, $x \succeq y$ iff $y \preceq x$. Similarly, $\succ, \prec$ are the irreflexive restrictions of $\succeq, \preceq$ respectively. Note that according to this ordering, $0.1 \succeq 0.2$.

<u>DEF 2</u>: If A is an atom, and $\mu \in \mathcal{T}$ (resp. $[0,1]$), then $A : \mu$ is an *annotated (resp. well-annotated) atom*.

<u>DEF 3</u>: If $A_0 : \mu_0, \ldots, A_k : \mu_k$ are well-annotated atoms, then

$$A_0 : \mu_0 \Leftarrow A_1 : \mu_1 \& \ldots \& A_k : \mu_k$$

is a *q-clause*.

<u>DEF 4</u>: A *quantitative logic program* (QLP) is a finite set of q-clauses.

An interpretation I may be considered to be a mapping from the Herbrand base B_Q of a QLP Q to $\mathcal{T}$. We now give a paraconsistent (cf.[3,4, 5,6,7]) model-theoretic semantics for QLPs.

<u>DEF 5</u>: *[Satisfaction]* (The symbol $\models$ denotes satisfaction.)

(1) I *satisfies* a ground annotated atom $A : \mu$ iff $I(A) \succeq \mu$.

(2) I satisfies a conjunction $A_1 : \mu_1 \& \ldots \& A_k : \mu_k$ of ground annotated atoms iff $I \models A_i : \mu_i$ for all $i = 1, \ldots, n$.

(3) I satisfies the ground q-clause

$$A_0 : \mu_0 \Leftarrow A_1 : \mu_1 \& \ldots \& A_k : \mu_k$$

iff one of the following conditions holds:

(a) I satisfies $A_0 : \mu_0$.

(b) I does not satisfy $A_1 : \mu_1 \& \ldots \& A_k : \mu_k$.

(4) I satisfies the existentially quantified formula $(\exists)F$ iff $I \models F(\vec{t}/\vec{x})$ where $\vec{t}$ and $\vec{x}$ are tuples of variable-free terms and variable symbols respectively, and $F(\vec{t}/\vec{x})$ denotes the simultaneous replacement of all free occurrences in F of a variable symbol x_i in $\vec{x}$ by the corresponding term t_i in $\vec{t}$.

(5) I satisfies the universally closed formula $(\forall x)F$ iff $I \models F(\vec{t}/\vec{x})$ for every $\vec{t}/\vec{x}$.

This completes the definition of the models of a QLP. Throughout this paper, we will use the words interpretation (resp. model) to refer to an Herbrand interpretation (resp. Herbrand model). Observe that all q-clauses are implicitly universally quantified. If F is a formula, we use the notation $(\exists)F$ and $(\forall)F$ to denote the existential closure and universal closure, respectively, of F (cf. Lloyd [10]). The above definition can be easily extended to handle disjunctions of annotated atoms, but this is not required for the purposes of this paper. Also, it was shown in [12, Theorem 14] that the device of annotations is powerful enough to make the use of negation unnecessary; consequently, we do not consider negated atoms in this paper. The notion of unification is extended to annotated atoms as follows: $A : \mu$ and $B : \rho$ are unifiable iff A and B are unifiable. (Note that we are not defining the result of unifying two atomic formulas yet). Similarly, applying the substitution θ to $A : \mu$ results in $A\theta : \mu$.

<u>DEF 6</u>: The mapping T_Q from Herbrand interpretations to Herbrand interpretations is defined as:

$T_Q(I)(A) = \sqcup\{\mu | A \Leftarrow B_1 : \psi_1 \& \ldots \& B_k : \psi_k$ is the ground instance of a q-clause in Q and $I \models B_1 : \psi_1 \& \ldots \& B_k : \psi_k\}$. ($\sqcup$ and $\sqcap$ will be used to denote least upper bounds and greatest lower bounds respectively under the $\preceq$ ordering). The $\preceq$ ordering on truth values is extended to interpretations in the natural way, i.e. $I_1 \preceq I_2$ iff $I_1(A) \preceq I_2(A)$ for all $A \in B_Q$.

Theorem 7: [12] T_Q is monotone. ∎

Theorem 8: [12] I is a model of the QLP Q iff $T_Q(I) \preceq I$. ∎

It follows from the above theorem that every QLP has a least model, and that this least model is identical to the least fixed-point of the T_Q operator. In addition, the least model is semi-computable.

<u>DEF 9</u>: The *upward (downward) iterations* of T_Q are defined as:

$$T_Q \uparrow 0 = \Delta \quad ; \quad T_Q \downarrow 0 = \nabla$$

$$T_Q \uparrow \alpha = T_Q(T_Q \uparrow (\alpha - 1)) \quad ; \quad T_Q \downarrow \alpha = T_Q(T_Q \downarrow (\alpha - 1))$$

$$T_Q \uparrow \lambda = \sqcup_{\alpha < \lambda} T_Q \uparrow \alpha \quad ; \quad T_Q \downarrow \lambda = \sqcap_{\alpha < \lambda} T_Q \downarrow \alpha$$

where Δ, ∇ are the interpretations that assign the truth value 0.5, $\top$ respectively to each ground atom $A \in B_Q$, and α, λ are successor and limit ordinals respectively.

Theorem 10: [12] $T_Q \uparrow \omega$ is the least model of Q. ∎

<u>DEF 11</u>: An interpretation I is *nice* iff $I(A) \neq \top$ for all $A \in B_Q$.

<u>DEF 12</u>: A QLP Q is said to be *nice* iff for every pair of q-clauses C_1, C_2 whose heads $A_1 : \mu_1, A_2 : \mu_2$ are unifiable, $\sqcup\{\mu_1, \mu_2\} \neq \top$.

Theorem 13: [12] If Q is a nice QLP, then $T_Q \uparrow \omega$ and $T_Q \downarrow \omega$ are nice models of Q. ∎

<u>DEF 14</u>: A model of the QLP Q is said to be *supported* iff for every $A \in B_Q$ such that $I(A) = \mu \neq 0.5$ there is a q-clause in Q having a ground instance of the form $A : \mu \Leftarrow B_1 : \psi_1 \& \ldots \& B_k : \psi_k$ such that $I \models B_1 : \psi_1 \& \ldots \& B_k : \psi_k$.

As only well-annotated atoms can occur in QLPs, every supported model is nice.

Theorem 15: [12] I is a supported model of Q iff I is a nice fixed-point of T_Q. ∎

<u>DEF 16</u>: A QLP Q is *well-covered* iff every variable symbol that occurs in the body of a q-clause also occurs in the head of that q-clause.

<u>DEF 17</u>: A QLP Q is *decent* iff $T_Q \downarrow \omega = \text{gfp}(T_P)$.

Theorem 18: [12] If Q is a nice, decent QLP, then $T_Q \downarrow \omega$ is the greatest supported model of Q. ∎

With this, we conclude our informal overview of the notations introduced in [12], as well as the declarative semantics investigated there.

3. Processing Existential Queries

<u>DEF 19</u>: If Q is a QLP and G an ex-query, then an *answer substitution* is a substitution for the variable symbols occurring in G.

<u>DEF 20</u>: If Q is a QLP and $\Leftarrow A : \mu$ an ex-query, then θ is a *correct* answer substitution iff

$$T_Q \uparrow \omega \models (\forall)(A\theta : \mu).$$

<u>DEF 21</u>: If Q is a QLP and $\Leftarrow A : \mu$ an ex-query, then the *and-or tree* $\Upsilon(Q, A : \mu)$ associated with Q is defined as:

(1) the root $\Re(\Upsilon(Q, A : \mu))$ of $\Upsilon(Q, A : \mu)$ is an or-node labelled $A : \mu$.

(2) if N is an or-node, then N is labelled by a single annotated atom.

(3) each and-node is labelled by a q-clause in Q and a substitution.

(4) descendants of an or-node are all and-nodes and vice-versa.

(5) if N is an or-node labelled by $A : \mu$ ($\mu \neq 0.5$), and if C is a q-clause in Q whose head $B : \rho$ unifies with $A : \mu$ (with m.g.u. θ), then there is a descendant of N labelled with C and θ. An or-node with no descendants is called an *uninformative node*.

(6) if N is an and-node labelled by the q-clause C and substitution θ, then for every annotated atom $B : \rho$ in the body of C, there is a descendant or-node labelled with $B\theta : \rho$. An and-node with no descendants is called a *success node*.

The label of a node N is denoted by $\ell(N)$. If N is an or-node, then $\ell(N)$ is an annotated atom. If N is an and-node, then $\ell(N) = (\ell(N)_1, \ell(N)_2)$ where $\ell(N)_1$ is the clause labelling N and $\ell(N)_2$ is the substitution labelling N.

Example 22. Consider the QLP Q given below and the query $\Leftarrow p(a, W) : 0.77$. The and-or tree $\Upsilon(Q, p(a, W) : 0.77)$ is shown in Figure 2.

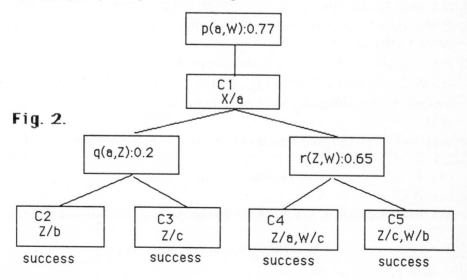

Fig. 2.

$$C1: \quad p(X,Y) : 0.8 \Leftarrow q(X,Z) : 0.2 \& r(Z,Y) : 0.65$$

$$C2: \quad q(a,b) : 0.4 \Leftarrow$$

$$C3: \quad q(a,c) : 0.1 \Leftarrow$$

$$C4: \quad r(a,c) : 0.6 \Leftarrow$$

$$C5: \quad r(c,b) : 0.9 \Leftarrow$$

Associated with each node in the AND/OR tree $\Upsilon(Q, A : \mu)$ is a set of pairs called DV-pairs.

<u>DEF 23</u>: If v is a truth value and σ a substitution, then (v, σ) is a *DV-pair*.

<u>DEF 24</u>: If Q is a QLP and $\Xi = \{(v_1, \sigma_1), \ldots, (v_m, \sigma_m)\}$ is the set of DV-pairs associated with the root of $\Upsilon(Q, A : \mu)$ where A is a possibly non-ground atom, then Ξ is *correct* iff for all $1 \leq i \leq m$, it is the case that

$$T_Q \uparrow \omega \models (\forall)(A\sigma_i : v_i)$$

If N is a node, then $DV(N)$ denotes the set of DV-pairs associated with N. It is well known that any substitution may be considered to be a conjunction of equations – we will, whenever, it is convenient, choose to regard a substitution in this light.

Algorithm 25: [DV-Pair Association Algorithm]

Input: A QLP Q and an ex-query $\Leftarrow A : \mu$, $\mu \neq \top$ such that $\Upsilon(Q, A : \mu)$ is finite.

(1) If N is an uninformative node, then $DV(N) = \{(0.5, \bowtie)\}$.

(2) If N is a success node labelled by the substitution θ and the unit clause $D : \delta$, then
$$DV(N) = \{(\delta, \theta)\}$$

(3) If N is an non-uninformative or-node labelled with $D : \delta$ whose descendant and-nodes are $N_1, \ldots, N_k$ then $DV(N) = \bigcup_{i=1}^{k} DV(N_i)'$ where $DV(N_i)'$ is the set of DV-pairs obtained by restricting all substitutions in the DV-pairs of $DV(N_i)$ to the variables of D.

(4) Suppose N is an and-node that is not a success node labelled with the clause

$$D : \delta \Leftarrow B_1 : \psi_1 \& \ldots \& B_k : \psi_k$$

and the substitution θ. Let N_i be the descendent of N labelled with $B_i\theta : \psi_i$. Let

$$T_i = \{(v, s) | (v, s) \in DV(N_i) \ \& \ v \succeq \psi_i\}$$

Then, $DV(N) = \{(\delta, \xi^1_{j_1} \xi^2_{j_2} \cdots \xi^k_{j_k})|$ for all $1 \leq i \leq k, \exists v_i$ such that $(v_i, \xi^i_{j_i}) \in T_i$ and $\xi^1_{j_1} \cup \xi^2_{j_2} \cup \cdots \cup \xi^k_{j_k}$ is solvable $\}$. (If $\xi^i_{j_i} = \bowtie$ for some $1 \leq i \leq k$, then the system of equations $\xi^1_{j_1} \cup \xi^2_{j_2} \cup \cdots \cup \xi^k_{j_k}$ is unsolvable).

End Algorithm 25

Intuitively, the "substitution" $\bowtie$ is a symbol denoting a failed attempt at unification. We observe that given a query $A : \mu$ and a QLP Q, the tree $\Upsilon(Q, A : \mu)$ is unique (upto isomorphism); moreover, there is exactly one way of associating sets of DV-pairs with the nodes of this tree. In addition, it is easily seen that if $\Upsilon(Q, A : \mu)$ is finite, then the set of DV-pairs associated with any node N is finite.

Example 26: Consider the QLP Q below, and the query $p(X) : 0.8$:

$$p(X) : 0.9 \Leftarrow q(X) : 0.2 \ \& \ r(X) : 0.6$$

$$p(X) : 0.3 \Leftarrow q(X) : 0.2 \ \& \ r(X) : 0.2$$

$$r(a) : 0.7 \Leftarrow$$

$$r(b) : 0.1 \Leftarrow$$

$$q(a) : 0.1 \Leftarrow$$

$$q(b) : 0.2 \Leftarrow$$

Then the set of DV-pairs associated with the root of $\Upsilon(Q, p(x) : 0.8)$ is:

$$\{(0.9, X = a), (0.3, X = b)\}$$

To answer the query, we need to find a substitution θ such that $T_Q \uparrow \omega(p(X)\theta) \succeq 0.8$. There is one DV-pair in the set of DV-pairs associated with the root of $\Upsilon(Q, p(x) : 0.8)$ that has the truth value 0.9 which is greater than 0.8 (in the $\succeq$ ordering). Thus, the answer to this query is "YES", together with the correct answer substitution $\{X/a\}$.

For the sake of convenience, we define the *height* of a finite and-or tree to be the maximum number of or-nodes (excluding the root) along any path of the tree. The height of the tree Γ is denoted by $h(\Gamma)$.

Theorem 27: [Soundness] If Q is a QLP and $\Leftarrow A : \mu$ an ex-query such that the and-or tree $\Upsilon(Q, A : \mu)$ is finite and such that

$$\{(v_1, \sigma_1), \ldots, (v_k, \sigma_k)\}$$

is the set of DV-pairs associated with $\Re(\Upsilon(Q, A : \mu))$, then

$$T_Q \uparrow \omega \models (\forall)(A\sigma_i : v_i)$$

for all $1 \leq i \leq k$.

Proof: By induction on the height of $\Upsilon(Q, A : \mu)$.

Base Case: $[h(\Upsilon(Q, A : \mu)) = 0]$ Then either the root $\Re(\Upsilon(Q, A : \mu))$ is an uninformative node, or its immediate descendants are all success nodes.

Subcase 1: If $\Re(\Upsilon(Q, A : \mu))$ is an uninformative node, then $v_i = 0.5$ for all $1 \leq i \leq k$. It is trivially true, therefore, that $T_Q \uparrow \omega \models (\forall)A\sigma : 0.5$ for any substitution σ.

Subcase 2: Suppose $D_1, \ldots, D_m$ are the immediate descendent nodes of $\Re(\Upsilon(Q, A : \mu))$; then each D_i must be a success node labelled with $(A_i : \rho_i \Leftarrow, \theta_i)$. Then each $v_i = \rho_i$ and $\sigma_i = \theta_i'$ for all $1 \leq i \leq k$ (where θ_i' is the restriction of θ_i to the variables of A). Thus, as $T_Q \uparrow \omega$ is a model of Q,

$\quad T_Q \uparrow \omega \models (\forall)A_i : \rho_i$, i.e.

$\quad T_Q \uparrow \omega \models (\forall)A_i\theta_i' : \rho_i$, i.e.

$\quad T_Q \uparrow \omega \models (\forall)A : \rho_i$.

Inductive Case: Suppose the height of the $\Upsilon(Q, A : \mu)$ is $r + 1$. Let the tree be as shown in Fig. 3:

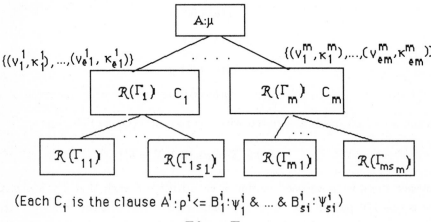

(Each C_i is the clause $A^i : \rho^i <= B_1^i : \psi_1^i \ \& \ \ldots \ \& \ B_{si}^i : \psi_{si}^i$)

Fig. 3

We want to show that

$$T_Q \uparrow \omega \models (\forall)(A\sigma_i : v_i)$$

for any $1 \leq i \leq k$. Let the set of DV-pairs associated with $\Re(\Gamma_i)$ be $\{(v_1^i, \kappa_1^i /), \ldots, (v_{e_i}^i, \kappa_{e_i}^i /)\}$. Clearly, $\sigma_i = \kappa_j^i /$ for some $1 \leq j \leq e_i$. But $\kappa_j^i /$ is the composition of substitutions

$$\tau_{x1}^{ij} \tau_{x2}^{i(j+1)} \ldots \tau_{xs_i}^{is_i}$$

where the set of DV-pairs associated with each $\Re(\Gamma_{ij})$ is $(w_{i1}^{ij}, \tau_{i1}^{ij}), \ldots, (w_{is_i}^{ij}, \tau_{is_i}^{ij})\}$. But the individual substitutions in the above composition are in the set of DV-pairs associated with the root of a tree of height less than or equal to r. Thus, by the induction hypothesis,

$$T_Q \uparrow \omega \models (\forall)[B_j^i \tau_{ix}^{ij} : w_{ix}^{ij}]$$

for $1 \leq x \leq s_i$. As (v_i, σ_i) is a DV-pair associated with the and node $\Re(\Gamma_i)$, (hence $v_i = \rho_i$) and as $\sigma_i = \kappa_i^j \prime$ for some j it must be the case that for all $1 \leq j \leq s_i$,

$$T_Q \uparrow \omega \models (\forall)[B_1^i \theta_i : \psi_1^i \& \ldots \& B_{s_i}^i \theta_i : \psi_{s_i}^i)\kappa_j^i]$$

As $T_Q \uparrow \omega$ is a model of Q,

$$T_Q \uparrow \omega \models (\forall)(A^i \theta_i \kappa_j^i : \rho^i)$$

But $A^i \theta_i = A$ and $\rho^i = v_i$. Thus,

$$T_Q \uparrow \omega \models (\forall) A \kappa_j^i : v_i$$

whence as $\sigma_i = \kappa_j^i \prime$ (the restriction of κ_i^j to the variables of A), it follows that

$$T_Q \uparrow \omega \models (\forall)(A\sigma_i : v_i))$$

The theorem follows. ∎

We have proved in [12, Theorem 42] that there is no loss of generality in assuming that annotated atoms of the form $A : 0.5$ do not occur either in a QLP or a goal. Unless explicitly stated otherwise, we shall make this assumption throughout the rest of this paper. Before we proceed to establish completeness results for existential queries, we state the completeness theorem for *ground* queries.

Theorem 28 [12] If Q is a well-covered QLP, and A is a ground atom such that

$$T_Q \uparrow \omega \models A : \mu$$

then $\nu(\Re(\Upsilon(Q, A : \mu))) \succeq T_Q \uparrow \omega(A)$. ∎

<u>DEF 29</u>: Let $\Xi = \{(v_1, \sigma_1), \ldots, (v_m, \sigma_m)\}$ be a set of DV-pairs. The *scope* of a variable symbol x is the set

$$\{x = t_i | x = t_i \in \sigma_j, 1 \leq j \leq m\}$$

<u>DEF 30</u>: A set of DV-pairs $\Xi = \{(v_1, \sigma_1), \ldots, (v_m, \sigma_m)\}$ is said to be *fully scoped* w.r.t. a variable symbol x and a set **TERM** of terms, if for any $t \in$ **TERM**, there is an equation $x = t_i$ in the scope of x such that t is an instance of t_i.

<u>DEF 31</u>: If Q is a QLP and G a goal, then a correct set $\Xi = \{(v_1, \sigma_1), \ldots, (v_m, \sigma_m)\}$ of DV-pairs is said to be *nullary* iff
(1) $m = 0$ or

(2) If x is a variable symbol occurring in G, then $\Xi = \{(v_1, \sigma_1), \ldots, (v_m, \sigma_m)\}$ is fully scoped w.r.t. x and the Herbrand Universe U_Q of Q.

We now attempt to extend the completeness theorem for ground queries to a completeness theorem for existential queries – we remove the condition that Q be well covered; however, we require that the QLP be function free.

Theorem 32: (Completeness) If Q is a function free QLP and $T_Q \uparrow \omega \models (\forall)A : \mu$ (where $\mu \neq \top$) and $\Upsilon(Q, A : \mu)$ is finite, then

(1) The set Ξ of DV-pairs associated with the root of $\Upsilon(Q, A : \mu)$ is nullary, and

(2) If σ is any substitution of ground terms for all of the variables of A, then there is a DV-pair (v_i, σ_i) in Ξ such that $v_i \succeq \mu$ and there is a substitution γ such that $\sigma = \sigma_i \cdot \gamma$. (i.e. γ is more general than σ).

Proof: As Q is function free, the set of ground instances of A is finite; hence, there is an integer n such that $T_Q \uparrow n \models (\forall)(A : \mu)$. The rest of the proof proceeds by induction on n – the induction hypothesis used is that the conjunction of (1) and (2) hold. We just use this to prove (1) – the proof for (2) is not very different and is omitted here.

Base Case: $[n = 1]$ Then $T_Q \uparrow 1 \models (\forall)(A\sigma : \mu)$. Thus, there is a finite multiset $\{C_1, \ldots, C_m\}$ of q-clauses in Q of the form

$$A'_i : \rho_i \Leftarrow$$

where $\rho_i \succeq \mu$ and such that $A'_i \theta_i = A$ for some most general substitution θ_i and such that $\sqcup\{\rho_i | 1 \leq i \leq m\} \succeq \mu$. Then the set of DV-pairs $\{(\rho_1, \theta'_1), \ldots, (\rho_m, \theta'_m)\}$ is nullary.

Inductive Case: The argument proceeds along similar lines – $\Upsilon(Q, A : \mu)$ is as shown in Fig. 4 below:

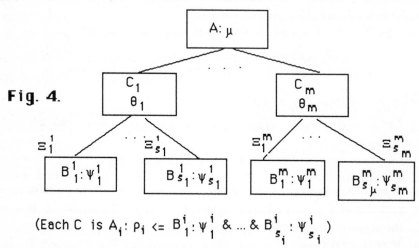

Fig. 4.

(Each C is $A_i : \rho_i <= B^i_1 : \psi^i_1$ & ... & $B^i_{s_i} : \psi^i_{s_i}$)

Suppose $T_Q \uparrow (n + 1) \models (\forall)(A : \mu)$. Let A' be any ground instance of A. Therefore,

$T_Q \uparrow (n+1) \models A' : \mu$, and we have to show that for some $(v, \sigma) \in \Xi$, $v \succeq \mu$ and A' is an instance of $A\sigma$. As $T_Q \uparrow (n+1) \models A' : \mu$, some clause C_i in Q has a ground instance of the form

$$(A_i : \rho_i \Leftarrow B_1^i : \psi_1^i \& \ldots \& B_{s_i}^i : \psi_{s_i}^i : \psi_{s_i}^i)\gamma$$

such that $\rho_i \succeq \mu$ and $\gamma = \theta\tau$ where θ is the mgu of A_i and A, $A_i\gamma = A'$, τ is some substitution and $T_Q \uparrow n \models (\forall)(B_1^i : \psi_1^i \& \ldots \& B_{s_i}^i : \psi_{s_i}^i)\gamma$. By the induction hypothesis (using (2) above), there is a substitution γ' more general than γ such that the set of DV-pairs associated with each node labelled $B_j^i\theta : \psi_j^i$ $(1 \leq j \leq s_i)$ is nullary; hence, the set Ξ_i of DV-pairs associated with the node labelled C_i and θ_i contains (ρ_i, γ'), whence Ξ contains (ρ_i, γ'). ∎

The restriction to function-free QLPs in the completeness result is not as great a restriction as one might imagine it to be. With this slight strengthening of the conditions imposed in [14,12], it is possible to process arbitrary existential queries (in [12,14], completeness results that were obtained were applicable only for ground queries to well-covered QLPs. Neither of these two conditions is applicable here). We now extend the notion of answer substitutions to SLDq-resolution (which was introduced in [12]). SLDq-resolution is a technique that is applicable to nice QLPs. It is similar to SLD-resolution, and hence is much more efficient (w.r.t. time) than the and-or tree construction method.

DEF 33: A *conjunctive existential query* (or cex-query, for short) is a query of the form $(\exists)(A_1 : \mu_1 \& \ldots \& A_k : \mu_k)$.

DEF 34: A *computation rule* is a function that selects from any conjunctive existential query $A_1 : \mu_1 \& \ldots \& A_k : \mu_k$, an annotated atom $A_i : \mu_i$.

DEF 35: If $\Leftarrow A_1 : \mu_1 \& \ldots \& A_k : \mu_k$ is a conjunctive existential query, and if C is the following q-clause in Q

$$A : \rho \Leftarrow B_1 : \psi_1 \& \ldots \& B_m : \psi_m$$

and if A_i and A are unifiable (with mgu θ) and if the computation rule R selects $A_i : \mu_i$ and if $\rho \succeq \mu_i$, then the *q-resolvent* of the above existential query and C with respect to R is the existential query

$$(\Leftarrow A_1 : \mu_1 \& \ldots \& A_{i-1} : \mu_{i-1} \& B_1 : \psi_1 \& \ldots \& B_m : \psi_m \& A_{i+1} : \mu_{i+1} \& \ldots \& A_k : \mu_k)\theta.$$

$A_i : \mu_i$ is called the annotated atom *q-resolved upon*.

DEF 36: An *SLDq-deduction* from the existential query G_0, the QLP Q and the computation rule R is a sequence

$$\langle G_0, C_1, \theta_1 \rangle, \ldots, \langle G_i, C_{i+1}, \theta_{i+1} \rangle, \ldots$$

where G_0 is the initial query, and G_{i+1} is derived by q-resolving G_i and a copy of the input q-clause C_{i+1} with respect to R. By a copy of C_{i+1}, we mean a renaming of the variables in C_{i+1} so that the renamed version of C_{i+1} contains no variables occurring in $G_0, \ldots, G_i, C_1, \ldots, C_i$.

DEF 37: An *SLDq-refutation* (of length n) from G_0 and the QLP Q using the computation rule R is an SLDq-deduction

$$\langle G_0, C_1, \theta_1 \rangle, \ldots \langle G_n, C_{n+1}, \theta_{n+1} \rangle$$

with G_{n+1} being the empty query. The restriction θ of $\theta_1 \cdots \theta_n$ to the variables of G_0 is called the *R-computed answer substitution*.

Theorem 38: [Soundness] If Q is a nice QLP then every R-computed answer substitution is correct.

Proof: Suppose G_0 is a cex-query and suppose

$$\langle G_0, C_1, \theta_1 \rangle, \ldots, \langle G_n, C_{n+1}, \theta_{n+1} \rangle$$

is an SLDq-refutation of G_0 from Q using the computation rule R. Let θ be the R-computed answer substitution, i.e. θ is the restriction of $\theta_1 \ldots \theta_n$ to the variables of G_0. We need only show that

$$T_Q \uparrow \omega \models (\forall) G_0 \theta$$

We will proceed by induction on the length of the SLDq-refutation.

Base Case [n=0] Then G_0 is a unit goal, i.e. of the form $\Leftarrow A : \mu$, and there is a unit q-clause C_1 in Q having the form

$$A' : \rho \Leftarrow$$

and $\rho \succeq \mu$. As $T_Q \uparrow \omega$ is a model of Q,, we have $T_Q \uparrow \omega \models (\forall) A' : \rho$,

whence $T_Q \uparrow \omega \models (\forall) A' \theta_1 : \rho$

whence $T_Q \uparrow \omega \models (\forall) A \theta_1 : \rho$ (as $A' \theta_1 = A \theta_1$)

whence $T_Q \uparrow \omega \models (\forall) A \theta_1 : \mu$ (as $\rho \succeq \mu$).

which completes this case of the proof.

Inductive Case: Suppose

$$\langle G_0, C_1, \theta_1 \rangle, \ldots, \langle G_{n+1}, C_{n+2}, \theta_{n+2} \rangle$$

is an SLDq-refutation of length $(n+1)$ of the cex-query G_0 [of the form $(\exists)(A_1 : \mu_1 \& \ldots \& A_k : \mu_k)$] from the nice QLP Q using the computation rule R. Suppose the q-resolvent G_1 of G_0 and C_1 via R is

$$(A_1 : \mu_1 \& \ldots \& A_{i-1} : \mu_{i-1} \& B_1 : \delta_1 \& \ldots \& B_m : \delta_m \& A_{i+1} : \mu_{i+1} \& \ldots \& A_k : \mu_k) \theta_1$$

where C_1 is

$$A' : \rho \Leftarrow B_1 : \delta_1 \& \ldots \& B_m : \delta_m$$

and $\rho \succeq \mu_i$ and θ_1 is the mgu of A and A'. Now, by the inductive hypothesis,

$$T_Q \uparrow \omega \models (\forall) G_1 \phi$$

where ϕ is the restriction of $\theta_2 \ldots \theta_n$ to the variable symbols of G_1. But the R-computed answer substitution θ of G_0 is simply the restriction of $\theta_1.\phi$ to the variables of G_0. We have, by the inductive hypothesis, that $T_Q \uparrow \omega \models$

$$(\forall)(A_1 : \mu_1 \& \ldots \& A_{i-1} : \mu_{i-1} \& B_1 : \delta_1 \& \ldots \& B_m : \delta_m \& A_{i+1} : \mu_{i+1} \& \ldots \& A_k : \mu_k)\theta_1.\phi$$

In particular,

$$T_Q \uparrow \omega \models (\forall)(B_1 : \delta_1 \& \ldots \& B_m : \delta_m)\theta_1.\phi$$

Thus, as $T_Q \uparrow \omega$ is a model of Q, and hence of C_1, we have $T_Q \uparrow \omega \models (\forall)(A'\theta_1.\phi : \rho)$, whence $T_Q \uparrow \omega \models (\forall)(A_i\theta_1.\phi : \mu_i)$ (as $A_i\theta_1 = A'\theta_1$), whence

$$T_Q \uparrow \omega \models (\forall)(A_1 : \mu_1 \& \ldots \& A_k : \mu_k)\theta_1.\phi$$

whence $T_Q \uparrow \omega \models (\forall)(A_1 : \mu_1 \& \ldots \& A_k : \mu_k)\theta$ where θ is the restriction of $\theta_1.\phi$ to the variables of G, which completes the proof. ∎

Theorem 39: [Completeness] If G_0 is a goal of the form $A_1 : \mu_1 \& \ldots \& A_k : \mu_k$, Q a nice, function-free QLP, and if $T_Q \uparrow \omega \models (\forall)(G_0\sigma)$ for some σ, then there exists an R-computed answer substitution θ (where θ is more general than σ) such that $T_Q \uparrow \omega \models (\forall)(G_0\theta)$.

Proof: We will prove by induction on n that for all natural numbers n, $T_Q \uparrow n \models (\forall)(G_0\sigma)$ implies the existence of an SLDq-refutation having θ as the R-computed answer substitution where θ is some substitution more general than σ.

Base Case: $[n = 1]$ If $T_Q \uparrow 1 \models (\forall)(G_0\sigma)$, then for each ground instance $A_i : \mu_i$, there is a q-clause in Q of the form

$$A_i' : \rho_i \Leftarrow$$

where $\rho_i \succeq \mu_i$ and $A_i\theta_i = A_i'$. Suppose $j_1, \ldots j_k$ is the order in which the literals of G_0 are selected by R, then the following diagram (Fig. 5) gives the SLDq-refutation of G_0 via R. Also, the restriction θ of $\theta_1 \ldots \theta_n$ to the variable symbols of G_0 is more general than σ. Thus, as $T_Q \uparrow \omega$ is a model of each of the clauses of the form $A_i' : \rho_i$, we have $T_Q \uparrow \omega \models (\forall)(G_0\theta)$.

Inductive Case Similar. ∎

With some ingenuity, it can be shown that the above completeness result can be made even stronger - viz. the restriction that Q be function-free can be removed. The proof proceeds along the same lines as the proof of the completeness of SLD-resolution given in [10]. We will just state the following lemmas without proof.

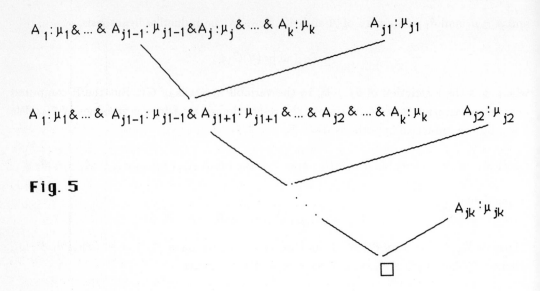

$$A_1:\mu_1\,\&\,\cdots\,\&\,A_{j1-1}:\mu_{j1-1}\,\&\,A_j:\mu_j\,\&\,\cdots\,\&\,A_k:\mu_k \qquad A_{j1}:\mu_{j1}$$

$$A_1:\mu_1\,\&\,\cdots\,\&\,A_{j1-1}:\mu_{j1-1}\,\&\,A_{j1+1}:\mu_{j1+1}\,\&\,\cdots\,\&\,A_{j2}\,\&\,\cdots\,\&\,A_k:\mu_k \qquad A_{j2}:\mu_{j2}$$

$$A_{jk}:\mu_{jk}$$

Fig. 5

$\square$

<u>DEF 40</u>: An *unrestricted* SLDq-refutation is an SLDq-refutation, except that we drop the requirement that the substitutions θ_i be most general unifiers. They need only be unifiers.

Lemma 41: [mgu Lemma] Suppose Q is a QLP and G a goal such that there is an unrestricted SLDq-refutation of G of length n. Then there is an SLDq-refutation of G from Q of the same length. In addition, if $\theta_1, \ldots, \theta_n$ are the unifiers from the unrestricted SLDq-refutation, and if $\theta'_1, \ldots, \theta'_n$ are the mgu's from the SLDq-refutation, then there is a substitution γ such that $\theta_1 \cdots \theta_n = \theta'_1 \cdots \theta'_n \gamma$.

Proof: Essentially the same as the proof of Lemma 8.1 of Lloyd [10]. ∎

Lemma 42: [Lifting Lemma] Let Q be a QLP, G a goal and θ a substitution. Suppose there exists an SLDq-refutation of $G\theta$ from Q. Then there is an SLDq-refutation of G from Q of the same length. Furthermore, if $\theta_1, \cdots, \theta_n$ are the mgu's from the SLDq-refutation of $G\theta$ and $\theta'_1, \ldots, \theta'_n$ are the mgu's from the SLDq-refutation of G from Q, then there is a substitution γ such that $\theta\theta_1 \cdots \theta_n = \theta'_1 \cdots \theta'_n \gamma$.

Proof: Essentially the same as the the proof of Lemma 8.2 of Lloyd [10]. ∎

Theorem 43: If Q is a nice QLP and $A \in B_Q$ is a ground atom such that $T_Q \uparrow \omega \models A : \psi$, then there is an SLDq-refutation of the query $A : \psi$ from Q.

Proof: Suppose $T_Q \uparrow \omega \models A : \psi$. Then, for some integer n, $T_Q \uparrow n \models A : \psi$. It can now be proved by induction on n that $A : \psi$ has an SLDq-refutation from Q.

Base Case: [$n = 1$] If $T_Q \uparrow 1 \models A : \psi$, then there is a q-clause in Q of the form

$$A' : \rho \Leftarrow$$

such that A, A' are unifiable and $\rho \succeq \psi$. Clearly, $A : \psi$ has an SLDq-refutation from Q.

Inductive Case: Suppose $T_Q \uparrow (n+1) \models A : \psi$. Then there is a ground instance of a q-clause of the form

$$A\prime : \rho \Leftarrow B_1 : \mu_1 \& \ldots \& B_m : \mu_m$$

such that $A\theta = A'\theta$ and $T_Q \uparrow n \models B_1\theta : \mu_1 \& \ldots \& B_m\theta : \mu_m$, where each $B_i\theta$ is ground. By the induction hypothesis, each $B_i\theta : \mu_i$ has a SLDq-refutation, and so these can be combined into an SLDq-refutation of $(B_1 \& \ldots \& B_m)\theta$ from Q. Thus, $A : \psi$ has an *unrestricted* SLDq-refutation from Q, and so by the mgu lemma, $A : \psi$ has an SLDq-refutation from Q. ∎

Theorem 44: [Strong Completeness] Let Q be a nice QLP and G a goal. Suppose $T_Q \uparrow \omega \models (\exists)G$. Then there is a computation rule R and an SLDq-refutation of G from Q via R.

Proof: Suppose $T_Q \uparrow \omega \models (\exists)G$, where G is the goal $(\exists)(A_1 : \psi_1 \& \ldots \& A_k : \psi_k)$. Hence, there is a ground instance $G\theta$ of G such that $T_Q \uparrow \omega \models G\theta$. By Theorem 43, $G\theta$ has an SLDq-refutation from Q. By the Lifting Lemma, there exists an SLDq-refutation of G from Q. ∎

Example 45. Consider the following nice QLP and its associated and/or tree..

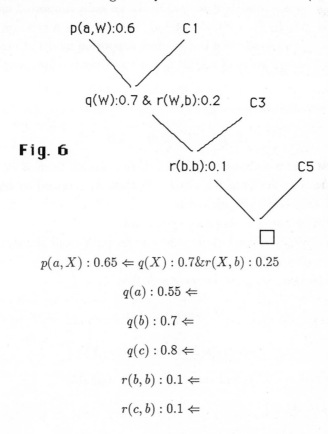

Fig. 6

$$p(a, X) : 0.65 \Leftarrow q(X) : 0.7 \& r(X, b) : 0.25$$

$$q(a) : 0.55 \Leftarrow$$

$$q(b) : 0.7 \Leftarrow$$

$$q(c) : 0.8 \Leftarrow$$

$$r(b, b) : 0.1 \Leftarrow$$

$$r(c, b) : 0.1 \Leftarrow$$

and the query $(\exists)(p(a, W) : 0.6)$. Fig. 6 shows an SLDq-refutation of this query from Q. We have thus shown that SLDq-resolution can be used compute answer substitutions in much the same way as SLD-resolution is used to compute answer substitutions in pure logic programming. What is also fascinating is that many of the properties of SLD-resolution generalize in a very simple and straightforward manner to the case of SLDq-refutations. Lemmas 41 and 42 are examples in this regard. A close inspection of the proofs of Lemmas 8.1 and 8.2 of [10] shows that they can be almost directly transcribed into proofs of Lemmas 41 and 42 respectively. In addition, the and-or tree construction procedure described earlier strengthens considerably the results obtained earlier in [12] and [14] where only ground queries could be handled, and only well-covered QLPs could be considered.

4. On Computing the Greatest Supported Model

Theorem 18 asserts that the greatest supported model of a decent, nice QLP is (partially) computable. It has been shown in [12, Theorem 56] that this leads to a form of default reasoning called s-consistency. Essentially, a ground query $A : \mu$ is s-consistent with the QLP Q iff some supported model of Q satisfies $A : \mu$. For decent, nice QLPs, this means that $A : \mu$ is s-consistent with Q iff the greatest supported model of Q (which, by Theorem 18 is just $T_Q \downarrow \omega$) satisfies $A : \mu$. There is thus a need for a computational procedure for (partially) computing the greatest supported model of nice, decent QLPs.
DEF 46: The *dependency graph* of a QLP Q is a (possibly infinite) graph on

$$\{A : \mu | A \in B_Q \text{ and } \mu \text{ is in } [0, 1]\}$$

such that if

$$A : \mu \Leftarrow B_1 : \psi_1 \& \dots \& B_k : \psi_k$$

is a ground instance of a q-clause in Q, then there is an arc from $A : \mu$ to $B_i : \mu_i$ for all $1 \leq i \leq k$. If there is an arc from $A : \rho$ to $B : \rho'$, then $A : \rho$ is said to *depend on* $B : \rho'$.
DEF 47: A QLP Q is said to be *cycle-free*
(1) if its dependency graph contains no cycles and
(2) for every $A \in B_Q$ and any μ_1, μ_2, there is no path from $A : \mu_1$ to $A : \mu_2$ in the dependency graph of Q.
Example 48. The QLP Q_1 given below is cyle-free

$$even(0) : 1 \Leftarrow$$

$$even(s(s(X))) : 1 \Leftarrow even(X) : 1$$

This is because $even(s(s(X)))$ and $even(X)$ are not unifiable.

Algorithm 49 $gsm(Q, A)$

INPUT: A nice, decent, cycle-free, well-covered QLP Q and a cr-query $A :? (A \in B_Q)$.

OUTPUT: A truth value μ such that $T_Q \downarrow \omega(A) = \mu$.

(1) compute $CL(Q, A)$, the set of q-clauses in Q whose head's atomic part unifies with A.

(2) compute $AN(Q, A)$, the multiset of annotations of heads of q-clauses in $CL(Q, A)$.

(3) **if** $CL(Q, A) = \emptyset$ **then** $gsm(Q, A) := 0.5$. **halt.**

(4) **else**

    (4.1) Select the maximal element ρ in $AN(Q, A)$ and a q-clause C in $CL(Q, A)$ whose head contains the annotation ρ. [Such a maximal element is guaranteed to exist as Q is nice]. $gsm(Q, A) := \rho$.

    (4.2) Suppose C is $B : \rho \Leftarrow B_1 : \psi_1 \& \ldots \& B_m : \psi_m$ and θ is the mgu of A and B. [As Q is well-covered, $(B_1 : \psi_1 \& \ldots \& B_m : \psi_m)\theta$ is ground.

    (4.2.1) If $(\forall i)[1 \leq i \leq m \& B_i : \psi_i$ does not depend on $A : \rho \Rightarrow gsm(Q, B_i) \succeq \psi_i]$, then **halt.**

    (4.2.2) **else** $CL(A) := CL(A) - \{C\}$;   Delete one copy of ρ from $AN(A)$.

    (4.3) goto (3).

end gsm.

End Algorithm 49

Once we focus attention on a particular QLP Q, the above algorithm computes the function

$$\lambda A.GSM(Q, A)$$

which we denote by ζ_Q; a function from B_Q to $[0, 1] \cup \{\top\}$. Thus, ζ_Q is an Herbrand interpretation.

Theorem 50: [Soundness] If Q is a QLP, then $T_Q \downarrow \omega \succeq \zeta_Q$.

Proof: It suffices to show that ζ_Q is a supported model of Q. Suppose $\zeta_Q(A) = \rho \neq 0.5$ for some $A \in B_Q$. Then $GSM(Q, A) = \rho$, i.e. there exists a q-clause C in $CL(Q, A)$ such that

$$A : \rho \Leftarrow B_1 : \psi_1 \& \ldots \& B_k : \psi_k$$

is a ground instance of C and such that $GSM(Q, B_i) \succeq \psi_i$ for all $1 \leq i \leq k$, i.e. $\zeta_Q(B_i) \succeq \psi_i$ for all $1 \leq i \leq k$, whence ζ_Q is supported. ∎

Theorem 51: [Completeness] If Q is a nice, decent, well covered, cycle free QLP and if $A \in B_Q$, then $\zeta_Q(A) \succeq T_Q \downarrow \omega(A)$.

Proof: The proof proceeds by induction on the number of times step (4.3) of the algorithm is reached.

Base Case: $[n = 0]$ In this case, $GSM(Q, A)$ is the maximal annotation in the heads of clauses whose head's atomic part unifies with A; hence $GSM(Q, A) \succeq T_Q \downarrow \omega(A)$.

Inductive Case: $[n+1]$ We need to show that at Step $(n+1)$, $GSM(Q,A) \succeq T_Q \downarrow \omega(A)$. At Step n $GSM(Q,B) \succeq T_Q \downarrow \omega(B)$ for all atoms $B \in B_Q$. To enter step $(n+1)$, $GSM(Q,B_i) \not\succeq \psi_i$ for some $1 \leq i \leq k$ where the selected clause C in G is:

$$A : \rho \Leftarrow B_1 : \psi_1 \& \ldots \& B_k : \psi_k$$

Thus, by the induction hypothesis, $T_Q \downarrow \omega(B_i) \not\succeq \psi_i$ for some $1 \leq i \leq k$. Thus, at Step $(n+1)$, the GSM algorithm selects the clause with the largest possible annotation, whence

$$GSM(Q,A) \succeq T_Q \downarrow \omega(A) \qquad \blacksquare$$

Example 52. Consider the nice, decent, well-covered QLP Q given below:

$$C1 : \qquad p(a) : 0.7 \Leftarrow p(a) : 0.7$$

$$C2 : \qquad p(b) : 0.8 \Leftarrow p(a) : 0.6$$

(1) Consider the query $p(a)$:?. Initially, $CL(Q,p(a)) = \{C1\}$ and $AN(Q,p(a)) = \{0.7\}$. The condition in step (4.2.1) of the *gsm* algorithm is satisfied (as there is no $1 \leq i \leq m$ distinct from $p(a)$ - here m=1), and the algorithm halts with success. $gsm(Q,p(a)) = 0.7$. (2) If the query $p(b)$ is posed, then $CL(Q,p(b)) = \{C2\}$ and $AN(Q,p(b)) = \{0.8\}$. The *gsm* algorithm is called recursively with arguments Q and $p(a)$. $gsm(Q,p(a)) = 0.7 \succeq 0.6$, and so the condition of (4.2.1) is satisfied. The algorithm halts, returning $gsm(Q,p(b)) = 0.8$. Note that in this example, the QLP Q is *not* cycle-free, but the *gsm* algorithm still halts giving the correct answer. We now give an example where s-consistency plays an important role.

Example 53. Consider a (very highly simplified) murder case in which the unfortunate victim is John. The facts may be established in terms of a QLP as follows:

C1:	$murdered(john) : 1 \Leftarrow$
C2:	$murderer(john, X) : 0.95 \Leftarrow motive(X) : 0.85 \quad \& \quad opportunity(X) : 0.95$
C3:	$murderer(john, X) : 0.98 \Leftarrow lying(X) : 0.99$
C4:	$motive(butler) : 0.90 \Leftarrow$
C5:	$motive(chauffeur) : 0.85 \Leftarrow$
C6:	$opportunity(butler) : 0.95 \Leftarrow$
C7:	$opportunity(chauffeur) : 0.99 \Leftarrow$
C8:	$lying(chauffeur) : 1.00 \Leftarrow$
C9:	$lying(butler) : 0.9 \Leftarrow$

The above q-clauses say, respectively, that

(1) John was definitely murdered.

(2) The murderer must have had a strong motive, as well as an excellent opportunity to commit the crime.

(3) Anyone who is lying automatically becomes a big suspect.

(4) The chauffeur had a good motive, though the butler had an even stronger motive.

(4) Both the chauffeur and the butler had a good opportunity to commit the murder, though the chauffeur had the better opportunity.

(5) The chauffeur has admitted to lying; however, some of the statements made by the butler sound very phony.

s-consistency allows us to ask questions like:

(a) Is it conceivable that the gardener committed the crime ?. The *gsm* algorithm will reply that there is no support for this belief.

(b) Is it conceivable that the butler committed the crime? The *gsm* algorithm says that it has 95 percent belief that he did.

(c) Is it conceivable that the chauffeur committed the crime ? The *gsm* algorithm says that this is even more likely, viz. it believes (99 percent) that the chauffeur did it.

5. Conclusions

If logic programming is to be a pragmatic tool for the construction of expert systems and the design of very large knowledge bases, it must have some means for dealing with uncertain information as well as inconsistent information. The framewok developed in [12] provides such a facility – moreover, it has the advantage of being semantically well founded.

We have extended the results of [12] in this paper by giving an operational semantics that is complete for quantitative logic programs that are neither nice nor well-covered. That this semantics is restricted to Herbrand interpretations only is not important – it is shown in [2] that such semantics can be easily extended to be applicable to interpretations based on pre-interpretations that satisfy certain conditions concerning the treatment of equality. In addition, we have been able to give an operational mechanism for computing the greatest supported model of nice, decent, well-covered QLPs. As is evident from Example 53, this has applications in reasoning about beliefs, and in processing questions like "Is it conceivable that *p* holds ?" where *p* is some proposition. Further work to strengthen the results on computing the GSM would involve relaxing the *well-coveredness* and *cycle-freedom* conditions. However, Theorem 18 does not allow the conditions of niceness and decency to be dropped, because doing so makes the greatest supported model non-recursively enumerable, and hence, uncomputable.

Acknowledgements. I have benefited from discussions on quantitative deduction with

Professor Howard Blair and Aida Batarekh. This work was supported by U.S. Air Force Contract F30602-85-C-0008.

REFERENCES

[1] Apt,K.,Blair,H.,Walker,A. "Towards a Theory of Declarative Knowledge", to appear, 1988.

[2] Blair, H.A., Brown, A.L., Subrahmanian, V.S. "A Logic Programming Semantics Scheme, Part I" , Logic Programming Research Group Tech. Report LPRG-TR-88-8, Syracuse University.

[3] Blair,H.A., Subrahmanian,V.S. "Paraconsistent Logic Programming" , *Proc. 7th Intl. Conf. on Foundations of Software Tech. & Theoretical Computer Science*, Lecture Notes in Computer Science Vol. 287, pps 340-360, Springer Verlag, Dec. 1987.

[4] Blair,H.A., Subrahmanian,V.S. "Foundations of Generally Horn Logic Programming" , in preparation.

[5] da Costa,N.C.A. "On the Theory of Inconsistent Formal Systems" , *Notre Dame J. of Formal Logic*, 15, pps 497-510, 1974.

[6] da Costa,N.C.A. "A Semantical Analysis of the Calculi C_n" , *Notre Dame J. of Formal Logic*, 18, pps 621-630, 1977.

[7] da Costa,N.C.A., Alves,E.H. "Relations between Paraconsistent Logic and Many Valued Logic" , *Bull. of the Section of Logic*, 10,pps 185-191, 1981.

[8] Fitting,M. "A Kripke-Kleene semantics for logic programs" , *J. of Logic Programming*, 2,4, pps 295-312, 1985.

[9] Fitting,M. "Notes on the Mathematical Aspects of Kripke's Theory of Truth" , *Notre Dame J. of Formal Logic*, 27, 1, pps 75-88, 1986.

[10] Lloyd,J.W. "Foundations of Logic Programming" , Springer.

[11] Lassez,J.-L.,Maher,M. "Optimal Fixed-points of Logic Programs" , *Theoret. Comput. Sci.*, 39, pps 115-125, 1985.

[12] Subrahmanian,V.S. "On the Semantics of Quantitative Logic Programs" , *Proc. 4th IEEE Symp. on Logic Prog.*, pps 173-182, San Francisco, Sep. 1987.

[13] Subrahmanian,V.S. "Towards a Theory of Evidential Reasoning in Logic Programming" , *Logic Colloquium '87*, Granada, Spain, July 1987.

[14] Van Emden,M.H. "Quantitative Deduction and its Fixpoint Theory" , *J. of Logic Prog.*, 4,1, pps 37-53, 1986.

An Environment For Automated Reasoning About Partial Functions *

David A. Basin

Department of Computer Science,
Cornell University, Ithaca, NY 14853

Abstract

We report on a new environment developed and implemented inside the Nuprl type theory that facilitates proving theorems about partial functions. It is the first such automated type-theoretic account of partiality. We demonstrate that such an environment can be used effectively for proving theorems about computability and for developing partial programs with correctness proofs. This extends the well-known proofs as programs paradigm to partial functions.

Key words and phrases. Automated program development, computability, constructivity, partial functions, tactics, theorem proving, type theory, unsolvability.

1   Introduction

Over the past 20 years, research by Martin-Löf [8], Constable [3], Coquand and Huet [6], and others has demonstrated that constructive type theory provides a useful foundation for theorem proving and program development. The Nuprl proof development system, developed at Cornell [4], has been used to demonstrate that type theory, in practice, provides a rich framework for theorem proving. However, current type theories are inadequate for reasoning about partial functions. Partial functions cannot be typed in full generality and must be approximated as total functions on subsets of their domain of convergence. This approach is problematic as the exact domain often cannot be represented. Also, these theories provide no means for abstract reasoning about termination, nor do they provide fixpoint induction rules. As a result, they are ill suited for proving theorems about partial computations and for developing partial programs.

---

*This research was supported in part by NSF grant DCR83-03327

Recent work by Constable and Smith has established a theoretical foundation for reasoning about partial objects in type theory. In [5], they present a method of extending the logic of Nuprl to reason about nontermination. A new type constructor, embodied in the bar operator $\overline{T}$ on type T, is added. $\overline{T}$ is inhabited by terms that represent computations of elements in T. An inhabitant of $\overline{T}$ may diverge, but if it terminates, it converges to an inhabitant of T. For example, the type $\text{int} \rightarrow \overline{\text{int}}$ corresponds to the standard notion of a partial function space; it is inhabited by functions that take an integer argument and, if they converge on their input, return an integer.

We demonstrate that this partial type theory, along with proof assisting tactics, can be implemented within Nuprl. Once implemented, the resultant partial type environment is surprisingly powerful. We simultaneously gain the ability to give concise proofs of theorems about computation and we can extend the "proofs as programs" paradigm to partial program development via the type-theoretic equivalent of partial correctness reasoning. Hence, we dramatically increase the power of Nuprl, both as a theorem proving system and as a program development system.

In the next section, we describe the environment. In the third section, we show the development of a recursion theoretic proof in this environment. In section 4, we present a new paradigm for program development and provide an example of its use. Finally, we draw conclusions from our research.

2 The Environment

A type theory may be specified by defining the terms of the theory, and an evaluation relation, along with rules defining types, type membership, and type equality. Constable and Smith [12] extend the Nuprl type theory to a partial type theory as follows: The terms of the partial type theory are the terms of the underlying theory, in our case Nuprl, plus a new term $\text{fix}(f, x. b)$. The evaluation relationship is augmented to reflect the redex/contractum pair shown below.

$$\text{fix}(f, x. b)(a), \quad b[\text{fix}(f, x. b), a/f, x]$$

The rules for defining types, type membership, and type equality, are augmented with rules for reasoning about the partial types. A complete list of rules for the partial type theory may be found in [12]. Some representative ones, which we present refinement style[1], are given in figure 1.[2] BarCTotality provides one way

---

[1] In the presentation of a rule, the first line contains the goal to which the rule is applied. The ">>" symbol is the printable equivalent of the logical turnstile, and H represents a (possibly empty) list of hypotheses. After the goal, indented lines contain the subgoals that result from the application of the refinement rule. The reader is referred to [4] for a complete description of Nuprl and refinement style theorem proving.

[2] As the partial type theory is still in development, the final rules may differ.

1. $H \gg t$ in $\overline{T}$ by `BarCTotality`
   $H \gg t$ in T

2. $H \gg t$ in! T by `BarInIntro`
   $H \gg t$ in T

3. $H \gg \texttt{fix}(f, x.b)$ in $x : A \texttt{->} \overline{B}$ by `BarFix` $\texttt{U}i$
   $H, f : (x : A \texttt{->} \overline{B}), x : A \gg b$ in $\overline{B}$
   $H \gg A$ in $\texttt{U}i$
   $H, x : A \gg B$ in $\overline{\texttt{U}i}$

Figure 1: Sample Partial Type Rules

of demonstrating that t is in a type $\overline{T}$, namely by showing that t is total, i.e., t in T. Constable and Smith's theory comes with a termination predicate **in!** that allows reasoning about termination. If t converges in the type $\overline{T}$ then t **in!** T, meaning t is in T. By rule 2, when we can prove that t inhabits T, we can then prove t **in!** T. Perhaps the most interesting rule is `BarFix` which allows us to type partial functions. Given a functional $\lambda f.\lambda x.b$, where f inhabits $A \to \overline{B}$ and x inhabits A, then if we can demonstrate that the body b inhabits $\overline{B}$ and that A and $\overline{B}$ are types, then, by `BarFix`, the fixpoint $\texttt{fix}(f, x.b)$ inhabits $A \to \overline{B}$. This is similar to rules used in typing recursive programs that state that a recursive function is well typed when its body is well typed under the assumption that its arguments and recursive calls are appropriately typed. It is by this rule that partial functions, such as unbounded search, may be typed. We will see that this rule is also useful for proving goals via fixpoint induction.

Rather than modifying the Nuprl source code, these new terms and rules are added by appropriately endowing a Nuprl library with assumptions and tactics. For example, as any term in the λ-calculus is a Nuprl term, we define the new fixpoint term $\texttt{fix}(f, x.b)$ as $Y(\lambda f.\lambda x.b)$, where Y is the fixpoint combinator $\lambda f.((\lambda x.f(xx))(\lambda x.f(xx)))$. This defined term now reduces by ordinary β-reduction and is equal to $\texttt{fix}(f, x.b)$ in any type that either inhabits. The addition of new equality and membership rules is more complex. For each new rule, a tactic is created whose application mimics that of a new primitive inference rule. We suppress details as future versions of Nuprl will contain a context mechanism for implementing new rules.

With the refinement rules implemented, the environment is completed with the addition of proof assisting tactics. These tactics partially automate the theorem proving process and thereby relieve the user from many tedious proof details. The most important tactic is a partial type `Autotactic` which is automatically applied to subgoals that arise after each refinement step. This tactic contains modules that perform type checking, propositional and arithmetic reasoning, and backchaining.

We extend this tactic to automatically prove the well-formedness of most partial type expressions and perform simple kinds of reasoning about partial types. Another important tactic is `FixInd` which helps prove goals via fixpoint induction. We shall discuss this tactic in section 4.

3 Partial Object Proofs

Our initial motivation in implementing the partial type theory was to design an environment in which we can prove abstract versions of recursion-theoretic theorems. It was our hope that, by making the Nuprl type theory more expressive, we could then prove interesting facts about recursive and recursively enumerable sets, complete sets, reducibilities, and unsolvable problems. Moreover, we hoped that the resulting proofs would be comprehensible so that those who wished to learn recursion theory could use the environment and the proofs as an interactive textbook.

Our experience to date with the environment has been positive. We have used it to prove the undecidability of the halting problem, Rice's theorem, and have shown that not all recursively enumerable sets are recursive. Some of these proofs involve subtle reasoning and reductions of previously proved problems. We have also begun using the environment to explore a theory of abstract fixpoint algebras in which we prove theorems of recursion theory. In this abstract setting, it is easy to err, and the environment has proved valuable as a proof checker.

In this section, we present one of the proofs developed in the environment, a proof of the undecidability of the halting problem. This proof is interesting for several reasons. First, it demonstrates that the partial type theory is expressive enough to construct partial functions and reason about their termination properties. Second, it demonstrates the power of partial type tactics in theorem proving. General purpose tactics automatically type the diagonalizing combinator, prove all well-formedness subgoals, and manipulate fixpoint terms. Finally, the proof is direct. With the exception of some equality reasoning, the refinement steps are concise and the amount of reasoning compares favorably with that found in many textbook proofs.

The undecidability of the halting problem is informally: there does not exist a total procedure h such that, when given a computation x, h decides if x terminates. The problem's formalization depends on how one represents computations (e.g., partial recursive functions, λ-calculus terms) and our formalization takes place within the partial type theory. Here, computations that yield integer values are inhabitants of $\overline{\text{int}}$. A total decision procedure that takes a computation and returns a value inhabits $\overline{\text{int}} \rightarrow \text{int}$. And the convergence of a computation x in $\overline{\text{int}}$ is expressed by the convergence predicate `in!`, i.e., x `in!` `int`. Thus, we formalize the

```
* DEF bot
⊥ == fix(f,x.f(x))(0)
* DEF d
d == fix(f,x.int_eq(h(f(x));1;⊥;1))
```

Figure 2: Nuprl Definitions For Halting Problem.

theorem as:

>> ¬(∃h:$\overline{int}$→int.∀x:$\overline{int}$.x in! int ⇔ h(x) = 1 in int).

Our proof of this statement is similar to traditional proofs found in such textbooks as [10]. It does not, however, rely on an indexing of the partial recursive functions. Informally, our proof is as follows: Assume that a decision procedure h exists. To derive a contradiction, define a diagonal function d as

$$d(x) = \begin{cases} \bot & \text{if } h(d(x)) = 1 \\ 1 & \text{otherwise.} \end{cases}$$

Our Nuprl definition for d is given in figure 2. Now consider the value of $h(d(1))$. There are two cases, and as h is total, we can decide which holds. The first case is when $h(d(1)) = 1$. Then, by definition, $d(1) = \texttt{int\_eq}(h(d(1)); 1; \bot; 1)$ which evaulates to $\bot$ when $h(d(1)) = 1$ and 1 otherwise. Thus, by direct computation $d(1) = \bot$ and, by the definition of h, $h(d(1)) \neq 1$ as $\bot$ diverges. Hence, $h(d(1)) = 1 \Rightarrow h(d(1)) \neq 1$ yields a contradiction. The other case, $h(d(1)) \neq 1$, is argued similarly.

A linearized proof tree of the actual theorem is given in figure 3. Here, an expression's level of indentation indicates its position in the tree. Goals are preceded by >> and their hypotheses are the numbered lines found above at lesser indentation levels. Goals are followed by refinement rules and subgoals, if any, occur below at one deeper level of indentation.

Through the use of general purpose tactics, the logical structure of the proof closely follows our informal explanation. Initially, we apply Intro and elim refinement rules, which set up a proof by contradiction by assuming the existence of the decision procedure h. Then, using a tactic Seq[3], we add d to the hypothesis list and use a partial type tactic ReduceFixConcl to unwind the fixpoint combinator and perform the β-reductions necessary for equality reasoning. The Decide tactic sets up the case analysis. In the case $h(d(1)) = 1$, we cut in and prove that $d(1) = \bot$.

---

[3]Seq is short for sequence and is used like cut in logic to add new facts to the hypothesis list. Any sequenced fact must be proved. However, this is often automatically done by Autotactic. For a complete description of all tactics found here, see [7].

```
>> ¬(∃h:int->int. ∀x:int. x in! int <=> h(x) = 1 in int)
| BY (Intro ...)
| 1. ∃h:int->int. ∀x:int. x in! int <=> h(x) = 1 in int
|- >> void
| | BY elim 1 new h
| | 2. h:int->int
| | 3. ∀x:int. x in! int <=> h(x) = 1 in int
| |- >> void
| | | BY (Seq ['d in int->int'; 'd(1)=int_eq(h(d(1));1;⊥;1) in int'] ...)
| | | 4. d in int->int
| | |- >> d(1)=int_eq(h(d(1));1;⊥;1) in int
| | | | BY (ReduceFixConcl 'z' 'z(1)=int_eq(h(d(1));1;⊥;1) in int' ...)
| | | 4. d in int->int
| | | 5. d(1)=int_eq(h(d(1));1;⊥;1) in int
| | |- >> void
| | | | BY (Decide 'h(d(1))=1 in int' ...)
| | | 6. h(d(1))=1 in int
| | |- >> void
| | | | | BY (Seq ['int_eq(h(d(1));1;⊥;1)=⊥ in int';'d(1)=⊥ in int'] ...) THEN Try
| | | | |     (ReduceDecisionTerm 1 true ...)
| | | | | 7. int_eq(h(d(1));1;⊥;1)=⊥ in int
| | | | | 8. d(1)=⊥ in int
| | | | |- >> void
| | | | | | BY (EOn '⊥' 3 ...) THEN (OnLastHyp Elim ...)
| | | | | | 9. ⊥ in! int->h(⊥)=1 in int
| | | | | | 10. h(⊥)=1 in int->⊥ in! int
| | | | | |- >> h(⊥)=1 in int
| | | | | | | BY (SubstForInHyp 'd(1)=⊥ in int' 6 ...)
| | | | | | 9. ⊥ in! int->h(⊥)=1 in int
| | | | | | 10. h(⊥)=1 in int->⊥ in! int
| | | | | | 11. ⊥ in! int
| | | | | |- >> void
| | | | | | | BY (NoBotConv [11] ...)
| | | | 6. ¬(h(d(1))=1 in int)
| | | |- >> void
| | | | | BY (Seq ['int_eq(h(d(1));1;⊥;1) = 1 in int'; 'd(1) = 1 in int'] ...) THEN Try
| | | | |     (ReduceDecisionTerm 1 false ...)
| | | | | 7. int_eq(h(d(1));1;⊥;1) = 1 in int
| | | | | 8. d(1) = 1 in int
| | | | |- >> void
| | | | | | BY (EOn '1' 3 ...) THEN OnNthLastHyp 2 Elim THEN Try (BarInIntro ...)
| | | | | | 9. 1 in! int->h(1)=1 in int
| | | | | | 10. h(1)=1 in int->1 in! int)
| | | | | | 11. h(1)=1 in int
| | | | | |- >> void
| | | | | | | BY (SubstForInHyp 'd(1) = 1 in int' 6 ...)
```

Figure 3: Proof Of The Undecidability Of The Halting Problem.

Then, we introduce $\perp$ into hypothesis 3, which yields that $(h(\perp) = 1) \Rightarrow (\perp$ in! int). This allows us to substitute $d(1) = \perp$ into the hypothesis $h(d(1)) = 1$ and conclude that $\perp$ in! int. But as $\perp$ diverges, the tactic NoBotConv yields a contradiction, i.e., a proof that the empty type void is inhabited. Again, the other case is proven similarly.

It is interesting to compare our proof with the Boyer and Moore proof [2], the first machine verified proof of the unsolvability of the halting problem. They took four days to create their proof outline, which spans five pages of text, and contains 29 definitions and theorems, the final theorem being the unsolvability of the halting problem. Their proof took 75 minutes to verify on a DEC 2060. On the other hand, our type theory comes equipped with much of the computational machinery they needed to build. As a result, our proof development took under an hour. It consists of only two definitions, d and $\perp$, and no preliminary lemmas. Twelve refinement steps were required, each using primitive refinement rules or general purpose tactics, and each reflecting a step in our informal proof. The proof was verified by a Symbolics 3670 Lisp Machine in under 20 seconds.

4 Partial Program Development

As our partial type theory is constructive, proofs contain information on building witnesses. For example, the constructive content of a proof of

$$\gg \forall x : T. \exists y : T'. R(x,y) \tag{*}$$

is a function that for each x in type T yields a pair: a y in T' and a proof of R(x,y). This program can be automatically extracted from a proof with the Nuprl term_of operator and executed with the Nuprl evaluator. Thus, the type theory provides a natural way of interpreting proofs as programs.

In the base type theory, all extracted terms normalize to canonical terms. This is the type theoretic analogue of total correctness: a proof P of goal G in Nuprl is a demonstration that term_of(P) terminates and inhabits G. Hence, in the base type theory, we may develop programs by giving total correctness specifications in the form of $(*)$, where R(x,y) uses the propositions-as-types principle to express the desired specification as a type. This approach to program development has been used by Howe [7] and others to develop saddleback search, quicksort, and integer factoring programs.

Total correctness is a strong requirement. As the termination of some programs is difficult to prove, one needs the weaker requirement of partial correctness. Analogously, when given a proof P of goal G, we call term_of(P) partially correct with respect to its specification G if when it converges, it converges to a value inhabiting G. By extending the Nuprl type theory to a partial type theory, we extend our ability to automatically synthesize programs meeting total correctness specifications to those meeting partial correctness specifications. This follows from the semantics given to the partial types. They guarantee that if an object in $\overline{T}$ converges, then it inhabits the underlying type T. Hence, by placing a bar over all or part of the goal, we attain such a specification. For example, if we are trying to extract a partial function in $T \rightarrow T'$, then instead of stating a specification in the form of (*), our new goal becomes

$$\gg \forall \texttt{x:T}. \overline{\exists \texttt{y:T'.R(x,y)}}.$$

The extracted object from a proof of the above will be a function that takes an x in T, and, if the function converges, produces a witness y in T' such that $\texttt{R(x,y)}$.

As an example, consider a simple program that returns the integer square root y of an integer x. Our partial correctness specification for this program is the following.

$$\gg \forall \texttt{x:int}. \overline{\exists \texttt{y:int where y*y = x}} \qquad (**)$$

Such a program could be developed in a total type theory only by altering the specification; for example, finding the largest y such that $y * y \leq x$. However, in our extended theory, we can prove the partial specification and extract a program from it that computes the square root of integers that are perfect squares.

In figure 4, we give our proof of (**). The highlight of this proof is the way in which fixpoint induction is used to prove partial type inhabitation. Here, the BarFix refinement rule provides a not necessarily well-founded induction principle which can be used to prove the existence of a partial function meeting our correctness specification. A tactic FixInd is used that applies BarFix to the goal and manipulates the resulting subgoals to set up the inductive proof.

The proof begins with the tactic Intro which moves x, the integer whose square root we seek, into the hypothesis list. The goal is now of the form $\overline{P}$ where P is

$$\exists \texttt{y:int where y*y = x}.$$

To make an induction argument go through, we cut in a different hypothesis than $\overline{P}$, namely $\forall \texttt{z:int}.\overline{P}$. Under the propositions as types interpretation, this new proposition is the function type $\texttt{z:int} \rightarrow \overline{P}$ and, by BarFix, this type is inhabited by a fixpoint term whenever $(\texttt{z:int} \rightarrow \overline{P}) \rightarrow (\texttt{z:int} \rightarrow \overline{P})$ is inhabited. But the inhabitation of the latter type is equivalent (by two applications of Intro) to demonstrating that $\overline{P}$ is inhabited given the hypotheses $\texttt{z:int} \rightarrow \overline{P}$ and $\texttt{z:int}$. Thus, by introducing the fixpoint term, FixInd sets up a proof that $\overline{P}$ is inhabited, giving us a function in $\texttt{z:int} \rightarrow \overline{P}$ and z in int. This construction is subtle and, if we are not careful, Autotactic will use these two new hypotheses to prove $\overline{P}$ by introducing $\bot$, a diverging term that trivially satisfies our specification. Instead, we prove $\overline{P}$ by case analysis on $z * z = x$. When $z * z = x$, we explicitly introduce z as our witness for $\overline{P}$. Using BarCTotality (rule 1, figure 1), Autotactic proves that z inhabits T by hypothesis 4. In the case $z * z \neq x$, we use the function in $\texttt{z:int} \rightarrow \overline{P}$, which

```
>> ∀x:int. ∃y:int where y*y = x
|  BY (Intro ...)
|  1. x:int
|- >> ∃y:int where y*y = x
|  |  BY Seq ['∀z:int. ∃y:int where y*y = x']
|  |- >> ∀z:int. ∃y:int where y*y = x
|  |  |  BY FixInd
|  |  |  2. ∀z:int. ∃y:int where y*y = x
|  |  |  3. z:int
|  |  |- >> ∃y:int where y*y = x
|  |  |  |  BY (Decide 'z*z = x' ...)
|  |  |  |  4. z*z = x
|  |  |  |- >> ∃y:int where y*y = x
|  |  |  |  |  BY (ExplicitI 'z' + BarCTotality)
|  |  |  |  4. ¬(z*z = x)
|  |  |  |- >> ∃y:int where y*y = x
|  |  |  |  |  BY (EOn 'z+1' 2 ...)
|  |  2. ∀z:int. ∃y:int where y*y = x
|  |- >> ∃y:int where y*y = x
|  |  |  BY (EOn '0' 2 ...)
```

Figure 4: Square Root Proof.

we are inductively defining, on $z + 1$. The proof is completed by applying to 0 the function we constructed, proving $\overline{P}$ and providing a starting point for our square root search.

The extracted program from our proof is given below.[4]

$$\lambda x.(\mathtt{fix}(f, z.\mathtt{int\_eq}(z * z; x; z; f(z + 1)))(0))$$

This function, when given an integer x, starts at $z = 0$ and iterates through the natural numbers until it finds a z such that $z * z = x$. For perfect squares, it terminates with their non-negative square root, otherwise it diverges.

5 Conclusion

We have used the environment to develop proofs of interesting theorems in recursion theory and as a tool to explore new theories of computation. In these settings, the environment has proved valuable in preventing faulty reasoning and in leading to the development of readable and natural proofs. We have also used it to develop partial programs which cannot be naturally developed in any total type theory. The proofs generating these programs are concise, as the user provides the main algorithmic ideas and leaves Autotactic to fill in the details. These results demonstrate that partial type environments serve as important tools for both abstract theorem proving and partial program development.

---

[4]Several inner redices are β-reduced to clarify program structure.

Acknowledgements

I would like to thank Robert Constable and Scott Smith for laying the foundations of the Nuprl partial type theory, Doug Howe for his abundant assistance in using the Nuprl system, and Stuart Allen and Nax Mendler for helpful discussions.

References

[1] Robert S. Boyer and J. Strother Moore. *A Computational Logic*. Academic Press, 1979.

[2] Robert S. Boyer and J. Strother Moore. A mechanical proof of the unsolvability of the halting problem. *Journal of the Association for Computing Machinery*, July 1984.

[3] R.L. Constable. Constructive mathematics and automatic program writers. In *Proceedings of IFIP Congress*, Ljubljana, 1971.

[4] R.L. Constable et al. *Implementing Mathematics with the Nuprl Proof Development System*. Prentice Hall, 1986.

[5] R.L. Constable and S.F. Smith. Partial objects in constructive type theory. In *Symposium on Logic in Computer Science*, Computer Society Press of the IEEE, 1987.

[6] Thierry Coquand and Gérard Huet. A theory of constructions. 1984. Unpublished manuscript.

[7] Douglas J. Howe. *Automating Reasoning in an Implementation of Constructive Type Theory*. PhD thesis, Cornell, 1987.

[8] Per Martin-Löf. Constructive mathematics and computer programming. In *Sixth International Congress for Logic, Methodology, and Philosophy of Science*, pages 153–175, North Holland, Amsterdam, 1982.

[9] Lawrence Paulson. Lessons learned from LCF: a survey of natural deduction proofs. *Comp. J.*, 28(5), 1985.

[10] H. Rogers, Jr. *Theory of Recursive Functions and Effective Computability*. McGraw-Hill, 1967.

[11] N. Shankar. *Towards Mechanical Metamathematics*. Technical Report 43, University of Texas at Austin, 1984.

[12] S.F. Smith. *The Structure Of Computation in Type Theory*. PhD thesis, Cornell, 1988. In preparation.

The Use of Explicit Plans to Guide Inductive Proofs [*]

Alan Bundy

Department of Artificial Intelligence
University of Edinburgh
Edinburgh, EH1 1HN, Scotland

Abstract

We propose the use of explicit proof plans to guide the search for a proof in automatic theorem proving. By representing proof plans as the specifications of LCF-like tactics, [Gordon *et al* 79], and by recording these specifications in a sorted meta-logic, we are able to reason about the conjectures to be proved and the methods available to prove them. In this way we can build proof plans of wide generality, formally account for and predict their successes and failures, apply them flexibly, recover from their failures, and learn them from example proofs.

We illustrate this technique by building a proof plan based on a simple subset of the implicit proof plan embedded in the Boyer-Moore theorem prover, [Boyer & Moore 79].

Space restrictions have forced us to omit many of the details of our work. These are included in a longer version of this paper which is available from: The Documentation Secretary, Department of Artificial Intelligence, University of Edinburgh, Forrest Hill, Edinburgh EH1 2QL, Scotland.

Key words and phrases. Proof plans, inductive proofs, theorem proving, automatic programming, formal methods, planning.

1   Introduction

In this paper we propose a new technique for guiding an automatic theorem prover in its search for a proof, namely the use of *explicit proof plans*. This proposal was motivated by a current research project in the mathematical reasoning group at Edinburgh to develop automatic search control for the NuPRL program synthesis system, [Constable *et al* 86], and it was inspired by an earlier project of the group on the use of *meta-level inference* to guide an equation solving system, PRESS, [Bundy & Welham 81].

NuPRL can prove theorems by mathematical induction. In fact, inductive proofs are required for the synthesis of recursive programs: the type of induction used determining the type of recursion synthesised. In logic and functional programs, recursion is used in place of the imperative program constructs of iteration, *eg* while, until, do, etc. We are thus particularly interested in inductive proofs. The best work to date on the guidance of inductive proofs is that by Boyer and Moore, [Boyer & Moore 79]. Figure 1 contains a simple example of the kind of inductive proof found by their theorem prover. Hence, we have been adapting the techniques embedded in the Boyer-Moore theorem prover to the NuPRL environment, [Stevens 87].

In order to adapt the Boyer-Moore work we first need to understand why it works. Their program contains a large amount of heuristic information which is highly successful in guiding inductive proofs. However, the descriptions of these heuristics in [Boyer & Moore 79] are not always clear about why they are successful nor why they are applied in a particular order. Some heuristics are not appropriate in the NuPRL system, or require modification to make them appropriate. Thus it is necessary to rationally reconstruct the Boyer-Moore work in order to apply it to another system.

But we want to go further than this. We want to give a formal account of the Boyer-Moore heuristics, from which we can predict the circumstances in which they will succeed and fail, and with which we can explain their structure and order. We also want to apply the heuristics in a flexible

---

[*]I am grateful for many long conversations with other members of the mathematical reasoning group, from which many of the ideas in this paper emerged. In particular, I would like to thank Frank van Harmelen, Jane Hesketh and Andrew Stevens for feedback on this paper. The research reported in this paper was supported by SERC grant GR/D/44874 and Alvey/SERC grant GR/D/44270.

way and to learn new ones from example proofs. To achieve these goals we intend to represent the Boyer-Moore heuristics in an explicit proof plan based on the ideas of meta-level inference.

2   Explicit Proof Plans

We are confident that we can find such an explicit proof plan for a number of reasons. We believe that human mathematicians can draw on an armoury of such proof plans when trying to prove theorems. It is our intuition that we do this when proving theorems, and the same intuition is reported by other experienced mathematicians. One can identify such proof plans by collecting similar proofs into families having a similar structure, *eg* those proved by 'diagonalization' arguments. Many inductive proofs seem to have such a similar structure (see section 4 below). Within such families one can distinguish 'standard' from 'interesting' steps. The standard ones are those that are in line with the plan and the interesting ones are those that depart from it. The Boyer-Moore program proves a large number of theorems by induction using the same heuristics. These proofs all seem to belong to the same family.

The properties we desire of the proof plans that we seek are as follows:

- **Usefulness:** The plan should guide the search for a proof to a successful conclusion.
- **Generality:** The plan should succeed in a large number of cases.
- **Expectancy:** The use of the plan should carry some expectation of success, *ie* we ought to have some story to tell about why the plan often succeeds, and to be able to use this to predict when it will succeed and when it will fail.
- **Uncertainty:** On the other hand, success cannot be guaranteed. We will want to use plans in undecidable areas. If our ability to predict its success or failure was always perfect then the plan would constitute a decision procedure — which is not possible.
- **Patchability:** It should be possible to patch a failed plan by providing alternative steps.
- **Learnability:** It should be possible automatically to learn new proof plans.

The above properties argue for an *explicit* representation of proof plans with which one can reason. The reasoning would be used to account for the probable success of the plan under certain conditions (expectancy), and to replan dynamically when the plan fails (patchability). The explicit representation would enable plans to be learnt (learnability). The uncertainty property is discussed in section 7.

We have chosen to represent our plans in a sorted meta-logic. This gives an explicit representation to reason with, and also allows the plans to be very general (generality), in contrast to plans which are merely sequences of object-level rule applications.

We now turn to a detailed investigation of the Boyer-Moore heuristics in an attempt to extract from them the explicit meta-level proof plan that we require.

3   A Typical Inductive Proof

In order to investigate the Boyer-Moore heuristics, it will be instructive to study a typical proof of the kind that these heuristics can construct. Figure 1 is such a proof: the associativity of $+$ over the natural numbers. The first line is a statement of the theorem. Each subsequent line is obtained by rewriting a subexpression in the line above it. The subexpression to be rewritten is underlined and the subexpression which replaces it is overlined. The (recursive) definition of $+$ is given in the small box.

The proof is by backwards reasoning from the statement of the conjecture. The first step is to apply the standard arithmetic induction schema to the theorem: replacing x by 0 in the base case and by $s(x)$ in the induction conclusion of the step case. The equations constituting the recursive definition of $+$ are then applied: the base equation to the base case and the step equation to the step case. Two applications of the base equation rewrite the base case to an equation between two identical expressions, which reduces to *true*. Three applications of the step equation raise the occurrences of the successor function, s, from their innermost positions around the xs to being the outermost functions of the induction conclusion. The two arguments of the successor functions are identical to the two arguments of $=$ in the induction hypothesis. The induction hypothesis is then used to substitute one of these arguments for the other in the induction conclusion, and the induction hypothesis is dropped. The two arguments of the successor functions are now identical and reduce to *true*.

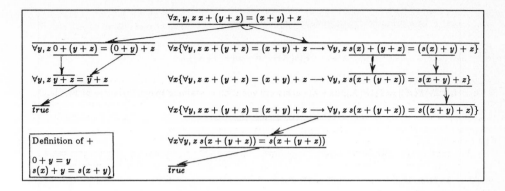

Figure 1: Proof of the Associativity of +

4 Simplified Boyer-Moore Proof Plan

We can pick out the general aspects of the proof in figure 1, and the above explanation of it, by displaying the schematic proof of figure 2. This schematic proof captures the spirit of the Boyer-Moore heuristics in a very simplistic way. Some of the extensions required to capture the full power of their theorem prover are discussed in the longer version of this paper.

In figure 2 capital letters indicate meta-variables. For instance, X and Y range over variables, F ranges over functions, and A, B_i, T_i, etc, range over terms. The difference between $T(X)$ and $T[X]$ is that the X in the round brackets signifies all occurrences of X in T whereas the X in the square brackets signifies some particular occurrence of X. Thus $T(Y)$ implies that all occurrences of X are replaced by Y, whereas $T[Y]$ implies that only one occurrence is replaced. In both cases the function or term may also contain variables other than X or Y. Note that the round bracket notation is unsound if the normal rules for substitution are applied to it. This is discussed further in the longer version of this paper.

Each arc is labelled with the name of the step that justifies the rewriting. Following LCF, [Gordon *et al* 79], we call these steps *tactics*. A $\sqrt{}$ sign beside a tactic indicates that it is guaranteed to succeed, whereas a ? indicates that it might fail. As in LCF, a tactic will be implemented as a program whose effect is to apply the appropriate rewritings to make the proof steps illustrated. However, whereas the primitive tactics provided in LCF apply only a small sequence of steps, we are also interested in designing tactics that will automatically complete a whole proof, or a substantial part of it, *ie* we are also interested in proof strategies.

Just as in the associativity proof of figure 1, the first tactic is to apply *induction*. Note that the induction scheme used, corresponds to the recursive scheme used to define F and that the induction variable to which it is applied is X, the variable in the recursive argument position of F. The major Boyer-Moore heuristic is to generalize this link between induction and recursion to most commonly occurring recursive data-structures and forms of recursion over them. The idea is to use the occurrence of recursive functions in the conjecture to suggest what induction scheme to use (one corresponding to the recursive structure of the function) and what variable(s) to induce on (those that occur in the recursive argument position(s) of the function). See [Stevens 87] for a more detailed analysis and rational reconstruction of this heuristic.

The equations that recursively define F are then applied to the base and step cases of the resulting formula, using the tactics *take-out* and *ripple-out*, respectively. The base case is simplified by this, but not solved as in the associativity proof. In the step case the occurrences of s are raised from their innermost positions to the outermost positions in the induction conclusion. The *ripple-out*[1] tactic does this using repeated applications of the step case of the recursive definition. This application of *ripple-out* is not guaranteed to succeed because the terms T_1, T_2 and B might not be of the right form. For a further description of this tactic and a definition of what form these terms must take for its success to be guaranteed, see figure 3. If *ripple-out* does succeed then *fertilization*[2] is

---

[1]The analogy is to a series of waves that carry the s from one place to another.

[2]The name is taken from Boyer and Moore. In the analogy the induction hypothesis is the sperm that fertilizes

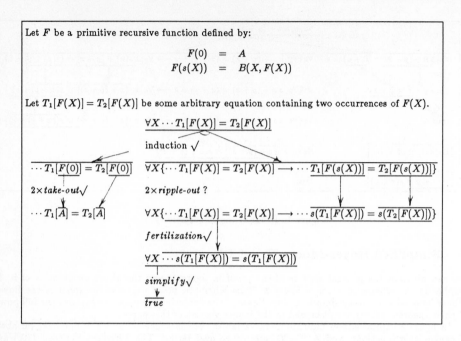

Let F be a primitive recursive function defined by:

$$F(0) = A$$
$$F(s(X)) = B(X, F(X))$$

Let $T_1[F(X)] = T_2[F(X)]$ be some arbitrary equation containing two occurrences of $F(X)$.

$$\forall X \cdots T_1[F(X)] = T_2[F(X)]$$

induction $\sqrt{}$

$$\overline{\cdots T_1[F(0)] = T_2[F(0)]} \qquad \forall X\{\cdots T_1[F(X)] = T_2[F(X)] \longrightarrow \cdots \overline{T_1[F(s(X))] = T_2[F(s(X))]}\}$$

$2 \times take\text{-}out \sqrt{} \qquad\qquad 2 \times ripple\text{-}out\ ?$

$$\cdots T_1[A] = T_2[A] \qquad \forall X\{\cdots T_1[F(X)] = T_2[F(X)] \longrightarrow \cdots \overline{s(T_1[F(X)])} = \overline{s(T_2[F(X)])}\}$$

$fertilization \sqrt{}$

$$\forall X \overline{\cdots s(T_1[F(X)]) = s(T_1[F(X)])}$$

$simplify \sqrt{}$

$$\overline{true}$$

Figure 2: Simplified Boyer-Moore Proof Plan

guaranteed to succeed. It substitutes the T_1 term for the T_2 term in the induction conclusion and the step case reduces to *true* via an application of some simplifying rules like the reflexivity axiom.

Note that, unlike the associativity proof in figure 1, the final step of the general proof does not solve the problem. However, the general proof does exchange the original conjecture for a sub-goal from which all occurrences of the function F have been eliminated. This can be seen as the aim of the general proof. Repeated applications of it will cause recursively defined functions to be systematically eliminated from the current sub-goals and replaced by the functions by which they are defined. If defined functions are arranged in a hierarchy with each defined function ordered above those by which it is defined, and a highest function is eliminated on each round, then a set of sub-goals will eventually be generated in which only primitive (*ie* non-recursive) functions occur. The proof of these will not require induction.

In the sub-proof describing the *ripple-out* tactic given in figure 3, the terms T_1 and T_2 take the form of a nested chain of recursively defined functions, F_i, where each F_i appears in the recursive position of F_{i+1} and the definitions of the F_i are all very simple. The step equations merely ripple the occurrences of s out once. By applying these step equations repeatedly the occurrences of s are rippled out from their innermost to the outermost position. Following Darlington we call a single application of the step equation an *unfold*[3].

This version of *ripple-out* is very simple and special purpose. To make it more general, we need to extend it not only to constructor functions other than s, but also enable it to supplement the use of unfolding step equations with the application of lemmas of a similar syntactic form. This latter extension is discussed in section 7.

5   The Specification of Tactics with Methods

In LCF or NuPRL tactics can be implemented as ML programs which will guide the application of rewrite rules to control the search for a proof. We have begun just such an implementation of

---

the step conclusion by making it provable.

[3] The analogy is with unfolding a piece of paper and taking out the present at the end.

For $1 \leq i \leq n$ let F_i be a primitive recursive function defined by:

$$F_i(0) = A_i$$
$$F_i(s(X)) = s(F_i(X))$$

Let $T_1[F(X)]$ or $T_2[F(X)]$ from the basic plan take the form $F_n(...F_2(F_1(X))...)$.

$$F_n(...F_2(F_1(s(X)))...)$$

$$unfold \sqrt{}$$

$$F_n(...F_2(\overline{s(F_1(X))})...)$$

recursively *ripple-out* $\sqrt{}$

$$\overline{s(F_n(...F_2(F_1(X))...))}$$

Figure 3: The Ripple-Out Sub-Plan

the Boyer-Moore heuristics, [Stevens 87]. However, such an implementation would not meet all the required properties of a proof plan. In particular, we require the ability to reason about the tactics in order to construct a proof plan for a problem and to replan when an existing proof plan fails. In order to conduct this reason we need to represent the conditions under which a tactic is applicable and the effect that it has if it succeeds, *ie* we need a *specification* of the tactic. Below we propose such a specification, which we plan to implement within the NuPRL or a similar framework.

Our specification formalism was adapted from that used in the LP system, [Silver 84], which was an extension of PRESS. The LP formalism was itself based on that of STRIPS. However, note that, unlike the plans formed in STRIPS-type plan formation, our plans will contain subroutines and recursion.

Following PRESS, we call the specification of a tactic, a *method*. A method is a frame containing information about the preconditions and effects of a tactic. A list of the slots in the frame and a description of the contents of each of them is given in table 1. Figure 4 is an example method for the *ripple-out* tactic. Methods for the other tactics in the simplified Boyer-Moore proof plan are given in the long version of this paper. Each of the slots contains a formula of our sorted meta-logic describing syntactic properties of the goal formulae before and after the tactic is applied. The meta-logical terms used in this paper are defined in table 2 and the sorts are defined in table 3. Definitions of the terms and sorts used in the remaining methods are given in the long version of this paper.

The description of the precondition is split between the input slot and the preconditions slot. The input slot contains a pattern which must match the before formula and the precondition contains additional information about the before formula which cannot be captured in this pattern. Similarly, the description of the effect is split between the output and the effects slots. The tradeoffs between representing information schematically in the input and output slots and representing it linguistically in the preconditions and effects slots, is discussed in the longer version of this paper.

A method represents an assertion in the meta-logic, namely that if a goal formula matches the input pattern and if the preconditions are true of it then the tactic is applicable. Furthermore, if the tactic application is successful then the resulting formula will match the output pattern and the effects will be true of it. The additional condition that the tactic application be successful means that the method is only a partial specification of the tactic. This is the key to the realisation of the uncertainty property and is discussed further in section 7 below.

The *ripple-out* tactic uses the repeated application of the unfold tactic to move the successor function from an innermost to an outermost position. *ripple-out* is illustrated in figures 1 and 3. The specification of the simple *ripple-out* tactic is given in table 4. The preconditions slot specifies that the input must be a nested sequence of simple recursive functions whose innermost argument

---

- **Name** - the name of the method. (We have followed the convention of using the tactic name for the method, augmented with additional arguments where necessary.)
- **Declarations** - a list of quantifier and sort declarations for meta-variables global to all the slots except the Tactics slot.
- **Input** - a schematic representation of the goal formula before the tactic applies.
- **Output** - a schematic representation of the goal formula after the tactic applies.
- **Preconditions** - a linguistic representation of further conditions required for the tactic to be applicable.
- **Effects** - a linguistic representation of additional effects of the method, including properties of the output and relationships between the input and output. They hold if the tactic applies.
- **Tactic** - a program for applying object-level rules of inference. This program is written in a subset of the same sorted meta-logic as the other slot values. This subset consists of applications of the object-level rules of inference and calls to sub-tactics. The tactic program serves also to specify the sub-tactics of this tactic and hence the sub-methods of this method. Meta-variables in this slot are local to each formula that constitutes the program[4].

Table 1: The Slots of a Method

has s as its dominant function. The output slot gives a pattern asserting that the output will an s whose argument is the input expression with the innermost s removed. No further effects information is required in this case. The tactic slot contains a recursively defined program, *ripple-out*, which takes a position and a formula and repeatedly applies *unfold* from that position to the outermost position. A specification of an extended version of *ripple-out* is given in section 7.

6 The Use of Proof Plans

In this formalism a proof plan is the method for one of the top-level tactics, *ie* it is the specification of a strategy for controlling a whole proof, or a large part of one. This super-method is so constructed that the preconditions of each of its sub-methods are either implied by its preconditions or by the effects of earlier sub-methods. Similarly, its effects are implied by the effects of its sub-methods. If the preconditions of a method are satisfied then its tactic is applicable. If the tactic application succeeds[5] then its effects are satisfied. The original conjecture should satisfy the preconditions of the plan; the effects of the plan should imply that the conjecture has been proved.

We can formalize this argument by associating with each method a formula of the form:

$$\forall O \in OSort.\ declarations(M)\{preconditions(M, input(M)) \wedge name(M, input(M)) = O$$
$$\longrightarrow effects(M, input(M), O) \wedge output(M) \equiv O\}$$

where M is a method name, $Slotname(M, ...)$ means the contents of slot $Slotname$ of method M applied to additional arguments ..., and $\equiv$ means syntactic identity. This formula can be read as asserting that if the input of a method satisfies the preconditions and if the tactic succeeds when applied to this input then the tactic's output matches the output slot and satisfies the effects of the method. We will call it the *expectancy* formula of the method, because it formalises our expectation that the method will do what it is intended to do.

Given, as an axiom, the expectancy formula for each of the sub-methods of a plan, axioms consisting of each of the tactic definitions, and various other axioms defining the meta-level terms of table 2, we can then prove as a theorem the expectancy formula for the super-method. I have carried out this programme for earlier versions of the methods described in section 5, *ie* given the expectancy formula for the method *unfold* I have proved the expectancy formula for the method *ripple-out*, then given them for: *induction*, *take-out*, *fertilization* and *simplify*, I have proved one for *basic-plan* (see [Bundy 87] for details). The structure of each proof is:

---
[5]The failure of tactics is discussed in section 7.

- $exp\text{-}at(Exp, Posn)$ is the sub-expression in expression Exp at position $Posn$. Positions are list of numbers which define an occurrence of one expression within another. For instance, [2,1] is the position of the 2nd argument of the 1st argument, eg the x in $f(g(2, x), 3)$. Note that the order of the list of numbers is the reverse of the usual convention. This simplifies some of the formulae in the sequel. We adopt the convention that 0 denotes the function symbol itself, so that [0,1] is the position of g in $f(g(2, x), 3)$.
- $single\text{-}occ(SubExp, Posn, SupExp)$ means that $SupExp$ contains precisely one occurrence of $SubExp$ and that this is at position $Posn$.
- $replace(Posn, NewExp, SupExp)$ is the expression obtained from $SupExp$ by replacing the sub-expression at position, $Posn$ with $NewExp$.
- $simp\text{-}rec(F, N)$ means F is a primitive recursive function whose Nth argument is the recursion argument, and whose step equation is of the simple form $F(s(X)) = s(F(X))$, where X is the Nth argument of $F(X)$
- $app(L_1, L_2)$ is the result of appending list L_1 to list L_2.
- $[Hd|Tl]$ is the list obtained from putting a new element Hd on the front on the list Tl.

Table 2: The Terms Used in the Meta-Logic

- $exprs$ is the set of all expressions.
- $terms$ is the set of all terms.
- $nums$ is the set of all natural numbers
- $posns$ is the set of all lists of natural numbers.

Table 3: The Sorts Used in the Meta-Logic

super-method preconditions $\longrightarrow$ sub-method preconditions
$\longrightarrow$ sub-method effects
$\longrightarrow$ super-method effects

The steps going between super- and sub-methods require the tactic definition for the appropriate link. The other step is by assumption. Such theorems prove that if the conjecture satisfies the preconditions of a plan and each of the sub-tactics succeed then the resulting formula will satisfy the effects of the plan.

The methods of section 5 were hand-coded to represent a rational reconstruction of a simple version of the implicit proof in the Boyer-Moore theorem prover. We are also interested in the use of the techniques of plan formation and/or automatic program synthesis to construct such proof plans automatically. Of relevance here is the work of Silver, [Silver 84], who developed the technique of *Precondition Analysis* for learning proof plans from examples in the domain of equation solving. Our representation of method is based on that of Precondition Analysis. Desimone has been extending this technique by removing some technical limitations which made it inapplicable to general proofs, [Desimone 87]. Precondition Analysis is also capable of learning new methods, *ie* the specifications of unknown tactics.

Also relevant is the work of Knoblock and Constable, [Knoblock & Constable 86], who have shown how NuPRL can be applied to the synthesis of its own tactics. We aim to explore this self-application of NuPRL to the generation of new tactics from the methods which specify them: methods which may have been learnt from example proofs using Precondition Analysis. Our representation of methods seems to be compatible with the specifications used by Knoblock and Constable.

7 How a Tactic may Fail

So far all the tactics that we have specified have been guaranteed to succeed provided their preconditions are met. Thus plans formed from them are guaranteed to succeed. However, as discussed in our

Name	$ripple\text{-}out(Posn)$
Declarations	$\forall Posn \in posns,\ \forall S\,Exp \in exprs,\ \forall Exp \in exprs.$
Input	$S\,Exp$
Output	$s(Exp)$
Preconditions	$S\,Exp = replace(Posn, s(exp\text{-}at(Exp, Posn)), Exp)) \wedge$
	$\{\forall Front \in posns, \forall Back \in posns, \forall N \in nums$
	$app(Front, [N|Back]) = Posn \longrightarrow simp\text{-}rec(exp\text{-}at(Exp, [0|Back]), N)\}$
Effects	nil
Tactic	$ripple\text{-}out([\,], Expr) = Expr$
	$ripple\text{-}out([Hd|Tl], Expr) = ripple\text{-}out(Tl, unfold([Hd|Tl], Expr))$

Table 4: The Ripple Out Method

list of desired properties of a proof plan, we cannot in general expect proof plans to be guaranteed successful, particularly in undecidable areas. Thus we must expect tactics to fail sometimes.

An example of a tactic that can sometimes fail is the *ext-ripple-out* tactic specified by the method in table 5. This method is similar to the one for the *ripple-out* tactic except that:

- we no longer require a nested sequence of simple recursive functions in the input, but only a single primitive recursive one;
- the output is no longer of the very simple form, $s(Exp)$, but is some expression containing a single occurrence of Exp;
- the tactic is defined using an extended version of $unfold$, called $wavelet$;
- and there is an extra equation to deal with the case that the rippling out process peters out benignly.

The idea of *wavelet*, for which we have not given a method, is that it can apply not just the step equations of recursive definitions, but any rewrite rule of the syntactic form:

$$G(B_1(X)) \Rightarrow B_2(X, G(X)) \tag{1}$$

eg[6]

$$even(X + Y) \Rightarrow even(X) \wedge even(Y) \tag{2}$$

where G is *even*, $B_1(X)$ is $X + Y$, $B_2(X, Z)$ is $Z + even(Y)$[7] and *even* is defined by:

$$even(0)$$
$$\neg even(s(0))$$
$$even(s(s(x))) \longleftrightarrow even(x)$$

Note that if $B_2(X, Z) \equiv Z$ then the rippling out will peter out benignly. The step equation for *even* is an example of such a rule.

$$even(s(s(X))) \Rightarrow even(X)$$

ext-ripple-out is able to ripple out expressions that the simple version cannot cope with. For instance, consider the expression:

$$even(s(x) \times y)$$

where $\times$ is defined by:

$$0 \times y = 0$$
$$s(x) \times y = x \times y + y$$

---

[6] Recall that we are reasoning backwards. The logical implication runs in the reverse direction.

[7] Recall also that our notation allows the term meta-variables, $eg\ B_1$, to contain variables, $eg\ Y$, other than those explicitly mentioned.

Name	$ext\text{-}ripple\text{-}out(app(Posn_2, Posn_1), Exp)$
Declarations	$\forall Posn_1 \in posns, \forall Posn_2 \in posns, \forall Exp \in exprs,$
	$\forall S\,Exp \in exprs, \forall S\,Exp_2 \in exprs.$
Input	$S\,Exp$
Output	$S\,Exp2$
Preconditions	$\exists B \in terms\ S\,Exp =$
	$replace(Posn_1, replace(Posn_2, exp\text{-}at(Exp, Posn_1), B), Exp))$
Effects	$\exists Posn_3\ single\text{-}occ(Exp, Posn_3, S\,Exp2)$
Tactic	$ext\text{-}ripple\text{-}out([], Expr, S\,Expr) = S\,Expr$
	$ext\text{-}ripple\text{-}out(Psn, Expr, Expr) = Expr$
	$wavelet(InPsn, S\,Expr, OutPsn, OutExpr)$
	$\longrightarrow$
	$ext\text{-}ripple\text{-}out(InPsn, Expr, S\,Expr) = ext\text{-}ripple\text{-}out(OutPsn, Expr, OutExpr)$

Table 5: The Extended Ripple-Out Method

After one application of *wavelet* using the step equation of $\times$ we get the expression:

$$even(x \times y + y)$$

We can now use *wavelet* to apply rule 2 to get:

$$even(x \times y) \wedge even(y)$$

which contains $even(x \times y)$ as required.

However, the following expression also fits the preconditions of the tactic:

$$even(s(x))$$

but there is no rewrite rule of form 1 rule that matches this expression, so *wavelet* will fail, causing the failure of *ext-ripple-out*. Thus *ext-ripple-out* might fail even though its preconditions are satisfied because an appropriate rewrite rule is missing. This is a typical way in which tactics fail. This meets the uncertainty property of proof plans: a proof plan might fail even though its preconditions are satisfied because one of its sub-tactics fails.

One could argue that the preconditions of methods should be strengthened so that they implied the success of the tactic. However, note that, in practice, this would amount to running the tactic 'unofficially' in the precondition, to see if it succeeded, before running it 'officially'.

8 The Required Properties are Satisfied

In this section we return to the desired properties of proof plans given in section 2 and see that each of them has been met by our proposals.

- **Usefulness:** As the tactics run they will each perform a part of the object-level proof, so the plan guides the proof search.
- **Generality:** The proof plan formalism is not restricted to describing a sequence of object-level rule applications. Meta-level specifications can describe a large set of rules. The powerful tactic language can combine these by sub-routining, recursion, conditionals, etc.
- **Expectancy:** If the conjecture meets the preconditions of the plan and each tactic succeeds then the effects of the plan will be true and the conjecture will be proved. Thus the plan is expected to succeed.
- **Uncertainty:** However, a tactic may fail, causing failure of the plan, so the plan is not guaranteed to succeed.
- **Patchability:** Since the preconditions and effects of a failing tactic are known, plan formation and/or program synthesis techniques may be (re)used to patch the gap in the plan with a subplan. This could be done automatically and dynamically enabling the theorem prover to recover from a failed plan without having to throw away those parts of the current proof that did succeed.
- **Learnability:** An extended version of Silver's Precondition Analysis might be used to learn proof plans from example proofs. New methods might also be learnt by this technique and the tactics corresponding to these synthesised by a self-referential use of NuPRL.

9 Conclusion

In this paper we have explored the use of proof plans to guide the search for a proof in automatic theorem proving. We have advocated the explicit representation of proof plans in a sorted meta-logic. We have developed a formalism for representing such proof plans as the specifications of LCF-like tactics. This proposal has been illustrated by developing a proof plan for inductive proofs based on the work of Boyer and Moore and others. We have developed tactics for running this proof plan and methods which specify each of them.

The domain of inductive proofs has proved a productive one since there is a rich store of heuristic knowledge available on how to guide such proofs. We have used this heuristic knowledge in the design of our tactics and methods. Our formalism has enabled us to explain why this heuristic knowledge is successful (when it is) and why it fails (when it does). In fact, we can give formal proofs that certain preconditions are sufficient for success, albeit in a very simple case. We have thus provided an analysis of the Boyer-Moore theorem prover which is serving as a good basis for extending and improving their ideas and for transporting them to a different system (NuPRL).

Our explicit representation suggests techniques for the dynamic construction of proof plans. We hope it will be possible to use these to recover from failure by constructing an alternative sub-plan to fill the gap left by a failed tactic. We are also exploring the use of these techniques to learn new proof plans from examples of successful proofs.

A major goal is the extension of the simple plans described above to incorporate some of the extensions described in the long version of this paper. This involves the identification of new meta-level relations, properties and functions, *eg* to describe a *wavelet* rule, and their incorporation in the meta-logic. We also intend to implement the ideas described in this paper by extending our version of the NuPRL system to use these proof plans for guiding search, recovering from failure and learning from examples.

References

[Boyer & Moore 79]   R.S. Boyer and J.S. Moore. *A Computational Logic*. Academic Press, 1979. ACM monograph series.

[Bundy & Welham 81]   A. Bundy and B. Welham. Using meta-level inference for selective application of multiple rewrite rules in algebraic manipulation. *Artificial Intelligence*, 16(2):189–212, 1981. Also available as DAI Research Paper 121.

[Bundy 87]   A. Bundy. *The derivation of tactic specifications*. Blue Book Note 356, Department of Artificial Intelligence, March 1987.

[Constable *et al* 86] R.L. Constable, S.F. Allen, H.M. Bromley, *et al*. *Implementing Mathematics with the Nuprl Proof Development System*. Prentice Hall, 1986.

[Desimone 87]   R.V. Desimone. Learning control knowledge within an explanation-based learning framework. In I. Bratko and N. Lavrač, editors, *Progress in Machine Learning – Proceedings of 2nd European Working Session on Learning, EWSL-87, Bled, Yugoslavia*, Sigma Press, May 1987. Also available as DAI Research Paper 321.

[Gordon *et al* 79]   M.J. Gordon, A.J. Milner, and C.P. Wadsworth. *Edinburgh LCF - A mechanised logic of computation*. Volume 78 of *Lecture Notes in Computer Science*, Springer-Verlag, 1979.

[Knoblock & Constable 86] T. B. Knoblock and R.L. Constable. Formalized metareasoning in type theory. In *Proceedings of LICS*, pages 237–248, IEEE, 1986.

[Silver 84]   B. Silver. Precondition analysis: learning control information. In *Machine Learning 2*, Tioga Publishing Company, 1984.

[Stevens 87]   A. Stevens. *A Rational Reconstruction of Boyer-Moore Recursion Analysis*. Research Paper forthcoming, Dept. of Artificial Intelligence, Edinburgh, 1987. submitted to ECAI-88.

LOGICALC: an environment for interactive proof development*

D. Duchier and D. McDermott
Department of Computer Science
Yale University

Abstract

LOGICALC is a system for interactive proof development. It lives in a graph editor environment based on DUCK's "*walk mode*" [10]. The user grows a tree of goals and plans by invoking *plan generators*. Goals are sequents of the form $A \Rightarrow p$, where A is an assumption set. Each plan has a validation that indicates how to conclude a proof of its parent goal from the proofs of its steps. When every step of a plan has been solved, their proofs are combined according to the validation and the resulting proof is handed to the parent goal. The logic is based on skolemization and unification. The system addresses subsumption within the goal graph, and automatic generalization of proofs.

Introduction

We present a system for interactive proof development. The project started out of the theorem proving demands of our spatial reasoning research. It was further justified by our interest in circumscription and the sad fact that most arguments of circumscription theory are quite tedious to work one's way through.

The requirements for the system were that it should find an acceptable compromise between being formal and being helpful, that it should lend itself, not only to proof checking, but to proof discovery as well, and, eventually, to alternative forms of automation, and that it should produce, not simply yes/no answers, but complete and self contained proof objects.

Our system is called LOGICALC, and its essential features are as follows:

---

*this research was supported by grant DAAA-15-87-K-0001 from BRL/DARPA

- It offers a goal/plan based approach to incremental proof refinement. Each goal may have more than one plan attached to it. Thus, the user can keep track of several alternative proof attempts at the same time.

- Goals are sequents of the restricted form $A \Rightarrow p$; where p is the conclusion to be derived, and A is the assumption set (represented as a data pool [9,10], for efficient access).

- Assumptions and goals are skolemized, so that quantifier-manipulation rules are unnecessary and the logic can be based on unification. A conclusion containing skolem terms and such that its proof does not depend on assumptions about them is appropriately generalized.

- The system avoids duplicating goals and maintains links between stronger and weaker goals. Thus, answers can be shared and appropriately propagated. As a consequence, what we think of as the *proof tree* is really a graph.

- The logic is first-order, but lambda-expressions can appear as terms. Appropriate support is provided in terms of inference rules, and the unifier can cope with them to a limited extent. [not discussed in this paper]

- The user can move from node to node in the graph; where a node may be a goal, a plan, an answer, a proof, or an assumption. Depending on the type of a node, the information displayed and the available commands will vary.

1 Overview

LOGICALC generalizes *backward chaining*. The generalization is just as powerful as resolution, and, indeed, resembles it in some ways, such as the use of skolemization for handling quantifiers. The details are quite different however, and the system is mostly based on *natural deduction*, although some of its techniques are related to McAllester's [8].

In a *backward chaining* approach, a prover works on *goals*, finding implications whose consequents unify with the current goal, and producing *subgoals* from their antecedents. This process bottoms out when a goal unifies with a *fact*. We generalize this paradigm in the following ways:

1. We provide several other methods of generating subgoals, such as reasoning by enumeration of cases, equality substitution, and proof by contradiction. A general method of generating subgoals is called a *plan generator* (cf. tactics in LCF [12]).

2. Goals and conclusions are represented by *sequents* [5] of the form $A \Rightarrow p$, where A is a set of assumptions. We say that the sequent $A \Rightarrow p$ is true iff p logically follows from A. Typically, A is a large data base, including an entire

theory plus local assumptions; LOGICALC provides mechanisms for managing it efficiently.

LOGICALC is an interactive system. The user grows a structure of goals and subgoals by issuing commands to try various plan generators. He can move through the tree and work concurrently on different approaches to a proof, or different parts of the same proof attempt.

A trivial example will help give the reader a feel for the system. Given $a < b$, $b = c$, $c < d$ and an axiom expressing the transitivity of "$<$," prove $a < d$. We express the givens thus:

```
(AXIOM A<B (< A B)) (AXIOM B=C (= B C)) (AXIOM C<D (< C D))
(AXIOM TRANS-< (FORALL (X Y Z) (IF (AND (< X Y) (< Y Z)) (< X Z))))
```

We see that LOGICALC expects formulae to be typed in a *lispish* notation. After these formulae have been read, they are processed and transformed into patterns (skolemized—see section 2), then stored in the global data base. The data base now contains respectively: `(< A B)`, `(= B C)`, `(< C D)`, and `(IF (AND (< ?X ?Y) (< ?Y ?Z)) (< ?X ?Z))`. We show LOGICALC's output both in `typewriter` font and **bold face** font (same as on the screen). User input is shown in *italics*.

```
> (LOGICALC)
Type the goal to be worked on: (< A D)
GOAL26 (top goal)
    Prove
        (< A D)
    (no assumptions)
    (no answers)
    (no proof plans)
    (no parasites, no more general goals, no less general goals)
```

After we have invoked LOGICALC and stated the theorem to be proven, the system displays the top goal of the proof. The user should now invoke a plan generator to produce an appropriate plan for this goal. Obviously, we want to do a little back chaining using the axiom of transitivity. The proper plan generator is *modus-ponens* (named after the inference rule that validates its plans).

w> *PLAN MODUS-PONENS TRANS-<*

To invoke a plan generator, the user should type **plan**, followed by the name of the plan generator, followed by whatever parameters it requires. *Modus ponens* expects, as input, the name of an assertion to be used as the major premise. The system responds with:

```
produced 1 fate
1 (plan):
      PLAN w> Plan MODUS-PONENS TRANS-<
      1  (< A ?Y) (1 answer)
```

```
    2  (< ?Y D) (1 answer)
Move to it? YES
```

We are informed that a plan has been produced, with two steps. Both already have answers. The reason is that LOGICALC will always try to find obvious answers to newly created goals; for instance, by matching them against assertions in the data base, or by stealing answers from goals that have related patterns. The system asks whether we want to see the new plan, and we agree.

Deductive plan for GOAL26
 Documentation: "w> Plan MODUS-PONENS TRANS-<"
 Supergoal:
   (< A D)
 Steps:
 |- 0 (=> (IF (AND (< ?X ?Y) (< ?Y ?Z)) (< ?X ?Z)))
    1 (< A ?Y) (no plans, 1 answer)
    2 (< ?Y D) (no plans, 1 answer)
 Validation:
   (MODUS-PONENS (0 (AND-INTRO (1 2) ())) ())
 (no successors)

Axioms and assumptions used in plans appear as pseudo-steps numbered from 0 downward, and are displayed prefixed with |-. Pseudo-step 0 is really axiom TRANS-<. The *validation* indicates how the conclusion will be drawn: first, the conclusions from steps 1 and 2 will be fed to the *and-intro* inference rule, thus concluding their conjunction, then *modus ponens* will be used with pseudo-step 0 (TRANS-<) as major premise and the newly concluded conjunction as minor premise. We decide to solve step 1 first. Typing 1 *moves* us to step 1.

```
w> 1
GOAL27
    Find: (Y) in:
       (< A ?Y)
    (no assumptions)
    Answers:
       1  ?Y        = B
    Supergoals:
       1  (< A D)
    (no proof plans)
    (no parasites, no more general goals, 1 less general goal)
```

We see that the system has already found an answer for us. The answer assigns the value B to variable ?Y. This is obviously the answer we want, so we decide to *move* to it.

```
w> ANSWER 1

Answer to goal:
    1  (goal): (< A ?Y)
```

```
        ?Y         = B
led to:
   2 (successor plan):
         (< B D)
```

An answer to a goal consists of a substitution assigning values to variables appearing in the goal, and of a proof in the form of a tree of sequents and validations, in the obvious form. When a step has been solved, a successor plan is produced by removing said step from the list of pending steps, and instantiating the remaining steps with the bindings provided by the answer. Creating a successor plan does not alter the original plan. The successor plan is simply appropriately inserted in the proof graph. A given plan may have many successors because its steps may have many unrelated answers.

The display reminds us what goal this is an answer to, how it *binds* the goal's variables, and finally what it led to, which is a successor plan. We chose to examine the successor plan:

w>*2*
Deductive plan for GOAL26
```
   Supergoal:
      (< A D)
   Steps:
   |- 0 (=> (IF (AND (< ?X ?Y) (< ?Y ?Z)) (< ?X ?Z)))
      [1  (< A ?Y)]
       2  (< B D) (no plans, no answers)
   Validation:
      (MODUS-PONENS (0 (AND-INTRO (1 2) ())) ())
   (no successors)
```

The solved step is now displayed between square brackets. The second step has been instantiated according to the answer's bindings (e.g. ?Y ← B). We move to it:

w>*2*
GOAL29
```
   Prove
      (< B D)
   (no assumptions)
   (no answers)
   Supergoals:
      1  (< A D)
   (no proof plans)
   (no parasites, 1 more general goal, no less general goals)
```

Now, we must substitute C for B in the goal. Plan generator *equality* will let us do that.

w>*PLAN EQUALITY B –USE B=C*

produced 1 fate

```
1 (conclusion): Yes
Move to it? YES
```

Answer to top goal:
```
   1  (goal): (< A D) Yes
```

Equality proposed a plan with (< C D) as sole step. The system found an obvious answer to that step (namely axiom C<D) and consequently produced a proof for the initial theorem. The answer binds no variables in the goal.[1] Such an answer we call a *yes answer*. We move to the answer's proof.

```
w>PROOF
Deductive proof
   Conclusion:
      (< A D)
   (no local assumptions)
   validation:
      (MODUS-PONENS (0 (AND-INTRO (1 2) ())) ())
   follows from:
      1   (=> (IF (AND (< ?X ?Y) (< ?Y ?Z)) (< ?X ?Z)))
      2   (=> (AND (< A B) (< B D)))
```

We could manually inspect (*walk* through) the proof, but it is more instructive to have LOGICALC display it in a *natural deduction* format:

```
w>DUMP
```

```
Axioms:
C<D         (< C D)
B=C         (= B C)
A<B         (< A B)
TRANS-<     (IF (AND (< ?X ?Y) (< ?Y ?Z)) (< ?X ?Z))
   1  (< B D)
      By (EQUALITY B=C C<D)
      Parameters: ((1))
   2  (< A D)
      By (MODUS-PONENS TRANS-< (AND A<B 1))
```

LOGICALC first displays the axioms used in the proof, then outputs the lines of the proof. Line 1 says that (< B D) can be inferred by *equality* using B=C and C<D as premises. Line 2 says that (< A D) can be concluded by *modus-ponens* with TRANS-< as major premise and the conjunction of A<B and line 1 as minor premise.

2   Syntax and Skolemization

LOGICALC is built on top of DUCK [10], which is built on top of NISP [11], which is built on top of LISP. DUCK provides a set of tools for maintaining a predicate calcu-

---

[1]hardly surprising since there are none.

lus database, including skolemization, unification, backward and forward chaining, and reason maintainance. It expects formulae to be typed in a lispish notation. Connectives and quantifiers have the following general syntax:

$$\neg p \longrightarrow \text{(NOT } p\text{)} \qquad\qquad p_1 \wedge \ldots \wedge p_n \longrightarrow \text{(AND } p_1 \ldots p_n\text{)}$$
$$p \supset q \longrightarrow \text{(IF } p \ q\text{)} \qquad\qquad p_1 \vee \ldots \vee p_n \longrightarrow \text{(OR } p_1 \ldots p_n\text{)}$$
$$(\forall x_1) \ldots (\forall x_n) \, P(x_1, \ldots, x_n) \longrightarrow \text{(FORALL } (x_1 \ \ldots \ x_n) \ (\text{P } x_1 \ \ldots \ x_n))$$
$$(\exists x_1) \ldots (\exists x_n) \, P(x_1, \ldots, x_n) \longrightarrow \text{(EXISTS } (x_1 \ \ldots \ x_n) \ (\text{P } x_1 \ \ldots \ x_n))$$
$$\lambda \, x_1, \ldots, x_n \, . \, t(x_1, \ldots, x_n) \longrightarrow \text{(LAMBDA } (x_1 \ \ldots \ x_n) \ (t \ x_1 \ \ldots \ x_n))$$

A variable binding list (1st argument of FORALL, EXISTS, or LAMBDA) may also contain type declarations. For instance, we can express the fact that, for any number, a lesser number can be found, by:

```
(AXIOM NO-FLOOR-<
    (FORALL (number X) (EXISTS (number BELOW) (< BELOW X))))
```

Type declarations play a double role. Firstly, they are used for issues of type checking related to well-formedness [10]. Secondly, they are inserted as additional constraints in the propositions where they appear.

In LOGICALC, propositions—whether assertions or goals—are represented by patterns with no quantifiers. This makes it possible to base the logic on unification. We rely on DUCK's *skolemization* facility to effect the transformation from input representation to internal representation. Axiom NO-FLOOR-< will really be skolemized as:

```
(IF (IS NUMBER ?X) (AND (IS NUMBER !.BELOW(?X)) (< !.BELOW(?X) ?X)))
```

where ?X is a *match* variable and !.BELOW(?X) is a skolem term that depends on ?X.

3 The Logic

LOGICALC combines *sequent logic* with unification. Traditionally, a sequent has been defined to be of the form $p_1 \wedge \ldots \wedge p_n \Rightarrow q_1 \vee \ldots \vee q_m$, which makes it possible to treat both sides of the sequent symmetrically as sets: $A \Rightarrow B$. We elected to restrict a sequent to be of the form $A \Rightarrow p$, where p is a single formula. Such a sequent is intended to convey the idea that p logically follows from assumption set A. We felt this was a simpler, more natural representation, whose implementation was straightforward in terms of DUCK's concepts and tools.[2]

Note that our decision does not restrict the power of the language, as any traditional sequent can be transformed into another one, isomorphic to a LOGICALC

---

[2]an assumption set is represented by a datapool.

sequent, by repeated application of the following inference rule: $A \Rightarrow B, p \vdash A, \neg p \Rightarrow B$ until only one formula remains in B.

Our logic requires only one axiom: $A, p \Rightarrow p$, which says that any proposition follows from a set of assumptions that contains it. Below, we list these inference rules which have been implemented so far:

Identity:	$A \Rightarrow p \vdash A \Rightarrow p$	Substitution:	$A \Rightarrow p \vdash A \Rightarrow \sigma(p)$
Assumption-Intro:	$A \Rightarrow p \vdash A, q \Rightarrow p$	Taut-Trans:[3]	$A \Rightarrow \sigma(a) \vdash A \Rightarrow \sigma(b)$
Modus-Ponens:	$A \Rightarrow p \supset q, A \Rightarrow p \vdash A \Rightarrow q$	Modus-Tollens:	$A \Rightarrow p \supset q, A \Rightarrow \neg q \vdash A \Rightarrow \neg p$
If-Intro:	$A, p \Rightarrow q \vdash A \Rightarrow p \supset q$	Not-Intro:	$A, p \Rightarrow q, A, p \Rightarrow \neg q \vdash A \Rightarrow \neg p$
Not-Elim:	$A \Rightarrow q, A \Rightarrow \neg q \vdash A \Rightarrow p$	Symmetry:[4]	$A \Rightarrow p(x, y) \vdash A \Rightarrow p(y, x)$
And-Intro:	$A \Rightarrow p_1, \ldots, A \Rightarrow p_n \vdash A \Rightarrow p_1 \wedge \ldots \wedge p_n$		
Reduction:	$A \Rightarrow p[(\% (\lambda \bar{x} \ t) \ \bar{a})] \vdash A \Rightarrow p[t[\bar{a}/\bar{x}]]$		
Abstraction:	$A \Rightarrow p[t] \vdash A \Rightarrow p[(\% (\lambda \bar{x} \ t[\bar{x}/\bar{a}]) \ \bar{a})]$		
Equality:[5]	$A \Rightarrow e_1 = e_2, A \Rightarrow p[e_1] \vdash A \Rightarrow p[e_2]$		

4 Resolution and Find Mode

Typing "`plan` *plangen parameters...*" invokes plan generator *plangen* on the current goal, with the given *parameters*. One particularly useful plan generator is *resolution*, which attempts to *detach* the goal from the assumption set. This is *not* the resolution rule.

We say that a proposition p can be *detached* from assertion a if p occurs as a subterm in a and a can be syntactically rewritten $r \wedge p$, or $q \supset p$, or $r \wedge (q \supset p)$. We also say that p is detachable from a *modulo* q: if q can be proven, then p can be inferred.

For instance, suppose we wish to prove p_3, and the assumption set contains assertion FOO of the form $p_1 \supset (p_2 \supset p_3)$. "`plan resolution foo`" will produce a plan with the following steps and validation:

```
|- 0  (IF P1 (IF P2 P3))
    1   P1
    2   P2
(MODUS-PONENS ((MODUS-PONENS (0 1) ()) 2) ())
```

LOGICALC makes it possible to combine the power of *resolution* with that of other plan generators. Whenever a plan generator expects an assertion of a certain form as a parameter, LOGICALC allows the user to name an assertion from which it can be detached instead. For instance, if the goal is $p(e_1)$ and assertion b is $r \supset (q \supset e_1 = e_2)$, typing "`plan equality` e_1 `-use` b" will produce a plan with the 3 steps $p(e_2)$, q, r; the last 2 steps are necessary to detach the equality.

---

[3]paramodulation

[4]$a \supset b$ is a tautology of propositional logic, $\sigma(a)$ and $\sigma(b)$ are wff of first order logic

[5]where p is symmetric

When invoking a plan generator that accepts assertions as parameters, the user will frequently be unsure which assertions he should name. Typing -find in place of the assertion parameter (e.g. "plan equality e_1 -use -find") will invoke *find mode* recursively.

Find mode lets the user browse through all the possible ways of detaching the pattern sought (e.g. "$e_1 = ...$" above) from the database, and select those he deems interesting. When *find mode* is exited, this information is returned to the plan generator. If p is detachable from a modulo q, then p will be used by the plan generator just as if it had been an assertion, but the plan produced will include q as a subgoal and the validation will indicate how to infer p from a and a proof of q.

5 Issues of Subsumption

It is often useful to think of the goal/structure as being a tree. In reality, similar subgoals may appear in different "branches" of the "tree," and we would rather not have to duplicate the work required to prove them. It would be clearly better if they could share their answers.

When LOGICALC creates a new goal, it compares it to those already existing. If the new goal is found to be identical (modulo renaming of variables) to an old goal, a *parasite* is created instead, pointing to the old goal (the host). A parasite shares all answers with its host. We see that the "tree" really has cross-over links.

It is possible to generalize this scheme, and share (adapted) answers among goals which are not necessarily equivalent, but nonetheless related. The general rule is:

an answer to $H \Rightarrow \sigma(p)$ is a valid answer to $H' \Rightarrow p$, where $H \subseteq H'$

LOGICALC implements this scheme incompletely. In particular, it is infeasible in practice to determine that $H \subseteq H'$. Since assumption sets are implemented as datapools [10], we simply check for hierarchical inclusion.

LOGICALC automatically generalizes conclusions. Consider a proof of $P(a)$ that depends on no assumption about a. Clearly, the proof can be generalized to a proof of $(\forall x) P(x)$. Such a mechanism can help remedy the overspecializing effects of skolemization: in order to prove $(\forall x) P(x)$, skolemization will have us prove (P !.X). If we can produce a proof of the preceding expression that makes no assumption about !.X,[6] the conclusion we really want to draw is (P ?X), which is the quantified expression skolemized as an assertion.

LOGICALC will automatically generalize a proof with respect to those skolem terms about which the proof makes no assumption (see also McAllester [8]).

---

[6]and it shouldn't, since !.X is brand new

Conclusion

Some of the theorems we have tried LOGICALC on include showing that a continuous function admits no jumps (taken from Bledsoe), and the circumscription proof outlined in Kautz's *Logic of Persistence*. We are currently working on adding autonomous proof-search mechanisms to the system.

References

[1] R. Boyer and J. Moore. *A Computational Logic*. *ACM Monograph Series*, Academic Press, 1979.

[2] J. Buines-Rozas. *GOAL: a goal oriented command language for interactive proof construction*. PhD thesis, Stanford University Department of Computer Science, 1979.

[3] R. Constable et al. *Implementing Mathematics with the Nuprl Proof Development System*. Prentice-Hall, 1986.

[4] T. Coquand and G. Huet. *Constructions: A higher Order Proof System for Mechanizing Mathematics*. Technical Report 401, INRIA, France, May 1985.

[5] K. Hintikka. Two papers on symbolic logic. *Acta Philosophica Fennica*, 8:1–115, 1955.

[6] J. Ketonen and J. Weening. *EKL—an interactive proof checker user's reference manual*. Technical Report 1006, Stanford University, 1984.

[7] E. Lusk and R. Overbeek. *The Automated Reasoning System ITP*. Technical Report 27, Argonne National Laboratory, 1984.

[8] D. McAllester. *ONTIC: A Knowledge Representation System for Matematics*. PhD thesis, Department of Electrical Engineering and Computer Science, MIT, May 1987.

[9] D. McDermott. Contexts and data dependencies: a synthesis. *IEEE Transactions on Pattern Analysis and Machine Intelligence*, 5(3):237–246, May 1979.

[10] D. McDermott. *The DUCK manual*. Technical Report 399, Yale University Department of Computer Science, June 1985.

[11] D. McDermott. *The NISP manual*. Technical Report 274, Yale University Department of Computer Science, June 1983.

[12] R. Milner. *Edinburgh LCF*. *Lecture Notes in Computer Science*, Springer-Verlag, 1979.

Implementing Verification Strategies
in the KIV-System

M. Heisel, W. Reif, W. Stephan

Universität Karlsruhe

Institut für Logik Komplexität und Deduktionsysteme

Postfach 6980

D - 7500 Karlsruhe

Abstract

We describe by two examples, BURSTALL's method for proving total correctness assertions and GRIES's method for program development, how verification strategies can be implemented in the KIV System. This system is based on Dynamic Logic and uses a metalanguage to program the generation of proofs. Strategies which are implemented in such an environment are always sound with respect to the basic logic. They can easily be extended and may be freely combined. We use the first example to demonstrate that parts of such strategies can be carried out automatically. The second example shows that program development is possible in the system.

1 Introduction

In this paper we describe by two examples how verification strategies can be implemented in an environment (the Karlsruhe Interactive Verifier) [HHRS 86] which uses Dynamic Logic (DL) as a logical basis and a functional metalanguage to program the generation of proofs. We will study BURSTALL´s method for proving total correctness assertions ("Program Proving by Hand Simulation and a little Induction") [Bu 79] and the program development method presented by GRIES in his book "The Science of Programming" [Gr 81]. The main purpose of this paper is to outline the general aspects of the KIV System as a tool to implement various verification strategies. Here we are not interested in the methods themselves and we will not discuss the advantages or disadvantages of either approch. The presented results are a small first step towards a verification system which comprises many different strategies that can be modified easily and combined freely.

Implementation of program proving methods like those mentioned above usually proceeds in two stages. First one has to work out the method from a logical point of view. For example in Chapter 10 of his book GRIES proves a theorem about alternative commands which is then used as a justification of his strategy to develop alternative commands presented in Chapter 14. In general these logical considerations are subject to errors to a degree that depends on the formal rigor one is willing to apply. Indeed our first rough ideas about (a DL version of) BURSTALL's method contained some substantial errors which became apparent during the actual implementation. In a second stage the method as it has been been worked out on paper is usually implemented using some programming language like LISP or PROLOG. In this stage additional errors may creep in. Instead we use a high-level programming environment which has built in a powerful logical formalism. Logical considerations and the actual implementation are no longer different tasks: methods for program proving are implemented by defining appropriate derived rules (together with validations of these rules), by using these additional rules to program more general inference steps which, in spirit of the ML-approach, are called tactics and by finally combining tactics into strategies. For example GRIES's theorem mentioned above is incorporated in the IF-strategy. It is carried out in a completely formal manner (for a special instance) when the strategy is called. In general all deduction steps are ultimately reduced to steps in the basic logical system.

There are many other systems which follow similar ideas, for example AUTOMATH [Br 80], Edinburgh LCF [GMW 79] and NUPRL [CKB 85]. The KIV System differs from these approaches in that it is designed for an application in the area of program proving. Some special features of the system, the use of Dynamic Logic, our notion of proof-trees, the use of metavariables and schematic proofs will be discussed below.

In the next section we present our version of DL and give a short overview on the Proof Programming Language PPL. Section 3 describes the general concepts like derived rules, tactics and strategies. Section 4 contains some details of the implementation of BURSTALL's Simulation-and-Induction-Strategy (SI). We use this example to discuss the problem of carrying out parts of strategies automatically. One advantage of our approach is that strategies can easily be altered or extended. We will mention some extensions of the SI strategy which go far beyond the original method. In Section 5 we demonstrate that our system is powerful enough to allow strategies where program proving and program development go hand-in-hand. GRIES's method serves as a prominent example.

2 The Logic and the Metalanguage of the KIV System

2.1 The Logic Underlying the KIV System

One major design aim of the KIV system has been to integrate all kinds of verification activities into one homogeneous software system. For this aim it is necessary to incorporate a number of different logical formalisms dealing with quite different types of statements about programs into a unified and more general logical framework. At least in the field of deterministic programs, Dynamic Logic (DL) has turned out to be a suitable logical basis for such a unified approach to program verification.

Dynamic Logic extends first-order logic by formulas $[\alpha]\varphi$, where α is an imperative program and φ again is a formula. $[\alpha]\varphi$ is read "if α terminates, φ holds". $<\alpha>\varphi$ is an abbreviation for $\neg[\alpha]\neg\varphi$ and has to be read " α terminates, **and** φ holds". For a survey of dynamic logic see [Ha 84]. The programming language considered here is made up of **skip, abort,** assignments $x := \tau$, compositions $\alpha;\beta$, conditionals **if ε then α else β fi** and **if ε then α fi**, while-loops **while ε do α od**, and iterations α^i.

With these extensions of the first-order language many interesting statements (which are subjects of well known verification strategies) can be formulated in DL: $<\alpha>\varphi$ stands for the weakest precondition of a deterministic program α with respect to φ, $\varphi \rightarrow [\alpha]\psi$ for a partial correctness assertion, $\varphi \rightarrow <\alpha>\psi$

for a total correctness assertion. The formula $[\alpha]\varphi \to [\beta]\psi$ stands for a program transformation, and $\exists i < (\textbf{if } \epsilon \textbf{ then } \alpha \textbf{ else skip fi})^i > \psi$ for an intermittent assertion of a loop **while ε do α od**.

We use a calculus for uninterpreted reasoning (see GOLDBLATT [Go 82]) to gain the flexibility that is necessary to overcome the restriction to a single verification technique e. g. invariants in HOARE's logic or in HAREL's calculus for DL [Ha 79]. These particular verification techniques then can be obtained in form of derived rules in our calculus (for examples see Sections 4 and 5).

Actually our calculus is a sequent calculus [Re 84], [Ri 78]. We write $\Gamma \Rightarrow \psi$ for sequents where Γ, a list of formulas, is called the antecedent and the formula ψ the succedent.

2.2 The Metalanguage PPL

To cope with the higher complexity of proofs in calculi for uninterpreted reasoning, the logic is embedded in a functional programming language to construct proofs, PPL [HRS 86]. PPL allows sound extensions of the basic logic and is used to implement various proof-searching strategies.

The central data structure of PPL are proof-trees: the proven sequent is the root of the tree, the axioms and hypotheses are its leaves, and $f_1,\ldots,f_n$ are sons of node (sequent) f iff f can be derived from $f_1,\ldots,f_n$ by application of some rule. The sequents may contain *metavariables* for any syntactical category like formulas or programs. Thus, we are able to carry out schematic proofs in PPL (a profitable feature, exploited in Section 5).

The basic rule schemes of the calculus are elementary proof-trees. Proof-trees consisting only of a single hypothesis are generated by **mkstree** ("make simple tree"). The operation **mktree** can be used to define arbitrary new rules (called user-defined or derived rules). However, to guarantee soundness, the user has to supply a validation for each of them. A validation is either a proof-tree verifying the rule directly or a PPL-function which can be used as a soundness check at any time the user wants. This function may succeed to verify the rule for some instantiations of the metavariables and fail for others.

New proof trees can be generated by operations called infer $(\tau, [i_1,..,i_n], [\tau_1,.., \tau_n])$ and refine (σ_1, i, σ_2). The infer operation performs a forward proof step by using the conclusions of n proof trees $\tau_1,\ldots,\tau_n$ as premisses of a proof tree τ yielding a new conclusion. The refine operation performs a backward proof step by replacing one premise of a proof tree σ_1 by a proof tree σ_2 yielding the premisses of σ_2 as new subgoals. Both operations use matching: the proof trees τ and σ_2 can be considered as generalized inference rules and thus be instantiated by applying a matcher Θ to some of the metavariables. Θ satisfies the following conditions: the i_k-th premise of $\Theta (\tau)$ is equal to the conclusion of τ_k for k=1,...,n (infer) or the conclusion of $\Theta (\sigma_2)$ is equal to the i-th premise of σ_1 (refine).

PPL allows an arbitrary combination of forward and backward proof steps. The control structures of the language are the usual ones including conditional, recursion, failure, and alternative. The last two, the **or**-construct together with **fail**, are used to implement proof search by backtracking. Input and output is done by **read** or **before** respectively. e_1 **before** e_2 outputs e_1 before evaluating e_2.

3 The Concepts to Implement Your Favourite Strategy

A verification method usually is more than just a particular logical system. It also induces a characteristic discipline how to use it. In PPL we reflect this distinction between the basic building blocks and the heuristics of a verification method in terms of derived rules, tactics, and strategies. The implementation of a method roughly proceeds as follows: (i) formalize the method in DL, and (ii) isolate the basic building blocks in terms of derived rules and tactics (see below). At this stage the system can already be used as a proof-checker. (iii) Finally implement the heuristics controlling the application of tactics. At this level any degree of automation is possible.

3.1 Derived Rules

The central concept of PPL, to embed specific logics in a more general one, is that of derived rules. In BURSTALL's method [Bu 74], for example, we need the rule scheme (see [HRS 87])

$$\Gamma \Rightarrow \exists i < (\text{ if } \varepsilon \text{ then } \alpha \text{ fi }) i > (\neg \varepsilon \wedge \varphi)$$

$$\Gamma \Rightarrow < \text{ while } \varepsilon \text{ do } \alpha \text{ od } > \varphi$$

to deal with while-loops (the rule connects the loop with its iteration steps). In PPL it can be defined by the operation **mktree** where the schematic part and the validation are separated. Thus proofs become modular objects. Although the rule contains metavariables, in this case we may execute (without failure) the validation at once, since the rule holds for all instantiations and our system is able to carry out proofs schematically. In other cases we may have to delay the validation (in order to avoid failure) until the rule becomes more and more instantiated. This means that the system may also be used to reason about not fully specified objects (e. g. in program development).

3.2 Tactics and Strategies

Using these derived rules it is not too difficult to program the basic building blocks of a verification method in PPL (e.g. the simulation steps in BURSTALL's method, see Section 4). These PPL programs are tactics in the sense of ML in the LCF system [GMW 79]. Presented with some goal, a tactic computes a number of subgoals that logically imply this goal, provided all the remaining delayed validations will be executed successfully at some time. A derived rule can be regarded as a special case of a tactic where the result can be expressed schematically.

We call a PPL program a strategy rather than a tactic if it decides where to apply which tactic. Several strategies may be combined freely to new strategies leading to heuristics at a high level of abstraction. Since strategies may, of course, fail they must be embedded in functions controlling failure by backtracking (using the alternative construct **or**). By explicitly programmed failure, the user may incorporate knowledge, e. g. about dead ends in the search space.

4 The SI Strategy

4.1 Tactics for Burstall's Method

SI ($\underline{S}$imulation and $\underline{I}$nduction) is a goal-directed proof strategy for total correctness assertions $\Gamma \Rightarrow$ $<\alpha>\psi$. Using a normalization tactic $\Gamma \Rightarrow <\alpha>\psi$ is transformed to $\Gamma \Rightarrow <\alpha_1> <\beta> \psi$ where α_1 is the first instruction of α and β the rest of α. The central idea of SI is to use symbolic execution of programs. Therefore we will have tactics for each form α_1 can take. We describe two of them. A more detailed discussion of these tactics is contained in [HRS 87].

ASSIGN: $\quad \Gamma \Rightarrow <x := \tau> <\beta> \psi$ is reduced to $\Gamma_x{}^y, x= \tau_x{}^y \Rightarrow <\beta> \psi$

$\qquad\qquad$ where y is a fresh variable

Note that this is just a description of the input-output behaviour of the PPL-function ASSIGN_TAC. During execution of ASSIGN_TAC the new antecedent $\Gamma_x{}^y, x= \tau_x{}^y$ is actually computed. If we allow program formulas to occur in Γ this is not a completely trivial task.

One of the tactics for while-commands expects as arguments two assertions φ and ψ_1 such that, intuitively speaking, from each "state" satisfying φ a "state" satisfying ψ_1 can be reached by some number i of iterations. This proposition is proven through induction on some (input-) variable which is supplied to this tactic as a third argument. We denote the ordering used in the induction proof by $<<$.

WHILE$_2$: $\quad$ Given φ, ψ_1 and u, $\quad \Gamma \Rightarrow$ <while ε do α_2 od> $<\beta> \psi$ $\quad$ is reduced to

$\qquad$ (1) $\varphi(u), \text{IND}(u) \Rightarrow \exists\, i <(\text{if } \varepsilon \text{ then } \alpha_2 \text{ fi})^i > \psi_1(u)$,

$\qquad$ (2) $\Gamma \Rightarrow \Theta(\varphi)$,

$\qquad$ (3) $\Gamma', \Theta(\psi_1) \Rightarrow \exists\, i <(\text{if } \varepsilon \text{ then } \alpha_2 \text{ fi})^i > (\neg\varepsilon \wedge <\beta> \psi)$

$\qquad$ where $\text{IND}(u)$ is the induction hypothesis:

$\qquad$ $\text{IND}(u) \equiv \forall\, u', \underline{x} \; ((u'<<u \wedge \varphi(u)) \rightarrow \exists\, i <(\text{if } \varepsilon \text{ then } \alpha_2 \text{ fi})^i > \psi_1(u))$ with

$\qquad$ $\underline{x}$ being the vector of the remaining variables that occur in $\varphi, \psi_1, \varepsilon$ and α_2. Θ is a substitution of metavariables for $\underline{x}$ and u. Γ' is a subset of Γ, where α_2 does not contain assignments to free variables of Γ'.

4.2 Towards a semi-automatic SI-Strategy

By adding some general tactics, for example for case analysis and simplifications, the user may now generate formal proofs in a way that is very close to informal proofs which are carried out by hand. Although this is a valuable result, (see [CBK 85]) we would like to go further.

In [HRS 87] we presented a completely interactive proof strategy combining all these tactics . This strategy offers the user some bookkeeping facilities. He is told what open subgoal should be tackled next. In cases where it seems to be hopeless to continue the proof he may type in "fail" and the system will lead him back to the last reasonable choice point. (This can be implemented using the rich backtracking facilities of PPL.) He still has to supply the induction hypothesis for while-loops, to prove all verification conditions generated by the tactics, and to decide what SI-tactic should be applied next. Note that in many situations more than one tactic is applicable. For example we might use the induction hypothesis or we might try to "execute" the body of a while-loop.

As an improvement of the above strategy, we have programmed an automatic simplification strategy (SIMPS). We use this strategy as a heuristic to decide without user-interaction what tactic we are going to apply. SIMPS is for example applied to the second subgoal of the while-tactic given above. If it succeeds the metavariables occurring in subgoal 3 become instantiated (using an infer step) and the strategy continues with subgoal 1 and (the modified) subgoal 3. If SIMPS fails on the second subgoal this is interpreted as a first guess that WHILE2_TAC is not applicable and the strategy tries to apply the next tactic (possibly the execute-tactic). SIMPS is also used to decide how to treat conditionals: in a given "state" we might be able to decide the test or we might have to consider both branches.

A central problem is to detect dead ends automatically. We have used SIMPS to program some simple heuristics for that task. If, for example, during the "execution" of the body of a while-loop we reach the situation ..., stack=s,... $\Rightarrow$ <stack := pop(stack)> <β> ψ ,where s is an input-variable and the antecedent cannot be simplified any further, the execution of SIMPS after ASSIGN_TAC will lead to a failure. This is due to the heuristic that as pop is "known" as a selector function and as there is no further information about s it is very unlikely that "executing" the assignment above will lead to a proof. In this case we have to use backtracking. As a result of the backtracking process we may abandon the "execution" of the body and try some other tactic. A sequent of the form s$\neq$emptystack , Γ $\Rightarrow$ <α_1><β> ψ will be simplified to $\Gamma_s^{\text{push}(s', y)}$ $\Rightarrow$ <α_1> <β> $\psi_s^{\text{push}(s', y)}$ where s´and y are fresh variables. Note that the condition s$\neq$emptystack may have been generated by "executing" a conditional.

The limitations of this heuristically guided strategy are obvious: it works only in situation where a decision can be made "at first sight". It is of no use in situations where we have to take into account deep mathematical results. However these situations do not occur frequently in the examples we have studied.

As a next and more ambitious step, one could think of an automatic generation of the induction hypothesis as it is done in the system of BOYER and MOORE [BM 79]. The main purpose of this section was to demonstrate that an increasing degree of automatization can be achieved in an environment like the KIV System by combining a given set of tactics which represent the logic of a method with "soft techniques" as are the heuristics described above.

4.3 Extensions

So far BURSTALL's original method has been extended by allowing the pre- and postconditions to contain (a certain kind of) program formulas. [HRS 87] contains a detailed example where this extension is used. To make use of program formulas occurring in an antecedent one has to add a new tactic, the so called *antecedent-tactic*, which is a generalization of the tactic for applying the inductive hypothesis. We are working on the problem to extend the SI-strategy to prove formulas <α_{rec}> φ $\rightarrow$ <α_{it}>ψ , where α_{rec} is a recursive program. To that end some new tactics dealing with the recursive program in the antecedents have to be developed. All other tactics (and heuristics) can be taken over from the SI method. It seems to be a promising idea to develop a collection of different verification methods in this step by step manner.

5 GRIES's Program Development Method

5.1 Informal Description

In his book "The Science of Programming" [Gr 81] GRIES describes in detail an informal method for the development of programs that uses invariants for the development of loops. The programming notation used is that of DIJKSTRA's guarded commands [Di 76].

The task is to find a program C satisfying the specification $Q \rightarrow wp(C,R)$, i.e., if C is started in a state where precondition Q holds then C always terminates in a state satisfying the postcondition R. Among others, the method uses the following strategy:

"To invent a guarded command, find a command C whose execution will establish postcondition R in at least some cases; find a boolean b satisfying $b \rightarrow wp(C,R)$; and put them together to form $b \rightarrow C$. Continue to invent guarded commands until the precondition of the construct implies that at least one guard is true."

5.2 Implementation of the Method in the KIV System

It is well known that the wp-semantics of guarded commands cannot be expressed directly in DL because guards may overlap. We therefore encode guarded commands (as well as **do** ... **od** commands) by programs which use "oracles". The statement which is proven during the development process says that for all such "oracles" our program is totally correct. Thereby we express the fact that any deterministic version can be chosen. Due to space limitations we do not present details of this encoding. Instead we use a simplified version where **if** $\varepsilon_1 \rightarrow \alpha_1 \Box \ldots \Box \varepsilon_n \rightarrow \alpha_n$ **fi** is encoded by IF ≡ **if** ε_1 **then** α_1 **else if** ε_2 **then** α_2 **else** ... **if** ε_n **then** α_n **else abort fi...fi**. and **do** $\varepsilon_1 \rightarrow \alpha_1 \Box \ldots \Box \varepsilon_n \rightarrow \alpha_n$ **od** by **while** $\varepsilon_1 \vee ... \vee \varepsilon_n$ **do** IF **od**. All the details of this formalism are hidden in the various tactics, for example the if-tactic (the subgoals generated by this tactic are stronger than necessary for the simplified version). The user is given the impression that he is using a structural editor for guarded commands.

We now give some tactics that are used by the PPL-program implementing the above method. Symbols beginning with "$" denote metavariables, whereas other symbols act as placeholders for formulas that may contain metavariables. The task of the tactics is to create new goals for recursive calls of GRIES's strategy.

Supplied with Γ, Δ, φ, where Γ is supposed to contain Δ, IF_TAC creates the following proof tree:

$$\Gamma, \neg\$b \Rightarrow\ <\$C2>\varphi \qquad\qquad \Gamma, \$b \Rightarrow\ <\$C1>\varphi$$

$$\Gamma \Rightarrow\ <\text{if }\$b\text{ then }\$C1\text{ else }\$C2\text{ fi}>\varphi$$

The validation that must be attached to the proof tree is simple.

Supplied with Γ, φ, ϕ and τ WHILE_TAC(Γ,φ,ϕ,τ) creates the following proof tree:

$$\phi, \neg\$b \Rightarrow \varphi \qquad \phi, \$b \Rightarrow \tau>0 \qquad \tau=t, t>0, \$b, \phi \Rightarrow\ <\$C>\phi \wedge \tau<t \qquad \Gamma \Rightarrow\ <\$C1>\phi$$

$$\Gamma \Rightarrow\ <\$C1;\text{while }\$b\text{ do }\$C\text{ od}>\varphi$$

The most important premise is the third where we have to show that the loop body decreases the bound function while maintaining the invariant. For this tactic the validation is more difficult than for IF_TAC. In fact we have to use tactics from SI. This is an example where the validation has to be postponed until the program is fully developed.

We now come to the interactive strategy. As a first step towards automation we added the possibility of backtracking. At any time the user may type "backtrack" which causes the system to return to the last call of the strategy. A next step could be to automatically find situations where backtracking is necessary. Only then can we begin to replace user interaction by automatic strategies

GRIES_STRAT (seq) =
print("The given specification is ", seq, "what kind of program do you want to develop?
        skip, assignments, conditional, loop, compound or backtrack?") **before**
let inp = read **and** q = ..antecedent of seq **and** r = ..postcondition of succedent of seq **in**
cond(inp = "backtrack" $\mapsto$ **fail**
        t $\mapsto$ (**cond**(inp = "skip" $\mapsto$ mkstree(q $\Rightarrow$ r)
                inp = "assignments" $\mapsto$ print("please give assignments") **before**
                                    let asg = read **in** EXECUTE_STRAT(q $\Rightarrow$ <asg>r),
                inp = "conditional" $\mapsto$ IF_STRAT(seq,q) ,
                inp = "loop" $\mapsto$ print("please give invariant and bound function") **before**
                                    let p = read **and** bf = read **in** WHILE_STRAT(seq,p,bf) ,
                inp = "compound" $\mapsto$ print("please give intermediate assertion") **before**
                                    let x = read **in** COMP_STRAT(seq,x)))
        or GRIES_STRAT(seq))

As an example for the substrategies of GRIES_STRAT, we present IF_STRAT which is a strategy for developing an arbitrary number of guarded commands. To ease understanding, we also give an informal description of the strategy which should be read simultaneously with the PPL-program.

First of all, a schema t1 for a conditional is generated by IF_TAC. Then GRIES_STRAT is given the second premise of the schema (Γ, \$b $\Rightarrow$ < \$C1 >φ) as a specification. The result t2 is a proof tree whose root looks like the given specification, but with \$C1 instantiated. Thus we have developed the "command"-part of a guarded command. The user has to give the corresponding guard b. t3 is just a copy of the root of t2, but \$b is instantiated with b. Then a refine-operation is used to instantiate \$b with b everywhere in t2 yielding t4. An infer-operation is used to attach the root of t4 to the second premise of t1 thereby instantiating t1 and yielding t5. Thus, in the schema created by IF_TAC \$b and \$C1 are instantiated and the second premise has been replaced by a proof tree while \$C2 is still open.

To develop a program for \$C2, we call IF_STRAT with the first premise of t5 as a specification. If we have enough guarded commands we type "enough" and ENDIF_TAC generates the program **abort** and eliminates the program from the formula. Otherwise, if we need another guarded command, we start the same procedure again. In other words, to prove the first premise of the schema for the conditional we have to build a proof tree t6 that has the same shape as the proof tree for the whole conditional. The result of IF_STRAT is a tree where t6 is attached to the first premise of t5 thereby instantiating t5.

IF_STRAT(seq,q) =   (* seq = q1 ⇒ < $C >r and q1 = q,¬b1,..,¬bn (n≥0) *)
   let inp = print("one more branch or enough or backtrack?", seq, Q) **before** read **and**
      q1 = ..antecedent of seq **and** r = ..postcondition of succedent of seq
  in cond(inp = "backtrack" ↦ **fail** ,

          inp = "enough" ↦ ENDIF_TAC(q1,r) ,

           t ↦ **letrec** t1 = IF_TAC(q1,q,r),   (* schema for conditional *)
                t2 = GRIES_STRAT(prem(t1,2)),   (*develop program for then-branch*)
                b =print("please give condition $b") **before** read
                t3 = mkstree(q,b ⇒ suc(concl(t2))),
                t4 = refine(t3,1,t2),   (*instantiate developed program with condition*)
                t5 = infer(t1, [2], [t4]),   (*instantiate schema with developed program*)
                t6 = IF_STRAT(prem(t5,1),q)   (*develop conditional for else-branch*)
        in infer(t5,[1],[t6]))   (*instantiate else-branch *)

5.3   Example

Let us illustrate how our strategy works by an example taken from [Gr 81]. The task is to develop a program that determines the maximum of two integers. The specification is: true ⇒ <$C> z≥x ∧ z≥y ∧ (z=x ∨ z=y). We decide do develop a conditional and the following proof tree is generated:

true, ¬$b ⇒ < $C2 > z≥x ∧ z≥y ∧ (z=x ∨ z=y)   true, $b ⇒ < $C1 > z≥x ∧ z≥y ∧ (z=x ∨ z=y)

               true ⇒ < **if** $b **then** $C1 **else** $C2 **fi**> z≥x ∧ z≥y ∧ (z=x ∨ z=y)

The command z:= x establishes the postcondition if x≥y. Therefore we choose z:= x for $C1 and x≥y for $b when proving the second premise. After instantiation we have

(*) true, ¬x≥y ⇒ < $C2 > R   true, x≥y ⇒ <z:= x> R
                                            (T)

        true ⇒ < **if** x≥y **then** z:= x **else** $C2 **fi** > R

where R ≡ z≥x ∧ z≥y ∧ (z=x ∨ z=y). Now IF-STRAT is called recursively with the arguments true, ¬x≥y ⇒ < $C2 > R for seq and true for q. This time we choose the command z:= y and the guard y≥x. Then we type "enough" and get the value of the recursive call:

true ⇒ x≥y ∨ y≥x
  |

true, ¬x≥y, ¬y≥x ⇒ <**abort** > R   true, y≥x ⇒ <z:= y> R

      true, ¬x≥y ⇒ < **if** y≥x **then** z:= y **else** **abort** **fi**> R

The root of this tree matches the first premise (*) of T, so that the infer-operation yields as the final result of IF_STRAT the following tree, where <z:= y> R and <z:= x> R have been reduced by

EXECUTE-STRAT to true, x≥y ⇒ x≥y   and   true, y≥x ⇒ y≥x respectively.

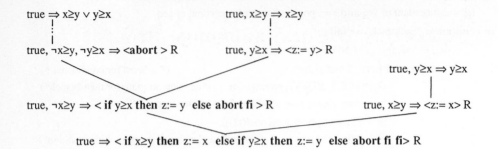

true ⇒ < if x≥y then z:= x else if y≥x then z:= y else abort fi fi> R

Hence we have developed and simultaneously proven correct a program, written in guarded-command-notation as: **if** x≥y → z:= x ∏ y≥x → z:= y **fi**

References

[BM 79]   Boyer, R.S./ Moore, J.S. A Computational Logic. Academic Press, New York 1979

[Br 80]   de Bruin, N.G. A Survey of the Project AUTOMATH. Essays in Combinatory Logic, Lambda Calculus, and Formalism, J.P. Selding and J.R. Hindley (eds.) Academic Press (1980), pp 589-606

[Bu 74]   Burstall, R.M. Program Proving as Hand Simulation with a little Induction. Information Processing 74, North-Holland Publishing Company (1974)

[CKB 85]   Constable,R./Knoblock,T./Bates,J. Writing Programs That Construct Proofs. Journal of Automated Reasoning, Vol.1, No.3, pp 285 - 326 (1985)

[Di 76]   Dijkstra, E.W. A Discipline of Programming. Prentice-Hall (1976)

[GMW 79]   Gordon,M/Milner,R./Wadsworth,C. Edinburgh LCF. Springer LNCS 78 (1979)

[Go 82]   Goldblatt, R. Axiomatising the Logic of Computer Programming. Springer LNCS 130 (1982)

[Gr 81]   Gries, D. The Science of Programming, Springer-Verlag (1981)

[Ha 79]   Harel, D. First Order Dynamic Logic. Springer LNCS 68 (1979)

[Ha 84]   Harel, D. Dynamic Logic. Handbook of Philosophical Logic, D. Gabbay and F. Guenthner (eds.), Reidel (1984), Vol. 2, 496-604

[HHRS 86]   Hähnle, R./Heisel, M./Reif, W./Stephan, W. An Interactive Verification System Based on Dynamic Logic. Proc. 8-th International Conference on Automated Deduction, J.Siekmann (ed), Springer LNCS 230 (1986), 306-315

[HRS 86]   Heisel,M./Reif, W./Stephan, W. A Functional Language to Construct Proofs. Interner Bericht 1/86, Fakultät für Informatik, Universität Karlsruhe (1986)

[HRS 87]   Heisel,M./Reif, W./Stephan, W. Program Verification by Symbolic Execution and In duction. Proc. 11-th German Workshop on Artificial Intelligence, K. Morik (ed), Informatik Fachberichte 152, Springer-Verlag (1987)

[Re 84]   Reif, W. Vollständigkeit einer modifizierten Goldblatt-Logik und Approximation der Omegaregel durch Induktion. Diplomarbeit, Fakultät für Informatik, Universität Karlsruhe (1984)

[Ri 78]   Richter, M. M. Logikkalküle, Teubner (1978)

Checking Natural Language Proofs *

Donald Simon

Department of Computer Science
University of Texas at Austin
Austin, Texas 78712

Abstract

Proofs in natural language contain much information useful for automatic
proof checking that is usually lost in translation to a formal language. This
paper describes a system which checks English language proofs in elementary
number theory that uses such information to guide the theorem prover. The
proof connector follows the argument presented in the proof and asks a theorem
prover to make the same deductions that the human reader of the proof is
assumed to make. This system has the added advantage of spotting fallacious
proofs of correct theorems. A more powerful theorem prover might prove the
theorem by ignoring a faulty proof.

1   Introduction

There are two basic approaches to automatically checking proofs: The first is to
formalize a proof by hand and then have the machine check the formal proof. The
second is to write a system that understands the natural language text and uses
both formal and informal knowledge. The first approach is the traditional approach;
Automath [3], THM [2], Ontic [5], to name a few, all require that the argument be in
formal terms. However, when replacing natural language formulations with formal
ones, information that is useful to the human reader is often lost. This information
could also be useful to the machine checker. On the other hand, the cost of pulling
this knowledge out of the NL text may be high. An open question is whether or not
the costs of processing the NL text to sift out this information outweighs, in terms
of simplifying the proving process, the costs of doing so.

---

*This work supported by National Science Foundation Grant CCR-8613 706.

For example, consider the differences between the statements, "Q is true and R is true," and "Q is true and hence R is true," There are three ways to interpret the first statement:

1. Prove Q, independently prove R, and then use both in the proof.

2. Prove Q, use Q to prove R, and then use R in the proof.

3. Prove Q, use Q to prove R, and then use both in the proof.

However, the second statement only has the latter two interpretations. With little overhead, the search space is made smaller by considering the adverb.

As a second example, consider a proof that starts with the clause, "If P is true." This could be the start of a proof by cases or a proof by contradiction. But if the proof begins, "If P were true," then it must be a proof by contradiction, since the subjunctive mood is used.

A second problem in automatic proof checking arises in the difference between a theorem prover and a true proof checker. In the first case, the object is to find a proof of some proposition regardless of the method. In the second, an argument is given, and the question is posed: does this argument prove the theorem? It is quite possible that a theorem is true, but a given proof for it is fallacious. If the theorem proving element of the system is too good, it may be able to prove the theorem ignoring the proof entirely. Does this mean the proof is correct? Of course not. Therefore, to actually do proof checking, we must be able to follow the given argument.

This paper will discuss a system that takes NL proofs in elementary number theory, and checks them by converting them into formal equivalents. The proofs that the system is targeted for are taken from [4]. In the remainder of this paper, we will look at an example of an NL proof, and the procedural knowledge incorporated in the system for checking this proof.

2 System Overview

Since one of our tenets is that the machine should have all the information that the human has, including typographical information, the decision was made to have the input be a source file to a text-formatting system, in our case, a LaTeX source file. The source proof is passed through a lexical analyzer, which returns a list of tokens for each sentence. Each list is passed to a sentence parser which applies syntactic and semantic information to parse the NL elements in the token list into an intermediate language. The proof connector takes this intermediate form and

Theorem: $a \in \mathbf{Z} \wedge b \in \mathbf{Z} \wedge a > 0 \Rightarrow \exists!(q,r), b = q * a + r \wedge 0 \leq r \wedge r < a$

Proof: first, we show that formula($\exists(q,r), (b = q * a + r \wedge 0 \leq r \wedge r < a)$) break
consider formula($D = \mathrm{set}(b - u * a, u \in \mathbf{Z})$) break
for formula($u = \mathrm{if}(b \geq 0, 0, b)$), formula($b - u * a \geq 0$),
 so that formula($\exists x, x \in D \wedge x \geq 0$) break
therefore formula(exists_smallest($x, x \in D \wedge x \geq 0$)) break
take formula($r = \mathrm{smallest}(x, x \in D \wedge x \geq 0)$)
 and take formula($q = \mathrm{the}(u, b - u * a = r \wedge u \in \mathbf{Z})$),
 i.e., let formula($q = \mathrm{largest}(q, q \in \mathbf{Z} \wedge b - q * a \geq 0)$) break
then, formula($r = b - q * a \cap b - q * a \geq 0$),
 whereas formula($r - a = b - (q + 1) * a \cap b - (q + 1) * a < 0$) break
hence formula($\exists(q,r), (b = q * a + r \wedge 0 \leq r \wedge r < a)$) break
to show the uniqueness of term((q,r)),
 assume that formula($b = q' * a + r' \wedge 0 \leq r' \wedge r' < a$) break
then if formula($q' < q$),
 we have formula($b - q' * a = r' \cap r' \geq b - (q - 1) * a$
$\cap b - (q - 1) * a = r + a \cap r + a \geq a$)
 and this contradicts formula($r' < a$) break
hence formula($\neg(q' < q)$) break
similarly, we show formula($q \geq q'$) break
therefore formula($q = q'$) and consequently formula($r = r'$) $\triangle$

Figure 1: Output from the Sentence Parser

follows the proof, calling a theorem prover to make the appropriate intermediate deductions. A complete, formal proof is returned by the proof connector.

The lexical analyzer and the sentence parser translate each sentence of the input proof into a list which contains the terms and formulas included in the sentence, and the other information, such as the verbs (if not translatable into a formula, e.g. non-operational verbs such as "consider," or "define"), and the adverbs. A slight oversimplification of this process would be to say that the mathematical terms and formulas that are represented in NL are transformed into their formal equivalents and the rest of the sentence is left as is. The parsed sentences are passed along to the proof connector.

As an example, we will consider Theorem 1-1 from [4]. The proof-checking system is given a LaTeX source file for the theorem and its proof. The output from the sentence parser appears in Figure 1. Segments of English text that represent mathematical terms and formulas have been replaced mechanically with formal equivalents and two references have been replaced by the appropriate referents by hand.

The proof connector follows the logic of the proof and calls a theorem prover to make the same deductions that a human reader of the proof is expected to do. The theorem prover answers the problems posed to it and returns the proofs of the intermediate steps. These subproofs are connected together into the proof of the main theorem that is claimed to exist.

3   Proof Connector

The proof connector is our main concern in this paper. It is an augmented phrase structure grammar. The grammar has at hand the sentences, but also the current goal to be proved (originally the theorem), a stack of facts (proof steps, definitions, hypotheses, and lemmas) that are currently in force, and the current context of the proof. The context includes such things as whether the proof is part of the basis or induction step of an induction proof, whether it is part of a proof of cases, etc. The grammar rules return the proof of the goal, or occasionally, in the cases context, the proof of something else. The proof connector will backtrack if a path fails to lead to a proof.

The context plays two roles in the proof connector. First, it serves to resolve references, and second, when the proof connector comes to the end of a context on the proof, it may choose to prove some result automatically. For example, a proof by cases of proposition P may state the following: "If Q then R, otherwise S," where it is easily seen that both R and S imply P, although this is never stated. The word "otherwise" in addition to denoting the start of the second case (and implying its hypothesis, namely not Q), also serves to end the first case. One of the proof rules states that if we come to the end of the case, see if the goal can be proved by what's on the stack.

This example, however, could be interpreted differently. The text that follows the line may read: "In either case, we have T, hence we conclude P." For this reason, we also have the proof rule, that a case may prove something other than the goal, and this proof is the one that is returned. This is the only case (so far) where a rule returns with a proof of some proposition other than the goal.

The term "stack" is really a misnomer. For example, in proving the conjunction P and Q, the proof may go as follows: "Prove R. Define A. Prove S about A and use R and S to prove P, then prove T about R and use T to prove B." After proving the first conjunct P, we would like to pop everything off the stack that was used to prove P, for reasons of efficiency in the theorem prover. But clearly, we cannot pop the definition of A from the stack since it is used in the proof of B. However, we may or may not want to pop the proof of R off the stack, depending on whether or not R is used to show that A is well defined. In general, the stack will be popped, and then if a definition that has been removed becomes necessary later in the proof, the connector will backtrack, and not pop the definition from the stack.

Let us now consider the fragment of the proof grammar in Figures 2 and 4 that is capable of following the informal proof of Theorem 1-1. (This grammar fragment is a simplification of the rules that actually occur in our system.)

The rules are written in an extended Prolog. In this version, we allow second order variables to occur inside quantified expressions as long as all the arguments of the second order expression are indicial variables of the quantifier and the expression will always be bound to a ground term when the clause is called. With these restrictions, there will be at most one valid unifier for the expression.

Several procedures require explanation:

deduce_formula(Q, Context, Facts, Qproof) is a call to the theorem prover to prove the proposition Q from the stack of Facts. Qproof, a proof of Q, is returned. The context is also passed in the event that Q contains unfilled references.

deduce_existence(X, Facts, Proof) proves that X exists.

form_proof(P, Proofs, Proof) creates a proof of P from a list of proofs.

When we run the proof grammar on the first part of the proof for Theorem 1-1, the calls in Figures 5 – 6 are made to the theorem prover. The listing indicates whether or not the theorem prover can prove the proposition. In the actual system, the proof of the proposition is returned if one is found.

As we have claimed, the proof connecter follows the logic of the proof and asks the theorem prover to make the same deductions that the reader of the proof would make.

4   Theorem Prover

The theorem prover is the heart of the deduction mechanism. In this section, we discuss the requirements of the theorem prover and describe our first attempts at constructing a suitable one. The first requirement is that the theorem prover be sufficiently powerful to prove the deductions that are left to the reader. However, the proof connector may take the wrong path and ask the theorem prover to prove either a non-theorem or a theorem without sufficient lemmas. For this reason, the theorem prover must fail fairly quickly when it is given a non-theorem or a theorem that's too hard to prove.

The second requirement of the theorem prover is that when it is asked to prove that P follows from Q, it returns a proof in which P follows from Q in a meaningful way. What "a meaningful way" means is currently undefined.

The theorem prover can make use of the knowledge that the facts that it is given are in a stack. In other words, the first facts are more current and may be more

prove(P ⇒ Q, Context, Facts, Proof) −→
  prove(Q, Context, [hyp(P),.. Facts], Proof1),
  { form_proof(P ⇒ Q, [suppose(P), Proof1], Proof) }.

prove(one(X,P), Context, Facts, Proof) −→
  prove(some(X,P) and unq(X,P), Context, Facts, Proof1),
  { form_proof(one(X,P), [Proof1], Proof) }.

prove(P and Q, Context, Facts, Proof) −→
  [first], prove(P, Context, Facts, Pproof), [break],
  prove(Q, Context, Facts, Qproof),
  { form_proof(P and Q, [Pproof, Qproof], Proof) }.

prove(P, Context, Facts, Proof) −→
  state_goal(P), prove(P, Context, Facts, Proof).

prove(unq(X, P(X)), Context, Facts, Proof) −→
  suppose(P(Y)),
  prove1(X = Y, Context, [hyp(P(Y)), hyp(P(X)),.. Facts], Proof1),
  { form_proof(unq(X, P(X)), [Proof1], Proof) }.

prove(P, Context, Facts, Proof) −→
  state_definition(X, Y1, Y2),
  prove_well_defined(X, Y1, Y2, Context, Facts, Def, Newfacts),
  prove(P, Context, Newfacts, Proof1),
  { form_proof(P, [Def, Proof1], Proof) }.

prove(P, Context, Facts, Proof) −→
  then,[formula(Q)],
  { deduce_formula(Q, Context, Facts, Qproof) },

prove(P, Context, [Step,.. Facts], Proof) −→
  hence,[formula(Q)],
  { deduce_formula(Q, Context, [Step,.. Facts], Qproof) },
  prove1(P, Context, [Qproof,.. Facts], Proof).

prove(P, Context, Facts, Proof) −→
  then, suppose(Hyp),
  prove(P, context(case(Hyp), Context), [hyp(Hyp),.. Facts], proof(Q, Qsteps)),
  { form_proof(Hyp ⇒ Q, [suppose(Hyp), proof(Q, Qsteps)], Proof1) },
  prove1(P, Context, [Proof1,.. Facts], Proof).

prove(P, Context, Facts, Proof) −→
  [this,contradicts,formula(Q)],
  { deduce_formula(Q, Context, Facts, Qproof),
  deduce_formula(not(Q), Context, Facts, NotQproof),
  form_proof(contradiction, [Qproof, NotQproof], Proof) }.

prove(P, Context, Facts, Proof) −→
  similarly,[formula(Q)],
  { form_proof(Q, [similarly], Proof1) },
  prove1(P, Context, [Proof1,.. Facts], Proof).

Figure 2: Grammar to parse Theorem 1–1

prove1(P, _, [proof(P, Psteps),.. Facts], proof(P, Psteps)) ⟶ [].

prove1(P, Context, Facts, Proof, [qed], [qed]) :-
  deduce_formula(P, Context, Facts, Proof).

prove1(P, Context, Facts, Proof) ⟶
  [break], prove(P, Context, Facts, Proof).

prove1(P, Context, Facts, Proof) ⟶
  [and], prove(P, Context, Facts, Proof).

prove1(P, Context, [Step,.. Facts], Proof) ⟶
  [so,that,formula(Q)],
  { deduce_formula(Q, Context, [Step,.. Facts], Qproof) },
  prove1(P, Context, [Qproof,.. Facts], Proof).

prove1(P, Context, Facts, Proof) ⟶
  [whereas,formula(Q)],
  { deduce_formula(Q, Context, Facts, Qproof) },
  prove1(P, Context, [Qproof,.. Facts], Proof).

state_goal(unq(X,_)) ⟶ [to,show,the,uniqueness,of,term(X)].

state_goal(P) ⟶ [we,show,that,formula(P),break].

state_definition(X, Y1, Y2) ⟶
  let,[formula(X = Y1)],
  more_definition(X, Y2).

state_definition(X, Y, empty) ⟶ [for,formula(X = Y)].

more_definition(X, Y2) ⟶ [ie], state_definition(X, Y2, empty).

more_definition(X, empty) ⟶ [and].

more_definition(X, empty) ⟶ [break].

prove_well_defined(X, Y1, empty, _, Facts, Def, [Def,.. Newfacts]) ⟶
  !, prove_existence(Y1, Facts, Newfacts, Proof),
  { Def = def(X = Y1, Proof) }.

prove_well_defined(X, Y1, Y2, Context, Facts, Def, [Proof3, Def,.. Newfacts]) ⟶
  prove_existence(Y1, Facts, Newfacts, Proof1),
  { deduce_formula(Y1 = Y2, Context, Newfacts, Proof2),
  form_proof(X = Y2, [def(X = Y1, Proof1), Proof2], Proof3),
  Def = def(X = Y1, Proof1) }.

prove_existence(Y, [proof(exists(Y), Steps),.. Facts], Facts, proof(exists(Y), Steps)) ⟶ [].

prove_existence(Y, Facts, Facts, Proof) ⟶ { deduce_existence(Y, Facts, Proof) }.

Figure 3: Grammar to parse Theorem 1–1

let $\longrightarrow$ [consider].
let $\longrightarrow$ [take].
let $\longrightarrow$ [let].

hence $\longrightarrow$ [so,that].
hence $\longrightarrow$ [therefore].
hence $\longrightarrow$ [then].
hence $\longrightarrow$ [hence].

suppose(P) $\longrightarrow$ [assume,that,formula(P)].
suppose(P) $\longrightarrow$ [if,formula(P)].

then $\longrightarrow$ [then].
then $\longrightarrow$ [we,have].
then $\longrightarrow$ [consequently].
then $\longrightarrow$ [].

Figure 4: Grammar to parse Theorem 1–1

useful in finding a proof than latter facts.

The prover we are currently working on relies on this. If it is trying to prove that P follows from Q, where Q is on the top of the stack of facts, it will use rewrite rules to rewrite Q into P. If that fails, the prover will select one rule and rewrite P into P' and then loop back and try to rewrite Q into P'. The rewrite rules chosen are based on the predicate and functions found in Q and P. This idea is similar to gazing [6]. For certain classes of formulas, a decision procedure may be used instead. For example, if the prover is trying to prove a ground inequality, a SUPINF prover is called [1]. Currently, the fallback position of the prover is to ask the user if the proposition is true in case of failure.

5 Current Status

The system we are building is specifically designed to handle one book, [4].

The current status of our system is as follows:

- The lexical analyzer and formula parser handles all proofs in the book. There are 71 theorems in the book of which 65 have proofs.

- The sentence parser is divided into two parts: a syntactic component which contains all the knowledge about the English syntax found in the text, and a semantic component, which contains knowledge about the meaning of the objects in the text. The syntactic component is complete for the book, about 700 sentences. The semantic component in an earlier version could handle 15 proofs and is currently being integrated into the new system and expanded.

- The proof grammar can parse about 10 proofs. Most pronomial references must be filled in by hand before the proof grammar runs.

- The theorem prover is under construction.

1. Does $\forall u, u \in \mathbf{Z} \Rightarrow b - u * a \in \mathbf{Z}$ follow from:
$a \in \mathbf{Z} \wedge b \in \mathbf{Z} \wedge a > 0$? Yes.

2. Does $b - u * a \geq 0$ follow from:
$u := \text{if}(b \geq 0, 0, b)$ $\qquad\qquad\qquad D := b - u * a, u \in \mathbf{Z}$
$a \in \mathbf{Z} \wedge b \in \mathbf{Z} \wedge a > 0$? Yes.

3. Does $\exists x, x \in D \wedge x \geq 0$ follow from:
$b - u * a \geq 0$ *** $\qquad\qquad\qquad u := \text{if}(b \geq 0, 0, b)$
$D := b - u * a, u \in \mathbf{Z}$ $\qquad\qquad\quad a \in \mathbf{Z} \wedge b \in \mathbf{Z} \wedge a > 0$? Yes.

4. Does $\exists \text{smallest}(x, x \in D \wedge x \geq 0)$ follow from:
$\exists x, x \in D \wedge x \geq 0$ *** $\qquad\qquad\quad u := \text{if}(b \geq 0, 0, b)$
$D := b - u * a, u \in \mathbf{Z}$ $\qquad\qquad\quad a \in \mathbf{Z} \wedge b \in \mathbf{Z} \wedge a > 0$? Yes.

5. Does $\exists! u, b - u * a = r \wedge u \in \mathbf{Z}$ follow from:
$r := \text{smallest}(x, x \in D \wedge x \geq 0)$ $\qquad u := \text{if}(b \geq 0, 0, b)$
$D := b - u * a, u \in \mathbf{Z}$ $\qquad\qquad\quad a \in \mathbf{Z} \wedge b \in \mathbf{Z} \wedge a > 0$? Yes.

6. Does $\text{the}(u, b - u * a = r \wedge u \in \mathbf{Z}) = \text{largest}(q, q \in \mathbf{Z} \wedge b - q * a \geq 0)$ follow from:
$r := \text{smallest}(x, x \in D \wedge x \geq 0)$ $\qquad u := \text{if}(b \geq 0, 0, b)$
$D := b - u * a, u \in \mathbf{Z}$ $\qquad\qquad\quad a \in \mathbf{Z} \wedge b \in \mathbf{Z} \wedge a > 0$? Yes.

7. Does $r = b - q * a$ follow from:
$q = \text{largest}(q, q \in \mathbf{Z} \wedge b - q * a \geq 0)$ *** $\quad q := \text{the}(u, b - u * a = r \wedge u \in \mathbf{Z})$
$r := \text{smallest}(x, x \in D \wedge x \geq 0)$ $\qquad\quad u := \text{if}(b \geq 0, 0, b)$
$D := b - u * a, u \in \mathbf{Z}$ $\qquad\qquad\qquad\quad a \in \mathbf{Z} \wedge b \in \mathbf{Z} \wedge a > 0$? Yes.

8. Does $b - q * a \geq 0$ follow from:
$q = \text{largest}(q, q \in \mathbf{Z} \wedge b - q * a \geq 0)$ *** $\quad q := \text{the}(u, b - u * a = r \wedge u \in \mathbf{Z})$
$r := \text{smallest}(x, x \in D \wedge x \geq 0)$ $\qquad\quad u := \text{if}(b \geq 0, 0, b)$
$D := b - u * a, u \in \mathbf{Z}$ $\qquad\qquad\qquad\quad a \in \mathbf{Z} \wedge b \in \mathbf{Z} \wedge a > 0$? Yes.

9. Does $r - a = b - (q + 1) * a$ follow from:
$r \geq 0$ *** $\qquad\qquad\qquad\qquad\qquad q = \text{largest}(q, q \in \mathbf{Z} \wedge b - q * a \geq 0)$ ***
$q := \text{the}(u, b - u * a = r \wedge u \in \mathbf{Z})$ $\qquad r := \text{smallest}(x, x \in D \wedge x \geq 0)$
$u := \text{if}(b \geq 0, 0, b)$ $\qquad\qquad\qquad\quad D := b - u * a, u \in \mathbf{Z}$
$a \in \mathbf{Z} \wedge b \in \mathbf{Z} \wedge a > 0$? Yes.

10. Does $b - (q + 1) * a < 0$ follow from:
$r \geq 0$ *** $\qquad\qquad\qquad\qquad\qquad q = \text{largest}(q, q \in \mathbf{Z} \wedge b - q * a \geq 0)$ ***
$q := \text{the}(u, b - u * a = r \wedge u \in \mathbf{Z})$ $\qquad r := \text{smallest}(x, x \in D \wedge x \geq 0)$
$u := \text{if}(b \geq 0, 0, b)$ $\qquad\qquad\qquad\quad D := b - u * a, u \in \mathbf{Z}$
$a \in \mathbf{Z} \wedge b \in \mathbf{Z} \wedge a > 0$? Yes.

Figure 5: Calls to the theorem prover

11. Does $\exists (q, r) b = q * a + r \wedge 0 \le r \wedge r < a$ follow from:

$r - a < 0$ *** $\qquad\qquad\qquad\qquad r \ge 0$ ***

$q = \text{largest}(q, q \in \mathbf{Z} \wedge b - q * a \ge 0)$ *** $\qquad q := \text{the}(u, b - u * a = r \wedge u \in \mathbf{Z})$

$r := \text{smallest}(x, x \in D \wedge x \ge 0)$ $\qquad\qquad u := \text{if}(b \ge 0, 0, b)$

$D := b - u * a, u \in \mathbf{Z}$ $\qquad\qquad\qquad a \in \mathbf{Z} \wedge b \in \mathbf{Z} \wedge a > 0?$ Yes.

Figure 6: Calls to the theorem prover

6 Conclusions

As the examples in this paper demonstrate, it is useful to consider the English text when checking a proof. The proof checking system is guided through the proof just as the human reader is, substantially reducing the time necessary to prove the theorem. One problem that still remains stems from the fact that any NL formulation of a proof is inherent informal and hence there will always be the question of whether or not the formal version of the proof is equivalent to the informal presentation. However, the simplicity of the proof grammar rules lead us to believe that the proof checking system will not stray from the intent of the author of the NL proof.

I would like to thank Dr. W. W. Bledsoe and Larry Hines for their help in the preparation of this paper.

References

[1] Bledsoe, W. W., "A New Method for Proving Certain Presburger Formulas," Fourth IJCAI, Tbilisi, USSR, (September 3–8, 1975).

[2] Boyer, R. S., and Moore, J S., *A Computational Logic*, Academic Press, (1979).

[3] de Bruijn, N. G., "Automath, a language for mathematics," Eindhoven University of Technology, T.H. –Report 68–WSK–05. AUT 1.

[4] LeVeque, W. J., *Elementary Theory of Numbers*, Addison-Wesley, Reading, Mass., (1962).

[5] McAllester, David, "Ontic: A Knowledge Representation System for Mathematics," MIT AI Lab Technical Report 979, (July 1987).

[6] Plummer, D., "GAZING: A Technique for Controlling Rewrite Rules," Phd Thesis, University of Edinburgh, Department of Artificial Intelligence, (December 1987).

Consistency of Rule-based Expert Systems

Marc Bezem

Centre for Mathematics and Computer Science
P.O. Box 4079, 1009 AB Amsterdam, The Netherlands

Consistency of a knowledge-based system has become a topic of growing concern. Every notion of consistency presupposes a notion of semantics. We present a theoretical framework in which both the semantics and the consistency of a knowledge base can be studied. This framework is based on first order flat many-sorted predicate logic and is sufficiently rich to capture an interesting class of rule-based expert systems and deductive databases. We analyse the feasibility of the consistency test and prove that this test is feasible for knowledge bases in Horn format without quantification.

1980 Mathematics Subject Classification: 68T30, 68T15.
1987 CR Categories: I.2.3, I.2.4, F.4.1.
Key Words & Phrases: knowledge-based systems, rule-based expert systems, knowledge representation, consistency.
Note: The work in this document was conducted as part of the PRISMA project, a joint effort with Philips Research Eindhoven, partially supported by the Dutch "Stimulerings-projectteam Informatica-onderzoek" (SPIN).

INTRODUCTION

The plan of this paper is as follows. First the reader is introduced to the knowledge representation used in rule-based expert systems. We shall indicate some semantical problems in relation to this knowledge representation. Then we explain in an informal way how many-sorted predicate logic comes in. In the next section we describe syntax and semantics of many-sorted predicate logic. We assume some knowledge of first order logic. Thereafter we shall be able to characterize rule-based expert systems as first order theories. The Tarski semantics solves the semantical problems mentioned above. Furthermore we shall derive several results on decidability and consistency of rule-based expert systems. Unfortunately, some natural equality and ordering axioms are not in Horn format (see [Re] for a discussion on the domain closure axiom). Hence testing consistency with a standard theorem prover would be very inefficient. In the last section we describe a technical device, a certain kind of null value, which allows feasible consistency testing in the presence of equality and ordering axioms, which are not in Horn format. This kind of null value, being quite different from null values as described in [IL], appears to be new. We shall focus our attention on rule-based expert systems, but the techniques can also be applied to deductive databases.

CONTENTS

1. Rule-based expert systems.
2. Many-sorted predicate logic.
3. Rule-based expert systems as many-sorted theories.
4. Testing consistency.
5. Conclusion.

1. RULE-BASED EXPERT SYSTEMS

1.1. In rule-based expert systems shells such as EMYCIN [BS] or DELFI2 [L], knowledge about some specific domain can be expressed in *facts* and in *rules* of the form

$$IF < antecedent > THEN < consequent >.$$

Facts are so-called object-attribute-value triples, or $<o,a,v>$ triples for short. The antecedent of a rule is a conjunction of disjunctions of conditions, and conditions are definite statements, such as *same, notsame* and *less than*, about $<o,a,v>$ triples. We restrict ourselves to rules having as consequent a conjunction of conclusions of the form *conclude* $<o,a,v>$. In most cases so-called certainty factors are associated with the facts and the rules. Certainty factors range from 1.00 (definitely true) to -1.00 (definitely false). The certainty factor of a fact expresses a measure of certainty about that fact, whereas the certainty factor of a rule scales the measure of certainty about the consequent with respect to the measure of certainty about the antecedent. In DELFI2 an *object tree* (called *context tree* in MYCIN) is used to state properties of and relations between different objects, which cannot be expressed by the rules. The nodes of this tree are objects, labeled by their attributes and respective domains of values. The path from a node to the root of this tree constitutes the context of that node. In other words: the objects occurring in the subtree of a node are sub-objects of the object belonging to that node. Furthermore it is stated in the object tree whether an attribute is singlevalued or multivalued.

The interpretation of the knowledge in rule-based expert systems is more operational than declarative: *same* $<o,a,v>$ is true if and only if $<o,a,v>$ occurs as fact (with certainty factor > 0.2), *conclude* $<o,a,v>$ has the effect that $<o,a,v>$ is added as fact (with appropriate certainty factor), and if the antecedent of a rule evaluates to true, then that rule may be *fired*, i.e. all conclusions occurring in the consequent are executed. We remark that *same* $<o,a,v>$ and *conclude* $<o,a,v>$ have the same declarative meaning as the fact $<o,a,v>$, i.e. attribute *a* of object *o* has value *v*.

1.2. Consider the following real-life example extracted from HEPAR, an expert system for the diagnosis of liver and biliary disease, built with DELFI2.

> IF same $<patient,complaint,colicky\_pain>$
>
> THEN conclude $<patient,pain,colicky>$ (1.00)

> IF same $<patient,abd\_pain,yes>$ AND
>
> same $<pain,character,continuous>$
>
> THEN conclude $<patient,pain,colicky>$ (-1.00)

> IF same $<patient,complaint,abdominal\_pain>$ OR
>
> same $<patient,pain,colicky>$
>
> THEN conclude $<patient,abd\_pain,yes>$ (1.00)

These three rules (from a rule base consisting of over 400 rules) show two *objects*, patient and pain, four *attributes*, namely complaint, pain and abd_pain of patient and character of pain, as well as several *values*.

1.3. A first observation, which can be made on the three rules above, is their inconsistency in the presence of the facts $<patient,complaint,colicky\_pain>$ (1.00) and $<pain,character,continuous>$ (1.00). The inference engine reasons backwards, using a simple loop check to prevent infinite looping.

Depending on the presence of the fact $<patient,complaint,abdominal\_pain>$ (1.00), the inference engine did or did not hit upon the contradiction $<patient,pain,colicky>$ (1.00 and -1.00). In *both* cases the contradiction was completely ignored.

A second observation is the following. Five items refer to pain: the values colicky_pain and abdominal_pain, the attributes pain and abd_pain of the object patient, and the object pain, which is a sub-object of patient, as stated by the object tree of HEPAR. The interrelations between these items do not seem to be expressible by the formalism.

These observations show a defect of the knowledge representation used in rule-based expert systems, namely the lack of a clear semantics.

1.4. Basically our approach amounts to interpreting $<o,a,v>$ by $a(o,v)$ in the multivalued, and by $a(o) = v$ in the singlevalued case. Here o and v are constants for elements of a domain O of objects and a domain V of values. In the multivalued case a denotes a *relation*, i.e. a subset of $O \times V$, and in the singlevalued case a *function* from O to V. If o is a sub-object of o', then o' is added as argument of a.

1.5. The following examples show how to extend our interpretation of $<o,a,v>$ triples to atoms.

> *conclude* $<patient,complaint,abdominal\_pain>$

becomes

> *complaint* $(patient,abdominal\_pain)$

> *less_than* $<patient,temperature, 36.8>$

becomes

> *temperature* $(patient) < 36.8$

> *same* $<pain,character,continuous>$

becomes

> *character* $(patient,pain) = continuous$

Note that the fact that pain is a sub-object of patient is expressed by adding *patient* as an argument of the function character. Another example, based on the rules of 1.2, is to be found in 4.6.

We feel that the existing formalisms for handling uncertainty are unsatisfactory. As we do not have a good alternative we leave the subject aside.

1.6. Under the interpretation described above, a rule-based expert system becomes a theory in first order many-sorted predicate logic, in short: a many-sorted theory. To keep this paper self-contained we give a short, introductory description of the syntax and semantics of many-sorted predicate logic in the next section. In Section 3 we shall characterize expert systems as many-sorted theories of a certain type. This approach has the following advantages:
- The declarative semantics of the expert system becomes perfectly clear, being the Tarski semantics of the associated many-sorted theory.
- Logical concepts such as decidability, consistency etc. get a clear meaning in relation to the expert system.
- Theorem proving techniques for testing consistency, such as resolution, become available for the expert system.

The choice for many-sorted instead of one-sorted logic is motivated as follows:
- It is natural to make a distinction between numerical data (ordered by $<$) and symbolic data (usually unordered).
- Subdividing a set of constants into sorts and typing the predicate and function symbols considerably reduces the number of well-formed formulas, which is important for debugging large knowledge bases.
- The many-sorted approach also imposes restrictions on unification, which considerably reduces the resolution search space (see for example [W]). We shall not make use of this advantage in the present paper.

2. Many-Sorted Predicate Logic

2.1. The syntax of many-sorted predicate logic extends the syntax of ordinary, one-sorted, predicate logic by having a set of *sorts* Σ, instead of just one sort. Moreover we have *variables* x_i^σ and *constants* c_i^σ for all sorts $\sigma \in \Sigma$. Furthermore we have *function symbols* $f_i^{\sigma_1 \times \cdots \times \sigma_m \to \sigma_0}$ $(m>0)$, where the notion of *type* $\sigma_1 \times \cdots \times \sigma_m \to \sigma_0$ replaces the notion of *arity* from the one-sorted case. We also have *proposition symbols* p_i and *predicate symbols* $P_i^{\sigma_1 \times \cdots \times \sigma_m}$ $(m>0)$ of type $\sigma_1 \times \cdots \times \sigma_m$. *Terms* are formed from variables and constants by function application (respecting the sorts). *Atoms* are either proposition symbols or the application of a predicate symbol to terms of appropriate sorts. With the help of propositional connectives and quantifiers, atoms are combined into *formulas*. The sets Σ, CONS, FUNC, PROP and PRED of, respectively, sorts, constants, function symbols, proposition symbols and predicate symbols, form together the *similarity type* of a specific many-sorted predicate calculus. For practical reasons we assume that the similarity type is finite.

2.2. A *many-sorted structure* $\mathfrak{M}$ consists of:
(a) A non-empty set A_σ for each $\sigma \in \Sigma$, called the *domain* of sort σ.
(b) For each constant c_i^σ an element $\bar{c}_i \in A_\sigma$.
(c) For each function symbol $f_i^{\sigma_1 \times \cdots \times \sigma_m \to \sigma_0}$ a mapping $\bar{f}_i : A_{\sigma_1} \times \cdots \times A_{\sigma_m} \to A_{\sigma_0}$.
(d) For each proposition symbol p_i a truth value $\bar{p}_i$.
(e) For each predicate symbol $P_i^{\sigma_1 \times \cdots \times \sigma_m}$ a mapping $\bar{P}_i : A_{\sigma_1} \times \cdots \times A_{\sigma_m} \to \{TRUE, FALSE\}$.

2.3. An *assignment* in $\mathfrak{M}$ is a mapping a assigning to each variable x_i^σ an element $a(x_i^\sigma)$ of A_σ.

2.4. The *interpretation* in $\mathfrak{M}$ of a term t under an assignment a, denoted by $I_a^{\mathfrak{M}}(t)$ or $\bar{t}$ for short, is inductively defined as follows:
(a) $\overline{x_i^\sigma} = a(x_i^\sigma)$
(b) $\overline{c_i^\sigma} = \bar{c}_i$
(c) $\overline{f_i^{\sigma_1 \times \cdots \times \sigma_m \to \sigma_0}(t_1,...,t_m)} = \bar{f}_i(\bar{t}_1, \ldots, \bar{t}_m)$
   The *truth value* in $\mathfrak{M}$ of an atom $P_i^{\sigma_1 \times \cdots \times \sigma_m}(t_1,...,t_m)$ under an assignment a is given by $\bar{P}_i(\bar{t}_1, \ldots, \bar{t}_m)$.

2.5. The *truth value* in $\mathfrak{M}$ of a formula F under an assignment a, denoted by $\mathcal{V}_a^{\mathfrak{M}}(F)$, is inductively defined as follows:
(a) If F is an atom, then $\mathcal{V}_a^{\mathfrak{M}}(F)$ is given by 2.4.
(b) $\mathcal{V}_a^{\mathfrak{M}}$ respects the truth tables of the propositional connectives.
(c) $\mathcal{V}_a^{\mathfrak{M}}(\forall x_i^\sigma F) = TRUE$ if and only if for all assignments a', which differ at most on x_i^σ from a, we have $\mathcal{V}_{a'}^{\mathfrak{M}}(F) = TRUE$.
(d) $\mathcal{V}_a^{\mathfrak{M}}(\exists x_i^\sigma F) = TRUE$ if and only if there exists an assignment a', which differs at most on x_i^σ from a, such that $\mathcal{V}_{a'}^{\mathfrak{M}}(F) = TRUE$.

2.6. A formula F is *true* in $\mathfrak{M}$, denoted by $\vDash_{\mathfrak{M}} F$, if $\mathcal{V}_a^{\mathfrak{M}}(F) = TRUE$ for all assignments a.

2.7. A *sentence* (or *closed* formula) is a formula without free variables (i.e. variables which are not bound by a quantifier). It will be clear that for sentences S the truth value $\mathcal{V}_a^{\mathfrak{M}}(S)$ does not depend on the assignment a. As a consequence we have either $\vDash_{\mathfrak{M}} S$ or $\vDash_{\mathfrak{M}} \neg S$, for every sentence S. Let SENT denote the set of sentences.

2.8. Let $\Gamma \subset$ SENT. $\mathfrak{M}$ is called a *model* for Γ, denoted by $\vDash_{\mathfrak{M}} \Gamma$, if $\vDash_{\mathfrak{M}} S$ for all $S \in \Gamma$.

2.9. $S \in$ SENT is called a *(semantical) consequence* of (or *implied* by) $\Gamma \subset$ SENT if for all many-sorted structures $\mathfrak{M}$ we have: if $\vDash_{\mathfrak{M}} \Gamma$, then $\vDash_{\mathfrak{M}} S$. This will be denoted by $\Gamma \vDash S$ (or $\vDash S$ if Γ is empty). Furthermore we define the *theory* of Γ as the set $Th(\Gamma) = \{ S \in \text{SENT} \mid \Gamma \vDash S \}$.

2.10. $\Gamma \subset$ SENT is called *consistent* if Γ has a model. $Th(\Gamma)$ is called *decidable* if there exists a mechanical decision procedure to decide whether a given sentence S is a semantical consequence of Γ or not.

2.11. Two many-sorted structures are called *elementarily equivalent* if exactly the same sentences are true in both structures.

2.12. REMARKS.

2.12.1. We refrain from giving an axiomatization of many-sorted predicate logic since our main concern will be model theory. Most textbooks on mathematical logic provide a complete axiomatization of ordinary (one-sorted) predicate logic. It suffices to generalize the quantifier rules in order to obtain an axiomatization of flat many-sorted predicate logic.

2.12.2. Of course, one-sorted predicate logic is a special case of many-sorted predicate logic. As a consequence, the latter is as undecidable as the former. More precisely: $\vDash S$ is undecidable, provided that the similarity type is rich enough (CHURCH, TURING, 1936, see also [M, 16.58]).

2.12.3. Conversely, many-sorted predicate logic can be embedded in one-sorted predicate logic by adding unary predicate symbols $S^\sigma(x)$, expressing that x is of sort σ, and replacing inductively in formulas $\forall x_i^\sigma F$ (resp. $\exists x_i^\sigma F$) by $\forall x (S^\sigma(x) \to F)$ (resp. $\exists x (S^\sigma(x) \wedge F)$). Let A' be the one-sorted sentence obtained from $A \in$ SENT in this way. It can be proved (see [M]) that $\vDash A$ if and only if $\Gamma \vDash A'$, where $\Gamma = \{ \exists x S^\sigma(x) \mid \sigma \in \Sigma \}$ expresses the fact that the domains are non-empty. This embedding allows us to generalize immediately many results on one-sorted predicate calculus to the many-sorted case (e.g. the compactness theorem). We shall not make use of this possibility in the present paper.

3. RULE-BASED EXPERT SYSTEMS AS MANY-SORTED THEORIES

3.0. We propose the following terminology for certain kinds of many-sorted theories:
- Indexed propositional expert systems.
- Universally quantified expert systems.

3.1. An *indexed propositional expert system* is a many-sorted theory axiomatized by:
(a) *Explicit axioms* (the rule base and the fact base), which are boolean combinations of atoms of the form $P^{\sigma_1 \times \cdots \times \sigma_m}(c,...,c')$ or of the form $f^{(\sigma_1 \times \cdots \times \sigma_m) \to \sigma_0}(c,...,c') =_{\sigma_0} c''$ (resp. $<_{\sigma_0} c''$, $>_{\sigma_0} c''$), with constants $c,...,c',c''$ of appropriate sorts. Such atoms (here and below called *constant-atoms*, or *c-atoms* for short) may be viewed as indexed propositions, which explains the name. Note that we conform to the convention to write $=$, $<$ and $>$ as infix predicates.
(b) *Implicit axioms* for equality of each sort and ordering of each sort for which an ordering is

appropriate. The axioms for $=_\sigma$, equality of sort σ, are (loosely omitting sort super- and subscripts):

$\forall x (x = x)$,

$\forall x_1, x_2 (x_1 = x_2 \rightarrow x_2 = x_1)$,

$\forall x_1, x_2, x_3 ((x_1 = x_2 \wedge x_2 = x_3) \rightarrow x_1 = x_3)$,

$\forall x_1, x_2 ((x_1 = x_2 \wedge F(x_1)) \rightarrow F(x_2))$ for formulas F,

$\neg c_i = c_j$ for $0 \leqslant i \neq j \leqslant n$,

$\forall x (x = c_0 \vee \cdots \vee x = c_n)$, the domain closure axiom.

These axioms express that $=$ is a congruence relation on a finite domain, where every element has exactly one name. Let EQ denote the set of equality axioms for all sorts σ, then we have by definition that either $EQ \vDash c_i = c_j$ or $EQ \vDash \neg c_i = c_j$ for all i,j. The axioms for $<$ and $>$ are:

$\forall x_1, x_2, x_3 ((x_1 < x_2 \wedge x_2 < x_3) \rightarrow x_1 < x_3)$,

$\forall x (\neg x < x)$,

$\forall x_1, x_2 (x_1 < x_2 \vee x_1 = x_2 \vee x_2 < x_1)$ for sorts which are totally ordered,

$\forall x_1, x_2 (x_1 < x_2 \leftrightarrow x_2 > x_1)$,

a subset Δ of $\{c_i < c_j \mid 0 \leqslant i \neq j \leqslant n\} \cup \{\neg c_i < c_j \mid 0 \leqslant i, j \leqslant n\}$ (see below).

These axioms express that $<$ is a transitive, irreflexive and (possibly) total ordering with inverse $>$. Let O denote the set of ordering axioms. We require that Δ is such that either $O \vDash c_i < c_j$ or $O \vDash \neg c_i < c_j$, for all i,j. It follows in particular that $EQ \cup O$ is consistent.

The idea behind the implicit axioms is that $=$ and $<$ are provided by the system and have a fixed meaning, whereas the other predicates are user-defined.

3.2. A *universally quantified expert system* differs from an indexed propositional one by allowing not only constants, but also variables in the explicit axioms. All explicit axioms are assumed to be universally closed.

3.3. Among the theories that do not fall under 3.1 and 3.2 are many theorem provers. A theorem prover might be an undecidable theory. The theoretical observations below show that, from a logical point of view, expert systems as defined in 3.1 and 3.2 are very simple, decidable theories.

3.4. LEMMA. *For every $S \in$ SENT there exists a boolean combination S' of closed atoms such that $EQ \vDash S \leftrightarrow S'$.*

PROOF. Replace inductively every subformula $\forall x F(x)$ by $F(c_0) \wedge \cdots \wedge F(c_n)$ and $\exists x F(x)$ by $F(c_0) \vee \cdots \vee F(c_n)$ where $\forall x (x = c_0 \vee \cdots \vee x = c_n) \in EQ$. $\square$

3.5. LEMMA. *For every closed atom A there exists a boolean combination A' of c-atoms such that $EQ \vDash A \leftrightarrow A'$.*

PROOF by giving a typical example. Let A be for instance $P(f(f'(c^{\sigma_0}), c^{\sigma_1}), c^{\sigma_2})$ with f of type $\sigma_2 \times \sigma_1 \rightarrow \sigma_0$ and f' of type $\sigma_0 \rightarrow \sigma_2$. Let $A^{\exists}$ be the sentence $\exists x^{\sigma_2} \exists y^{\sigma_0} [f'(c^{\sigma_0}) = x^{\sigma_2} \wedge f(x^{\sigma_2}, c^{\sigma_1}) = y^{\sigma_0} \wedge P(y^{\sigma_0}, c^{\sigma_2})]$. Now apply Lemma 3.4 to $A^{\exists}$ and obtain A'. $\square$

3.6. Both lemmas above can cause combinatorial explosions. Therefore they are only of theoretical use. They tell us that, in the presence of EQ, boolean combinations of c-atoms have the same expressive power as full many-sorted predicate logic.

3.7. LEMMA. *If $EQ \subset \Gamma \subset$ SENT, then every model of Γ is elementarily equivalent to a model whose domains consist of exactly the interpretations of all constants.*

PROOF. Let $\mathfrak{M}$ be a model of Γ with $EQ \subset \Gamma \subset$ SENT. By EQ the interpretations of the equality predicates in $\mathfrak{M}$ are equivalence relations which are congruences with respect to the interpretation of all other predicate and function symbols in $\mathfrak{M}$. It follows that $\mathfrak{M}$ and $\mathfrak{M}/_=$, the quotient structure of $\mathfrak{M}$ modulo equality, are elementarily equivalent. By the axioms $\forall x (x = c_0 \lor \cdots \lor x = c_n)$ and $\neg c_i = c_j$ $(i \neq j)$ from EQ, it follows that the domains of $\mathfrak{M}/_=$ consist of exactly the interpretations of all constants. $\square$

3.8. LEMMA. *If $EQ \subset \Gamma \subset$ SENT, then $Th(\Gamma)$ is decidable.*

PROOF. Let Γ be such that $EQ \subset \Gamma \subset$ SENT. Then we can apply Lemma 3.7 and observe that there are just finitely many non-isomorphic $\mathfrak{M}/_='s$. In other words: up to dividing out $=$ and identifying objects with the same name there are just finitely many different models of Γ. Moreover all models of Γ are finite. Hence we can, for any given $S \in$ SENT, test in finite time whether S holds in all models of Γ or not. $\square$

4. TESTING CONSISTENCY

4.1. Definition 2.10 and Lemma 3.7 suggest the following procedure for testing the consistency of theories Γ with $EQ \subset \Gamma$: generate all many-sorted structures whose domains consist of exactly the interpretations of all constants, and test each time whether the many-sorted structure is a model of Γ or not. This procedure is in general not feasible. A first step towards a feasible consistency test is the alternative characterization of consistency for indexed propositional expert systems, described in the following paragraphs.

Let Γ be an axiomatization of an indexed propositional expert system (see 3.1). Let $P_1,...,P_m$ be a list of all c-atoms of the form $P(c,...,c')$ occurring in Γ, and let $t_1,...,t_n$ list all terms $f(c,...,c')$ occurring in Γ. The idea behind the following construction is that it is not necessary to have an entire many-sorted structure to be able to interpret Γ.

Let $A_\sigma = \{c_0^\sigma,...,c_n^\sigma\}$ be the set of all constants of sort $\sigma \in \Sigma$. Equality of sort σ is interpreted by syntactical identity on A_σ. Then all equality axioms from EQ are satisfied. The ordering (if any) on A_σ is induced by O, i.e. $c_i < c_j$ if and only if $O \vDash c_i < c_j$. We need the following notions:
- A *truth valuation* of $P_1,...,P_m$ is an assignment of either *TRUE* or *FALSE* to each P_i $(1 \leq i \leq m)$.
- A *valuation* of $t_1,...,t_n$ is an assignment of a unique element of the appropriate domain A_σ to each t_j $(1 \leq j \leq n)$. It will be clear that a truth valuation of $P_1,...,P_m$ and a valuation of $t_1,...,t_n$ suffice for an interpretation of Γ. Each model of Γ yields a truth valuation of $P_1,...,P_m$ and a valuation of $t_1,...,t_n$. Conversely, each truth valuation of $P_1,...,P_m$ and valuation of $t_1,...,t_n$ for which every explicit axiom of Γ is true, can be extended in an arbitrary way to a many-sorted structure $\mathfrak{M} \vDash \Gamma$. Thus we have the following

THEOREM. *Let conditions be as above. Then we have: Γ is consistent if and only if there exists a truth valuation of $P_1,...,P_m$ and a valuation of $t_1,...,t_n$ for which every explicit axiom of Γ is true.*

4.2. Theorem 4.1 suggests a simple algorithm for testing the consistency of an indexed propositional expert system Γ: generate all valuations and truth valuations and test each time whether the explicit axioms of Γ are satisfied or not. This clearly leads to combinatorial explosions. A second step towards a feasible consistency test is imposing language restrictions on our expert systems.

A common language restriction is Horn format. A *clause* is a finite disjunction of atoms and negations of atoms (so called positive and negative *literals*). A *conjunctive normal form* is a finite conjunction of clauses. A *Horn clause* is a clause containing at most one positive literal. A *unit clause* is a clause containing one positive literal. The connection between Horn clauses and production rules is easily seen by the equivalence of $(A_1 \wedge \cdots \wedge A_k) \to B$ and $A_1^- \vee \cdots \vee A_k^- \vee B^+$, where the superscripts $+$ and $-$ denote whether a literal occurs positively or negatively. However, the implicit axioms $\forall x(x = c_0 \vee \cdots \vee x = c_n)$ and $\forall x_1, x_2 (x_1 < x_2 \vee x_1 = x_2 \vee x_2 < x_1)$ are not Horn clauses. As a consequence we can only require the explicit axioms to be Horn clauses. This is not sufficient for feasible consistency checking, as will be demonstrated in the next paragraph. The idea is to reduce the satisfiability problem for propositional logic, which is known to be NP-complete (see [GJ]), to the consistency problem of indexed propositional expert systems, all whose explicit axioms are Horn clauses.

Let σ be a sort with exactly two constants: c_0^σ and c_1^σ. Then we have $\forall x(x = c_0 \vee x = c_1) \in EQ$. For any propositional atom A, let f_A be a function symbol of type $\sigma \to \sigma$. It is not difficult to see that

$$\{\neg f_A(c_0) = c_0 \vee A, f_A(c_0) = c_0 \vee \neg A\} \bigcup EQ \vDash (\neg f_A(c_0) = c_1) \leftrightarrow A.$$

So the *positive* literal A is equivalent to the *negative* literal $\neg f_A(c_0) = c_1$ in any Γ containing the two Horn clauses $\neg f_A(c_0) = c_0 \vee A$ and $f_A(c_0) = c_0 \vee \neg A$ and the equality axioms EQ. Let C be any propositional conjunctive normal form. Let Γ_C be the indexed propositional expert system with explicit axioms $\neg f_A(c_0) = c_0 \vee A$ and $f_A(c_0) = c_0 \vee \neg A$ for all positive literals A occurring in C. Then Γ_C is consistent and satisfies $\Gamma_C \vDash C \leftrightarrow H_C$, where H_C is the set (conjunction) of Horn clauses obtained from C by replacing all positive literals A by their equivalent negative literal. Moreover we have that C is satisfiable if and only if $\Gamma_C \bigcup H_C$ is consistent. This yields a polynomial reduction of the satisfiability problem for propositional logic to the consistency problem of indexed propositional expert systems, all whose explicit axioms are Horn clauses. A similar reduction could be established using $\forall x_1, x_2 (x_1 < x_2 \vee x_1 = x_2 \vee x_2 < x_1)$ instead of $\forall x(x = c_0 \vee x = c_1)$.

4.3. In the previous subsection we showed that testing consistency is NP-hard without further language restrictions. The problem is to specify language restrictions, which are strong enough to guarantee a feasible consistency test, and still allow enough expressivity for some given application. This third and final step towards a feasible consistency test will be achieved in the following theorem by the introduction of constants *undefined*, which are different from (wrt. $=$) and incomparable with (wrt. $<$) any other constant.

THEOREM. *Let Γ be the axiomatization of an indexed propositional expert system satisfying the following conditions:*
(1) All explicit axioms of Γ are Horn clauses.
(2) For every sort $\sigma \in \Sigma$ there exists at least one constant, which does not occur in the explicit axioms of Γ. Such a constant, say c_0^σ, will be denoted by undefined$^\sigma$. So we have:
$\forall x(x = undefined \vee x = c_1 \vee \cdots \vee x = c_n) \in EQ$ and $\neg undefined = c_i \in EQ$ for all $1 \leq i \leq n$.
(3) For every sort $\sigma \in \Sigma$ the ordering (if any) of sort σ is partial with respect to undefined$^\sigma$. More precisely: we require that $O \vDash \neg undefined < c_i$ and $O \vDash \neg undefined > c_i$ for all $1 \leq i \leq n$.
(4) The orderings $<$ and $>$ do not occur in the positive literals occurring in the explicit axioms of Γ. Then the consistency of Γ can be tested in polynomial time.

PROOF. List the explicit axioms of Γ as follows (the super- and subscripted capitals denote c-atoms):

$$A_1^+$$
.
.
.
$$A_p^+$$
$$B_{1,1}^- \vee \cdots \vee B_{1,m_1}^-$$
.
.
.
$$B_{q,1}^- \vee \cdots \vee B_{q,m_q}^-$$
$$C_{1,1}^- \vee \cdots \vee C_{1,n_1}^- \vee D_1^+$$
.
.
.
$$C_{r,1}^- \vee \cdots \vee C_{r,n_r}^- \vee D_r^+$$

Now apply the following well-known algorithm, which is in fact a special case of hyper-resolution (see [R]). By "implied" below we mean "implied in the presence of $EQ \bigcup O$".

> WHILE cancellations possible AND no clause empty
>
> DO
>
> > cancel all clauses that contain a literal $B_{i,j}^-$, $C_{i,j}^-$ or D_k^+ which is implied by a unit clause;
> >
> > cancel all literals $B_{i,j}^-$ and $C_{i,j}^-$ whose complement is implied by a unit clause
> >
> > (and possibly get new unit clauses D_k^+!)
>
> OD;
>
> IF empty clause occurs OR unit clauses $f(c,...,c') = c_i$ and $f(c,...,c') = c_j$ occur with $i \neq j$
>
> THEN Γ is inconsistent
>
> ELSE Γ is consistent

Note that the WHILE-loop terminates since the number of c-atoms involved is strictly decreasing. By the cancellations the explicit axioms of Γ are transformed into an equivalent set of Horn clauses having the property that no unit clause implies or refutes any negative literal. Since we have either $EQ \vDash c_i = c_j$ or $EQ \vDash \neg c_i = c_j$ and either $O \vDash c_i < c_j$ or $O \vDash \neg c_i < c_j$, it follows that no term $f(c,...,c')$ occurs both in a unit clause and in a negative literal. As a consequence the consistency of Γ in the ELSE-part above can be seen by applying Theorem 4.1 with the following truth valuation of $P_1,...,P_m$ and valuation of $t_1,...,t_n$:

$$P_i = \begin{cases} TRUE \text{ if } P_i \text{ occurs as unit clause} \\ FALSE \text{ otherwise} \end{cases}$$

$$t_j = \begin{cases} c_i \text{ if } t_j = c_i \text{ occurs as unit clause} \\ undefined \text{ otherwise} \end{cases}$$

The algorithm is clearly quadratic in the number of occurrences of literals. To get the consistency of Γ in P we tacitly assumed that the primitive operations used in the algorithm above are in P. $\square$

REMARKS.

4.3.1. As follows by close inspection of the proof above, it would suffice to require the following weakening of condition (2): for every sort σ for which a function symbol f of type $\ldots \to \sigma$ occurs in a negative literal occurring in the explicit axioms of Γ, there exists at least one constant which does not occur on the right-hand side of an equation occurring in such a literal. However, we think it is more systematic to require condition (2) as it stands.

4.3.2. Condition (4) can not be missed, which can be seen as follows. Assume for some sort σ we have exactly three constants different from *undefined*, which are totally ordered by $c_1 < c_2 < c_3$. Then the unit clause $f(c) > c_1$ is equivalent to $f(c) = c_2 \lor f(c) = c_3$, which enables a similar construction as in the third paragraph of 4.2.

4.3.3. In view of condition (1), occurrences of *notsame* $<o,a,v>$ in the antecedent of a rule can be problematic. This problem can be overcome by postponing the consistency test until these occurrences evaluate to either *TRUE* or *FALSE*. Then the knowledge base can be transformed into an equivalent knowledge base satisfying (1).

4.3.4. In our opinion the domain closure axiom is realistic in many applications (apart from the fact that every computer is a finite automaton). The lemmas 3.4-3.8 essentially depend on it. However, Theorem 4.3 remains valid when the domain closure axioms are removed from EQ. For, detected contradictions do not depend on the domain closure axioms and, conversely, consistency *with* the domain closure axiom trivially implies consistency *without* it.

4.4. Let us briefly discuss the semantical consequences of the conditions (2) and (3) from the previous theorem, since they may slightly deviate from the intended meaning of the knowledge base. It is possible that the consistency of Γ essentially depends on the valuation $f(c,\ldots,c') = undefined$, i.e. that any valuation $f(c,\ldots,c') = c_i$ $(1 \leq i \leq n)$ would not yield a model for Γ. One could say that $f(c,\ldots,c') = undefined$ possibly saves the expert system from inconsistencies by preventing production rules with occurrences of $f(c,\ldots,c')$ in the antecedent from firing. Since such rules have obviously not been used in the inference, this may be considered an advantage. On the other hand, this may be considered a disadvantage in cases where $f(c,\ldots,c') = undefined$ is not realistic (e.g. *temperature* (*patient*) $= undefined$). In these cases we suggest to add the appropriate unit clause $f(c,\ldots,c') = c_i$ and to test consistency again.

4.5. It is not difficult to generalize the algorithm of 4.3 to universally quantified expert systems, although some care has to be taken in quantifying x in clauses containing literals of the form $\neg f(c,\ldots,c') = x$. In these cases only restricted quantification of the form $\forall x \neq undefined$ is allowed. Of course, hyper-resolution can become very inefficient (from $P(c_0), P(c_1), \neg P(x_1) \lor \cdots \lor \neg P(x_n) \lor Q(x_1,\ldots,x_n)$, for instance, 2^n instances of Q are generated), so we suggest limited use of variables (or, preferably, the use of a more efficient algorithm).

4.6. EXAMPLE. We demonstrate the consistency test in action on the case of 1.2. Taking into account the object tree (sub-objects, single/multivalued attributes) the three rules yield the following clauses in shorthand:

$$\neg C(p,cp) \lor P(p,c),$$

$$\neg f_{abdp}(p) = yes \lor \neg f_{char}(p,pa) = co \lor \neg P(p,c),$$

$$\neg C(p,ap) \lor f_{abdp}(p) = yes, \neg P(p,c) \lor f_{abdp}(p) = yes.$$

When we add the unitclauses $C(p,cp)$ and $f_{char}(p,pa) = co$, the empty clause is easily derived and the algorithm decides to inconsistency. If we only add $C(p,cp)$, then the algorithm decides to consistency on the basis of the following (truth-)valuation:

$$C(p,cp) = P(p,c) = TRUE, \quad C(p,ap) = FALSE,$$

$$f_{abdp}(p) = yes, \quad f_{char}(p,pa) = undefined.$$

5. CONCLUSION

We argued how to interpret a rule-based expert system as a many-sorted theory. Then the Tarski semantics yields a well-defined notion of consistency, and theorem proving techniques such as resolution can be used for testing consistency. The relevance for expert systems lies in the fact that inconsistencies make the conclusions of a knowledge-based system highly unreliable. For, what reason do we have to believe a conclusion if its negation could also be derived? We provided means for *detecting* inconsistencies, but did not discuss how to *deal* with them. Note that we considered an expert system as a fixed theory, whereas in practice the theory grows during the interaction with a user answering questions posed by the system (in this way the inconsistency of 1.2 was obtained). In our opinion only relatively consistent answers are to be accepted by the system. Starting with a consistent knowledge base we thus maintain consistency as an invariant of the interaction.

REFERENCES

[BS] B.G. BUCHANAN, E.H. SHORTLIFFE, *Rule-based expert systems: the Mycin experiments of the Stanford Heuristic Programming Project.* Addison-Wesley, Reading, Massachusetts (1984).

[GJ] M.R. GAREY, D.S. JOHNSON, *Computers and intractability: a guide to the theory of NP-completeness.* Freeman, San Francisco, California (1979).

[IL] T. IMIELINSKI, W. LIPSKI JR., *Incomplete information in relational databases.* Journal of the ACM 31, 4, pp. 761-791 (1984).

[L] P.J.F. LUCAS, *Knowledge representation and inference in rule-based expert systems.* Report CS-R8613, Centre for Mathematics and Computer Science, Amsterdam (1986).

[M] J.D. MONK, *Mathematical Logic*, Springer-Verlag, Berlin (1976).

[R] J.A. ROBINSON, *Automatic deduction with hyper-resolution.* International Journal of Computer Mathematics 1, pp. 227-234 (1965).

[Re] R. REITER, *Equality and domain closure in first order databases.* Journal of the ACM 27, 2, pp. 235-249 (1980).

[W] C. WALTHER, *A many-sorted calculus based on resolution and paramodulation.* Pitman, London (1987).

A MECHANIZABLE INDUCTION PRINCIPLE
FOR EQUATIONAL SPECIFICATIONS

Hantao Zhang [1,†], *Deepak Kapur* [2,†]*, and Mukkai S. Krishnamoorthy* [1]

[1]*Department of Computer Science*   [2]*Department of Computer Science*
Rensselaer Polytechnic Institute   *State University of New York at Albany*
Troy, NY 12180, U.S.A.   *Albany NY 12222, U.S.A.*

Abstract. Automating proofs of properties of functions defined on inductively constructed data structures is important in many computer science and artificial intelligence applications, in particular in program verification and specification systems. A new induction principle based on a constructor model of a data structure is developed. This principle along with a given function definition as a set of equations is used to construct automatically an induction scheme suitable for proving inductive properties of the function. The proposed induction principle thus gives different induction schema for different function definitions, just as Boyer and Moore's prover does. A novel feature of this approach is that it can also be used for proving properties by induction for data structures such as integers, finite sets, whose values cannot be freely constructed, i.e., constructors for such data structures are related to each other. This method has been implemented in RRL, a rewrite-rule based theorem prover. More than a hundred theorems in number theory including the unique prime factorization theorem, have been proved using the method.

1. INTRODUCTION

Computation structures and algorithms invariably involve data structures which can be recursively constructed using a set of constructors; well known examples of such data structures are natural numbers, lists, sequences, and trees. Functions on these data structures are often defined using recursive or iterative methods. In order to reason about the behavior of such functions, it is often necessary to reason inductively on the values of a data structure. Typically, this is done by defining a total measure function, such as size or length, on the data structure; the range of such a function is the set of natural numbers. Induction on the values of the data structure is then indirectly performed using induction on the natural numbers.

In this paper, an induction principle derived from a constructor model of a data structure is proposed. A constructor model is obtained from the way the values of the data structure are recursively built using its constructors (a constructor model is similar to Henkin model or Herbrand model built from a set of functions identified as constructors). The mechanization of the induction principle is discussed in the context of equational

† Partially supported by the National Science Foundation Grant no. CCR-8408461.

specifications in which functions are defined by equations. The basic idea was inspired by the way Boyer and Moore [Boyer&Moore 79] use their induction principle (and not by their induction principle itself). As in Boyer and Moore's approach, our induction principle can be combined with a function definition to generate an induction scheme; it turns out to be convenient for proving properties by induction for this function. Different induction schema are thus used for different functions determined by their definitions.

The notion of a *cover set* is introduced which is computed from a function definition given as a set of equations. For a unary function, a cover set is a finite set of terms which covers all the elements of the constructor model; for a n-ary function, a cover set is a finite set of n-tuples of terms which covers all n-tuples defined over the constructor model serving as inputs to the function. From the cover set computed from a function definition, a generalized version of the structural induction scheme is developed to prove properties of the function by induction.

The proposed method is general; it works also for data structures whose values cannot be freely constructed using constructors, that is, relations between constructors are accepted. Examples of such data structures are integers, rationals, finite sets, etc. It also works under certain circumstance when some functions are incompletely defined. Furthermore, the method can be automated and has been implemented in *RRL*, a *Rewrite Rule Laboratory*, a theorem prover based on rewriting techniques being developed at General Electric Corporate Research and Development and Rensselaer Polytechnic Institute. The method has been used to prove over a hundred theorems about lists and number theory, including the unique prime factorization theorem. The focus in this paper is on proving equations by induction from other equations, even though we believe that the method is general enough to prove any formula by induction.

Before discussing the technical details of the proposed method and related work on methods for proving properties by induction and their automation, we first motivate the approach with a simple example.

1.1. An Example

Consider, for example, a system E of two equations defining the function '+' (addition) on natural numbers which can be freely generated using the constructors $\{ 0, suc \}$:

$$E: \quad (1.1) \quad 0 + x = x, \qquad (1.2) \quad suc(x) + y = suc(x + y),$$

we can prove that

$$(1.3) \quad suc(0) + suc(0) = suc(suc(0))$$

by *equational reasoning* (i.e., inference rules for equality such as replacement of an expression by an equivalent expression and substitution of any expression for a universally quantified variable). However, the following equation cannot be proved by equational reasoning.

$$(1.4) \quad x + (y + z) = (x + y) + z.$$

It should be obvious that for each value of x, y, and z, (1.4) follows from (1.1) and (1.2), i.e., when ground terms (terms without variables) constructed from 0 and suc are substituted for x, y, and z in (1.4), all resulting equations can be reduced to identity using (1.1) and (1.2). This suggests that (1.4) may be proved by induction. Using Peano's induction principle for natural numbers on the variable x, we need to prove:

Basis case: $0+(y+z) = (0+y)+z$,

Inductive case: $x+(y+z) = (x+y)+z$ implies $suc(x)+(y+z) = (suc(x)+y)+z$.

In this way, Peano's induction principle provides a method to consider, in a finitistic manner, infinitely many cases for x. In the literature, a generalization of Peano's induction principle to arbitrary data structures has come to be known as the structural induction principle [Burstall 69].

Now let us consider a definition of the *gcd* function on natural numbers:

(1.5) $gcd(x,0) = x$,   (1.7) $gcd(x+y,y) = gcd(x,y)$,

(1.6) $gcd(0,x) = x$,   (1.8) $gcd(x,x+y) = gcd(x,y)$.

It can be shown that the above definition completely defines *gcd* on all pairs of natural numbers, with the equations (1.1) and (1.2) together. The goal is to prove that *gcd* function is commutative, i.e.,

(1.9) $gcd(x,y) = gcd(y,x)$.

This equation also cannot be proved by equational reasoning, but it can be proved by induction as it holds for every value of x and y. If we try to prove it by Peano's induction method on the variable x, we must show that:

(1.10) $gcd(0,y) = gcd(y,0)$,

(1.11) $gcd(x,y) = gcd(y,x)$ implies $gcd(suc(x),y) = gcd(y,suc(x))$.

The equation (1.10) follows from (1.6). However, (1.11) cannot be proved from (1.5) to (1.8) by equational reasoning. Even if we had used the induction scheme such as that

$gcd(x_1,y) = gcd(y,x_1)$ and $gcd(x_2,y) = gcd(y,x_2)$

imply $gcd(x_1+x_2,y) = gcd(y,x_1+x_2)$,

the equation (1.9) cannot be proved.

The method presented in this paper will be able to prove (1.9) from (1.1), (1.2) and (1.5) to (1.8). The definition of *gcd* itself suggests an induction scheme which considers all the infinitely many cases. For the tuple $<s, t>$, where s and t are arbitrary natural numbers, the following finite set of pairs of terms covers all the infinitely many cases:

$\{ <x, 0>, <0, y>, <x+y, y>, <x, x+y> \}$.

These pairs come from the definition of *gcd* (the left-hand sides of (1.5) to (1.8)). The non-recursive equation(s) in a definition suggests the basis step of an induction proof; while the recursive equation in the definition suggests the inductive step of the proof. For this example, corresponding to the above four pairs of terms, the following four formulae are generated:

(1.12) $gcd(x,0) = gcd(0,x)$,

(1.13) $gcd(0,y) = gcd(y,0)$,

(1.14) $gcd(x,y) = gcd(y,x)$ implies $gcd(x+y,y) = gcd(y,x+y)$,

(1.15) $gcd(x,y) = gcd(y,x)$ implies $gcd(x,x+y) = gcd(x+y,x)$.

And, each one of them can be easily proved using (1.5) to (1.8). This paper explains how and why (1.12) to (1.15) can be formulated in this way.

We should also mention that the induction principle presented in [Boyer&Moore, 79] can also be used to prove (1.9). However the axioms (or function definitions) input to their prover are in the form of lisp-like functions. For example, the *gcd* function is defined as:

$gcd(x,y) = $ *if (x = 0) then y elseif (y = 0) then x*
    elseif (x > y) then gcd(x-y, y) else gcd(x, y-x).

After establishing that <x, y> is strictly greater than <x-y, y> and <x, y-x> by a well-founded ordering, a proof scheme for (1.9), namely, $P(x,y) = (gcd(x,y) = gcd(y,x))$, can be generated in Boyer and Moore's prover as follows.

[1] $(x = 0)$ implies $P(x,y)$,

[2] [$not(x = 0) \wedge (y = 0)$] implies $P(x,y)$,

[3] [$not(x = 0) \wedge not(y = 0) \wedge (x>y) \wedge P(x-y,y)$] implies $P(x,y)$,

[4] [$not(x = 0) \wedge not(y = 0) \wedge not(x>y) \wedge P(x,y-x)$] implies $P(x,y)$.

They are all provable by expanding function calls and by equational reasoning.

Following Boyer and Moore's approach, we develop a method for generating an induction scheme from function definitions given in equational form. Furthermore, our method can handle incompletely defined functions and data structures with relations among constructors.

1.2. Related Work

McCarthy [McCarthy 63], and Burstall [Burstall 69] recognized the importance of inductive reasoning in program verification and computer science, and he proposed a structural induction principle for recursively defined data structures. Since then, proofs by induction and methods for automation of proofs by induction have been extensively studied ([Wegbreit&Spitzen 76], [Aubin 76], [Brotz 74]). Particular mention should be made of Boyer and Moore's impressive work and their theorem prover for proving inductive theorems [Boyer&Moore 79].

Musser [Musser 80] discovered the use of the Knuth-Bendix completion procedure for proving equations by induction from an equational specification of data types. Since the classical induction principle is not explicitly invoked in this method, this method has been called the inductionless induction method. This method has since been refined by [Goguen 80], [Huet&Hullot 80], [Dershowitz 83]. Kapur and Musser [Kapur&Musser 84] developed a model-theoretic (algebraic) framework and showed that the inductionless induction method can apply to a larger set of theorems than those that are valid in the initial model; they call it the *proof-by-consistency* method to contrast it with the commonly used *proof-by-contradiction* method. Paul [Paul 84] also studied theoretical issues related to this method. Jouannaud and Kounalis [Jouannaud&Kounalis 86] as well as Kapur, Narendran and Zhang [Kapur et al 86] developed extensions to consider relations over constructors. Kapur and Musser [Kapur&Musser 86] developed an extension of the proof-by-consistency method for ambiguous systems where the completeness assumption can be relaxed.

2. DEFINITIONS

We assume that the reader is familiar with some of the terminology and concepts commonly used in equational specifications and rewriting systems, such as sorts, variables, well-typed terms, equations, substitutions, congruence relation, equational congruence, etc. (see [Huet&Oppen 80]).

2.1. Terms, Equations, Substitutions and Congruences

Given a finite set F of function symbols and a set X of variables, we shall use $T(F,X)$ to denote all well-typed terms built on F and X. A term without variables is a *ground term*. The set of all ground terms is denoted by $GT(F)$. Whenever required, terms of a sort s are identified as $T(F,X)_s$. An equation is a pair $<m, n>$ of terms of the same sort, written as $m = n$.

A substitution is called *ground* if for $\sigma(x) \neq x$, $\sigma(x)$ is a ground term. A substitution σ is extended to $\sigma: T(F,X) \rightarrow T(F,X)$ in a natural way. A substitution can also be extended to a mapping from equations to equations. We shall use $\sigma(e)$ to denote $(\sigma(l) = \sigma(r))$ for an equation $e = (l = r)$. A relation R is said to be *stable under substitution* if for any terms t_1, t_2 and any substitution σ, $t_1 R t_2$ implies $\sigma(t_1) R \sigma(t_2)$.

Convention: Throughout the paper, we use $x, y, z, x', x_1, x_2, \cdots$ for variables, $t, m, n, t', t_1, t_2, \cdots$ for terms and $\sigma, \sigma', \cdots$ for substitutions. E will be a set of equations and $=_E$ denotes the equational congruence generated by E.

To prove a given equation $m = n$ by *equational reasoning* or *equational deduction* from E, is to prove that $m =_E n$; in this case, $m = n$ is said to be an *equational theorem deduced from* E. Let $EQ(E)$ denote the set of all equational theorems deduced from E, namely

$$EQ(E) = \{ m = n \mid m =_E n \}.$$

Convention: The congruence classes of $GT(F)_s$ (resp. $GT(F)$, $T(F,X)$, etc.) modulo $=_E$ will be denoted by $GT(F,E)_s$ (resp. $GT(F,E)$, $T(F,X,E)$). Each element in $GT(F,E)_s$ is denoted by $[t]_{=_E}$ for $t \in GT(F)_s$, or by $[t]$ if E is understood from the text. Obviously,

$$[t] = \{ t' \mid t' =_E t \wedge t' \in GT(F)_s \}.$$

2.2. Constructors and Specifications

Definition 2.1 (equational specification, constructors). An algebraic specification (or axiomatization) $SP = (S, F_c, F_d, E)$ where S is a set of sorts, F_c and F_d are two disjoint sets of function symbols and E is a set of equations over $T(F_c \cup F_d, X)$. F_c is called the set of *constructors* of SP, and F_d the set of *defined function symbols*. SP is also called an *equational system*, or a *specification*.

Let $EQ(SP)$ stand for the equational theory associated with an equational specification SP. That is, $EQ(SP) = EQ(E)$ for $SP = (S, F_c, F_d, E)$.

Example 2.2: Let us define addition and predecessor on the natural numbers and choose $\{0, suc\}$ as the constructors. The specification $SP = (S, F_c, F_d, E)$ of the addition and predecessor functions over the natural numbers, is given by

$$S = \{num\}, \qquad F_c = \{0, suc\}, \qquad F_d = \{pre, +\},$$
$$E = \{pre(0) = 0, \ pre(suc(x)) = x, \ 0{+}x = x, \ suc(x){+}y = suc(x{+}y)\}.$$

On the other hand, if the user intends to define addition on the integers, we have the following specification:

Example 2.3:

$$S = \{num\}, \qquad F_c = \{0, suc, pre\}, \qquad F_d = \{+\},$$
$$E = \{suc(pre(x)) = x, \ pre(suc(x)) = x,$$
$$0{+}x = x, \ suc(x){+}y = suc(x{+}y), pre(x){+}y = pre(x{+}y)\}.$$

We shall use $T(F_c, X)$ to denote the set of all terms that contains only constructors and/or variables and call them *constructor terms*. Consequently, the set of all ground constructor terms is denoted by $GT(F_c)$. A substitution σ is called a *constructor substitution* if $\sigma(x)$ is a constructor term for any variable x.

For each sort in a specification, we require its constructor set to be not empty. The set of defined functions may be empty.

Definition 2.4 (constructor model). Given a specification $SP = (S, F_c, F_d, E)$, the constructor model (D^c, F^c) of SP consists of a set D^c of domains and a set F^c of functions, such that

$$D^c = \{GT(F_c, E)_s \mid s \in S\}$$

where $GT(F_c, E)_s$ denotes the set of ground constructor terms of sort s modulo the relation $=_E$, and

$$F^c = \{f^c \mid f \in F \text{ and } f^c([t_1], ..., [t_n]) = [f(t_1, ..., t_n)] \text{ for } t_i \in GT(F)\}$$

where $F = F_c \cup F_d$ and $[t]$ denotes the equivalence class of the term t under $=_E$.

There is a close relation between the constructor model and the initial model of a specification [Goguen et al 78]. That is, the constructor model is a subalgebra of the initial model with respect to constructors. Since we focus on the problem how to prove theorems by induction automatically, a full treatment of the relation between the constructor model and initial model is the subject of another paper.

2.3. Inductive Theorems

We first define the class $IND(E)$ of inductive theorems. In the next section, we will examine different induction principles which could be used to prove the theorems of $IND(E)$.

Definition 2.5 (inductive theorem) Given $SP = (S, F_c, F_d, E)$, an equation e is an *inductive theorem* of SP if for all ground constructor substitutions $\sigma: X \to GT(F_c)$, $\sigma(e) \in EQ(SP)$.

Given $SP = (S, F_c, F_d, E)$, let

$$IND(SP) = \{ e \mid e \text{ is an inductive theorem of } SP \}$$

and we shall use $E \underset{ind}{\vdash} e$ to denote $e \in IND(SP)$ and use $=_{IND}$ for the relation induced by $IND(SP)$. That is, $t_1 =_{IND} t_2$ iff $<t_1, t_2> \in IND(SP)$. As a convention we will write $IND(E)$ for $IND(SP)$ when S, F_c and F_d are understood from the context.

Note that our notion of "inductive theorem" is related to the constructors F_c since we limit substitutions to the ground constructor terms.

Inference rules are intimately tied with theorems and models. Theorems are derived using inference rules and models provide the sound basis for inference rules in the sense that the theorems derived from the inference rules are valid in these models. As we defined a constructor model and inductive theorems, a natural question is about the relationship between those two notions. We claim below that inductive theorems are theorems of the constructor model.

An equation $t = t'$ is said *true* in a model A, if for all homomorphisms $\phi: T(F, X) \to A$, we have $\phi(t) = \phi(t')$ in A.

Theorem 2.6: Given $SP = (S, F_c, F_d, E)$ and an equation e, e is in $IND(E)$ iff e is true in the constructor model of SP.

3. COVER-SET INDUCTION PRINCIPLE

3.1. Criteria of Induction Principles

Directly from the definition of inductive theorems, we have the following induction principle for the class $IND(E)$:

Constructor Induction Principle:

Given (i) e is an equation and $x_1,...,x_k$ are some variables of e;

(ii) $E \underset{ind}{\vdash} e[x_1/t_1,...,x_k/t_k]$, for all $t_i \in GT(F_c)_{s_i}$, where t_i and x_i have the same sort for $1 \le i \le k$.

Then $E \underset{ind}{\vdash} e$.

We call it a *constructor* induction principle to emphasize that only constructor ground terms are substituted for variables. To use the above induction principle, we must establish infinitely many premises as there are usually infinitely many ground terms. Hence it is not practical.

The following structural induction principle was proposed by Burstall [Burstall 69].

Structural Induction Principle:

Given (i) e is an equation and x is a variable of sort s in e;

(ii) $E \cup \{e[x/x_{i_1}],...,e[x/x_{i_k}]\} \underset{ind}{\vdash} e[x/f(x_1,...,x_k)]$ for each

$f:s_1,...,s_k \to s \in F_c$, where x_i are variables of sort s_i for $1 \le i \le k$ and $x_{i_1}, ..., x_{i_k}$ are those x_i of sort $s_i = s$.

Then $E \underset{ind}{\vdash} e$.

The use of structural induction to prove programs correct is elegantly described in [Burstall 69]. Even though the above principle can be generalized to multiple variables, it is still not powerful enough to prove some theorems in $IND(E)$. For example, the structural induction principle failed in proving the commutativity of *gcd* as illustrated in the introduction.

We would like an induction principle to be both practical and powerful. The constructor induction principle is not practical. The structural induction principle is not powerful enough. The induction principle presented in [Boyer&Moore 79, pp.33-35] overcomes both of the two inconveniences; however, it requires the axioms in the form of lisp-like functions. Inspired by Boyer and Moore's approach, we develop a method for generating an induction scheme from a function definition in the equational form. In the following, after introducing the notion of a cover set, we present a generalized version of the structural induction principle, which is suitable for equational systems and is as powerful as Boyer and Moore's induction method.

3.2. Cover Set of Ground Constructor Terms

Since there are usually infinitely many ground constructor terms for each sort, we would like to choose a finite set of terms which "describes" every ground constructor term of that sort. Instead of proving that an equation is valid for all ground constructor instances, we prove that an equation is valid for all instances taken from that finite set.

Such a finite set of terms is called a *cover set*.

Definition 3.1 (cover function and cover set). Let GT_s be the set of all ground constructor terms of the sort s and M be a finite set of terms of sort s.

(1) We say a mapping Ψ, from M to the power set of GT_s, is a *cover function* of the sort s upon M if

(a) (completeness) for any term t in GT_s, there exists a term t_1 in M such that $t \in \Psi(t_1)$, and there exists a substitution σ such that $t =_E \sigma(t_1)$;

(b) (minimality) $\Psi(t_1) \neq \varnothing$ for $t_1 \in M$;

(c) (uniqueness) t_1 in (a) is unique.

$\Psi(t_1)$ is the set of terms *covered* by t_1 for $t_1 \in M$.

(2) M is said to be a *cover set* of s, if M possesses a cover function Ψ.

By definition, $\Psi(t)$ are pairwise disjoint subsets of GT_s. That is, for any $t_1, t_2 \in M$, $t_1 \neq t_2$ implies $\Psi(t_1) \cap \Psi(t_2) = \varnothing$. This condition is needed for the soundness of the induction scheme discussed later. A cover set may contain non-constructor terms. If M contains only constructor terms and there is no relation between constructors, then the relation $=_E$ in (a) above could be replaced by the identity relation.

There are infinitely many cover sets for a sort s. For example, any set containing a variable is a cover set. The set $\{ f(x_1,...,x_n) \mid f : s_1,...,s_n \to s \in F_c \}$ is also a cover set. Following are examples:

sort	cover set	Ψ
bool	{ true, false }	$\Psi(true) = \{ true \}$, $\Psi(false) = \{ false \}$
num	{ 0, suc(x) }	$\Psi(0) = \{ 0 \}$, $\Psi(suc(x)) = \{ suc^i(0) \mid i > 0 \}$
list	{ nil, cons(x, y) }	$\Psi(nil) = \{ nil \}$, $\Psi(cons(x, y)) = \{$ non-empty lists $\}$

Reduction ordering [Dershowitz 87]. A *reduction ordering* is a transitive monotonic relation on terms, which is terminating (or *Noetherian*) and stable under substitution. A reduction ordering respecting a congruence relation $=_E$ induces an ordering on congruence classes of $GT(F)$ induced by $=_E$. This ordering on congruence classes of $GT(F)$ can then be used to define an ordering on terms $T(F,X)$. The inductive inequivalence defined below captures this relation on terms.

Definition 3.2 (inductive inequivalence). Given a reduction ordering $>$, a set E of equations and a set G of ground terms, a term t_1 is said to be *inductively greater than* t_2 with respect to G if for any ground substitution σ and for any term $t \in G$, $t =_E \sigma(t_1)$ implies $t > \sigma(t_2)$. This is denoted by $t_1 >^i t_2$ with respect to G.

3.3. A New Induction Principle

From now on, let $SubT(t,G)$ denote some subterms of t such that for any term $t' \in SubT(t,G)$, t and t' are of the same sort and $t >^i t'$ with respect to G.

Cover-Set Induction Principle

Given (i) M is a cover set of s with the cover function Ψ;

(ii) e is an equation and x a variable of sort s in e;

(iii) $E \underset{ind}{\vdash} e[x/t_2]$ implies $E \underset{ind}{\vdash} e[x/t_1]$ for all $t_1 \in M$,

where $t_2 \in SubT(t_1, \Psi(t_1))$.

Then $E \underset{ind}{\vdash} e$.

Proof: If $E \underset{ind}{\vdash} e$ is false, then there exists a ground constructor substitution σ such that $\sigma(e) \notin EQ(E)$. Let σ be such a substitution that $t = \sigma(x)$ is minimal by a given reduction ordering. Then there exists a term $t_1 \in M$ and a substitution θ such that $t \in \Psi(t_1)$ and $t =_E \theta(t_1)$.

By the assumption, there exists a subterm $t_2 \in SubT(t_1, \Psi(t_1))$ such that $E \underset{ind}{\vdash} e[x/t_2]$ implies $E \underset{ind}{\vdash} e[x/t_1]$, or $E \underset{ind}{\vdash} e[x/\theta(t_2)]$ implies $E \underset{ind}{\vdash} e[x/\theta(t_1)]$ and $t =_E \theta(t_1) > \theta(t_2)$. We cannot have $E \underset{ind}{\vdash} e[x/t_1]$ since $E \underset{ind}{\vdash} e[x/\theta(t_1)]$ is false; so $E \underset{ind}{\vdash} e[x/\theta(t_2)]$ is false for some t_2 such that $t_1 >^i t_2$ with respect to $\Psi(t_1)$. That is a contradiction to the assumption that t is minimal. $\square$

Let $M = \{f(x_1, \ldots, x_n) \mid f : s_1, \ldots, s_n \to s \in F_c, x_i \text{ is of sort } s_i\}$. Then M is certainly a cover set of $GT(F_c)$ if there is no relation on constructors. (As shown later in an example, this is not a cover set when there are relations on constructors.) Using M in the above induction principle, we get the structural induction principle. Hence the cover-set induction principle is a generalization of the structural induction principle.

Like the structural induction principle, the number of premises to be established by the cover-set principle is finite. The next subsection illustrates the use of the cover-set induction principle on three simple examples.

3.4. Examples

One of the hardest problems in applying the cover-set induction principle is to find suitable cover sets. As proposed in [Boyer&Moore 79], the function symbols in the conjecture and their definitions offer an insight into the problem. We start with a simple example taken from [Boyer&Moore 79]. Here is the specification of *append* on the data structure *list*:

Example 3.3:

$S = \{num, list\}$, $\quad F_c = \{0, suc, nil, cons\}$, $\quad F_d = \{append\}$,

$E = \{$ [a] $append(nil, x) = x$,

$\quad\quad$ [b] $append(cons(x_1, x_2), y) = cons(x_1, append(x_2, y))$ $\}$.

Suppose we were tying to prove some conjecture involving a term $append(x,y)$. Since $append$ is defined using the terms $append(nil,y)$ and $append(cons(x_1,x_2),y)$, that suggests us that a cover set for the sort $list$ is $\{nil,cons(x_1,x_2)\}$ and the only subterm of $cons(x_1,x_2)$ of sort $list$ is x_2. It is easy to check that $cons(x_1,x_2)$ is inductively greater than x_2 since $cons$ is a free constructor.

The use of the induction principle is demonstrated in the following proof.

Theorem: $append(append(x,y),z) = append(x,append(y,z))$

Proof: Let $P(x) = \{all\ y,z\ [append(append(x,y),z) = append(x,append(y,z))]\}$, and we choose $\{nil,\ cons(x_1,x_2)\}$ as a cover set of sort $list$.

By the cover-set induction principle, the induction can be done following the scheme:

[1] $P(nil)$
[2] $P(x_2)$ implies $P(cons(x_1,x_2))$

The proof of [1] is trivial and the proof of [2] is as follows:

$Premises([2]) = \{append(append(x2,y),z) = append(x2,append(y,z))\}$,
$Left([2]) = append(append(cons(x_1,x_2),y),z)$

$\qquad = append(cons(x_1,append(x_2,y)),z)$ (by [b])
$\qquad = cons(x_1,append(append(x_2,y),z)$ (by [b])
$\qquad = cons(x_1,append(x_2,append(y,z)))$ (by $Premises([2])$)
$\qquad = append(cons(x_1,x_2),append(y,z))$ (by [b])
$\qquad = Right([2])$

Thus [2] holds, too.
So $E \underset{ind}{\vdash} P(x)$. □

Using the arguments of the definition of a function f as a cover set requires that the function f be completely defined. However, this does not mean that every function appearing in the conjecture must be completely defined. This can be illustrated by the following example.

Example 3.4: Now we add two new functions $append1$ and $length$ into Example 3.3 as follows, in which $length$ is not defined on nil.

$F_d' = F_d \cup \{\ append1:list,num \rightarrow list \quad length:list \rightarrow num\ \}$
$E' = E \cup \{\ append1(nil,x) = cons(x,nil),$
$\qquad\qquad append1(cons(y,z),x) = cons(y,append1(z,x)),$
$\qquad\qquad length(cons(y,z)) = suc(length(z))\}$

Theorem: $length(append1(y,x)) = suc(length(y))$.

Proof: There are two defined function symbols, *length* and *append* 1, appearing in the equation. As *length* is not completely defined, we choose *append* 1 as a candidate, whose definition will be used to formulate an induction scheme. Let $P(y) = [all\ x\ length(append\,1(y,x)) = suc(length(y))]$ and $\{nil, cons(y_1,y_2)\}$ be a cover set of sort *list*. By the cover-set induction principle, the induction can be done following the scheme:

[1] $P(nil)$,
[2] $P(y_1)$ implies $P(cons(y_1,y_2))$.

Both formulae are provable by equational reasoning. □

The cover-set principle also works on systems when there are relations among constructors. Note that such constructors are not allowed by the shell mechanism for introducing new data structures in Boyer and Moore's approach; constructors in their approach are assumed to be free.

Example 3.5: Let S, F and E be given as follows:

$S = \{int\}$, $F_c = \{0, suc, pre\}$, $F_d = \{minus\}$,
$E = \{\,pre(suc(x)) = x$, $suc(pre(x)) = x$,
           $minus(0) = 0$, $minus(pre(x)) = suc(minus(x))$,
           $minus(suc(x)) = pre(minus(x))\}$.

Theorem: $minus(minus(x)) = x$.

Proof: Let $P(x) = [minus(minus(x)) = x]$ and $\{0, pre(x), suc(x)\}$ be a cover set of sort *integer* with the cover function Ψ defined as follows:

$\Psi(0) = \{0\}$,
$\Psi(suc(x)) = \{suc^i(0) \mid i > 0\}$,
$\Psi(pre(x)) = \{pre^i(0) \mid i > 0\}$.

By the cover-set induction principle, the induction can be done following the scheme:

[1] $P(0)$,
[2] $P(x)$ implies $P(pre(x))$,
[3] $P(x)$ implies $P(suc(x))$.

Each of these formulae can be proved by equational reasoning. □

The above examples illustrated the power of the cover-set induction. That is, it can handle equational systems with incomplete functions or with relations among constructors.

The "inductive inequivalence" condition introduced in definition 3.2 above for $SubT(t,\Psi(t))$ is necessary for the soundness of the cover-set principle as the following example illustrates.

Example 3.6: Let S, F and E be given as follows:

$S = \{num\}$, $F_c = \{0, suc\}$, $F_d = \{f\}$,
$E = \{\,suc(suc(x)) = x$, $f(suc(x)) = f(x)\}$.

Since $GT_{num} = \{\ 0,\ suc(0)\ \}$. If we choose $\{\ suc(x)\ \}$ as the cover set, i.e., $M = \{$ $suc(x)\ \}$, and define $\Psi(suc(x)) = GT_{num}$, it is easy to check that M satisfies the conditions (a) and (b) of Definition 3.1. The only strict subterm of $suc(x)$ is x; but, $suc(x)$ is not inductively greater than x since for $\sigma = [x/suc(0)]$, $\sigma(suc(x)) =_E 0$ is not greater than $suc(0)$. If we used M as a cover set in the cover-set principle and did not require that elements in $SubT(t,\Psi(t))$ for each t in M be inductively smaller than t, we could prove any formula $P(y)$ which contains a subterm $f(y)$, by the scheme

$$P(x) \text{ implies } P(suc(x)).$$

For example, $P(y)$ might be $f(y) = 0$ (or $f(y) = suc(0)$). Note that neither $f(y) = 0$ nor $f(y) = suc(0)$ is in $IND(E)$.

3.5. Extension of the Cover-Set Principle

A cover-set as defined above is a finite set of terms; the cover-set induction principle based upon it is not powerful enough to prove many inductive properties, especially properties of some n-ary functions, $n > 1$, for example, the commutativity of gcd. In this subsection, we extend the notion of cover sets to a set of n-tuples of terms.

Roughly speaking, a cover set M of $s_1 \times \cdots \times s_n$ is a set of n-tuple terms over $s_1 \times \cdots \times s_n$, where s_i are sort names, such that for each $<t_1,\ldots,t_n>$ in $GT_{s_1} \times \cdots \times GT_{s_n}$, there exists a n-tuple $<t_1',\ldots,t_n'>$ in M and a substitution σ such that $t_i =_E \sigma(t_i')$ for $1 \le i \le n$. The formal definition is as follows:

Definition 3.7 (cover function and cover set revised). Given a finite set M of n-tuples of terms (taken from $T_{s_1} \times T_{s_2} \times \cdots \times T_{s_n}$),

(1) We say a mapping Ψ, from M to the power set of $GT_{s_1} \times GT_{s_2} \times \cdots \times GT_{s_n}$, is a *cover function* of $s_1 \times s_2 \times \cdots \times s_n$ upon M if

(a) (completeness) for any $A = <t_1,t_2,\ldots,t_k>$ in $GT_{s_1} \times GT_{s_2} \times \cdots \times GT_{s_k}$, there exists $B = <m_1,m_2,\ldots,m_k> \in M$ such that $A \in \Psi(B)$ and there exists a substitution σ such that $t_i =_E \sigma(m_i)$ for i = 1 to k;

(b) (minimality) $\Psi(B) \ne \varnothing$ for $B \in M$;

(c) (uniqueness) B in (a) is unique.

(2) M is said to be a *cover set* of $s_1 \times s_2 \times \cdots \times s_n$ if M possesses a cover function defined above.

The inductive inequivalence for tuples can be defined with respect to an n-tuple of sets of ground terms $<G_1,\ldots,G_k>$ as: $<m_1,m_2,\ldots,m_k>$ is inductively greater than $<t_1,t_2,\ldots,t_k>$ if there exists j, $1 \le j \le k$, such that $m_i = t_i$ for $1 \le i < j$ and $m_j >^i t_j$ with respect to G_j. In this ordering, components in tuples are compared from left-to-right. In fact, any well-founded inductive inequivalence ordering on tuples induced by an inductive inequivalence ordering on terms can be used.

For example, a cover set of $num \times num$ is
$$\{<x, 0>, <0, suc(x)>, <suc(x), suc(y)>\}.$$

The cover-set induction principle can be extended by using the new definition of a cover set. It is analogous to the one presented before; the reader should not have any difficulty in formulating the new principle and in giving a soundness proof.

Example 3.8 Let us look at the gcd example again. The axioms are given in Section 1.1, the equations (1.1) to (1.8). For this system, we can choose the cover set as
$$M = \{<0,y>, \quad <x,0>, \quad <x+y,y>, \quad <x,x+y>\}.$$
The cover function Ψ is given by

$\Psi(<0, y>) = \{<0, suc^i(0)> \mid i>0\},$

$\Psi(<x, 0>) = \{<suc^i(0), 0> \mid i \geq 0\},$

$\Psi(<x+y, y>) = \{<suc^i(0), suc^j(0)> \mid i \geq j > 0\},$

$\Psi(<x, x+y>) = \{<suc^j(0), suc^i(0)> \mid i > j > 0\}.$

Now let us prove $P(x,y) = [gcd(x,y) = gcd(y,x)]$. By the cover-set induction principle, the induction scheme is as follows:

[1] $P(0,y)$, [3] $P(x,y)$ implies $P(x+y,y)$,

[2] $P(x,0)$, [4] $P(x,y)$ implies $P(x,x+y)$.

These four formulae are identical to (1.12) to (1.15) after expanding P. Each can be proved by equational reasoning, so $P(x,y)$ belongs to $IND(E)$.

3.6. Some Experimental Results

The cover-set principle can also be extended to conditional equations systems or to equational systems with interpreted functions (interpreted functions are themselves assumed to be axiomatized by a finite set of equations) in a straightforward way. We have implemented such a version of the cover-set induction principle in *RRL*, a *Rewrite Rule Laboratory*, a theorem proving environment based on rewriting techniques, being developed at General Electric Corporate Research and Development and Rensselaer Polytechnic Institute. Experimented with this method have been performed on about a hundred problem in the number theory, including the unique prime factorization theorem. These preliminary results show that the cover-set induction principle is powerful and convenient for equational systems and/or Prolog-like systems.

The proof of the unique prime factorization theorem follows closely the proof given in [Boyer&Moore 79] with minor changes. It involves 3 sorts (bool, natural number and list), 23 functions (not including boolean operators) and 54 lemmas. It takes 4 minutes on a Symbolics LM 3670 to compile all axioms and to prove all lemmas. All the axioms and the lemmas are in the (conditional) equational form and can be found in Appendices A and B.

Many techniques described in [Boyer&Moore 79] are also implemented to support the cover-set induction principle. These include "rewriting," "generalization" and "irrelevance eliminating." However, the technique "definition expansion" is no longer needed by our system, since definitions (or axioms) and lemmas are all transformed into rewrite rules and are not distinguishable in the rewriting. The technique "cross-fertilization" is not used in our system, either.

The cover-set induction principle is one of the inference rules used in proving these lemmas. Since we do not have enough space to present other inference rules here, we do not go into details on these experiments or on how to implement the cover-set principle in a rewriting paradigm; this is discussed in the forthcoming thesis of the first author. The following is just an outline of how to use the cover-set principle to prove theorems.

To use the cover-set principle in RRL, the user must input to RRL the declaration of the arities of functions and their definitions. The definitions are in the equational form and will be considered as axioms. The system turns equations into rewrite rules by choosing the greater term (under the reduction ordering) as the left-hand side, so that the rewrite rules have the termination property. Then the system checks whether each function is completely defined; if it is, the left-hand sides of the rewrite rules made from its definition can be used as a cover set when that function is involved in an equation to be proved. (The current algorithm in RRL for checking whether a function is completely defined fails to prove the completeness of the functions *quotient*, *remainder*, *gcd* and *primefactors* used in the proof of unique prime factorization theorem. Hence, in a strict sense, our proof of the unique prime factorization theorem is not totally *automatic*.)

When a lemma is proved as a part of the inductive proof being developed, a heuristic function is used to find a term in the lemma. That term must have a non-constructor as its root and have at least one variable as an argument of the root function symbol. Then the left-hand sides of the definition of the root function will be used as a cover set and an induction scheme is automatically constructed guided by the cover set induction principle. This induction scheme produces a set of equations and these equations will be simplified by other inference rules, such as equational reasoning. If all of the equations are simplified to identity, the lemma is proved and can be saved in the system to prove other lemmas. If one of the equations cannot be proved, the system either reports a failure or applies again the cover-set principle on it.

3.7. Comparison with Boyer and Moore's Prover

As we mentioned before, the most significant idea we learned from Boyer and Moore's prover is that an induction scheme can be constructed from the definition of a function appearing in the equation to be proved. That is also the basic common point in the use of the two induction principles.

The major difference between the cover-set principle and Boyer and Moore's induction principle is that our axioms are in equational form while theirs are in lisp-like functional form. Besides this, Boyer and Moore's prover does not allow relations between constructors and does not accept incompletely specified functions. Our principle does not have such a requirement as examples 4.4 and 4.5 illustrated.

Comparing with Boyer and Moore's induction principle, the soundness of the cover-set principle also relies on well-founded orderings. The difference is that the well-founded ordering used in Boyer and Moore's approach is a semantic ordering. For example, x is greater than $sub1(x)$ for the destructor $sub1$, even though x is a subterm of $sub1(x)$. For the cover-set principle, the inductive ordering is defined via reduction ordering, which can be constructed by many syntactic criteria (See [Dershowitz 87] for a survey).

Boyer and Moore's prover has been used to prove many impressive theorems (see [Boyer&Moore 86] for an overview), while our experiments are thus far very limited. However, even from our limited experience, the method proposed in this paper appears to have considerable potential. For example, in the proof of the unique prime factorization theorem, some lemmas used by Boyer and Moore's prover, including most of the bridge-lemmas, were not needed by our prover. This is because rewrite rules in RRL are always kept in normal form. Some of the lemmas used in the proof by Boyer and Moore's prover can be simplified to identity by normalization; so there is no need to keep them in the system. We noticed that about a fifth of the lemmas between #274 and #325 given in [Boyer&Moore 79, pp.360-366] can be simplified in this way by our prover.

We plan to study heuristics used in Boyer and Moore's prover, and to combine techniques used in the inductionless induction approaches. In [Kapur et al 86], we introduced a method based on test sets for proving inductive properties using the inductionless induction approach. A test set of a complete rewriting system is a finite set of terms describing all the irreducible ground constructor terms of the rewriting system. The concept of a cover set introduced in this paper appears similar to the concept of a test set. We plan to investigate relationship between test sets and cover sets.

Acknowledgement: We would like to thank David Musser for helpful comments on an earlier draft of this paper.

REFERENCES

[Aubin 76] Aubin, J., *Mechanizing structural induction*. Ph.D. Thesis, University of Edinburgh, Edinburgh, 1976.

[Boyer&Moore 79] Boyer, R.S. and Moore, J S., *A computational logic*. (Academic Press, New York, 1979).

[Boyer&Moore 86] Boyer, R.S. and Moore, J S., "Overview of a theorem-prover for a computational logic," in: Proc. *8th Intl. Conf. on Automated Deduction (CADE-8)*, Oxford, U.K., 1986, LNCS , Springer-Verlag, NY.

[Brotz 74] Brotz, D., *Proving theorems by mathematical induction*. Ph.D. Thesis, Computer Science Dept., Stanford University, Stanford 1976.

[Burstall 69] Burstall, R., "Proving properties of programs by structural induction," *Computer Journal* 12(1), 41-48, 1969.

[Dershowitz 83] Dershowitz, N., *Applications of the Knuth-Bendix Completion Procedure*. Laboratory Operation, Aerosapce Corporation, Aerospace Report No. ATR-83(8478)-2, 15 May, 1983.

[Dershowitz 87] Dershowitz, N., "Termination of rewriting," *J. of Symbolic Computation* 3, 1987, 69-116.

[Goguen 80] Goguen, J.A., "How to prove algebraic inductive hypotheses without induction," *Proc. of the Fifth Conference on Automated Deduction*, 1980.

[Goguen et al 78] Goguen, J.A., Thatcher, J.W. and Wagner, E.W., "Initial algebra approach to the specification, correctness, and implementation of abstract data types," in: R.T. Yeh (ed.), *Data Structuring, Current Trends in Programming Methodology*, 4 (Prentice-Hall, Englewood Cliffs, NJ, 1978).

[Guttag 75] Guttag, J., *The Specification and Application to Programming of Abstract Data Types*. Department of Computer Science, Univ. of Toronto, Ph.D. Thesis, CSRG-59, 1975.

[Guttag&Horning 78] Guttag, J.V. and Horning, J.J., "The algebraic specification of abstract data types," *Acta Informatica* 10(1), 1978, 27-52.

[Hsiang&Dershowitz 83] Hsiang, J. and Dershowitz, N., "Rewrite methods for clausal and non-clausal theorem proving," in: *Proc. Tenth EATCS, Inter. Collo. on Automata, Languages, and Programming*, Barcelona, Spain, 1983.

[Huet 80] Huet, G., "Confluent Reductions: Abstract Properties and Applications to Term Rewriting Systems," *JACM* 27(4), October 1980.

[Huet&Hullot 80] Huet, G. and Hullot, J.M., "Proofs by induction in equational theories with constructors," in: *21st IEEE Symposium on Foundations of Computer Science*, Syracuse, NY. 1980, 96-107.

[Huet&Oppen 80] Huet, G. and Oppen, D., "Equations and rewrite rules: a survey," in: R. Book (ed.), *Formal Languages: Perspectives and Open Problems*, (Academic Press, New York, 1980).

[Jouannaud&Kounalis 86] Jouannaud, J.-P., and Kounalis, E., "Proofs by Induction in Equational Theories Without Constructors," in: *Proc. of Logic in Computer Science Conference*, Cambridge, MA, 1986.

[Kanamori&Fujita 86] Kanamori, T., Fujita, H., "Formulation of induction formulas in verification of Prolog programs," *Proc. of 8th Intl Conf. on Automated Deduction (CADE-8)*, Oxford, U.K., 1986.

[Kapur&Musser 84] Kapur, D., and Musser, D.R., "Proof by Consistency," *Proc. of an NSF Workshop on the Rewrite Rule Laboratory*, Sept. 4-6, 1983. Schenectady, G.E. R&D Center Report GEN84008, April 1984. (also in *Artificial Intelligence* 31, 1987, 125-57).

[Kapur&Musser 86] Kapur, D., and Musser, D.R., "Inductive reasoning with incomplete specifications," in: *Proc. of Logic in Computer Science Conference*, Cambridge, MA, 1986.

[Kapur et al 85] Kapur, D., Narendran, P., and Zhang, H., "On Sufficient Completeness and Related Properties of Term Rewriting Systems," Unpublished Manuscript, General Electric R&D Center, Schenectady, NY, Oct. 1985. To appear in *Acta Informatica*.

[Kapur et al 86] Kapur, D., Narendran, P., and Zhang, H., "Proof by induction using test sets," *Proc. of 8th Intl Conf. on Automated Deduction (CADE-8)*, Oxford, U.K., 1986.

[Kapur&Sivakumar 83] Kapur, D. and Sivakumar, G., "Experiments with and Architecture of RRL, a Rewrite Rule Laboratory," Proc. of *An NSF Workshop on the Rewrite Rule Lab.*, Sept. 1983. General Electric R&D Center Report 84GEN008, 33-56, April 1984.

[Kapur et al 86(2)] Kapur, D., Sivakumar, G., and Zhang, H., "RRL: A Rewrite Rule Laboratory," *Proc. of 8th Intl Conf. on Automated Deduction (CADE-8)*, Oxford, U.K., 1986.

[Kirchner 84] Kirchner, H., "A General Inductive Algorithm and Application to Abstract Data Types," Proc. *7th Intl. Conf. on Automated Deduction (CADE-7)*, LNCS 170, Springer-Verlag, May 1984.

[Knuth&Bendix 70] Knuth, D., and Bendix, P., "Simple Word Problems in Universal Algebras," in: Leech (ed.) *Computational Problems in Abstract Algebra*, Pergamon Press, 1970, 263-297.

[Lankford 81] Lankford, D.S., *A simple explanation of inductionless induction.* MTP-14, Louisiana Tech University, Ruston, LA, 1981.

[McCarthy 63] McCarthy, John, "A basis for a mathematical theory of computation," *Computer Programming and Formal Systems*, P. Braffort and d. Hirschberg [ed.], Norht-Holland, Amsterdam, 1963, 33-70.

[Musser 80] Musser, D.R., "On Proving Inductive Properties of Abstract Data Types," *Proc. 7th Principles of Programming Languages*, Las Vegas, Jan. 1980.

[Musser&Kapur 82] Musser, D.R., and Kapur, D., "Rewrite Rule Theory and Abstract Data Type Analysis," *EUROCAM 1982* LNCS 144 (ed. Calmet), Springer-Verlag, 77-90, April 1982.

[Paul 84] Paul, E., "Proof by induction in equational theories with relations between constructors," in: B. Courcelle (ed.), *Ninth Colloquium on Trees in Algebra and Programming*, Bordeaux, France, 1984, 211-215.

[Wegbreit&Spitzen 76] Wegbreit, B., and Spitzen, J.M., "Proving properties of complex data structures," *JACM* 23(2), 1976, 389-396.

Appendix A: Axioms Used in Proving
Unique Prime Factorization Problem

[0 : num]

[suc : num -> num]

[+ : num, num -> num]

x + 0 == x

x + suc(y) == suc(x + y)

[* : num, num -> num]

x * 0 == 0

x * suc(y) == x + x * y

[< : num, num -> bool]

0 < suc(x) == true

x < 0 == false

suc(x) < suc(y) == x < y

[sub1 : num -> num]

sub1(0) == 0

sub1(suc(x)) == x

[- : num, num -> num]

0 - x == 0

x - 0 == x

suc(x) - suc(y) == x - y

[quotient : num, num -> num]

quotient(x, 0) == 0

quotient(x, y) == 0 if x < y

quotient(y + x, y) == suc(quotient(x, y))

    if not(0 = y)

[remainder : num, num -> num]

remainder(x, 0) == x

remainder(x, y) == x if x < y

remainder(y + x, y) == remainder(x, y)

[divides : num, num -> bool]

divides(x, y) == remainder(y,x) = 0

[>= : num, num -> bool]

x >= x == true

0 >= suc(y) == false

suc(x) >= y == x >= y if not(suc(x) = y)

[gcd : num, num -> num]

gcd(x, 0) == x

gcd(0, y) == y

gcd(x + y, y) == gcd(x, y)

gcd(x, x + y) == gcd(x, y)

[nil : list]

[cons : univ, list -> list]

[append : list, list -> list]

append(nil, y) == y

append(cons(x, y), z) ==

    cons(x, append(y, z))

[member : univ, list -> bool]

member(x, nil) == false

member(x, cons(x, z)) == true

member(x, cons(y, z)) == member(x, z)

    if not(x = y)

[delete : univ, list -> list]

delete(x, nil) == nil

delete(x, cons(x, y)) == y

delete(x, cons(y, z)) == cons(y, delete(x, z))

    if not(x = y)

[perm : list, list -> bool]

perm(nil, nil) == true

perm(nil, cons(x, y)) == false

perm(cons(x, y), nil) == false

perm(cons(x, y), z) == member(x, z) and

    perm(y, delete(x, z))

[prime1 : num, num -> bool]

prime1(x, 0) == false

prime1(x, suc(0)) == true

prime1(x, suc(y)) == not(divides(suc(y), x))

    and prime1(x,y) if not(y = 0)

[prime : num -> bool]

prime(0) == false

prime(suc(x)) == prime1(suc(x), x)

[primelist : list -> bool]

primelist(nil) == true

primelist(cons(x, y)) == prime(x) and

    primelist(y)

[timelist : list -> num]

timelist(nil) == suc(0)

timelist(cons(x, y)) == x * timelist(y)

[primefactors : num -> list]

primefactors(0) == nil

primefactors(suc(0)) == nil

primefactors(x * y) == append(primefactors(x),

    primefactors(y)) if

    not(x = 0) and not(y = 0)

Appendix B: Lemmas Proved and Used in Proving

Unique Prime Factorization Problem

1. quotient(0, y) == 0
2. remainder(0, y) == 0
3. x + y == y + x
4. y * quotient(y + x, y) ==
    y * suc(quotient(x, y))
5. (remainder(x, y) + (y * quotient(x, y))) == x
6. (y * quotient(x, y)) == x if divides(y, x)
7. (x * (y + z)) == ((x * y) + (x * z))
8. (x * y) * z == x * (y * z)
9. x * y == y * x
10. 0 < x == not(x = 0)
11. x < suc(0) == x = 0
12. (x + y) = 0 == (x = 0) and (y = 0)
13. (x + y) = y == (x = 0)
14. (x + z) = (y + z) == (x = y)
15. (x * y) = 0 == (x = 0) or (y = 0)
16. (x * y) = x == (y = suc(0)) if not(x = 0)
17. (x * y) = x == (y = suc(0)) or (x = 0)
18. (x * y) = suc(0) == (x = suc(0)) and (y = suc(0))
19. quotient((x * y), x) == y if not((x = 0))
20. remainder((y * x), x) == 0
21. remainder((y * z), x) == 0 if
        (remainder(z, x) = 0) and not(x = 0)
22. remainder(x+y,z) == remainder(x, z) if
        remainder(y, z) = 0
23. (quotient(x, y) < x) == true if
        (not(x = 0)) and (not(y = 0)) and
        (not(y = suc(0)))
24. divides(x, timelist(y)) == true if member(x, y)
25. delete(x, y) == y if not member(x, y)
26. primelist(delete(x, y)) == true if primelist(y)
27. y < suc(y) == true
28. (x = suc(0)) == false if prime(x)
29. (x = 0) == false if prime(x)
30. quotient(z + y, x) ==
        quotient(z, x) + quotient(y, x) if
        divides(x, z) and not(x = 0)
31. quotient(z * y, x) == y * quotient(z, x) if
        divides(x, z) and not(x = 0)
32. timelist(delete(x, y)) ==
        quotient(timelist(y), x) if
        (not((x = 0)) and member(x, y))
33. timelist(x) = 0 == false if primelist(x)

34. timelist(primefactors(x)) == x if not(x = 0)
35. primelist(primefactors(x)) == true if not(x = 0)
36. 0 >= u == u = 0
37. prime1(w*z, u) == false if
        not(z = suc(0)) and not(z = 0) and
        (u >= z) and not(u = suc(0))
38. suc(x) < y == true if (x < y) and not(suc(x) = y)
39. u >= z == not(u < z)
40. (u * y) < suc(y) == false if
        not(u = 0) and not(u = suc(0)) and
        not(y = 0)
41. sub1(x) < y == x < suc(y) if not(x = 0)
42. prime1(x, sub1(x)) == false if
        not(z = suc(0)) and not(z = x) and
        not(x = 0) and not(x = suc(0)) and
        divides(z, x)
43. remainder(x, y) = 0 == false if
        prime(x) and not(y = suc(0)) and not(x = y)
44. gcd(x, y) == gcd(y, x)
45. gcd((x * z), (y * z)) == (z * gcd(x, y))
46. gcd(x*y, z) = y == false if
        (remainder(z, x) = 0) and
        not(remainder(y, x) = 0)
47. remainder(y * z, x) = 0 == false if
        (gcd(x*y, y*z) = y) and not(y = 0) and
        not(remainder(y, x) = 0)
48. gcd(x, suc(0)) == suc(0) if not((x = 0))
49. gcd(x, y) == suc(0) if
        (remainder(x, gcd(x, y)) = 0) and
        not(x = 0) and not(x = suc(0)) and
        prime1(x, sub1(x)) and not(gcd(x, y) = x)
50. remainder(x, gcd(x, y)) == 0
51. gcd(x, y) = x == false if
        not(remainder(y, x) = 0)
52. remainder(y * z, x) = 0 == false if
        prime(x) and not(divides(x, y)) and
        not(divides(x, z))
53. member(x, y) == true if
        prime(x) and primelist(y) and
        divides(x, timelist(y))
54. perm(x, y) == true if
        primelist(x) and primelist(y) and
        timelist(x) = timelist(y)

FINDING CANONICAL REWRITING SYSTEMS EQUIVALENT TO A FINITE SET OF GROUND EQUATIONS IN POLYNOMIAL TIME*

Jean Gallier[1], Paliath Narendran[2], David Plaisted[3],
Stan Raatz[4], and Wayne Snyder[1]

[1]University of Pennsylvania
Philadelphia, Pa 19104
[2]General Electric Company
Corporate Research and Development
Schenectady, N.Y. 12345
[3]University of North Carolina
Chapel Hill, N.C. 27514
[4]Rutgers University
New Brunswick, N.J.

Abstract. In this paper, it is shown that there is an algorithm which, given any finite set E of ground equations, produces a reduced canonical rewriting system R equivalent to E in polynomial time. This algorithm based on congruence closure performs simplification steps guided by a total simplification ordering on ground terms, and it runs in time $O(n^3)$.

1 Introduction

In this paper, it is shown that there is an algorithm which, given any finite set E of ground equations, produces a reduced canonical rewriting system[1] R equivalent to E in polynomial time.

It has been known for some time that because total reduction orderings on ground terms exist, Knuth-Bendix type completion procedures do not fail on input sets consisting of *ground equations* and terminate with a canonical system equivalent to the input set. This has been noted by Dershowitz [5] who attributes the result to Lankford [16]. The precise reason is that newly formed equations can always be oriented (because a reduction ordering total on ground terms can be used). Actually, if one examines carefully the inference rules describing the Knuth-Bendix completion procedure (Knuth and Bendix [13]) in the formalism of Bachmair, Dershowitz, and Plaisted [1,2], one will notice that because the rules are ground, the inference rule yielding critical pairs never applies, but instead the simplification rules apply. From this and the fact that newly formed equations can always be oriented, it is easy to see that the completion procedure always halts with success. However, the complexity of such a procedure is unclear, and to the best of our knowledge, no polynomial-time algorithm has been presented. In this paper, we give such an

---

* This research was partially supported by the National Science Foundation under Grant No DCR-86-07156.

[1] A canonical system is a confluent and Noetherian system. The term reduced is defined in the next section.

algorithm based on congruence closure (Kozen [14,15], Nelson and Oppen [17], Downey, Sethi, and Tarjan [6]) that runs in time $O(n^3)$.

The basic intuition behind the algorithm is the following. Let $\prec$ be a reduction ordering total on ground terms. Given a finite set E of ground equations, we run a congruence closure algorithm on E, obtaining its congruence closure in the form of a partition Π. Recall that the equivalence classes of Π consist of the sets of subterms occurring in E that are congruent modulo E. Let $C_1, \ldots, C_n$ be the nontrivial equivalence classes on Π.[2] For each class C_i, we can form a set of rules as follows: let ρ_i be the least element of C_i (w.r.t. $\prec$), let S_i be the set of all rules of the form $\lambda \to \rho_i$, where $\lambda \in C_i$, $\lambda \succ \rho_i$. Now, the union of the sets of rules just constructed is almost the answer. The problem is that the sets S_i may not be reduced. In order to reduce them, some simplification steps must be performed. But care must be exercised in performing these simplifications, because naive simplification strategies can take exponential time (see example 5.2). Fortunately, it is possible to carry out these simplifications in polynomial time. Roughly speaking, the trick is to choose the classes C_i to form the sets of rules S_i in an order such that the next class selected is the one containing the least element belonging to nontrivial classes. In this fashion, we minimize the number of simplification steps to be applied. What happens in this algorithm is that congruence closure is performed only once at the beginning, and that every time a new set of rules S_i is produced (as the result of picking the right class), S_i is simplified to a canonical set and it is also used to simplify the current partition and the current set of rules formed so far. For every such step, at least the selected class C_i collapses to a trivial class, and so this process is guaranteed to stop in a number of steps bounded by the number of nontrivial classes in the original congruence closure. Another very useful fact in the ground case is that given a set R of ground rewrite rules that is Noetherian, if R is reduced then it is canonical. Hence, our algorithm produces a Noetherian and reduced set of rules. The correctness of this algorithm is nontrivial, and we give a rigorous proof.

2 Preliminaries

We review briefly the concepts that will be needed in this paper. As much as possible, we tried to use notation and terminology consistent with Huet and Oppen [11] and Gallier [7]. Given a ranked alphabet (or signature) Σ, the set of ground terms on Σ is denoted by T_Σ. In this paper, only *finite* ranked alphabets will be considered. Given a term t and a tree address α in t, t/α denotes the subterm of t rooted at α. Given two terms $s, t \in T_\Sigma$ and a tree address α in s, the term $s[\alpha \leftarrow t]$ is the result of replacing the subterm rooted at α in s by t.

Definition 2.1 Let $\Longrightarrow \subseteq A \times A$ be a binary relation on a set A. The transitive closure and the transitive and reflexive closure of $\Longrightarrow$ are denoted by $\Longrightarrow^+$ and $\Longrightarrow^*$ respectively. The *converse* (or *inverse*) of the relation $\Longrightarrow$ is denoted by $\Longrightarrow^{-1}$ or $\Longleftarrow$. We say that $\Longrightarrow$ is *Noetherian* or *well founded* iff there are no infinite sequences $\langle a_0, a_1, \ldots, a_n, a_{n+1}, \ldots \rangle$ of elements of A such that $a_n \Longrightarrow a_{n+1}$ for all $n \geq 0$.

---

[2] that is, those containing at least two elements

Definition 2.2 A partial order $\preceq$ on a set A is a binary relation $\preceq \subseteq A \times A$ that is reflexive, transitive, and antisymmetric. We let $\succeq = \preceq^{-1}$. Given a partial order $\preceq$ on a set A, the strict ordering $\prec$ associated with $\preceq$ is defined such that $s \prec t$ iff $s \preceq t$ and $s \neq t$. We let $\succ = \prec^{-1}$. A strict ordering $\prec$ on a set A is *well founded* iff $\succ$ is well founded according to definition 2.1.

Definition 2.3 A strict ordering $\prec$ on (ground) terms is *monotonic* iff for every two terms s, t and for every function symbol f, if $s \prec t$ then $f(\ldots, s, \ldots) \prec f(\ldots, t, \ldots)$. The strict ordering $\prec$ has the *subterm property* iff $s \prec f(\ldots, s, \ldots)$ for every term $f(\ldots, s, \ldots)$ (since we are considering symbols having a fixed rank, the deletion property is superfluous, as noted in Dershowitz [4]). A *simplification ordering* $\prec$ is a strict ordering that is monotonic and has the subterm property. It is shown in Dershowitz [4] that for finite ranked alphabets, any simplification ordering is well founded, and that there exist total simplification orderings on ground terms.

Note that if a strict ordering $\prec$ is total, monotonic, and well founded, we must have $s \prec f(\ldots, s, \ldots)$ for every s, since otherwise, by monotonicity, we would have an infinite decreasing chain. From this, we have immediately that for a finite ranked alphabet Σ, a total monotonic ordering $\prec$ is well founded iff it is a simplification ordering (as noted in Dershowitz [4]). Such an ordering will be called a *total simplification ordering* on ground terms.

Definition 2.4 Let $E \subseteq T_\Sigma \times T_\Sigma$ be a binary relation on ground terms. We define the relation $\longrightarrow_E$ over T_Σ as follows. Given any two terms $t_1, t_2 \in T_\Sigma$, then $t_1 \longrightarrow_E t_2$ iff there is some pair $(s, t) \in E$ and some tree address α in t_1 such that

$$t_1/\alpha = s, \quad \text{and} \quad t_2 = t_1[\alpha \leftarrow t].$$

When $t_1 \longrightarrow_E t_2$, we say that t_1 *rewrites* to t_2, or that we have a *rewrite step*. The subterm t_1/α of t_1 is called a *redex of t_1 at β*, or simply a *redex*. When a pair (s, t) is used in a rewrite step, we also call it a *rewrite rule* (or *rule*), and use the notation $s \rightarrow t$ to emphasize its use as a rewrite rule. The idea is that the pair is used *oriented* from left to right.

Let $\overset{*}{\longrightarrow}_E$ be the reflexive and transitive closure of $\longrightarrow_E$. The relation $\longleftrightarrow_E$ is defined as follows: for every pair of terms $s, t \in T_\Sigma$,

$$s \longleftrightarrow_E t \quad \text{iff} \quad s \longrightarrow_E t \quad \text{or} \quad t \longrightarrow_E s.$$

In such a rewrite step, the pair (s, t) is used as a two-way rewrite rule (that is, non-oriented). In such a case, we denote the pair as $s \doteq t$ and call it an *equation*. Let $\overset{*}{\longleftrightarrow}_E$ be the transitive and reflexive closure of $\longleftrightarrow_E$. It is well known that $\overset{*}{\longleftrightarrow}_E$ is the smallest congruence on T_Σ containing E. When we want to fully specify a rewrite step, we use the notation $t_1 \longrightarrow_{[\alpha, s \rightarrow t]} t_2$.

Definition 2.5 Given a set R of ground rewrite rules and a total simplification ordering $\prec$, we say that R is *compatible with* $\prec$ iff $r \prec l$ for every $l \rightarrow r \in R$.

Given a set R of ground rewrite rules, we say that R *is reduced* iff

(1) No lefthand side of any rewrite rule $l \rightarrow r \in R$ is reducible by any rewrite rule in $R - \{l \rightarrow r\}$;

(2) No righthand side of any rewrite rule $l \to r \in R$ is reducible by any rewrite rule in R.

Definition 2.6 Let $\longrightarrow \subseteq T_\Sigma \times T_\Sigma$ be a binary relation on T_Σ. We say that $\longrightarrow$ is *locally confluent* iff for all $t, t_1, t_2 \in T_\Sigma$, if $t \longrightarrow t_1$ and $t \longrightarrow t_2$, then there is some $t_3 \in T_\Sigma$ such that $t_1 \longrightarrow^* t_3$ and $t_2 \longrightarrow^* t_3$. We say that $\longrightarrow$ is *confluent* iff for all $t, t_1, t_2 \in T_\Sigma$, if $t \longrightarrow^* t_1$ and $t \longrightarrow^* t_2$, then there is some $t_3 \in T_\Sigma$ such that $t_1 \longrightarrow^* t_3$ and $t_2 \longrightarrow^* t_3$.

It is well known (Huet [10]) that a Noetherian relation is confluent iff it is locally confluent. We say that a set of rewrite rules (or equations) R is Noetherian, locally confluent, or confluent iff the relation $\longrightarrow_R$ associated with R given in definition 2.4 has the corresponding property. We say that R is *canonical* iff it is Noetherian and confluent. Note that since a reduced set of ground rewrite rules has no critical pairs, by [10], it is locally confluent. A reduced set of ground rewrite rules compatible with $\succ$ is also Noetherian because $r \prec l$ for every rule $l \to r$, and $\prec$ is a simplification ordering. Hence, by [10], such a set is confluent.

3 The Algorithm

We shall define a sequence of triples $\langle \mathcal{E}_i, \Pi_i, \mathcal{R}_i \rangle$ where $\mathcal{E}_i$ is a finite set of ground equations, Π_i is a partition (associated with $\mathcal{E}_i$), and $\mathcal{R}_i$ is a set of ground rewrite rules. Given a triple $\langle \mathcal{E}_i, \Pi_i, \mathcal{R}_i \rangle$, we let $\mathcal{T}_i$ be the set of all subterms of terms occurring in equations in $\mathcal{E}_i$ or in rewrite rules in $\mathcal{R}_i$. The algorithm below makes use of the *congruence closure* of a finite set of ground equations (Kozen [14,15], Nelson and Oppen [17], Downey, Sethi, and Tarjan [6]). Congruence closures are represented by their associated partition Π. Given an equivalence relation represented by its partition Π, the equivalence class of t is denoted by $[t]_\Pi$, or $[t]$. Recall that s, t are in the same equivalence class of Π iff s and t are subterms of the terms occurring in E and $s \overset{*}{\longleftrightarrow}_E t$ (for details, see Gallier [7]). The congruence closure algorithm will only be run once on E to obtain Π_0, but the partition Π_i may change due to further steps (simplification steps). Note that for the purpose of defining the algorithm, it is sufficient to deal with pairs $\langle \Pi_i, \mathcal{R}_i \rangle$, but the component $\mathcal{E}_i$ is necessary for the proof of correctness, and this is why the method is presented in terms of triples.

begin algorithm

Initially, we set $\mathcal{E}_0 = E$, $\mathcal{R}_0 = \emptyset$, and run a congruence closure algorithm on the ground set E to obtain Π_0. $i := 0$;

while Π_i has some nontrivial equivalence class[3] **do** {Simplification steps}

Let ρ_{i+1} be the smallest element[4] of the set

$$\bigcup_{C \in \Pi_i, |C| \geq 2} C$$

of terms belonging to nontrivial classes in Π_i.[5] Let C_{i+1} be the nontrivial class that contains

---

[3] that is, a class containing at least two elements, in which case $\mathcal{E}_i$ has at least one nontrivial equation.

[4] in the ordering $\prec$

[5] where $|C|$ denotes the cardinality of the set C

ρ_{i+1}, and write $C_{i+1} = \{\rho_{i+1}, \lambda^1_{i+1}, \ldots, \lambda^{k_{i+1}}_{i+1}\}$, where $k_{i+1} \geq 1$, since C_{i+1} is nontrivial. Let $\mathcal{S}_{i+1} = \{\lambda^1_{i+1} \to \rho_{i+1}, \ldots, \lambda^{k_{i+1}}_{i+1} \to \rho_{i+1}\}$.[6]

{Next, we use the rewrite rules in $\mathcal{S}_{i+1}$ to simplify the rewrite rules in $\mathcal{R}_i \cup \mathcal{S}_{i+1}$, the partition Π_i, and the equations in $\mathcal{E}_i$.}[7]

To get $\mathcal{R}_{i+1}$, first, we get a canonical system equivalent to $\mathcal{S}_{i+1}$. For this, for every lefthand side λ of a rule in $\mathcal{S}_{i+1}$, replace every maximal redex of λ of the form λ^j by ρ, where $\lambda^j \to \rho \in \mathcal{S}_{i+1} - \{\lambda \to \rho\}$.[8] Let $\mathcal{S}'_{i+1}$ be the set of simplified rules.[9] Also, let $\mathcal{R}'_{i+1}$ be the set obtained by simplifying the lefthand sides of rules in $\mathcal{R}_i$ using $\mathcal{S}_{i+1}$ (reducing maximal redexes only), and let

$$\mathcal{R}_{i+1} = \mathcal{R}'_{i+1} \cup \mathcal{S}'_{i+1}.$$

Finally, use $\mathcal{S}_{i+1}$ to simplify all terms in Π_i and $\mathcal{E}_i$, using the simplification process described earlier to obtain Π_{i+1} and $\mathcal{E}_{i+1}$.

$i := i + 1$

endwhile

{All classes of Π_i are trivial, and the set $\mathcal{R}_i$ is a canonical system equivalent to E.}[10]

end algorithm

We will have to justify the fact that when the lefthand side of a rule $l \to r \in \mathcal{R}_i$ is simplified to a rule $l' \to r$, it is still true that $r \prec l'$ holds, and that righthand sides are never simplified.

Note that during the step where $\mathcal{S}_{i+1}$ is used to simplify all terms in Π_i and $\mathcal{E}_i$, the class C_{i+1} is simplified to the trivial class $\{\rho_{i+1}\}$.

We claim that any sequence defined by the above procedure terminates, and that the set $\mathcal{R}_i$ obtained in the last step is a reduced canonical system equivalent to the original set E of equations. For this, we will need a number of lemmas, but first, the method is illustrated in the following example.

Example 3.1 Let $\mathcal{E}_0 = E$ be the following set of ground equations:

$$E = \{f^3 a \doteq a,$$
$$f^5 a \doteq a,$$
$$a \doteq d,$$
$$gha \doteq a,$$
$$gma \doteq a,$$
$$ha \doteq c,$$
$$mgc \doteq b\}.$$

---

[6] For simplicity of notation, we occasionally omit the subscript $i+1$.

[7] This is one of the crucial steps that ensures a polynomial time algorithm.

[8] By a maximal redex of λ, we mean a redex of λ that is not a proper subterm of any other redex of λ. The simplified term is irreducible w.r.t $\mathcal{S}_{i+1}$, so these replacements are only done once, and they can be done in parallel because they apply to independent subterms of λ.

[9] We shall prove that this is a canonical system equivalent to $\mathcal{S}_{i+1}$.

[10] $\mathcal{E}_i$ must consist entirely of trivial equations, that is, equations of the form $s \doteq s$.

Let $\prec$ be a total simplification ordering such that, $d \prec c \prec b \prec a \prec f \prec g \prec h \prec m$. After computing the congruence closure for E, we have the initial partition

$$\Pi_0 = \{\{d, a, fa, f^2a, f^3a, f^4a, f^5a, gc, gha, gma\},$$
$$\{c, ha\},$$
$$\{b, mgc, ma\}\}.$$

The class $\{d, a, fa, f^2a, f^3a, f^4a, f^5a, gc, gha, gma\}$ is selected, since d is the least term. We have

$$S_1 = \{a \rightarrow d,$$
$$fa \rightarrow d,$$
$$f^2a \rightarrow d,$$
$$f^3a \rightarrow d,$$
$$f^4a \rightarrow d,$$
$$f^5a \rightarrow d,$$
$$gc \rightarrow d,$$
$$gha \rightarrow d,$$
$$gma \rightarrow d\}.$$

After simplification, we obtain the reduced system

$$\mathcal{R}_1 = \{a \rightarrow d,$$
$$fd \rightarrow d,$$
$$gc \rightarrow d,$$
$$ghd \rightarrow d,$$
$$gmd \rightarrow d\}.$$

The partition Π_0 simplifies to

$$\Pi_1 = \{\{d\},$$
$$\{c, hd\},$$
$$\{b, md\}\},$$

and $\mathcal{E}_0$ to

$$\mathcal{E}_1 = \{d \doteq d,$$
$$hd \doteq c,$$
$$md \doteq b\}.$$

The next class selected is $\{c, hd\}$. We have

$$S_2 = \{hd \rightarrow c\}.$$

After simplification, we have

$$\mathcal{R}_2 = \{a \rightarrow d,$$
$$fd \rightarrow d,$$
$$gc \rightarrow d,$$
$$gmd \rightarrow d,$$
$$hd \rightarrow c\},$$

$$\Pi_2 = \{\{d\},$$
$$\{c\},$$
$$\{b, md\}\},$$

and

$$\mathcal{E}_2 = \{d \doteq d,$$
$$c \doteq c,$$
$$md \doteq b\}.$$

Finally, the class $\{b, md\}$ is selected, we have

$$\mathcal{S}_3 = \{md \rightarrow b\},$$

and after simplification, we have

$$\mathcal{R}_3 = \{a \rightarrow d,$$
$$fd \rightarrow d,$$
$$gc \rightarrow d,$$
$$gb \rightarrow d,$$
$$hd \rightarrow c,$$
$$md \rightarrow b\},$$

$$\Pi_3 = \{\{d\},$$
$$\{c\},$$
$$\{b\}\},$$

and

$$\mathcal{E}_3 = \{d \doteq d,$$
$$c \doteq c,$$
$$b \doteq b\}.$$

The reduced canonical system equivalent to E is $\mathcal{R}_3$.

4 Correctness and Termination of the Method

We now prove that the method described in section 3 terminates and produces a reduced canonical system equivalent to the original ground set E.

Definition 4.1 A set S of ground rewrite rules is *right uniform* iff $r = r'$ for all $l \to r$ and $l' \to r' \in S$.

Definition 4.2 Let S be a right uniform set of rules. The relation $\rightarrowtail_S$ is defined such that $\rightarrowtail_S$ is like rewriting using S, except restricted to maximal redexes. Formally, given any two ground terms s, t,

$$s \rightarrowtail_S t \quad \text{iff} \quad t = s[\beta \leftarrow r],$$

where s/β is a maximal redex of s such that $s/\beta \to r \in S$ (a maximal redex is a redex that is not a proper subterm of any other redex). Let $\rightarrowtail_S^+$ be the transitive closure of $\rightarrowtail_S$, and $\rightarrowtail_S^*$ its reflexive and transitive closure.

Lemma 4.3 Let S be a set of ground rewrite rules compatible with $\prec$ such that

(1) S is right uniform

(2) Whenever $\lambda \to r \in S$ and $\lambda' \overset{*}{\longleftrightarrow}_S \lambda$ where λ' is a subterm of the lefthand side of some rule in S ($\lambda' = l/\beta$ for some rule $l \to r \in S$), then $\lambda' \to r \in S$.

Let $s \to r$ be any rule in S and let $S_1 = S - \{s \to r\}$. Then, the following statements hold. (i) In any simplification sequence $s \rightarrowtail_{S_1}^+ s'$, there cannot be two steps applied at addresses β_1 and β_2 in that order such that β_1 is an ancestor of β_2. (ii) In any simplification sequence $s \rightarrowtail_{S_1}^+ s'$, there cannot be two steps applied at addresses β_1 and β_2 in that order such that β_2 is an ancestor of β_1.

In order to keep the paper sufficiently short, the proof is omitted. From lemma 4.3, we obtain the following corollary.

Corollary 4.4 If S is a set of rewrite rules compatible with a total simplification ordering $\prec$ and satisfying conditions (1) and (2) of lemma 4.3, then for every ground term u, there is a unique ground term v irreducible (w.r.t. S) such that $u \rightarrowtail_S^+ v$, and v is obtained from u by replacing all maximal (independent) redexes of u by r, the common righthand side of all rules in S.

Lemma 4.5 Let S be a set of ground rewrite rules compatible with $\prec$ and satisfying conditions (1) and (2) of lemma 4.3. Let

$$S' = \{l' \to r \mid l \rightarrowtail_S^* l', \ l' \succ r, \ l \to r \in S\},$$

where l' is the normal form of l w.r.t. $\rightarrowtail_S$. Then S' is a reduced canonical system equivalent to S.

Again, in order to keep the paper sufficiently short, the proof is omitted.

Lemma 4.6 For every $i \geq 0$,

$$\rho_{i+1} \prec \rho_{i+2} = min(\bigcup_{C \in \Pi_{i+1}, |C| \geq 2} C).$$

The proof is also omitted. From lemma 4.6, we obtain the following corollary.

Lemma 4.7 For every $l \to r \in \mathcal{R}_i$, r is never simplified by any rule in $\mathcal{S}_{i+1}$, and if l simplifies to l', then $r \prec l'$.

Proof. By lemma 4.6, $\rho_{i+1} \prec \rho_{i+2}$ for all $i \geq 0$. Since the simplication rules in $\mathcal{S}_{i+1}$ are of the form $\lambda \to \rho_{i+1}$ and the set of righthand sides of rules in $\mathcal{R}_i$ is $\{\rho_1, \ldots, \rho_i\}$, the result is clear. $\square$

Lemma 4.8 (1) The sequence $\langle \mathcal{E}_i, \Pi_i, \mathcal{R}_i \rangle$ is finite and its length m is bounded by the number of nontrivial equivalence classes in Π_0.

(2) Π_i is the partition associated with the congruence closure of $\mathcal{E}_i$ for every i, where $0 \leq i \leq m$.

Proof. (1) follows from the fact that when Π_{i+1} is derived from Π_i, the equivalence class C_{i+1} (in Π_i) of ρ_{i+1} collapses to the trivial class $\{\rho_{i+1}\}$. (2) is shown by induction on i. The details are straightforward and are left to the reader (we use the fact that both Π_i and $\mathcal{E}_i$ are simplified by the set $\mathcal{S}_{i+1}$). $\square$

Lemma 4.9 Let m be the length of the sequence $\langle \mathcal{E}_i, \Pi_i, \mathcal{R}_i \rangle$. Then,

$$\overset{*}{\longleftrightarrow}_E = \overset{*}{\longleftrightarrow}_{\mathcal{E}_i \cup \mathcal{R}_i},$$

for all i, $0 \leq i \leq m$.

Proof. The proof is by induction on i. It is rather straighforward and is omitted. $\square$

We now prove the following crucial lemma.

Lemma 4.10 The system $\mathcal{R}_m$ is reduced.

Proof. For every i, $1 \leq i \leq m$, the set $\{C'_1, \ldots, C'_{n_i}\}$ is defined as follows: $\{C'_1, \ldots, C'_{n_i}\}$ consists of all classes in the set $\{C_1, \ldots, C_i\}$ of nontrivial equivalence classes selected by the algorithm, and all singletons of the form $\{u\}$, where u is some subterm of a term in one of the classes $C_1, \ldots, C_i$ and $u \notin C_k$ for every k, $1 \leq k \leq i$. Given any set C'_j, its representative ρ'_j is defined such that $\rho'_j = \rho_k$, the representative chosen by the algorithm if $C'_j = C_k$ is a nontrivial class, else $\rho'_j = u$, the single element in the set $C'_j = \{u\}$. We also order the set $\{C'_1, \ldots, C'_{n_i}\}$ to form the sequence $\langle C'_1, \ldots, C'_{n_i} \rangle$ as follows: C'_k precedes C'_l iff $\rho'_k \prec \rho'_l$.

We say that a term u in Π_i is *simplified (at stage i)* if, either u is not in any of the classes $C'_1, \ldots, C'_{n_i}$, or $u = \rho'_k$ for some k, $1 \leq k \leq n_i$. For $i = 0$, we define $\{C'_1, \ldots, C'_{n_i}\}$ as the empty set and $\langle C'_1, \ldots, C'_{n_i} \rangle$ as the empty sequence. We shall prove the following claim by induction on i.

Claim. For every i, $0 \leq i \leq m - 1$, the following properties hold:

(a) If $l \to r$ is a rule in $\mathcal{R}_{i+1}$ then all proper subterms of l and all subterms of r are simplified, and every proper subterm of any term $u \in \Pi_{i+1} - \mathcal{R}_{i+1}$ is simplified.

(b) If u is a representative of some C'_l, $1 \leq l \leq n_i$, then every proper subterm of u is also the representative of some C'_k, $1 \leq k < l$.

Proof of claim. The claim is true for $i = 0$ since $\{C'_1, \ldots, C'_{n_i}\}$ is the empty set. The induction step is established as follows. Observe that the new class $C_{i+1} = C'_{n_{i+1}}$ chosen by the algorithm has the property that all proper subterms of the representative $\rho_{i+1} = \rho'_{n_{i+1}}$ are previously chosen representatives. This is because since $\prec$ has the subterm property, $u \prec \rho'_{n_{i+1}}$ for every proper subterm u of $\rho'_{n_{i+1}}$. Then, because $\rho'_{n_{i+1}}$ is chosen minimal, we must have $u \in C'_k$ for some $k < n_{i+1}$ and the induction hypothesis applies. Thus property (b) holds. To prove (a), we simply note three facts. (1) Righthand sides of rules in S_{i+1} or $\mathcal{R}_{i+1}$ are representatives of $C_1, \ldots, C_{i+1}$, since by lemma 4.7, only lefthand sides of rules in $\mathcal{R}_i$ are simplified. (2) When the lefthand side l of a rule in S_{i+1} or $\mathcal{R}_i$ is simplified, either the rule disappears, or some proper subterm u of l is replaced by ρ_{i+1}. But then, every proper subterm u of l is either a subterm of $\rho_{i+1} = \rho'_{n_{i+1}}$, or by the induction hypothesis a subterm of ρ'_k for some k, $1 \leq k \leq n_i$, or u is not in any of the classes $C'_1, \ldots, C'_{n_i}$. This shows that u is either the representative of one of the classes C'_k, $1 \leq k \leq n_{i+1}$, or that $u \notin C'_k$ for every k, $1 \leq k \leq n_{i+1}$. Hence, u is simplified at stage $i + 1$. (3) The same property applies to proper subterms of terms in $\Pi_{i+1} - \mathcal{R}_{i+1}$. This proves (a) and concludes the proof of the claim. $\square$

We now apply the claim to the rules in $\mathcal{R}_m$. Therefore, every subterm u of a term in a rule from $\mathcal{R}_m$ is the representative of some equivalence class in $C'_1, \ldots, C'_{n_m}$ or $u \notin C'_k$ for every k, $1 \leq k \leq n_m$, except possibly for lefthand sides of rules. Thus, every subterm u of a term in a rule from $\mathcal{R}_m$ is the representative of some equivalence class in $C'_1, \ldots, C'_{n_m}$ or belongs to some trivial class of Π_m, except possibly for lefthand sides of rules. This means that no rewrite rule in $\mathcal{R}_m$ can be used to further simplify $\mathcal{R}_m$, except possibly to simplify a lefthand side at the top level. Assume that some rule $l_2 \to r_2$ in $\mathcal{R}_m$ simplifies the lefthand side of some rule $l_1 \to r_1$ in $\mathcal{R}_m$. Then, l_2 and l_1 must be identical, so l_1, l_2, r_1, and r_2 are all in the same equivalence class. Since r_1 and r_2 are both the representative of this class, we have $r_1 = r_2$. However, by definition, a rule does not reduce itself. Thus, $\mathcal{R}_m$ is reduced. $\square$

We finally have our main result.

Theorem 4.11 Given a finite set E of ground equations, the procedure of section 3 terminates with a reduced canonical system $\mathcal{R}_m$ equivalent to E.

Proof. The termination of the procedure is shown in lemma 4.8. By lemma 4.9, $\xleftrightarrow{*}_E = \xleftrightarrow{*}_{\mathcal{E}_m \cup \mathcal{R}_m}$, where m is the index of the final triple $\langle \mathcal{E}_m, \Pi_m, \mathcal{R}_m \rangle$. However, for this last triple, $\mathcal{E}_m$ consists of trivial equations, and so, $\xleftrightarrow{*}_E = \xleftrightarrow{*}_{\mathcal{R}_m}$. Finally, by lemma 4.10, $\mathcal{R}_m$ is reduced. Because $\mathcal{R}_m$ is reduced and ground, there are no critical pairs, and since $\mathcal{R}_m$ is also Noetherian, it is confluent. $\square$

5 Complexity of the Procedure

In this section, we analyze the complexity of the procedure.

Lemma 5.1 The algorithm of section 3 runs in time $O(n^3)$, where n measures the size of E.

Proof. First, observe that the number of subterms of terms occurring in E is $O(n)$, where n measures the size of E (say the length of the string obtained by concatenating the equations in E written in prefix notation). The number m of nontrivial equivalence classes of Π_0 is bounded by $\lfloor n/2 \rfloor$. Every term t is simplified using rules in S_{i+1} by replacing maximal (independent) subterms of t by ρ_{i+1}. If a DAG structure with sharing of common subterms is used, for each round, the simplification of all terms in Π_i and $\mathcal{R}_i$ by rules in S_{i+1} can be performed in $O(n)$. Hence, the complexity of the simplifications for m rounds is $O(n^2)$. The contribution of the congruence closure is $O(n^2)$. Finally, we need to make sure that there are total simplification orderings such that the least element of a set of k ground terms can be determined in time $O(k^2)$. However, this is not difficult to achieve. For example, one can use a recursive path ordering where sequences of subtrees are compared using a lexicographic ordering. Hence, the complexity of the comparisons for m rounds is $O(n^3)$. Therefore, the complexity of the algorithm is $O(n^3)$. $\square$

Note that the dominant factor in the time complexity of the procedure is the process of finding least elements with respect to a simplification ordering. If this can be reduced, the time complexity of the algorithm will also be reduced. The following example shows that a naive approach to simplification can lead to an exponential-time complexity.

Example 5.2 Given any integer $k > 1$, consider the following set E of equations:

$$gf^k c \doteq fg^k c$$
$$\ldots \doteq \ldots$$
$$gffffc \doteq fggggc$$
$$gfffc \doteq fgggc$$
$$gffc \doteq fggc$$
$$gfc \doteq fgc$$
$$gc \doteq fc.$$

The set of nontrivial classes of the partition Π_0 obtained after computing the congruence closure of E is

$$\{\{fc, gc\},$$
$$\{ffc, fgc, gfc, ggc\},$$
$$\{fffc, fggc, gffc, gggc\},$$
$$\ldots$$
$$\{f^k c, fg^{k-1}c, gf^{k-1}c, g^k c\},$$
$$\{fg^k c, gf^k c\}\}.$$

It is clear that there are total simplification orderings induced by the total order on the function symbols such that $c \prec f \prec g$. The order in which the classes are selected by our algorithm amounts to simplifying E bottom-up. First, the set $\mathcal{R}_1$ of rewrite rules associated with the class $\{fc, gc\}$ is

$$\{gc \to fc\},$$

and we replace all occurrences of gc in Π_0 by fc, obtaining the partition Π_1 whose nontrivial blocks are

$$\{\{ffc, gfc\},$$
$$\{fffc, fgfc, gffc, ggfc\},$$
$$\cdots$$
$$\{f^k c, fg^{k-2}fc, gf^{k-1}c, g^{k-1}fc\},$$
$$\{gf^k c, fg^{k-1}fc\}\}.$$

Next, $\mathcal{R}_2$ is the set

$$gfc \to ffc$$
$$gc \to fc,$$

and we replace all occurrences of gfc in Π_1 by ffc, obtaining the partition Π_2 whose nontrivial blocks are

$$\{\{fffc, gffc\},$$
$$\cdots$$
$$\{f^k c, fg^{k-3}ffc, gf^{k-1}c, g^{k-2}ffc\},$$
$$\{gf^k c, fg^{k-2}ffc\}\}.$$

Proceeding in this fashion, we obtain the reduced system $\mathcal{R}_{k+1}$

$$gf^k c \to f^{k+1} c$$
$$\cdots \to \cdots$$
$$gfffc \to ffffc$$
$$gffc \to fffc$$
$$gfc \to ffc$$
$$gc \to fc$$

in time $O(k^2)$. On the other hand, if we don't compute the congruence closure of E but simply transform E into the following set R of rewrite rules

$$gf^k c \to fg^k c$$
$$\cdots \to \cdots$$

$$gfffc \to fggggc$$
$$gfffc \to fgggc$$
$$gffc \to fggc$$
$$gfc \to fgc$$
$$gc \to fc,$$

and simplify R from the top-down, this takes exponential time in k. Indeed, in order to simplify $gf^kc \to fg^kc$ to $gf^kc \to f^{k+1}c$, $2^k - 1$ steps are required. This is shown by proving by induction that g^kc simplifies to f^kc in $2^k - 1$ steps. For $k = 1$, this is obvious using the last rule $gc \to fc$. Assuming inductively that $g^{k-1}c$ simplifies to $f^{k-1}c$ in $2^{k-1} - 1$ steps, then

$$g^kc = gg^{k-1}c \Rightarrow^* gf^{k-1}c$$

in $2^{k-1} - 1$ steps,

$$gf^{k-1}c \Rightarrow fg^{k-1}c$$

using the second rule, and

$$fg^{k-1}c \Rightarrow^* ff^{k-1}c$$

in $2^{k-1} - 1$ steps again. The total number of steps is $2^{k-1} - 1 + 1 + 2^{k-1} - 1 = 2^k - 1$, as claimed. Hence, it will take

$$2^k - 1 + 2^{k-1} - 1 + \ldots + 2^2 - 1 + 2^1 - 1 = 2^{k+1} - (k+2)$$

steps to reduce R top-down.

6 Relation to Other Work

In this section we clarify the relationship between our work and the work of Dauchet et al. [3], Otto and Squier [18], and Kapur and Narendran [12], and clear up some possibly confusing points. Dauchet et al. [3] prove that it is decidable whether a set of ground rewrite rules is confluent. The algorithm is fairly involved and its complexity is not clear, but it is unlikely that it runs in polynomial time. This is not in contradiction with our result. In fact, this decidability result has no bearing on our problem. Indeed, since our goal is to find a canonical system equivalent to the input system R, the orientation of the rules in R is irrelevant, and we are free to reorient the rules so that we have a Noetherian system. Having oriented the rules in R properly, we force confluence by interreducing the rules using our algorithm. Hence, we don't care whether the original set is confluent or not. Of course, Dauchet et al. must accept the original orientation of the rules in R and they cannot change it. It is somewhat amusing to think that it might be faster to apply our algorithm to get a reduced canonical system than to test whether the given rules are confluent! Whether our work can be helpful for giving an alternate confluence test is another story, but we have not explored this path.

Both Otto and Squier [18] and Kapur and Narendran [12] show that there exist finite Thue systems with a decidable word problem for which no equivalent finite canonical system exists.

Otto and Squier actually prove this result for finitely presented monoids with a decidable word problem. At first glance, this may seem to contradict our result. Indeed, strings are ground terms after all! However, we are forgetting that the free monoid over an alphabet Σ satisfies the *associativity* axiom

$$\forall x \forall y \forall z [x \cdot (y \cdot z) = (x \cdot y) \cdot z],$$

which is *not* equivalent to any finite set of ground equations. In fact, the associativity axiom is equivalent to *infinitely many* ground equations, all ground instances of the form $u \cdot (v \cdot w) = (u \cdot v) \cdot w$ obtained by substituting arbitrary strings $u, v, w \in \Sigma^*$ for the variables x, y, z. This explains the apparent contradiction. Our algorithm deals with a *finite* set of ground equations on the *initial* Σ-algebra T_Σ, where Σ is a finite ranked alphabet. The free monoid Σ^* is isomorphic to the *quotient* $T_\Delta / \equiv$ of the initial algebra T_Δ on the ranked alphabet $\Delta = \Sigma \cup \{\cdot, \epsilon\}$ (where $\cdot$ is a binary symbol, ϵ a constant, and every letter in Σ is a constant) by the least stable congruence $\equiv$ containing the set of (nonground) equations

$$\{x \cdot (y \cdot z) = (x \cdot y) \cdot z, \; x \cdot \epsilon = x, \; \epsilon \cdot x = x\}.$$

This is not the free Δ-algebra.

In principle, our algorithm can deal with a finite set E of nonground equations provided that there is a known bound k on the number of instances of equations used, but then the running time of our algorithm is $O(k^3)$, where k has nothing to do with the number of equations in the input set E.

7 Conclusion

An algorithm that produces a (reduced) canonical system equivalent to a set of ground equations has been presented and proved correct. This algorithm calls the congruence closure algorithm only once and performs simplification steps carefully. The present version of the algorithm runs in time $O(n^3)$. It is possible that using more sophisticated data structures the running time of the algorithm can be improved, but in this paper we are more concerned with correctness, and the issue of efficiency is left for further research. It is worth noting that this algorithm is at the heart of the decision procedure showing that rigid unification (first introduced in Gallier, Raatz, and Snyder [8]) is NP-complete, a result to appear in a forthcoming paper [8]. The algorithm of this paper seems attractive in applications where it is useful to compile a set of ground equations into a canonical set of rules efficiently, but this remains to be explored.

8 References

[1]   Bachmair, L. "Proof Methods for Equational Theories", Ph.D thesis, University of Illinois, Urbana Champaign, Illinois (1987).

[2]   Bachmair, L., Dershowitz, N., and Plaisted, D., "Completion without Failure," Proceedings of CREAS, Lakeway, Texas (May 1987), also submitted for publication.

[3] Dauchet, M., Tison, S., Heuillard, T., and Lescanne, P., "Decidability of the Confluence of Ground Term Rewriting Systems," *LICS'87*, Ithaca, New York (1987) 353-359.

[4] Dershowitz, N,. "Termination of Rewriting," *Journal of Symbolic Computation* 3 (1987) 69-116.

[5] Dershowitz, N,. "Completion and its Applications," Proceedings of CREAS, Lakeway, Texas (May 1987).

[6] Downey, Peter J., Sethi, Ravi, and Tarjan, Endre R. "Variations on the Common Subexpressions Problem." *J.ACM* 27(4) (1980) 758-771.

[7] Gallier, J.H. *Logic for Computer Science: Foundations of Automatic Theorem Proving*, Harper and Row, New York (1986).

[8] Gallier, J.H., Raatz, S., and Snyder, W., "Theorem Proving using Rigid E-Unification: Equational Matings," *LICS'87*, Ithaca, New York (1987) 338-346.

[9] Gallier, J.H., Narendran, P., Plaisted, D., and Snyder, W., "Rigid E-Unification is NP-complete," *LICS'88*, Edinburgh, Scottland (July 1988)

[10] Huet, G., "Confluent Reductions: Abstract Properties and Applications to Term Rewriting Systems," *J.ACM* 27:4 (1980) 797-821.

[11] Huet, G. and Oppen, D. C., "Equations and Rewrite Rules: A Survey," in *Formal Languages: Perspectives and Open Problems*, R.V.Book, ed., Academic Press, New York (1982).

[12] Kapur, D., and Narendran, P., "A Finite Thue System With Decidable Word Problem and Without Equivalent Finite Canonical System," *Theoret. Comp. Sci.* 35 (1985) 337-344.

[13] Knuth, D.E. and Bendix, P.B., "Simple Word Problems in Univeral Algebras," in *Computational Problems in Abstract Algebra*, Leech, J., ed., Pergamon Press (1970).

[14] Kozen, Dexter. Complexity of Finitely Presented Algebras, Technical Report TR 76-294, Department of Computer Science, Cornell University, Ithaca, NY (1976).

[15] Kozen, Dexter. Complexity of Finitely Presented Algebras, *9th STOC Symposium*, Boulder Colorado, 164-177 (May 1977)

[16] Lankford, D.S., "Canonical Inference," Report ATP-32, University of Texas (1975),

[17] Nelson Greg, and Oppen, Derek C. Fast Decision Procedures Based on Congruence Closure. *J. ACM* 27(2) (1980) 356-364.

[18] Otto, F., and Squier, C., "The Word Problem for Finitely Presented Monoids and Finite Canonical Rewriting Systems," *RTA'87*, Bordeaux, France (1987) 74-82.

Towards Efficient "Knowledge-Based" Automated Theorem Proving for Non-Standard Logics

Michael A. McRobbie, Robert K. Meyer and *Paul B. Thistlewaite*

Automated Reasoning Project
Australian National University
G.P.O. Box 4
Canberra, A.C.T., 2601
Australia

UUCP: {uunet,hplabs,ubc-vision,ukc,mcvax,nttlab}!munnari!arp.anu.oz!xxx
[where xxx = mam (*McRobbie*), rkm (*Meyer*) or pbt (*Thistlewaite*)]

0. Abstract

In this paper we give an introduction to a technique for greatly increasing the efficiency of automated theorem provers for non-standard logics. This technique takes advantage of the fact that while most important non-standard logics do not have finite *characteristic* models (in the sense that truth tables are a finite characteristic model for classical propositional logic), they do have finite models. These models validate *all* the theorems of a given logic, though *some* non-theorems as well. They invalidate the rest of the non-theorems. Hence this technique involves using the models to direct a search by an automated theorem prover for a proof by filtering out or pruning the items of the search space which the models invalidate.

1. Introduction

There has recently been, if not an explosion, then certainly a minor detonation of interest among researchers in artificial intelligence (AI) and computer science in logics that have been so far mainly studied by logicians and philosophers, in some cases for millenia.

These are logics such as deontic, epistemic, intuitionistic, modal, paraconsistent, relevant and temporal logics all of which are usually collectively called non-standard or non-classical logics. Other logics which may also be included in this class but which were invented primarily with the needs of AI and computer science in mind include dynamic logic, linear logic and non-monotonic logic.

Here is not the place to survey all the research that has been done in this area (most of it very recent) as it now forms quite a substantial corpus. However anyone familiar with the appropriate parts of the literature (for example [14] and [19]), could not help but be aware of it. But for the reader unfamiliar with this territory, we suggest a feel for it might be gleaned from [40] and [42]. Further references may also be found in these books. Another useful reference is the recent special issue of **Logique et Analyse** devoted to this topic. (See [9].) For the reader unfamiliar with the wider territory of non-standard logics, the broad outlines of the landscape are comprehensively described in [10].

Given the nature of much of this research, many people have correctly pointed to the importance of developing efficient (automated) theorem proving systems for these logics. However so far very little work has in fact been done on implementing such systems. Where such systems have been implemented either directly or indirectly via a theorem prover for classical first-order logic, we believe it is fair to say that they have usually not been able to prove anything more difficult than elementary theorems of these logics. (The system KRIPKE described in detail in [40] may be an exception.)

In this paper we give an introductory description of a technique for greatly increasing the efficiency of automated theorem provers for these logics. In brief this technique takes advantage of the fact that while most important non-standard logics do not have finite *characteristic* models (in the sense that truth tables are a finite characteristic model for classical propositional logic), they do have finite models. These models validate *all* the theorems of a given logic, though *some* non-theorems as well. They invalidate the rest of the non-theorems. The earliest known example of such a logic is propositional intuitionistic logic, and that it had these properties was proved in [13]. In fact it is argued in [33] that the only really interesting logics are logics such as these.

Hence this technique involves using finite models of a given logic to direct or partially control a search for a proof by a theorem prover. This is done by restricting the search to a sub-space of the search space generated just by objects (appropriately translated) of the underlying proof theory of the theorem prover that are valid in the finite models specified. The rest of the search space generated by the objects of this proof theory that the models invalidate is never searched. Consequently we call this technique the *model pruning technique*.

For this reason we (somewhat cautiously) describe this technique as knowledge-based since the search for a proof is largely directed by the models which may be thought of as encoding sophisticated logical information about the structure of the logic. It is of course now part of the accepted wisdom of AI that if searches in symbolic computation are not to fall prey to combinatorial explosion, they must incorporate domain-specific knowledge in such a way so as to give direction to the search.

However two important facts should be noted about this technique. First, as will become clear, it can be used in theorem proving essentially only for *propositional* logics or for formal systems that can be interpreted as propositional logics. (We leave for another day the question of how this technique could be integrated together with a theorem prover for first-order logic.) Second, it cannot be used in resolution-based theorem proving in a straightforward way.

As regards the first of these facts, we simply believe that in implementing theorem provers for non-standard logics, it is important to get their propositional fragments right first since these fragments are in general more complex than classical propositional logic, and in some cases even undecidable. As to the second fact, we shall return to it in Section 6. Suffice it to say that we regard neither of these points as in any way detracting from the importance of this technique.

We conclude this Section by noting that though we arrived at the idea independently, the model pruning technique is broadly speaking, not a particularly new idea. In fact it is claimed in [36] that the general idea goes back to the very birth of AI as it was proposed by Minsky at the 1956 Dartmouth Conference.

In the AI literature the use of a related technique in theorem proving for geometry was advocated and described in [12]. The viability of using a technique like this in the context of first order theories was noted in [31] and [36], where typically relational models are used. However in all these cases, it has been claimed that the computational overheads involved in implementing these techniques are *prohibitively* high. Whether or not this is true, it is certainly not true in the case of implementations of the model pruning technique for theorem provers for non-standard logics. The computational overheads of this technique are high but not prohibitively so. (We discuss this question in Section 4.) Without it efficient automated theorem proving for many non-standard logics may well be very difficult.

In the logical literature, an embryonic form of our technique may be extracted from [30], 264-269. A rudimentary form of it is also pointed to in [7] where, in the context of a discussion of the Gentzen decision procedure for the intuitionistic propositional logic, Dummett suggests that in actually using this decision procedure to construct a proof search tree, the construction could be considerably simplified by using truth tables to test the application of rules for validity. (See [20] for more on this topic.)

2. Models

Our presentation of the formal theory of models is based essentially on that of [15], though most of our terminology comes from [7].

In what follows we assume a background (propositional) language L that is relativised to the logic or logics under consideration. Where necessary, we take a logic to be specified by its Hilbert-style (i.e. axiomatic) formulation, which in turn fixes the background language. The background *language* L is a structure $<E,C,F>$ where,

(i) E is the set of *elementary objects* of L, e.g. *propositional variables, propositional constants*;

(ii) C is a finite set $\{c_1,...,c_n\}$ the members of which are the *connectives* of L, such that for each $c_i \in C$, c_i is of fixed degree;

(iii) F is the set of *formulas* of L and is the smallest set such that $E \subseteq F$ and which is closed under all the members of C. We let $A,B,E,A_1, ...$ be schematic variables ranging over these formulas.

A *model* $\mathcal{M}$ for L is a structure $<V,D,O>$ where,

(i) V is a finite non-null set of *values*, of non-trivial cardinality, i.e. greater than 1, and whose members are usually represented by the integers 0,1,2,... ;

(ii) D is a non-null set of *designated values* such that $D \subseteq V$ - conversely the members of the set V-D are known as *undesignated values*;

(iii) O is a set of *operators* $\{o_1,...,o_n\}$ correlated one-one with members of C in L such that if o_j is correlated with c_i and c_i is of degree d, then $o_j : V^d \to V$.

An *assignment* for L relative to some fixed $\mathcal{M}$ is a function $f:E \to V$. This function can then be extended to a homomorphism, called a *valuation*, that maps all members of F into V. A formula A *holds* in $\mathcal{M}$ with respect to an assignment if that assignment maps A to a member of D. If A holds for all such assignments, A is said to be *valid* in $\mathcal{M}$. If A does not hold for at least one such assignment, A is said to be *invalid* in $\mathcal{M}$ and such an assignment is called a *refutation*.

Given a logic **X**, $\mathcal{M}$ is said to be a *model* for **X** iff all the axioms of **X** are valid in $\mathcal{M}$ and if the rules of **X** preserve this property, i.e. if all the *theorems* of **X** are valid in $\mathcal{M}$. If *exactly* the theorems of **X** are valid in $\mathcal{M}$, then $\mathcal{M}$ is said to be a *characteristic model* for **X**.

If this chain of definitions is not clear, the following Venn diagram may be useful for illustrating the crucial case of interest to us in this paper where $\mathcal{M}$ is a model for **X**. Note that when B-C={ } in this diagram, this is the case where $\mathcal{M}$ is characteristic for **X**.

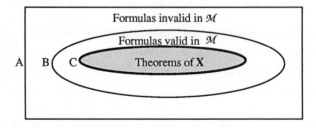

Figure 1.

All the preceding represents the usual set-theoretic mystification that is *de rigueur* in a technical paper. However it obscures the fact that models are very simple structures. Let us consider a well known example. If we let our logic be the classical propositional calculus **K** and let the set C of L be $\{\sim,\to,\&,\vee\}$, then an example of a model for **K** which we shall call 2 is just the following structure:

$<\{0,1\},\{1\},\{o_1,o_2,o_3,o_4\}>$

where we take the operators o_1 to o_4 to be defined just as follows:

$o_1(1)=0$ and $o_1(0)=1$
$o_2(1,1)=o_2(0,1)=o_2(0,0)=1$ otherwise $o_2(1,0)=0$
$o_3(1,1)=1$ otherwise $o_3(1,0)=o_3(0,1)=o_3(0,0)=0$
$o_4(1,1)=o_4(1,0)=o_4(0,1)=1$ otherwise $o_4(0,0)=0$

Of course to show that _2_ is a model for **K**, it is necessary to show that all the theorems of **K** are valid in it. However we leave this task to the reader should he or she actually find it necessary to do so. Naturally the reader will have recognized _2_ as the most famous model of all - truth tables, and as everyone knows _2_ is not only a model for **K**, but also a characteristic one.

In what follows we shall adopt some standard methods for displaying models. First, to designate the operations in a model we shall use the same symbols as we use to designate the set of connectives correlated with these operations. In practise this never presents any problems, as context always resolves any ambiguities. Second, as an abbreviatory device, we shall represent operations by tables (or _matrices_ as they are better known) as is normal, and the particular way we choose to do this follows [1]. Third, where possible we further abbreviate these tables by collapsing them into simple algebraic structures such as lattices, which we represent by Hasse diagrams. Fourth, designated values are distinguished by starring them.

Therefore we can now represent _2_ as follows:

~	
0	1
*1	0

→	0	1
0	1	1
*1	0	1

*1 and 0 are joined in a two element lattice (Hasse diagram).

where the tables for the operations & and ∨ can be reconstructed from the _meet_ and _join_ operations on this two element lattice.

However in the world of non-standard logic _2_ is almost totally useless, as its powers of logical discrimination are severely limited (if not crippled). To take just one topical example, it can do very little to help us separate out some of the theorems of **K** that are not theorems of the intuitionistic propositional logic **J**. A model that does do this is the following model, which we shall call G, and which essentially comes from [13]. (Note that G contains _2_ as a _submodel_, where submodel is defined in the obvious way.)

$$\begin{array}{c|c}\sim & \\\hline 0 & 2\\ 1 & 0\\ *2 & 0\end{array}\qquad\begin{array}{c|ccc}\rightarrow & 0\ 1\ 2\\\hline 0 & 2\ 2\ 2\\ 1 & 0\ 2\ 2\\ *2 & 0\ 1\ 2\end{array}$$

This model is in fact the smallest model for **J** that is not a model for **K**, since such well known non-theorems of **J** as the Law of the Excluded Middle, A∨~A, and Peirce's Law, A→B→A→A are invalid in it. (A fuller discussion of this model, the class of models it belongs to and the role it plays in an efficient implementation of Gentzen's decision procedure for **J** is to be found in [20].)

A more complex example of a model is the following, called $\mathcal{M}_0$ (Note again that it contains 2 as a submodel.)

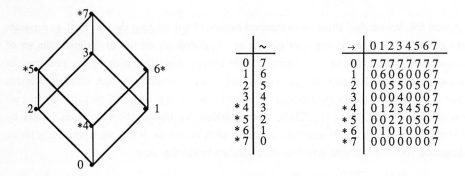

$$\begin{array}{c|c}\sim & \\\hline 0 & 7\\ 1 & 6\\ 2 & 5\\ 3 & 4\\ *4 & 3\\ *5 & 2\\ *6 & 1\\ *7 & 0\end{array}\qquad\begin{array}{c|cccccccc}\rightarrow & 0\ 1\ 2\ 3\ 4\ 5\ 6\ 7\\\hline 0 & 7\ 7\ 7\ 7\ 7\ 7\ 7\ 7\\ 1 & 0\ 6\ 0\ 6\ 0\ 0\ 6\ 7\\ 2 & 0\ 0\ 5\ 5\ 0\ 5\ 0\ 7\\ 3 & 0\ 0\ 0\ 4\ 0\ 0\ 0\ 7\\ *4 & 0\ 1\ 2\ 3\ 4\ 5\ 6\ 7\\ *5 & 0\ 0\ 2\ 2\ 0\ 5\ 0\ 7\\ *6 & 0\ 1\ 0\ 1\ 0\ 0\ 6\ 7\\ *7 & 0\ 0\ 0\ 0\ 0\ 0\ 0\ 7\end{array}$$

Again the tables for the operations & and ∨ can be reconstructed from the *meet* and *join* operations on this eight element lattice. This model has the interesting property that if an implicational formula A→B is valid in it, A→B share a variable, i.e. there is a propositional variable that occurs at least once in both A and B. So it can separate out implications whose truth rests purely on the truth or falsity of their antecedents or consequents, and implications, a necessary condition for whose truth is that their antecedents and consequents must "share content", in this case through the medium of a shared propositional variable. For more on this model and the important role that it has played in investigations into the concept of implication see [1].

Of course many readers will be aware that the study of models such as these is an extremely large area of research and forms part of the study of *many-valued logics*. Here models are also known as *matrix models*, *many-valued models*, and so on. To the reader unfamiliar with this subject we highly recommend [32] and the excellent bibliography contained therein which is essentially complete until the end of 1965. This bibliography, supplemented by [43] for the period 1966 to 1974, as well as the recent survey article by Urquhart in [10], will give the reader an idea of the enormity of this subject.

3. **Proof Theory and the Model Pruning Technique**

We begin this section by discussing in the abstract an important property that a proof method for a logic must have, in order for the model pruning technique to be applicable to it.

In general there are many formal methods possible for proving the theoremhood or validity of statements in a logic. Of course underlying each of these methods is always a formal system that specifies exactly how the method is to work, as well as a theorem which establishes that the method is sound and complete with respect to this logic, i.e. a theorem which establishes that using this method, all and only the theorems of the logic can be proved, or to put it another way, that the theorems of the formal system are exactly those of the logic. Such a formal system is usually called a *formulation* of the logic in question.

In the case of classical logic there are many such formulations known and a number of these are listed in Figure 2, roughly in the order in which they were discovered. For each of these, a reference is listed where further details concerning them may be found. However in the case of other logics it is important to note that with the obvious exception of 1, it is not always known how to formulate them in such a wide variety of ways.

1.	*Hilbert axiomatic formulations*	Just about any good logic textbook.
2.	*Gentzen/Jaskowski natural deduction formulations*	Prawitz [29]
3.	*Gentzen sequent formulations*	Kleene [16]
4.	*Beth semantic tableau formulations*	Beth [2]
5.	*Schütte disjunctive formulations*	Schütte [35]
6.	*Smullyan analytic tableau formulations*	Smullyan [38]
7.	*Robinson resolution formulations*	Gallier [11]
8.	*Kowalski connection graph formulations*	Kowalski [17] (though see Eisinger [8])
9.	*McRobbie/Belnap proof tableau formulations*	McRobbie and Belnap [22]
10.	*Andrews/Bibel connection/mating formulations*	Bibel [3]

Figure 2.

All of these formulations can be automated and most of them have. (See [4].) All of these formulations can be refined in various ways and most of them have - in some cases almost *ad nauseum*. (For a very good survey of such refinements in the case of 7 see [18].)

Of course the question may be quite rightly raised as to exactly how distinct some of these formulations are. For example, 6 can be seen as a more elegant version of 4, 5 as an epicycle on 3, 8 and 10 as epicycles on 7, and 9 as a proof theoretic refinement of 6 (and in the case of classical logic, a trivial one). We certainly would not want to disagree too strongly with these views, but merely note in passing that from a suitably lofty and abstract viewpoint, nearly all of these methods are the same. However operationally speaking, the difference between them remains considerable.

But let us now consider these formulations more abstractly. They can be very broadly divided into two classes - *compositional* and *decompositional*. Formulations 1, 2, 7, 8 and 10 fall into the former class while formulations 3, 4, 5, 6 and 9 could be said to fall into the latter.

In very rough terms, a compositional formulation of a logic is one in which the theoremhood of a formula is determined by starting with axioms (*or their equivalents*) which are by definition theorems and the choice of which is determined by the logical properties of this formula, and then applying the rules of the formulation to these axioms to produce new theorems, and so on recursively until a proof of the formula is arrived at. (Of course there is in general no guarantee that this process will always lead to such a proof in a finite amount of time.) In a compositional formulation of a logic, the objects generated in a search for a proof of a theorem are themselves *always* theorems.

Conversely, a decompositional formulation of a logic is one in which the theoremhood of a formula is determined by recursively decomposing the formula (possibly normal-formed in some way) according to the rules of the formulation until axioms (*or their equivalents*) are reached. (Again there is in general no guarantee that this process will always lead to such a proof in a finite amount of time.) Even if this formula is a theorem, objects may be generated in the process of decomposing it that are not! *And it is here where the crucial difference between the two classes lies.*

We do not pretend that this is a precise distinction nor that our categorization with respect to it is uncontentious. Its purpose is simply to draw attention to the fact that the way in which theoremhood is determined in the formulations in these two classes differs fundamentally. It will also become clear in what follows that the model pruning technique can be used with decompositional formulations of logics but *not* in general with compositional formulations.

Given this discussion, our classification of 3 as a decompositional formulation may raise some eyebrows, since such formulations are normally thought of as compositional. However in practice they are almost always used in a "decompositional mode" (so to speak) and in fact can easily be formulated in just this way, which is exactly how we want to think of them.

We can best illustrate the difference in formulations by considering a Gentzen sequent formulation (or just sequent formulation) of a logic. It is not necessary to specify the details of this logic, nor of its sequent formulation. For our purposes it will suffice if the set of theorems of the logic are contained in those of classical logic, though nothing in the following discussion rests on this.

In such formulations objects of the form $\alpha \Rightarrow \beta$, called *sequents*, are studied. Here α and β are finite collections of formulas - in this case we will let them be multisets, though they could also be sets, sequences or some other kind of collection of formulas. (In what follows we will let small Greek letters range over such collections.) We will simplify matters by assuming that β contains one and only one member, though the membership of α is unrestricted and can be null. Hence the sequents actually studied in this formulation are of the form $\alpha \Rightarrow B$. Where α contains the (not necessarily distinct) formulas $A_1,...,A_n$, we interpret this sequent as the formula $A_1 \rightarrow. \: ... \: \rightarrow.A_n \rightarrow B$. In displaying formulas we adopt the bracketing conventions of [1] in order to reduce bracketing to a minimum.

We further assume that our sequent formulation contains the following two rules common to the sequent formulation of many logics, including for instance the pure implicational fragments of **J** and the relevant logic **R** which we call **J$_\rightarrow$** and **R$_\rightarrow$**, while we call their sequent formulations **LJ$_\rightarrow$** and **LR$_\rightarrow$**. (For further details about these logics and the nomenclature system we use to identify them, see [1], Part I, §1 to §3 and *et passim*.)

Implication on the Left (I$\Rightarrow$)   *Implication on the Right* ($\Rightarrow$I)

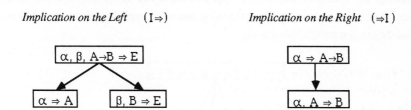

In representing these rules and instances of their applications, we reverse the usual order in which they are stated to reflect the fact that, as we have already noted, proofs in sequent systems begin with theorems and terminate with axioms which we shall take to be all instances of the sequent $A \Rightarrow A$. We call a sequent to which a rule is applied the upper sequent and the sequent(s) that result from this rule application the lower sequent(s). Following [6], we call the formula $A \rightarrow B$ the principal constituent of these rules and the multisets α and β their parametric constituents.

Now let us consider informally what is involved in finding a proof of the following formula in this formulation.

(1) $(A \rightarrow.B \rightarrow B) \rightarrow.(E \rightarrow E \rightarrow B) \rightarrow.(B \rightarrow.A \rightarrow B \rightarrow A) \rightarrow.B \rightarrow A$

In order to prove this, we must first be able to prove the sequent,

(2) $\Rightarrow (A \rightarrow.B \rightarrow B) \rightarrow.(E \rightarrow E \rightarrow B) \rightarrow.(B \rightarrow.A \rightarrow B \rightarrow A) \rightarrow.B \rightarrow A$

which in turn means constructing a tree of sequents using only the rules of this formulation which commences with (2) and all of whose branches terminate in axioms. Four applications of the rule ⇒I generate the following sequent:

(3) A→.B→B, E→E→B, B→.A→B→A, B ⇒ A

But now we are faced with a large number of ways in which to apply I⇒ to this sequent. There are three implicational formulas which can be the principal constituents of an application of this rule, and for each such formula it is easy to show that there are eight possible ways in which this rule can be applied corresponding to all the ways in which the parametric constituents can be distributed among the lower sequents - a total of 24 in all.

Recall that the point of this proof theoretic interlude is to see how in a decompositional formulation of a logic like the one that we are looking at, objects (in this case sequents) can be generated in the search for a proof that are not (under translation) themselves theorems of the logic. So now we look at a number of applications of I⇒ to (3) with B→.A→B→A as the principal constituent. In Figure 3 we represent this situation using an AND/OR tree (for which see [28]), where the AND-links between AND-linked nodes (which correspond to these rule applications) are indicated by thick lines and they in turn are OR-linked to (3).

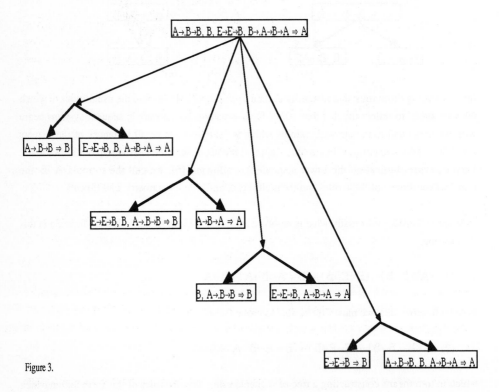

Figure 3.

Now let us consider the lower sequents generated by these rule applications taken from left to right in Figure 3. In the first case the sequent $A \rightarrow . B \rightarrow B \Rightarrow B$ corresponds on our interpretation to the formula $(A \rightarrow . B \rightarrow B) \rightarrow A$ which can easily be seen not to be a theorem of **K** (for example by assigning A and B the values 0 and 1 respectively, which makes it invalid in 2). So given that we have assumed all the theorems of our logic are theorems of **K**, in searching for a proof of (2) we do not need to decompose either of these sequents any further, since this cannot lead to a proof of (3) and hence of (2). However as the reader can easily verify, both the formulas answering to the lower sequents of the second rule application are valid in 2.

And the informal procedure we have just outlined is the essence of the model pruning technique. As we have just seen, in a decomposition formulation of a logic, objects can be generated by the application of rules which are not theorems of the underlying logic. Consequently in searching for a proof of a formula, an appropriately chosen model (or class of models) is used to stop the application of rules which generate such objects by testing these objects, appropriately translated if necessary, for validity in this model (or class of models). In what follows it will simplify matters a little if we simply identify a sequent with the formula to which it corresponds on our interpretation. This then allows us to talk about a sequent's being valid or invalid in a model $\mathcal{M}$. We call a rule application that leads to the generation of at least one sequent that is invalid in $\mathcal{M}$ an $\mathcal{M}$-*invalid* rule application or, in general, a *bad rule application*. Otherwise we say it is $\mathcal{M}$-*valid*.

It is crucial to note here that this procedure will not stop all bad rule applications from being made in the search for a proof unless the model being used is characteristic for the underlying logic. Unless it is, it should be clear from our discussion in Section 2 that there will always be some cases where a sequent generated by a rule is valid in the model but is not a theorem of the underlying logic. However, as we shall see, judiciously chosen models can drastically reduce the number of such bad applications.

Now let us look at a more sophisticated case. Let us assume that the underlying logic is $J_\rightarrow$. How do we proceed in searching for a proof of (2) in $LJ_\rightarrow$? Clearly if the first rule application in Figure 3 is 2-invalid, it is also $\mathcal{G}$-invalid since 2 is a sub-model of $\mathcal{G}$. What about the second rule application? Here the sequent $A \rightarrow B \rightarrow A \Rightarrow A$ corresponds to the formula $A \rightarrow B \rightarrow A \rightarrow A$ which we noted in Section 2 is known as *Peirce's law* and which though a theorem of **K** is famously not one of $J_\rightarrow$. This fact can be verified by assigning A and B the values 1 and 0 respectively which makes this formula invalid in $\mathcal{G}$ and hence which makes this rule application $\mathcal{G}$-invalid.

However both sequents generated by the third rule application in Figure 3 are valid in $\mathcal{G}$ as the reader can again easily if not quickly verify. Thus applications of $I \Rightarrow$ and $\Rightarrow I$ together with the further $LJ_\rightarrow$ rule of *weakening* (for a statement of which see e.g. [1], 51-57) lead to a $LJ_\rightarrow$ proof of (2).

So now let us look at an even more sophisticated example. Let us assume that the underlying logic is $R_\rightarrow$. Again given that the first two rule applications are $\mathcal{G}$-invalid, they will also be $\mathcal{M}_\mathcal{G}$-invalid since it is easy to show that with respect to the *positive* connectives of **J** and **R**, $\mathcal{G}$ is a sub-model of $\mathcal{M}_0$. But what about the third rule application? Here the formula $B\to.(A\to.B\to B)\to B$ which corresponds to the sequent B, $A\to.B\to B \Rightarrow B$, is proved axiomatically in $\mathbf{J}_\to$ using the axiom $A\to.B\to A$ and the sequent in turn is proved using the weakening rule in $\mathbf{LJ}_\to$. However as this axiom and this rule are paradigmatically not an axiom or rule of $\mathbf{R}_\to$ or $\mathbf{LR}_\to$, it will come as no surprise to hear that $B\to.(A\to.B\to B)\to B$ is invalid in $\mathcal{M}_0$, e.g. assign A and B the values 0 and 4 respectively. Hence the third rule application is $\mathcal{M}_0$-invalid.

However it can be easily though labouriously shown that the fourth and final rule application is $\mathcal{M}_0$-valid, and it turns out that further applications of just the rules $I\Rightarrow$ and $\Rightarrow I$ to the lower sequents of this rule application will lead to an $\mathbf{LR}_\to$ proof of (2).

The efficacy of these models in directing a search for a proof in the various sequent formulations we have discussed by providing a means for identifying bad rule applications in such searches, can be seen from Figure 4.

Total Number of Possible Applications of $I\Rightarrow$ to (3)	$\mathcal{Z}$-Valid Applications	$\mathcal{G}$-Valid Applications	$\mathcal{M}_0$-Valid Applications
24	9	5	3

Figure 4.

We note here that the reader interested in checking our claims about the validity of the rule applications in the various models we have displayed, especially $\mathcal{M}_0$ will find this a very lengthy, tedious, and error-prone exercise. However the process of testing a sequent for validity in a model, *model testing* for short, is easily automated. Hence the reader interested in exploring the model testing technique further, should either write their own program to do this or simply obtain a copy of the beautiful interactive model testing program TESTER due to Belnap and Isner. (See [1], 86-87.) A UNIX version of this program will soon be available from the Automated Reasoning Project at the Australian National University.

We now summarise a little more formally the model pruning technique in order to stress its general applicability. In order to use it as component of a theorem prover for some logic **X**, a decompositional formulation of this logic is necessary. A theorem prover for **X** based on this formulation takes as input a formula A (possibly normal-formed in some way) and recursively constructs a proof search tree with A as its origin by applying the rules of this formulation in all possible $\mathcal{M}$-valid ways, where $\mathcal{M}$ is a model for **X**. Again more than one model of **X** can be employed here. If a proof of A exists, the proof search tree will contain it as a *subtree* (where subtree is defined in the obvious way). However there is of course no general guarantee that such a proof search tree will be finite, and hence no general guarantee that a proof of A can be found in a finite amount of time. Further in constructing a proof search tree, no one search strategy is uniquely motivated. Rather specific properties of a logic or implementational issues might dictate a depth or breadth first strategy, or more likely a bidirectional strategy (in the sense of [28]).

We conclude this Section by noting that other kinds of models, e.g. algebraic models and Kripke structures, can also be utilized in this technique. However it is not clear that any computational advantages will flow from doing so as the various kinds of models for a logic are usually starightforwardly equivalent.

4. The Model Pruning Technique and Implementation Issues

During a conversation in which the first author was discussing the model pruning technique with Larry Wos, Wos recalled that some 20 years ago a group theorist proposed a technique similar to the model pruning technique for reducing the search space in theorem proving in group theory. Wos said that in reviewing this work he commented that the computational overheads involved would be prohibitively expensive. He queried whether this might not also be so for the model pruning technique.

In practice this technique certainly turns out to be computationally expensive but it is definitely not prohibitively so. We make this claim on the basis of our experiences with the implementation of a number of theorem provers for a variety of non-standard logics such as intuitionistic logics, modal logics and especially relevant logics which tend to be considerably harder than the previous two kinds.

What we have found is that theorem provers for these logics are incapable of proving anything other than the simplest theorems unless they also incorporate the model pruning technique in some form. However if this is done, such theorem provers are capable of proving extremely difficult theorems of these logics within reasonable time constraints. In fact we tentatively venture the claim that the theorem proving system KRIPKE mentioned in Section 1 which makes extensive use of the model pruning technique, has solved the hardest problems yet solved by a theorem prover for non-standard logics. We present these problems as a challenge to the automated theorem proving community in [24]. Our reseach in this area commenced in the late 1970s and much, though certainly not all of it, is announced or described in [20], [23], [24], [25], [26], [39], [41] and culminating in [40].

We have noted then, that though the model pruning technique can contribute much to the efficiency of theorem provers in non-standard logics, it is computationally expensive. In particular its implementation means including a procedure in the theorem proving program that tests every rule application for validity in a model or class of models. Our experience shows that depending on the average number of distinct variables in a sequent generated by such a rule application and depending on the size of the model(s), theorem provers using the model pruning technique can spend between 50 and 90% of their time just engaged in model testing.

Obviously the more time a theorem prover actually spends looking for a proof the better. Thus it is very important to reduce if possible the amount of time a theorem prover using the model pruning technique spends engaged in model testing, without of course compromising (too much) the logical basis of the technique itself. In what follows we note four ways in which this can be done (all of which have in some form been utilized in KRIPKE), and a further three ways in which it could be done (the first two being presently under investigation).

• *Improvements to the Testing Algorithm.* The simplest algorithm for model testing is one that makes all possible assignments of values to variables in the sequents and then evaluates them according to the model. However improvements can be made to this algorithm that increase the speed of model testing by eliminating some redundant testing.

• *Reprogramming the Model Testing Procedures.* The crucial procedures in the theorem prover that actually carry out the model testing can be written in a faster lower level language, e.g. ASSEMBLER.

• *Refuting Strength of Models.* In choosing models, clearly the ones to choose are those that are best at invalidating non-theorems of the logic under consideration, a property of the model we call its *refuting strength*. The more bad rule applications a model can identify the better, since more of the proof search tree can then be constructed in a finite amount of time hence increasing the chances of finding a proof of a theorem. But how do we measure refuting strength? This is a hard and complex problem that we wrestle with at length in Chapter 3 of [40] without making much progress. However some idea of the refuting strength of models can always be gained empirically, e.g. by running a theorem prover a number of times on a set of test examples, each time changing the model(s), and then comparing the run times.

• *Dynamic Change of Models.* Even if a precise measure of the refuting strength of a model is found, it is unlikely that it will be an absolute measure. Hence some models might be better at invalidating certain kinds of formulas than others, but not universely so. Thus the model pruning technique can be implemented in such a way that the theorem prover monitors the kinds of sequents being generated by rule applications and changes the models being used depending on whether or not they have certain logical properties and on other factors. (It should also be mentioned that the monitoring of the effect of certain models has in turn led to refinements and simplifications of the rules themselves.)

• *Statistical and Neural Net Based Model Testing.* Statistical techniques could be used to identify a relatively small subset of value assignments that have the greatest possibility of invalidating sequents. All model testing would then be done only with respect to this set. There may also be benefits in identifying this subset using neural net techniques. For instance a neural net implemented on a neurocomputer could monitor the model testing being carried out by a theorem prover and progressively "learn" what the assignments were most likely to lead to the invalidation of sequents. If necessary the testing of all assignments of values could still be carried out but in an order determined by the neural net that aims to maximize the possibility of finding a refutation as soon as possible.

• *Supercomputer and Multiprocessor Based Model Testing.* It is possible to vectorize the model testing procedures and execute them separately on a vector processing supercomputer. This would increase the speed of a theorem prover by as much as an order of magnitude. Model testing is clearly also capable of parallelization and consequent execution on a multiprocessor. Again the potential speedups here are very large.

• *Model Testing Chip.* Should theorem provers for non-standard logics find widespread application and should the model pruning technique become an integral part of these, this might justify the building of special purpose model testing co-processor chips. (We note here that one of the referees of this paper suggested that this point is superfluous since most theorem provers have "rigid" algorithms that could be handled this way. This is certainly true. However our point is rather that the regular and extremely elementary nature of the computation involved in model testing suggests that such chips would be simple and inexpensive to design and build. This is also the view of a number of electrical engineers we have consulted.)

5. How to Generate Models

Obviously in order to use the model pruning technique one must be able to efficiently generate models for logics. A detailed treatment of this problem takes us well beyond the topic of this paper but a few words concerning it seem in order.

Computationally all the models for a logic up to a certain size can be generated by the "simple" method of enumerating *all* the models up to this size, and as each one is enumerated, testing to see whether or not it is a model for the logic in question. This is known as the *Test and Change Method.* But as the reader would no doubt expect, this method is unfeasible for all except the smallest models. For example if we wanted to find all models for **R** with up to 10 values, this would mean using this naive method, enumerating n models where,

$$n = 2^4 + 3^9 + ... + 9^{81} + 10^{100}$$

However fundamental research on this problem by Brady, Belnap, Meyer, Pritchard and Slaney, and more recently Malkin and Martin, has lead to algorithms that can generate classes of models such as these in a quite reasonable amount of time. In essesse these algorithms use properties of a logic to reduce the number of models that must be enumerated using the naive Test and Change Method. Interestingly this is something like the dual of the model pruning technique.

The *locus classicus* for research in this area is still [37], which surveys much of this work. A number of new programs for efficiently generating classes of models such as these have been developed recently at the Automated Reasoning Project at the Australian National University and are discussed in [27]. Research is also being carried out with a view to finding efficient methods for storing, managing and retrieving classes of models for various logics once they have been generated, using modern database management techniques, with a view to making data such as this generally available to those interested.

6. The Model Pruning Technique and Resolution Theorem Proving

We will conclude this paper by briefly discussing why there appears to be no straightforward way of using the model pruning technique in resolution theorem proving *in its most simple and elementary form*. However we expect it can be used with some of the more sophisticated refinements of resolution such as those discussed in [5], Chapter 6. But this is a topic for another paper which we hope one day to write.

This discussion will also serve to clarify our possibly controversial classification of formulations of type 7 in Figure 2 as compositional formulations. We will assume the reader's familiarity with this style of theorem proving.

The following set of clauses is generated by (1):

(4) $\{\sim A, B, A \vee \sim B, B \vee E, B \vee \sim E, \sim A \vee B \vee \sim B\}$

and the following is a resolution proof (though certainly not the shortest) which shows that (1) is a theorem of **K**.

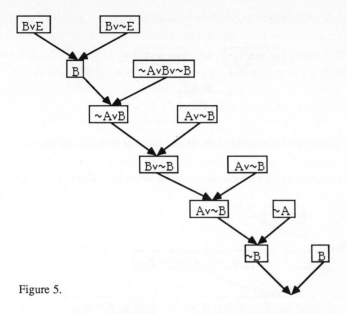

Figure 5.

Now all resolution proofs can be translated into proofs in a sequent formulation of **K**, specifically into what is called a left-handed sequent formulation of **K** where objects studied are all of the form $\alpha \Rightarrow$. We call this formulation $L_l\mathbf{K}$. (For such formulations see, e.g. [22].) If $\alpha \Rightarrow$ is a theorem of $L_l\mathbf{K}$ and α contains the (not necessarily distinct) formulas $A_1,...,A_n$, then we interpret this sequent as saying that $\sim(A_1 \& ... \& A_n)$ is a theorem of **K**.

The essence of this translation is as follows.

First negate all clauses in the resolution proof and then replace each one by the set of normal-formed clauses that are equivalent to this negated formula. Then to each node between and including a tip in this proof and its root, add a copy of the unnegated clause that originally occured at this tip. It is now not hard to show by a simple inductive argument that this translation converts all resolution proofs into $L_l\mathbf{K}$ proofs. In particular all the tips in such proofs are of the general form,

(5) $A_1{}^*, ... , A_n{}^*, A_1 v...v A_n \Rightarrow$

(where each A_i is a literal and each corresponding $A_i{}^*$ is its negation, possibly normal formed by the removal of double negations) which is trivially provable in $L_l\mathbf{K}$, and the only other rule used is *cut*, which we take in the following form (where for the purposes of this Section we take $L_l\mathbf{K}$ in its compositional guise):

214

Cut

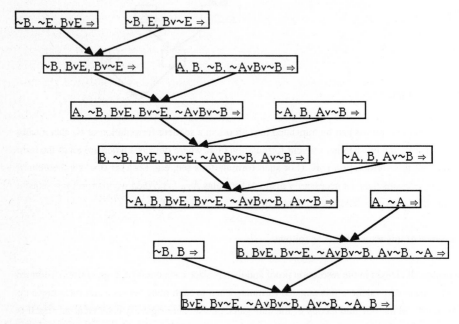

In fact we can straightforwardly formulate L$_I$K with (5) taken as an additional axiom.

Hence the resolution proof in Figure 5 translates into the following L$_I$K proof .

Figure 6

Hopefully it should now be clear that the clauses generated in a resolution proof *always* translate as provable sequents in a corresponding sequent formulation and thus, that the model pruning technique can never be used in the version we have described in this paper since all such sequents (appropriately interpreted as indicated previously) will *always* be valid in *any* model for the underlying logic. (For an extended treatment of the topic of this section see [21]. For a similar analysis of resolution proofs from the point of view of sequent formulations, see [11].

7. Acknowledgements

The people we wish to thank for helpful conversations on the subject matter of this paper, for comments on earlier drafts of it or for directing us to parts of the literature of which we were not aware, reads like a Who´s Who of automated theorem proving and computational logic: Bundy, Eisinger, Gabbay, Lusk, Ohlbach, Overbeek, Rasiowa, Alan Robinson, Siekmann, Slaney, Stanton, Wallen and Wos. Earlier versions of this paper were presented as parts of larger papers to the Austrian Society for Cybernetic Studies, the Universities of Edinburgh, Kaiserslautern, Karlsruhe, and Leige and at Imperial College and the Sorbonne. It benefitted from discussions at all these places. This paper was written while the first author held a Visiting Professorship in the Department of Computer Science, University of Kaiserslautern, West Germany, funded under Sonderforschungsbereich 314. He is extremely grateful to Professor Jörg Siekmann for having made this possible.

8. References

[1] Anderson, A.R. and Belnap, N.D. Jr. **Entailment: The Logic of Relevance and Necessity**, Vol. 1, Princeton U.P., Princeton, New Jersey, 1975.

[2] Beth, E.W. **Formal Methods: An Introduction to Symbolic Logic and to the Study of Effective Operations in Arithmetic and Logic**, Reidel, Dordrecht, 1962.

[3] Bibel, W. "Mating in Matrices", **Communications of the ACM**, 26 (1983), 844-852.

[4] Bledsoe, W.W. "Non-Resolution Theorem Proving", **Artificial Intelligence**, 9 (1977), 1-35.

[5] Bundy, A. **The Computer Modelling of Mathematical Reasoning**, Academic Press, London, 1983.

[6] Curry, H.B. **Foundations of Mathematical Logic**, McGraw-Hill, New York, 1963. (Corrected reprint, Dover, 1977.)

[7] Dummett, M. **Elements of Intuitionism**, Oxford U.P., Oxford, 1977.

[8] Eisinger, N. "What You Always Wanted to Know About Clause Graph Resolution", 316-336, in **Proceedings of the 8th International Conference on Automated Deduction**, ed. J. Siekmann, **LNCS**, Vol. 230, Springer, Berlin, 1986.

[9] Farinas del Cerro, L. and Orlowska, M. (eds.) **Automated Reasoning in Non-Classical Logic**, Special double issue of **Logique et Analyse**, 28 (1985), 115-294.

[10] Gabbay, D.M. and Guenthner, F. (eds.) **Handbook of Philosophical Logic**, Vol. 1, Reidel, Dordrecht, 1983. (Vol.2 was published in 1984 and Vol. 3 in 1985.)

[11] Gallier, J.H. **Logic for Computer Science: Foundations of Automatic Theorem Proving**, Harper and Row, New York, 1986.

[12] Gelernter, H. Hanson, J.R. and Loveland, D.W. "Empirical Explorations of the Geometry Theorem-Proving Machine", 153-163, **Computers and Thought**, eds. E. Feigenbaum and J. Feldman, McGraw-Hill, New York, 1963.

[13] Gödel, K. "Zum intuitionistischen Aussagenkalkül", **Anzeiger der Akademie der Wissenschaften Wien, mathematisch naturwissenschaftliche Klasse**, 69 (1932), 65-66.

[14] Halpern, J.Y. (ed.) **Theoretical Aspects of Reasoning About Knowledge: Proceedings of the 1986 Conference**, Morgan Kaufmann, Los Altos, California, 1986.

[15] Harrop, R. "Some Structure Results for Propositional Calculus", **Journal of Symbolic Logic**, 30 (1965), 271-292.

[16] Kleene, S. C. **Introduction to Metamathematics**, North-Holland, Amsterdam, 1952.

[17] Kowalski, R.E. **Logic for Problem Solving**, North-Holland, Amsterdam, 1979.

[18] Loveland, D.W. **Automated Theorem Proving: A Logical Basis**, North-Holland, Amsterdam 1978.

[19] McDermott, J. (ed.) **IJCAI 87: Proceedings of the Tenth International Joint Conference on Artificial Intelligence**, Vols. 1 and 2, Morgan Kaufmann, Los Altos, 1987.

[20] McGovern, G. and McRobbie, M.A."On an Efficient Computer Implementation of A Decision Procedure for Intuitionistic Logic Using Matrix Models", Manuscript, 1987.

[21] McRobbie, M.A. "What is Relevant Resolution? A Simple Tutorial", Manuscript, 1985. (Presented at the **1985 Conference of the Australasian Association for Logic**.)

[22] McRobbie, M.A. and Belnap, N.D. Jr. "Relevant Analytic Tableaux", **Studia Logica**, 38 (1979), 187-200.

[23] McRobbie, M.A., Meyer, R.K. and Thistlewaite, P.B. "Computer-Aided Investigations into the Decision Problem for Relevant Logics: The Search for a Free Associative Operation", **Australian Computer Science Communications**, 5 (1982), 236-267. **(Proceedings of the 6th Australian Computer Science Conference**, ed. L.M. Goldschlager, Basser Department of Computer Science, University of Sydney.)

[24] McRobbie, M.A., Meyer, R.K. and Thistlewaite, P.B. "A Challange", Manuscript, 1987.

[25] McRobbie, M.A., Thistlewaite, P.B. and Meyer, R.K. "A Mechanized Decision Procedure for Non-Classical Logic: The Program KRIPKE", **Bulletin of the Section of Logic**, Polish Academy of Sciences, 9 (1980), 189-192.

[26] McRobbie, M.A., Thistlewaite, P.B. and Meyer, R.K. "A Mechanized Decision Procedure for Non-Classical Logic - the Program KRIPKE", (Abstract), **Journal of Symbolic Logic**, 47 (1982), 717.

[27] Malkin, P.K. and Martin, E.P. "Logical Matrix Generation and Testing", see these **Proceedings**.

[28] Nilsson, N.J. **Principles of Artificial Intelligence**, Springer, Berlin, 1980.

[29] Prawitz, D. **Natural Deduction: A Proof Theoretic Study**, Almqvist and Wiksell, Stockholm 1985.

[30] Rasiowa, H. and Sikorski, R. **The Mathematics of Metamathematics**, Polish Scientific Publishers (PWN), Warsaw, 1963. (3rd Edition, 1970.)

[31] Reiter, R. "A Semantically Guided Deductive System for Automated Theorem Proving", **IEEE Transactions on Computers**, C-25 (1976), 328-334.

[32] Rescher, N. **Many-Valued Logic**, McGraw-Hill, New York, 1969.

[33] Routley, R. and Wolf, R.G. "No Rational Logic Has a Finite Characteristic Matrix", **Logique et Analyse**, 17 (1974), 317-321.

[34] Sandford, M.M. **Using Sophisticated Models in Resolution Theorem Proving**, **LNCS**, Vol. 90, Springer, Berlin, 1980.

[35] Schütte, K. **Proof Theory**, trans, J. Crossley, Springer, Berlin, 1977.

[36] Siekmann, J."Geschichte und Anwendungen", in **Deduktionssysteme: Automatisierung des logischen Denken**, eds. K.H. Bläsius and H.-J. Bürckert, Oldenbourg, Munich and Vienna, 1987.

[37] Slaney, J.K. **Computers and Relevant Logics: A Project in Computing Matrix Model Structures for Propositional Logics**, Ph.D. Thesis, Australian National University, 1980.

[38] Smullyan, R.M. **First-Order Logic**, Springer, Berlin, 1968.

[39] Thistlewaite, P.B, McRobbie, M.A. and Meyer, R.K. "The KRIPKE Automated Theorem Proving System", 705-706 in **Proceedings of the 8th International Conference on Automated Deduction**, ed J. Siekmann, **LNCS,** Vol. 230, Springer, Berlin, 1987.

[40] Thistlewaite, P.B, McRobbie, M.A. and Meyer, R.K. **Automated Theorem Proving in Non-Classical Logic**, Research Notes in Theoretical Computer Science, Pitman, London, and Wiley, New York, 1987.

[41] Thistlewaite, P.B., Meyer, R.K. and McRobbie, M.A. "Advanced Theorem Proving for Relevant Logics", **Logique et Analyse**, 28 (1985), 233-258.

[42] Turner, R. **Logics for Artificial Intelligence**, Ellis Horwood, Chichester 1984.

[43] Wolf, R.G. "A Survey of Many-Valued Logic", 167-323 in **Modern Uses of Multiple-Valued Logic**, ed. J.M. Dunn and G. Epstein, Reidel, Dordrecht, 1977.

Propositional Temporal Interval Logic is PSPACE Complete

A.A. Aaby[t] and K.T. Narayana[t]

ABSTRACT

We define a notion of $\pi\alpha$ equivalence of two execution sequences, where π is the set of variables shared between the two sequences and α is a set of variables disjoint from π appearing in only one of them. We call the set of variables α as auxiliary variables. We extend the notion of $\pi\alpha$ equivalence to formulas in temporal logics, and there by to classes of temporal logics. Under such a notion, we provide sound and complete translation scheme from Propositional Temporal Interval Logic(PTIL) to Linear Time Propositional Temporal Logic (PTL). We do so via the introduction of a chop operator into PTL. The PTIL that we consider is of Swartz, Melliar-Smith variety[13]. The translations that we give are Polynomial in space and time. Together with the results of Sistla and Clarke[14], we conclude that the satisfiability problem for PTIL is PSpace. Known decision procedures for PTIL are exponential in space[9]. The translations provide a means with which synchronization skeletons could be synthesized from specifications given in PTIL. We have constructed a prolog based prototype implementation of the synthesizer.

1 Introduction

Temporal Logics[11] have been found to be extremely suitable for the specification of concurrent systems. The logics have modalities which speak about the present, the future and the continuous future. Further, Propositional Temporal Logics satisfy the *finite model property*. As a result, synchronization skeletons, which in most part are finite state systems, can be synthesized once a specification of the behaviours of the concurrent system is given in the logic. A variety of logics like Linear Time Temporal Logic(PTL), and Branching Time Temporal Logic(BTL) have been investigated in the literature for their expressive powers (see survey paper by Pnueli[10]) and for the complexity of the decision procedures[2, 3, 8, 14].

From a specification point of view, both PTL and BTL are low level in the sense that the programmer has to keep track of an extraordinary amount of detail while giving a specification of a behaviour. It is to alleviate this problem, Swartz, Melliar-Smith and Vogt advocated Propositional Temporal Interval Logic(PTIL) as a formalism for the specification of concurrent systems. Essentially, the formalism provides abstract operators for constructing intervals of interest and for specifying that a property holds in a given interval. The logic is event based with an event being designated by a *two state* interval. Events are captured as state transitions with a formula characterizing the event being *false* in the first state of the interval, and being *true* in the second state of the interval. The syntax and the semantics of PTIL (given in section 6) facilitates the construction of a very complex set of interval formulas. Decision procedures for satisfiability of a PTIL formula must then address this inherent complexity of the logic. The same can be recast as a synthesis procedure when the side effect of the procedure is a transition diagram yielding the set of all execution sequences in which the formula is satisfiable. If the generated transition diagram is empty, then the formula is unsatisfiable and therefore the negation of the formula is a theorem. In order to motivate the reader, we highlight here some issues in the synthesis problem. Suppose we are given a PTIL formula of the form

---

[t] Department of Computer Science, Whitmore Laboratory, The Pennsylvania State University, University Park, Pa 16802

$$[\![a \Rightarrow b]\!][\![c \Rightarrow d]\!]f.$$

We briefly describe its semantics. An event as stated earlier is a two state interval. Assume that a , b , c , and d are all propositions. Then event a is said to occur in a given execution sequence if we can find a two state interval in that execution sequence such that in the first state of that interval a is false and in the second state a is true. If any of the events a, b, c, and d do not occur in an execution sequence, then that execution sequence is a model of the above formula. If all the events a, b, c, and d occur say in that order in a computation sequence, then formula f needs to be satisfied in the state in which the proposition c makes a transition to a true value from its false value in the preceding state. Such an occurrence of the event c, and the set of states characterized by f must occur prior to the occurrence of the next event d within the interval spawned by the occurrence of the events a and b. The synthesis problem concerns itself with the synthesis of the event terms a, b, c, and d and further the formula f. Once they have been synthesized, the constructions must ensure that the events a, b, c, and d are properly placed. Further, we have to guarantee that f is satisfied in the finite set of states of the interval characterizing the occurrence of events c and d. Note that this formulation is in addition to the vacuous truth of the formula in the case when any of the events a, b, c, and d do not occur. Thus, appropriate placement of the event structures forms a crucial problem in any synthesis procedure.

A natural question that arises then is whether in the light of the complexity of the synthesis problem, is PTIL worth pursuing? There is evidence to the positive. We can expend some computing cycles as long as we can guarantee that the programmer is relieved of low level detail in the specification. Further, we show that low level details, if any, can be incorporated into the specification by a translation scheme from PTIL into PTL.

Thus, we provide a PSPACE decision procedure for PTIL of the Schwartz,Melliar-Smith and Vogt[13] variety. Decision procedures for PTIL have been investigated by Plaisted[9] by first defining a low level temporal logic and then providing translation of PTIL formulas into that low level logic. Plaisted's decision procedure has exponential space complexity. He conjectured in his paper[9] that PTIL is in PSPACE. We provide translations from PTIL formulas to until formulas. These translations are computable in PSPACE. Since Sistla and Clarke[14] have shown that there is a decision procedure for until formulas which is PSPACE complete, we establish a PSPACE complete decision procedure for PTIL.

Our approach is as follows. First we consider linear time propositional temporal logic(PTL) with the until operator(PTL(Until)). We augment PTL(Until) with the chop operator to yield Choppy logic (our logic should be contrasted with that of Rosner and Pnueli[12] in that their chop operator is a fusion operator, while our chop operator is a concatenation operator). Then we introduce the notion of $\pi\alpha$ equivalence of temporal formulas. The notion of $\pi\alpha$-equivalence of temporal formulas allows for the selective introduction of auxiliary variables into one of the formulas. Thus, we provide translations converting formulas in Choppy Logic into $\pi\alpha$-equivalent formulas in PTL(Until). The translations are computable in PSPACE(formula length). Then we provide translations converting formulas of PTIL into $\pi\alpha$ equivalent formulas in Choppy Logic. Again the translations involve the selective use of auxiliary variables. Theses translations are sound with respect to $\pi\alpha$-equivalence and complete. The translations are again in PSPACE(formula length).

The paper is organized as follows. Section 2 gives the syntax and semantics of PTL. Section 3 introduces Chop operator and its semantics. Section 4 introduces the notions of $\pi\alpha$-equivalence of execution sequences and its extension to formulas. Section 5 gives $\pi\alpha$- equivalent translations from Choppy Logic to PTL and further establishes that the translations are polynomial in space to the length of the formula. Section 6 gives the syntax and semantics of PTIL. Section 7 formulates $\pi\alpha$-equivalent translations from PTIL to Choppy Logic. Section 8 establishes that the translations are Polynomial in space to the length of the formula. Section 9 gives an overview of a prototype implementation. Section 10 concludes the paper.

2 Syntax and Semantics of PTL(until)

Syntax

Let $\Pi = \{P,...\}$ be a finite set of atomic propositions and N the set of the negations of atomic propositions. Let f be a typical formula. Then the set of formulas of the logic is inductively defined as follows:

$$f ::= P \mid \neg f \mid f \wedge f \mid f \vee f \mid \bigcirc f \mid \diamond f \mid \square f \mid f \, U f$$

Semantics

Let Σ be a finite set of states. Let M be a function from Σ to 2^{Π}. That is, for $s_i \in \Sigma$, $M(s_i)$ is a subset of Π. Let $\sigma = s_0 s_1 s_2 \cdots , s_i \in \Sigma$, be a computation sequence, possibly infinite in length. Let $\vDash$ be a relation between σ, M, i and j, positions in σ, and f, a formula. Then, the triple, $<\sigma,M,i,j>$, is called an interpretation and $<\sigma,M,i,j> \vDash f$ is read as f is valid in the interpretation, $<\sigma,M,i,j>$. For a given context both σ and M are fixed. Therefore, they are made implicit in the semantic definitions. Let P designate an arbitrary atomic proposition and, a and b designate arbitrary formulas. The semantics of formulas are given by structural induction as follows:

$<i,j> \vDash P$ iff $P \in M(s_i)$,

$<i,j> \vDash \neg a$ iff not $<i,j> \vDash a$,

$<i,j> \vDash a \vee b$ iff $<i,j> \vDash a$ or $<i,j> \vDash b$,

$<i,j> \vDash a \wedge b$ iff $<i,j> \vDash a$ and $<i,j> \vDash b$,

$<i,j> \vDash \bigcirc a$ iff i<j and $<i+1,j> \vDash a$,

$<i,j> \vDash \diamond a$ iff $\exists k(i \le k \le j \wedge <k,j> \vDash a)$,

$<i,j> \vDash \square a$ iff $\forall k(i \le k \le j \rightarrow <k,j> \vDash a)$,

$<i,j> \vDash a \, U b$ iff $\exists k(i \le k \le j \wedge <k,j> \vDash b \wedge \forall l(i \le l < k \rightarrow <l,j> \vDash a))$,

3 Syntax and Semantics of Choppy Logic

Syntax
Extend the syntax of PTL(until) with: $f \, Cf$.

Semantics
Extend the semantic definitions of PTL(until) with:

$<i,j> \vDash a \, Cb$ iff $\exists k(i \le k < j \wedge <i,k> \vDash a \wedge <k+1,j> \vDash b)$.

4 $\pi\alpha$-Equivalence

Definition:

Let $\sigma = s_0 s_1 \cdots$ and $\tau = t_0 t_1 \cdots$ be state sequences. Then, σ and τ are $\pi\alpha$-*equivalent* ($\sigma \equiv_{\pi\alpha} \tau$) iff *there are two disjoint sets of variables*, π and α, where π is a set of variables shared between σ and τ, α is a set of auxiliary variables of τ, and there is a monotonic function $f \in [Nat \rightarrow Nat]$ such that $f(0)=0$, $\forall i \, \exists k i < k \wedge f(i) < f(k)$, and

$$\forall\, a \in \pi \forall p \in \alpha \forall i,j\,[a \in s_i = a \in t_{f(i)} \wedge [f(i) \le j < f(i+1) \to a \in s_i = a \in t_j \wedge p \in t_{f(i)} = p \in t_j]].$$

This definition states that two state sequences are $\pi\alpha$-equivalent, if in the second of the two sequences duplicate states are mapped to a single state of the first sequence. This is an equivalence relation. The reflexive property is established by the identity function. The transitivity property is established by the composition of functions. The symmetric property is established by choosing π as before, letting α be empty and defining an inverse function g as follows.

$$g(j) = \text{if } \exists\, i : f(i) = j \text{ then } \min\{i \mid f(i) = j\} \text{ else } \max\{i \mid f(i) < j\}$$

Proposition: The function g is monotonic.

Proof: The proof is performed by induction. Since $f(0) = 0$ and $\min\{i \mid f(i)=0\}=0$, $g(0) = 0$. We must show that $g(j) \le g(j+1)$. There are two cases.

Assume that there is a i such that $f(i) = j$. Then $g(j) = \min\{i \mid f(i)=j\}$ by the definition of g. Let k be that minimum i. If $f(k+1) = f(k) = j$ then $g(j+1) = g(j) = k$ by the definition of g. If $f(k+1) > f(k)$ and $f(k+1) = 1$ then either $g(j+1) = 1$ and $1 > g(j)$ or $g(j+1) = \max\{i \mid f(i) < j+1\} = k$ and therefore, $g(j+1) \ge g(j)$.

Assume that there is no i such that $f(i) = j$. Then $g(j) = \max\{i \mid f(i)<j\}$ by the definition of g. Let k be that maximum i. If there is an i such that $f(i) = j+1$, then since f is monotonic, $\max\{i \mid f(i)<j\} < \min\{i \mid f(i)=j+1\}$ and $g(j) < g(j+1)$. If no such i exists then $g(j+1) = g(j)$. ‖‖‖

Now we must show that

$$\forall\, a \in \pi \forall p \in \alpha[a \in t_i = a \in s_{g(i)} = a \in s_j \wedge p \in s_{g(i)} = p \in s_j].$$

Since α is empty, the conditions on variables of α are satisfied. Assume $g(j) < g(j+1)$. Since $\sigma \equiv_{\pi\alpha} \tau$, for each i such that $f(i) = j$, the condition on the variables of π holds.

Definition:

Let S and T be sets of state sequences then S and T are $\pi\alpha$-*equivalent* iff

$$\forall\, s \in S\, \exists\, t \in T : (s \equiv_{\pi\alpha} t) \wedge \forall\, t \in T\, \exists\, s \in S : (s \equiv_{\pi\alpha} t).$$

This definition states that two sets of state sequences are $\pi\alpha$-equivalent iff for each state sequence in one set, there is a corresponding state sequence in the other set such that the two sequences are $\pi\alpha$-equivalent. As before, this is an equivalence relation. The reflexive property is established by the identity function. The transitivity property is established by the composition of functions. The symmetric property is established by defining the inverse function as before.

Since temporal formulas specify sets of state sequences, the definition may be extended to temporal formula as well. A direct extension of $\pi\alpha$-equivalence to formulas is possible. However under a direct extension, if two formulas are $\pi\alpha$-equivalent, their negations might not be $\pi\alpha$-equivalent. Note that (as a trivial example) the formulas $a\, U\, b$ and $(a \wedge p)\, U\, b$ are certainly equivalent under the constraint that $\Box(p \to true)$. The negations of the formulas may be written as $(\bigcirc \Box \neg b) \vee ((\neg b)\, U\, (\neg b \wedge \neg a))$ and $(\bigcirc \Box \neg b) \vee ((\neg b)\, U\, (\neg b \wedge (\neg a \vee \neg p)))$ respectively. The negations are not equivalent since it is possible for both $a\, U\, b$ and $\neg((a \wedge p)\, U\, b)$ to be satisfied by the execution sequence defined by $\bigcirc(a \wedge \neg b) \wedge \bigcirc b$. This example motivates the following definitions. Equivalent formulas will consist of two parts, a base formula and a constraint formula. The structure of a constraint formula is given in the next definition.

Definition:

$C(\pi\alpha)$ is a constraint formula iff it is a formula of the form: $\square \wedge (p_i \to B_i)$ where p_i is either an auxiliary variable or the negation of an auxiliary variable and B_i is a logical formula.

A constraint formula defines what must hold whenever an auxiliary variable (or the negation of an auxiliary variable) is satisfied in a given state. The negation of the conjunct of a base formula and a constraint formula cannot follow the standard rules for negation. In the previous example, the negation of the constraint formula is $\diamondsuit(p \wedge false)$ which is false. The negation of the base formula must be constrained instead by $\square(\neg p \to false)$. This motivates the definition which follows.

Definition:

If $C(\pi\alpha))$ is a constraint formula, then $\neg(g \wedge C(\pi\alpha))$ is the formula $\neg g \wedge \bar{C}(\pi\alpha))$ where $\bar{C}(\pi\alpha)$ is formed from $C(\pi\alpha)$ by replacing p_i with $\neg p_i$ and by replacing B_i with $\neg B_i$.

Definition:

Let f, g be temporal formula where π is a set of variables shared between f and g, α is a set of variables appearing in g and disjoint from π. Then, f and g are $\pi\alpha$-*equivalent* iff for every σ there is a τ such that

$$\{\sigma \vDash f \text{ then } [(\sigma \equiv_{\pi\alpha} \tau) \text{ and } \tau \vDash g]\} \text{ and } \{\sigma \vDash \neg f \to [\sigma \equiv_{\pi\alpha} \tau \wedge \tau \vDash \neg g]\} \quad \wedge$$
$$\{[\sigma \vDash g] \text{ then } [\sigma \equiv_{\pi\alpha} \tau \text{ and } \tau \vDash f]\} \text{ and } \{[\sigma \vDash \neg g] \text{ then } [\sigma \equiv_{\pi\alpha} \tau \text{ and } \tau \vDash \neg f]\}$$

The definition states that temporal formulas are $\pi\alpha$-equivalent whenever there are corresponding $\pi\alpha$-equivalent execution sequences. As before, this is an equivalence relation. The reflexive property is established by the identity function. The transitivity property is established by the composition of functions. The symmetric property is established by defining the inverse function as before.

5 Translations from Choppy Logic to PTL(until)

Lemma 1: (a) The chop operator may be expressed in until logic and (b) the complexity(length) of the expression is polynomial.
Proof: Let L be a formula of propositional logic(does not contain temporal operators), then the translation rules for choppy formulas are:
Standard cases

$$L\,Cf = L \wedge \bigcirc \diamondsuit f$$
$$(f \vee g)Ch = (f\,Ch) \vee (g\,Ch)$$
$$(\bigcirc f)Cg = \bigcirc(f\,Cg)$$
$$(\diamondsuit f)Cg = \diamondsuit(f\,Cg)$$
$$(\square L)Cg = L \wedge [\bigcirc(L\,\mathbf{U}\,g)] \text{ where L does not contain temporal operators}$$
$$(f\,Cg)Ch = f\,C(g\,Ch)$$

Special cases
Stage one (where $\square[(p \to true) \wedge (\neg p \to true)]$)

$$(f \wedge g)Ch \equiv_{\pi\alpha} [(f \wedge \square \neg p)C(p \wedge h)] \wedge [(g \wedge \square \neg p)C(p \wedge h)]$$
$$(\square f)Cg \equiv_{\pi\alpha} [((f \wedge \square \neg p)Cp)] \wedge [((f \wedge \square \neg p)Cp)\,\mathbf{U}\,(p \wedge g)]$$

where f contains temporal operators

$$(f\,\mathbf{U}\,g)Ch \equiv_{\pi\alpha} [(f \wedge \square \neg p)C(h \wedge p)]\,\mathbf{U}\,[(g \wedge \square \neg p)C(h \wedge p)]$$

Stage two(h is $p \wedge Q$ for some Q)

$(L \wedge \Box \neg p)Ch = [L \wedge \neg p] \wedge [\neg p \ \mathbf{U} h]$ where L does not contain temporal operators.

$((f \vee g) \wedge \Box \neg p)Ch = [(f \wedge \Box \neg p)Ch] \vee [(g \wedge \Box \neg p)Ch]$

$(f \wedge g \wedge \Box \neg p)Ch = [(f \wedge \Box \neg p)Ch] \wedge [(g \wedge \Box \neg p)Ch]$

$(\bigcirc f \wedge \Box \neg p)Ch = \neg p \wedge \bigcirc [(f \wedge \Box \neg p)Ch]$

$(\diamond f \wedge \Box \neg p)Ch = \neg p \ \mathbf{U} [(f \wedge \Box \neg p)Ch]$

$(\Box f \wedge \Box \neg p)Ch = [((f \wedge \Box \neg p)Cp)] \wedge [((f \wedge \Box \neg p)Cp) \ \mathbf{U} h]$

$[(f \ \mathbf{U} g) \wedge \Box \neg p]Ch = [(f \wedge \Box \neg p)Cp] \ \mathbf{U} [(g \wedge \Box \neg p)Ch]$

$((f Cg) \wedge \Box \neg p)Ch = (f \wedge \Box \neg p)C(g \wedge \Box \neg p)Ch$

To establish part **a** of the lemma the rules must be shown to be complete and sound. The rules are complete since there is a rule for each type of Choppy formula. This includes the cases which introduce an auxiliary variable, that is, the stage two rules are complete after taking into account DeMorgan's Laws and the propositional tautologies. Soundness is easily established by an appropriate manipulation of the semantic definitions and the use of $\pi \alpha$ equivalence.

To establish part **b** of the lemma, we must define a notion of formula length and show that the translation rules increase the length of a choppy formula by at most a polynomial. We define two notions of length. The first is the length of the formula defined as the number of symbols it contains. Let $|F|$ denote this concept of the length of the formula F, formally,

$|P| = 1$

$|\neg a| = |a| + 1$

$|a \vee b| = |a| + |b| + 1$

$|a \wedge b| = |a| + |b| + 1$

$|\bigcirc a| = |a| + 1$

$|\diamond a| = |a| + 1$

$|\Box a| = |a| + 1$

$|a \ \mathbf{U} b| = |a| + |b| + 1$

$|a \ Cb| = |a| + |b| + 1$

The second notion of length is the length of the translated formula. Let $|F|_T$ denote the length of the formula which results when the translations are applied to the chop formulas in F. Then $|F|_T$ is defined as:

$|P|_T = |P|$

$|\neg a|_T = |a|_T + 1$

$|a \vee b|_T = |a|_T + |b|_T + 1$

$|a \wedge b|_T = |a|_T + |b|_T + 1$

$|\bigcirc a|_T = |a|_T + 1$

$|\diamond a|_T = |a|_T + 1$

$|\Box a|_T = |a|_T + 1$

$|a \ \mathbf{U} b|_T = |a|_T + |b|_T + 1$

Clearly, $|F| = |F|_T$ when F does not contain the chop operator. Therefore, we must show that the chop operator increases the length polynomially. This is established by induction on the length of the first argument of the chop operator. Assume that the lemma holds for shorter formulas, then since

$|LCf|_T = |L \wedge \bigcirc \diamond f|_T = |L| + |f|_T + 3$

$|(f \vee g)Ch|_T = |f Ch|_T + |g Ch|_T + 1$

$|(\bigcirc f)Cg|_T = |f Cg|_T + 1$

$|(\diamond f)Cg|_T = |f Cg|_T + 1$

$|(\Box L)Cg|_T = |L| + |L \ \mathbf{U} g|_T + 2$

$|(f Cg)Ch|_T = |f C(g Ch)|_T$

the lemma holds for these cases.

The next three cases introduce auxiliary variables. We perform induction on the length of the first argument of the chop operators in the translated formula. That is, we do induction on the right hand sides (the length of the first argument to the chop operator). Assume that the lemma holds for shorter formulas, then since

$|(f \wedge g)Ch|_T = |(f \wedge \square \neg p)C(p \wedge h)|_T + |(g \wedge \square \neg p)C(p \wedge h)|_T$

$|(\square f)Cg|_T = |((f \wedge \square \neg p)Cp)|_T + |((f \wedge \square \neg p)Cp)U(p \wedge g)|_T + 1$

    where f contains temporal operators

$|(f U g)Ch|_T = |(f \wedge \square \neg p)C(h \wedge p)]U[(g \wedge \square \neg p)C(h \wedge p)|_T$

and

$|(L \wedge \square \neg p)Ch|_T = |L| + |h|_T + 7$

$|((f \vee g) \wedge \square \neg p)Ch|_T = |(f \wedge \square \neg p)Ch|_T + |(g \wedge \square \neg p)Ch|_T + 1$

$|(f \wedge g \wedge \square \neg p)Ch|_T = |(f \wedge \square \neg p)Ch|_T + |(g \wedge \square \neg p)Ch|_T + 1$

$|(\bigcirc f \wedge \square \neg p)Ch|_T = |(f \wedge \square \neg p)Ch|_T + 4$

$|(\diamondsuit f \wedge \square \neg p)Ch|_T = |(f \wedge \square \neg p)Ch|_T + 3$

$|(\square f \wedge \square \neg p)Ch|_T = |((f \wedge \square \neg p)Cp)|_T + |(f \wedge \square \neg p)Cp|_T + |h|_T + 2$

$|[(f U g) \wedge \square \neg p]Ch|_T = |(f \wedge \square \neg p)Cp|_T + |(g \wedge \square \neg p)Ch|_T + 1$

$|((f Cg) \wedge \square \neg p)Ch|_T = |(f \wedge \square \neg p)C(g \wedge \square \neg p)Ch|_T$

the lemma holds.

The use of auxiliary variables has a disadvantage in that each occurrence of a chop formula requires a different auxiliary variable. Two cases where this is unacceptable are when a chop formula is in the scope of an always or an until operator. In these cases there is the potential for an exponential number of auxiliary variables, one for each state. There are two solutions. The first is to use the following rule:

$$\frac{f(P), f(Q)}{f(P)},$$

where P and Q are auxiliary variables. The rule states that two instances of a formula may be replaced with a single instance. This allows us to reduce the number of auxiliary variables to a number which is at most the size of the closure of a formula[6]. The second solution is to provide recursive definitions for the three cases which introduce auxiliary variables. This approach requires a base case. The formulas follow.

$(L \wedge b)Cc = L \wedge (bCc)$

$(a \wedge b)Cc = L \wedge (fCc)$ where $L \wedge f = a \wedge b$ by a tautology.

$(\square a)Cb = (a \wedge (\neg \bigcirc true \vee \bigcirc \square a))Cb$

$(a U b)Cc = (b \vee a \wedge \bigcirc (a U b))Cc$

These require a least fixed point operator for the solution.

6 Syntax and Semantics of PTIL

Syntax

Extend the syntax of Choppy Logic with:

$\llbracket f \rrbracket f \mid \llbracket begin(t) \rrbracket f \mid \llbracket end(t) \rrbracket f \mid \llbracket t \Rightarrow t \rrbracket f \mid \llbracket t \Rightarrow \rrbracket f \mid \llbracket \Rightarrow t \rrbracket f \mid \llbracket t \Leftarrow t \rrbracket f \mid \llbracket t \Leftarrow \rrbracket f \mid \llbracket \Leftarrow t \rrbracket f$.

The set of interval terms is defined as follows:

$$t ::= f \mid begin(t) \mid end(t) \mid (t \Rightarrow t) \mid (t \Leftarrow t).$$

Semantics

Let P designate an arbitrary atomic proposition and a and b designate arbitrary formulas, $\perp$ designate the empty interval, and I and J designate interval terms then extend the semantic definitions of Choppy Logic with:

$<i,j> \models [\![I]\!] a$ iff $F(I,<i,j>,f) \models a$,

$\perp \models a$.

To define function **F**, the functions **changeset**, **first** and **last** must be defined. The function, **changeset**, given a formula and an interval, returns a set of intervals where each interval represents two successive states, the first where the formula is unsatisfied, and the second where the formula is satisfied.

$\text{changeset}(a,<i,j>) = \{ <k-1,k> \mid k \in <i,j>, <k-1,j> \models \neg a, <k,j> \models a \}$

Note that the formula may have just become satisfied in the current state, i, in that case, the interval $<i-1,i>$ is in the set returned by the function. The functions **first** and **last** return, respectively, the initial and final indices of a given interval.

$\text{first}(<i,j>) = i$

$\text{last}(<i,j>) = j$

$\text{last}(<i,\infty>) = \perp$

Definition of **F**

F: interval term $\times$ interval $\times$ direction $\rightarrow$ interval

f stands for forward search direction

b stands for backward search direction

d stands for direction parameter

$F(a,<i,j>,f) = \min(\text{changeset}(a,<i,j>))$

$F(a,<i,j>,b) = \max(\text{changeset}(a,<i,j>))$

$F(\text{begin}(I),<i,j>,d) = <\text{first}(F(I,<i,j>,d)),\text{first}(F(I,<i,j>,d))>$

$F(\text{end}(I), <i,j>,d) = <\text{last}(F(I,<i,j>,d)),\text{last}(F(I,<i,j>,d))>$

$F(I\Rightarrow ,<i,j>,d) = <\text{last}(F(I,<i,j>,d)), j>$

$F(I\Leftarrow ,<i,j>,d) = <\text{last}(F(I,<i,j>,b)), j>$

$F(\Rightarrow J,<i,j>,d) = <i, \text{last}(F(J,<i,j>,f))>$

$F(\Leftarrow J,<i,j>,d) = <i, \text{last}(F(J,<i,j>,d))>$

$F(I\Rightarrow J,<i,j>,d) = F(\Rightarrow J, F(I\Rightarrow ,<i,j>,d),f)$

$F(I\Leftarrow J,<i,j>,d) = F(\Leftarrow J, F(I\Leftarrow ,<i,j>,d),f)$

where min and max are defined on sets of disjoint intervals, and both return $\perp$ if changeset is empty and further, max returns $\perp$ if changeset is of infinite size. The expression $I \Rightarrow J$ selects the next occurrence of the event I in the outer context and then the next following event J. The expression $I \Leftarrow J$ selects the last occurrence of the event I in the outer context and then the next following event J. The begin and end operators construct the beginning and end points of an interval respectively. They are useful for the construction of intervals which begin just prior to an event such as $[\![a \Rightarrow begin(b)]\!] \Box c$, that is, c must hold up to but not including the event **b** (the state in which b becomes true).

7 Translation semantics for the interval operator

The semantics of PTIL refer to the previous state and also allows for the possibility that an event sequence does not occur. To simplify the translation rules, we introduce last time and weak until operators by way of definitions. Further, we introduce a propositional constant to denote the initial state of the execution sequence.

$f \wedge \bigcirc true = \bigcirc \ominus f$

$a \, \mathbb{U} \, b = (\Box(a \wedge \neg b)) \vee (a \, U \, b)$.

$init = \neg \ominus true$

The translation of a temporal interval logic formula to a propositional temporal logic formula requires the construction of the sequence of events which determine the left-hand side of the interval, the construction of the sequence of events which determine the right-hand side of the interval and the correct placement of the nested formula within the sequence of events. We provide definitions to construct the left-hand sequence of events and a function, E which constructs the right-hand sequence of events and which ties the nested formula to both the left and right-hand sequence of events.

Lemma 2: The interval operator may be expressed in until logic.
Proof:
All of the PTIL formulas may be expressed in one of the five forms. The five forms are:

$$[\![I \Rightarrow \ ]\!] \ where \ I \ is \ an \ event$$
$$[\![begin(I) \Rightarrow \ ]\!]F \ where \ I \ is \ an \ event$$
$$[\![\Rightarrow \ J]\!]F$$
$$[\![I \Leftarrow \ ]\!]F \ where \ I \ is \ an \ event$$
$$[\![begin(I) \Leftarrow \ ]\!]F \ where \ I \ is \ an \ event$$

The following properties of PTIL formulas provide for this reduced form.

Properties of Temporal Interval Logic Formulas

Single state intervals
$[\![begin(I)]\!]F = [\![begin(I) \Rightarrow \ ]\!][\![\Rightarrow \ begin(I)]\!]F$ where I is an event.
$[\![end(I)]\!]F = [\![I \Rightarrow \ ]\!][\![\Rightarrow \ I]\!]F$ where I is an event.
$[\![end(I \Rightarrow \ J)]\!]F = [\![I \Rightarrow \ ]\!][\![end(J)]\!]F$
$[\![end(I \Leftarrow \ J)]\!]F = [\![I \Leftarrow \ ]\!][\![end(J)]\!]F$

Double state interval
$[\![I]\!]F = [\![begin(I) \Rightarrow \ ]\!][\![\Rightarrow \ I]\!]F$ where I is an event

General intervals (From $[\![I \Rightarrow \ J]\!]F$)
$[\![I \Rightarrow \ J]\!]F = [\![I \Rightarrow \ ]\!][\![\Rightarrow \ J]\!]F$
$[\![(I \Rightarrow \ J) \Rightarrow \ ]\!]F = [\![I \Rightarrow \ ]\!][\![J \Rightarrow \ ]\!]F$
$[\![(I \Leftarrow \ J) \Rightarrow \ ]\!]F = [\![I \Leftarrow \ ]\!][\![J \Rightarrow \ ]\!]F$
$[\![end(I) \Rightarrow \ ]\!]F = [\![I \Rightarrow \ ]\!]F$
$[\![begin(I \Rightarrow \ J) \Rightarrow \ ]\!]F = [\![begin(I) \Rightarrow \ ]\!][(\neg[\![J]\!]true) \vee (F \wedge \neg[\![J]\!]false)]$
$[\![begin(I \Leftarrow \ J) \Rightarrow \ ]\!]F = [\![begin(I) \Leftarrow \ ]\!][(\neg[\![J]\!]true) \vee (F \wedge \neg[\![J]\!]false)]$
$[\![\Rightarrow \ begin(I \Rightarrow \ J)]\!]F = [\neg[\![I \Rightarrow \ J]\!]true] \vee [(\neg[\![I \Rightarrow \ J]\!]false) \wedge [\![\Rightarrow \ begin(I)]\!]F]$
$[\![\Rightarrow \ begin(I \Leftarrow \ J)]\!]F = [\neg[\![I \Leftarrow \ J]\!]true] \vee [(\neg[\![I \Leftarrow \ ]\!]false) \wedge [\![\Rightarrow \ begin(I \Leftarrow \ )]\!]F]$
$[\![\Rightarrow \ end(J)]\!]F = [\![\Rightarrow \ J]\!]F$

The following cases require translations to PTL(until) which are given later:
$[\![I \Rightarrow \ ]\!]F$ I is an event
$[\![begin(I) \Rightarrow \ ]\!]F$ I is an event
$[\![\Rightarrow \ J]\!]F$ with special subcases
$\quad [\![\Rightarrow \ begin(I)]\!]F$
$\quad [\![\Rightarrow \ (I \Rightarrow \ J)]\!]F$
$\quad [\![\Rightarrow \ (I \Leftarrow \ J)]\!]F$

General intervals (From $[\![I \Leftarrow \ J]\!]F$)
$[\![I \Leftarrow \ J]\!]F = [\![I \Leftarrow \ ]\!][\![\Rightarrow \ J]\!]F$
$[\![(I \Rightarrow \ J) \Leftarrow \ ]\!]F = [\![I \Leftarrow \ ]\!][\![J \Rightarrow \ ]\!]F$
$[\![(I \Leftarrow \ J) \Leftarrow \ ]\!]F = [\![I \Leftarrow \ ]\!][\![J \Rightarrow \ ]\!]F$
$[\![\Leftarrow \ J]\!]F = [\![\Rightarrow \ J]\!]F$
$[\![end(I) \Leftarrow \ ]\!]F = [\![I \Leftarrow \ ]\!]F$
$[\![begin(I \Rightarrow \ J) \Leftarrow \ ]\!]F = [\![begin(I) \Leftarrow \ ]\!][(\neg[\![J]\!]true) \vee (F \wedge \neg[\![J]\!]false)]$

$[\![begin\,(I \Leftarrow\ J) \Leftarrow\ ]\!]F = [\![begin\,(I) \Leftarrow\ ]\!][(\neg[\![J]\!]true) \vee (F \wedge \neg[\![J]\!]false)]$

The following cases require translations to PTL(until) which are given later:

$[\![I \Leftarrow\ ]\!]F$ I is an event

$[\![begin\,(I) \Leftarrow\ ]\!]F$. I is an event

Properties of begin and end

begin(begin(I)) = begin(I)

end(begin(I)) = begin(I)

end(end(I)) = end(I)

begin(end(I)) = end(I)

Cases not covered by the previous formulas

$[\![I \Rightarrow\ ]\!]F$ where I is an event

$[\![begin\,(I) \Rightarrow\ ]\!]F$ where I is an event

$[\![I \Leftarrow\ ]\!]F$ where I is an event

$[\![begin\,(I) \Leftarrow\ ]\!]F$ where I is an event

$[\![\Rightarrow\ J]\!]F$

The translations for these cases

Note that the semantics of PTIL require that the previous state be included in the search for the occurrence of an event therefore, if $T([\![I]\!]F)$ is the translation of $[\![I]\!]F$ then the actual formula to be used is $(init \wedge T([\![I]\!]F)) \vee \ominus T([\![I]\!]F)$. For each rule, we present informal arguments in pictorial form to justify the soundness of the rule. Let F' be the translation of the PTIL formula F into PTL(until). Then, the rules for each of the five cases are:

$[\![I \Rightarrow\ ]\!]F = I \cup (\neg I \cup (\neg I \wedge \bigcirc (I \wedge F')))$.

$\leftarrow I \rightarrow$	$\leftarrow \neg I \rightarrow$	$\neg I$	I
			F'

Each column represents either a single state or if the column contains entries of the form $\leftarrow I \rightarrow$, then the column represents a sequence of states(possibly empty).

$[\![I \Leftarrow\ ]\!]F = true \cup (\neg I \cup (\neg I \wedge \bigcirc (I \wedge F' \wedge \bigcirc \square (I \vee \bigcirc \neg I))))$

$\leftarrow true \rightarrow$	$\leftarrow \neg I \rightarrow$	$\neg I$	I	$\square (I \vee \bigcirc \neg I)$
			F'	

$[\![begin\,(I) \Rightarrow\ ]\!]F = I \cup (\neg I \cup (\neg I \wedge F' \wedge \bigcirc I))$

$\leftarrow I \rightarrow$	$\leftarrow \neg I \rightarrow$	$\neg I$	I
		F'	

$[\![begin\,(I) \Leftarrow\ ]\!]F = true \cup (\neg I \wedge F' \wedge \bigcirc (I \wedge \square (I \vee \bigcirc \neg I)))$

$\leftarrow true \rightarrow$	$\neg I$	$I \wedge \square (I \vee \bigcirc \neg I)$
	F'	

The remaining form is $[\![\Rightarrow\ J]\!]F$. For treatment of formulas of this kind, we introduce an auxiliary variable Q to mark the states of an interval; $\neg Q$ for the interior of the interval, and Q for the state following the interval. This permits us to replace $[\![\Rightarrow\ I]\!]f$ with $(f \wedge \square \neg Q)CQ$, and a formula which constructs the event I. The translations of the last form are:

$[\![\Rightarrow\ J]\!]F = \neg E(f,n,J,\_,done) \vee [((F' \wedge \square \neg Q)CQ) \wedge E(f,q,J,Q,done)]$

where $\square [(Q \rightarrow true) \wedge (\neg Q \rightarrow true)]$ or J does not occur.

←J→	←¬J→	¬J	J
←¬Q→	←¬Q→	¬Q ¬Q	Q
$(F' \wedge \Box \neg Q)CQ$			

E function: the event sequence formula.

E(D,M,I,Q,N) = FI, formula which constructs the interval term
        D direction: f-forward, b-backward
        M mode: q-Q on to mark end of interval, n-Q not wanted
        I interval term
        Q auxiliary variable which marks the end of the interval
        N continuation point: done-no continuation, Next- continuation formula
   note: the two modes allow the construction of two different temporal
        formulas, one in which not Q holds until the end point at which time
        Q holds and one which does not contain the auxiliary variable

$E(f,n,I,\_,done) = I \, U(\neg I \, U(\neg I \wedge \bigcirc I))$

←I→	←¬I→	¬I	I

$E(f,n,I,\_,Next) = I \, U(\neg I \, U(\neg I \wedge Next \wedge \bigcirc I))$

←I→	←¬I→	¬I	I
		Next	

$E(f,q,I,Q,done) = (I \wedge \neg Q) \, U((\neg I \wedge \neg Q) \, U(\neg I \wedge \neg Q \wedge \bigcirc (I \wedge \neg Q \wedge \bigcirc Q)))$

←I→	←¬I→	¬I	I	
←¬Q→	←¬Q→	¬Q	¬Q	Q

$E(f,q,I,Q,Next) = (I \wedge \neg Q) \, U((\neg I \wedge \neg Q) \, U(\neg I \wedge \neg Q \wedge \bigcirc I))$

←I→	←¬I→	¬I	I
←¬Q→	←¬Q→	¬Q	
		Next	

In the next two rules, I and J may be terms. Therefore, we simply illustrate the effect of the rule. The first lines indicate the construction of the event I, the second lines the construction of the event J(FJ), the third lines the value of Q and the fourth the placement of the next events in the sequence.
$E(D,M,I \Rightarrow J,Q,Next) = E(D,M,I,Q,FJ)$ where FJ is $E(f,M,J,Q,Next)$.

←I→	←¬I→	¬I	I		
		←J→	←¬J→	¬J	J
←¬Q→	←¬Q→	←¬Q→	←¬Q→	¬Q	
				Next	

$E(D,M,I \Leftarrow J,Q,Next) = E(b,M,I,Q,FJ)$ where FJ is $E(f,M,J,Q,Next)$.

←true→	¬I	$I \wedge \square(I \vee \bigcirc \neg I)$		
	←J→	←¬J→	¬J	J
←¬Q→	←¬Q→	←¬Q→	←¬Q→	
			Next	

$E(b,n,I,\_,done) = \diamond(\neg I \wedge \bigcirc(I \wedge \square(I \vee \bigcirc \neg I))))$

←true→	¬I	$I \wedge \square(I \vee \bigcirc \neg I)$

$E(b,n,I,\_,Next) = \diamond(\neg I \wedge Next \wedge \bigcirc(I \wedge \square(I \vee \bigcirc \neg I))))$

true	¬I	$I \wedge \square(I \vee \bigcirc \neg I)$
	Next	

$E(b,q,I,Q,done) = (\neg Q)\,\mathbf{U}(\neg I \wedge \neg Q \wedge \bigcirc(I \wedge \square(I \vee \bigcirc \neg I) \wedge \neg Q \wedge \bigcirc Q)))$

	¬I	$I \wedge \square(I \vee \bigcirc \neg I)$	
←¬Q→	¬Q	¬Q	Q

$E(b,q,I,Q,Next) = (\neg Q)\,\mathbf{U}(\neg I \wedge \neg Q \wedge Next \wedge \bigcirc(I \wedge \square(I \vee \bigcirc \neg I)))$

	¬I	$I \wedge \square(I \vee \bigcirc \neg I)$
←¬Q→	¬Q	
	Next	

$E(f,q,begin(I),Q,done) = (I \wedge \neg Q)\,\mathbf{U}((\neg I \wedge \neg Q)\,\mathbf{U}(\neg I \wedge \neg Q \wedge \bigcirc(I \wedge Q)))$

←I→	←¬I→	¬I	I
←¬Q→	←¬Q→	¬Q	Q

$E(f,q,begin(I),Q,Next) = (I \wedge \neg Q)\,\mathbf{U}((\neg I \wedge \neg Q)\,\mathbf{U}(\neg I \wedge \neg Q \wedge (init \wedge Next \vee \ominus Next) \wedge \bigcirc I))$

←I→	←¬I→	¬I	I
←¬Q→	←¬Q→	¬Q	
		init∧Next or	
		⊕ Next	

$E(b,n,begin(I),\_,Next) = \diamond(\neg I \wedge Next \wedge \bigcirc(I \wedge \bigcirc \square(I \vee \bigcirc \neg I)))$

←true→	¬I	$I \wedge \square(I \vee \bigcirc \neg I)$
	Next	

$E(b,q,begin(I),Q,done) = \neg Q\,\mathbf{U}(\neg I \wedge \neg Q \wedge \bigcirc(I \wedge \square(I \vee \bigcirc \neg I)))$

←true→	¬I	$I \wedge \square(I \vee \bigcirc \neg I)$
←¬Q→	¬Q	Q

$$E(b,q,begin(I),Q,Next) = \neg Q \,\mathbf{U}\,(\neg I \wedge \neg Q \wedge (init \wedge Next \vee \ominus Next) \wedge \bigcirc(I \wedge \square(I \vee \bigcirc \neg I)))$$

←true→	¬I	$I \wedge \square(I \vee \bigcirc \neg I)$
←¬Q→	¬Q	
	$init \wedge Next$ or	
	$\ominus Next$	

$E(D,n,begin(I \Rightarrow J),\_,Next) = E(D,n,begin(I),\_,FJ)$ where FJ is $E(f,n,J,\_,Next)$

←I→	←¬I→	¬I	I		
		←J→	←¬J→	¬J	J
					Next

$E(D,q,begin(I \Rightarrow J),Q,done) = E(D,q,begin(I),Q,done) \wedge E(D,n,begin(I),\_,FJ)$ where FJ is $E(f,n,J,\_,done)$

←I→	←¬I→	¬I	I		
		←J→	←¬J→	¬J	J
←¬Q→	←¬Q→	¬Q	Q		

$E(D,q,begin(I \Rightarrow J),Q,Next) = E(D,q,begin(I),Q,done) \wedge E(f,n,begin(I),\_,FJ)$ where FJ is $E(f,n,J,\_,Next)$

←I→	←¬I→	¬I	I		
		←J→	←¬J→	¬J	J
					Next
←¬Q→	←¬Q→	¬Q	Q		

$E(f,M,begin(I \Leftarrow J),Q,Next) = E(b,M,begin(I \Rightarrow J),Q,Next)$

$E(b,M,begin(I \Leftarrow J),Q,done) = E(b,M,begin(I \Rightarrow J),Q,done)$

The negation of each of the five cases is given here. Recall that the negation of an interval formula holds only if the interval occurs and the nested formula is false.

$\neg[[I \Rightarrow]]F = I \,\mathbf{U}\,(\neg I \,\mathbf{U}\,(\neg I \wedge \neg F' \wedge I))$

$\neg[[begin(I) \Rightarrow]]F = I \,\mathbf{U}\,[(I \wedge \neg F' \wedge \bigcirc(\neg I \wedge \bigcirc I)) \vee (\neg I \,\mathbf{U}\,(\neg I \wedge \neg F \wedge \bigcirc(\neg I \wedge \bigcirc I)))]$

$\neg[[I \Leftarrow]]F = true \,\mathbf{U}\,(\neg I \,\mathbf{U}\,(\neg I \wedge \neg F' \wedge \bigcirc(I \wedge \bigcirc \square(I \vee \bigcirc \neg I))))$

$\neg[[begin(I) \Leftarrow]]F = true \,\mathbf{U}\,(\neg I \wedge \neg F' \wedge \bigcirc(I \wedge all(I \vee \bigcirc \neg I)))$

$\neg[[\Rightarrow J]]F = [((\neg F' \wedge \square \neg Q)\mathbf{C}Q) \wedge E(f,q,J,Q,done,true)]$

8 Complexity

Theorem: Interval logic is PSPACE complete.

Proof:

To show that PTIL is PSPACE complete it suffices to note that PTL(until) is PSPACE complete[14] and that PTL(until) logic is contained in PTIL. We now show that translations from PTIL to PTL(until) are of polynomial complexity. This shows that PTIL is also PSPACE complete.

The proof is by induction on the length of a formula. Assume that the translations from PTIL to PTL(until) are of polynomial complexity for shorter formulas. The definition of the length of a formula is extended to temporal interval formulas by:

$|[I]F| = |I| + |F|$ where

$|begin(t)| = |t| + 1$

$|end(t)| = |t| + 1$

$|I \Rightarrow\ |=|I|+1$
$|\Rightarrow\ J|=|J|+1$
$|I \Rightarrow\ J|=|I|+|J|+1$
$|I \Leftarrow\ |=|I|+1$
$|\Leftarrow\ J|=|J|+1$
$|I \Leftarrow\ J|=|I|+|J|+1$

The extension of the $|\ |_T$ operator is implicitly given in the statement of the complexity of the rule and we choose to use the simplier notation of the length operator to also represent the $|\ |_T$ operator in the remainder of the paper. The complexity of each translation rule follows the order in which the rules were given in section 7.

$|I| + |F| + 1 \rightarrow 4|I| + |F'| + 7$
$|I| + |F| + 1 \rightarrow 5|I| + |F'| + 14$
$|I| + |F| + 1 \rightarrow 4|I| + |F'| + 7$
$|I| + |F| + 1 \rightarrow 4|I| + |F'| + 11$
$|I| + |F| + 1 \rightarrow |E(f,n,J,\_,done)| + |E(f,q,J,Q,done)| + |F'| + 6$
$|E(f,n,I,\_,done)| = 4|I| + 6$
$|E(f,n,I,\_,Next)| = 4|I| +\ |Next| + 7$
$|E(f,q,I,Q,done)| = 4|I| + 20$
$|E(f,q,I,Q,Next)| = 4|I| + 15$
$|E(D,M,I \Rightarrow J,Q,Next)| = |E(D,M,I,Q,FJ)| + |E(f,M,J,Q,Next)|$
$|E(b,n,I,\_,done)| = 4|I| + 9$
$|E(b,n,I,\_,Next)| = 4|I| +\ |Next| + 10$
$|E(b,q,I,Q,done)| = 4|I| + 20$
$|E(b,q,I,Q,Next)| = 4|I| +\ |Next| + 14$
$|E(f,q,begin(I),Q,done)| = 4|I| + 17$
$|E(f,q,begin(I),Q,Next)| = 4|I| + 2|Next| + 20$
$|E(b,n,begin(I),\_,Next)| = 4|I| +\ |Next| + 11$
$|E(b,q,begin(I),Q,done)| = 4|I| + 14$
$|E(b,q,begin(I),Q,Next)| = 4|I| + 2|Next| + 19$
$|E(D,n,begin(I \Rightarrow J),\_,Next)| = |E(D,n,begin(I),\_,FJ)| + |E(f,n,J,\_,Next)|$
$|E(D,q,begin(I \Rightarrow J),Q,done)| = |E(D,q,begin(I),Q,done)| + |E(D,n,begin(I),\_,FJ)| +$
$\qquad\qquad\qquad\qquad |E(f,n,J,\_,done)| + 1$
$|E(D,q,begin(I \Rightarrow J),Q,Next)| = |E(D,q,begin(I),Q,done)| + |E(f,n,begin(I),\_,FJ)| +$
$\qquad\qquad\qquad\qquad |E(f,n,J,\_,Next)| + 1$

Where necessary, apply the induction hypothesis. In each case the resulting formula is polynomially related to the length of the given formula. Therefore, we have established the theorem.

9 Implementation of the Decision Procedure for TIL

We have constructed a synthesizer for PTL(until)[7, 8, 12] and have extended it to include the chop operator and the interval operator. The synthesizer is written in Prolog and utilizes the pattern matching capabilities of Prolog to simplify the construction of the transition diagrams. Direct implementations are provided for $\bigcirc, \ominus, \diamondsuit, \square, U, \mathbb{U}$ and C. The equivalences presented earlier are used to implement the interval operator. The auxiliary variables required by the equivalences are internally generated.

For each formula, the implementation obtains an ω-graph representation. An ω-graph is a directed graph whose nodes correspond to formulas and whose edges are labeled with atomic formulas. The implementation allows for incremental construction of the ω-graph. The ω-graph of a formula is constructed from the ω-graphs of its subformulas[5]. We can also designate a set of ω-graphs as a library. This library is utilized to prevent the repeated construction of an ω-graph for common sub-formulas. Some formulas in the library may be distinguished as model formulas. If a given formula unifies with a model formula, then an ω-graph for the given formula may be constructed from the ω-graph for the model formula by substitution. The graph of an unsatisfiable formula is empty. The transitions constructed by the implementation have the form, $t(s_i, s_j, cond)$, where the first two entries are states and cond is the set of propositional variables or their negations which must hold for that transition.

9.1 ω-graph semantics for PTL

The development here is similar to the tableau methods[7, 12] for PTL. A natural semantics for PTL is presented where the semantics of a formula is defined in terms of a set of its subformulas.

Syntax

Let $\Pi = \{P,...\}$ be a set of atomic propositions and N the set of the negations of atomic propositions. The set of formulas is defined as follows:

$$f ::= \text{ term} \mid P \mid \neg f \mid f \wedge f \mid f \vee f \mid \bigcirc f \mid \diamond f \mid \Box f \mid f \text{ U} f \mid f \text{ C} f.$$

Where term is a propositional constant with the following property:

$$term \wedge \bigcirc true \equiv false.$$

Semantics

The closure of a formula is used to provide a bound on the size of the ω-graph corresponding to a given formula. The closure of a formula, cl(f), is the smallest set satisfying the following:

$f \in cl(f)$

$a \vee b \in cl(f) \rightarrow a,b \in cl(f)$

$a \wedge b \in cl(f) \rightarrow a,b \in cl(f)$

$\bigcirc a \in cl(f) \rightarrow a \in cl(f)$

$\neg \bigcirc a \in cl(f) \rightarrow term, \bigcirc \neg a \in cl(f)$

$\diamond a \in cl(f) \rightarrow a, \bigcirc \diamond a \in cl(f)$

$\Box a \in cl(f) \rightarrow a \wedge term, \bigcirc \Box a \in cl(f)$

$a \text{ U} b \in cl(f) \rightarrow a,b, \bigcirc a \text{ U} b \in cl(f)$

$\neg(a \text{ U} b) \in cl(f) \rightarrow \Box \neg b, \neg b \text{ U}(\neg b \wedge \neg a) \in cl(f)$

$a \text{ C} b \in cl(f)$ then

        $a \equiv L \rightarrow L, \bigcirc \diamond b \in cl(f)$

        $a \equiv L \wedge term \rightarrow L, \bigcirc b \in cl(f)$

        $a \equiv L \wedge \bigcirc g \rightarrow L, \bigcirc(g \text{ C} b) \in cl(f)$

        $a \equiv (g \vee h) \rightarrow g \text{ C} b, h \text{ C} b \in cl(f)$

        $a \equiv \neg(L \text{ C} b) \rightarrow \neg L, L \wedge \bigcirc \diamond (b \text{ C} B) \in cl(f)$

        $a \equiv \neg((L \wedge term) \text{ C} b) \rightarrow \neg L, L \wedge \neg \bigcirc (b \text{ C} B) \in cl(f)$

        $a \equiv not((L \wedge \bigcirc g) \text{ C} b) \rightarrow \neg L, \neg \bigcirc (g \text{ C} b) \in cl(f)$

        $a \equiv Y \rightarrow Y \text{ C} B \in cl(f)$

$\neg(a \text{ C} b) \in cl(f)$ then

        $a \equiv term \rightarrow \neg \bigcirc B \in cl(f)$

        $a \equiv (L \wedge term) \rightarrow \neg L \vee (L \wedge \neg \bigcirc C) \in cl(f)$

        $a \equiv L \rightarrow \neg L \vee (L \wedge \neg \bigcirc \diamond B) \in cl(f)$

        $a \equiv \diamond A \rightarrow \Box \neg (A \text{ C} B) \in cl(f)$

        $a \equiv (A \vee B) \rightarrow \neg((A \text{ C} C) \vee (B \text{ C} C)) \in cl(f)$

        $a \equiv Y \rightarrow \neg(Y \text{ C} B) \in cl(f)$

where L is a formula which contains neither temporal operators nor the propositional constant, *term* and Y is the formula on the right-hand side of the following equivalences corresponding to a left-hand side of the equivalence of which a syntactically matches. The equivalences are as follows:

$(A \vee B) \wedge C \equiv (A \wedge C) \vee (B \wedge C)$

$$(A \wedge B) \wedge C \equiv A \wedge (B \wedge C)$$
$$\bigcirc A \wedge \bigcirc B \equiv \bigcirc (A \wedge B)$$
$$A \wedge B \equiv B \wedge A$$
$$\diamond A \equiv A \vee \bigcirc \diamond A$$
$$\square A \equiv A \wedge (term \vee \bigcirc \square A)$$
$$A \ U B \equiv B \vee (A \wedge \bigcirc (A \ U B))$$
$$(A \vee B) CC \equiv (A CC) \vee (B CC)$$
$$(\bigcirc A) CB \equiv \bigcirc (A CB)$$
$$(\diamond A) CB \equiv \diamond (A CB)$$
$$(L \wedge A) CB \equiv L \wedge (A CB)$$
$$(A CB) CC \equiv A C (B CC)$$
$$L CB \equiv L \wedge \bigcirc \diamond B$$
$$(L \wedge term) CB \equiv L \wedge \bigcirc B$$
$$(L \wedge \bigcirc A) CB \equiv L \wedge \bigcirc (A CB)$$
$$a \equiv Y \rightarrow Y CB \in \mathbf{cl}(f)$$
$$\neg\neg A \equiv A$$
$$\neg (A \vee B) \equiv \neg A \wedge \neg B$$
$$\neg (A \wedge B) \equiv \neg A \vee \neg B$$
$$\neg \bigcirc A \equiv term \vee \bigcirc \neg A$$
$$\neg \diamond A \equiv \square \neg A$$
$$\neg \square A \equiv \diamond \neg A$$
$$\neg (A \ U B) \equiv (\square \neg B) \vee ((\neg B) \ U \neg (A \vee B))$$

The size of $\mathbf{cl}(f)$ is bounded by $4|f|$[7]. Let $\Sigma(f)$ be the set of subsets of $\mathbf{cl}(f)$. The size of $\Sigma(f)$ is $2^{4|f|}$. An element of $\Sigma(f)$ is written as a triple, (F,N,P), where F, N and P are conjunctions of sub-formulas of f. The triple designates a transition between two states. F is the initial state. N is the successor state. P is designates the assignments to atomic propositions which hold in the initial state. The rules for computing the closure of a formula enable us to decompose a given formula into two sets of subformulas. One set designating the properties of the successor state and the other set designating the assignments to the atomic proposition in the current state. For example, the formula, $a \ U b$, may be decomposed into the sets $\{\bigcirc (a \ U b)\}$ and $\{a\}$ or the sets ϕ and $\{b\}$ where a and b are atomic propositions. We define the semantics of a formula, f, as a set of subsets of $\Sigma(f)$. Such a set is representable as a graph. In this context, a graph is a set of transitions where each transition names two states. We let a formula designate the state name, a conjunction of next-time formulas designate the name of the next state and atomic propositions designate the conditions applicable to that transition. For example, the formula, $a \ U b$, the graph is representable by the two transitions, $(a \ U b, a \ U b, a)$ and $(a \ U b, \phi, b)$. Note that the next-time operator is dropped from the next-time formulas.

The symbol Glf is designates the graph of f. The set Glf is defined as follows:

$$Glterm = \{(term, \phi, \{term\})\}$$
$$GlP = \{(P, \phi, \{P\})\}$$
$$Gl\neg P = \{(\neg P, \phi, \{\neg P\})\}$$
$$Gl(a \vee b) = \{(a \vee b, N, A) | (a, N, A) \in (G \ la) \ or \ (b, N, A) \in (G \ lb)\} \cup \bigcup_{\alpha} G \ l\alpha$$

$$where \ \alpha \in \{N | (a, N, C) \in G \ la \ or \ (b, N, C) \in G \ lb\}$$
$$Gl(a \wedge b) = \{(a \wedge b, Na \wedge Nb, A \cup B) | (a, Na, A) \in (G \ la), (b, Nb, B) \in (G \ lb)\} \cup \bigcup_{\alpha} G \ l\alpha$$

$$where \ \alpha \in \{Na \wedge Nb | (a, Na, A) \in (G \ la), (b, Nb, B) \in (G \ lb)\}$$
$$G \ l(\bigcirc a) = \{(\bigcirc a, a, \phi)\} \cup G \ la$$
$$G \ l(\ominus a) = \{(\ominus a, a, \ominus a)\} \cup G \ la$$
$$G \ l(\neg \bigcirc a) = \{(\neg \bigcirc a, \phi, term), (not \bigcirc a, \neg a, \phi)\} \cup G \ l\neg a$$
$$G \ l(\neg \bigcirc a) = \{(\neg \bigcirc a, \phi, term), (not \bigcirc a, \neg a, \phi)\} \cup G \ l\neg a$$

$$G \mid (\diamond a) = \{(\diamond a, N, C) \mid [(a, N, C) \in G \mid a, C = A] \; or \; [N = \diamond a, C = \diamond a]\} \cup \underset{\alpha}{\cup} G \mid \alpha$$

$\qquad$ where $\alpha \in \{N \mid (a, N, A) \in G \mid a\}$

$$G \mid (\Box a) = \{(a, \phi, C \wedge term) \mid [(a, \phi, C) \in G \mid a\} \cup \{(a, N \wedge \Box a, C) \mid [(a, N, C) \in G \mid a\} \cup G \mid N \wedge \Box a$$

$\qquad$ where $\phi \cup \Box a = \Box a$

$$G \mid (a \; U \; b) = \{(a \; U \; b, N, B) \mid (b, N, B) \in G \mid b\} \cup (\underset{\alpha}{\cup} G \mid \alpha) \cup (\underset{\beta}{\cup} G \mid \beta)$$

$\qquad$ where $\alpha \in \{N \wedge (a \; U \; b) \mid (a, N, A) \in G \mid a\}$

$\qquad$ where $\beta \in \{\beta \mid (b, N, B) \in G \mid b\}$

$$G \mid (a \; \Cup \; b) = \{(a \; \Cup \; b, N, B) \mid (b, N, B) \in G \mid b\} \cup (\underset{\alpha}{\cup} G \mid \alpha) \cup (\underset{\beta}{\cup} G \mid \beta)$$

$\qquad$ where $\alpha \in \{N \wedge (a \; \Cup \; b) \mid (a, N, A) \in G \mid a\}$

$\qquad$ where $\beta \in \{\beta \mid (b, N, B) \in G \mid b\}$

$\qquad$ note: this until does not require eventuality checking

$$G \mid (\neg(a \; U \; b)) = G \mid ((\neg b) \; \Cup \; \neg(a \vee b))$$

$$G \mid (\neg(a \; \Cup \; b)) = G \mid ((\neg b) \; U \; \neg(a \vee b))$$

$$G \mid ((L \wedge term) C b) = \{((L \wedge term) C b, b, L)\} \cup G \mid b$$

$$G \mid (L C b) = \{(L C b, \diamond b, L)\} \cup G \mid \diamond b$$

$$G \mid (a C b) = \{(a C b, N, A)\} \cup \underset{\alpha}{\cup} G \mid \alpha$$

$\qquad$ where $\alpha \in \{N C b \mid (a, N, A) \in G \mid a\}$

$G \mid (\neg(a \; C b))$ is constructed by utilizing the equivalences given earlier.

Not all transitions permitted by this construction are consistent. Nor are all eventualities necessarily satisfied. Inconsistent transitions are easy to recognize and to delete. A transition, t(M,N,C) is consistent iff, for each x in P, both x and $\neg x$ are not in C. There are three types of formulas in which eventualities must be checked. They are formulas which contain either eventuality operator, the until operator or the chop operator. The satisfaction of eventualities is checked by partitioning the graph into strongly connected components, then deleting any component in which there is an eventuality condition and no path to a component in which the eventuality condition is satisfiable.

9.2 Specification and Synthesis of Hardware

We give the example of the synthesis of hardware aribiter from PTIL specifications. The example highlights some of the performance issues associated with the synthesizer.

9.2.1 Hardware Arbiter

The hardware arbiter coordinates the access to a resource by several users. Dill and Clarke give a specification of a hardware arbiter in CTL[4], a variant of branching time logic. The specifications in PTIL of the arbiter given here are based upon their discription of the arbiter and includes in addition the specification of linear waiting. The protocol, which involves the exchange of signals, is described as follows:

1. If the user i maintains the request UR_i, then the arbiter will eventually seek the transfer of parameters from that user by raising the request T_{R_i} to the transfer module corresponding to user i.

$$\forall i ([\![UR_i \Rightarrow \;]\!] \diamond T_{R_i})$$

Hp1. It is assumed that if the arbiter seeks the transfer of parameters from a user i (say) by raising the signal T_{R_i}, then the user will eventually complete the transfer of the parameters by raising the signal TA_i.

$$\forall i ([\![T_{R_i} \Rightarrow \;]\!] \diamond TA_i)$$

2. After the transfer of parameters by a user i designated by the raising of TA_i, the arbiter will eventually make a request to the resource module by raising the signal RS.

$$\forall i(\llbracket TA_i \Rightarrow \rrbracket \diamond RS)$$

Hp2. It is assumed that upon the receipt of a request RS from the arbiter module, the resource module will eventually grant the resource to the arbiter by raising the signal RA.

$$\llbracket RS \Rightarrow \rrbracket \diamond RA$$

3. Upon the receipt of the acknowledgment RA from the resource module, the arbiter module will eventually grant access to the resource by raising the signal UA_i corresponding to the raised signal T_{R_i}.

$$\forall i(\llbracket RA \Rightarrow \rrbracket T_{R_i} \rightarrow \diamond UA_i)$$

4. It is assumed that after the utilization of the granted resource by user i, the user releases the resource by eventually lowering UR_i.

$$\forall i(\llbracket UA_i \Rightarrow \rrbracket \diamond \neg UR_i)$$

5. The arbiter module eventually lowers its request to the transfer module, the resource module and its acknowledgment to the user module i upon the lowering of UR_i.

$$\forall i(\llbracket \neg UR_i \wedge UA_i \Rightarrow UR_i \rrbracket^* \llbracket \Rightarrow \neg T_{R_i} \rrbracket^* \llbracket \Rightarrow \neg RS \rrbracket \diamond \neg UA_i)$$

Hp3. It is assumed that upon lowering T_{R_i}, User i should eventually lowers TA_i before the next time T_{R_i} is raised by the arbiter module.

$$\llbracket \neg T_{R_i} \Rightarrow T_{R_i} \rrbracket \diamond \neg TA_i$$

Hp4. It is assumed that upon lowering of the request RS by the arbiter module, the resource module will eventually lower the signal RA before the next time the request RS is raised by the arbiter module.

$$\llbracket \neg RS \Rightarrow RS \rrbracket \diamond \neg RA$$

6.*Mutual exclusion*: The system should not select two users simultaneously for granting access to the resource.

$$\forall i \neq j \square \neg (T_{R_i} \wedge T_{R_j})$$

7.*Selection*: The system selects one of the pending requests nondeterministically for granting access to the resource.

$$\forall i \neq j(UR_i \wedge UR_j \rightarrow (*\llbracket T_{R_i} \Rightarrow T_{R_i} \rrbracket \vee *\llbracket T_{R_j} \Rightarrow T_{R_j} \rrbracket))$$

8. *Linear Waiting*: Between two successive selections of a request from a user for access to the resource, every other user whose request for access to the resource is pending is selected. That is no user need to wait more than every other requesting user has his turn at most once.

$$\forall i \neq j(\llbracket UR_i \Rightarrow \rrbracket^* \llbracket UR_j \wedge \neg T_{R_j} \wedge T_{R_i} \Rightarrow \neg UR_i \rrbracket \rightarrow$$

$$\forall i \neq j(\llbracket UR_i \Rightarrow \rrbracket^* \llbracket UR_j \wedge \neg T_{R_j} \wedge T_{R_i} \Rightarrow \neg UR_i \rrbracket(\llbracket \Rightarrow T_{R_j} \rrbracket \square \neg UR_i \vee \llbracket \Rightarrow T_{R_j} \rrbracket \diamond T_{R_j}))$$

The synthesis of the formula proceeds by translating each PTIL formula to a corresponding PTL formula and then synthesizing the PTL formula. Table 9.1 lists the time required (in seconds) to construct the corresponding ω-graph of each formula. The global graph of the specification is the pointwise combining of the graphs of individual formulas while simultaneously rejecting inconsistent computation sequences. The synthesis of the ω-graph is given in [1]

9.3 Performance Issues and Needed Improvements

There are several areas in which the synthesizer needs improvement. The first is the simplification of the generated ω-graph. For example in a generated ω-graph two transitions t(m,n,[a,b]) , and t(m,n,[a, not b]) from the state m to the state n may be replaced with a single transition t(m,n,[a]). Similarly two transitions t(m,n,[a]), and t(m,n,[a, b]) may be replaced with the transition t(m,n,[a]). In both examples, the value of b is irrelevant to the transition. Currently the synthesizer detects situations of the second type but not situations of the first type. Note that the above concerns redundant transitions. Another aspect of inefficiency is the generation of redundant states. The simplest situation is when two states have identical outgoing transitions. The synthesizer checks for this case.

Formula	Time
A1	1.04
A2	1.12
A3	1.40
A4	1.10
A5	1237.12
A6	0.16
A7	48.86
A8	3306.78
H1	1.14
H2	0.92
H3	40.58
H4	30.96

Table 9.1: Time (in seconds) to Construct Graph

There are also ways of optimizing on the number of states by merging states under a weaker notion than the notion of identical outgoing transitions. While some of the optimizations have been implemented, the system requires further pruning. The second area in which the synthesizer needs improvement is the storage of the subgraphs. Consider table 9.1 with the times for the construction of the ω-graphs for the axioms of the arbiter. The times for the construction of the axioms 5 and 8 appear to be excessive. The time is highly dependent on the size of the database of transitions. Both axioms generate a large number of subformulas. Many of the subformulas are used just once in the construction of the graph of some other formula. If the graphs for these subformulas were eliminated, the data base would be smaller. This results in reduced search thereby permitting a reduced construction time. Currently efforts are underway to tune the system.

10 Conclusions

We proved that the decision procedures for PTIL is PSPACE Complete. This is a positive result. The result shows that under an appropriate notion a syntactic characterization of PTIL can be given in terms of PTL and vice versa. The approach of $\pi\alpha$- equivalence is interesting in the sense that we can establish equivalences of classes of temporal logics under that notion.

A prototype implementation has been carried out in Prolog. Elements of hardware have been synthesized using the implementation.

11 Acknowledgements

We thank Piotr Berman for his encouragement and critical remarks on an earlier draft outline of the paper.

References

1. Aaby,A.A and Narayana,K.T, Synthesis of Hardware from Propositional Temporal Interval Logic , Computer Science Research Report, 1988.

2. Ben-Ari,M., Manna,Z. and Pnueli,A., The Temporal Logic of Branching Time, *Acta Informatica 20*, (1983), pp. 207-226.

3. Clarke,E.M. and Emerson,E.A., Using Branching Time Temporal Logic to Synthesize Synchronization Skeletons, *Science of Computer Programming 2*, (1982), pp. 241-266.

4. Dill,D.L and Clarke,E.M, Automatic Verification of Asynchronous Circuits Using Temporal Logic, *1985 Chappel Hill Conference on VLSI*, , May 1985.

5. Fujita,M., Tanaka,H. and Moto-oka,T., Specifying Hardware in Temporal Logic and Efficient Synthesis of State-Diagrams Using Prolog, *Proceedings of the International Conference on Fifth Generation Computer Systems*, , 1984.

6. Lichtenstein,O and Pnueli,A, Checking That Finite State Concurrent Programs Satisfy Their Linear Specifications, *12th ACM Symp. on Prin. of Prog. Lang.*, , January 1985, pp. 97-107.

7. Lichtenstein,O, Pnueli,A and Zuck,L, The Glory of the Past, *Proc. Logics of Programs*, New-York, June 1985.

8. Manna,Z. and Wolper,P., Synthesis of Communicating Processes from Temporal Logic Specifications, *Trans. Prog. Lang and Systems 6*, 1 (January 1984), pp. 68-93.

9. Plaisted,D,A., A Low Level Language for Obtaining Decision Procedures for Classes of Temporal Logics, in *Logics of Programs*, vol. LNCS 164, Springer-Verlag, 1983, pp. 403-420.

10. Pnueli,A, Application of Temporal Logic to the Specification and Verification of Reactive Systems: A Survey of Current Trends, *Lecture Notes in Computer Science 224*, (1986), pp. 510-584, Springer-Verlag.

11. Pnueli,A., The Temporal Logic of Programs, *IEEE Annual Symp. on Foundations of Computer Science*, , November 1977.

12. Rosner,R. and Pnueli,A., A Choppy Logic, in *Symposium on Logic in Computer Science*, Cambridge,Massachusetts, 1986, pp. 306-313.

13. Schwartz,R.L., Melliar-Smith,P.M. and Vogt,F.H., An Interval Logic for Higher Level Temporal Reasoning, *Conference on Principles of Distributed Computing*, , 1983.

14. Sistla,A.P. and Clarke,E.M., The Complexity of Propositional Linear Temporal Logics, *J. ACM 32*, 3 (July 1985), pp. 733-749.

Computational Metatheory in Nuprl *

Douglas J. Howe

Department of Computer Science
Cornell University
Ithaca, NY 14853-7501

Abstract

This paper describes an implementation within Nuprl of mechanisms that support the use of Nuprl's type theory as a language for constructing theorem-proving procedures. The main component of the implementation is a large library of definitions, theorems and proofs. This library may be regarded as the beginning of a book of formal mathematics; it contains the formal development and explanation of a useful subset of Nuprl's metatheory, and of a mechanism for translating results established about this embedded metatheory to the object level. Nuprl's rich type theory, besides permitting the internal development of this partial reflection mechanism, allows us to make abstractions that drastically reduce the burden of establishing the correctness of new theorem-proving procedures. Our library includes a formally verified term-rewriting system.

Key words and phrases. Theorem proving, tactics, type theory, reflection, formal metamathematics, constructive mathematics.

1   Introduction

Most theorem-proving systems that allow the user to soundly extend the inference mechanism with new theorem-proving procedures use one of two approaches. The first approach is to require that new procedures be proven correct in a formalized metatheory [5,2,14,11]. The second is to provide a tactic mechanism [7,4], which permits arbitrary new procedures, but which requires each application of such a procedure to generate a proof in terms of the primitive inference rules. It appears that only the second approach has been used in significant applications. In this paper we

---

*This research was supported in part by NSF grant CCR-8616552.

show how Nuprl [4], which incorporates a tactic mechanism, can be "boot-strapped" to encompass both approaches. We show how Nuprl's powerful type theory can be used to develop a completely internal account of a portion of its metatheory, and to make abstractions that significantly reduce the burden of formally verifying new theorem proving procedures. This partial reflection mechanism, together with some applications, has been implemented in Nuprl, resulting in a completely formal account of an extension to Nuprl that allows new theorem-proving procedures to be programmed in the type theory. This implementation provides some evidence for the practical potential of the approach.

As a simple example of the limitations of the tactic mechanism, consider equations involving an associative commutative operator ".". One way to check that an equation

$$a_1 \cdot a_2 \cdot \ldots \cdot a_m = b_1 \cdot b_2 \cdot \ldots \cdot b_n$$

holds is to sort the sequences $a_1, \ldots, a_m$ and $b_1, \ldots, b_n$, with respect to some ordering on terms, and check that the results are identical. A tactic that emulated this informal procedure would have to chain together appropriate instances of lemmas (*e.g.*, for the associativity and commutativity of ·) and rules (*e.g.*, the substitution rule). This indicates two major problems. First, the tactic writer must be continually concerned with generating Nuprl proofs, and this can increase the intellectual effort involved in constructing theorem-proving procedures. Secondly, there is a major efficiency problem. In the example, even though it is known in advance that the sorting algorithm is sufficient to establish equality, every time the tactic is called it must "re-justify" the algorithm.

The mechanisms we have implemented allow procedures such as the one just described to be used directly. The main part of our implementation consists of a large collection of Nuprl definitions, theorems and proofs. This *library* may be regarded as the beginning of a book of formal mathematics; it contains the formal development and explanation of a useful subset of Nuprl's meta-theory, and of a mechanism for translating results established about this embedded meta-theory to the object level. The most important application of this is to the automation of reasoning: one can write Nuprl programs that directly encode such meta-level procedures as the sorting-based algorithm given above, prove that the program is correct, and then apply it in the construction of Nuprl proofs. Applications of such programs can be done in a manner consistent with Nuprl's proof development paradigms.

The core of our library contains the development of the partial reflection mechanism. Natural representations for a certain subclass of the terms of the type theory (roughly, the quantifier-free terms) and for contexts (*i.e.*, definitions and hypotheses) are constructed. Theorems are then proven in Nuprl that connect the representations to the objects represented. This allows the results of computations involving representations to be used in making judgments about the represented objects. The remainder of the library contains the development of several applications. These include a term rewriting system and a procedure involving the algorithm discussed above.

The basic idea of the reflection mechanism is simple. Using Nuprl's recursive-type constructor, we define a type Term0[1] that represents a certain subset of the terms of Nuprl's theory. We then define what it means for a term (*i.e.*, a member of Term0) to be well-formed with respect to an environment, which is a Nuprl object which associates atoms with Nuprl objects. Environments are used to represent definitions and hypotheses. Let Term(α) be the type of all members of Term0 that are well-formed with respect to α. We can demonstrate the existence of Nuprl functions type and val such that for any α,

$$\text{type}(\alpha) \quad \text{in} \quad \text{Term}(\alpha)\text{->SET}$$

and

$$\text{val}(\alpha) \quad \text{in} \quad \text{t:Term}(\alpha) \text{ -> type}(\alpha)(\text{t}),$$

where a member of SET is a type together with an equality relation on that type.

To see how the above can be put to use, consider a simplified example. Suppose that we have constructed a function f of type

$$\text{Term0 -> Term0},$$

and that we can show that it preserves equality in some environment α, *i.e.*, that for any term t in Term(α),

$$\text{type}(\alpha)(t) = \text{type}(\alpha)(f(t)) \text{ in SET}$$

and

$$\text{val}(\alpha)(t) = \text{val}(\alpha)(f(t)) \text{ in type}(\alpha)(t)$$

(*i.e.*, for the second equality, the equivalence relation from type(α)(t) is satisfied). For example, f might be based on the sorting procedure mentioned earlier. Suppose that we are in the course of proving a Nuprl theorem, that the current goal is of the form >> T (where >> is Nuprl's turnstile), and that we want to "apply" f to T. The first step is to *lift* the goal. To do this, we apply a special tactic which takes as an argument the environment α. This tactic first computes a member t of Term(α) such that the application val(α)(t) computes to a term identical to T, and then it generates the subgoal

$$\text{>> val}(\alpha)(t)$$

(assuming that the environment α is appropriate). We now apply a tactic which, given f, has the effect of computing $f(t)$ as far as possible, obtaining a value $t' \in$ Term(α), and producing a subgoal

$$\text{>> val}(\alpha)(t').$$

---

[1] We will use typewriter typeface for terms (that possibly contain definition instances) of Nuprl's type theory. Italicized identifiers within these terms will be meta-variables.

At this point, we can proceed by applying other rewriting functions, or by simply computing the conclusion and obtaining an "unlifted" goal $\gg T'$ to which we can apply other tactics.

An important aspect to this work is that the partial reflection mechanism involves no extensions to the logic; the connection between the embedded meta-theory and the object theory is established completely within Nuprl. This means, first, that the soundness of the mechanism has been formally verified. Secondly, there is considerable flexibility in the use of the mechanism. All of its components are present in the Nuprl library, so one can use them for new kinds of applications without having to do any metatheoretic justification. Rewriting functions are just an example of what can be done; many other procedures that have proven useful in theorem proving, congruence closure for example, can be soundly added to Nuprl. Thirdly, the mechanism provides a basis for stating new kinds of theorems. For example, one can formalize in a straightforward way the statement of a familiar theorem of analysis: *if $f(x)$ is built only from x, $+$, $\cdot$, $\ldots$, then f is continuous.*

Finally, and perhaps most importantly for the practicability of this approach, it is possible to make abstractions that drastically reduce the burden of verifying the correctness of new procedures. Since contexts (environments) are represented, one can prove the correctness of a procedure once for a whole class of applications. For example, a simplification procedure for rings could be proven correct for any environment where the values of certain atoms satisfy the ring axioms, and then be immediately applied to any particular context involving a ring. A general procedure such as congruence closure could be proved once and for all to be correct in any environment whatsoever. Another kind of abstraction is indicated by the rewriting example to be given later. Functions such as **Repeat** are proved to map rewriting functions to rewriting functions; these kinds of combinators can often reduce the proof of correctness for a new rewriting function to a simple typechecking task that can be dealt with automatically.

One of the main motivations of this work is to provide tools that will be of use in formalizing Bishop-style constructive real analysis [1]. The design of the reflection mechanism incorporates some of the notions of Bishop's set theory. However, our work is more generally applicable, since Nuprl's type theory is capable of directly expressing problems from many different areas. It should be of use in other areas of computational mathematics (*e.g.*, in correct-program development). Using the techniques developed in AUTOMATH [6] and LF [8], it can be applied to reasoning in a wide variety of other logics (constructive in character or not). For example, any first-order logic can be embedded within Nuprl; in such a case, the reflection mechanism will encompass the quantifier-free portion of the logic.

In the next section is a brief description of some of the relevant components of Nuprl. This is followed by a concrete example of term-rewriting. The following section gives some details about what has been implemented. Finally, there is a discussion of some related work, and some concluding remarks are made. For a more complete account of the work described in this paper, and for a complete listing of the library that was constructed, see [9].

2 Nuprl

Nuprl [4] is a system that has been developed at Cornell by a team of researchers, and is intended to provide an environment for the solution of formal problems, especially those where computational aspects are important. One of the main problem-solving paradigms that the system supports is that of *proofs-as-programs*, where a program specification takes the form of a mathematical proposition implicitly asserting the existence of the desired programs. Such a proposition can be formally proved, with computer assistance, in a manner resembling conventional mathematical arguments. The system can extract from the proof the implicit computational content, which is a program that is guaranteed to meet its specification. With this paradigm, the construction of correct programs can be viewed as an applied branch of formal constructive mathematics.

The logical basis of Nuprl is a constructive type theory that is a descendent of a type theory of Martin-Löf [12]. The rules of Nuprl deal with *sequents*, which are objects of the form

$$x_1 : H_1, \ x_2 : H_2, \ \ldots, \ x_n : H_n \ \gg \ A.$$

Sequents, in the context of a proof, are also called *goals*. The terms H_i are referred to as *hypotheses* or *assumptions*, and A is called the *conclusion* of the sequent. Such a sequent is said to be true when, roughly, there is a procedure that, under the assumption that H_i is a type for $1 \leq i \leq n$, takes members x_i of the types H_i, and produces a member of the type A. An important point about the Nuprl rules is that they allow top-down construction of this procedure. They allow one to *refine* a goal, obtaining subgoal sequents such that a procedure for the goal can be computed from procedures for the subgoals.

The Nuprl type theory has a large set of type constructors. They can be roughly described as follows. The type Int is the type of all integers. Atom is the type of all character strings (delimited by double quotes). The disjoint union $A \mid B$ has members inl(a) and inr(b) for $a \in A$ and $b \in B$. The dependent product $x : A \# B$ (written $A \# B$ when x does not occur free in B) has members $< a, b >$ for $a \in A$ and $b \in B[a/x]$. The dependent function space $x : A -> B$ (written $A -> B$ when x does not occur free in B) has as members all functions $\lambda x.b$ such that for all $a \in A$, $b[a/x] \in B[a/x]$. The type A list has as members all lists whose elements are in A. The type $\{x : A \mid B\}$ is the collection of all members of A such that $B[a/x]$ has a member. The type $a = b$ in A (for $a, b \in A$) has the single member axiom if a and b are equal as members of A, and has no members otherwise. The members of the recursive type $\text{rec}(T.A)$ are the members of $A[\text{rec}(T.A)/T]$. Finally, there is a cumulative hierarchy of universes U1, U2, ..., where each Ui is obtained via closing under the other type constructors.

Associated with each type constructor are forms for making use of members of the type; $e.g.$, projection for pairs, an induction form for integers, and application for functions. These forms give rise to a system of computation, and the proceeding discussion of membership, to be accurate, must be amended to reflect the fact that $b \in A$ whenever $a \in A$ and b computes to a. The usual connectives and quantifiers of predicate calculus can be defined in terms of the above type constructors using the *propositions-as-types* correspondence.

Proofs in Nuprl are tree-structured objects where each node has associated with it a sequent and a refinement rule. The children of a node are the subgoals (with their subproofs) which result from the application of the refinement rule of the node to the sequent. A refinement rule can be either a primitive inference rule, or a tactic (written in the ML language [7]). If it is a tactic, then there is a hidden proof tree whose leaves are the children of the refinement.

Nuprl has a large number of inference rules; we only point out one which is of special significance for our work. The *evaluation rule* allows one to invoke Nuprl's (reasonably efficient) evaluator on either a hypothesis or the conclusion of a sequent. The term is given as input to the evaluator, is computed as far as possible, and is replaced in the sequent by the result of the computation. This rule is used to execute theorem-proving procedures that are based on the reflection mechanism.

Interactions with Nuprl are centred on the *library*. A library contains an ordered collection of definitions, theorems, and other objects. New objects are constructed using special-purpose window-oriented editors. The text editor, together with the definition facility, permit very readable notations for objects of Nuprl's type theory. The proof editor is used to construct proofs in a top-down manner, and to view previously constructed proofs. Proofs are retained by the system, and can be later referred to either by the user, or by tactics.

3   An Example

We will now look at an example developed in our library, where a simple rewriting function is applied to an equation involving addition and negation over the rational numbers. The environment used is α_Q, which contains "bindings" for rational arithmetic. For example, it associates to the Nuprl atom "Q" the set ($i.e.$, member of Set) Q of rational numbers, and to the atom "Q_plus" the addition function. The rewriting functions for this environment are collected into the type $\text{Rewrite}(\alpha\_Q)$. The requirement for being a rewriting function over α_Q is actually stronger than indicated earlier; the criteria given there must hold for any α which extends α_Q.

The member of `Rewrite(`α`_Q)` we will apply is

```
Repeat( TopDown( rewrite[x;y]( -(x+y) -> -x+-y ) ) )
```

(this function is named `norm_wrt_Q_neg`). This does just what one would expect: given a term, it repeatedly, in top-down passes, rewrites subterms of the form `-(`$x+y$`)` to `-x+-y`. The functions `Repeat` and `TopDown` both have type

```
Rewrite(α) -> Rewrite(α)
```

for any α. The argument to `TopDown` is by definition the application of a function of three arguments, which must be: a list of atoms that indicate what constants are to be interpreted as variables for the purpose of matching; and two members of `Term0` which form the left and right sides of a rewrite-rule. Nuprl definitions are used to make members of `Term0` appear as close as possible to the terms they represent. The term `x+y` that appears in the definition of `norm_wrt_Q_neg` is, when all definitions are expanded,

```
inl( < "Q_plus", <inl(<"x",nil>) . inl(<"y",nil>) . nil > ).
```

The sequent we apply our function to is

```
>> -(w+x+y*z+z) = -w+-x+-(y*z)+-z in Q
```

(the hypotheses declaring the variables to be rationals have been, and will be, omitted). The first step is to apply the tactic

```
(LiftUsing ['α_Q'] ...),
```

(where the "..." indicates the use of a Nuprl definition that applies a general purpose tactic called the *autotactic*) which generates the single subgoal

```
α:Env, (...)
>> ↓( val( α, -(w+x+y*z+z) = -w+-x+-y*z+-z in Q ) )
```

($\downarrow$ is an information-hiding operator that will not be discussed here—see [4] for an explanation). The notation "(...)" indicates that the display of a hypothesis has been suppressed by the system. This elided hypothesis contains (among other information) the fact that α is equal to an environment formed by extending α_Q by entries for representatives of the variables `w`, `x`, `y` and `z`. The next step is to apply the tactic

```
(RewriteConcl 'norm_wrt_Q_neg' ...),
```

which generates a single subgoal with conclusion

```
↓( val( α,-w+-x+-(y*z)+-z = -w+-x+-(y*z)+-z in Q ) ).
```

The proof is completed using an equality procedure to be described later.

4   The Implementation

In this section is a presentation of the Nuprl library containing the partial reflection mechanism and applications. The library contains more than 1300 objects, and is by far the largest development undertaken so far using Nuprl. An ideal description of the library would involve the use of the Nuprl system itself; the system is designed not only for the construction of formal arguments, but also for their presentation. The best that can be done here is to explain some of the basic definitions and theorems. The next three subsections are devoted to the library objects and tactics that establish the reflection mechanism. The last two subsections describe some implemented examples of the use of the mechanism.

Of the 1300 objects in the library, about 800 are theorems, 400 are definitions, and 100 are ML objects. The library can be roughly divided according to topic (in order): basics; lists; sets; the representation of Nuprl terms; association lists and type environments; function environments and combined environments; booleans and "partial" booleans; well-formedness and evaluation of terms; the monotonicity of certain functions with respect to environment extension, and operations on environments; term rewriting; and other applications. The library divisions corresponding to the first three topics are of somewhat lesser interest, and will be discussed very briefly in this section.

Before proceeding, we need to establish a few notational conventions. Typewriter typeface will be used exclusively to denote Nuprl terms or names of Nuprl objects. In the context of a Nuprl term, a variable which is in italics will denote a meta-variable. Thus x+y denotes a Nuprl term where x and y are *Nuprl* variables, whereas in $x+y$, x and y range over Nuprl terms. We will be somewhat sloppy in the use of meta-variables, using them, for example, to denote parameters of Nuprl definitions, but the context should make clear what the intention is.

The presentation below is somewhat in the style of conventional mathematics. Usually, explicit references to the existence of Nuprl objects will not be made. Unless stated otherwise, when a definition is made, there is a corresponding definition in the library; and when it is asserted that a certain statement, written in Nuprl syntax, is true, or can be proved, then there is a corresponding theorem in the library.

The "basics" section of the library contains mostly simple definitions, such as for logical concepts and definitions pertaining to the integers. There are also definitions that provide alternate notations for simple combinations of base Nuprl terms. For example, we define a "let" construct in the usual way:

$$\text{let } x = t \text{ in } t' \;\equiv\; (\lambda\, x.t')(t),$$

and use it (together with the definitions for projection from pairs) in the definition let2:

$$\text{let } x,y = p \text{ in } t \;\equiv\; \text{let } x = p.1 \text{ in let } y = p.2 \text{ in } t.$$

The "lists" section of the library contains a fairly substantial development of list theory.

In order to account for equivalence relations that are not built into the type theory[2] (*e.g.*, equality of real numbers), we define a *set* to be a pair consisting of a type together with an equivalence relation on the type. We define a hierarchy of collections of sets that parallels the universe hierarchy: a member of Set(i) is a pair consisting of a type from Ui together with an equivalence relation on that type. For S a set, define $|S|$ to be the type component of S, and for a and b members of $|S|$, define $a=b$ in S to be the application of the equivalence relation of S to a and b. We also define some particular sets that will be used later in the library: Set and SET are Set(6) and Set(7), respectively, and

$$\text{Prop} \quad \equiv \quad \text{<U6, } \lambda \text{ P,Q. P <=> Q>.}$$

The choices of 6 and 7 are somewhat arbitrary; we only need to guarantee that the levels are high enough to accommodate anticipated applications.

It is a simple matter to define some basic set constructors. We will overload our notations and use S_1->S_2 to denote the *set* of functions from S_1 to S_2, and S_1#S_2 for the product of S_1 and S_2. Also, #(L) will denote the product of the sets in the list L. We will not discuss here our treatment of Bishop's notion of *subset* [1].

4.1   Representation

In this section we discuss first the representation in Nuprl of a simple class of Nuprl terms, and then the representation of contexts (*i.e.*, the library and hypotheses).

The type Term0 of "raw" terms, which do not necessarily represent an object of the Nuprl theory, is defined via the recursive type

```
rec(T. (Atom#T list) | (T#T#Atom) | (Atom#T#T) | Int ).
```

Members of this type, which will be referred to as *meta-terms*, or just *terms* when no confusion can result, can be thought of as trees with four kinds of nodes. The meaning of the different kinds of nodes is suggested by the notations for the injections:

$$
\begin{aligned}
f(l) &\equiv \quad \text{inl(<}f,l\text{>)} \\
x = y \text{ in } A &\equiv \quad \text{inr(inl(<}x,y,A\text{>))} \\
y\{i \; x\} &\equiv \quad \text{inr(inr(inl(<}i,x,y\text{>)))} \\
n &\equiv \quad \text{inr(inr(inr(}n\text{))).}
\end{aligned}
$$

The above are all members of Term0 whenever f, A, i are in Atom, x and y are in Term0, l is a list of members of Term0, and n is an integer. These kinds of nodes will

---

[2]Nuprl has a quotient-type constructor, but it cannot be used for our purposes (see [4]).

be referred to, respectively, as function-application nodes, equality-nodes, i-nodes, and integer nodes. The first will represent terms which are function applications; the second, terms which are equalities (in the set sense); and the last, terms which are integers. The inclusion of i-nodes is motivated by Bishop's notion of *subset* [1]; i-nodes will be ignored in the rest of this paper.

Contexts are represented with association lists (or "a-lists"), of type

$$\texttt{(Atom \# } A \texttt{) list}$$

for A a type. Such lists are used to associate Nuprl objects with the atoms that appear in meta-terms. In dealing with a-lists, and in the term-rewriting system described later, extensive use is made of *failure*. For A a type, define $?A$ to be the type $A \,|\, \texttt{True}$ (where $\texttt{True}$ is a single-element type). A value $\texttt{inl}(a)$ of $?A$ is denoted by $\texttt{s}(a)$ ($\texttt{s}$ for "success"), and the unique value of the form $\texttt{inr}(a)$ is denoted by $\texttt{fail}$. In later sections, the distinction between a and $\texttt{s}(a)$ will often be glossed. The library has many objects related to failure and a-lists.

A basic definition for a-lists is of a-list application, written $l\{A\}(a)$. If A is a type, if l is in $\texttt{Atom\#}A$ $\texttt{list}$, and if a is in $\texttt{Atom}$, then $l\{A\}(a)$ is in $?A$. Evaluation of $l\{A\}(a)$ either results in failure, or returns some b such that $<a,b>$ is a member of l.

Contexts are represented as pairs of a-lists, the first component of which is a type environment, and the second of which is a function environment. Type environments associate atoms with members of $\texttt{Set}$ and are members of the type:

$$\texttt{TEnv} \;\equiv\; \texttt{(Atom \# S:Set \# ?triv\_eq(S)) list} .$$

The predicate $\texttt{type\_atom}$ is defined so that $\texttt{type\_atom}(\gamma, a)$ is true if and only if either a is the atom $\texttt{"Prop"}$ or a is bound in the type environment γ. Also, $\texttt{type\_atom}(\gamma, a)$ has the property that if, during the course of a proof, γ and a are sufficiently concrete (*e.g.*, if they do not contain any free variables), it can be evaluated to either $\texttt{True}$ or $\texttt{False}$. Define

$$\texttt{AtomicMType}(\gamma) \;\equiv\; \{\ \texttt{a: Atom} \mid \texttt{type\_atom}(\gamma, \texttt{a})\ \}$$

(M is for "meta"). For members a of this type, define $\gamma(a)$ to be $\texttt{<Prop,fail>}$ if a is $\texttt{"Prop"}$, otherwise the value obtained from looking up a in γ. We define

$$\texttt{MType}(\gamma) \;\equiv\; \{\ \texttt{t: Atom list \# Atom} \mid \texttt{all\_type\_atoms}(\gamma, \texttt{t})\ \} .$$

This is the type of "meta-types" for functions; the first component of such a meta-type is a list of atoms that represent a function's argument types, and the second component represents the result type.

We can now define evaluation for members of $\texttt{MType}(\gamma)$. If mt is such a member, and $mt.1$ (*i.e.*, the first component of the pair mt) is non-empty, then we can evaluate *mt*'s "domain type":

```
dom_val(γ, mt)  ≡
    let l,b = mt in
    #((map λa. val(γ,a) on l to SET list)).
```

Evaluation of function meta-types is defined as:

```
val(γ, mt)  ≡
let l,b = mt in
if null(l) then val(γ,b) else dom_val(γ, mt) -> val(γ,b).
```

A value $val(\alpha, x)$ will often be referred to as the *meaning* or *value* of x in α.

Function environments associate an atom, called a *meta-function*, with a meta-type, with a value that is a member of the value of the meta-type, and with some additional information about the value. More specifically, for γ a type environment, define

```
FEnv(γ)  ≡
Atom # mt:MType(γ) # f:|val(γ,mt)| # val_kind(γ,mt,f)  list.
```

The type $val\_kind(\gamma, mt, f)$ can be ignored here. Definitions analogous to those made for type environments are made for function environments. Thus define function-environment application, the predicate `fun_atom`, and the type $MFun(\alpha)$ of meta-functions. For f in $MFun(\alpha)$, $mtype(\alpha, f)$ is the meta-type assigned to f by $\alpha.2$, and $val(\alpha, f)$ is the value assigned to f.

We can now define the type whose members represent contexts:

```
Env  ≡  γ:TEnv # FEnv(γ).
```

Members of `Env` will simply be called *environments*, and the variable α will always denote an environment. For most of the definitions presented so far that take a type environment as an argument, there is a new version which takes an environment as an argument. In what follows, only these new versions will be referred to, so the same notation will be used.

4.2 Well-Formedness and Evaluation

This section presents the core of the partial reflection mechanism. We formalize a notion of well-formedness for members of `Term0`, and construct an evaluation function for well-formed terms.

Roughly, a meta-term t is well-formed in an environment α if "the meta-types match up". For example, if t is a function-application $f(l)$, then we require that the domain component of the meta-type associated with f in α "matches" the list of terms l; *i.e.*, that they have the same length as lists, and that the atomic meta-types of f's domain "match" the atomic meta-types computed for the terms in l. A simple definition of "match" for atomic meta-types would be equality as atoms. However, since each f can have only one meta-type associated with it, this definition would preclude us from representing such simple terms as x+y, where x and y are Nuprl variables declared to be in a sub-type of the integers, and where + is addition over the integers. Also, it would preclude us from dealing with simple partial functions, *i.e.*, from representing terms like $g(x)$ where the domain type of g is a subtype of the type of x and x satisfies the predicate of the subtype. This leads us to consider a notion of matching of atomic meta-types that extends the simple version by allowing the associated values to stand in a subtype relationship. Consider the case of a meta-term $f(t)$, where f has meta-type domain the singleton list $[A]$, t has the meta-type B, and the value in α of A is a subtype of the value of B. Then for $f(t)$ to be well-formed we require that the *value* of t satisfy the subtype predicate associated with A.

Thus well-formedness depends on evaluation, which in turn is only defined on well-formed terms, and so we need to deal with these notions in a mutually recursive fashion. In particular, we first define the function which computes meta-types of terms, the evaluation function, and the well-formedness predicate, and then prove that they are well-defined (*i.e.*, have the appropriate types) simultaneously by induction.

Several functions below are defined by recursion. We will give the recursive definitions informally, although the formal versions are obtained directly.

The meta-type of a well-formed term is simple to compute; define $\mathtt{mtype}(\alpha,t)$ to be: if t is $f(l)$, then $\mathtt{mtype}(\alpha,f).2$; if t is $u=v$ in A, then "Prop"; if t is n then "Int". Also, define $\mathtt{type}(\alpha,t)$ to be the value in α of t's metatype. Evaluation of terms is defined using an auxiliary function that applies a function g to a list l and produces a tuple (denoted by $g\{\alpha\}(l)$) of the results. Thus we define $\mathtt{val}(\alpha,t)$ to be: if t is $f(l)$, then

$$\mathtt{if}\ \mathtt{null(mtype}(\alpha,f).1)\ \mathtt{then}\ \mathtt{val}(\alpha,f)$$
$$\mathtt{else}\ \mathtt{val}(\alpha,f)\ (g\{\alpha\}(l))$$

where $g(t)$ is $\mathtt{val}(\alpha,t)$; if t is $u=v$ in A, then

$$\mathtt{val}(\alpha,u)\mathtt{=val}(\alpha,v)\ \mathtt{in}\ \mathtt{val}(\alpha,A);$$

and if t is n, then n.

Due to space limitations, we will not give a detailed definition of the well-formedness predicate. Instead, we will just informally describe the criteria for a function-application $f(l)$ to be well-formed in an environment α. First, f must be bound as a function in α, and the domain part of its meta-type must be a list r with the same length as l. Secondly, the terms in l must be well-formed in α. Finally, for each t in l and corresponding a in r, a and $\mathtt{mtype}(\alpha,t)$ must "match" in the sense outlined earlier in this section. Note that this last criterion involves $\mathtt{val}(\alpha,t)$.

The assertion that a term t is well-formed in an environment α is written $\mathtt{wf@}(\alpha,t)$. The central theorem of the partial reflection mechanism is the following:

$$\forall\alpha\!:\!\mathtt{Env}.\ \forall t\!:\!\mathtt{Term0}.\ \mathtt{wf@}(\alpha,\mathtt{t})\ \mathtt{in}\ \mathtt{U}\ \&$$
$$\downarrow(\mathtt{wf@}(\alpha,\mathtt{t}))\ \texttt{=>}\ (\mathtt{mtype}(\alpha,\mathtt{t})\ \mathtt{in}\ \mathtt{AtomicMType}(\alpha)$$
$$\&\ \mathtt{val}(\alpha,\mathtt{t})\ \mathtt{in}\ |\mathtt{type}(\alpha,\mathtt{t})|)$$

(where $\mathtt{U}$ is a particular universe—we use membership in a universe to assert that a term is a well-formed type). Finally, define $\mathtt{Term}(\alpha)$ to be the set of all meta-terms that are well-formed in α.

For the reflection method to be useable, it is necessary that the lifting procedure outlined in the first section of this chapter be reasonably fast. Since each lifting of a sequent will require establishing that some meta-terms are well-formed, the well-formedness predicate $\mathtt{wf@}$ is not suitable, imposing excessive proof obligations. Instead, we use a characterization $\mathtt{wf}$ of the predicate that embodies an algorithm for the "decidable part" of well-formedness. To prove $\mathtt{wf}(\alpha,t)$ for concrete α and t, we apply Nuprl's evaluation rule, which simplifies the formula to a conjunction of propositions that assert that some terms satisfy some subtype predicates. In many cases arising in practice, the simplified formula will be $\mathtt{True}$ (or $\mathtt{False}$).

The characterization $\mathtt{wf}$ is accomplished using the "partial booleans". Define

$$\mathtt{Bool}\ \equiv\ \mathtt{True}\vee\mathtt{True}\qquad\text{and}\qquad\mathtt{PBool}\ \equiv\ \mathtt{Bool}\vee\mathtt{U},$$

so a partial boolean is either a boolean or a proposition. We define $\mathtt{PBool}$ analogues of the propositional connectives, and of some of the other basic predicates. The characterization is built essentially by replacing components of $\mathtt{wf@}$ by their $\mathtt{PBool}$ analogues.

We will usually want our assertions about the correctness of theorem-proving procedures to respect environment extension. For example, we may have constructed a function f that normalizes certain expressions and that is correct in an environment α_Q that associates certain atoms with the operations of rational arithmetic. Clearly we will want to do more than just apply f to terms that are well-formed in α_Q; at the very least, we will want to apply f to a term like $\mathtt{x+y}$, where $\mathtt{x}$ and $\mathtt{y}$ are variables declared in some sequent. This necessitates a notion of *sub-environment*. The environment α_1 is a sub-environment of α_2, written $\alpha_1\subset\alpha_2$, if $\alpha_1.1$ and $\alpha_1.2$ are sub-a-lists of $\alpha_2.1$ and $\alpha_2.2$, respectively. Most of the functions that have been defined so far that take an environment as an argument are *monotonic* with respect to environment extension. For example, if $\alpha_1\subset\alpha_2$, and if t is well-formed in α_1, then t is well-formed in α_2 and its values in the two environments are equal. A substantial portion of the library is devoted to monotonicity theorems.

4.3 Lifting

Suppose that we wish to use some procedure that operates on meta-terms to prove a sequent

$$A_1, \ A_2, \ \ldots, \ A_n \ \gg \ A_{n+1}.$$

We first *lift* the sequent using the tactic invocation LiftUsing *envs*, where *envs* is an ML list of Nuprl terms that are environments, call them $\alpha_1, \alpha_2, \ldots, \alpha_m$. All but one of the subgoals generated by this tactic will in general be proved by the autotactic (a general purpose tactic that runs after all other top-level tactic invocations). The remaining subgoal will have the form

$$\alpha : \text{Env}, \ (\ldots), \ \text{val}(\alpha, A_1'), \ \text{val}(\alpha, A_2'), \ \ldots, \ \text{val}(\alpha, A_n')$$
$$\gg \ \text{val}(\alpha, A_{n+1}'),$$

where $(\ldots)$ is an elided hypothesis, and each A_i' is a member of Term0 that represents A_i. By the use of definitions, A_i' will usually appear to the user to be exactly the same as A_i. The preceding description of lifting contains two simplifications. First, the hypotheses shown may be interspersed with hypotheses that have declared variables; the lifting procedure ignores such hypotheses. Secondly, some of the A_i may remain in the hypothesis list, not being replaced by $\text{val}(\alpha, A_i')$. This happens when the representation computed for A_i is trivial.

The first step in lifting a sequent is to apply a procedure that uses the α_j's to compute the representations A_i' in Term0 for the terms A_i, $1 \leq i \leq n + 1$. This procedure also produces a list of environment additions for those subterms that are unanalyzable with respect to the α_j's; each such subterm s will be represented by a new metafunction constant whose associated value will be s. The new variable α is added to the sequent to be lifted, as well as the new hypothesis $(\ldots)$. This elided hypothesis contains the following assertions: that the α_j are consistent (*i.e.*, that they agree on the intersection of their domains); that α is equal to the combination (call it $\overline{\alpha}$) of the α_j and the environment additions; that $\alpha_j \subset \alpha$ for each j; and that each A_i' is a well-formed meta-term that has meta-type "Prop". Of course, to justify the addition of this hypothesis, the lifting tactic must prove each of the assertions. This is accomplished by the use of simplification (using the evaluation rules) and of lemmas (to handle some details concerning the fact that some substitutions of $\overline{\alpha}$ for the variable α must be done before simplification). Most of these assertions will be proved automatically. The final stage of the lifting procedure is to replace each A_i by $\text{val}(\alpha, A_i')$; the tactic only needs to use evaluation in order to justify this.

The elided hypothesis contains most of the information necessary to prove the applicability of meta-procedures. For example, if f is in Rewrite(α'), then to prove that f is applicable all that needs to be shown is that $\alpha' \subset \alpha$. An important point about this new hypothesis is that it is easy to maintain. For example, rewriting functions preserve meta-types and values, so when a rewrite is applied to a sequent, we can quickly update the hypothesis to reflect the changes in the sequent. Thus, the work of applying a rewrite to a lifted sequent is dominated by the work of doing the actual rewriting.

4.4 Term Rewriting

The most important application of the reflection mechanism is to equality reasoning. In this section is a description of the implementation in Nuprl of a verified term-rewriting system. This implementation consists of a collection of definitions, theorems, and tactics that aid the construction of term-rewriting (*i.e.*, equality-preserving) functions. Using this collection, one can easily turn equational lemmas into rewrite rules, and concisely express a wide variety of rewriting strategies. Proving the correctness of rewriting strategies generally reduces to simple typechecking.

4.4.1 The Type of Rewriting Functions

Informally, a function f of type `Term0->?Term0` is a rewrite with respect to an environment α if for any term t that is well-formed in some extension α' of α, $f(t)$ either fails or computes to a term t' such that: (1) t' is well-formed in α'; (2) t' and t have the same meta-type, call it A; and (3) the meanings of t and t' are equal in the set which is the meaning of A. Define `Rewrite`(α) to be the collection of all such f. Note that monotonicity (with respect to environment extension) is built into the definition of `Rewrite`, so that if f is in `Rewrite`(α), and $\alpha \subset \alpha'$, then f is in `Rewrite`(α').

4.4.2 Basic Rewriting Functions

The basic units of conventional term rewriting are *rewrite rules*. These are ordered pairs $\langle t, t' \rangle$ of terms containing variables, where the variables of t' all occur in t. A term u rewrites to a term u' via the rule $\langle t, t' \rangle$ if there is a substitution σ for the free variables of t such that u' is obtained by replacing an occurrence of $\sigma(t)$ in u by $\sigma(t')$. In our setting, we will view a rewrite rule as a member of `Rewrite`(α); corresponding to $\langle t, t' \rangle$ will be the function which, given u, if there is a substitution σ such that $\sigma(t) = u$, returns $\sigma(t')$, or else fails. These rules will usually be obtained by "lifting" a theorem that is a universally quantified equation.

We first need to develop a theory of substitution for meta-terms. The *variables* of a meta-term t are defined with respect to a list l of identifiers, and are the subterms of t that are applications of a member of l to no arguments. A *substitution* is a member of the type `(Atom # Term0) list`. The application of a substitution *subst* to a meta-term t is written $subst(t)$. A matching function is extracted from the proof of the specification

> ∀vars:Atom list. ∀t1,t2:Term0.
> ?(∃s:Atom#Term0 list where ... & s(t1)=t2 in Term0)

(where we have elided an uninteresting conjunct).

The rewrite associated with a "lifted" equation is simply defined in terms of the matching function (call it `match`). For *vars* a list of atoms, and for u, v terms, the function is

> rewrite{*vars*}(u->v) ≡
> λt. let s(subst) = match(u,t,vars) in s(subst(v)): ?Term0.

(recall that `s(a)` denotes a successful computation). To prove that

$$\text{rewrite}\{vars\}(u\text{->}v) \text{ in Rewrite}(\alpha)$$

for particular u and v, one uses a special tactic that makes use of a collection of related lemmas. This tactic often completes the proof without user assistance.

There are two simple basic rewrites that are used frequently. The first is the trivial rewrite,

$$\text{Id} \quad \equiv \quad \lambda\text{t. s(t)}.$$

The second is `true_eq`, which rewrites an equality meta-term $a=b$ in A to `True` when a and b are "syntactically" equal.

4.4.3 Building Other Rewriting Functions

The approach to term rewriting of Paulson [13] is easily adapted to our setting. His approach involves using analogues of the LCF *tacticals* [7] as combinators for building rewrites, where his rewrites produce LCF proofs of equalities. Using these combinators, we can concisely express a variety of rewriting strategies (Paulson used his rewriting package to prove the correctness of a unification algorithm [13]). Furthermore, since we can prove that each combinator is correct (*i.e.*, it combines members Rewrite(α) into a member of Rewrite(α)), proving that a complex rewriting function is correct often involves only simple typechecking.

If f and g are in Rewrite(α), then so are the following.

- f THEN g. This rewriting function, when applied to a term t, fails if $f(t)$ fails, otherwise its value is $g(f(t))$.

- f ORELSE g. When applied to t, this has value $f(t)$ if $f(t)$ succeeds, otherwise it has value $g(t)$.

- **Progress** f. This fails if $f(t)$ succeeds and is equal (as a member of Term0) to t, otherwise it has value $f(t)$.

- **Try** f. If $f(t)$ fails then the value is t, otherwise it is $f(t)$.

- **Sub** f. This applies f to the immediate subterms of t, failing if f failed on any of the subterms.

- **Repeat** f. This is defined below.

In Nuprl, all well-typed functions are total. However, many of the procedures, rewrites in particular, that we will want to write will naturally involve unrestricted recursion, and in such cases we will not be interested in proving termination. For example, many theorem-proving procedures perform some kind of search, and searching continues as long as some criteria for progress are satisfied. In such a case, it will often be difficult or impossible to find a tractable condition on the inputs to the procedure that will guarantee termination. It appears that the effort to incorporate partial functions into the Nuprl type theory has been successful [3], but a final set of rules has not yet been formulated and implemented.

For our purposes, there is a simple-minded solution to this problem which should work in all cases of practical interest. We simply use integer recursion with a (very) large integer. Thus, for A a type, a a member of A, and f a function of type $A\text{->}A$, we can define $\texttt{fix}\{A\}(a,f)$ to be an approximation to the fixed-point of f. One difference between this and a true fixed-point operator is that we must take into account the effectively bogus base case a. Using this definition it is easy to define **Repeat**:

$$\text{Repeat}(f) \quad \equiv \quad \texttt{fix}\{\texttt{Term0->?Term0}\}(\texttt{Id}, \lambda\texttt{g. (f THEN g) ORELSE Id}).$$

We can also directly define, using Id as the "base case", a fixed-point operator **rewrite_letrec** for rewriting functions that satisfies

$$\forall\alpha\texttt{:Env. } \forall\texttt{F:Rewrite}(\alpha)\texttt{->Rewrite}(\alpha).$$
$$\texttt{rewrite\_letrec(F) in Rewrite}(\alpha).$$

As an example of the use of our combinators, we define a rewrite that does a bottom-up rewriting using f:

$$\text{BotUp}(f) \quad \equiv \quad \texttt{letrec g = (Sub(g) THEN Try}(f)).$$

Normalization with respect to f could be accomplished with

$$\texttt{Repeat (Progress (BotUp}(f))).$$

A rewriting function is applied using a special tactic. The subgoals generated by the tactic are the rewritten sequent, and a subgoal to prove that the rewriting function is well-typed. This latter subgoal often requires only simple typechecking that can be handled automatically.

4.5 Other Applications

There are also two smaller implemented applications in our library. First, as an example of a procedure that works on an entire lifted sequent, there is an equality procedure that uses equalities from the lifted hypotheses, and decides whether the equality in the conclusion follows via symmetry, transitivity, and reflexivity. Secondly, there is an implementation of a version of the sorting algorithm described earlier. This algorithm normalizes expressions over a commutative monoid (a set together with a commutative associative binary operation over that set and an identity element). Given a term of the form $a_1 \cdot a_2 \cdot \ldots a_n$, where $\cdot$ is the binary operation of the monoid, and where no a_i is of the form $b \cdot c$, the algorithm "explodes" the term into the list $[a_1; \ldots; a_n]$, removes any a_i which is the identity element of the monoid, sorts the list, according to a lexicographic ordering on terms induced by an ordering on atoms taken from the current environment, and finally "implodes" the list into a right-associated term.

5 Comparison with Other Work

The idea of using a formalized metatheory to provide for user-defined extensions of an inference system has been around for some time. Some of the early work was done by Davis and Schwartz [5] and by Weyrauch [14].

The work that is closest to our own is that of Boyer and Moore on *metafunctions* [2]. Their metafunctions are similar to our rewriting functions. One of the points which most sharply distinguishes our work is that in our case the entire mechanism is constructed within the theory. An important implication is that we can abstract over environments, drastically reducing the burden of proving the correctness of theorem proving procedures. Also, we do not have, as Boyer and Moore do, one fixed way of applying proven facts about the embedded meta-theory. The user is free to extend the mechanisms; for example, to provide for the construction and application of conditional rewrites. Another difference is that the higher-order nature of the Nuprl type theory allows us to construct useful functions such as the rewrite combinators. Finally, the represented class of objects is much richer than what can be represented in the quantifier-free logic of Boyer and Moore.

Constable and Knoblock [11,10] have undertaken an independent program to unify the meta-language and object theory of Nuprl. They have shown how to represent the proof structure and tactic mechanism of Nuprl in Nuprl's type theory. This is largely complementary to the work described here. Given an implementation of their approach, in order to obtain the same functionality provided by our library, virtually all of our work would have to be duplicated, although in a more convenient setting. They do not deal with representing contexts, and do not impose the kind of structure on represented terms that allows direct construction of such procedures as rewriting functions.

6   Conclusion

There are some questions concerning the practicability of this whole reflection-based approach that cannot yet be fully answered. Some small examples have been developed, but conclusive answers will not be possible until the approach is tested in a major application. It is hoped that it will be a useful tool for the implementation of a significant portion of Bishop's book on constructive analysis [1].

One of these questions is whether encodings in Nuprl of reasoning procedures are adequately efficient. The data types these procedures operate over are not significantly different from those used in the implementation of Nuprl, so this question reduces to the open question of whether Nuprl programs in general can be made efficient. In the example given earlier involving rational arithmetic, the actual rewriting (the normalization of the term that is the application of the rewriting function to its argument) took about ten seconds. The normalization was done using Nuprl's evaluator, which employs a naive call-by-need algorithm, and with no optimization of extracted programs.

Another question is whether lifting a sequent and maintaining a lifted sequent are sufficiently easy. Currently, these operations are somewhat slow. For instance, in the example, the lifting step and the rewriting step each took a total of about two to three minutes. The main reason for this problem has to do with certain gross inefficiencies associated with Nuprl's term structure and definition mechanism. Solutions are known and awaiting implementation.

The most difficult and important question is whether it is feasible to formally verify in Nuprl significant procedures for automating reasoning. The term rewriting system gives some positive evidence, as do some of the other examples outlined above that point out the utility of the means for abstraction available in Nuprl.

References

[1] Errett Bishop. *Foundations of Constructive Analysis*. McGraw-Hill, New York, 1967.

[2] R. S. Boyer and J Strother Moore. Metafunctions: proving them correct and using them efficiently as new proof procedures. In R. S. Boyer and J Strother Moore, editors, *The Correctness Problem in Computer Science*, chapter 3, Academic Press, 1981.

[3] Robert L. Constable and Scott F. Smith. Partial objects in constructive type

theory. In *Proceedings of the Second Annual Symposium on Logic in Computer Science*, IEEE, 1987.

[4] Robert L. Constable, et al. *Implementing Mathematics with the Nuprl Proof Development System*. Prentice-Hall, Englewood Cliffs, New Jersey, 1986.

[5] Martin Davis and Jacob T. Schwartz. Metamathematical extensibility for theorem verifiers and proof-checkers. *Computers and Mathematics with Applications*, 5:217–230, 1979.

[6] N. G. de Bruijn. The mathematical language AUTOMATH, its usage and some of its extensions. In *Symposium on Automatic Demonstration, Lecture Notes in Mathematics vol. 125*, pages 29–61, Springer-Verlag, New York, 1970.

[7] Michael J. Gordon, Robin Milner, and Christopher P. Wadsworth. *Edinburgh LCF: A Mechanized Logic of Computation*. Volume 78 of *Lecture Notes in Computer Science*, Springer-Verlag, 1979.

[8] Robert Harper, Furio Honsell, and Gordon Plotkin. A framework for defining logics. In *The Second Annual Symposium on Logic in Computer Science*, IEEE, 1987.

[9] Douglas J. Howe. *Automating Reasoning in an Implementation of Constructive Type Theory*. PhD thesis, Cornell University, 1988.

[10] Todd B. Knoblock. *Metamathematical Extensibility in Type Theory*. PhD thesis, Cornell University, 1987.

[11] Todd B. Knoblock and Robert L. Constable. Formalized metareasoning in type theory. In *Proceedings of the First Annual Symposium on Logic in Computer Science*, IEEE, 1986.

[12] Per Martin-Löf. Constructive mathematics and computer programming. In *Sixth International Congress for Logic, Methodology, and Philosophy of Science*, pages 153–175, North Holland, Amsterdam, 1982.

[13] Lawrence C. Paulson. A higher-order implementation of rewriting. *Science of Computer Programming*, 3:119–149, 1983.

[14] Richard W. Weyhrauch. Prolegomena to a theory of formal reasoning. *Artificial Intelligence*, 13:133–170, 1980.

Type Inference in Prolog *

H. AZZOUNE

Laboratoire de Genie Informatique
IMAG-Campus B.P. 53 X
38041 Grenoble Cedex France

Abstract

This paper presents a type inference system for Prolog. The idea is to describe sets of terms for which the predicates may succeed. We calculate ordered sets of terms containing some variables called "type variables" (corresponding to variables appearing in simultaneously heads and bodies of clauses), and we generate inequations (from the bodies of clauses) on these type variables. The inequations are then solved in order to instanciate type variables by their domains, obtained from this resolution. The resolution of inequations is done by unification and strategy resolution simulation, with a particular treatment of recursive calls.

Key words. Prolog, Type, Type Inference.

1. Motivations

We think *"types"* can play a natural and useful role in Prolog. There are several directions for applications of type inference results. They provide useful information not only to programmers but also to infer characteristics of execution-time behaviors from program texts; such a task is called *Program Analysis*. Prolog currently lacks any forms of type checking, being designed as a language with a single type (Herbrand Universe). In practice, Prolog programs are very error-prone, small syntactical errors doesn't produce compile-time errors. The results are indeed different (undesired) computations. The fact that it focus type errors can only be detected at run-time. In some situation, we have affirmative answers but the results are false. A reason is the lack of type declarations or a type inference mechanism. The results of the type

---

(*) This work was supported by a CNET contract

inference allow us to detect some miscoding in writing programs; in changing terms by anothers, these errors can appear in the results of the types infered. Actual implementations are done with few optimizations using the structures of arguments. More optimizations can be made using the structure of arguments (propagate this structure to the rest of variables depending on it,...). The results of type inference can be also used for debugging, by detecting difference between the model intended in programmer's mind and the types infered. Faillure situations can be detected statically: clauses which cannot contribute to successful resolution, and goals which cannot succeed. We determine all the solutions and their cardinalities in the case of some non-recursive and non-polymorphic predicates. The *well-typing* notion is introduced, to check the coherence of all predicate calls [9]. Other applications can be made using these results.

In this paper we present a type inference system for Prolog Programs. The idea is to describe sets of terms (or schemas of terms) for which the predicates may succeed. For all terms not described by these sets, the calls cannot succeed. The computation of these sets is done by application of rules on the program. We compute ordered sets of terms containing some variables called *type variables*, (corresponding to variables appearing simultaneously in heads and bodies of clauses). We generate some inequations on these type variables (from the bodies of clauses), which describe all the possible configurations of calls. The generation and the resolution of the inequations are done by simulation of the resolution strategy with a particular treatment on recursive calls. This resolution of inequations is substantially equivalent to Mishra resolution [8] or to the sequential approximation used in [5]. These sets of types can contain terms for which the predicates fail. We call such sets of terms, *the difference sets*. We try to minimize these sets. Ideally these sets should be empty, our aim is to minimize them.

After summarizing some notations, we define a notion of type element in section 2. We give rules of type inference system in section 3. The section 4, presents abstracts of the first works on the types in Prolog.

2. Basic notations

In this section we present some notations used in the rest of the paper. In the following, we assume familiarity with the basic terminologies of first order logic. We also assume knowledge about Herbrand Universe, unification, resolution strategy of Prolog, (for more details see [1,6,13]). In the examples, we follow the syntax of DEC-10 Prolog [11]. As syntactical variables, we use X,Y,A,B, ... for type variables, p,q,r, ... for predicates, T_1,T_2, ... for components of predicate types, and T_{i1}, T_{i2}, ... for components of T_i. A T_{ij} is the type of the component i at the predicate definition j. Before the inference system is applied, it is necessary to transform the program, such that each clause has a distinct set of variables. We neglect the effect of the extra logical predicates ! (cut), assert and retract. Negative subgoals are ignored.

Type elements

The notion of type element allows us to represent all the ground terms manipulated in a program. The type elements manipulated by our inference system are given by the following definition.

Definition: A type element can be:

* An empty set noted by $\varnothing$. This type element is the type of variables of undefined predicates, or variables with many occurences in a clause body, whose intersection of domains in different occurences is empty. For example the variable X in the following clause p(..X..):-X=a,...,X=b.has the type $\varnothing$. If a variable takes as type the element $\varnothing$, we eliminate the correponding combination call, the corresponding branch in the search tree is a failure branch.
* A constant c. Example: 10, -2, -10, ...
* An atom a. Example: [], a, b, ...
* A structure $f(t_1,..,t_n)$ where f is a functor and each t_i is a type element for i=1,n.
  Example: f([],a) , s(s(0)), [a|b], ...
* A symbol any. This symbol is an abreviation of a type set. Any element of Herbrand Universe can be represented by this symbol. It represents all the Herbrand Universe. We have no restriction for the values of variables belonging to the type any, but a variable repeated in the clause takes the same value. For Example: the variable K in the clause append([],K,K). has the type any.
* A set $\{ t_1, ..., t_m \}$, where each t_i is a type element. Example: {a, b, {c, s(0)}, [], 5}.
* An expression $r_i ! (t_1, ..., f(r_i), ...)$ (recursion on r_i) which represents the smallest set of terms $S = \{ t_1, ..., f(x), ...$ such that $x \in S \}$. It is equivalent to the grammatical rule $A \rightarrow t_1 / ... / f(A) /...$ This operator ! allows us to describe infinite sets of terms. It captures a non-trivial subset of Prolog terms (for recursive predicates). This same notation was introduced by Mihsra [8]. When we manipulate $f(r_i)$ the symbol r_i could be replaced by one term of the expression $(t_1, ..., f(r_i), ...)$ as much as it is necessary.

In practical programming context, it will be possible to declare some pre-defined primitive types, and a number of pre-defined predicates with their types using the previous definition. We can define primitive types for example, integer, atom,....

3. Inference system

In this section we present our inference system. The inference is done in three steps. In the first one, we compute ordered sets of type containing possibly type variables and we generate the inequations. In the second step the inequations are normalized by application of a function DECOMP. The third step concerns the resolution of inequations.

Definition

We define a type T_p of a n-ary predicate p with m definitions (heads of clauses), as a tuple $< T_1, ..., T_n >$. A component T_i is the ordered set $T_{i1} \vee T_{i2} \vee ... \vee T_{im}$ of types of arguments of each predicate definition in the program (we take the lexical apparition order of clauses).

A type T_{ij} of a predicate argument in one definition is a superset (domain) of the success set of ground terms taken by the argument itself.

3.1. Type calculation

In the first step, we compute ordered sets (with variables) corresponding to the predicate type. This computation is done by application of the following inference rules. The results are the types of predicates eventualy containing type variables. In the following, T(t) " *denote the type of t*".

1- If an argument t is a constant c, then T(t) = c.

2- If an argument t is an atom a, then T(t) = a.

3- If an argument t is a variable appearing only in a clause head, then T(t) = any.

4- If an argument t is a variable X appearing the in head and body of a clause, then
   T(t) = X where X is a type variable (to be instanciated by its domain after the generation and the resolution of inequations).

5- If an argument t is a structure $f(t_1, ..., t_n)$ where f is a functor and t_i a term, then
   $T(t) = f(t'_1, ..., t'_n)$ where t'_i is the type of term t_i for i=1,n.

6- If an argument t is a tuple $< a_1, ..., a_n >$, then $T(t) = <t_1, ..., t_n >$, t_i is the type of a_i.

7- For an atomic formula pa, head of a clause, T(p) = t where t is the type of a.

8- For all the definitions of predicate p given by $p_1, ..., p_m$, the type of predicate p is
   the ordered disjunction $T_1 \vee T_2 \vee ...\vee T_m$ where T_i is the type of the definition i.

9- The relation $\vee$ is defined by:

$$T_1 = <T_{11}, T_{12}, ..., T_{1n}>$$
$$T_2 = <T_{21}, T_{22}, ..., T_{2n}>$$
$$... \quad ... \quad ...$$
$$T_m = <T_{m1}, T_{m2}, ..., T_{mn}>$$

$$T_1 \vee T_2 \vee ... \vee T_m = < T_{11} \vee T_{21} \vee ... \vee T_{m1}, \; ... \; , T_{1n} \vee T_{2n} \vee ... \vee T_{mn} >$$

Example: Compute the type (with type variables) of the predicate append.

$$append([],K,K).$$
$$append([X|L],M,[X|N]):-append(L,M,N).$$

$$T(append) = T(append_1) \ v \ T(append_2)$$

$T(append_1) = T(append([],K,K))$	by rule (8)
$= <T([]),T(K),T(K)>$	by rules (6,7)
$= <[],T(K),T(K)>$	by rule (2)
$= <[],any,any>$	by rule (3)

$T(append_2) = T(append([X|L],M,[X|N]))$	by rule (8)
$= <T([X|L]),T(M),T([X|N])>$	by rules (6,7)
$= <[T(X)|T(L)],T(M),[T(X)|T(N)]>$	by rule (5)
$= <[any|T(L)],T(M),[any|T(N)]>$	by rule (3)
$= <[any|L],M,[any|N]>$	by rule (4)

The variables L, M and N are type variables, to be instanciated by their domains obtained after the generation and the resolution of the inequations.

$T(append) = T(append_1) \ v \ T(append_2)$	by rule (8)

$= <[],any,any> \ v \ <[any|L],M,[any|N]>$	
$= <[] \ v \ [any|L] \ , \ any \ v \ M \ , \ any \ v \ [any|N]>$	by rule (9)

where each set is an ordered set. This order corresponds to the lexical apparition order of clauses (the component i correspond to the clause i).

3.2. The generation of Inequations

The type sets calculated in the previous step may contain variables, called *type variables* corresponding to variables appearing in simultaneously heads and bodies of clauses (rule 4). These variables can take the type "any" with true results, but these domains are very large and we said that our aim is to minimize the domains of these variables such that they contain the success sets. Till now, we have treated the heads of clauses. This section concerns the treatment of the bodies of the clauses. We treat the bodies of clauses to instanciate these variables by their domains. We produce some systems of inequations corresponding to all the possibilities of calls of predicates. The generation of inequations is done by resolution strategy and unification simulation, with a particular treatment on recursive calls. The unification simulation is done in the following way. In the call $p(..., t_1, ...)$ of a definition clause $p(...,t_2,...):-...,$ the unification of t_1 with t_2 produces an inequation $t_1 \leq t_2$ (the type of term t_1 is enclosed in the type of term t_2). We take equality and inclusion but not only equality, because t_2 can contain a set of type element on account of possibly bactrack of body calls of the clause, and because some approximations would be made on the recursive calls. So the call $p(t_1, t_2, ..., t_n)$ of a definition $p(t'_1, t'_2, ..., t'_n)$ produces the system of inequations (the symbol "," should be read as "and"):

$$t_1 \leq t'_1, t_2 \leq t'_2, ... , t_n \leq t'_n.$$

The inequations are generated by application of the following inference rules:

1- A call of the definition i, $p(t'_1, ..., t'_n)$ by $p(t_1, ..., t_n)$, produces the system:

$$s_i = p(t_1 \leq t'_1, ..., t_n \leq t'_n), \text{ body of the definition i}$$

2- Consider that a definition i of predicate p is the clause $p:-p_1, ..., p_n$. For the body $p_1, ..., p_n$ with m_i definitions for predicate p_i, we have Πm_i (product of m_i) combinations of possible calls.

$$i_1 = 1, m_1 \ (i_2 = 1, m_2 \ (...(i_n = 1, m_n (\text{one combination } p_{i1}, ..., p_{in}))...)$$

3- For one combination $p_1, ..., p_n$ we produce an expression:

$$S_j = s_1, s_2, ... , s_n$$

where s_i is the system generated by rule (1) from the goal p_i. We eliminate the expression S_j if one variable has the empty type element $\emptyset$ or the system is impossible (impossible unificationn or impossible decomposition). This impossibility corresponds to a failure branch in the search tree. This failure situation can be detected for exemple by an unfolding algorithm [12].

4- For each combination of rule (2) of the definition i of predicate p, we produce:

$$SS_{pi} = p_i(S_1 + S_2 + ... + S_k) \text{ with } k \leq \Pi m_i.$$

where S_j is the expression for the combination j. All the expressions S_j has the same left hand side members in their inequations. These members correspond to the arguments of the body goals. (the operator + should be read as "or")

5- For all the r definitions of predicate p, we obtain:

$$I_p = p(SS_{p1} + SS_{p2} + ... + SS_{pr})$$

where SS_{pi} is the expression of the definition i obtained from the rule (4).

6- For all the programs with predicates p, q, r, ..., s we obtain the following expression to be resolved:

$$I = I_p , I_q , I_r , ... , I_s$$

where I_h is the expression for predicate h obtained from the rule (5).

3.3. Normalization

Definition: An inequation system has a normal form if all the left hand side members of its inequations are atomic or variables.

Generally an inequation has the form $r \leq s$ where r is a term. A term may be a constant, an atom, a variable or a structure $f(t_1, ..., t_k)$ where each t_i is a term. To obtain the values of variables appearing in the terms r, or to verify the possibility or not to unify r with s where r is a structure, we introduce a function **DECOMP**. This function decomposes an inequation which its left hand side being a structure. Some constraints are possibly added on the decomposed terms. These constraints will be verified after the calculation of the terms s. We propagate the new term structure to the other positions of these variables decomposed. The function DECOMP is defined by:

DECOMP: Inequation $\rightarrow$ Inequation(s) x Constraints

* DECOMP($t_1 \leq t_1$) = $t_1 \leq t_2$ x $\emptyset$ where t_1 is an atomic or a variable.
* DECOMP($f(_1, ..., t_n) \leq Var$) = DECOMP(t_i , V_i) i=1,n x $\{f(V_1,...,V_n) = Var\}$.
* DECOMP($f(t_1,...,t_n) \leq f(t'_1,...,t'_n)$) = DECOMP(t_i,t'_i) i=1,n x $\emptyset$.
* Otherwise Impossible.

In the second form, all the ground terms possibly taken by the variable Var could be separated into two classes. The first one is formed by n-ary terms of the form $f(\_,....,\_)$. The second one is a set σ formed by the rest of ground terms (all ground terms possibly taken by Var except ground n-ary terms of the form $f(\_,...,\_)$). If a ground term t taken by the variable Var belongs to the first class, then the unification of $f(t_1,...,t_n)$ with t may succeed. If the term t belongs to the second class σ, then the unification fails. So the other occurences of the variables Var are replaced only by the expression $f(V_1,...,V_n)$ (terms of the form $f(\_,...,\_)$) and all terms belonging to the set σ are failure situations (impossible unifcation of $f(t_1,...,t_n)$ with them).

Example:

      DECOMP($[A] \leq [a]$) = $A \leq a$, $[] \leq []$ = $A \leq a$ with no constraint

      DECOMP($a \leq f(..)$)   impossible.

      DECOMP($[A] \leq B$, $C \leq B$) = $A \leq B'$, $C \leq [B']$ with B = [B']
      DECOMP($f(A,B) \leq C$) = $A \leq C_1$ & $B \leq C_2$ with C = $f(C_1,C_2)$

In the last example, the variable C takes either a term of the form $f(\_,\_)$ and the unification may succeed, or a term unlike $f(\_,\_)$ and the unification fails. The ground terms possibly taken by the variable C are separated into two classes. The first one is the class of terms of the form $f(\_,\_)$, and the other one σ is formed by the rest of ground terms. The terms belonging to σ have no role for calculating the domains of variables A and B, because the values of this form taken by C cannot succeed (impossible unification). The terms unlike $f(\_,\_)$ are eliminated because they cannot unify with $f(A,B)$. To unify $f(A,B)$ and C, we create a structure $f(C_1,C_2)$ and we memorize it. The variable C is replaced in its other occurences by only terms of the form $f(\_,\_)$ which correspond to f(C1,C2).

The function **DECOMP** is applied to a system I. We obtain an another system I' (possibly I, if all the left hand side members of inequations are atomic or variables), with possibly some constraints to be verified after the terms of constraints are decomposed. The decomposed variables are replaced in the other occurences by their new expressions.

3.4. The resolution of inequaltions

In this section the algorithms for the resolution of the systems of inequations are given. We show how the algorithms avoid some situations of loops. Required informations are memorized allowing to eliminate these loops.

* For the type calculation of predicate p (to obtain the domains of type variables), we calculate its type for each one of its definition (We resolve all the expressions of rule (5)), we apply the relation **v** (rule (9) of the type calculation) to the results.
* To compute the type of one definition (to obtain variable domains for one definition), we compute the type for each one of its call combination. (We resolve all the expressions of rules (2) and (4)).
* To compute the type of one call combination, we calculate types of all the calls (we resolve all the expressions of rule (3)), with a particular treatment on recursive calls. An inequation system corresponding to a predicate q is resolved after this predicate calls a clause with no body or it makes a recursive call (it calls a goal for which we are calculating the type).
* When we have resolved a system of one call combination, we apply the condition algorithm to the inequations of this combination.

* When we have resolved all the combinations of one definition, we compute a domain of each inequation. For the inequation i, we take the set formed by the right hand side members of inequations i of each combination, in order to obtain $X_i \leq \{ t_1,...,t_k \}$. To obtain the expressions of one component T_{ij}, we produce as much expressions as there are multiple definitions of variables contained in this component T_{ij}.

This treatment corresponds to the search tree traversal with a particular treatment on the recursive calls. The traversal of a search tree with no recursive calls is finite, so this treatment of resolution terminates. When we have a recursive call q, in calculating the predicate type p, we suppose that type of q is $<T_1 , ... , T_n>$. We take the components T_{qij} corresponding to the definition j of the predicate q. At the end of the calculation of type predicate p and the resolution of inequations, we obtain expressions:

$$T_{ij} = \psi(Y_k^*)$$

where Y_k may be (with eventually a constraint) a constant, an atom, a free variable V, a set $\{Y_1,.....,Y_n\}$, a type component T_{mn}, a structure $f(Y_1^*)$ (where f is a functor), or a variable X belonging to a constraint (generated by DECOMP). For one type component T_{ij}, we may have multiple expressions. These expressions are resolved like this:

* We take one expression $T_{ij}= \{T_{ij1}, T_{ij2}, ...\}$,
(**) We push T_{ij} on Rec_steack initially empty and we handle its right hand side.
(***) If the term Tijk is:
    * a constant then assert it.
    * an atom then assert it.
    * a free variable then assert it.
    * a set $\{T_{ij1}...,T_{ijk}\}$ then treat each component by (***).
    * a variable T_{kl} **then** if $T_{kl} \in$ Rec_steack
                then assert r_{kl}, we have recursion on T_{kl}.
                else * we push T_{kl} on Rec_steack.
                    * we compute all the expressions T_{kl} by (***).
                    Compute the possible intersection of T_{kl} [1].
                    We create as much T_{ij} as results of T_{kl}.
                *f(...), then compute its arguments by (***).
                    Create as much T_{ij} as results f(...).
                *A variable X belonging to a constraint. We decompose the corresponding constraint by applying the decomposition algorithm and we resolve its results by **.
                    We create as much T_{ij} as results of X.

---

(1) intersection of T_{ij} and T_{kl} is computable if they don't contain any components T_{rs}.

At the end of this resolution, the free variables are replaced by the type element any. We obtain possibly more than one expressions for each components T_{ij}, we compute the final expression by an intersection algorithm. Each expression has the form:

$$T_{ij} = \{T_{ij1},...,T_{ijn}\} \text{ where}$$
$$T_{ijk} = \text{constant / atom / any / } f(T_{ijk},...,T_{ijk}) / r_m \ ! \ g(...r_m..) / \{T_{ijk},...,T_{ijk}\}.$$

Each T_{ijk} may have an associated constraint generated by a condition algorithm. When we have calculated the intersection, we use the verification algorithm to verify the constraint generated by the condition algorithm.

Algorithms

We describe now the different algorithms used in the resolution of inequations. Examples are given to show how these algorithms may loop, so it is necessary to memorize informations allowing to eliminate them.

Condition algorithm:

* Eliminate all the true inequations from the system of one clause.
* Separate the rest of inequations into two classes:
  a. Those whose the left hand side member belongs to the set of head variables.
  b. The rest of inequations.
* The class (b) forms a condition C_i on the solution of the class (a).
* Replace inequations $r \leq s$ of class (a) by $r \leq \text{constraint}(C_i, s)$.
* We verify this condition after that resolution of T_{ij} terminates.

Example.

Apply this algorithm to the following system with $\{A, B\}$ as set of variables appearing in the head of the clause.

$$p(A \leq a, [] \leq \{[], T_{rs}\}, a \leq T_{ij}, C \leq V, C \leq [], C \leq T_{kl}, B \leq b)$$

The application of the algorithm gives as class (a) the following system:

$$A \leq a(C_1), B \leq b(C_1)$$

and as class (b) the constraint C_1 given by: (to be verified after that T_{ij} and T_{kl})

$$a \leq T_{ij}, C \leq [], C \leq T_{kl}$$

Decomposition algorithm:

To determine the expressions of a variable V belonging to a constraint C generated by the function DECOMP, we take the expression T_{ij} corresponding to the decomposed term, we memorize the couple (C, T_{ij}) and we try to decompose the constraint C with the expressions T_{ij}. If we act on a memorized couple, we assign to the variable V the type any. If we want to decompose a constraint with a type component T_{kl}, we decompose it with all its expressions. If we want to decompose a constraint with a variable belonging to an another constraint, we memorize the couple and we decompose this last one and we decompose the results with the first. The other cases of decomposition of type elements are simple. Notice that without memorizing, the algorithm may loop.

Example:

Compute the expression of the component $T_{11} = X$ with the resolved system $X \leq T'_{11}$ and the constraint $T_{11} = [T'_{11}]$. The application of the decomposition algorithm gives $T_{11} = $ any.

Intersection algorithm..

When all expressions of one T_{ij} are resolved, we compute the intersection of these expressions to obtain the more general one which corresponds to the type of component T_{ij}. The only case where the intersection algorithm may loop, is the case of recursion. When we calculate the intersection of two recursive expressions $r_i ! (...)$ and $r_j ! (...)$, the couple set $\{r_i, r_j\}$ is memorized and when we handle a memorized set, we know that we have a recursion at this level of intersection. The other combinations of intersection of type elements are simple.

Example:

The intersection of expressions $r_1 ! (s(0) , s(r_1))$ and $r_2 ! (s(0) , s(s(r_2)))$ gives $r_{12} ! (s(s(0)) , s(s(r_{12})))$.

Algorithm verification

To verify the possibility or not of a system, we have introduced conditions corresponding to inequations whose their left hand side member don't appear in the set of variables of the clause head, and we didn't verify them.[*] , or that we are multiple

---

(*) because the right hand side member contains a type component T_{lm}

occurences of a variable and this variable didn't appear in the set of head variables. To verify a condition C_i, we memorize it and we try to verify the expressions of this condition using the resolute expressions T_{ij} themselves may contain conditions. If we handle a memorized constraint, we suppose that it is true. The fact to memorize the conditions allows us to eliminate loops in the algorithm.

Example:

Verify the constraint $C_1 = \{ a \leq T_{11} \}$ in the expression $T_{11} = a(C_1)$ (the term a is a solution if C_1 is true). The application of the verifcation algorithm gives T_{11}=a.

3.5. Example of type inference:

Compute the type of predicate rev given by the following program.

```
rev([],[]).
rev([A|B],C):-rev(B,D),append(D,[A],C).
append([],K,K).
append([X|L],M,[X|N]):-append(L,M,N).
```

In application of first inference rules (for type calculation), we obtain:

$$T_{rev} = <T_1 , T_2> = <T_{11} \vee T_{12} , T_{21} \vee T_{22}> = < [] \vee [A|B] , [] \vee C >.$$

with A, B, C as type variables, to instanciate by their domains computed from the systems of inequations.

In the generation of inequations, we take an execution of predicate rev in the general sense. The first clause has no body, then no inequations from it (no type variables). In the second clause, we have a body with 2 goals with 2 definitions for each one. We have 2*2=4 possible combinations of unification of rev and append with their definitions.

1) The unification of rev and append with their first definitions (which have no body), produces the system:

$$\begin{array}{ll}
\text{rev } B \leq [] \quad \text{no body} & B \leq [] \\
\quad D \leq [] & D \leq [] \\
\end{array}$$

$$\text{DECOMP}$$

$$\begin{array}{ll}
\text{append } D \leq [] \quad \text{no body} & D \leq [] \\
\quad [A] \leq K_1 & A \leq K'_1 \\
\quad C \leq K_1 & C \leq [K'_1] \\
& \text{with constraint } K_1 = [K'_1]
\end{array}$$

The intersection of the second and third inequations is non-empty. The domains of the variables A, B, and C are solution with no-constraints.

2)The unification of rev with its first definition (which has no body), and append with its second definition (it has a body), produces a system:

rev $B \leq []$ no body
$\quad D \leq []$

$\qquad\qquad\qquad\qquad\qquad\qquad\qquad$ impossible (second and third inequation)

append $D \leq [X|L]$ body (append$_2$)
$\quad [A] \leq M$
$\quad C \leq [X|N]$

In this system, the variable D cannot have a solution because the intersection of domains of the second and third inequations would be empty. The system is impossible and the corresponding branch in the search tree is to eliminate. This situation could be detected by an unfolding preprocessor.

3)rev is unified with its second definition (which has a body) and append is unified with its first definition (which has no body), produces the system:

rev $B \leq [A|B]$ body(rev$_2$))$\qquad\qquad$ rev $B \leq [A|B]$ body(rev2)
$\quad D \leq C$ $\qquad\qquad\qquad\qquad\qquad\qquad D \leq C$

$\qquad\qquad\qquad\qquad$ DECOMP

append $D \leq []$ no body$\qquad\qquad$ append $D \leq []$ no body
$\quad [A] \leq K$ $\qquad\qquad\qquad\qquad\qquad A \leq K'$
$\quad C \leq K$ $\qquad\qquad\qquad\qquad\qquad C \leq [K']$

$\qquad\qquad\qquad\qquad\qquad$ with constraint $K = [K']$

We are computing the type of predicate rev, and for this computation we need the type of this same predicate from a body call rev (recursive call). We suppose that the type of this predicate is equal to $<T_{11} \vee T_{12}, T_{21} \vee T_{22}>$, we take its second components $< T_{12}, T_{22} >$ corresponding to the second clause defintion, then the system has no calls now. It becomes:

rev $B \leq T_{12}$ no body $\qquad\qquad$ rev $B \leq T_{12}$ no body
$\quad D \leq T_{22}$ $\qquad\qquad\qquad\qquad\qquad D \leq T_{22}$

append $D \leq []$ no body $\qquad\qquad$ append $D \leq []$ no body
$\quad A \leq K'$ $\qquad\qquad\qquad\qquad\qquad A \leq K'$
$\quad C \leq [K']$ $\qquad\qquad\qquad\qquad\qquad C \leq [K']$

In this example we can know that the second component T_{22} cannot contains the term [], then if the intersection of the second and the third inequations is empty, then we eliminate this system. It corresponds to a failure branch in the search tree, which can be detected by an unfolding preprocessor. If we don't apply unfolding to detect this situation, when we take the solution of this system, we manipulate it with a constraint "$T_{22} \cap []$" that will be false after that T_{22} is computed, then we eliminate it.

4) rev is unified with its second definition (which has a body) and append with its second definition (which has a body) produces a system:

rev $B \leq [A|B]$ a body(rev$_2$)   rev $B \leq T_{12}$ no body
    $D \leq C$ $B \leq T_{22}$

append $D \leq [X|L]$ a body (append$_2$) $D \leq [X|L]$ body(append$_2$)
       $[A] \leq M$ $[A] \leq M$
       $C \leq [X|N]$ $C \leq [X|N]$

In this system we have a call for predicate append, then we must compute its type, and we replace the result in this system. We compute now the types of second definition of predicate append. After application of first inference rule, we find:

$$T_{32} = [X|L]$$
$$T_{42} = M$$
$$T_{52} = [X|N]$$

with 2 systems to resolve (the body of the second clause has 1 call with 2 definitions):

5) append(L,M,N) is unified with its first definition (which has no body), produces a system:

append $L \leq []$ no body $L \leq []$ no body
        $M \leq K_2$ $\rightarrow$ $M \leq K_2$
        $N \leq K_2$ $N \leq K_2$

6) append(L,M,N) is unified with its second definition (which has a body append), produces the system:

append $L \leq [X|L]$ a body (append$_2$)
       $M \leq M$
       $N \leq [X|N]$

to compute the predicate type append, we need the type of this same predicate. We suppose that its type is $< T_{31} \vee T_{32} , T_{41} \vee T_{42} , T_{51} \vee T_{52} >$, we take its second definition $< T_{32} , T_{42} , T_{52} >$ and we obtain:

$$L \leq T_{32} \text{ no body}$$
$$M \leq T_{42}$$
$$N \leq T_{52}$$

All the variables L, M and N appear the clause head, the application of the condition algorithm give an empty set as class (b). The domains of the 2 systems are solution with no constraint.

All systems of append are resolved, and the domains of variables L, M, and N are computed:

$$L \leq \{ [] , T_{32} \}$$
$$M \leq \{ K_2 , T_{42} \}$$
$$N \leq \{ K_2 , T_{52} \}$$

All the variables L, M, and N belong to the call append([X|L],M,[X|N])

$$T_{32} \leq [X|\{[],T_{32}\}]$$
$$T_{42} \leq \{K_2,T_{42}\}$$
$$T_{52} \leq [X|\{K_2,T_{52}\}]$$

We replace these results in the 4th system, (one inequation for each variable):

$B \leq T_{12}$	$B \leq T_{12}$
$D \leq T_{22}$	$D \leq T_{22}$
$D \leq [X|\{[],T_{32}\}]$	$D \leq [X|\{[],T_{32}\}]$
$[A] \leq \{K_2 , T_{42}\}$	$A \leq \{K'_2,T'_{42}\}$
$C \leq [X|\{K_2,T_{52}\}]$	$C \leq [X|\{[K'_2],T_{52}\}]$

$$K_2 = [K'_2] \text{ and } T_{42} = [T'_{42}]$$

We replace K_2 by $[K'_2]$ in the rest of expressions (the only values taken by K_2 has the form [_]. we obtain $T_{42} = \{ [K'_2] , [T'_{42}] \}$ and $T_{52} = [X|\{[K'_2],T_{52}\}]$. The variable D doesn't appear in the type variables set, and it appear in more than one occurence in the system. The application of the condition algorithm gives as class (b) (constraint C_1) the inequations {$D \leq T_{22}$, $D \leq [X|\{[],T_{32}\}]$} to be verified after that T_{22} and T_{32} will be calculated. We obtain as domains:

$$B \leq T_{12}(C_1)$$
$$A \leq \{K'_2, T'_{42}\}(C_1)$$
$$C \leq [X|\{[K'_2], T_{52}\}](C_1)$$

The domains of variables B, D and C computed from the first and fourth systems.

$$B \leq \{[], T_{12}(C_1) \}$$
$$A \leq \{K'_1, \{k'_2,T'_{42}\}(C_1) \}$$
$$C \leq \{[K'_1], \{[X|\{[K'_2],T_{52}\}]\}(C_1) \}$$

We replace the variables A, B, and C by their doimains in T_{12}, T_{22} (we have one solution only for each variable, we obtain one expression for each expression T_{ij}):

$$T_{11} = []$$
$$T_{12} = [A|B] = [\{any, \{any, T'_{42}\}(C_1)\} \mid \{[], T_{12}\}(C_1)]$$

$$T_{21} = []$$
$$T_{22} = C = \{[any], [any \mid \{[any], T_{52}\}](C_1)\}$$

We compute now the expressions T'_{42} and T_{52} from their new expressions. We find that $T'_{42} = any$ and $T_{52} = r_1 \mathbin{!} [any \mid \{[any], r_1\}]$. We replace these expressions in T_{12} and T_{22} and we obtain:

$$T_{12} = r_2 \mathbin{!} [\{any, \{any, any\}(C_1)\} \mid \{[], r_2\}(C_1)]$$
$$T_{22} = \{[any], [any \mid \{[any], r_1 \mathbin{!} [any \mid \{[any], r_1\}]\}]\}(C_1)\}$$

To verify the constraint C_1, we find that the intersection of T_{22} with $[X \mid \{[], T_{32}\}]$ is non-empty, then the constraint is true.

$$T_{12} = r_2 \mathbin{!} [any \mid \{[], r_2\}] = list(any)$$
$$T_{22} = \{[any], [any \mid \{[any], r_2 \mathbin{!} [any \mid \{[any], [any \mid r_2]\}]\}]\}\} = list(any)$$

$$T(rev) = < [] \lor List(any) , [] \lor List(any) >$$

We compute the type of predicate append, like this last example.

$$T_{31} = []$$
$$T_{32} = [X|L] = [X \mid \{[], T_{32}\}] = r_3 \mathbin{!} [any \mid \{[], r_3\}]$$
$$T_{41} = K_3 = any$$
$$T_{42} = M = \{K_3 , T_{42}\} = \{any\}$$
$$T_{51} = K_3 = any$$
$$T_{52} = [X \mid \{K_3 , T_{52}\}] = r_4 \mathbin{!} [any \mid \{any, r_4\}]$$

$$T(append) = < [] \lor r_3 \mathbin{!} [any \mid \{[], r_3\}] , any \lor any , any \lor r_4 \mathbin{!} [any \mid \{any, r_4\}] >$$

Example: Compute the type of predicate p given by the following program.

$$p(a).$$
$$p(f(X)):-p(X).$$
$$p(g(Y)):-p(Y).$$

$T_p = <T_1, T_2, T_3>$ with (after computation):

$T_1 = a$ with no inequations

$T_2 = f(X)$ with inequations $X \leq a + X \leq T_2 + X \leq T_3 \rightarrow X \leq \{a, T_2, T_3\}$

$T_3 = g(Y)$ with inequations $Y \leq a + Y \leq T_2 + Y \leq T_3 \rightarrow Y \leq \{a, T_2, T_3\}$

After resolution we find:

$T_1 = a$

$T_2 = r_1 \, ! \, f(\{a, r_1, r_2 \, ! \, g(\{a, r_1, r_2\}) \}))$

$T_3 = r_3 \, ! \, g(\{a, r_4 \, ! \, f(\{a, r_4, r_3\}), r_3 \}))$

Remark:

In the first example, we remark that the type of predicate append is computed twice. Firstly at computing the type of predicate rev, and secondly for the computation of the type of predicate append. At this time we are working to develop an incremental method. Problems arise for the propagation of constraints from the function **DECOMP**.

4. Related works

Most of the investigations of type inference have been made for functional programs. A few works are done for Prolog with different point of views. In the literature, we find two approachs.

4.1. Explicit Approach

Bruynooghe [3] proposed to introduce types to Prolog in order to enhance reliability and readability. He suggests to add redundant information about the flow of data through clauses and gives a method to analyse the consistency between this additionnal information and the text of Prolog clauses. This information consists of type declarations. Each procedure is writen with declarations stating the number and the types of the arguments and the intended patterns of the call (mode). These informations are required from the user, and a normal Prolog interpreter can discard them. He associates a set of meaningful values (a type) with each argument in a procedure.

Nilsson [10] introduces the notion of data domains of a program, which is reminiscent of the mathematical notion of many-sorted term-algebra. He considers the compilation of Prolog into target programs of procedural high level language (Pascal). The crucial point in his approach is the introduction of a notion of data types in Prolog, termed domains, to emphasize the abstract form of the type concept, and to avoid confusion with the target language data types. The domain (type) concept is introduced for obtaining a division of the Herbrand Universe of ground terms of the program, in order to distinguish various usages of data in the program. This concept is useful through the static constraints being imposed on the source program. He requires some informations from the programmer. A Prolog program is accompanied by a number of domain declarations. These domains are similar to the notion of sorts of a many-sorted term-algebra.

Mycroft and Okeefe [9] describe a polymorphic type schema for Prolog programs which makes static type checking possible. The additions to the language are type declarations. They consider type specifications as restrictions on arguments to predicates and functors. They proposed a schema of types given by a grammar which are essentially the same as those which occur in ML [4]. They introduce the notion of well-typing and show that well-typed programs do not go wrong. They think that a new compiler for typed Prolog could improve the speed of compiled clauses, and the greatest gain is that of programmer time provided by early detection of errors.

We think that explicit approach makes the program unreadable for every one except their authors. The declarations cause an explosion of the program size. It is very interesting to have a system that can compute the predicate types automaticaly, (that correspond to inference approach) and can use the information given by the user.

4.2. Inference approach

Mishra [8] proposes a type inference system for non-polymorphic predicates. His system infer types for these predicates without explicit definition of data types or type declarations. He uses some regular expressions for using algorithms on regular languages. This approach is characterized by the slogan "ill-typed program cannot succeed". The type of a predicate is defined as a set of all terms for which the predicate may succed. For any term not described by the type of predicate, the predicate cannot succeed. He takes inspiration from Milner's work on ML [4]. Regular trees are used to represent the predicates types. He computes the type of non-polymorphic predicates as regular trees and generates inequations. Some transformations on the inequation are applied to obtain ground regular trees for which the language accepted by an automate (for this ground regular tree), is regular. Algorithms for determining the intersection of two ground regular trees, determining if one ground regular tree is a subset of another,....are derived.

Example: Compute the type of predicate f_father.

$$father(a,b).$$
$$father(b,c).$$
$$f\_father(X,Y):-father(X,Z),father(Z,Y).$$

The type of predicate f_father is $T(f\_father)=T_1 \# T_2$ with the inequations:

$$T_1 \leq a + b$$
$$Z \leq b + c$$
$$Z \leq a + b$$
$$T_2 \leq b + c$$

He computes the intersection of the second and third inequations by algorithms derived from their regular expressions counterparts. This intersection is non-empty, then the first and the last inequations are the solutions.

$T(f\_father) = a + b \times b + c$ with the denotation $\{ (a,b) , (a,c) , (b,b) , (b,c) \}$

The type infered by our system is <a,c> corresponding in this case to the solution.

Kanamori and Horiuchi [5] present a type inference method. The idea is to describe a superset of the success set by associating a type substitution (an assignment of sets of ground terms to variables) with each head of definite clauses. The computation of the superset is done by sequentiel approximation, which is substantially equivalent to solving inequations. They introduce explicitly type construct to define data structures (like List for example). The idea, is to describe an Herbrand interpretation I covering success set under a restriction J in terms of types. This is performed by defining an appropriate transformation T' satisfying the following theorem.

Theorem: Let I and J be Herbrand interpretation. I is said to cover success set under a restriction J when it contains the intersection of the minimum Herbrand model M_0 and J. An Herbrand interpretation J is said to be closed with respect to a conjunction of definite clauses P, when for any ground instance of definite clause in P such that the head is in J, any ground atom in the body is also in J. ($M_0 \cap J$ is computable within it). If J is closed with respect to P, $T(I) \subseteq T'(I)$ and $T'(I) \cap J \subseteq I$, then I covers success set under J.

4.3. Discussions

We have presented only two works on type inference in Prolog, because in our knowledge these are the first works on type inference. We remark that our system applies for monomorphic and polymorphic predicates. The Mishra's type inference system [8] is monomorphic, it is not applicable for polymorphic predicate. Our system infers the types of predicates automaticaly with no required declaration, but it can use informations given by user. The Kanamori's type inference system needs explicit declarations of type constructs.

5. Conclusion

We have presented a type inference method for Prolog programs. This type inference method is an element of our static verifcation system for Prolog under development. We will relate our work to that one of abstract interpretation [2] and on the hybrid interpretation method of Kawamura and Kanamori.

Bibiography

[1] Apt, K.R. and Van Emden M.H., "Contributions to the Theory of Logic Programming", *Journal of ACM, Vol 29 No 3* pp 841-862 1982.

[2] Bruynooghe M., "Abstract Interpretation: Towards the Global Optmization of Prolog Programs", *Proceedings 1987 Symposium on Logic Programming* August 31- September 1987 San Francisco IEEE Society Press pp 192-204.

[3] Bruynooghe M., "Adding redundancy to obtain more reliable and more readable Prolog Programs", *Proc. First Int. Logic Prog. Conf* pp 129-133 1982.

[4] Gordon, M.J.C.,Milner,A.J.R.G., Morris,L.,Newey,M.and Wadsworth,C., "A Metalanguage for interactive proof in LCF", *in Proc. 5th ACM Symposium Principles of Programming Language*, Tucson AZ, 1978.

[5] Kanamori T. and Horiuchi K., "Type Inference in Prolog and its Applications", *ICOT TR 95*, December 1984.

[6] Kowalski R.A., "Logic for Problem Solving", *Elsevier Science Publishing* , 1979 North Holland.

[7] Mellish C.S., "An Automatic Generation of Mode Declaration for Prolog Programs", *D.A.I. Research Paper No 163*, University of Edinburgh 1981.

[8] Mishra P., "Towards a Theory of Types in Prolog", *Proc 1984 Inter. Symp. on Logic Programmig*, pp289-298.

[9] Mycroft A. and O'Keefe R.A., "A polymorphic Type System for Prolog", *Logic Programming Workshop* 1983.

[10] Nilsson J.F., "On the Compilation of a Domain-Based Prolog", *Information Processing* pp 293-298 1983.

[11] Pereira L.M., Pereira F.C.N. and Warren D.H.D., "User's Guide to DEC system-10 Prolog", *Occasional Paper 15, Dept of Artificial Intelligence*, Edinburgh 1979.

[12] Tamaki H. and Sato T., "Unfold/Fold Transformation of Logic Programs", *Second International Logic Programming Conference*, Uppsala 1984.

[13] Van Emden, M.H. and Kowalski, R.A., "The Semantics of Predicate Logic as Programming", *Journal of ACM Vol 23 No 4*, pp733-742 1976.

Procedural Interpretation of Non-Horn Logic Programs

Jack Minker[1,2]   Arcot Rajasekar[1]

Department of Computer Science[1]

and

Institute for Advanced Computer Studies[2]

University of Maryland

College Park, Maryland 20742

Abstract

Procedural interpretation in logic programming consists of two parts : answering positive queries and answering negative queries. Answering positive queries can be done using a general theorem prover. To answer negative queries it is necessary, in most practical cases, to augment a theorem prover with default rules. Such an approach has been taken with Horn clause logic programs in the definition of SLDNF-resolution. We describe a similar proof procedure for non-Horn programs and define an inference system, SLINF-resolution, for this purpose. SLI-resolution is used as the main inference mechanism and a weaker form of the generalized closed world assumption, called the Support-for-Negation Rule, is used as the default rule for answering negative queries.

Keywords and phrases: generalized closed world assumption, logic programming, negation, non-horn programs, procedural interpretation, support-for-negation.

1   Introduction

Procedural interpretation in logic programming consists of two parts : answering positive queries and answering negative queries. Answering positive queries can be achieved using a general theorem prover, while answering negative queries is generally more complex. The inference procedure can also be made efficient for answering positive queries by using some characteristics of the program under consideration as done for Horn logic programs (SLD-resolution) [Hil74]. But, in the case of negative queries, a theorem prover does not lead to an efficient procedure since it requires negative information to be represented explicitly as part of the logic program. Such an explicit representation is not feasible in many applications such as deductive databases and artificial intelligence where the amount of negative information may overwhelm the system. The solution to this problem has been to use default rules to implicitly infer negated facts from the system. Theorem provers augmented with such default rules offer attractive inference systems for logic programs. Such an approach has been taken with Horn programs in the definition of SLDNF-resolution [Cla78]. We describe a similar proof procedure for non-Horn programs and define an inference system, SLINF-resolution, for this purpose.

SLD-resolution (SL-resolution for Definite programs) [Hil74] is a sound and complete proof procedure system for definite Horn programs. SLD-resolution is efficient when compared to a general theorem prover such as SL-resolution [KK71], since it uses the definiteness (one positive literal) characteristic of Horn programs and need not perform ancestry resolution or factoring. However, for general Horn programs and for answering negative queries in Horn programs SLD-resolution is insufficient and is augmented with a default rule called *Negation-As-Failure* (NAF) [Cla78], which states that a ground literal $\neg A$ can be inferred when a proof for A using SLD-resolution fails finitely. NAF is a weaker form of another rule of negation called the *Closed World Assumption* (CWA) [Rei78], which allows one to infer $\neg A$ from a program if A is not provable from the program. This definition of negation is not fully computable in many cases [She88] and hence the weaker definition of finite failure in NAF is used. SLDNF-resolution [Cla78] is the procedural implementation of the negation-as-failure rule in the framework of SLD-resolution.

For non-Horn programs ancestry resolution is necessary if one uses SL-resolution. Any sound and complete resolution based theorem prover could be of used to answer a positive query. In this paper, we describe LUST-resolution, developed by Minker and Zanon [MZ82], and argue that it is an appropriate proof procedure for non-Horn programs in answering positive queries. We rename LUST-resolution as SLI-resolution (SL-resolution for Indefinite programs) and show that it is an extension of SLD-resolution. SLI-resolution shares with SLD-resolution the property of arbitrary literal selection and reduces to SLD-resolution when used with Horn programs.

For answering negative queries we augment SLI-resolution with a default rule. For non-Horn programs Minker [Min82] proposed a *Generalized Closed World Assumption* (GCWA), which states that a ground literal $\neg A$ can be inferred from a logic program P if for all positive clauses K such that $A \vee K$ is provable from P then K is also provable from P. The GCWA shares with the CWA the problem of computability. There are two difficulties involved with the GCWA : first, $P \vdash A \vee K$ or $P \vdash K$ might not be decidable for some K and second, since K is a clause if $P \vdash A \vee K$ we have $P \vdash K \vee K$ and $P \vdash K \vee K'$, for all possible positive clause K', making the computability of the GCWA difficult. The first drawback is either due to infinite loops or due to a derivation having an unbounded term property [Van88]. The second difficulty is caused by clauses not having distinct literals. We address the infinite loop and the non-distinct literal problems in our approach. We translate the generalized closed world assumption into a *Support-for-Negation* (SN) property for individual ground atoms and define a set of clauses which are necessary to be proved to infer the negation of the atoms. The set SN for a ground atom consists of all K's where $A \vee K$ are provable from the program. We tighten the definition of the support-for-negation property by replacing the provability of $A \vee K$ as the derivability, using an SLINF-derivation, which is a restricted form of SLI-derivation. We define a default rule for negation, called Support-for-Negation Rule (SN Rule), based on the SN property. This approach is quite similar to the tighter definition of NAF from the CWA. The new definition of SN takes care of the problems of infinite loops and non-distinct literals because of the admissibility condition in the SLINF-derivation, but the unbounded term size problem is not taken care of[1]. We describe a query answering procedure based on the Support-for-Negation property and show that it is sound for answering negative queries in non-Horn programs. We also show that the procedure reduces to NAF when applied to Horn programs. Przymusinski [Prz86] provides an algorithm for answering queries using the GCWA. Although not specifically stated, the Przymusinski's algorithm implements the SN rule. The algorithm defined by Przymusninski is based on MILO-resolution, a modification of OL-resolution [CL73]. The algorithm presented in this paper differs from the Przymusinski's algorithm in two ways. First it allows more freedom in the selection of literals during resolution. Second, it is based on a theory which is an extension of the negation theory of Horn programs. The Przymusinski algorithm is based on the theory of circumscription.

The next section presents the proof procedures for Horn programs. Section 3 describes SLI-resolution and defines the concept of Support-for-Negation and develops the query answering procedure for handling negation. Section 4 compares the Horn and non-Horn procedures.

2 Horn programs

In this section we discuss inference procedures for Horn programs. There are several resolution-based procedures which are sound and complete for answering queries from Horn and non-Horn programs. Kowalski and Kuehner [KK71] introduced a modified form of linear resolution for logic programs called SL-resolution – Linear Resolution with Selection Function, which is closely related to model-elimination developed by Loveland [Lov78]. The selection function in SL-resolution is restricted to certain literals. Minker and Zannon [MZ82] defined a complete and sound modification to SL-resolution called LUST-resolution (Linear Resolution with Unrestricted Selection Function based on Trees) for logic programs which has no restrictive selection function. For Horn programs, Hill [Hil74] presented a complete and sound linear resolution procedure called LUSH resolution – Linear Resolution with Unrestricted

---

[1]The unbounded term size property is undecidable [Van88]

Selection function for Horn programs – for answering positive queries. The advantage of this procedure is that one can arbitrarily select literals in a clause on which to expand, and neither ancestry resolution nor factoring are required. Apt and van Emden [AvE82] renamed LUSH-resolution to be SLD-resolution – **SL**-resolution for Definite Programs.

2.1 SLD-resolution

Horn programs are a subset of logic programs and have clauses of the form :
$$A \leftarrow B_1, \ldots, B_n \quad n \geq 0,$$
where $A, B_1, \ldots, B_n$ are atomic and all variables are universally quantified. In *general Horn programs* the B_i's are literals. SLD-resolution [AvE82] is a unification-based replacement procedure, for answering positive queries in Horn programs, which develops a progression of sub-goals to reach a refutation. A goal in an SLD-derivation is of the form $\leftarrow A_1, \ldots, A_n, n \geq 0$, where the A_i's are atoms. The formal definition of an SLD-derivation is given below:

Definition 1 *Let P be a Horn logic program, G be a goal $= \leftarrow A_1, \ldots, A_m, \ldots A_n$. An SLD-derivation from P with top-goal G consists of a (finite or infinite) sequence of goals $G_0 = G, G_1, \ldots$, such that for all $i \geq 0$, G_{i+1} is obtained from G_i as follows:*
(1) A_m is an atom in G_i and is called the selected atom
(2) $A \leftarrow B_1, \ldots, B_q$ is a program clause in P
(3) $A_m\theta = A\theta$ where θ is a substitution (most general)
(4) G_{i+1} is the goal $\leftarrow (A_1, \ldots, A_{m-1}, B_1, \ldots, B_q, A_{m+1}, \ldots, A_k)\theta$ *[]*

Definition 2 *An SLD-refutation from P with top-goal (or query) G is a finite SLD-derivation of the null clause $\Box$ from P with top-goal G. If $G_n = \Box$, we say the SLD-refutation has length n.*   *[]*

SLD-resolution is the system using SLD-derivation as the inference mechanism. SLD-resolution is sound and complete for Horn programs in answering positive queries. In the next subsection we discuss negation in Horn programs. For that we need the notion of an SLD-tree. An SLD-tree is a tree of goal clauses. The root node of an SLD-tree is the top-goal (or query). For each subgoal in the tree all the subgoals which can be derived using SLD-derivation from the goal and the program in one-step are attached as children to the goal. A branch in the tree fails when the SLD-derivation fails at the leaf node. A thorough treatment of SLD-trees can be found in [Llo84].

2.2 CWA, NAF and SLDNF-resolution

The *Closed World Assumption* (CWA) [Rei78] is a default rule developed by Reiter
to answer negative queries in Horn programs. For a Horn program P and a ground
atom A, the rule can be stated as follows

$$\frac{A \text{ is not provable from } P}{\neg A}$$

The rule may be read as follows. If the numerator is satisfied then one can conclude
the denominator. In general, the set of negative literals which can be inferred
under the CWA is not recursively enumerable [She88]. Clark [Cla78] proposed a
weaker form of negation based on finitely failed SLD-trees, called *Negation-As-
Failure* (NAF):

$$\frac{A \text{ has a finitely failed } SLD-tree}{\neg A}$$

Clark [Cla78] augmented the SLD-resolution procedure with the NAF rule and introduced SLDNF-resolution as an appropriate inference system for answering negative
queries in Horn programs. We provide the formal definition of SLDNF-resolution below. A general goal in an SLDNF-derivation is of the form $\leftarrow L_1, \ldots, L_n, \quad n \geq 0$,
where the L_i's are literals.

Definition 3 *Let P be a Horn logic program, G be a general goal. An SLDNF-
derivation from P with top-goal G consists of a (finite or infinite) sequence of general
goals $G_0 = G, G_1, \ldots,$ such that for all $i \geq 0$, G_{i+1} is obtained from G_i as follows:
Suppose G_i is $\leftarrow L_1, \ldots, L_m, \ldots, L_n$ and L_m is selected for derivation :
(a) Suppose L_m is a positive literal then
    (1) $A \leftarrow B_1, \ldots, B_q$ is a program clause in P
    (2) $L_m\theta = A\theta$ where θ is a substitution (most general)
    (3) G_{i+1} is the goal $\leftarrow (L_1, \ldots, L_{m-1}, B_1, \ldots, B_q, L_{m+1}, \ldots, L_k)\theta$
(b) Suppose L_m is a negative ground literal and $(L_m = \neg A)$ then
    (1) Attempt to construct a finitely failed SLD-tree with $\leftarrow A$ as top-goal (query)
    (2) If $\leftarrow A$ succeeds then goal G_i fails
    (3) If $\leftarrow A$ fails finitely then
        G_{i+1} is the goal $\leftarrow L_1, \ldots, L_{m-1}, L_{m+1}, \ldots, L_k$* []

An SLDNF-refutation from P with general top-goal (or query) G is a finite SLDNF-
derivation of the null clause □. Clark [Cla78] showed that SLDNF-resolution is
sound with respect to program completion which states that a Horn clause can be
viewed not only as an IF-condition for proving the atom at the head of the clause
but also as an ONLY-IF-condition, effectively achieved by changing the direction
of the implication in a clause. SLDNF-resolution has two disadvantages. First , it
may flounder, that is, it may reach a goal where all literals are negative and non-
ground. Second, it does not detect infinite loops in the derivation and the derivation

does not fail. The problem of unbounded term size in the derivation is yet another problem but is undecidable.

In the next section we describe the counterparts for SLD-resolution, CWA, NAF and SLDNF-resolution which apply to non-Horn programs.

3 Non-Horn Programs

A *non-Horn program* consists of clauses of the form :
$$A_1, \ldots, A_m \leftarrow B_1, \ldots, B_n \quad n \geq 0, \ m > 0,$$
where $A_1, \ldots, A_m, B_1, \ldots, B_n$ are atomic and all the variables are universally quantified. In *general non-Horn programs* the B_i's are literals. Horn programs are a subset of non-Horn programs.

3.1 SLI-resolution

We describe the proof-procedure for answering positive queries in non-Horn programs. Since, we are dealing with non-Horn programs, ancestry resolution and factoring are necessary. LUST-resolution developed in [MZ82] is the basis for our proof procedure for answering queries in non-Horn programs. We use it for the following reasons :

1. LUST-resolution is sound and complete for theorem proving [MZ82].

2. As in SLD-resolution, it allows an arbitrary literal to be selected in a clause.

3. It provides a convenient basis for developing a procedure for answering negative queries.

4. When restricted to Horn clauses, LUST-resolution reduces to SLD-resolution.

We rename LUST-resolution as SLI-resolution – **SL**-resolution for **I**ndefinite clauses, to follow the nomenclature used by Apt and van Emden when they renamed LUSH-resolution as SLD-resolution. SLI-resolution is described below.

SLI-resolution is a complete and sound inference system for theorem proving. It is a modification of SL-resolution that allows expansion of arbitrary literals. The inference system SLI is a renaming by [MR87] of the LUST-resolution based inference system defined in [MZ82].

SLI-resolution is defined using trees as the basic representation. Each node in the tree is a literal and there are two types of literals : a marked literal or an A-literal and an unmarked literal called a B-literal. A non-terminal node is always an A-literal whereas a terminal literal can be either an A- or a B-literal. An SLI-derivation operates on admissible and minimal t-clauses. A t-clause is a special representation of a clause and embeds the information about the ancestry of each literal.

Definition 4 *([MZ82]) A t-clause C, is an ordered pair $<C,m>$ where*

- **C** *is a labeled tree whose root is labeled with the distinguished symbol ε, and whose other nodes are labeled with literals; and*

- **m** *is a marking relation on the node such that every non-terminal node is marked.* []

A t-clause can also be viewed as a pre-order representation of a resolution tree. It is a well-parenthesized expression such that every open parenthesis is followed by a marked literal. A non-Horn program is represented by t-clauses which have only one A-literal (a reserved literal ε^*). A goal clause is also a t-clause.

Example 1 *$(\varepsilon^* \, a \, b \, \neg c)$ is a t-clause representation of $a \lor b \leftarrow c$*
$(\varepsilon^* \, a \, (c^* \, d \, (e^* \, b) \, f) \, e)$ is another example of a t-clause. []

A literal is marked if it has been selected in an SLI-derivation. During derivation, an unmarked literal in the goal t-clause is selected and marked. This literal can be either positive or negative. The selected literal is unified with a complementary literal in the program clause. The resolvent is attached as a subtree to the literal in the goal clause. The t-clause is then made admissible and minimal by performing factoring, ancestry-resolution and truncation. The notions of factoring, ancestry-resolution and truncation are similar to that in SL-resolution [KK71]. Program and goal clauses are represented in the form $(\varepsilon^* L_1 \ldots L_n)$, where ε is a special symbol, $*$ is a marking and the L_is are literals.

We next give a formal definition for the SLI-derivation. First, we define two sets of literals which are used during resolution.
$\gamma_L = \{$M: where M is a B-literal, and M is one node off the path from the root of L$\}$
$\delta_L = \{$N: where N is an A-literal, and N is on the path between the root and L$\}$
A t-clause is said to satisfy the admissibility condition (AC) if for every occurrence of every B-literal L in it the following conditions hold:
(i) No two literals from γ_L and L have the same atom.
(ii) No two literals from δ_L and L have the same atom.
A t-clause is said to satisfy the minimality condition (MC) if there is no A-literal which is a terminal node.
γ_L and δ_L are used while performing factoring and ancestry respectively. AC and MC make sure that the factoring and ancestry are performed as soon as possible. Now we have the framework for describing an SLI-resolution. We modify the definition given in [MZ82] and provide a definition similar to that of SLD-resolution.

Definition 5 *Consider a t-clause C_0. Then C_n is a $\underline{tranfac - derivation}$ (truncation, ancestry and factoring) of C_0 when there is a sequence of t-clauses $C_0, C_1, \ldots, C_n$*

such that for all i, $0 \leq i \leq n$, C_{i+1} *is obtained from* C_i *by either t-factoring, t-ancestry, or t-truncation.*

C_{i+1} *is obtained from* C_i *by* $t - factoring$ *iff*

    (1) C_i *is* $(\alpha_1 \ L \ \alpha_2 \ M \ \alpha_3)$ *or* C_i *is* $(\alpha_1 \ M \ \alpha_2 \ L \ \alpha_3)$;

    (2) L *and* M *have the same sign and unify with mgu* θ ;

    (3) L *is in* γ_M(*i.e.,* L *is in an higher level of the tree*);

    (4) C_{i+1} *is* $(\alpha_1\theta \ L\theta^* \ \alpha_2\theta \ \alpha_3\theta)$ *or* C_{i+1} *is* $(\alpha_1\theta \ \alpha_2\theta \ L\alpha_3\theta)$

C_{i+1} *is obtained from* C_i *by* $t - ancestry$ *iff*

    (1) C_i *is* $(\alpha_1 \ (L \ \alpha_2 \ (\alpha_3 \ M \ \alpha_4) \ \alpha_5) \ \alpha_6)$;

    (2) L *and* M *are complementary and unify with mgu* θ ;

    (3) L *is in* δ_M;

    (4) C_{i+1} *is* $(\alpha_1\theta \ (L\theta^* \ \alpha_2\theta \ (\alpha_3\theta \ \alpha_4\theta) \ \alpha_5\theta) \ \alpha_6\theta)$;

C_{i+1} *is obtained from* C_i *by* $t - truncation$ *iff*

    either C_i *is* $(\alpha \ (L^*) \ \beta)$ *and* C_{i+1} *is* $(\alpha \ \beta)$.

    or C_i *is* (ε^*) *and* C_{i+1} *is* $\square$.   *[]*

Definition 6 *Consider a t-clause* C_i *to be* $(\varepsilon^* \ \alpha_1 \ L \ \beta_1)$ *and a program-clause* B_{i+1} *to be* $(\varepsilon^* \ \alpha_2 \ M \ \beta_2)$ *and* L *to be an arbitrary literal selected for expansion. Then* C_{i+1} *is* <u>*derived*</u> *from* C_i *and* B_j *if the following conditions hold :*

    (a) L *and* M *are complementary and unify with mgu* θ'_{i+1} ;

    (b) C'_{i+1} *is* $(\varepsilon^* \ \alpha_1\theta \ (L\theta^* \ \alpha_2\theta \ \beta_2\theta)\beta_1\theta)$

    (c) C_{i+1} *is a tranfac-derivation of* C'_{i+1}

    (d) C_{i+1} *satisfies the admissibility and minimality conditions*   *[]*

Definition 7 *An* <u>*SLI* − *derivation*</u> *of a t-clause* E *from a non-Horn program* P *with top t-clause* $(\varepsilon^* \ \neg C)$ *is a sequence of t-clauses* $(C_1, \ldots, C_n)$ *such that:*

- C_1 *is a tranfac-derivation of* C, *and* C_n *is* E;

- *For all* i, $1 \leq i \leq n$ C_{i+1} *is derived from* C_i *and a program clause* B_{i+1} *in* P *[]*

Definition 8 *An* <u>*SLI* − *refutation*</u> *from a program* P *with top t-clause* $(\varepsilon \ \neg C)$ *is an SLI-derivation of the null clause* $\square$. *Then we write* $P \vdash_{SLI} C$.   *[]*

The example given next illustrates the SLI-refutation procedure.

Example 2 *Consider the program*

(1) $(\varepsilon^* \, a \, \neg c \, \neg d \, \neg e)$
(2) $(\varepsilon^* \, \neg d \, c)$
(3) $(\varepsilon^* \, f \, e \, \neg g)$
(4) $(\varepsilon^* \, a \, \neg f)$
(5) $(\varepsilon^* \, d)$
(6) $(\varepsilon^* \, g)\}$

and the goal clause

(7) $(\varepsilon^* \, \neg a)$

We show that there is an SLI-refutation.

(7)	$(\varepsilon^* \, \underline{\neg a})$	*goal clause*
(8)	$(\varepsilon^* \, (\neg a^* \, \neg c \, \neg d \, \underline{\neg e}))$	*derivation (7,1)*
(9)	$(\varepsilon^* \, (\neg a^* \, \underline{\neg c} \, \neg d \, (\neg e^* \, f \, \neg g)))$	*derivation (8,3)*
(10)	$(\varepsilon^* \, (\neg a^* \, (\neg c^* \, \underline{\neg d}) \, \neg d \, (\neg e^* \, f \, \neg g)))$	*derivation (9,2)*
(11)	$(\varepsilon^* \, (\neg a^* \, (\underline{\neg c^*}) \, \neg d \, (\neg e^* \, f \, \neg g)))$	*t-factoring (10)*
(12)	$(\varepsilon^* \, (\neg a^* \, \neg d \, (\neg e^* \, \underline{f} \, \neg g)))$	*t-truncation (11)*
(13)	$(\varepsilon^* \, (\neg a^* \, \neg d \, (\neg e^* \, (f^* \, \underline{a}) \, \neg g)))$	*derivation (12,4)*
(14)	$(\varepsilon^* \, (\neg a^* \, \neg d \, (\neg e^* \, (\underline{f^*}) \, \neg g)))$	*t-ancestry (13)*
(15)	$(\varepsilon^* \, (\neg a^* \, \underline{\neg d} \, (\neg e^* \, \neg g)))$	*t-truncation (14)*
(16)	$(\varepsilon^* \, (\neg a^* \, (\underline{\neg d^*}) \, (\neg e^* \, \neg g)))$	*derivation (15,5)*
(17)	$(\varepsilon^* \, (\neg a^* \, (\neg e^* \, \underline{\neg g})))$	*t-truncation (16)*
(18)	$(\varepsilon^* \, (\neg a^* \, (\neg e^* \, (\neg g^*))))$	*derivation (17,6)*
	$\square$	*t-truncations (18)*

[]

Minker and Zanon [MZ82] show that SLI-resolution is complete and sound for theorem proving with arbitrary clauses.

Theorem 1 *([MZ82]) Let S be a set of input t-clauses. Then, C is deducible from S by SLI-refutation iff C is a logical consequence of S* *[]*

3.2 Negation in non-Horn programs

First, we describe the generalized closed world assumption (GCWA) [Min82] which is the basis for answering negative queries in our approach. We also develop the concept of Support-for-Negation which is used in the development of the query answering procedure.

3.2.1 GCWA and Support-for-Negation

Minker [Min82] developed a consistent rule for inferring negation in non-Horn programs called the *Generalized Closed World Assumption* (GCWA). He provided a proof-theoretic and a model-theoretic definition for the GCWA and showed that they are equivalent. [MR87] presented a fixpoint definition for the GCWA. The original definition in [Min82] was restricted to function-free non-Horn programs but the restriction is not needed as shown in Shepherdson [She88].

Definition 9 *([Min82])*
Let P be a non-Horn logic program (function-free) and C a ground atom, then $\neg C$ can be inferred from P iff $C \notin E$ where E is a set of ground atoms defined as
$$E = \{A \mid A \text{ is a ground atom and } P \vdash A \vee K, K \text{ is a positive (possibly null)}$$
$$\text{clause and } K \text{ is not provable from } P \}.$$
GCWA(P) is the set of atoms whose negation can be inferred from the program using the GCWA []

The set GCWA(P) like that of CWA(P) is not fully computable. There are two reasons for this. The first is that $P \vdash A \vee K$ may not be decidable for some K. The second difficulty is that if $P \vdash A \vee K$ then we have $P \vdash A \vee K \vee K$ which will give rise to an infinite set to be proved before concluding the negation of A. The decidability problem arises because of infinite loops and unbounded term property in some derivation. The second difficulty is due to the clause K containing replicated literals. This problem can be solved if we restrict K to having only distinct literals. We solve the problem of non-distinct literals and infinite loops by giving a weaker definition for negation in non-Horn programs. The definition is based on the Support-for-Negation property, described below.

Next we introduce the *Support-for-Negation* property. The concept of Support-for-Negation stems from the definition of the generalized closed world assumption.

Definition 10 *Given a logic program P and a ground atom A, Support-for-Negation of A, is defined as :*
$$SN(A) = \{K \mid K \text{ is a ground positive (possibly null) clause and } P \vdash A \vee K\} \quad []$$

From the definition for GCWA, we can see that $\neg A$ can be assumed if all the clauses in SN(A) are logical consequences of P. This definition is next modified such that the problems of infinite loops and non-distinct literals are taken care of. The new definition is based on SLI-resolution.

Definition 11 *A $\underline{B - clause}$ is the clause formed as a disjunct of the B-literals in a t-clause. A B-clause is positive when it contains no negative B-literals.* []

Definition 12 *Given a logic program P and a ground atom A, Support-for-Negation of A (using SLI-resolution) is defined as :*
$SNSLI(A) = \{K \mid K$ *is a positive B-clause derived using an SLI-derivation from P and top-clause $(\varepsilon^* \neg A)$ }* []

The next result follows from the above definitions.

Corollary 1 *Given a non-Horn program P, a ground atom A in P is in GCWA(P) iff all the clauses in SNSLI(A) are logical consequences of P* []

We have given two definitions for support-for-negation. The one based on provability defines a complete support-for-negation set (SN), which has been reduced using derivability from an SLI-derivation (SNSLI). The second definition takes care of the problem of non-distinct literals because it is based on derivability and not on provability. The problem of infinite loops is solved since SLI-derivation detects such loops because of the admissibility condition and fails the derivation. The support-for-negation set as defined by SNSLI(A) can be very large and any reduction in the size of this set would be useful. Next, we give a definition for such a reduced set. We define a subset of SNSLI and show that this subset is sufficient for inferring negation in a program. We show that if all clauses in the subset are provable from the program, it implies that all clauses in the SNSLI set are also provable from the program. So, to infer the negation of a ground atom using the GCWA, we have to show that this subset (rather than the whole set SNSLI) is a logical consequence of the program. The definition of this subset is procedural and uses a modified SLI-derivation called SLINF-derivation. We use this definition of support-for-negation to develop a query answering procedure.

3.2.2 SLINF-resolution

SLINF-derivation is a variation of an SLI-derivation such that only negative literals in a t-clause C_i, $i = 1, 2, \ldots$ can take part in the application of a derivation. The rationale for this restriction is that our interest is in deriving positive clauses from the top-clause. When we have an SLI-derivation with top-clause $\neg C$, the clause formed using the B-literals at any step of the derivation is a logical consequence of the program and $\neg C$. So, restricting the t-extension rule from selecting positive literals derives positive clauses wherever possible, while retaining completeness (there may be cases where a t-clause may have some negative B-literal which cannot resolve with any input clause. In such cases we do not reach a refutation in the SLINF-derivation). The modified derivation is defined below:

Definition 13 *Consider a t-clause C_i to be $(\varepsilon^* \alpha_1 L \beta_1)$ and a program-clause B_{i+1} to be $(\varepsilon^* \alpha_2 M \beta_2)$ and L to be an arbitrary literal selected for expansion. Then*

C_{i+1} is $\underline{neg-derived}$ from C_i and B_j if the following conditions hold :

    (a) L is a negative literal, L and M are complementary and unify with mgu θ'_{i+1}
    (b) C'_{i+1} is $(\varepsilon^* \; \alpha_1\theta \; (L\theta^* \; \alpha_2\theta \; \beta_2\theta)\beta_1\theta)$
    (c) C_{i+1} is a tranfac-derivation of C'_{i+1}
    (d) C_{i+1} satisfies the admissibility and minimality conditions   []

Definition 14 An $\underline{SLINF-derivation}$ of a t-clause E from a non-Horn program P with top t-clause C is a sequence of t-clauses $(C_1, \ldots, C_n)$ such that:

- C_1 is a tranfac-derivation of C, and C_n is E;

- For all i, $1 \le i \le n$ C_{i+1} is neg-derived from C_i and a program clause B_{i+1} in P

Definition 15 An $\underline{SLINF-refutation}$ from the program P with top t-clause C is an SLI-derivation of a positive B-clause C' from the top t-clause C and program P.

    Then we write $P \bigcup \{\neg C\} \vdash_{SLINF} C'$ []

The modified definition of the support-for-negation set can be stated as follows :

Definition 16 Given a logic program P and a ground atom A, Support-for-Negation of A (using SLINF), is defined as :
$SNSLINF(A) = \{K \mid K$ is derived by an SLINF-Refutation from P with top-clause $(\varepsilon^* \; \neg A) \}$ []

    The set of clauses defined by SNSLINF is contained in or equal to that of SNSLI. That is, if a clause is derivable using an SLINF-derivation it is also derivable using an SLI-derivation. We show that these two sets are logically equivalent under program P. We do this by showing that the clauses in SNSLI(A) are logical consequences of $SNSLINF(A) \cup P$ and that SNSLINF(A) is a subset of SNSLI(P). The advantage, as noted earlier, is that the reduced size of the support-for-negation set makes it easier to compute negation. Note that this reduced set need not be the optimal support-for-negation set since two clauses K and $K \vee K'$ can be in the set SNSLINF. $K \vee K'$ is a redundant clause in the set, since the provability of K from the program also implies the provability of $K \vee K'$. The equivalence of these two sets are given by the two theorems below whose proofs are given in [MR87].

Theorem 2 Given a program P and a ground atom A,
    $P \bigcup SNSLINF(A) \Longrightarrow_{SLI} SNSLI(A)$ ·   []

Theorem 3 Given a program P and a ground atom A,
    $SNSLINF(A) \subseteq SNSLI(A)$ []

From the two theorems we see that if a clause K in SNSLI(A) is provable from program P, there exists a clause K' in SNSLINF(A) which is provable from P. Also if all clauses in SNSLINF(A) are provable from P then all clauses in SNSLI(A) are provable from P.

Example 3 *Let* $P = \{p \vee q, \ r \leftarrow q\}$
$\quad SNSLI(p) = \{q, \ r\}$
$\quad SNSLINF(p) = \{q\}$
We can see that $SNSLINF(p) \subseteq SNSLI(p)$ *and*
$\quad SNSLINF(p) \bigcup P \vdash_{SLI} SNSLI(p)$ $\qquad\qquad\qquad\qquad\qquad\quad$ *[]*

We can now state a new rule for negation in non-Horn programs. We call this rule Support-for-Negation Rule (SN Rule).

$$\frac{\forall K, \ K \in SNSLINF(A), \ P \vdash_{SLI} K}{\neg A}$$

The soundness of this rule with respect to the GCWA follows from Corollary 1 and Theorems 1 & 2.

3.2.3 Query Answering Algorithm

In this section we provide a procedure which can be used for answering queries. We use SLI-resolution for answering positive queries and the SN Rule for negations. For a negative query $\neg A$ the algorithm derives a positive B-clause K from the top t-clause $(\varepsilon^* \ \neg A)$ using an SLINF-derivation. If K is provable from the program then the algorithm tries another SLINF-refutation and a new K. If at any time a K is not provable (finitely) then the algorithm fails the query. If there are no more SLINF-refutations the query succeeds. The algorithm is stated for unit queries and is easily extended to a conjunction of literals.

Procedure 1 *QUERY ANSWERING ALGORITHM (QAA)*
$\quad$ *Procedure: (Unit Query: L, $\qquad$ Program: P)*

$\qquad$ *Positive Query : L is atomic and L = A*
$\qquad\qquad$ *If there is an SLI-refutation from P and top-clause $(\varepsilon^* \ \neg A)$*
$\qquad\qquad\qquad\qquad\qquad\qquad\qquad\qquad$ *then SUCCEED L*
$\qquad\qquad\qquad\qquad\qquad\qquad\qquad\qquad$ *else FAIL L*
$\qquad$ *Negative Query : L is ground and L = $\neg A$*
$\qquad\qquad$ *1. Construct an SLINF-refutation from P and top-clause $(\varepsilon^* \ \neg A)$*
$\qquad\qquad\quad$ *which is distinct from the previous SLINF-refutations.*
$\qquad\qquad\quad$ *Let K be the positive B-clause derived*

If $P \vdash_{SLI} K$ then Step 2
else FAIL L
2. If no more distinct SLINF-refutations can be constructed
then SUCCEED L
else Step 1

[]

Example 4 *Let $P = \{p \vee q \leftarrow m, \ r \vee m, \ s \vee m, \ q \vee r\}$ and let the query be $\neg p$*

1. Step 1 in QAA constructs an SLINF-refutation and ends in a B-clause $q \vee r$

2. Since $P \vdash_{SLI} q \vee r$ we go to Step 2 in QAA

3. Since another distinct SLI-refutation is possible we go to Step 1 of QAA

4. Step 1 in QAA constructs an SLINF-refutation which is distinct from the one in step 1 of this example and ends in a B-clause $q \vee s$

5. Since $P \nvdash_{SLI} q \vee s$ the query FAILS and we cannot conclude $\neg p$. []

4 Comparison of Horn and non-Horn Procedures

4.1 SLD-resolution and SLI-resolution

From the definitions of SLD and SLI resolutions we can see that apart from the *tranfac-derivation* (truncation,ancestry and factoring) step both the definitions are similar. The tranfac-derivation step is needed because we are dealing with non-Horn programs. If the program under consideration is Horn then the tranfac derivation step can be simplified to just t-truncation and the two derivations would be the same. In an SLD-derivation the t-truncation step is built in as an elimination of a query atom when unifying with an assertion. In an SLI-derivation the elimination is explicitly done as repeated applications of t-truncation. The main reason for the similarity between SLI and SLD derivations comes from the the fact that both use SL-derivation as the basis of the inference method and both have the property that they allow arbitrary literal selection. Hence, SLI and SLD behave identically when operating on Horn programs. SLI-derivation has an advantage over SLD-derivation since it uses the admissibility condition as a loop-checker and fails infinite loops in the derivation. But this advantage is bought at the cost of time and space. For Horn implementation of SLI-resolution the admissibility condition can be an optional check done for some literals identified by the user or a pre-compiler. The idea is similar to the optional 'occur-check' suggested for unification [Pla84]. When operating on Horn programs , SLI and SLINF derivations are equivalent since there are no negative literals involved.

4.2 SN-Rule and NAF

The reduction of GCWA to CWA when dealing with Horn programs has been well established [Min82]. Here we compare the two weaker rules of negation, SN and NAF. We show that the SN Rule reduces to NAF Rule when applied to Horn programs. When dealing with Horn programs and a ground negative query $\neg A$, SNSLINF(A) is either empty or contains the empty clause or is undecidable because of some infinite derivation. When SNSLINF(A) is empty, it is because all SLINF-derivations end in B-clauses which contain negative literals. This situation is the same as having a finitely failed SLD-tree for A. When SNSLINF(A) has the empty clause, it is the same as having an SLI-refutation for A. In this case an SLD-refutation will also have a refutation and fail the negative query. Hence, in both decidable cases the SN Rule reduces to the NAF Rule when operating on Horn programs. But there is one difference, the difference comes from the 'loop-check' capability of SLINF-resolution.

5 Conclusion

We have presented SLI-resolution as a viable proof procedure for answering positive queries in non-Horn programs. We have shown that SLI-resolution is an extension to SLD-resolution and reduces to SLD-resolution for Horn clauses. We have given a default rule of negation, Support-for-Negation Rule, for non-Horn programs. This definition is a weaker form of the GCWA. We have described a modified form of SLI-resolution called SLINF-resolution and used it to define the new rule of negation. We have also developed a query answering algorithm for handling unit queries in non-Horn programs based on the SN Rule. We have also shown that the SN-Rule reduces to NAF Rule when dealing with Horn programs.

Acknowledgements

We wish to express our appreciation to the National Science Foundation for their support of our work under grant number IRI-86-09170 and the Army Research Office under grant number DAAG-29-85-K-0-177. We wish to thank Jorge Lobo for his useful discussions during the development of the paper.

References

[AvE82] K.R. Apt and M.H. van Emden. Contributions to the Theory of Logic Programming. *J.ACM*, 29(3):841–862, 1982.

[CL73]   C. L. Chang and R. C. T. Lee. *Symbolic Logic and Mechanical Theorem Proving*. Academic Press, New York, 1973.

[Cla78] K. L. Clark. Negation as Failure. In H. Gallaire J. Minker, editor, *Logic and Data Bases*, pages 293–322, Plenum Press, New York, 1978.

[Hil74] R. Hill. *LUSH Resolution and its Completeness*. Technical Report, University of Edinburgh School of Artificial Intelligence, August 1974.

[KK71]   R. A. Kowalski and D. Kuehner. Linear Resolution with Selection Function. *Artificial Intelligence*, 2:227–260, 1971.

[Llo84] J.W. Lloyd. *Foundations of Logic Programming*. Springer–Verlag, 1984.

[Lov78] D.W. Loveland. *Automated Theorem Proving: A Logical Basis*. North–Holland Publishing Co., 1978.

[Min82] J. Minker. *On Indefinite Databases and the Closed World Assumption*. Springer-Verlag, 1982.

[MR87]   J. Minker and A. Rajasekar. A Fixpoint Semantics for Non-Horn Logic Programs. July 1987. Submitted to Journal of Logic Programming.

[MZ82]   J. Minker and G. Zanon. An extension to linear resolution with selection function. In *Information Processing Letters*, pages 191–194, June 1982.

[Pla84] D.A. Plaisted. An Efficient Bug Location Algorithm. In *Proc. Second Int. Conf. on Logic Programming*, Uppsala, 1984.

[Prz86] T. Przymusinski. Query answering in circumscriptive and closed world theories. *Proc. Amer. Assoc. for Artificial Intelligence '86*, 186–190, August 1986.

[Rei78] R. Reiter. On closed world data bases. In H. Gallaire J. Minker, editor, *Logic and Data Bases*, pages 55–76, Plenum Press, New York, 1978.

[She88] J.C. Shepherdson. Negation in logic programming. In J. Minker, editor, *Foundations of Deductive Databases and Logic Programming*, Morgan-Kaufman Pub., 1988.

[Van88] A. Van Gelder. Negation as failure using tight derivations for general logic programs. In J. Minker, editor, *Foundations of Deductive Databases and Logic Programming*, Morgan-Kaufmann, 1988.

Recursive Query Answering with Non-Horn Clauses

Shan Chi and Lawrence J. Henschen

Department of Electrical Engineering and Computer Science
Northwestern University
Evanston, Illinois 60208

Abstract

In this paper, an algorithm is presented for answering recursive queries under the Generalized Closed World Assumption (GCWA) in a database with positive non-Horn ground clauses. It is proved that the algorithm generates all the answers under the GCWA. We consider only the transitive closure recursions in which only one base relation is involved. The set of ground clauses is stored as one relation so that a modified join operator can be applied to the clauses. The relation can be visualized as a directed graph with each tuple representing an edge. A query (only closed queries are considered in this paper) is answered by extracting the paths from the relation, forming the negative clauses from these paths, and sending the negative clauses together with the non-Horn clauses to a theorem prover. We proved that if the empty clause is derived then the answer to the query is *true*. This algorithm is efficient for two reasons: 1) facts (positive unit ground clauses) are processed by the modified join operator, and 2) the theorem prover handles only the non-Horn ground clauses and some negative ground clauses.

1   Introduction

The efficient processing of recursive queries in relational database systems has been studied intensively in recent years. To the authors' knowledge, the recursions studied are exclusively based upon Horn databases. In this paper, we consider the recursions with non-Horn ground clauses.

When databases contain incomplete data, some incomplete data can be represented by disjunctions of facts. For example, we may know someone's blood type is either A or O but do not have the complete information to tell exactly which. Then instead of leaving a null value [4] in the blood type field, we may record *A or O*. Or putting it into a clause form, we get the non-Horn clause *Blood-Type(John,A)* ∨ *Blood-Type(John,O)*. This representation carries more information then the null value representation. For example, if we are asked "Is John's blood type B?" we may safely answer *no*. While in the null value representation, we can only answer *don't know*.

One problem posed by a non-Horn database is related to the negative information implied by the database. The Closed World Assumption (CWA)[12] is commonly used to derive negative information in Horn databases. In [12], Reiter also pointed out that the CWA may cause inconsistency in a non-Horn database. Minker[9] then proposed the Generalized Closed World Assumption (GCWA) for deriving negative information in non-Horn databases. Under the GCWA, we can assume $\sim A$ only if A does not occur in any minimal model of the database. We adopt the GCWA in our query processing system.

Grant and Minker[5] proposed a method for processing queries under the GCWA. In this method, all the ground instances and database models are generated. Yahya and Henschen[14] extended the GCWA to the extended GCWA (EGCWA) such that negative clauses can also be assumed. They developed a deductive approach to query answering under the EGCWA. In this approach a large group of clauses need to be proved at query time. Bossu and Siegel [2] proposed an algorithm for answering queries based upon the so-called *subimplication*. To answer the query, their algorithm generates the characteristic formulas using a theorem prover. All these approaches seem to be too inefficient for two reasons: 1) not enough preprocessing is done before the query processing, and 2) a large set of clauses are generated.

Henschen and Park[8] proposed the compilation approach to query processing in non-Horn databases. In this approach, the database is divided into two sets: one, called the RDB, stores unit ground clauses and the other, called the CDB, stores other clauses. The CDB is processed at compile time and the RDB clauses (or tuples) are retrieved at query time. One advantage of this approach comes from the applications of relational operators, like select and join, to the RDB clauses. The other comes from the elimination of many redundant clauses at the compilation stage by applying the reduction theorems they developed. However, storing all the non-Horn clauses in the CDB also causes some disadvantages: 1) updating a non-Horn clause in the CDB requires the recompilation of the CDB, and 2) too many clauses may be generated in the compilation stage due to the existence of non-Horn clauses.

Other than one example of recursive query given in [8], these papers [2,8,5,14] did not deal with recursions explicitly. On the other hand, the recursive query processing techniques developed so far [1,3,6,7,10,11] considered only Horn databases. The reason for not considering non-Horn databases is a natural one—the data model these techniques are based upon is the relation model and relations can be directly mapped to Horn clauses in first-order theory. With non-Horn clauses in the database, query answering usually requires a theorem prover because the relational operators like select and join can not be applied to non-Horn clauses. Another reason for not considering non-Horn databases is a practical one—in most database applications today, only definite information (represented as unit ground clauses) is involved. However we feel that in the future a deductive database will be coupled with other programs, such as computer vision program and radar signal analysis program, through which incomplete information might be gathered. Then a non-Horn database provides a much better modeling of the world of interest than a Horn database with null values does.

In this paper, we present an efficient algorithm that processes recursive queries in a non-Horn database. Even though the database and the recursive pattern considered are very simple, the theorems can be extended to more complicated databases with other types of linear recursions.

2 Preliminaries

A database DB can be considered as the union of two disjoint sets of clauses, IDB and EDB, representing the intensional database and the extensional database, respectively. The EDB consists of all the ground clauses and IDB consists of other clauses. We further divide the EDB into two disjoint sets, $DEDB$ and $IEDB$, where $DEDB$ contains only unit ground Horn clauses and $IEDB$ contains other EDB clauses. Clauses in $DEDB$ are said to be *definite* and those in $IEDB$ are *indefinite*. A literal is definite (indefinite) if it belongs to an definite (indefinite) clause. We assume that a clause in DB is not subsumed by any other clause in DB. Under this assumption, a literal in EDB must be either definite or indefinite but not both (otherwise the definite clause containing the literal subsumes the indefinite clause containing the same literal). The database we considered contains no function symbol and the IDB contains only the following two rules:

$$\text{rule 1} \;=\; R(x,y){\sim}A(x,y)$$
$$\text{rule 2} \;=\; R(x,z){\sim}R(x,y){\sim}A(y,z)$$

Note that the $\vee$ (or) connective between literals are omitted. The EDB contains only A literals. We are interested in getting the answer under the GCWA for the closed query $?R(a,b)$. (Though the algorithm can be easily extended to answer open queries, we will not discuss it in this paper.) The definition of the answers to a query under the GCWA is based upon the concept of minimal clauses.

Definition A clause C is minimal in a set S of clauses if C is not subsumed by any other clause derivable from S.

In the sequel, unless otherwise specified a clause is said to be minimal if the database DB does not derive any of its subclauses.

Definition (GCWA) The ground unit clause ${\sim}A$ can be assumed to be true if DB does not derive any minimal positive clause containing A.

Definition The answer to the closed query $?R(a,b)$ is *true* if $DB \vdash R(a,b)$. The answer is *possible* if DB derives a non-unit minimal positive clause containing $R(a,b)$. The answer is *false* otherwise.

Example 2.1 The following shows the database DB, including two IDB clauses C_1

and C_2, one *IEDB* clause C_3, and one *DEDB* clause C_4. Some clauses derived from *DB* (C_5 through C_8) are also shown. Note that we use numbers to represent constants in the database.

$$
\begin{aligned}
C_1 &= R(x,y){\sim}A(x,y) \\
C_2 &= R(x,z){\sim}R(x,y){\sim}A(y,z) \\
C_3 &= A(a,1)A(1,2) \\
C_4 &= A(1,3) \\
[C_1,C_{3.1}]\ C_5 &= R(a,1)A(1,2) \\
[C_2,C_{5.1},C_{3.2}]\ C_6 &= R(a,2)A(a,1)A(1,2) \\
[C_2,C_{5.1},C_4]\ C_7 &= R(a,3)A(1,2) \\
[C_1,C_4]\ C_8 &= R(1,3)
\end{aligned}
$$

where the clause numbers enclosed in [and] are the clauses involved in the resolution and the number after the decimal point indicates the literal resolved upon. For query $?R(a,1)$, we should answer *possible* because *DB* derives the minimal positive clause C_5. Similarly, the answer to $?R(a,3)$ is *possible* (from C_7). However, for query $?R(a,2)$, the answer should be *false* under the GCWA because no minimal positive clause containing $R(a,2)$ is derivable from *DB*. (Note C_6 contains $R(a,2)$ but is not minimal because it is subsumed by C_3.) For query $?R(1,3)$, the answer should be *true* because $R(1,3)$ is derivable from *DB* (see C_8).

A direct approach to query answering is generating all minimal positive clauses. This approach requires the level saturation with a theorem prover and the subsumption tests. In the above example, to make sure C_5 is minimal, one must check all other clauses derivable from *DB* to see whether any of them subsumes C_5. In a real database, the number of clauses is usually too large for this approach to be practical.

The algorithm proposed in this paper uses the modified join operator to facilitate the query processing. Therefore all ground clauses must be stored as relations. Unit ground clauses can be directly stored as tuples in relational databases. In order to store the non-Horn ground clauses as tuples, we need to break the clauses into individual literals . We also need to label the literals from each non-Horn clause so that they can be re-assembled into that clause. Labeling literals also allows us to distinguish the definite literals (representing Horn clauses) from indefinite literals (those from a positive non-Horn clause).

Labeling the literals can be implemented in many different ways. One straightforward implementation is to add one field to each tuple holding the clause identifier containing the literal. All definite tuples should contain a special identifier in this field to indicate their definiteness. Indefinite clauses can then be re-assembled by joining on this field. Inverted files can be created to speed up the retrievals. Also the definite tuples and indefinite tuples can be stored in two separate tables so that no extra field is needed for the definite table. In any case, these implementation details should be transparent to us. In our model, all A literals (definite or indefinite) are stored in relation A.

Theorem 2.1 (Generic Rule) *$IDB \vdash R(x_1, x_n) \sim A(x_1, x_2) \ldots \sim A(x_{n-1}, x_n)$ for any $n > 1$.*

Theorem 2.1 suggests that a sequence of literals $A(a_1, a_2), A(a_2, a_3), \ldots, A(a_{n-1}, a_n)$ be used for deriving a clause containing $R(a_1, a_n)$. This leads us to the following definition.

Definition A set of tuples $\{(a_1, a_2), (a_2, a_3), \ldots, (a_{n-1}, a_n)\}, n > 1$ in a relation, abbreviated as $[a_1, a_2, a_3, \ldots, a_n]$, is called a *path* from a_1 to a_n. A path is non-empty if not otherwise specified. If the path $[a_1, a_2, \ldots, a_n]$ is in relation A, then $A(a_1, a_2), A(a_2, a_3), \ldots, A(a_{n-1}, a_n)$ are called the *literals in the path*. A path is said to be *definite* if all the literals in that path are definite; otherwise, it is *indefinite*. In the sequel, all paths refer to the paths in the relation A. A tuple (a, b) in relation A is said to be *induced* by a clause C if C contains $A(a, b)$. The clause C is called the *inducing* clause.

From the definition, we may notice that a definite tuple (a, b) in relation A implies the clause $A(a, b)$ is in *EDB*. Also an indefinite tuple (a, b) in relation A implies the existence of at least one inducing clause $A(a, b)C$ in *EDB* where C is the subclause containing other positive ground A literals. Therefore, the existence of a path implies the existence of all its tuples in the relation, which in turn implies the existence of all the inducing clauses.

Example 2.2 In this example, we shall give an overview of the algorithm presented in this paper before the algorithm is formally derived. Assume the database contains the following clauses:

$$
\begin{array}{llll}
C_1 & = & R(x, y) \sim A(x, y) & \qquad C_8 & = & A(6, 9) \\
C_2 & = & R(x, z) \sim R(x, y) \sim A(y, z) & \qquad C_9 & = & A(8, 10) \\
C_3 & = & A(a, 1) & \qquad C_{10} & = & A(1, 11)A(11, 12) \\
C_4 & = & A(a, 3) & \qquad C_{11} & = & A(a, 2)A(1, 4)A(3, 6) \\
C_5 & = & A(2, 5) & \qquad C_{12} & = & A(a, 2)A(4, 8)A(7, 10) \\
C_6 & = & A(4, 7) & \qquad C_{13} & = & A(1, 4)A(9, 8)A(6, 10) \\
C_7 & = & A(5, 8) &
\end{array}
$$

where C_1 and C_2 form the *IDB*, C_3 through C_9 form the *DEDB*, and C_{10} through C_{13} form the *IEDB*. The paths in relation A can be easily seen in the following graph:

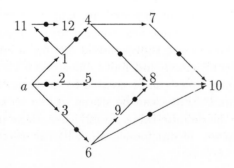

Each literal $A(b,c)$ in DB is represented as an edge from b to c in the graph. Note that the edges representing indefinite literals are marked with dots. $[2,5,8,10]$ is a definite path in relation A. It implies the existence of the definite literals $A(2,5)$, $A(5,8)$, and $A(8,10)$ in DB. By resolving these literals with the generic rule, DB derives $R(2,10)$. $[3,6,9]$ is an example of indefinite paths. It is indefinite because one of the literals in the path, $A(3,6)$ is indefinite. $A(3,6)$ is induced by the indefinite clause $C_{11} = A(a,2)A(1,4)A(3,6)$. $A(6,9)$ is a definite literal (from C_7) in the path. By resolving C_{11} and C_7 with the generic rule, DB derives the minimal clause $R(3,9)A(a,2)A(1,4)$. To give an overview of the algorithm, we assume the query to be $?R(a,10)$. The algorithm will extract the indefinite literals in the paths from a to 10 as follows:

$\{A(1,4), A(7,10)\}$	From path $[a,1,4,7,10]$
$\{A(1,4), A(4,8)\}$	From path $[a,1,4,8,10]$
$\{A(a,2)\}$	From path $[a,2,5,8,10]$
$\{A(3,6), A(9,8)\}$	From path $[a,3,6,9,8,10]$
$\{A(3,6), A(6,10)\}$	From path $[a,3,6,10]$

These sets of literals are then transformed into negative clauses:

$$
\begin{aligned}
C_{14} &= {\sim}A(1,4){\sim}A(7,10) \\
C_{15} &= {\sim}A(1,4){\sim}A(4,8) \\
C_{16} &= {\sim}A(a,2) \\
C_{17} &= {\sim}A(3,6){\sim}A(9,8) \\
C_{18} &= {\sim}A(3,6){\sim}A(6,10)
\end{aligned}
$$

The clauses in $IEDB$ (C_{10} through C_{13}) and these negative clauses are sent to a theorem prover. Theorem 3.11 shows that if the empty clause is derived then $DB \vdash R(a,10)$. Therefore, $?R(a,10)$ should be answered $true$. One derivation of the empty clause (by hyperresolutions[13]) are shown below:

$$
\begin{aligned}
[C_{11}, C_{12}, C_{14}]\ C_{19} &= A(a,2)A(3,6)A(4,8) \\
[C_{11}, C_{15}, C_{19}]\ C_{20} &= A(a,2)A(3,6) \\
[C_{13}, C_{17}, C_{20}]\ C_{21} &= A(a,2)A(1,4)A(6,10) \\
[C_{11}, C_{18}, C_{21}]\ C_{22} &= A(a,2)A(1,4) \\
[C_{12}, C_{14}, C_{22}]\ C_{23} &= A(a,2)A(4,8) \\
[C_{15}, C_{22}, C_{23}]\ C_{24} &= A(a,2) \\
[C_{16}, C_{24}]\ C_{25} &= \square
\end{aligned}
$$

Note that in the derivation, only the indefinite ground clauses from $IEDB$ and the negative ground clauses derived from the paths are involved. Also note that $R(a,10)$ is definite even though no single definite path from a to 10 exists. In effect, there exists a deduction in which various indefinite literals get merged. Theorem 3.11 shows that this merging can be discovered by a theorem proving strategy that separates resolution over the IDB from resolution over the EDB, the latter being simple enough to do at query-processing time.

3   Paths and Minimal Models

In this section, we shall develop the theorems that the algorithm is based upon. We shall use L to represent a literal, C to represent a clause, and LC to represent a clause containing the literal L and the subclause C. L and C with subscripts are used to represent different literals and clauses.

Theorem 3.1 *Let C be a positive ground clause. If all the minimal models of a set S of clauses satisfy C then S derives a clause subsuming C.*

Proof   If S is inconsistent then S derives the empty clause $\square$ that subsumes C. Now suppose S is consistent. If all the minimal models of S satisfy C then $S \cup \sim C$ is inconsistent, therefore $S \cup \sim C \vdash \square$. Since $\sim C$ contains only unit clauses, we may modify the refutation tree of the derivation so that all the resolutions with the clauses in $\sim C$ are grouped to the root of the tree (the formal proof is omitted). Let $\sim L_1, \ldots \sim L_n$ be the clauses from $\sim C$ occurring in the modified refutation tree (with L_1 being the first clause from $\sim C$ resolved). At least one such clause exists otherwise S would have been inconsistent. Let C' be the clause resolving with L_1 in the modified refutation tree. Note that C' resolves with $\sim L_1, \ldots \sim L_n$ to derive the empty clause. The derivation of the empty clause is possible only if $L_1 \ldots L_n$ is a ground instance of C'. In that case, C' subsumes $L_1 \ldots L_n$, a subclause of C. Therefore C' subsumes C.   Q.E.D.

In Theorem 3.1, if S is the *DB* defined in this paper, and every minimal of *DB* satisfies the positive unit ground clause C then $DB \vdash C$ because all the positive clauses derivable from *DB* are ground and the only ground clause subsuming C must be C itself.

Theorem 3.2 *Let S be a set of clauses and M be a minimal model of S. Then a positive literal L is in M if and only if there exists a ground instance LC of a clause in S such that M falsifies C.*

Proof   (if) If M is a minimal model of S then M must satisfy all the ground instances of the clauses in S. In particular, M must satisfy LC. If M falsifies C then M must satisfy L in order to satisfy LC. Therefore L must be in M.

(only if) Let $LC_1, LC_2, \ldots, LC_n$ be all the ground instances containing L. At least one such instance exists, and among these instances at least one C_i is falsified by M for otherwise $M - \{L\}$ is also a model, conflicting with the minimality of M. Q.E.D.

The above theorem can be extended so that two or more literals in the minimal model are considered at the same time.

Theorem 3.3 *Let S be a set of clauses, and M be a minimal model of S. Then the positive literals $L_1, \ldots, L_n$ are in M if and only if there exist the ground instances $L_1 C_1, \ldots, L_n C_n$ of the clauses in S such that $C_1 C_2 \ldots C_n$ is falsified by M.*

Definition   An interpretation M of DB is said to *satisfy* a path if all the literals in the path are in M; otherwise, it is said to *falsify* the path.

Theorem 3.4 $R(a_1, a_n)$ *is in a minimal model M of DB if and only if M satisfies a path from a_1 to a_n.*

Proof (if) Let the path satisfied by M be $[a_1, a_2, \dots, a_n]$. Consider one instance of the generic rule: $R(a_1, a_n) \sim A(a_1, a_2) \dots \sim A(a_{n-1}, a_n)$. M must satisfy this instance since the rule is derived from DB. However, all the negative literals in this instance are falsified by M because $A(a_1, a_2), \dots, A(a_{n-1}, a_n)$ are in M. By Theorem 3.2 $R(a_1, a_n)$ must be in M for the instance to be satisfied.

(only if) If M is a minimal model of DB then it must satisfy all the ground instances of rule 1 and 2. For any constant a_i, the ground instances of the clauses in DB containing $R(a_1, a_i)$ can be classified into the following two types:

$$\text{type 1} \;=\; R(a_1, a_i) \sim A(a_1, a_i)$$
$$\text{type 2} \;=\; R(a_1, a_i) \sim R(a_1, x) \sim A(x, a_i)$$

where x can be replaced with any constant in DB. By Theorem 3.2, if $R(a_1, a_i) \in M$ then there must exist a ground instance $R(a_1, a_i)C_i$ such that C_i is falsified by M. If $R(a_1, a_i)C_i$ is of type 1 then $\sim A(a_1, a_i)$ must be falsified by M, i.e., $A(a_1, a_i) \in M$. If $R(a_1, a_i)C_i$ is of type 2 then there exists some a_{i+1} such that $\sim R(a_1, a_{i+1}) \sim A(a_{i+1}, a_i)$ is falsified by M, or equivalently, $R(a_1, a_{i+1}), A(a_{i+1}, a_i) \in M$. Note that the choice of a_{i+1} is nondeterministic when there are two or more alternatives.

By applying the above argument to a_n we have: if $A(a_1, a_n) \notin M$ then $\exists a_{n+1}$ such that $R(a_1, a_{n+1}), A(a_{n+1}, a_n) \in M$. Furthermore, if $A(a_1, a_{n+1}) \notin M$, then $\exists a_{n+2}$ such that $R(a_1, a_{n+2}), A(a_{n+2}, a_n) \in M$. By repeating this deduction process, we may obtain a sequence of ground instances of rule 2 whose negative literals are all falsified by M:

$$C_n \;=\; R(a_1, a_n) \sim R(a_1, a_{n+1}) \sim A(a_{n+1}, a_n)$$
$$C_{n+1} \;=\; R(a_1, a_{n+1}) \sim R(a_1, a_{n+2}) \sim A(a_{n+2}, a_{n+1})$$
$$\cdots$$
$$C_{n+k-1} \;=\; R(a_1, a_{n+k-1}) \sim R(a_1, a_{n+k}) \sim A(a_{n+k}, a_{n+k-1})$$

Note that the choice of each a_{i+1} must be such that for any $j \leq i, a_{i+1} \neq a_j$ to avoid cycles in the deduction.   If such a choice does not exist, then $M - \{R(a_1, a_n), R(a_1, a_{n+1}), \dots, R(a_1, a_i)\}$ is also a model (because all C_i's are still satisfied for their negative R literals), which conflicts with the minimality of M. Therefore, the sequence of clauses can be chosen such that each clause brings in a new constant. As the number of constants in DB is finite, the sequence can not run indefinitely. For some $k \geq 0$, there must be a $A(a_1, a_{n+k}) \in M$ terminating the deduction. Since all the negative literals of clauses in this sequence are falsified by M, the literals $A(a_1, a_{n+k})$, $A(a_{n+k}, a_{n+k-1}), \dots, A(a_{n+1}, a_n)$ must be in M. These literals form a path from a_1 to a_n satisfied by M.   Q.E.D.

Example 3.1 Let *EDB* contain the following four clauses:

$$\begin{aligned}
C_1 &= A(1,2)\ A(4,6) \\
C_2 &= A(2,3)\ A(5,6) \\
C_3 &= A(4,6)\ A(5,6) \\
C_4 &= A(4,5)
\end{aligned}$$

Then *DB* has the following three minimal models:

$$\begin{aligned}
M_1 &= \{A(4,5), A(4,6), A(5,6), R(4,5), R(4,6), R(5,6)\} \\
M_2 &= \{A(4,5), A(4,6), A(2,3), R(4,5), R(4,6), R(2,3)\} \\
M_3 &= \{A(4,5), A(5,6), A(1,2), R(4,5), R(5,6), R(1,2), R(4,6)\}
\end{aligned}$$

Note that there exists an indefinite path $[1, 2, 3]$ (due to the indefinite literals $A(1,2)$ and $A(2,3)$). However, because there is no minimal model satisfying this path, $R(1,3)$ shall not occur in any minimal model. On the other hand, $R(4,6) \in M_3$ implies a path from 4 to 6 satisfied by M. We can see that $A(4,5), A(5,6) \in M_3$ form this path. In the proof of Theorem 3.4, a nondeterministic process is shown for finding the path in a minimal model. Let us follow that process: Since $R(4,6) \in M_3$ and $A(4,6) \notin M_3$, we should look for some constant b such that both $R(4,b)$ and $A(b,6)$ are in M_3. 5 is the only choice in this case. (In the proof, it is shown that such a constant always exist.) Continuing the process, we found $A(4,5) \in M_3$ which terminates the deduction. And the A literals produced in this process, $A(5,6)$ and $A(4,5)$, form the path.

Theorem 3.5 *If a minimal model M of DB satisfies a path from a_1 to a_n then DB derives a positive ground clause $R(a_1, a_n)C$ such that M satisfies $R(a_1, a_n)$ and falsifies C.*

Proof Let the path from a_1 to a_n be $[a_1, a_2, \ldots, a_n]$. By definition this path implies the existence of the inducing clauses in *EDB*:

$$A(a_1, a_2)C_1$$
$$A(a_2, a_3)C_2$$
$$\ldots$$
$$A(a_{n-1}, a_n)C_{n-1}$$

Note that all C_i's are positive or empty because we do not allow negative literals in *EDB*. From Theorem 3.3, we know we can always choose the set of clauses such that $C_1 C_2 \ldots C_n$ is falsified by M. Resolving these clauses with the generic rule derives $R(a_1, a_n)C_1 \ldots C_{n-1}$. Let $C = C_1 \ldots C_{n-1}$. Then $DB \vdash R(a_1, a_n)C$, where C is falsified by M, and $R(a_1, a_n)$ is satisfied by M from Theorem 3.4.   Q.E.D.

Example 3.2 Use the *DB* defined in Example 3.1. M_1 satisfies the path $[4, 5, 6]$ (because $A(4, 5), A(5, 6) \in M_1$). Therefore, $DB \vdash R(4, 6)C$ for some positive ground clause C falsified by M_1. The following resolutions will derive such a clause:

$$[C_4, C_5] \, C_7 \; = \; R(4, 5)$$
$$[C_7, C_2, C_6] \, C_8 \; = \; R(4, 6)A(2, 3)$$

It can be seen that M_1 satisfies $R(4, 6)$ and falsifies $A(2, 3)$.

Theorem 3.6 *Let S be a set of clauses. Suppose S derives the positive ground clause LC where L is a positive literal and C is the subclause such that there is a minimal model M of S satisfying L and falsifying C. Then S derives a minimal positive ground clause containing L.*

Proof If LC is not minimal then S derives a subclause of LC that is minimal. If this minimal clause does not contain L then it must be a subclause of C, therefore must be falsified by M, conflicting with the fact that it is derivable from S. Therefore this minimal clause contains L. Q.E.D.

Example 3.3 Let S be the *DB* defined in Example 3.1. From Example 3.2 we know that $DB \vdash R(4, 6)A(2, 3)$, and the minimal model M_1 of *DB* satisfies $R(4, 6)$ and falsifies $A(2, 3)$. From Theorem 3.6, *DB* derives a minimal positive clause containing $R(4, 6)$. The following resolutions derive such a minimal positive clause:

$$[C_4, C_5] \, C_7 \; = \; R(4, 5)$$
$$[C_{3.1}, C_5] \, C_8 \; = \; R(4, 6)A(5, 6)$$
$$[C_7, C_{8.2}, C_6] \, C_9 \; = \; R(4, 6)$$

Theorem 3.7 *DB derives a minimal positive ground clause containing $R(a_1, a_n)$ if and only if there is at least one minimal model of DB satisfying a path from a_1 to a_n.*

Proof (if) Let M be a minimal model of *DB* satisfying a path from a_1 to a_n. By Theorem 3.5, *DB* derives a positive ground clause $R(a_1, a_n)C$ such that M satisfies $R(a_1, a_n)$ and falsifies C. From Theorem 3.6, we conclude that *DB* derives a minimal positive ground clause containing $R(a_1, a_n)$.

(only if) Let $R(a_1, a_n)C$ be a minimal positive ground clause derivable from *DB*. If $R(a_1, a_n)$ is not in any minimal model, then all the minimal models of *DB* must satisfy C (in order to satisfy $R(a_1, a_n)C$). From Theorem 3.1, we know *DB* must derive a clause subsuming C, which conflicts with the minimality of $R(a_1, a_n)C$. Therefore, there must exist a minimal model M of *DB* in which $R(a_1, a_n)$ occurs. From Theorem 3.4, we know that there is a path from a_1 to a_n satisfied by M. Q.E.D.

In Example 3.1, it can be seen that there exists a path from 1 to 3 while there is no minimal model satisfying this path. Therefore, DB does not derive any minimal clause containing $R(1,3)$. (Note that even though DB does derive a clause $R(1,3)A(4,6)A(5,6)$, the clause is not minimal because it is subsumed by C_3.) Another path $[4,5,6]$ is satisfied by M_3 and it is shown in Example 3.3 that the minimal clause $R(4,6)$ is derivable from DB.

Theorem 3.8 $DB \vdash R(a_1, a_n)$ *if and only if for each minimal model M of DB there exists a path from a_1 to a_n satisfied by M.*

Proof (if) From Theorem 3.4, we know that if a minimal model satisfies a path from a_1 to a_n then $R(a_1, a_n)$ is in that model. Now if every minimal model of DB satisfies a path from a_1 to a_n then $R(a_1, a_n)$ is in all the minimal models of DB. Therefore, $DB \vdash R(a_1, a_n)$.

(only if) For any minimal model M of DB, if $DB \vdash R(a_1, a_n)$ then $R(a_1, a_n) \in M$. From Theorem 3.4, we know that M satisfies a path from a_1 to a_n. Q.E.D.

It should be pointed out that the path satisfied by a minimal model of DB need not be the same as the paths satisfied by other minimal models of DB. In Example 3.1, we have the following three minimal models:

$$
\begin{aligned}
M_1 &= \{A(4,5), A(4,6), A(5,6), R(4,5), R(4,6), R(5,6)\} \\
M_2 &= \{A(4,5), A(4,6), A(2,3), R(4,5), R(4,6), R(2,3)\} \\
M_3 &= \{A(4,5), A(5,6), A(1,2), R(4,5), R(5,6), R(1,2), R(4,6)\}
\end{aligned}
$$

As shown in Example 3.3, $DB \vdash R(4,6)$. The above theorem says every minimal model of DB must satisfy a path from 4 to 6. It can be seen that M_1 and M_3 satisfy the path $[4,5,6]$, while M_2 satisfies the path $[4,6]$.

The following theorem shows that only the indefinite literals in the minimal models need to be considered for testing whether a minimal model satisfies a path.

Theorem 3.9 *A minimal model satisfies a path if and only if all the indefinite literals in that path are in the minimal model.*

Proof Each definite literal must be in every minimal model. Therefore if all the indefinite literals in a path are in the minimal model M, then all the literals in the path are in M and the path will be satisfied by M. On the other hand, in order to satisfy a path, a minimal model must contain all the literals in that path, including those indefinite ones. Q.E.D.

We shall show that minimal models need not be generated. Instead, an optimized theorem prover can be used for answering queries. This theorem prover does not resolve the clauses in IDB and $DEDB$. It resolves the $IEDB$ clauses with the negations of paths defined below.

Definition The *negation* of a path containing the indefinite literals $L_1, L_2, \ldots, L_n$ is the negative clause $\sim L_1 \ldots \sim L_n$. The negation of a definite path is the empty clause.

Theorem 3.10 *A minimal model M of DB satisfies a path K if and only if M falsifies the negation of K.*

Proof (if) Let the negation of K be $\sim L_1 \ldots \sim L_n$. For M to falsify the negation, all the positive ground literals $L_1, \ldots, L_n$ must be in M. As M contains all the indefinite literals in K, by Theorem 3.9, M must satisfy K.

(only if) Let $L_1, \ldots, L_n$ be all the indefinite literals in K. If M satisfies K then $L_1, \ldots, L_n$ are all in M. Therefore, M falsifies the negation of $K, \sim L_1 \ldots \sim L_n$. Q.E.D.

Theorem 3.11 *Let S be the set (conjunction) of the negations of all the paths from a to b. $DB \vdash R(a, b)$ if and only if $IEDB \cup S \vdash \square$.*

Proof (if) If $IEDB \cup S \vdash \square$, then $IEDB \cup S$ is inconsistent and has no model. Therefore, every minimal model of DB must falsify S. As S is a conjunction of negative clauses, every minimal model of DB must falsify at least one negative clause in S. By Theorem 3.10, every minimal model of DB must satisfy at least one path. From Theorem 3.8 we may conclude $DB \vdash R(a, b)$.

(only if) By Theorem 3.8, if $DB \vdash R(a, b)$ then each minimal model of DB satisfies a path from a to b. Following Theorem 3.10, we know each minimal model of DB must falsify the negation of a path, and therefore must falsify S, the conjunction of the negations of all the paths from a to b. As all the atoms of S come from DB, we may conclude $DB \cup S$ is inconsistent, therefore, $DB \cup S \vdash \square$. In the derivation of the empty clause, IDB clauses (rule 1 and 2) must not be involved for otherwise, an R literal will be generated and can not be resolved away. Also the $DEDB$ clauses (definite literals) must not be involved because all the literals in S are indefinite literals. Therefore, $IEDB \cup S \vdash \square$. Q.E.D.

Theorem 3.12 *Let S be the set of the negations of all the paths from a to b. DB derives a minimal positive ground clause containing $R(a, b)$ if and only if in one hyperresolution step $IEDB \cup S$ derives a positive (or empty) clause not subsumed by any clause in $IEDB$.*

Proof (if) Suppose in one hyperresolution step, $IEDB \cup S$ derives a positive (possibly empty) clause C not subsumed by any clause in $IEDB$. Let the clauses involved in the resolution be $L_1 C_1, \ldots, L_n C_n$ and the negation of path $K, \sim L_1 \ldots \sim L_n$. Then $C = C_1 \ldots C_n$. Now we want to show that there exists a minimal model M of DB falsifying C. Suppose this is not true. Then all the minimal models of DB satisfy C, or equivalently, $DB \cup \sim C \vdash \square$. Note that $\sim C$ is a conjunction of negative ground literals. These literals are indefinite because all the literals in C come from indefinite clauses. In the derivation of the empty clause $\square$, we may use the set-of-support restriction with $\sim C$ being the set of support. Recall all literals of C are indefinite, so no negative clause from $\sim C$ can resolve with *DEDB*. Therefore the deduction of $\square$ must resolve with one clause from *IEDB*. Under this restriction, the only way to derive the empty clause is to have all the literals in an indefinite clause $C' \in IEDB$ resolved away by the negative literals in $\sim C$. All the literals in C' must be in C. In other words, C' subsumes C, which conflicts with the premise that C is not subsumed by any clause in *IEDB*. Therefore, there must exist a minimal model M falsifying C. By Theorem 3.3, all the literals $L_1, \ldots, L_n$ must be in M. It follows from Theorem 3.9 that M satisfies K. From Theorem 3.7, we may conclude that a minimal positive ground clause containing $R(a, b)$ is derivable from *DB*.

(only if) Theorem 3.7 shows that if a minimal positive ground clause containing $R(a, b)$ is derivable from *DB* then at least one minimal model of M satisfies a path from a to b. Let M be such a model and K be the path satisfied by M. Also let the negation of K be $\sim L_1 \ldots \sim L_n$. This negation is in S since K is a path from a to b. On the other hand, $L_1, \ldots, L_n$ must be all in M as M satisfies K. We may find the inducing clauses $L_1 C_1, \ldots, L_n C_n$ such that M falsifies all the subclauses $C_1, \ldots, C_n$ (Theorem 3.3). Resolving these inducing clauses with the negation of K, we get a new positive clause $C_1 \ldots C_n$. We know this clause is not subsumed by any clause in *IEDB* because it is falsified by M while all the clauses in *IEDB* are satisfied by M. Q.E.D.

Theorems 3.11 and 3.12 suggest an efficient way of testing whether *DB* derives a minimal positive clause containing $R(a, b)$ and whether that clause is unit. The algorithms will be discussed in the next section.

4   Algorithms

In this section, we shall introduce the algorithms for answering queries based upon the theorems developed in the previous section. The first two algorithms are for extracting the indefinite literals from the paths. The third algorithm answers the query according to the results from Theorems 3.11 and 3.12. Before introducing the algorithms we need to define the incident indefinite tuples.

Definition   The *incident indefinite tuples* of all the paths from a to b are the indefinite tuples (x, y) such that there is a (possibly empty) path from a to x and a definite (or empty) path from y to b. The set of incident indefinite tuples of all the paths from a to b is denoted as $IT_a(b)$.

For any constant b to which there are definite paths from a, we shall mark it by using a dummy incident tuple (a_0, b_0). Therefore, if $IT_a(b) = \{(a_0, b_0)\}$ then there is a definite path from a to b.

Algorithm 4.1   *Compute Incident Indefinite Tuples*

Description: Given the relation A and a constant a, this algorithm generates $IT_a(x)$ for all the constants x to which there is a path from a.

Input: Constant a and relation A.

Output: Set S containing the constants to which there is a path from a.

Subroutines: Procedure INSERT adds an element to a set. Function MERGE returns $\{(a_0, b_0)\}$ if one of the arguments is $\{(a_0, b_0)\}$; otherwise, it returns the union of the two arguments.

Variables: S is for collecting all the constants to which there are (definite or indefinite) paths from a. Each element x in S is associated with a set of indefinite incident tuples $x.it$. *FRONT* and *NEW-FRONT* are for propagating the newly explored constants. *New-it* is the newly generated set of indefinite incident tuples. This temporary variable is used for checking whether any new indefinite incident tuples are generated.

$S \leftarrow \emptyset$;
$FRONT \leftarrow \emptyset$;
for each $(a, y) \in A$ **do begin** (* Select with a *)
  if $y \notin S$ **then begin** INSERT(y, S); $y.it \leftarrow \emptyset$ **end**;
  if (a, y) is definite **then** INSERT$((a_0, b_0), y.it)$;
  if (a, y) is indefinite **then** INSERT$((a, y), y.it)$;
  end;

$FRONT \leftarrow S$;
while $FRONT \neq \emptyset$ **do begin**
  $NEW\text{-}FRONT \leftarrow \emptyset$;
  (* Modified Join *)
  for each $(x, y) \in A$, such that $x \in FRONT$ **do begin**
    if $y \notin S$ **then begin** INSERT(y, S); $y.it \leftarrow \emptyset$ **end**;
    if (x, y) is definite **then** $new\text{-}it \leftarrow$ MERGE$(x.it, y.it)$;
    if (x, y) is indefinite **then** $new\text{-}it \leftarrow$ MERGE$(\{(x, y)\}, y.it)$;
    if $new\text{-}it \neq y.it$ **then begin** INSERT$(y, NEW\text{-}FRONT)$; $y.it \leftarrow new\text{-}it$ **end**;
    end;
  $FRONT \leftarrow NEW\text{-}FRONT$;
  end;

At this point, the set S contains all the vertices reachable from a through a definite or an indefinite path. For any vertex $x \in S$, if $x.it = \{(a_0, b_0)\}$ then there is a definite path from a to x; otherwise, $x.it$ contains the incident indefinite tuples of x. This algorithm terminates because a constant is included into *NEW-FRONT* only if a new indefinite incident tuple is found while the number of indefinite tuples is finite.

Algorithm 4.2 *Generate the paths*

Description: Given the set S computed from Algorithm 4.1, this algorithm will generate all the paths (recording only indefinite literals) from a to each constant reachable from a.

Input: Relation A and the set S computed from Algorithm 4.1.

Output: The set S each of whose elements is associated with a set of paths from (a_0, b_0).

Subroutines: The notations σ and $\bowtie$ are the selection and the join operators in the relational algebra, respectively. If not otherwise specified, the joins will join the first attributes of the two relations. The number used in the selection condition indicates the attribute position. INSERT is for adding a tuple to a set of tuples. PATH-MERGE will merge two sets of paths such that the subsumed paths are deleted from the result. A path K_1 is subsumed by another path K_2 if all the tuples in K_2 are in K_1. The subsumption test in PATH-MERGE will prevent the algorithm from generating paths with cycles. The notation $A \times B$ is the cross product of two sets of paths such that for each path $K_1 \in A$ and a path $K_2 \in B$, the concatenation $K_1 K_2$ is in the product.

Variables: D stores the set of constants reachable from a through definite paths. I stores the set of indefinite tuples whose second attribute values are not in D. *FRONT* and *NEW-FRONT* are for propagating the newly reached indefinite tuples through new paths. *New-path* stores the newly merged set of paths. If it is equal to the old set of paths then a cycle is detected and the path should not be further explored.

$$D \leftarrow \sigma_{it=\{(a_0,b_0)\}} S;$$
$$I \leftarrow \sigma_{\text{indefinite} \wedge 2 \notin D} A;$$
for each $(x,y) \in I$, **do** $y.path \leftarrow \emptyset$;
$$FRONT \leftarrow D \bowtie I;$$
for each $(x,y) \in FRONT$ **do** $y.path \leftarrow \{(x,y)\}$;
while $FRONT \neq \emptyset$ **do begin**
  $NEW\text{-}FRONT \leftarrow \emptyset$;
  for each $(u,v) \in FRONT$, **do begin**
    $T \leftarrow (\sigma_{(u,v)\in it} S) \bowtie I$;
    for each $(x,y) \in T$, **do begin**
      $new\text{-}path \leftarrow \text{PATH-MERGE}(v.path \times \{(x,y)\}, y.path)$;
      if $new\text{-}path \neq y.path$ **then begin**
        $y.path \leftarrow new\text{-}path$;
        $\text{INSERT}((x,y), NEW\text{-}FRONT)$;
      end
    end

end;
    FRONT ← *NEW-FRONT*
end;

This algorithm always terminates because a tuple is added to NEW-FRONT only if a new acyclic path leads to this tuple while there are a finite number of acyclic paths. Given the set S with the associated indefinite incident tuples and the associated paths, we can answer a query $?R(a, y)$ for any constant y by using the following algorithm.

Algorithm 4.3 *Query Answering*

Description: Given a constant y and the set S output from Algorithm 4.2, this algorithm will generate an answer to the query $?R(a, y)$.

Subroutines: The projection operator π_2 is to project the second attribute of a relation. Also a theorem prover is called in this algorithm to determine the answers.

Variables: T is the set of constants from which there are definite paths to y. Each constant in T is associated with a set of paths from a to the constant.

if $y.it = \{(a_0, b_0)\}$ then answer *true*
else begin
    $T \leftarrow \pi_2 \ y.it$;
    NEGATION ← ∅;
    for each $v \in T$ do
        for each path $(a_1, b_1) \dots (a_n, b_n) \in v.path$ do
            Insert $\sim A(a_1, b_1) \dots \sim A(a_n, b_n)$ into NEGATION;
    Send *IEDB*∪ NEGATION to a theorem prover;
    if it derives □ then answer *true*
    else if it derives a positive clause not subsumed by any clause in *IEDB*
        then answer *possible*
    else answer *false*
end;

Example 4.1 We shall use the *DB* defined in Example 2.2. After applying Algorithm 4.1, the set S will contain the following elements with the associated indefinite incident tuples.

$$1.it = \{(a_0, b_0)\}$$
$$2.it = \{(a, 2)\}$$
$$3.it = \{(a_0, b_0)\}$$
$$4.it = \{(1, 4)\}$$

$$5.it = \{(a,2)\}$$
$$6.it = \{(3,6)\}$$
$$7.it = \{(1,4)\}$$
$$8.it = \{(4,8),(a,2),(9,8)\}$$
$$9.it = \{(3,6)\}$$
$$10.it = \{(7,10),(4,8),(a,2),(9,8),(6,10)\}$$
$$11.it = \{(1,11)\}$$
$$12.it = \{(11,12)\}$$

The set S and the associated indefinite incident tuples are then sent to Algorithm 4.2 to generate the paths as follows:

$$2.path = \{(a,2)\}$$
$$4.path = \{(1,4)\}$$
$$6.path = \{(3,6)\}$$
$$8.path = \{(1,4)(4,8),(3,6)(9,8)\}$$
$$10.path = \{(1,4)(7,10),(3,6)(6,10)\}$$
$$11.path = \{(1,11)\}$$
$$12.path = \{(1,11)(11,12)\}$$

Note that not all the constants are associated with paths. To answer the query $?R(a,10)$, by following Algorithm 4.3, we need to project the second constants of $10.it$ into T. There we get $T = \{2,8,10\}$. Then we can generate the negations of the paths in $2.path$, $8.path$, and $10.path$. The remaining steps for answering the query $?R(a,10)$ is shown in Example 2.2

Let us try another query: $?R(a,9)$. Following Algorithm 4.3, we get $T = \{6\}$. There is only one path in $6.path$ whose negation is $\sim A(3,6)$. We may resolve this negation with C_{10} to get a positive clause $A(a,2)A(1,4)$ which is not subsumed by any clause in $IEDB$. Since we can not derive the empty clause, the answer to the query $R(a,9)$ is *possible*.

If the query is $?R(a,12)$, there is again only one path whose negation is $\sim A(1,11)\sim A(11,12)$. It can be seen that $IEDB$ and this negative clause can not derive any new positive clause. Therefore, the answer is *false*.

5 Summary

We presented a new method for answering recursive queries in non-Horn databases. The modified join operator is used for processing the EDB while the relatively small $IEDB$ is processed by a theorem prover. A few examples of answering closed queries in the form $?R(a,b)$ are shown. However, answering an open query like $?R(a,y)$ is just a matter of collecting the paths to the constants in S (Algorithm 4.1), and sending their negations to the theorem prover. The performance of this method can be further optimized. In particular, the theorem prover can be well optimized because it deals only with positive and negative ground clauses. We did not get into these details in this paper. Most theorems need to be modified if the assumptions about the database are changed. Further research for more complicated recursive patterns need to be done.

References

[1] Bancilhon, F. and Ramakrishnan, R., "An amateure's introduction to recursive query processing strategies," *ACM SIGMOD Conference on Management of Data*, (1986).

[2] Bossu, G. and Siegel, P., "Saturation, nonmonotonic reasoning and the closed-world assumption," *Artificial Intelligence* **25**, (1985), pp. 13-63.

[3] Chang, C.L., "On evaluation of queries containing derived relations," *Advances in Data Base Theory* **1**, H. Gallaire and J. Minker, and J.M. Nicolas, Eds., Plenum Press, New York, (1981), pp. 235–260.

[4] Codd, E.F., "Extending the database relational model to capture more meaning," *ACM Transactions on Database Systems* **4**, 4, (December 1979), pp. 339–434.

[5] Grant, J. and Minker, J., "Answering queries in indefinite databases and the null value problems," University of Maryland, College Park, Maryland, (July 1981).

[6] Han, J., *Pattern-Based and Knowledge-Directed Query Compilation for Recursive Databases*, Ph.D. thesis, University of Wisconsin–Madison, (1985).

[7] Henschen, L.J. and Naqvi, S., "On compiling queries in recursive first-order databases," *JACM* **31**,1, (January 1984), pp. 47–85.

[8] Henschen, L.J. and Park, H., "Indefinite and GCWA inference in indefinite deductive databases," *Proc. AAAI-86*.

[9] Minker, J., "On indefinite database and the closed world assumption," *Lecture Notes in Computer Science* **138**, Springer-Verlag, (1982), pp. 292–308.

[10] Minker, J., and Nicolas J., "On Recursive Axioms in Relational Databases," *Tech. Rep.* No. 1119, University of Maryland, College Park, Maryland, (1981).

[11] Reiter, R., "Deductive question answering on relational data bases," *Logic and Databases*, H. Gallaire and J. Minker, Eds., Plenum Press, New York, (1978), pp. 149-177.

[12] Reiter, R., "Question answering on relational databases," *Logic and Databases*, H. Gallaire and J. Minker, Eds., Plenum Press, New York, (1978), pp. 55-76.

[13] Wos, L., Overbeek, R., Lusk, E. and Boyle, J. : *Automated Reasoning, Introduction and Applications*, Prentice-Hall, (1984).

[14] Yahya, A. and Henschen, L.J., "Deduction in non-Horn databases," *Journal of Automated Reasoning* **1**, No.2, (1985), pp., 141–160.

CASE INFERENCE IN RESOLUTION-BASED LANGUAGES

T. Wakayama
School of Computer and Information Science
313 Link Hall, Syracuse University, Syracuse, NY 13244

T.H. Payne
Department of Mathematics and Computer Science
University of California, Riverside, CA 92521

Abstract. Informally, *case inference* is a type of inference that inherently involves disjunctions in deriving definite consequences. We show that a difficulty with efficient implementation of case inference in resolution-based languages stems from the fact that case inference *always* requires derived clauses to be reused as side clauses: in general, the number of derived clauses is quite large, and storing all of them seems unacceptably inefficient in programming language settings. However, our results also show that in retrieving definite information, this use of derived clauses is necessary *only* when case inference is required. This in turn leads to our next finding that storing a relatively small class of derived clauses, which is characterized in terms of certain properties of case inference, is sufficient for proving all definite consequences. We then show that a conservative approximation of this class can be, in effect, precomputed for clause sets not containing purely negative clauses.

§1. Introduction

In logic programming and deductive databases, there has been a strong interest in developing languages that can reason properly in the presence of *indefinite* information, i.e., information expressed as a disjunction of atomic facts [2,3,4,8,9,10,14]. It has been known that the presence of this type of information in databases causes serious computational difficulties: the efficiency of SLD-resolution is no longer available, yet more general resolution schemes are not quite efficient enough to be the basis of programming languages.

One reason, among others, that it is still desirable to keep indefinite information in databases is that it could play, in certain circumstances as illustrated below, an indispensable role in deriving *definite* consequences. In this study, we refer to this type of inference as *case inference* and identify its basic yet sufficient computational mechanism in the context of linear resolution. We then briefly discuss the feasibility of adapting it into an efficient Prolog-like environment.

To see how case inference arises naturally in application, consider the following:

(1.1) Example.

$symptom(S)$.

$cause(C_1) \lor cause(C_2) \leftarrow symptom(S)$.

$treatment(T_0) \leftarrow cause(C_1)$.

$$treatment(T_1) \leftarrow cause(C_1).$$

$$treatment(T_0) \leftarrow cause(C_2).$$

$$treatment(T_2) \leftarrow cause(C_2).$$

In addition, suppose that the presence of the symptom S requires some immediate treatment, and that the doctors can not afford to wait for further test results which may identify the exact cause of the symptom S and allow them finer treatments (such as T_1 or T_2). The point of this example is that the present state of the knowledge base, although *incomplete*, is already mature enough to establish the validity of the treatment T_0, and that any inference procedure intended for such applications should be able to detect this situation.

However, this seemingly simple situation is by no means an easy one for resolution-based programming languages, e.g., try SLDNF-resolution on the above example with query $\exists x$ $treatment(x)$. The major reason for this is that, as we will see in §3, case inference *always* requires the use of derived clauses (i.e., center clauses in linear resolution including the top goal clause) as side clauses: in general, the number of derived clauses is quite large and their management is a major efficiency problem in linear resolution. However, our results also show (§3) that in retrieving definite information, the use of derived clauses is necessary *only* when case inference is required. This in turn leads to the finding (§4) that storing a relatively small class of derived clauses, characterized in terms of certain properties of case inference, is sufficient for proving all definite consequences. In §5, we briefly discuss how a conservative approximation of this class can be, in effect, precomputed for a set of definite and indefinite clauses.

Although any implementation of case inference in the context of linear resolution must have some way of reintroducing at least some derived clauses back to the refutation, storing them is not the only way. Loveland has shown, in his recent work on *near Horn Prolog*[8], that desired derived clauses can be, in effect, regenerated by reentering once-resolved subgoals. We call his approach the *reentry* approach, and ours the *reduction* approach, where reduction is, as we will see in the next section, a way of storing and utilizing derived clauses. The two approaches are quite different, and it is interesting to compare them. But a full comparison is beyond the scope of this presentation, and we will mention only a few points on this in later sections.

Throughout, we employ the following conventions:

- All upper case letters denote literals, either positive or negative.
- An A-clause is a clause that contains the literal A.
- If $\mathcal{F}$ is a formula then $\exists(\mathcal{F})$ and $\forall(\mathcal{F})$ denote the existential closure and the universal closure of $\mathcal{F}$, respectively.
- $\Gamma \models \mathcal{F}$ means that the formula $\mathcal{F}$ is a logical consequence of the set of axioms Γ in the sense of classical first order logic.

§2. Ordered Linear Resolution and Definite Answers

The form of linear resolution we study is OL-resolution[1], which is essentially the same as SL-resolution[5] and the model elimination method [7]. In this section, we introduce basic definitions on OL-resolution and establish a preliminary result on question-answering aspects of OL-resolution.

(2.1) Definition. A *program* is a consistent set of finitely many clauses. A *positive* program is a program in which every clause has at least one positive literal. A *Horn* program is a positive program in which every clause has exactly one positive literal. A program is *ground* if every clause in it is ground. **P** denotes a program, and ground(**P**) the set of all ground instances of clauses in **P** over the (unspecified) language of **P**.

(2.2) Definition. A *query* is a conjunction of literals with fixed order. (We sometimes speak of a query as a sequence). We assume that queries are existentially closed, and we usually omit the quantifiers. We call members of a query *subgoals* and allow some of them to be distinguished as *marked subgoals*. (We use [ ] to indicate marked subgoals). A query is said to be *reducible* if the right-most subgoal is unmarked and its negation is unifiable with some marked subgoal in the query.

The definition of query here is different from ordinary ones in two respects: the order of its literals matters, and some literals may be marked ones. This extra machinery is a way of storing derived clauses of previous resolution steps. For instance, a query $q = A\&[B]\&C\&\neg B$ records the fact that the negation of the initial segment of q up to the marked subgoal, namely, the clause $\neg A \vee \neg B$, was previously derived. Furthermore, marked subgoals serve only as a recording device and do not participate in resolution. This point will become clear in following definitions.

(2.3) Definition. A *resolution triple* is a triple of the form $< q, s, \theta >$ where q is a query, s is a clause or the *reduction symbol*, #, (intended to indicate that q is reducible), and θ is a substitution. A resolution triple $< q\&A, s, \theta >$ is said to *derive* a query p if one of the following two cases holds:

- s is a clause that contains a literal unifiable with A, and θ is their mgu. In this case, if s has the form, $A_0 \leftarrow A_1 \& \cdots \& A_n$, where A_0 is a literal unifiable with A, then p is obtained from the query $q\theta\&[A\theta]\&A_1\theta\& \cdots \&A_n\theta$ by successively dropping all marked subgoals in the right-most position.

- s is the reduction symbol, $q\&A$ is reducible, and θ is an appropriate mgu. In this case, p is obtained from the query $q\theta$ in the same way as above, and this derivation step is called a *reduction*.

(2.4) Definition. An *OL-derivation* from a query q in a program **P** is a sequence of resolution triples $\delta_1 =< q_1, s_1, \theta_1 >, \delta_2 =< q_2, s_2, \theta_2 >, \cdots$ such that,

- q_1 is q (We assume that the first query has no marked subgoals).

- if s_i is a clause, it is a variant of a clause in $\mathbf{P}\cup \{\neg q\}$. We call such s_i's *side clauses*. We usually represents side clauses in the form $A_0 \leftarrow A_1 \& \cdots \& A_n$, indicating that A_0 is the literal to be resolved upon.

- δ_i derives q_{i+1}.

(2.5) Definition. An *OL-proof* of a query q in **P** is a finite OL-derivation from q in **P** such that the last triple derives the empty query.

Note that reduction in OL-derivation is actually a way of utilizing derived clauses as side clauses. For instance, the derivation of $A\&[B]\&C$ from the triple $< A\&[B]\&C\&\neg B, \#, \{ \} >$ is really a derivation from the triple $< A\&[B]\&C\&\neg B, \neg B \leftarrow A, \{\} >$ followed by *merging-left* on

A and subsequent removal of $[\neg B]$, where merging-left refers to removal of all but the left-most identical subgoal in a query.

(2.6) Example. The following is an OL-proof of the query $\exists x \; treatment(x)$ in the program in example 1.1. (*cause*, *symptom*, and *treatment* are abbreviated as c, s, and t, respectively).

$$
\begin{array}{lclcl}
t(x) & ; & t(T_0) \leftarrow c(C_1) & ; & \{x/T_0\} \\
[t(T_0)]\&c(C_1) & ; & c(C_1) \leftarrow \neg c(C_2)\&s(S) & ; & \{\} \\
[t(T_0)]\&[c(C_1)]\&\neg c(C_2)\&s(S) & ; & s(S) & ; & \{\} \\
[t(T_0)]\&[c(C_1)]\&\neg c(C_2) & ; & \neg c(C_2) \leftarrow \neg t(T_0) & ; & \{\} \\
[t(T_0)]\&[c(C_1)]\&[\neg c(C_2)]\&\neg t(T_0) & ; & \# & ; & \{\}
\end{array}
$$

One aspect of OL-proof procedure not illustrated in the above example is that the negation of the original query may appear as a side clause. We will set up some terminology to discuss this point.

(2.7) Definition. Let Δ be an OL-proof of a query q in a program $\mathbf{P}$. The top occurrence of q or an occurrence of a variant of $\neg q$ as a side clause in Δ is called an *entry point* of Δ. A *single-entry* (*multi-entry*) OL-proof is an OL-proof that has a single entry point (multiple entry points).

(2.8) Definition. An *answer* is a nonempty finite set of substitutions which replace the same set of variables. If q is a query and $\{\theta_1,\cdots,\theta_n\}$ is an answer, then $q\{\theta_1,\cdots,\theta_n\}$ denotes $q\theta_1 \vee \cdots \vee q\theta_n$. An answer is *indefinite* if it has two or more elements; otherwise it is *definite* and we write θ for $\{\theta\}$. A query q *has* an answer Θ in $\mathbf{P}$ if $\mathbf{P} \models \forall(q\Theta)$.

(2.9) Definition. Let Δ be an single-entry OL-proof with the mgu's $\theta_1, \cdots, \theta_n$. Then the substitution obtained by restricting $\theta_1\cdots\theta_n$ to the variables of q is called the *answer computed* by Δ.

An interesting observation in example 2.6 is that the top subgoal can be reentered at the last triple instead of a reduction, i.e., multi-entry OL- proofs can facilitate case inference. In fact, Loveland's reentry approach is essentially a generalization of this reentry mechanism: in his reentry mechanism, subgoals other than the top one can be reentered as well. This freedom, of course, comes with a control decision as to which subgoals should be reentered, and this decision affects the completeness of the procedure as well as efficiency. Consequently, the approach requires some book-keeping of previously resolved subgoals. But, other than that, the approach requires no means of storing derived clauses, and, in particular, it is independent of the reduction mechanism.

On the other hand, our reduction approach relies entirely on the reduction mechanism for retrieving definite answers including those that require case inference, and leaves the reentry option only for indefinite answers. Our first proposition (2.12) gives a foundation for this approach, establishing that single-entry OL-proofs are sufficient for computing all definite answers. A preliminary definition and a lemma follow first.

(2.10) Definition. Let Δ be an OL-proof of a query q in a program $\mathbf{P}$ with mgu's $\theta_1,\cdots,\theta_n$. A *ground instance* of Δ is an OL-proof obtained from Δ by applying the composition $\theta_1\cdots\theta_n$ to every query and side clause in Δ, then replacing every θ_i by the empty substitution, and finally instantiating all remaining variables uniformly by constants. When it is desirable to place some restrictions on these constants, we will refer to them as *instantiation constants* in the ground instance of Δ.

The following version of lifting lemma adapted to the context of this study is useful. Its proof is essentially identical to that seen in [6].

(2.11) Lemma (Lifting Lemma). Let q be a query and θ a substitution. Suppose that $q\theta$ has a single-entry OL-proof Δ in a program $\mathbf{P}$. Then q has a single-entry OL-proof Δ' in $\mathbf{P}$ such that for some substitution γ, $\theta\theta_1\cdots\theta_n = \theta_1'\cdots\theta_n'\gamma$, where $\theta_1,\cdots,\theta_n$ are the mgu's in Δ and $\theta_1',\cdots,\theta_n'$ are the mgu's in Δ'.

(2.12) Proposition. Let $\mathbf{P}$ be a program and q a query. Then q has a definite answer θ' in $\mathbf{P}$ iff it has a single-entry OL-proof in $\mathbf{P}$ that computes θ where θ' is an instance of θ.

Proof. (soundness) Let Δ be a single-entry OL-proof of q in $\mathbf{P}$, and θ the answer computed by Δ. Let Δ' be a ground instance of Δ, and q' the first query in Δ'. We assume that every instantiation constant in Δ' is a constant new to the underlying language of $\mathbf{P}$. We then extend $\mathbf{P}$ to $\mathbf{P}'$ by extending the language of $\mathbf{P}$ to include all the instantiation constants in Δ'. In view of the Theorem on Constants[11], which states, in this context, that $\mathbf{P}' \models q'$ implies $\mathbf{P} \models \forall(q\theta)$, it suffices to show that $\mathbf{P}' \models q'$. But since Δ has a single entry point, every side clause in Δ' is an instance of a clause in $\mathbf{P}$ (and hence in $\mathbf{P}'$). By the soundness of OL-proof procedure, we have $\mathbf{P}' \models q'$.

(completeness) Suppose that q has a definite answer θ' in $\mathbf{P}$. Then $\mathbf{P} \models \forall(q\theta')$. We take a ground instance q' of $q\theta'$ by assigning to each variable in $q\theta'$ a constant not appearing in $\mathbf{P}$. We will show that q' has a single-entry OL-proof in $\mathbf{P}$. Let A be any member of q'. By the completeness of OL-proof procedure, we have an OL-proof Δ_A of A in $\mathbf{P}$. Since $[A]$ is present throughout Δ_A, we can assume that Δ_A has a single entry point. Combining these OL-proofs for all members of q', we obtain an OL-proof Δ' of q' in $\mathbf{P}$. Note that Δ' has a single entry point because $\neg q'$ is not in ground($\mathbf{P}$) by the consistency of $\mathbf{P}$. Since q' is ground, the answer computed by Δ' is empty. Now we will replace every constant employed to obtain q' from $q\theta'$ by a new variable throughout Δ'. Then we obtain a single-entry OL-proof Δ_1 of $q\theta'$ in $\mathbf{P}$. Note that the answer computed by Δ_1 is still empty. By applying the lifting lemma to Δ_1 we obtain a single-entry OL proof Δ_2 of q in $\mathbf{P}$ such that $\theta'\theta_1'\cdots\theta_n' = \theta_1\cdots\theta_n\sigma$, where $\theta_1',\cdots,\theta_n'$ are the mgu's in Δ_1 and $\theta_1,\cdots\theta_n$ are the mgu's in Δ_2. But the restriction of $\theta'\theta_1'\cdots\theta_n'$ to the variables of q is θ' since that of $\theta_1'\cdots\theta_n'$ is empty. Hence we conclude that θ' is an instance of the answer computed by Δ_2. ∎

§3. Case Inference and Derived Clauses

The following series of definitions are intended to capture the informal notion of case inference.

(3.1) Definition. If a ground clause has the form $A_0 \leftarrow A_1 \& \cdots \& A_n$ such that $A_1 \& \cdots \& A_n$ is a logical consequence of a program $\mathbf{P}$ then the clause is said to be *determinative* in $\mathbf{P}$; otherwise it is said to be *nondeterminative* in $\mathbf{P}$.

(3.2) Definition. A subgoal A is said to *directly require cases* in $\mathbf{P}$ if it has definite answers in $\mathbf{P}$, and for every ground instance A' of A which is a logical consequence of $\mathbf{P}$, every A'-clause in ground$(\mathbf{P})$ is nondeterminative in $\mathbf{P}$. A subgoal A is a *direct-case consequence (dc-consequence)* of $\mathbf{P}$ if for some instance A' of A, $\exists(A')$ directly requires cases in some subset of ground$(\mathbf{P})$.

(3.3) Definition. $\mathbf{P}$ is *case-dependent* if it has a dc-consequence; otherwise it is *case-free*.

(3.4) Example. Every Horn program is case-free.

(3.5) Definition. A subgoal A *requires cases* in $\mathbf{P}$ if it has definite answers in $\mathbf{P}$, and every subset of ground$(\mathbf{P})$ that entails some ground instance of A is case-dependent. A subgoal A is a *case-consequence* of $\mathbf{P}$ if for some instance A' of A, $\exists(A')$ requires cases in some subset of ground$(\mathbf{P})$.

We are now ready to formally study relationships between case inference and the use of derived clauses. The following syntactic property of dc-consequences is a key to the understanding of such relationships.

(3.6) Lemma. If a subgoal K is a dc-consequence of $\mathbf{P}$, then for some ground instance K' of K which is a logical consequence of $\mathbf{P}$, ground$(\mathbf{P})$ contains at least two distinct K'-clauses.

Proof. Suppose that K is a dc-consequence of $\mathbf{P}$. Then for some instance K' of K, $\exists(K')$ directly requires cases in some subset $\mathbf{P}'$ of ground$(\mathbf{P})$. Since K' has definite answers in $\underline{\mathbf{P}}'$, some ground instance K'' of K' is a logical consequence of $\mathbf{P}'$. But then every K''-clause in $\mathbf{P}'$ is nondeterminative in $\mathbf{P}'$. Hence $\mathbf{P}'$ contains at least two K''-clauses. ∎

(3.7) Definition. A ground program $\mathbf{P}$ is *minimal for* a ground literal A if A is a logical consequence of $\mathbf{P}$ but not a logical consequence of any proper subset of $\mathbf{P}$.

(3.8) Lemma. Let $\mathbf{P}$ be a ground program minimal for some ground literal, say L. Then for every ground literal A which is a logical consequence of $\mathbf{P}$ but not a dc-consequence of $\mathbf{P}$, $\mathbf{P}$ contains exactly one A-clause.

Proof. Let $\mathbf{P}_A$ be a subset of $\mathbf{P}$ minimal for A. Since A is not a dc-consequence of $\mathbf{P}$, A does not directly require cases in $\mathbf{P}_A$, i.e., some A-clause, say $A \leftarrow B_1 \& \cdots \& B_n$, in $\mathbf{P}_A$ is determinative in $\mathbf{P}_A$. By the A-minimality of $\mathbf{P}_A$, $\mathbf{P}_A$ contains no $\neg A$-clauses. Hence $B_1 \& \cdots \& B_n$ is a logical consequence of $\mathbf{P}_A - A(\mathbf{P}_A)$, where $A(\mathbf{P}_A)$ is the set of all A-clauses in $\mathbf{P}_A$. This in turn implies that L is a logical consequence of $\mathbf{P}' = \mathbf{P} - A(\mathbf{P}_A) \cup \{ A \leftarrow B_1 \& \cdots \& B_n \}$ since every model of $\mathbf{P}'$ is a model of $\mathbf{P}$. Finally, the L-minimality of $\mathbf{P}$ forces $\mathbf{P}' = \mathbf{P}$, i.e., $\mathbf{P}$ has exactly one A-clause. ∎

(3.9) Definition. Let $\Delta = <q_1, s_1, \theta_1>, \cdots, <q_n, s_n, \theta_n>$ be an OL-proof. Suppose that A, either marked or unmarked, is the k-th subgoal in a query q_i in Δ. Then the *continuations* of A in q_i are defined as follows:
- A itself is a continuation of A,
- the k-th member of q_{i+1} (if exists) is a continuation of A, and
- if a subgoal is a continuation of A then its continuation is a continuation of A.

Suppose further that q_i has exactly k members. Then the *subderivation* from A (the occurrence of A in q_i) in Δ is the sequence $<q'_i, s_i, \theta_i>, <q'_{i+1}, s_{i+1}, \theta_{i+1}>, \cdots, <q'_{i+k}, s_{i+k}, \theta_{i+k}>$ such that , for each $0 \leq j \leq k$, q'_{i+j} is the tail segment of q_{i+j} starting at the continuation of A, and, q_i, q_{i+1}, $\cdots$, q_{i+k} are all of the queries in Δ that contain a continuation of A. The subderivation from A in Δ is a *subproof* of A in Δ if every reduction during this subderivation takes place within the subderivation.

(3.10) Lemma. Let $\mathbf{P}$ be a ground program and A a ground literal. Then A is a logical consequence of a case-free subset of $\mathbf{P}$ iff A has a single-entry OL-proof in $\mathbf{P}$ that has no reductions.

Proof. Suppose that A is a logical consequence of a case-free subset of $\mathbf{P}$. Since every subset of a case-free set is case-free, there exists a case-free subset $\mathbf{P}_A$ of $\mathbf{P}$ minimal for A. We proceed by induction on cardinality of $\mathbf{P}_A$. By lemma 3.8, $\mathbf{P}_A$ has a unique A-clause , say $A \leftarrow B_1 \&$ $\cdots \& B_n$, which is determinative in $\mathbf{P}_A$. But then $B_1 \& \cdots \& B_n$ is a logical consequence of the case-free set $\mathbf{P}_A - \{ A \leftarrow B_1 \& \cdots \& B_n \}$. By induction, each B_i has a single-entry OL-proof Δ_i that has no reductions. We combine these OL-proofs linearly and obtain an OL-proof of A in $\mathbf{P}$ that has no reductions. This OL-proof of A has a single entry point because each Δ_i has no occurrences of $\neg A$ as side clauses.

Suppose that A has a single-entry OL-proof in $\mathbf{P}$ without reductions. Let Δ be such a proof that has a minimal number of side clauses. We claim that the set $\mathbf{P}'$ of side clauses in Δ is case-free. If not, $\mathbf{P}'$ has a ground dc-consequence, say K. By lemma 3.6, $\mathbf{P}'$ contains two or more K-clauses. Since every subgoal in Δ is a logical consequence of $\mathbf{P}'$, every K-clause in $\mathbf{P}'$ must appear in the form $K \leftarrow B_1 \& \cdots \& B_n$ as a side clause. This implies that Δ includes two or more subproofs of K. Replacing all of these subproofs by a shortest one, we obtain an OL-proof of A whose set of side clauses is strictly smaller than $\mathbf{P}'$. ∎

(3.11) Proposition. Suppose that a subgoal K has definite answers in a program $\mathbf{P}$. Then the subgoal requires cases in $\mathbf{P}$ iff every single-entry OL-proof of it in $\mathbf{P}$ has reductions.

Proof. Suppose that K does not require cases in $\mathbf{P}$. Then some ground instance K' of K is a logical consequence of some case-free subset $\mathbf{P}'$ of ground($\mathbf{P}$). Then by lemma 3.10, K' has a single-entry OL-proof Δ' in $\mathbf{P}'$ that has no reductions. Applying the lifting lemma (the classical version as seen in [1]) to Δ', we obtain an OL-proof of $\exists(K)$ that has no reductions. This OL-proof has a single entry point because $\mathbf{P}$ has no variants of the clause $\{\neg A\}$.

Now suppose that some single-entry OL-proof Δ of $\exists(K)$ has no reductions. Let Δ' be a ground instance of Δ, K' the first query in Δ', and $\mathbf{P}'$ the set of side clauses in Δ'. Then Δ' is a single entry OL-proof of K' in $\mathbf{P}'$ that has no reductions. By lemma 3.10, K' is a logical consequence of a case-free subset of $\mathbf{P}'$. But $\mathbf{P}'$ is a subset of ground($\mathbf{P}$) because Δ has a single-entry point. Hence we have found a case-free subset of ground($\mathbf{P}$) which entails a ground instance of K. ∎

§4. Reduction on Dc-consequences

When OL-proofs are lengthy, they often contain a large number of marked subgoals, yet, typically, only few of them actually participate in reduction. Most of them are just sitting there, doing nothing other than increasing the search space. In this section, we show that most of these subgoals can in fact be omitted, retaining only those that are characterized as dc-consequences. For this purpose, we consider a version of OL-proof, called OL*-proof, in which reduction occurs only on those marked subgoals which are dc-consequences.

(4.1) Definition. An OL-proof Δ in **P** is *canonical* if for every right-most subgoal A in Δ which is a logical consequence of **P**, the subderivation from A is a subproof of A in Δ.

(4.2) Lemma. Suppose that a ground program **P** is minimal for a ground literal A and that A has a canonical OL-proof in **P**. Then A has an OL*-proof in **P**.

Proof. Let Δ be a canonical OL-proof of A in **P**. Note that if a ground literal L is a logical consequence of **P** but not a dc-consequence of **P**, **P** has a unique L-clause(lemma3.8), which must then be determinative in **P**. But then the canonicity of Δ forces that no reductions occur on continuations of L. Hence, we proceed by induction on the number of reductions in Δ on marked subgoals which are not logical consequences of **P**. If it is zero, the canonical proof is already an OL*-proof. Suppose that the reduction on $[M]$ in the k-th triple is the earliest such occurrence. Let $A_1, \cdots, A_n, M$ be the *ancestor chain* of $[M]$, i.e.,

- M is the right-most subgoal in m-th triple, $m < k$, such that $[M]$ in k-th triple is its continuation.
- A_n is the right-most subgoal in j_n-th triple, where the side clause in this triple is the *parent clause* of M, i.e., the side clause through which M was introduced to the proof, and similarly for $A_{n-1}, \cdots, A_1$.
- A_1 is A.

We first show that some member A_i of the chain is a dc-consequence of **P**. Suppose not. We can assume, without loss of generality, that M's parent clause has the form $A_n \leftarrow B_L \& M \& B_R$. Since A_n is not a dc-consequence, we conclude, by lemma 3.8, that A_n is not a logical consequence of **P**. Repeating this procedrue, we reach the contradiction that A_1 is not a logical consequence of **P**.

In the k-th triple, we replace the reduction symbol by $\neg M \leftarrow B_L \& \neg A_n \& B_R$. Since B_R is right-most in $(j_n + 1)$-th triple, this subgoal completes its subderivation prior to the k-th triple, and hence it has no reductions on subgoals other than dc-consequences (by the choice of k). We can use this subderivation for the subgoal B_R in $(k + 1)$-th triple because the set of marked subgoals to the left of B_R in $(k + 1)$-th triple is a super set of that of B_R in $(j_n + 1)$-th triple. The right-most B_L in $(k + 1)$-th triple will be merged left to the occurrence of B_L to the left of $[M]$. Repeat this procedure until we introduce $\neg A_i$, the negation of the dc-consequence in the chain, into the proof, which will be resolved, of course, by means of reduction on $[A_i]$. We then obtain an OL-proof of A in **P** that has one less number of reductions on subgoals other than dc-consequences. By induction, we obtain an OL*-proof of A in **P**. ∎

(4.3) Proposition. A query q has a definite answer θ' in **P** iff it has a single-entry OL*-proof in **P** that computes θ where θ' is an instance of θ.

Proof. (Soundness) Immediate from proposition 2.12.

(Completeness) Suppose that q has a definite answer θ' in **P**. We take a ground instance q' of $q\theta'$ by assigning to each variable a constant not appearing in **P**. Let A be any member of q' and $\mathbf{P}_A$ a subset of ground(**P**) minimal for A. We can obtain a single-entry canonical OL-proof of A in $\mathbf{P}_A$ (a simple induction on cardinality of $\mathbf{P}_A$). We then obtain, by lemma 4.2, a single-entry OL*-proof of A in $\mathbf{P}_A$. By applying the lifting lemma (the classical version) to this proof, we get a single-entry OL*-proof of A in **P**. Combining these proofs for all members of q', we obtain a single-entry OL*-proof of q' in **P**. Now apply the lifting lemma as in the proof of proposition 2.12. ∎

If a marked subgoal in an OL-derivation in **P** is not a dc-consequence of **P**, none of its continuations is a dc-consequence of **P**. Hence, the proposition says that, at least for the purpose of computing definite answers, we can drop all those marked subgoals which are not dc-consequences of **P**. This, in general, significantly reduces length of derived queries and hence the size of the search space at each resolution step.

§5. Discussion

In order to take advantage of the above finding in actual implementation, we must identify, in advance, the set of those subgoals characterized as dc-consequences, or some reasonably good approximation of it. The following syntactic analysis gives one such approximation for positive programs.

(5.1) Definition. A *case-tree* of a positive program **P** is a tree such that:

- the root has n children $A_1, \cdots, A_n$, where $A_1 \vee \cdots \vee A_n$ is the *head* of an indefinite clause in **P**. (We assume that every clause in **P** is in the form $A_1 \vee \cdots \vee A_n \leftarrow B_1 \& \cdots \& B_m$, $m \geq 0$, where A_i's and B_i's are all positive, and we refer to the disjunction as the head of the clause).

- any other node, say N, has m children $B_1, \cdots, B_m$, if there is a clause in **P** such that its head is $B_1 \vee \cdots \vee B_m$, and its body contains a literal unifiable with N.

- no branch has multiple occurrences of the same literal.

Note that every case-tree of **P** is finite and **P** has finitely many of them.

(5.2) Definition. Let T be a case-tree of a positive program **P** with n branches $\beta_1, \cdots, \beta_n$. Let $\beta_i[j_i]$ denote the literal labeling the j_i-th node in β_i. If a set of literals $\{\beta_1[j_1], \cdots, \beta_n[j_n]\}$ is unifiable, we call $\beta_i[j_i]$, $1 \leq i \leq n$, *merging literals* of T (or of **P**).

We then have,

(5.3) Proposition. Every dc-consequence of a positive program **P** is unifiable with some merging literal of **P**.

The proof of the proposition requires an extensive analysis of case-trees and their relationships to OL-proofs (see [13]).

Given this proposition, what we hope to have is the following refinement of OL-resolution for positive programs:

- Given a positive program **P**, we first identify all merging literals. Then, at each resolution step, if A is the right-most subgoal, B the literal to be resolved upon in the side clause, and θ their mgu, we mark and retain $A\theta$ *only* when B is a merging literal.

However, this refinement does not immediately follow from the proposition, i.e., it is quite possible that A is a dc-consequence, yet B is *not* a merging literal. In such a case, should we still retain $[A\theta]$ in the refutation? Fortunately, this is not necessary (see also [13]), i.e., at each resolution step, the subgoal need not be checked against the set of all merging literals to see if it is unifiable with some of them; rather, it is the case that the corresponding side clause can carry all the information necessary to determine whether the subgoal should retain its continuations in the refutation.

Further implementation issues are not discussed in this paper, but the overall strategy is to look at application of propositions 4.3 and 5.3 as a refinement of *Prolog Technology Theorem Prover* (PTTP) for positive programs[12] (PTTP uses both reduction and reentry mechanisms, and it is first-order complete). We finally note that in the limited context of positive programs, the two propositions can be applied to other aspects of PTTP (e.g., use of contrapositives) to obtain further refinements.

Acknowledgements. Many thanks to Professor H. Blair, Professor E. Sibert, A. Batarekh, and V.S. Subrahmanian for their valuable suggestions.

References

[1] Chang,C.L., and Lee,R.C.T., *Symbolic Logic and Mechanical Theorem Proving*, Academic Press, New York, 1973.

[2] Grant,J., and Minker,J., Answering Queries in Indefinite Databases and the Null Value Problem. Technical Report 1374, University of Maryland, 1983.

[3] Gallaire,H., Minker,J., and Nicolas,J-M., Logic and Databases: A deductive Approach. *Computing Surveys*, Vol.16, No. 2, 1984.

[4] Imielinski,T., and Lipski,W., On Representing Incomplete Information in a Relational Database. *Proceedings of the 7th International Conference on Very Large Databases.* IEEE, New York, 1981.

[5] Kowalski,R., and Kuehner,D., Linear Resolution with Selection Function, *Artificial Intelligence 2*, 1971.

[6] Lloyd,J.W., *Foundations of Logic Programming*, Springer-Verlag, 1984.

[7] Loveland,D.W., Mechanical Theorem Proving by Model Elimination. *J. ACM*, 1968.

[8] Loveland,D.W., Near Horn Prolog, *4th International Conference in Logic Programming*, Melbourne, 1987.

[9] Minker,J., On Indefinite Databases and the Closed World Assumption. *Lecture Notes in Computer Science, No. 138*, Springer-Verlag, New York, 1982.

[10] Pryzymusinski,T., On the Semantics of Stratified Deductive Databases, *Foundations of Deductive Databases and Logic Programming*, J. Minker. ed., 1987.

[11] Shoenfield,J., *Mathematical Logic*. Addison-Wesley, Reading, Mass. 1967.

[12] Stickel,M.E., A Prolog Technology Theorem Prover: Implementation by an Extended Prolog Compiler. *Lecture Notes in Computer Science, No 230*, Springer- Verlag, New York, 1986.

[13] Wakayama,T., and Payne,T.H., Refining Linear Resolution for Clause Sets with Disjunctive Heads, in preparation.

[14] Yahya,A., and Henschen,L., Deduction in Non-Horn Databases, *J. of Automated Reasoning 1*, 1985.

Notes on Prolog Program Transformations, Prolog Style, and Efficient Compilation to The Warren Abstract Machine

Ralph M. Butler
Division of Computer and Information Science
Univ. of North Florida
Jacksonville, FL 32216

Rasiah Loganantharaj
Center for Advanced Computer Studies
P.O.Box 4330, USL
Lafayette, LA 70504-4330

Robert Olson
Mathematics and Computer Science Division
Argonne National Laboratory
Argonne, Illinois 60439

Abstract

Most of the efficient implementations of Prolog are based on Warren's abstract machine [1,4]. Evan Tick [3] has argued that a substantial percentage of execution time in such implementations is directly related to the creation and the use of choice points. In this paper, we discuss how the overhead of choice point creation and processing can be avoided without sacrificing the declarative meaning of Prolog programs.

1 Introduction

It is extremely difficult to sort out and justify all of the factors that contribute to a "good" programming style. In this short work we shall try to expose a number of issues that all impinge on exactly how one should write Prolog programs. In the process, we hope to clarify some of the interplay between declarative beauty, efficiency, and compilation.

We will spend our time commenting on exactly how one might code a rather simple little routine *intersection*(X, Y, Z) which computes Z as the intersection of X and Y (versions of this routine has appeared repeatedly in the literature of logic programming). Finally, we will demonstrate the operation of a program which we have developed to perform some of the suggested transformations at the Prolog-to-Prolog level.

2 Computing the Intersection of Two Sets

Let us suppose that we are representing a *set* as an ordered list which contains no duplicates. In this case, all of the following constitute acceptable definitions of "*intersect/3*" in the sense that they correctly compute the desired result:

Version 1:
```
intersection([],_,[]).
intersection([H|T],L,[H|T2) :- member(H,L), !, intersection(T,L,T2).
intersection([_|T],L,T2) :- intersection(T,L,T2).
```

Version 2:
```
intersection([],_,[]).
intersection([H|T],L,[H|T2]) :- member(H,L), intersection(T,L,T2).
intersection([H|T],L,T2) :- not member(H,L), intersection(T,L,T2).
```

Version 3:
```
intersection([],_,[]).
intersection([H|T],L,Answer) :-
    (member(H,L) - >
      (   intersection(T,L,T2), Answer = [H|T2] )
      ;
      (   intersection(T,L,Answer) )
    ).
```

Version 4:
```
intersection([],_,[]).
intersection([H|T],L,Answer) :- intersection(T,L,T2),
    (member(H,L) - >
      (   Answer = [H|T2] )
      ;
      (   Answer = T2 )
    ).
```

Version 5:
```
    intersection(L1,L2,Answer) :-
        (L1 = []) - >
            (   Answer = [] )
            ;
            (   L1 = [H|T],
                intersection(T,L2,T2),
                    (member(H,L2) - >
                    (   Answer = [H|T2] )
                    ;
                    (   Answer = T2 ))
            ).
```

There is no question that some of these versions are somewhat less desirable than others (indeed, we believe that some will come close to inspiring open hostility when brought up in polite conversations). We intend to explore exactly what there is about these routines that make them more or less desirable.

In order to fully explore these different routines, we shall develop several topics which only peripherally relate to issues of "taste". We will make a variety of assumptions, some of them involving rather specific details of Prolog implementations:

1. First, we assume that we will be discussing a Prolog implementation based on the Warren Abstract Machine [1,4]. Since Warren's description does not include a discussion of how to implement the "cut" or the programming construct $A- > B; C$ we will take the liberty of proposing some alternatives that have occurred to us.

2. We assume that a choice point is created for a routine when a unique clause cannot be discriminated by the first argument of the invocation (e.g., when version 1 of "intersection/3" is invoked with a nonempty list as the first argument, a choice point will be created).

3. We will take the liberty of extending the instruction set of the Warren Abstract Machine in ways that allow efficient implementation of the "if-then-else" programming construct.

2.1 Implementation of if-then-else

Evan Tick [3] has argued that a substantial percentage of execution time in current Prolog implementations is directly related to the creation and use of choice points. We have gradually come to the following conclusion:

The if-then-else is a truly fundamental programming construct that is implemented inefficiently by current technology. That is, the implementation of if-then-else via the creation of choice points is both unnecessary and extremely costly.

Let us gradually explore exactly how we arrived at this point of view.

First, let us consider one of the standard implementations of the programming construct: $A- > B; C$.

$A- > B; C : -A, !, B.$
$A- > B; C : -C.$

This clearly represents serious overhead. First, the creation of the choice point requires the storage of a substantial amount of "machine state", and then the use of *call/1* introduces the overhead of symbol table accesses.

Consider the use of this programming construct when the routines A, B, and C are known at compilation. In this case, the overhead of *call/1* can be avoided by just generating appropriate code in-line. For example, using the extensions we've implemented in the version of WAM implemented at Argonne National Laboratory [2] the code for $g1- > g2; g3$ would be

```
            save_chpt          Y0
            try_me_else        lab(1)
            call               g1
            set_chpt           Y0
            call               g2
            goto               lab(2)
    lab(1)
            trust_me_else_fail
            call               g3
    lab(2)
```

This still represents a fair amount of overhead, since the choice point is created. Let us suppose that *g1* in the above construct is just a "test" (e.g., I < 7, var(X), etc.). In this case, it really represents a situation where one of two mutually-exclusive branches should be taken. This is, in fact, the concept of if-then-else implemented in common procedural programming languages. To implement it efficiently in Prolog, we need an instruction like *test_and_branch g1,lab(1)* which causes a branch to lab(1), if the "simple test" *g1* fails. Here the "simple test" must be an built-in predicate that does not introduce a choice point and does not alter the "environment" on failure (e.g., a unification test would have the property that, on failure, any new bindings would be removed without the mechanism of backtracking). In this case, the in-line implementation of $g1- > g2; g3$ would be

```
          test_and_branch   g1,lab(1)
          call              g2
          goto              lab(2)
lab(1)
          call              g3
lab(2)
```

This is more like it! Here, there is no overhead for either the creation of the choice point nor the use of *call/1*.

Now, it seems to us, that the creation of a choice point should reflect "conditional reasoning", in the sense that one of a variety of alternatives are to be explored. It should not be the basic mechanism for implementing a construct as fundamental and simple as if-then-else.

Having advocated an efficient implementation of if-then-else as critically important, we would also like to state that we believe that it should be used only very rarely in Prolog code. This is due to the fact that it really does impair the declarative beauty of Prolog code. However, we hope to show in this document that, with some minimal restrictions in style, it is possible to easily create a translator that will transform "declarative" Prolog into "if-then-else" Prolog that can be efficiently compiled. This topic seems to us to be of substantial importance.

2.2   Version 1 of intersection/3

We believe that this version is the most commonly employed, of those presented. It has the virtue of not duplicating the invocation of *"member/2"* and is considered "efficient". It does cause the creation of a choice point, and it uses the *cut* in a way that destroys the logical reading of the clauses.

2.3   Version 2 of intersection/3

From our perspective, there is a certain beauty in this version. It can be read declaratively, and there is none of the procedural flavor that exists in the last three versions. On the other hand, it appears to be the least efficient of the three versions. Under existing implementations this routine will result in both the creation of a choice point and the duplication of the "member" test.

2.4   Version 3 of intersection/3

This version is significant (in our view) because

1. it does not result in the duplicate invocation of *"member/2"*,

2. it can be read declaratively (albeit with a procedural flavor), and

3. it is fairly straightforward to transform version 2 to this form, when it is known that *"member/2"* is a determinate test.

The last point is somewhat dubious, since *"member/2"* is normally not defined as a determinate check. Whether a single definition of *"member/2"* should be used (that is, an openly nondeterminate definition), or whether two versions – *"member_test/2"* and *"member_gen/2"* – should be used is a point widely debated. We feel that two separate intrinsics will probably be included in high-performance Prologs of the future solely to provide performance, without sacrificing declarative readability. In view of our comments on the implementation of "if-then-else" tests, it should be noted that if *"member_check/2"* is included as a "simple test", the generated code would result in neither the creation of a choice point nor in a duplicate "member" check.

2.5   Version 4 of intersection/3

This version represents a slight improvement over version 3, since the generated code is somewhat less. It should not result in improved performance (as far as we can determine), and the added compiler complexity required to automatically transform version 3 into this version seems to be less important than the transformation mentioned above (for transforming version 2 into version 3).

2.6   Version 5 of intersection/3

This version of the routine reflects fairly accurately the approach that might be taken using a standard procedural language. Given the indexing scheme proposed by Warren, it really buys nothing. That is, there seems little point in attempting to transform version 4 into this code, since Warren's indexing scheme already discriminates between the two clauses being "collapsed" into a single clause.

2.7   Comments on the intersection/3 Versions

From our perspective, it seems clear that the Prolog programmer would like to write version 2 of the routine. This is practical, assuming that two versions of the "member" check are provided (and that the compiler is informed that the version being utilized is determinate). Furthermore, if the *"member_test/2"* intrinsic is included as a built-in "simple test", the code generated should avoid the overhead of choice point creation/processing.

We have also done a similar study on exactly how one might code another simple routine $merge(X, Y, Z)$ which merges the contents of the ordered lists X and Y, producing the list Z. Because of space limitation, it is omitted in this paper.

3 Our Program Transformations

We are currently developing a system which performs Prolog-to-Prolog transformations on Prolog programs. It is treated as a preprocessor to our Prolog compiler so that if one desires, he may compile a Prolog program without any transformations being performed. It is often easier to debug the original code than code which has been transformed.

Below are some examples of the types of transformations which we are currently able to perform. Note that for each example, there are two types of transformations performed. The first type of transformation removes all non-variables from the head of a clause and replaces it with a variable which, in the body of the clause, is unified with the original value. These transformations are performed first, because it can be rather tricky to collapse clauses for a procedure if there are differences in the heads. Mode declarations are supplied for each predicate ("+" means input, "-" means output, and "?" means either - treated as output) to assist in determining where the unification with the original value should be performed. Input variables are unified at the front of the clause, and output variables at the end. It is not necessary to supply the mode declarations, but more collapsing is likely to occur if they are given. The second type of transformation performed is to locate complementary tests in a set of clauses and to alter them to the if-then-else construct.

The Intersection Example:

```
determ(member).
intersection(+L1,+L2,-L3).
intersection([],_,[]).
intersection([H|T],L,[H|T2]) :- member(H,L), intersection(T,L,T2).
intersection([H|T],L,L2) :- not member(H,L), intersection(T,L,L2).
```

is transformed into:

```
intersection(X0,X1,X2) :-
        X0=[X3|X4] − >
        (member(X3,X1) − > intersection(X4,X1,X5),
            X2=[X3|X5]
        ;
        intersection(X4,X1,X2)
        )
        ;
        X2=[].
```

Note that the member predicate is designated as determinate, or single solution.

The Merge Example:

```
merge(+L1,+L2,-L3).
merge([],L,L).
merge([H|T],[],[H|T]).
merge([H1|T1],[H2|T2],[H1|T]) :- H1 @=< H2,
        merge(T1,[H2|T2],T).
merge([H1|T1],[H2|T2],[H2|T]) :-
        H2 @< H1,
        merge([H1|T1],T2,T).
```

is transformed into:

```
merge(X0,X1,X1) :- X0=[].
merge(X0,X1,X2) :-
        X0=[X3|X4],
        (X1=[X5|X6] − >
        (X3 @=< X5 − >
            merge(X4,[X5|X6],X7), X2=[X3|X7]
            ;
            merge([X3|X4],X6,X7), X2=[X5|X7]
        )
        ;
        X2=[X3|X4]
    ).
```

Here, X <= Y and Y < X are recognized as complementary tests. The program similarly recognizes [] vs. [H|T] as complementary when used as the values unified with some variable.

Example of the Length Predicate

```
length(+List,-Length).
length([],0) :- !.
length([H|T],L) :- length(T,L1), L is L1 + 1.
```

is transformed into:

```
length(X0,X1) :-
    (X0 = [X2|X3] - >
        length(X3,X4), X1 is X4 + 1 )
    ;
    !, X1 = 0.
```

4   Summary of Our Evolving Outlook

We have touched on a number of issues that have become gradually clearer to us as we explored the use of compiler optimizations to improve the quality of code generated by our Prolog compiler. It appears to us that the overhead of choice point creation and processing can be avoided for most common situations that would be handled using an "if-then-else" in a common procedural language. This reduction of choice point processing will have, we believe, a substantial impact on the efficiency of generated code. Whether it is practical to advocate such a position, depends on a number of issues (such as mode declarations, a convenient means for declaring the appropriate routines free of side-effects, and which tests are determinate).

We believe that these program transformations support Prolog programmers in a coding style which accurately reflects the declarative meaning of a procedure without sacrificing any performance which would typically be gained by omitting complementary tests (e.g. member and not member). Further, we hope that these transformations represent the initial group of an entire set of such optimizations which may be made at the Prolog level.

5   Acknowledgements

The authors wish to acknowledge Dr. Ross Overbeek of Argonne National Laboratory for his guidance in this work. None of the work would have been possible without his ideas, insights, and support. He has been our most valuable resource.

References

[1] J. Gabriel, T. Lindholm, E. Lusk and R. A. Overbeek, "Tutorial on Warren abstract machine," Technical Report ANL-84-84, Argonne National Laboratory, Argonne, Illinois, October, 1984.

[2] E. L. Lusk, R. A. Overbeek, and R. Butler, "The Warren abstract machine for computational logic: Extension to support cut, the occur check, subsumption checks and unit databases, (in preparation).

[3] E. Tick, "Lisp and Prolog memory performance," Technical Report 86-291, Departments of Electrical Engineering and Computer Science, Stanford University, January, 1986.

[4] D. H. D. Warren, "An abstract prolog instruction set," SRI Technical Note 309, SRI International, October, 1984.

Exploitation of Parallelism in

Prototypical Deduction Problems

Ralph M. Butler

Nicholas T. Karonis

ABSTRACT

In this paper we consider two computations: the calculation of the elements in a finite semigroup and a theorem prover for propositional logic. As a first step towards exploitation of multiprocessors for more general deduction systems, we implemented portable programs to perform each of these tasks. The programs utilize both loosely-coupled multiprocessors (using message passing) and tightly-coupled multiprocessors using shared-memory. We believe that the basic algorithms developed here will carry over to the more general case of deduction systems for the predicate calculus and to rewrite systems as well. Indeed, the algorithms have been applied to a first-order deduction system developed by William McCune. In one experiment on that system we successfully derived a proof of Sam's lemma (a classic benchmark problem) in under 13 seconds using 12 processors on an Encore Multimax.

1. Introduction

One builds deduction systems to study what formulas can be derived from a set of initial formulas using a set of specified inference rules. In this paper, we wish to consider the question of how such systems might be structured to utilize multiprocessors. In examining this question, it is useful to first analyze the abstract structure employed by many deduction systems. Consider, for example, the specific case of asking whether or not the empty clause can be deduced using binary resolution from a set of clauses $S = \{C1, C2, C3, ... Cn\}$. The basic algorithm used in most systems for non-redundantly performing the computation is as follows:

```
basic_resolution_algorithm
{
(1)     Let Set_of_Support initially be the set S.

(2)     Let Usable_Formulas be empty.

        while (the empty clause has not been deduced and
               Set_of_Support is not empty)
        {
(3)         Select one element of Set_of_Support and call it the Given_Clause.

(4)         Move the Given_Clause from Set_of_Support to Usable_Formulas.

(5)         Compute the set of resolvents that can be formed using the Given_Clause and one other
               clause in Usable_Formulas (keeping only those that are not subsumed by clauses in
               Set_of_Support or Usable_Formulas). Add these generated clauses to Set_of_Support.
        }
}
```

If step (5) is performed by "pairing" the Given_Clause with each element of **Usable_Formulas** and attempting to compute resolvents from the pair, then it can be observed that the algorithm causes each possible pair of clauses to be examined exactly once.

Now consider the (seemingly unrelated) question of how to compute the set of words in a finite semigroup from an initial set of "generators" and the given semigroup operation. To emphasize the similarity of this computation with the one discussed above, let us call the set of generators $S = \{C1, C2, C3, ... Cn\}$. Here $C1$, $C2$, ... Cn are specific elements of the semigroup in question. One algorithm for computing the entire semigroup is as follows:

```
generate_semigroup
{
(1)     Let Set_of_Support initially be the set S.

(2)     Let Usable_Elements be empty.

        while (Set_of_Support is not empty)
        {
(3)          Select one element of Set_of_Support and call it
                the Given_Element.

(4)          Move the Given_Element from Set_of_Support to Usable_Elements.

(5)          Compute the set of elements that can be formed using the Given_Element
                and one other element in Usable_Elements (keeping only those that are not
                already present in Set_of_Support or Usable_Elements).  Add
                these generated elements to Set_of_Support.
        }
}
```

Here, we are obviously presenting the algorithm in a format designed to emphasize the similarity with the preceding computation. This general structure is, perhaps, worth abstracting to a form that includes almost arbitrary inferences rules (e.g., non-binary rules), and the restrictions imposed by the set-of-support strategy:

```
abstract_deduction_algorithm
{
(1)     Partition S into two sets.  Those with "support" put into Set_of_Support;
            those without put into Usable_Formulas.

        while (an "end condition" does not exists and Set_of_Support is not empty)
        {
(2)          Select one element of Set_of_Support and call it the Given_Formula.

(3)          Move the Given_Formula from Set_of_Support to Usable_Formulas.

(4)          Compute the set of inferred formulas that can be formed using the Given_Formula
                and other formulas in Usable_Formulas (keeping only those that are pass
                some specified criteria).  Add these generated formulas to Set_of_Support.
        }
}
```

While the above abstraction leaves an enormous number of critical issues unspecified, it does offer a framework for investigating several issues that will be central to the successful exploitation of multiprocessors.

In our investigation, we chose to examine two computational problems that conform to the abstract structure described above: computing the elements of a finite semigroup and attempting to show the unsatisfiability of a set of clauses in propositional logic using binary resolution and subsumption. In the following sections, we will describe these problems in detail, along with our approaches to implementing solutions on multiprocessors.

2. The Semigroup Problem

2.1. A Short Description of the Problems

In the semigroup problem that we studied, the program is given a set of words (n-tuples) and an operation table which defines how new words may be generated from existing ones. The operation is commonly referred to as *product*, although it is not a commutative operation as is the arithmetic product of integers. The product of two words in the semigroup is determined by computing the product of each of the *n* respective values in those two words. There is a finite set of values which each element of a word may assume. For example, in the semigroup problem described in [3] , there are five unique values for the elements of a word of the semigroup:

```
0,e,f,a,b
```

Below is an operation table defining the product of each pair of elements:

```
  | 0  e  f  a  b
__|_____
0 | 0  0  0  0  0
e | 0  e  0  a  0
f | 0  0  f  0  b
a | 0  0  a  0  e
b | 0  b  0  f  0
```

The generation of the entire semigroup proceeds by beginning with some starter set of words, and generating as many new words from that starter set as possible. Each generated word is compared to existing words to see if it already exists. If a generated word is new, it is added to the set of words to eventually have its product taken with all other words in the set. The set of all possible words for a given starter set is potentially quite large, but typically the set of unique words generated is much smaller than the set of all possible words. For example, in one particular 125-tuple problem, there were 5^{125} possible words that could have been generated, but the semigroup actually contained only 1435 unique words. Even in a problem that only generates a small number of words, however, there is often more work than a single processor can handle in a reasonable amount of time.

2.2. The Algorithm Used to Compute the Semigroup

For this problem, we chose to explore an implementation that was based on message-passing. That is, it does not assume that any of the coordinating processes share memory. We coded it portably, using the techniques described in[1] , which means that the program can be executed (without changing the source code) on a wide variety of hardware configurations. We conducted our experiments on a Sequent B8000, a Sequent B21000, and Encore Multimax, and a set of Sun workstations. This section presents both verbal and pseudocode descriptions of the algorithm.

2.2.1. Verbal Description of the Algorithm

A sequential algorithm to solve the semigroup problem is quite straightforward (as described in the introduction). Recall that the algorithm is supplied with an operation table and an initial set of unique words. The algorithm keeps an ordered list of unique words which is initialized as the **Set_of_Support**. Processing the list involves starting at the top of the list and processing each word in the list. Processing a word in the list involves computing the product of that word with each word in the list above it and finally with itself. Because the operation table is not commutative, each pair of words must be "multiplied" twice, switching the order of the operands. Each product is checked to see if it already exists in the list of words. If found to be unique, the product is placed at the end of the ordered list. If not unique, it is simply discarded. The list is processed in this manner until there are no more words in the list. The final list represents the complete set of words that can be generated from the original set of words with the given operation table.

The parallel version of this algorithm is also quite straightforward. Consider an environment where there exists one master process and one or more slave processes, where the master is responsible for handing out work and coordinating the activities of the slaves. The master is also responsible for collecting and organizing the slaves' results. In addition, the master must determine when there is no more work to do and to inform all the slaves to halt. The slaves are responsible for requesting work from the master, performing that work, and reporting their results to the master. The slaves continue this loop until they are notified that there is no more work to do and that they should halt.

During its initialization procedure, the master process must read the starter set of words and the operation table, create the slave processes, and send a copy of the words and table to each slave. The slaves receive this information and initialize their local copies of the data structures. At this point, the processes are all ready to begin computing the semigroup.

After initialization of all processes, the master process waits for requests for work from the slaves. It coordinates the allocation of work to the slaves by assuring that no two slaves receive the same piece of work to perform and also ensuring that no piece of work is left undone. The master also receives candidate new words from the slaves. If it receives a candidate that is indeed a new word, then it adds that word to its master list and sends a copy of that word to each slave (along with an indication of where that word is to be placed in the list).

The slave processes request work from the master process. The work may come in one of two forms: either a new word to add to the local list, or as an index of an existing word from which the slave is to compute products with other words. When a slave receives an index of a word, it multiplies that word with each word above it in the list. Each product is checked to verify that it does not already exist in the slave's local copy of the word list. If the word does exist locally, it is discarded and processing continues. If the word does not exist locally, the slave sends a copy of it to the master to determine if it is in the master list. If the master discovers that the word is new, a a copy is sent to all slaves to be added to their local lists.

It is important to understand the justification for having both a master list of words as well as a local copy at each slave. The major reason for having both lists is that, since each word tends to be generated many times during a single execution, the communication overhead would be staggering if each slave sent every word that it generated to the master. It is much cheaper in terms of communication requirements to have each slave discard as many of these generated words as possible. If a word is n determination as to whether that word is indeed a new one. Thus, each slave's list is a subset of the master list. The slave's list does not have to be completely up to date because the master makes all final determinations.

2.2.2. Pseudocode for Master Process

The master process maintains two critical data structures used to keep track of communication with slave processes. The *Ready_Slave_Queue* is used to record which slave processes are waiting for the master process to transmit work. Since each slave process keeps a local copy of the database of words, the master must transmit commands to the slaves whenever alterations are made to the master database. To accomplish this, the master process maintains a separate queue for each process. The queue will contain commands reflecting changes is the master database. These commands are of three forms:

1. An *add* command is of the form "add word W to your database", where W is some word that the master has added to the master database.

2. A *move* command is of the form "move the nth word to the *Usable_Words* list (where n is a parameter included in the command).

3. A *generate* command is of the form "generate all new words that can be formed from the nth word and words in the *Usable_Words* list.

All queued commands are transmitted to a slave, when the slave requests work. With these comments in mind, the pseudo-code for the master process is as follows:

```
semigroup_master_process:
{

    Initialize Set_of_Support with starter set of words.

    Let Usable_Words be empty.

    Create the slave processes.

    We keep a queue of slaves "waiting for work" called Ready_Slave_Queue.  Initially, all
        of the slaves are put in this queue (which simply records the fact that initially the master
```

```
      must send each of the slaves some work).

For each process, we keep a queue of "commands" to be sent to the slave process.  Initially,
   these queues must each contain one command per word in the initial Set_of_Support.  These
   commands will be "add word Wi to your local Set_of_Support" (where Wi is a specific word).

while (Set_of_Support is not empty or not all of the slaves are waiting for work)
{
      if (the Ready_Slave_Queue is not empty and Set_of_Support is not empty)
      {
         Select a process P from the Ready_Slave_Queue (removing it from the queue)

         Select W from Set_of_Support (moving it from Set_of_Support to Usable_Words).
            Queue commands to each slave of the form "move W to Usable_Words".

         Queue a command to P of the form "generate all new words that can be formed from W".

         Send to P all messages queued for P.  This may actually require multiple
            send operations, since there is a well-defined limit on the actual size of transmitted messages
      }
      else
      {
         Receive a message from one of the slaves.  This message may contain 0 or more new words.
            It may (or may not) also include a request for more work (i.e., it may be asking for
            another "generate" command).

         for each new word in the message
         {
            if (the word does not already exist in either Set_of_Support or Usable_Words)
            {
               Add the word to Set_of_Support.

               Queue "add" commands for each of the slave processes
            }

            if (the slave requested more work)
            {
               Add the slave to the Ready_Slave_Queue.
            }
         }
      }
}
      Send a "termination" message to each slave processss
}
```

When the master process terminates, *Usable_Words* will contain the entire set of words in the semigroup.

2.2.3. Pseudocode for Slave Processes

The slave processes all communicate with the master, updating their local databases and generating new words to be transmitted back to the master process. The logic for the slave processes is as follows:

```
semigroup_slave_process:
{
      Initialize Set_of_Support and Usable_Words to empty.

      Receive a message from the master process.  The message is either a "termination" message
         or a message containing a sequence of "add", "move", and "generate" commands.

      while (the message is not the "termination" message)
      {
         for each command in the message
         {
            if (the command is an "add" command)
            {
               Add the word included in the command to Set_of_Support.

            }
            else if (the command is a "move" command)
            {
               Move the designated word from Set_of_Support to Usable_Words
            }
            else the command must be a generate command
            {
               Generate New_Words equal to the set of new words that can be generated using
                  the designated word (from the "generate" command) and words in Usable_Words.
```

```
Send the words in New_Words back to the master.  This may take an arbitrary number
    of actual send operations, since the size of a physical message is strictly limited.
    The last physical message sent should include a "request for more work" to the master process.
        }
      }
    }
  }
```

2.3. Results

The above algorithm "works" in the sense that all of the required work is distributed evenly to the slaves, and none of the computation is duplicated. However, it should be noted that no attempt was made to "buffer" work to a slave. That is, no attempt was made to have the master send work to a slave before the slave had actually requested it. In situations in which the cost for message passing is low (e.g., in cases in which shared-memory is utilized), this should pose no problem. However, in cases in which the message-passing is slow (e.g., using a set of Sun workstations connected via a local area network), we anticipated some degradation.

We conducted experiments on semigroup problems of three sizes: 25-tuple, 36-tuple, and 125-tuple. Our solution to the 36-tuple problem was interesting from a mathematician's perspective because it was a previously unsolved problem. However, it did not run long enough using this technique to acquire interesting data about the effectiveness of parallelizing the algorithm. Thus, the 125-tuple problem (although it had been previously solved) was selected for collection of timing data to compare the relative speedup values of executions involving varying numbers of slaves. This problem was run on each of three different machine configurations, using 1, 2, and 3 slaves in each of the three environments. Because portability was one of our main goals, it is important to note that the exact same code was compiled in each target environment. The differences appeared only in the code generated by the parallel programming tools described above [1].

Table 1 gives the speedup values for the 125-tuple on each of three different machine configurations. The first two entries show relative speedups for two shared-memory machines, the Sequent Balance 8000 and the Encore Multimax. The third entry shows a single experiment which we ran using a combination of Sequents and Encores, i.e. the master process ran on one Sequent; one slave ran on a second Sequent and the other two slaves each ran alone on an Encore. For this third entry, the communication between the slaves and master was done via an Ethernet. The fourth and final entry shows the speedups gained by running on a set of Sun workstations connected via an Ethernet.

		Number of slaves		
Machine(s)		1	2	3
Sequent		267	143	89
Encore		208	105	70
Sequent & Encore		-	-	80
Suns		94	50	34

Table 1. Timings (minutes) for 125-Tuple Experiments

We see from the table that speedups are near linear. They are slightly less than linear, however, because as mentioned earlier, this type of algorithm tends to have substantial communication overhead during the first few minutes while the processes discover lots of new words. As time goes on, generated words are more and more likely to be in the slaves' local tables. As this occurs, the amount of communication required decreases dramatically to the point where there is very little communication except when a slave needs to request more work; very few new words are being sent up to the master or making their way back to down to the slaves.

The fact that communication overhead plays a role only during the early minutes of the computation makes the Sun workstation timings especially encouraging. The implication is that, for a relatively modest price, one can develop a serious parallel processing environment for research into areas such as automated deduction.

3. The Propositional Logic Theorem-Prover

The experiments reported above were encouraging, but one important phenomenon that occurs in realistic deduction systems was not present: no previously existing formulas (or, words, in the case of the semigroups problem) were deleted. This issue of deleting substantial numbers of formulas during a sizable execution must be handled for the algorithm to be viewed as suitable for implementation of deduction systems.

In addition, the algorithm given above uses only relatively large granularity multiprocessing (i.e., the units of work that get distributed are quite large). No attempt was made to utilize the capabilities offered by shared-memory multiprocessors.

Our next experiment involved the implementation of a theorem prover for propositional logic. Our goal in constructing the parallelized propositional theorem-prover was not to produce an extremely high-performance propositional logic theorem-prover. Rather, it was to create a prototype that would generalize to systems for the predicate calculus and generalized rewrite systems. Hence, we did not actively explore algorithms like the Davis-Putnam algorithm[2] , but rather chose to implement binary resolution with set-of-support and subsumption. We modified the algorithm given in the previous section in two ways:

1. The commands sent from the master process to the slaves were of three types in the previous algorithm: "add", "move", and "generate". We added a fourth type, "delete". This command was used when the master, during addition of a new formula, detected that it was desirable to delete an already existing formula (do to "back subsumption" in the case of the propositional theorem-prover).

2. Each slave process created a number of new processes which, together with the slave, shared memory. In this case, we chose to call the slave process a "local master", since it managed a "cluster" of processes. The created processes were then called "local slaves". A simple "dispatcher" was used to distribute work among the processes in a cluster.

The first change was relatively straightforward, although the implications were substantial (e.g., we now had an algorithm that had all of the functionality required to support distributed rewrite algorithms or general theorem-proving algorithms). The second alteration requires some explanation.

Consider the problem of forming all resolvents that can be produced using a single given clause and a set of "usable clauses". One way to do this would be to, via some suitable indexing, determine the subset of clauses in the "usable clause" set that had literals that complemented those in the given clause; call this subset the "clashable clauses" subset. We can define a "match" to be a 3-tuple (g,c,v), where g is the given clause, c is a clause from the "clashable clauses" subset, and v is the literal of the given clause that is to be "resolved upon". Corresponding to each match we have the following computational task:

1. Form the resolvent from the two clauses.

2. Determine whether it is subsumed by any clauses that are already kept in the local copy of the database.

3. If the clause appears to be "new", add it to the set of clauses to be returned to the master process.

Thus, each match represents a unit of work that is much finer grained than "compute all new resolvents using a specified given clause" (which is the basic unit of work transmitted from the master process to the local master processes), but is still large enough to be successfully distributed among the local slaves (assuming that synchronization overheads are low due to shared-memory and that there are a modest number of local slaves (we consider 1 to 15 local slaves to be appropriate, depending on the machine on which the cluster executes).

When a local master receives a unit of work from the master process, the set of matches is computed (this can be done rapidly using suitable indexing). These "matches" are added to the pool of work being dispatched to local slaves. The local master and local slaves take matches from the pool and process them, until all of the matches have been processed. At this point, the local master behaves just as the slave process in the semigroup example. That is, it transmits the new clauses back to the master process and waits for a new set of commands (or a termination message) to be returned from the master. This basic algorithm is described by the following psedo-code, which gives the general logic used by the master process, the local master processes, and the local slave processes:

```
master_process:
{

    Initialize Usable_Formulas to contain input formulas that are not members of the set-of-support.

    Initialize Set_of_Support with those input formulas that are members of the set-of-support.

    Create the local_master processes.

    We keep a queue of local_masters "waiting for work" called Ready_Local_Master_Queue.
        Initially, all of the local_masters are put in this queue (which simply records the fact that
        initially the master must send each of the local_masters some work).

    For each process, we keep a queue of "commands" to be sent to the local_master process.
        Initially, these queues must each contain one command per formula in the initial Set_of_Support.
        These commands will be "add formula Wi to your local Set_of_Support"
        (where Wi is a specific formula).

    while (Set_of_Support is not empty or not all of the local_masters are waiting for work)
    {
        if (the Ready_Local_Master_Queue is not empty and Set_of_Support is not empty)
        {
            Select a process P from the Ready_Local_Master_Queue (removing it from the queue)

            Select W from Set_of_Support (moving it from Set_of_Support to
                Usable_Formulas).  Queue commands to each local_master of the form
                "move W to Usable_Formulas".

            Queue a command to P of the form "generate all new formulas that can be formed from W".

            Send to P all messages queued for P.  This may actually require multiple send
                operations, since there is a well-defined limit on the actual size of transmitted messages.
        }
        else
        {
            Receive a message from one of the local_masters.  This message may contain 0 or more
                new formulas.  It may (or may not) also include a request for more work (i.e., it may be asking
                for another "generate" command).

            for each new formula in the message
            {
                if (the formula does not already exist in either Set_of_Support or Usable_Formulas)
                {
                    Add the formula to Set_of_Support.

                    Queue "add" commands for each of the local_master processes

                    Check to see whether or not the new formula subsumes any previously existing
                        formulas.  For each subsumed formula, delete the formula from the database and
                        queue a "delete" command for each of the local_masters.
                }

                if (the local_master requested more work)
                {
                    Add the local_master to the Ready_Local_Master_Queue.
                }
            }
        }
    }
    Send a "termination" message to each local_master processs
}

local_master_process:
{
    Initialize Set_of_Support and Usable_Formulas to empty.
```

Initialize the **Pool_of_Work** to empty.

Create the set of local slave processes, which aid the local master in computing new formulas from
    each given formula transmitted from the master process. The logic for the local slaves is given below.
    Essentially, they wait for work to show up in **Pool_of_Work**, and then help to generate
    new formulas to be sent back to the master.

Receive a message from the master process. The message is either a "termination" message or
    a message containing a sequence of "add", "move", and "generate" commands.

```
while (the message is not the "termination" message)
{
    for each command in the message
    {
        if (the command is an "add" command)
        {
            Add the formula included in the command to Set_of_Support.
        }
        if (the command is a "delete" command)
        {
            Delete the formula indicated in the command from the database.
        }
        else if (the command is a "move" command)
        {
            Move the designated formula from Set_of_Support to Usable_Formulas
        }
        else the command must be a generate command
        {
            Generate New_Formulas equal to the set of new formulas that can be generated
                using the Given_Formula (from the "generate" command) and formulas in
                Usable_Formulas.  This is accomplished by calling the routine gen_formulas.

            Send the formulas in New_Formulas back to the master.  This may take an
                arbitrary number of actual send operations, since the size of a physical
                message is strictly limited.  The last physical message sent should include a
                "request for more work" to the master process.
        }
    }
}
Signal "program end" which causes all of the local slaves waiting in the routine work to exit.
```

```
gen_formulas:
{
    Form the set of matches that can be composed from the Given_Formula and formulas
        in Usable_Formulas and add the set to Pool_of_Work.

    Invoke the routine work to aid the local slaves in computing new clauses.
}
```

```
local_slave:
{
    Invoke the routine work.  This routine is used by both the local master and all of
        the local slaves to cooperatively generate new clauses whenever a new "generate" command
        arrives from the master process.
}
```

```
work:
{
    Wait for either work to show up in the Pool_of_Work or a "program end" condition.

    while (a "match" has been acquired or
            (all of the "matches" for a single Given_Formula have been processed
             and this process is a slave))
    {
        if (a "match" has been acquired)
        {
            Form the derived formula.

            If (the formula is not subsumed by an existing formula)
            {
                Add the new formula to New_Formulas.
            }
        }
        Wait for either work to show up in the Pool_of_Work or a "program end" condition.
    }
}
```

The last routine, *work*, requires some explanation (a complete description of this pattern of logic is given in[1] , and the reader is directed to that work for all of the details). The local master invokes *work* to aid the local slaves in computing all of the new formulas derivable from a designated "given formula". When these have been computed, the local slaves remain looping in *work* waiting for more work or a "program end" condition. The local master must exit *work* to return the results to the master and acquire more commands to process.

We ran the program using a number of moderately demanding theorems. The timings that we got are reflected in the following table:

Number of Clusters	Processes per Cluster (including local master)		
	1	2	3
1	15.2	10.1	9.0
2	8.2	1.9	2.8
3	.7	.4	2.3

Table 2. Timings (minutes) for the same Problem Submitted to the Propositional Theorem Prover

These results also require some explanation. First, we have not yet determined what performance could be attained using a well-coded version of the algorithm for uniprocessors. We believe that it would run at most about 20-25% faster than our case using one cluster with one slave (we are in the process of verifying this belief).

Note the anomaly of super-linear speedups (e.g., compare the case of a single cluster with one process versus two clusters, each with two processes). This occurs due to the distinct orderings of clauses in the **Set_of_Support** list. Our algorithm is very sensitive to the ordering of clauses in the set-of-support. When an algorithm such as the one we are using is employed, different orderings may occur on distinct runs, resulting on wide variations on the execution time. It is worth noting that there are at least three distinct phenomena going on:

1. In the sequential run, suppose that *C1* and *C2* occur at the head of the set-of-support, and that *C1* is chosen as the first "given clause". If one of the clauses generated using *C1* subsumes *C2*, then *C2* will never become the "given clause". On the other hand, in the case in which multiple clusters are executing, *C1* and *C2* will probably become given clauses simultaneously, wasting execution time in comparison with sequential execution.

2. Supposing again that *C1* and *C2* are at the head of the set-of-support, and supposing that the set-of support is sorted on some *weighting[4]* function, then clauses generated by *C1* may become the given clause before *C2*. In the case of parallel execution, clauses generated by *C2* may subsume clauses generated by *C1* before they become the given clause.

3. The simple reordering of the generated clauses due to the use of the local slaves to aid in generating the set can lead to widely varying results.

We have no basis for commenting on how these phenomena might impact execution on a wide class of problems. The test case that we employed was a fairly contrived clause set designed to be particularly demanding (the run with a single local master and no local slaves generated over 4100 non-subsumed clauses, some containing 6 literals), and we are certainly aware that examination of a wider class of problems would be a natural next step.

4. Summary and Future Research

We have chosen prototypical deduction problems and shown that they can be parallelized. We have also shown that such parallel solutions can be made portable across parallel architectures that are based on either the shared-memory model or the message-passing model. Further, we have demonstrated that relatively inexpensive hardware may be utilized as the development environment for this type of research.

Having ported our code to a true message-passing environment (e.g. the set of Suns), we would next like to port it to one of the commercially available hypercubes. A project is currently underway to put the Argonne parallel programming tools on a hypercube. When that project is completed, we plan to use those tools to port the semigroup problem.

As mentioned earlier, this research was preliminary work designed to investigate the potential for parallelism in automated deduction systems. We believe that great potential has been realized, and thus our future research interests also include development of a complete, parallel automated deduction system.

5. Acknowledgements

This work would not have been possible without the help of Ewing Lusk, Robert McFadden, and Ross Overbeek. In particular, we would like to thank Argonne National Laboratory for providing us the opportunity to work on the machines in their Advanced Computing Research Facility.

References

1. Jim Boyle, Ralph Butler, Terry Disz, B. Glickfeld, E. L. Lusk, R. A. Overbeek, J. Patterson, and R. Stevens, *Portable Programs for Parallel Processors*, Holt, Rinehart, and Winston, New York, New York (1987).

2. M. Davis and H. Putnam, "A computing procedure for quantification theory," *Journal of the ACM* 7, pp. 201-215 (1960).

3. Ewing L. Lusk and Robert B. McFadden, "Using Automated Reasoning Tools: A Study of the Semigroup F2B2," *Semigroup Forum* 36(1), pp. 75-88 (1987).

4. J. McCharen, R. Overbeek, and L. Wos, "Complexity and related enhancements for automated theorem-proving programs," *Computers and Mathematics with Applications* 2, pp. 1-16 (1976).

A Decision Procedure for Unquantified Formulas of Graph Theory

Louise E. Moser

Department of Mathematics and Computer Science
California State University, Hayward
Hayward, CA 94542

Abstract

A procedure for deciding the validity of ground formulas in a theory of directed graphs is described. The procedure, which decides equality and containment relations for vertex, edge and graph terms, is based on the method of congruence closure and involves the formation of equivalence classes and their representatives. An interesting aspect of the procedure is its use of maximal, rather than minimal, normal forms as equivalence class representatives. The correctness, efficiency, and limitations of the procedure are discussed.

Key words and phrases. Directed graph, decision procedure, congruence closure, normal form, equivalence class representative.

1 Introduction

Decision procedures for theories that include equality [7,11] typically involve the formation of equivalence classes and representatives for those classes. For such decision procedures the two main questions that must be answered affirmatively are

- Does the procedure produce a unique representative for terms that are semantically equal?

- Does the procedure terminate?

The decision procedure for graph theory described here is likewise concerned with the formation of equivalence classes and their representatives. It is based on the method of congruence closure [7,14], and is designed to be interfaced with Shostak's procedure for deciding combinations of theories [12]. Unfortunately, Shostak's method of solving equations and eliminating variables cannot be applied directly to

This research was supported by the National Science Foundation under grant DCR-8408544 and by the National Security Agency, Office of Cryptographic Research, under grant MDA904-84-H-0009.

The author's current address is: Department of Computer Science, University of California, Santa Barbara, CA 93106.

our problem, because it applies to "algebraically solvable" theories and the theory of graphs considered here is not algebraically solvable. Although we can solve vertex and edge equations (and indeed we do), it is impossible to solve graph equations in our theory; thus, we had to devise a new method for deciding such formulas.

The theory to which our decision procedure applies includes the operators of union and intersection. The union operator allows graphs to be constructed from simple vertex and edge graphs. The theory does not include a complement or difference operator. The union and intersection operators are associative, commutative, distributive, and idempotent. These properties are handled in our procedure by means of normal forms, i.e., by rewriting intersections of unions as unions of intersections, ordering arguments, and eliminating redundant arguments and function symbols.

An equivalence class representative can usually be chosen to be any element of the class but, in decision procedures, a minimal normal form is often used. An interesting aspect of the procedure described here is the need to use maximal, rather than minimal, normal forms as equivalence class representatives. In general, unique minimal normal forms that allow the determination of equality and containment relations for graph terms do not exist. One can, of course, choose an order such that the maximal normal forms register as minimal, but such orders do not correspond to intuitive notions of complexity, or to the algorithm used.

In our decision procedure, the equivalence class representative for a graph term is the normal form of the union of all terms in the system equivalent to the given term, or to terms contained in the given term. It is obtained by expanding new terms using existing equations in the system and by expanding existing representatives using new equations. Formation of these unique maximal normal forms depends on the finiteness of the expressions involved. Because maximal forms result from combining terms that already exist, the cost of constructing these terms is not very great. The technique of using maximal forms as equivalence class representatives may generalize to other theories for which it is impossible to solve equations for minimal normal forms.

As described in [6,12], the decision procedure for graph theory presented here can be used in combination with decision procedures for other theories. Decision procedures for set theory are considered in [1,2], for Boolean algebra in [3], and for lattice theory in [4]. Graph theory is considered in [9] in the context of term rewriting systems and category theory. In [5] we establish the decidability of formulas in graph theory by providing a procedure based on a decision procedure for propositional calculus; the correctness of that procedure is proved using model theory.

A decision procedure for graph theory, when incorporated into a specification and verification system such as that described in [13], can provide support for a user, who needs to establish rigorous proofs about properties of graphs, but who is accustomed to dealing with graphs intuitively. The decision procedure for graph theory given here has been developed to provide the underlying mechanical support for verifying such graphical properties. With a decision procedure for graph theory, the system can mechanically verify basic properties of graphs that are tedious to check, and thus can free the user to work at a more abstract level. A prototype of our decision procedure for graph theory has been implemented in Pascal.

2   The Language

In this section, we describe the language to which our decision procedure applies. The language was chosen to allow the fundamental concepts of graph theory to be expressed in familiar terms, thereby making it easier for the user to formulate his or her definitions, lemmas, and theorems. No doubt there exist other languages of graph theory, which are not as rich and for which the cost of the decision process is not as great, but such languages are more cryptic and more difficult to use. The language we consider here is strongly typed and includes the following symbols:

> parentheses ( )
> the usual logical connectives $\neg, \wedge, \vee, \Rightarrow, \equiv$
> free variables for the types vertex, edge, and graph
> $\qquad v1, v2, v3, \ldots, e1, e2, e3, \ldots, g1, g2, g3, \ldots$
> the constant ϕ
> the unary operators H, T, VG, EG
> the binary operator E
> the nary operators U, I
> the binary relators $\in, \subseteq, =$

The argument of H (head) is an edge, and its value is a vertex. Similarly, for T (tail). The arguments of E (edge) are vertices, and its value is an edge. The argument of VG (vertex graph) is a vertex, and its value is a graph consisting of a single vertex. Similarly, for EG (edge graph). The arguments of U (union) are graphs, and its value is a graph. Similarly, for I (intersection). The first argument of $\in$ is a vertex or an edge, and its second argument is a graph. The arguments of $\subseteq$ are graphs, while the arguments of $=$ are either vertices, edges, or graphs.

Formulas are formed from these symbols in the usual manner. The theory of graphs we consider is the conventional theory, an axiomatization of which is given in [5].

3   Terminology

Definition. The *elementary graph* terms are (1) ϕ, (2) the graph variables, and (3) terms of the form $VG(v)$ and $EG(e)$.

3.1   Normal Forms

The normal forms of vertex terms are v, $H(e)$, and $T(e)$. The normal forms of edge terms are e and $E(t1,t2)$, where $t1$ and $t2$ are vertex variables or heads (tails) of edge variables, but not $E(H(e),T(e))$ because this can be simplified to e.

The normal forms of graph terms are the empty graph; edge graphs; vertex graphs; graph variables; polyadic intersections of edge graphs, vertex graphs, and graph variables; polyadic unions of intersections, edge graphs, vertex graphs, and graph variables. When written in normal form, the arguments of unions and intersections are in strictly decreasing order. The order of terms is the usual recursive lexicographic order, where the order of function symbols is

$$\operatorname{order}(U) > \operatorname{order}(I) > \operatorname{order}(EG) > \operatorname{order}(VG) > \operatorname{order}(H) > \operatorname{order}(T).$$

The recursive lexicographic order on terms is extended to equations, and thus provides a normal form for equations. An equation is in normal form if its left and right sides are in normal form and the higher-order term is on the left.

In rewriting terms in normal form, we use basic properties of the operators that are implied by the axioms, such as the associative, commutative, distributive, and idempotent properties of union and intersection. From these properties, it follows that the depth of nesting of function symbols in a term in normal form is at most five as, for example, in

Example. $U(I(EG(E(H(e),v)),g1),g2)$.

In the examples presented here, we assume that $\operatorname{order}(g1) > \operatorname{order}(g2) > \ldots$ We also assume that there are no other relations in the system involving the given terms other than the ones indicated. The terms in these examples are terms that arise in the midst of processing – they may be supplied by the user or they may be generated by the system.

Example. The normal form of $U(g2,U(I(g1,g2),g2))$ is $U(I(g1,g2),g2)$.

Note that two terms are equal if they have identical normal forms. However, two terms may be semantically equal and yet have different normal forms as, for example, $U(I(g1,g2),g2)$ and $g2$.

Definition. A term $t1$ is a *part* of the term $t2$ if and only if $t1$ and $t2$ are both unions (intersections) and each argument of $t1$ is an argument of $t2$ when $t1$ and $t2$ are in normal form.

Example. $U(g1,g3)$ is a part of $U(U(g1,g2),g3)$ since the normal form of $U(U(g1,g2),g3)$ is $U(g1,g2,g3)$ and each argument of $U(g1,g3)$ is an argument of $U(g1,g2,g3)$.

Note that a term that is a union or intersection is a part of itself. Also, note that if $t1$ and $t2$ are intersections and $t1$ is a part of $t2$, then $t2$ is a subgraph of $t1$.

Example. $I(g1,g2)$ is a part of $I(g1,g2,g3)$.

Definition. A term $t1$ *is contained in* the term $t2$, or $t2$ *contains* $t1$, if and only if either (1) $t2$ is a union and $t1$ is an argument of $t2$, (2) $t1$ and $t2$ are both unions and $t1$ is a part of $t2$, (3) $t1$ is an intersection and $t2$ is an argument of $t1$, or (4) $t1$ and $t2$ are both intersections and $t2$ is a part of $t1$.

3.2 Equivalence Class Representatives

In a decision procedure for graph theory, an equivalence class representative cannot be chosen as an arbitrary element of the class (or even the normal form of such an element), because such a choice will not necessarily result in the determination that semantically equal terms are equivalent. An equivalence class representative must in some sense be chosen as a "canonical form." However, unlike in algebraically solvable theories, an equivalence class representative for graph terms cannot be chosen to be a single variable or even a minimal expression. Consider the following example.

Example. If $U(g1,g2,g4) = g5$ and $U(g1,g3,g4) = g6$, then $U(g3,g5) = U(g2,g6)$.

In this example the variable $g5$ would naturally be chosen as the minimal normal form for $U(g1,g2,g4)$. Similarly, $g6$ would be chosen as the minimal normal form for $U(g1,g3,g4)$. However, if these are chosen as the equivalence class representatives, it is difficult to see how one would conclude that $U(g3,g5) = U(g2,g6)$.

The problem here is that we cannot solve equations for minimal normal forms. The difference and complement operators, which might allow us to solve equations involving graph terms, are not defined because of the dangling edges that result when a vertex is removed from a graph. Unless the incident edges are removed along with the vertex, the resulting object is not a well-defined graph.

As a result of considering examples such as this, we were led to devise a new method for establishing equality of graph terms. The method is based on the use of maximal, rather than minimal, normal forms as equivalence class representatives. Thus, we have

Definition. The *equivalence class representative of a union* t is the normal form of the union of (1) all arguments of t, (2) all unions in the system that are a part of t, (3) all intersections in the system that have an argument or a part that is an argument of t, and (4) all terms in the system equivalent to these terms.

The *equivalence class representative of an intersection* t is the normal form of a union that includes (1) an intersection of all terms in the system equivalent to arguments of t and (2) all terms in the system equivalent to intersections of which t is a part.

The *equivalence class representative of an elementary graph term* t is the normal form of a union that includes (1) t, (2) all intersections in the system that have t as an argument, (3) $VG(H(e))$ (or $VG(T(e))$) if $H(e)$ (or $T(e)$) occurs in the system and t is $EG(e)$, (4) $I(EG(e),t)$ if $EG(e)$ occurs in the system and t is $VG(H(e))$ (or $VG(T(e))$), and (5) all terms in the system equivalent to these terms.

Example. If $U(g1,g2,g3) = U(g6,g7)$ and $U(g3,g4) = g8$, then the equivalence class representative of $U(g1,g2,g3)$ and $U(g6,g7)$ is $U(g1,g2,g3,g6,g7)$, and the equivalence class representative of $U(g3,g4)$ and $g8$ is $U(g3,g4,g8)$. Furthermore, the equivalence class representative of $U(g1,g2,g3,g4,g5)$ is $U(g1,g2,g3,g4,g5,g6,g7,g8)$. This is found by expanding $U(g1,g2,g3,g4,g5)$ to $U(g1,g2,g3,g4,g5,U(g1,g2,g3,g6,g7),$ $U(g3,g4,g8))$ and then normalizing the latter term. The expansion takes place because $U(g1,g2,g3)$ and $U(g3,g4)$ are each a part of $U(g1,g2,g3,g4,g5)$, and $U(g1,g2,g3,g6,g7)$ and $U(g3,g4,g8)$ are their respective representatives.

To hold down the computational cost of our procedure, we avoid introducing unnecessary graph terms and consider only graph variables that occur in the formula. Thus, an equivalence class representative is not an absolute canonical form among terms in the Herbrand universe, but rather a canonical form relative to the universe of terms that have been constructed. If terms are processed in a different order, different equivalence class representatives may be formed. However, for a given formula, the same representative is generated for all terms that are equal, as shown in Section 6 below.

4 The Data Structures

The data structures used by the algorithm are similar to those used by Shostak [12] and include

- An arglist for each term that is a nonvariable expression
- A uselist for each equivalence class; an element of the uselist is a nonvariable expression for which there is an element in the equivalence class that is an argument of the expression
- A findfield that gives the equivalence class representative
- A ringfield for each term that gives access to the members of the equivalence class of that term
- A sigfield for each term that gives the signature which for a variable is the variable and for a nonvariable expression is the normal form of the term obtained by applying its function symbols to the representatives of their arguments.

For efficiency, only one instance of each term is kept in the system.

4.1 The Ring

Instead of the usual tree structure used in union-find programs, we use a ring to represent an equivalence class. The ring contains terms that are equivalent because of the equations and predicates that occur in the formula and because of the graph-theoretic rules that are built into the decision procedure. These include rules such as $H(E(v1,v2)) = v1$, $E(H(e),T(e)) = e$, $U(I(g1,g2),g2) = g2$, etc. A ring is used rather than a tree because in parts of the procedure it is necessary to access each of the terms in an equivalence class.

4.2 The Uselist

While the arglist of a function gives access to the arguments of the function, the uselist of a term gives access to the functions of which the term is an argument. The uselist is kept for the equivalence class, i.e., the representative, rather than for each term in the class. Graph terms that are unions or intersections are kept on the uselists of their arguments, but not on the uselists of terms contained in these terms. When two classes are merged, terms on their uselists are rewritten in terms of the new representative. The uselist is also used to deal with negations. In particular, a negated equality $\neg\ t1 = t2$ is kept on the uselists of its arguments $t1$ and $t2$.

Example. If $\neg\ I(g1,g2) = g1$, then $\neg\ g1 = g2$.

If $g1 = g2$, the equivalence class representative of $g1$ and $g2$ is $U(I(g1,g2),g1,g2)$. Hence, $I(g1,g2)$ is expanded to $I(U(I(g1,g2),g1,g2),U(I(g1,g2),g1,g2))$ which is normalized to $U(I(g1,g2),g1,g2)$. Thus, $I(g1,g2)$ and $g1$ have the same representative, and so they are equal. This is a contradiction because $\neg\ I(g1,g2) = g1$ occurs on the uselists of $g1$ and $I(g1,g2)$.

5   The Algorithm

Our procedure is a refutation procedure. It takes the negation of the formula and puts it into disjunctive normal form. It then determines if the negated formula is satisfiable and, therefore, if the original formula is valid. When an equation is encountered in a formula, or when one is constructed during the processing of the formula, it is pushed onto a stack for subsequent processing along with the other conjuncts. If one conjunct on the stack is unsatisfiable, then the negated formula is unsatisfiable, the original formula is valid, and the procedure terminates without emptying the stack. Otherwise, all the conjuncts on the stack are satisfiable, the original formula is invalid, and the procedure terminates with an empty stack.

If a case split occurs in which all the disjuncts are unsatisfiable, then the negated formula is unsatisfiable and the original formula is valid. If some disjunct is satisfiable, then the negated formula is satisfiable, the original formula is invalid, and the procedure returns immediately. A second stack is used to defer the processing of disjunctions. This is done to reduce the cost of the algorithm, which increases significantly when a disjunction occurs because a new context has to be created for each disjunct.

Each equation and predicate is processed by canonizing its arguments and then calling the graphsolver. The canonizer returns the equivalence class representative of a term. If the term is a nonvariable expression that has not previously been processed, it first computes the signature of the term and adds it to the uselist of its arguments. The newly computed signature then becomes the equivalence class representative. For a graph term, the signature is computed by creating the maximal expanded form and then normalizing it.

When an equation is popped from the stack, the procedure checks whether the two terms are already in the same equivalence class, and thus known to be equal. If not, the two equivalence classes are merged and a new equivalence class representative is formed. If the merged terms are graph terms, the procedure then replaces these terms by their representative in all equivalence class representatives that contain these terms. If the merged terms are vertex or edge terms, the procedure simply replaces the lower-order term by the higher-order term.

The graphsolver determines the equivalence class representatives for vertex, edge, and graph terms. For graph terms, the equivalence class representatives are the maximal normal forms described in Section 3.2. For vertex and edge terms, an equivalence class representative is simply the term of highest order in a class. The graphsolver uses the procedure graphsigma to rewrite terms in normal form. Graphsigma uses a depth-first strategy to rewrite a term in normal form, starting with the arguments of the term. If a term is equal to its own signature, the term itself is returned. The procedure graphsigma consists of mutually recursive procedures for vertex, edge, and graph terms.

The graphsolver accepts an equation or a predicate and replaces the predicate by an equivalent equation, as follows:

$v \in g$ is replaced by $U(VG(v),g) = g$
$e \in g$ is replaced by $U(EG(e),g) = g$
$g1 \subseteq g2$ is replaced by $U(g1,g2) = g2$.

It then compares the signatures of the left and right sides of the equation. If they are identical symbolic forms, it returns the predicate true. Otherwise, depending on the type of term on the left side of the equation, it calls specific solvers for vertex, edge, and graph equations. The graphsolver may have the side-effect of pushing lower-order equations onto the stack.

The solvers for vertex and edge terms, solvevertex and solveedge, solve equations, eliminate variables, and reduce equations to simpler forms, as in [12] for the theory of car, cdr, and cons. In the process of isolating and eliminating edge variables, new vertex variables may be introduced. The vertex and edge solvers use the fact that two edges are equal if and only if their heads and tails are equal. The solver for graph terms consists of create and replace procedures, which expand equivalence class representatives to maximal forms, and reduction procedures, which reduce existing equations to simpler equations that are a consequence of the existing equations. These procedures are described in more detail below.

5.1 Expansion Procedures

In our decision procedure, existing equations are used to expand new terms, and new equations are used to expand existing representatives in the formation of equivalence class representatives. This is done by means of the procedures create and replace. Create builds an equivalence class representative for a new term t. Replace then replaces t by its new representative in each equivalence class representative that contains t.

Example. If $U(g1,g3,g3,g4)$ and $U(g1,g5,g6)$ are on the uselist of $g1$ and the new equation $U(g1,g2,g3) = U(g5,g6)$ is processed, by create we have

$$\text{ring}$$
$$U(g1,g2,g3,g5,g6)$$
$$U(g1,g2,g3) \longleftrightarrow U(g5,g6)$$

Here, $U(g1,g2,g3,g4,g5)$ is the representative for the equivalence class containing $U(g1,g2,g3)$ and $U(g5,g6)$. Then, by replace we have

$$\text{uselist for } g1$$
$$\downarrow$$
$$U(g1,g2,g3,g4) = \quad U(g1,g2,g3,g4,g5,g6)$$
$$\downarrow$$
$$U(g1,g5,g6) \quad = \quad U(g1,g2,g3,g5,g6)$$
$$\downarrow \qquad \text{push these equations}$$

The equation $U(g1,g2,g3,g4) = U(g1,g2,g3,g4,g5,g6)$ is pushed onto the stack, because $U(g1,g2,g3)$ is contained in $U(g1,g2,g3,g4)$ and because the representative of $U(g1,g2,g3)$ is $U(g1,g2,g3,g5,g6)$. Similarly, $U(g1,g5,g6) = U(g1,g2,g3,g5,g6)$ is pushed onto the stack.

5.1.1 Create

Create builds an equivalence class representative for a new term t, as a maximal union of terms in the system that are equivalent to terms contained in t. For a union or intersection t, create first calls itself recursively for terms in the system contained in t and then builds the representative for t itself. Consequently, only the equivalence class representatives of terms contained in t need to be included in the maximal form and not each of the terms in their equivalence classes. If t is an intersection, create expands the arguments of t first and then creates a union of all terms in the system that are equivalent to terms contained in t. If t is an elementary term, create searches equivalence classes for $H(e)$, $T(e)$, and $EG(e)$. If t is $EG(e)$ and $H(e)$ occurs, then $VG(H(e))$ is included in the equivalence class representative. Similarly, if $T(e)$ occurs. If t is $VG(H(e))$ and $EG(e)$ occurs, then $I(EG(e),t)$ is included in the representative. Similarly, if $VG(T(e))$ occurs.

Example. $U(I(g1,g2,g3),I(g1,g2)) = I(g1,g2)$.

Since $I(g1,g2,g3)$ is contained in $I(g1,g2)$, create includes $I(g1,g2,g3)$ in the equivalence class representative of $I(g1,g2)$ to yield $U(I(g1,g2,g3),I(g1,g2))$.

5.1.2 Replace

In our procedure, when the equivalence classes of vertex (or edge) terms $t1$ and $t2$ are merged, the term $t2$ of higher order becomes the new representative, and $t1$ is replaced by $t2$ in all terms on the uselist of $t1$. This is similar to what is done in Shostak's procedure for algebraically solvable theories. On the other hand, when the equivalence class representatives of graph terms $t1$ and $t2$ are merged, our procedure creates a new equivalence class representative t and replaces $t1$ and $t2$ by this new representative in the terms in which they are contained. The new representative is the graph-theoretic union of the representatives of $t1$ and $t2$. In replacing $t1$ $(t2)$ by t, if $t1$ $(t2)$ is a union or elementary term, replace includes t

in the equivalence class representative of each term that contains $t1$ ($t2$). If $t1$ ($t2$) is an intersection, replace introduces for each term u that contains $t1$ ($t2$) a new equation that equates the equivalence class representative of u with the union of t and the representative of u. It then includes t in the representative of each union that has $t1$ ($t2$) as an argument.

Example. If $U(g1,g2,g4) = g5$ and $U(g1,g3,g4) = g6$, then $U(g3,g5) = U(g2,g6)$.

By create we have

$$U(g1,g2,g4,g5) \qquad\qquad U(g1,g3,g4,g6)$$

$$U(g1,g2,g4) \longleftrightarrow g5 \qquad\qquad U(g1,g3,g4) \longleftrightarrow g6$$

Then by replace we have

(a) $U(g1,g2,g3,g4,g5)$ (b) $U(g1,g2,g3,g4,g5,g6)$

$$\downarrow$$

$$U(g3,g5) \qquad\qquad U(g1,g2,g3,g4,g5) \longleftrightarrow U(g3,g5)$$

Since the representative of $g5$ is $U(g1,g2,g4,g5)$ and $g5$ is an argument of $U(g3,g5)$, $U(g1,g2,g4,g5)$ is included in the equivalence class representative of $U(g3,g5)$ which yields $U(g1,g2,g3,g4,g5)$ as in (a). Furthermore, since the representative of $U(g1,g3,g4)$ is $U(g1,g3,g4,g6)$ and $U(g1,g3,g4)$ is contained in $U(g1,g2,g3,g4,g5)$, $U(g1,g3,g4,g6)$ is included in the equivalence class representative of $U(g3,g5)$ which yields $U(g1,g2,g3,g4,g5,g6)$ as in (b).

Again by replace we have

(c) $U(g1,g2,g3,g4,g6)$ (d) $U(g1,g2,g3,g4,g5,g6)$

$$\downarrow$$

$$U(g2,g6) \qquad\qquad U(g1,g2,g3,g4,g6) \longleftrightarrow U(g2,g6)$$

Since the representative of $g6$ is $U(g1,g3,g4,g6)$ and $g6$ is an argument of $U(g2,g6)$, $U(g1,g3,g4,g6)$ is included in the equivalence class representative of $U(g2,g6)$ which yields $U(g1,g2,g3,g4,g6)$ as in (c). Furthermore, since the representative of $U(g1,g2,g4)$ is $U(g1,g2,g4,g5)$ and $U(g1,g2,g4)$ is contained in $U(g1,g2,g3,g4,g6)$, $U(g1,g2,g4,g5)$ is included in the equivalence class representative of $U(g2,g6)$ which yields $U(g1,g2,g3,g4,g5,g6)$ as in (d).

Thus, when $U(g1,g2,g3,g4,g6) = U(g1,g2,g3,g4,g5,g6)$ is popped from the stack, the two equivalence classes are merged, and $U(g3,g5)$ is shown to be equal to $U(g2,g6)$.

As the next example illustrates, the expansion process terminates when the new equivalence class representative is identical to the old one.

Example. $U(g1,g2) = g1$.

By create the equivalence class representative of $g1$ and $U(g1,g2)$ is $U(U(g1,g2),g1)$ which is normalized to $U(g1,g2)$. Since $U(g1,g2)$ is on the uselist of $g1$, it is expanded to $U(U(g1,g2),g1)$ by replace. This term is then normalized to $U(g1,g2)$. Thus, $U(g1,g2)$ is its own representative, and the process terminates.

5.2 Reduction Procedures

The procedure reduce, together with the procedures solvevertex and solveedge, are the reduction procedures. The reduce procedure reduces certain forms of graph equations to equations of lower order which are a consequence of the given equation, or to the predicate false. For example, $EG(f) = EG(e)$ is reduced to $f = e$, and $VG(v) = \phi$ is reduced to false. In some cases, a disjunction of equations results, as in the case of $I(EG(e),VG(u)) = VG(u)$, which is reduced to $H(e) = u \vee T(e) = u$. The processing of such disjunctions is deferred. Here is another example.

Example. If $I(VG(v),VG(u)) = VG(u)$, then $U(VG(v),VG(u)) = VG(u)$.

The reduce procedure reduces $I(VG(v),VG(u)) = VG(u)$ to $v = u$. When these terms are merged, the canonizer replaces u by v in all terms on the uselist of u. Thus, $VG(u)$ becomes $VG(v)$, and $U(VG(v),VG(u))$ becomes $U(VG(v),VG(v))$ which is normalized to $VG(v)$.

Note that in the procedure reduce it does not suffice to equate equivalence class representatives. We must search the equivalence classes and consider pairs of equal terms. However, we don't need to consider each term in the first class against every term in the second class; we need only consider one elementary term in the first class against each term in the second.

6 Correctness

We now give a justification of the correctness of the graphsolver procedure that decides the validity of formulas in graph theory.

Lemma. If $t1$ is a subgraph of $t2$, then the normalized expanded form of $t1$ is contained in the normalized expanded form of $t2$.

Proof. If $t1$ is a subgraph of $t2$, then $t1$ is contained in the expanded form of $t2$. For if $t1$ is a subgraph of $t2$ then, by the definition of containment and the replacement of $g1 \subseteq g2$ by $U(g1,g2) = g2$, some term x in the equivalence class of $t1$ is contained in some term y in the equivalence class of $t2$. In the expansion procedures, all terms x' that are in the system and are equivalent to x are included in the expanded form of y, which is the expanded form of $t2$, since y and $t2$ are in the same class. In particular, $x' = t1$ is contained in the expanded form of $t2$.

Furthermore, the expanded form of $t1$ includes not only all terms equivalent to $t1$, but all terms equivalent to terms contained in $t1$. Such terms are subgraphs of $t2$ and, therefore, by the note above are also contained in the expanded form of $t2$. It follows that the expanded form of $t1$ is contained in the expanded form of $t2$ which implies that the normalized expanded form of $t1$ is contained in the normalized expanded form of $t2$.

Theorem 1. If $t1$ and $t2$ are semantically equal, then the graphsolver procedure computes a unique normal form for $t1$ and $t2$.

Proof. If $t1$ and $t2$ are semantically equal, then $t1$ $(t2)$ is a subgraph of $t2$ $(t1)$. By the previous lemma, the equivalence class representative of $t1$ $(t2)$ is contained in the equivalence class representative of $t2$ $(t1)$. But these terms are contained in each other; hence, they have the same order and are therefore identical.

Theorem 2. The graphsolver procedure terminates.

Proof. In a given formula there are a finite number of equations and predicates. Each predicate is replaced by an equivalent equation. The reduction procedures may add equations to the system, but since these equations are of lower order, this process eventually terminates.

As equations are added to the system, equivalence class representatives are expanded and new terms are generated. But, only finitely many new normal forms can be generated, because these terms are built from a finite number of vertex, edge, and graph variables and because the depth of nesting of function symbols in a term in normal form is at most five. Thus, eventually the normal form of a newly expanded term will be identical to the normal form of the term just expanded, and the expansion process will terminate.

When no graph terms can be further expanded and no vertex or edge variables can be eliminated, the graphsolver procedure terminates, with a maximal normal form for each ring and a minimal number of rings.

7 Efficiency

Because of the complexity of the theory involved, the algorithm is necessarily expensive. If l = the number of vertex variables, m = the number of edge variables, and n = the number of graph variables in the formula, then there are at most $L = l + 2m$ vertex terms, $M = l^2 + m$ edge terms, $N1 = 2^{L+M+n} - (L + M + n + 1)$ intersections, $N2 = 2^{N1+L+M+n} - (N1 + L + M + n + 1)$ unions, and thus

$$N = N1 + N2 + L + M + n + 1 = 2^{2^{l^2+L+3m+n-1}}$$

graph terms, when these are represented in normal form. In the above calculations, 1 was added in for the empty graph ϕ.

The procedure described here eliminates edge variables at the expense of introducing new vertex variables. If $m1$ is the number of edge variables eliminated, then the number of edge variables in the system is $m2 = m - m1$ and the number of vertex variables in the system is $l2 = l + m1$, where l = the number of vertex variables and m = the number of edge variables that occur. Replacing l by $l2$ and

m by $m2$ in the previous formulas, we see that the set of normal forms of vertex, edge, and graph terms is finite (though large). Of course, not all these terms are actually generated by the procedure. If the procedure finds that a conjunction of the negated formula is unsatisfiable (or a disjunction is satisfiable), it terminates immediately without emptying the stack.

Our decision procedure for graph theory is designed to be interfaced with Shostak's procedure for deciding combinations of theories. Thus, the efficiency of our procedure is dependent upon that of Shostak's. While the complexity of his algorithm is unknown, in practice we have found it to be quite efficient. Furthermore, our decision procedure for graph theory is intended to be used in combination with decision procedures for other theories. In practice, relatively simple graph terms are embedded in quite complicated propositional formulas that may also involve terms from other complex theories, such as Presburger arithmetic [8]. Therefore, the cost of our procedure must be considered in relation to the cost of the procedures with which it is combined.

8 Limitations

Interesting theorems in graph theory almost invariably require the use of quantification and functions, neither of which is present in the unquantified theory to which our decision procedure applies. Without the ability to define functions and, in particular, predicates to represent higher-level abstractions, proofs of interesting theorems become very tedious. Of course, graph theory with both quantification and predicates is undecidable. A semi-decision procedure for the more complex theory, based for instance on resolution [10], would have a cost which is much higher than the cost of our procedure and may proceed indefinitely without ever terminating.

Practical mechanical theorem provers depend on extensive interaction with the user, who must necessarily have a good understanding of the proofs to be performed. By use of Skolemization and user-supplied instantiation of existentially quantified variables, proofs in the quantified theory with functions and predicates can be reduced to equivalent proofs of ground formulas, which can be decided by our procedure, together with decision procedures for propositional calculus, Presburger arithmetic, and equality of uninterpreted functions. The user, based on his or her understanding of the proof, is expected to be able to provide the appropriate instantiations of the existentially quantified variables, and will prefer to provide this extra information to the computer in return for a much faster and more predictable decision process.

Acknowledgment. I wish to thank Michael Melliar-Smith and Rob Shostak for helpful discussions concerning this work.

References

[1] A. Ferro, E. G. Omodeo, and J. T. Schwartz, "Decision procedures for elementary sublanguages of set theory. I. Multilevel syllogistic and some extensions," *Comm. Pure App. Math.* 33 (1980), pp. 599-608.

[2] A. Ferro, E. G. Omodeo, and J. T. Schwartz, "Decision procedures for some fragments of set theory," in: *Proc. 5th Conf. on Automated Deduction*, LNCS 87, (Springer-Verlag, Berlin-Heidelberg-New York, 1980), pp. 88-96.

[3] J. Hsiang, "Topics in automated theorem proving and program generation," *Artificial Intelligence* 25 (1985), pp. 255-300.

[4] J. C. C. McKinsey, "The decision problem for some classes of sentences without quantifiers," *J. Symb. Logic* 8, 3 (1943), pp. 61-76.

[5] L. E. Moser, "Decidability of formulas in graph theory," conditionally accepted by *Fundamenta Informaticae*.

[6] G. Nelson and D. C. Oppen, "Simplification by cooperating decision procedures," *ACM Trans. Prog. Lang. and Syst.* 1, 2 (1979), pp. 245-257.

[7] G. Nelson and D. C. Oppen, "Fast decision procedures based on congruence closure," *J. ACM* 27, 2 (1980), pp. 356-364.

[8] M. O. Rabin and M. J. Fisher, "Super-exponential complexity of theorem proving procedures," in: *Proc. AMS SIAM Symp. Appl. Math.*, (AMS, Providence, RI, 1973), pp. 27-42.

[9] J. C. Raoult, "On graph rewritings," *Theor. Comp. Sci.* 32 (1984), pp. 1-24.

[10] J. A. Robinson, "A machine oriented logic based on the resolution principle," *J. ACM* 12 (1965), pp. 23-41.

[11] R. E. Shostak, "An algorithm for reasoning about equality," *Comm. ACM*, 21, 7 (1978), pp. 583-585.

[12] R. E. Shostak, "Deciding combinations of theories," *J. ACM* 31, 1 (1984), pp. 1-12.

[13] R. E. Shostak, R. Schwartz, and P. M. Melliar-Smith, "STP: A mechanized logic for specification and verification," in: *Proc. 6th Conf. on Automated Deduction*, LNCS 138, (Springer-Verlag, Berlin-Heidelberg-New York, 1982), pp. 32-49.

[14] R. E. Tarjan, "Efficiency of a good but not linear set union algorithm," *J. ACM* 22, 2 (1975), pp. 215-225.

Adventures in Associative-Commutative Unification
(A Summary)

Patrick Lincoln and Jim Christian
MCC, Systems Languages
3500 W. Balcones Cntr. Dr.
Austin, TX, 78759
(512) 338-3717

Abstract

We have discovered an efficient algorithm for matching and unification in associative-commutative (AC) and associative-commutative-idempotent (ACI) equational theories. In most cases of AC unification and in all cases of ACI unification our method obviates the need for solving diophantine equations, and thus avoids one of the bottlenecks of other associative-commutative unification techniques. The algorithm efficiently utilizes powerful constraints to eliminate much of the search involved in generating valid substitutions. Moreover, it is able to generate solutions lazily, enabling its use in an SLD-resolution-based environment like Prolog. We have found the method to run much faster and use less space than other associative-commutative unification procedures on many commonly encountered AC problems.

1 Introduction

A number of computer science applications, including term rewriting, automatic theorem proving, software verification, and database retrieval require AC unification (also called "bag unification") or ACI unification ("set unification"). A complete unification algorithm for AC theories was developed several years ago by Mark Stickel [16]. Independently, Livesey and Siekmann published a similar algorithm for AC and ACI unification. Their procedures center around generating solutions to a linear diophantine equation, each coefficient of which represents the multiplicity of some subterm in one of the unificands. There are two nagging properties of this method. First, it requires generating a basis for the solution space of the diophantine equation. Second, there can be a large amount of search involved in actually generating solutions once a basis is discovered.

We have found an algorithm for dealing with associative-commutative theories without resorting to the solution of diophantine equations. By weakening the variable abstraction introduced by Stickel, most cases of AC and all cases of ACI unification can be solved by working only with equations in which all coefficients are unity. The basis of solutions of such an equation possesses a highly regular structure; this allows us to optimize the representation of the problem and avoid spending time finding a basis. We are able instead to begin generating unifiers almost immediately. In addition, our representation allows the incorporation of several simple but powerful constraints in a way that is much more natural and efficient than previous methods have allowed.

Our algorithm can solve AC matching problems (so-called "one-way unification"), most cases of AC unification, and all cases of ACI unification very efficiently – in most cases, several times faster than Stickel's algorithm. However, if repeated variables occur in one unificand, our algorithm may return redundant unifiers. If repeated variables occur in both unificands, our algorithm may not terminate. In these easily detected cases, it suffices to dispatch to some complete algorithm, like Stickel's; the overhead in making the decision to dispatch is negligible. Fortunately, these cases occur in only a few percent of many applications of AC unification like Knuth-Bendix completion [12]. In some applications like database query compilation, these cases never occur. Thus our procedure can achieve a significant improvement in average execution speed of AC unification. Furthermore, our procedure requires nominal space overhead for generating solutions, and is amenable to the lazy generation of solutions required in an SLD-resolution environment like that of Prolog. We give a summary of our full paper [2] in the text that follows.

2   History Of AC Unification

Mark Stickel was the first to develop a complete, terminating algorithm for AC unification; the algorithm was initially presented in 1975 [15]. Livesey and Siekmann published a similar algorithm in 1976 [13]. Most AC unification procedures in use today are essentially modifications of that of Stickel or of Livesey and Siekmann, but a few novel approaches have been proposed. Within the loose framework of Stickel's method there are two hard problems: generating a basis of solutions to linear homogeneous diophantine equations, and searching through all combinations of this basis for a solution to the given AC unification problem.

Since Gordan's study of diophantine equations in 1873, only in the last few years has there been any significant progress made regarding the generation of their bases. Fortenbacher, Huet, and Lankford have separately proposed a number of refinements to Gordan's basic method. Recently, Zhang has discovered a class of diophantine equations which can be quickly solved. However, no published algorithm has proven to be superior in all cases. [5,4,7,12,6,18].

The extraction of solutions to AC problems given a basis of solutions to the diophantine equation is also an area of concern. In the past few years Fortenbacher [4] has proposed a method of reducing the search space by eliminating certain basis elements. Claude Kirchner has recently developed an AC unification algorithm within the framework of the Martelli-Montanari unification procedure, but, like Stickel's, his method requires solving diophantine equations [11,14]. Also, Hullot invented an algorithm for AC unification which involves ordered partitions of multisets [9]. While his algorithm is faster than Stickel's, it does not seem to offer nearly the dramatic speed increases we have obtained with our procedure. We have not implemented Hullot's algorithm, but base our judgement on timing comparisons listed in his paper. In Germany, Büttner has developed a parallel algorithm for AC unification [1]. The method involves linear algebraic operations in multi-dimensional vector spaces, but he fails to provide the details necessary for a realistic comparison. Recently,

Kapur [10] has developed an algorithm based on Stickel's method that uses Zhang's equation solving technique. The survey by Huet and Oppen [8] summarizes results in related areas.

3 Our Method

There are two basic difficulties with previous algorithms. First, generation of a basis for a diophantine equation is an expensive operation. Second, given a basis, the search which must be performed to produce solutions can be very expensive. It is thus necessary to enforce several non-trivial constraints [4,17]. Fortenbacher [4] has described many obvious optimizations of Stickel's method, and Stickel himself has implemented quite impressive constraints on the generation of solutions which tame this search problem. As we shall soon see, our algorithm is able to exploit similar constraints in a very natural and efficient way.

3.1 Slaying the Diophantine Dragon

The preparation phase of our algorithm is very similar to previous approaches. First, both terms are put through a "flattening" operation which removes nested AC function symbols. This operation can be viewed more precisely as term reduction by root application of the rewrite rule $f(t_1, \ldots, f(s_1, \ldots, s_m), \ldots, t_n) \longrightarrow f(t_1, \ldots, s_1, \ldots, s_m, \ldots, t_n)$. Hence, the term $f(f(a, a), a, f(g(u), y, x))$ will be flattened to $f(a, a, a, g(u), y, x)$, while $f(a, f(b, g(c)), f(y, y), z)$ will be changed to $f(a, b, g(c), y, y, z)$. The flattened term $f(s_1, s_2, \ldots, s_n)$ is merely syntactic sugar for the right-associative normal form $f(s_1, f(s_2, \ldots, f(s_{n-1}, s_n) \ldots))$. The validity of the flattening operation is guaranteed by the associativity axiom, which implies that nesting of AC function symbols is largely irrelevant.

After flattening terms, the next step is to remove subterms which occur pairwise in both unificands. For instance, after deleting duplicate subterms from $f(a, a, a, g(u), y, x)$ and $f(a, b, g(c), y, y, z)$ we obtain the terms $f(a, a, g(u), x)$ and $f(b, g(c), y, z)$, specifically by removing one occurrence of a and one occurrence of y from each.

We suppose that both terms are sorted so that atomic constants are grouped together, followed by function terms, followed by variables. For instance, $f(a, g(u), a, y, a, x)$ would be sorted to produce $f(a, a, a, g(u), y, x)$. An important non-intuitive point is that compared to the time required to generate all unifiers, the time spent flattening and sorting terms is insignificant.

In the ACI case, repeated constants, terms, and variables are removed from each unificand. For example, $f(a, a, b, x, x)$ is simplified to $f(a, b, x)$. In the AC case, this step is omitted.

Now, our generalization step differs from others, in that we assign a distinct variable for *each* argument. Thus, while Stickel's algorithm would convert the $f(a, a, g, x)$ to $f(x_1, x_1, x_2, x_3)$, ours will produce $f(x_1, x_2, x_3, x_4)$. Effectively, we convert the problem of solving the unification problem $f(X_1, \ldots, X_m) = f(Y_1, \ldots, Y_n)$ into the equivalent conjunction of problems $f(x_1, \ldots, x_m) = f(y_1, \ldots, y_n) \wedge x_1 = X_1 \wedge \ldots \wedge x_m = X_m \wedge y_1 = Y_1 \wedge \ldots \wedge y_n = Y_n$, where the x_i and y_j are distinct variables.

		a	a	$g(u)$	x
		x_1	x_2	x_3	x_4
b	y_1	$z_{1,1}$	$z_{1,2}$	$z_{1,3}$	$z_{1,4}$
$g(c)$	y_2	$z_{2,1}$	$z_{2,2}$	$z_{2,3}$	$z_{2,4}$
y	y_3	$z_{3,1}$	$z_{3,2}$	$z_{3,3}$	$z_{3,4}$
z	y_4	$z_{4,1}$	$z_{4,2}$	$z_{4,3}$	$z_{4,4}$

	C	T	V
C	0	0	$\leftrightarrow$
T	0	$\oplus$	$\leftrightarrow$
V	$\updownarrow$	$\updownarrow$	any

Table 1: Matrix for a simple problem and some constraints

Notice that the diophantine equation corresponding to any pair of such generalized terms will have only unit coefficients. Such an equation has a few nice properties, namely given a diophantine equation of the form $x_1 + \ldots + x_m = y_1 + \ldots + y_n$, the minimal solution basis is that set of solutions such that, for each solution, exactly one x_i has value one, exactly one y_j has value one, and all other variables have value zero. Also, the number of basis solutions is nm.

Knowing that the basis has such a nice, regular structure, we need not explicitly generate it; for, given only the respective arities of the generalized unificands, we can immediately construct a two dimensional matrix, where each column is labeled with an x_i, and each row is labeled with one of the y_j. Each entry i,j in the matrix is a boolean value, that corresponds to a new variable, $z_{i,j}$, which represents the solution vector which assigns a one to x_j and y_i. Thus every true boolean value i,j in a solution matrix corresponds to one basis element of the solution of the diophantine equation. Any assignment of true and false to all the elements of a matrix represents a potential solution to the AC unification problem in the same way that any subset of the basis elements of the diophantine equation represents a potential solution to the same AC problem.

For instance, suppose we are given the (already flattened) unificands $f(a, a, g(u), x)$ and $f(b, g(c), y, z)$. Substituting new variables for each argument, we obtain $f(x_1, x_2, x_3, x_4)$ and $f(y_1, y_2, y_3, y_4)$. The associated solution matrix is displayed in Table 1

In our implementation, we do not create the entire n by m matrix; rather, we will utilize a more convenient and compact data structure. But for now, let us pretend that the matrix is represented as a simple 2-dimensional array. And as we will demonstrate below, the matrix representation is inherently amenable to constraining the search for unifiers.

3.2 Constraining Search

Remember that unificands are sorted in the preparation step of our algorithm. Hence, a given solution matrix comprises nine regions, illustrated in Table 1. In the table, C, T, and V stand, respectively, for atomic constants, functional terms, and variables. An entry in the lower left region of the matrix, for instance, corresponds to an assignment in the (unprepared) unificands of a constant in one and a variable in the other.

As Table 1 indicates, there are several constraints on the distribution of ones and zeros within a solution matrix. First, notice that there must be at least one non-zero entry in each row and column of a solution matrix, so that all variables in the generalized terms receive

		a	a	$g(u)$	x
		x_1	x_2	x_3	x_4
b	y_1	0	0	0	1
$g(c)$	y_2	0	0	1	0
y	y_3	0	0	0	1
z	y_4	1	1	0	0

Unifying substitution:

$$x \leftarrow f(b, y)$$
$$z \leftarrow f(a, a)$$
$$u \leftarrow c$$

Table 2: A solution to the matrix

an assignment. The upper left corner involves assignments to incompatible constants (since we have removed duplicate arguments from the unificands, no constants from one term can possibly unify with any constant from the other term). This part of any solution matrix, then, must consist only of zeros. Similarly, the C/T and T/C regions of a solution matrix must contain all zeros. The C/V region is constrained to have exactly a single one in each column, since any additional ones would cause the attempted unification of a functional term, say $f(z_{1,1}, z_{1,2})$, with a constant. Similarly, any T row or T column must contain exactly one one. Finally, the V/V region of a matrix can have any combination of ones and zeros which does not leave a whole row or column filled only with zeros.

In reality, the nine regions depicted in Table 1 are further partitioned to handle repeated variables and constants; for now, we will ignore this detail, and assume that all arguments within a unificand are distinct.

3.3   Generating Solutions

Once a unification problem has been cast into our matrix representation, it is not a difficult matter to find unifying substitutions. The approach is to determine a valid configuration of ones and zeros within the matrix, perform the indicated assignments to the variables in the generalized terms, and finally unify the arguments of the original unificands with their variable generalizations.

Consider the matrix in Table 1. We know that $z_{1,1}$ must be zero, since it falls within the C/C region of the matrix. Likewise, $z_{1,2}$, $z_{1,3}$, $z_{2,1}$, and $z_{2,2}$ must always be zero. In fact, the only possible position for the required one in the y_1 column is at $z_{1,4}$. Filling out the rest of the matrix, we arrive at the solution shown in Table 2; after assigning the nonzero $z_{i,j}$'s to the x and y variables, and then unifying the variables with the original unificand arguments, we obtain the substitution shown at the side of Table 2. Note that the step at which $g(u) \leftarrow g(c)$ was derived, and recursively solved to produce $u \leftarrow c$ has been omitted.

3.4   Lazy generation of solutions

Enumerating solutions essentially amounts to performing simple binary operations on regions of the matrix. For instance, in the variable-constant region of the martix, a binary rotate instruction and a few numeric comparisons usually suffice to generate the next solution from the current state.

	a	a
x	1	0
y	0	1

	a	a
x	0	1
y	1	0

Table 3: Redundant matrix configurations for $f(a, a) = f(x, y)$

3.5 Repeated terms

Until now, we have assumed that all arguments within a unificand are distinct. This will always be true for ACI unification, since the repeated terms are removed during preparation. However, this is not necessarily the case for AC unification. In practice, repeated terms occur infrequently; Lankford, for instance, has found that more than 90 percent of the unification problems encountered in some completion applications involve unificands with distinct arguments. Nevertheless, the ability to handle repeated arguments is certainly desirable.

Our algorithm can easily be adapted to handle repetitions in constants and functional terms in either or both unificands. Repeated variables are more difficult to handle. If they occur in a single unificand, our algorithm is complete and terminating, but may return redundant unifiers. If they occur in both unificands, the algorithm might generate subproblems at least as hard as the original, and thus not terminate. Stickel's algorithm can be employed whenever repeated variables are detected; the overhead involved in making this decision is negligible. Thus in the worst cases we do simple argument checking, and dispatch to Stickel's algorithm. We have several methods of minimizing the use of Stickel's algorithm but we have not yet discovered a straightforward, general method. In [2] we prove that our procedure does indeed terminate whenever variables are repeated in at most one of the unificands.

The set of unifiers returned by our algorithm is guaranteed to be complete. However, the set of unifiers may not be minimal. The set of unifiers returned by Stickel's algorithm is similarly not guaranteed to be minimal. If a minimal set of unifiers is required, it suffices to simply remove elements of the non-minimal set which are subsumed by other unifiers in the set.

Assuming no repeated variables in one term, our algorithm can handle arbitrary repetitions of constants and functional terms. But before disclosing the modification to our algorithm which facilitates handling of repeated arguments, we show with a simple example why the modification is needed. Suppose we wish to unify $f(a, a)$ with $f(x, y)$, which is a subproblem of the earlier example. Without alteration, our algorithm as so far stated will generate the two configurations shown in Table 3. While the matrix configurations are distinct, they represent identical unifying substitutions – namely $\{x \leftarrow a, y \leftarrow a\}$.

The solution to this problem is surprisingly simple. In short, whenever adjacent rows represent the same term, we require that the contents of the upper row, interpreted as a binary number, be greater than or equal to the contents of the lower row. A symmetric restriction is imposed on columns. See the full paper for a detailed explanation of these restrictions.

3.6   An Algorithm for Associative-Commutative Unification

Until now, we have concentrated almost exclusively on the matrix solution technique which lies at the heart of our AC unification algorithm. Following is a statement of the unification algorithm proper. This will serve, in the next section, as a basis for results involving the completeness and termination of our method. The algorithm is presented as four procedures: `AC-Unify`, `Unify-With-Set`, `Unify-Conjunction`, and `Matrix-Solve`.

Procedure `AC-Unify`: Given two terms x and y, return a complete set of unifiers for the equation $x =_{AC} y$.

Step 1   If x is a variable, then see if y is a functional term and x occurs in y. If both are true, return `fail`. Otherwise, return $\{\{x \leftarrow y\}\}$, unless $x = y$ — in that case, return the null substitution set $\{\{\}\}$.

Step 2   If y is a variable, then see if y occurs in x. If it does, return `fail`. Otherwise, return $\{\{y \leftarrow x\}\}$.

Step 3   If x and y are distinct constants, return `fail`.

Step 4   If x and y are the same constant, return $\{\{\}\}$.

Step 5   At this point, x and y are terms of the form $f(x_1, \ldots, x_m)$ and $g(y_1, \ldots, y_n)$. If $f \neq g$, return `fail`.

Step 6   If f is not an AC function symbol, and $m = n$, then call procedure `Unify-With-Set` with the substitution set $\{\{\}\}$ and the conjunction of equations $x_1 =_{AC} y_1 \wedge \ldots \wedge x_n =_{AC} y_n$, and return the result. If $m \neq n$, return `fail`.

Step 7   Flatten and sort x and y, and remove arguments common to both terms. Call the resulting terms $\hat{x}$ and $\hat{y}$, respectively. Assume $\hat{x} = f(x_1, \ldots, x_j)$ and $\hat{y} = f(y_1, \ldots, y_k)$. Set up the conjunction of equations $f(X_1, \ldots, X_j) =_{AC} f(Y_1, \ldots, Y_k) \wedge X_1 =_{AC} x_1 \wedge \ldots \wedge X_j =_{AC} x_j \wedge Y_1 =_{AC} y_1 \wedge \ldots \wedge Y_k =_{AC} y_k$, where the X_i and Y_i are new, distinct variables. Call this conjunction E.

Step 8   Let T be the result of applying `Matrix-Solve` to the conjunction E. If $T = $ `fail`, return `fail`.

Step 9   Call procedure `Unify-With-Set` with the set of substitutions T and the conjunction of equations $X_1 =_{AC} x_1 \wedge \ldots \wedge X_j =_{AC} x_j \wedge Y_1 =_{AC} y_1 \wedge \ldots \wedge Y_k =_{AC} y_k$, and return the result.

Procedure `Unify-With-Set`: Given a set of substitutions T and a conjunction of equations E, return $\bigcup_{\theta \in T} CSU(\theta E)$, where $CSU(X)$ is a complete set of unifiers for X.

Step 1   Let $S = \{\}$.

Step 2   For each $\theta \in T$, set S to $S \cup \{\bigcup_{\sigma_j \in Z} \{\theta \cup \sigma_j\}\}$, where Z is the result of applying procedure `Unify-Conjunction` to θE.

Step 3   Return S.

Procedure `Unify-Conjunction` Given a conjunction of equations $E = e_1 \wedge \ldots \wedge e_n$, return a complete set of unifiers for E.

Step 1   . Let V be the result of calling procedure `AC-Unify` with e_1. If $n = 1$, return V. If $V = $ `fail`, return `fail`.

Step 2   . Call procedure `Unify-With-Set` with the set of substitutions V and the conjunction $e_2 \wedge \ldots \wedge e_n$, and return the result.

Procedure Matrix-Solve Given a conjunction of equations $f(X_1, \ldots, X_m) =_{AC} f(Y_1, \ldots, Y_n) \wedge X_1 =_{AC} x_1 \wedge \ldots \wedge X_m =_{AC} x_m \wedge Y_1 =_{AC} y_1 \wedge \ldots \wedge Y_n =_{AC} y_n$, where the X_i and Y_i are distinct variables, determine a set of substitutions which will unify $f(X_1, \ldots, X_m)$ with $f(Y_1, \ldots, Y_n)$.

Step 1 Establish an m-by-n matrix M where row i (respectively column j) is headed by X_i (Y_j).

Step 2 Generate an assignment of 1s and 0s to the matrix, subject to the following constraints. If x_i (y_j) is a constant or functional term, then exactly a single 1 must occur in row i (column j). If x_i and y_j are both constants, or if one is a constant and the other is a functional term, then $M[i, j] = 0$. Also, there must be at least a single 1 in each row and column. Finally, if $x_i = x_{i+1}$ for some i, then row i interpreted as a binary number must be less than or equal to row $i + 1$ viewed as a binary number. (Symmetrically for y_j and y_{j+1}.)

Step 3 With each entry $M[i, j]$, associate a new variable $z_{i,j}$. For each row i (column j) construct the substitution $X_i \leftarrow f(z_{i,j_1}, \ldots, z_{i,j_k})$ where $M[i, j_l] = 1$, or $X_i \leftarrow z_{i,j_k}$ if $k = 1$. (symmetrically for Y_j).

Step 4 Repeat Step 2 and Step 3 until all possible assignments have been generated, recording each new substitution. If there is no valid assignment, return **fail**.

Step 5 Return the accumulated set of substitutions.

When there are repeated variables in both unificands, it is possible that our algorithm will not terminate. For example, in the unification of $f(x, x)$ with $f(y, y)$ one of the recursive subproblems generated is identical (up to variable renaming) to the original problem. However, as we prove in the full version of this paper [2] our algorithm is totally correct in other cases.

4 Benchmarks

Although we have not presented the details of our actual implementation, it should be obvious that more efficient data structures exist than an n by m boolean matrix. In particular, finding the next matrix configuration often reduces to one lisp *incf* and a few comparisons in our most optimized code. In order to quickly generate a unifier from a matrix configuration we utilize some auxiliary data structures.

The table on the next page reflects the time in seconds necessary to prepare unificands and to find and construct all AC unifiers. For each problem, timings were supplied by Kapur and Zhang (RRL), Stickel (SRI), and by ourselves (MCC). All data were collected on a Symbolics 3600 with IFU. As shown in the table, our algorithm is consistently three to five times faster than Stickel's and Kapur's.

These benchmarks do not include any problems with repeated variables, since in such cases, our algorithm would either return non-minimal sets of unifiers, or it would dispatch to Stickel's procedure. This is not as serious a concession as it might appear, since the most common cases of AC Unification are the ones without repeated variables. In fact, Lankford has found that less than 8 percent of uses of AC unification in applications like Knuth-Bendix completion have repetitions of anything, and less than three percent have repetitions on both sides [12]. Also, in the case of ACI, variables are never repeated.

Problem	♯	RRL	SRI	MCC
$xab = ucde$	2	0.020	0.018	0.005
$xab = uccd$	2	0.023	0.011	0.005
$xab = uccc$	2	0.018	0.008	0.004
$xab = uvcd$	12	0.045	0.047	0.013
$xab = uvcc$	12	0.055	0.032	0.014
$xab = uvwc$	30	0.113	0.096	0.034
$xab = uvwt$	56	0.202	0.171	0.079
$xaa = ucde$	2	0.028	0.013	0.005
$xaa = uccd$	2	0.023	0.009	0.004
$xaa = uccc$	2	0.021	0.006	0.005
$xaa = uvcd$	8	0.043	0.032	0.010
$xaa = uvcc$	8	0.035	0.020	0.011
$xaa = uvwc$	18	0.087	0.062	0.023
$xaa = uvwt$	32	0.192	0.114	0.051

Problem	♯ solns	RRL	SRI	MCC
$xya = ucde$	28	0.093	0.094	0.024
$xya = uccd$	20	0.068	0.050	0.018
$xya = uccc$	12	0.045	0.026	0.013
$xya = uvcd$	88	0.238	0.247	0.064
$xya = uvcc$	64	0.211	0.133	0.048
$xya = uvwc$	204	0.535	0.538	0.160
$xya = uvwt$	416	0.918	1.046	0.402
$xyz = ucde$	120	0.375	0.320	0.118
$xyz = uccd$	75	0.185	0.168	0.072
$xyz = uccc$	37	0.093	0.073	0.038
$xyz = uvcd$	336	0.832	0.840	0.269
$xyz = uvcc$	216	0.498	0.431	0.171
$xyz = uvwc$	870	2.050	2.102	0.729
$xyz = uvwt$	2161	5.183	5.030	1.994

5 Future Extensions

With simple modifications, our algorithm can apparently handle arbitrary combinations of associativity, commutativity, identity, and idempotence. We say "apparently" because we have not yet proven completeness or termination in all these cases, but preliminary findings have been encouraging. Of particular interest is unification with the single axiom of associativity, which corresponds to the word problem in free semigroups. The matrix representation of unifiers seems well suited to capturing this limited form of equational theory, but the details of this are beyond the scope of this report.

6 Conclusion

We have just described an algorithm which we believe to be the most efficient way of solving a large class of associative-commutative matching and unification problems. The algorithm obviates the need for solving diophantine equations, and it utilizes a matrix representation which conveniently enforces powerful search constraints. Compared to Stickel's and Kapur's procedures, our method often yields a significant improvement in speed. Certainly, applications of AC unification stand to benefit from our research.

We would like to thank Dallas Lankford for introducing us to his diophantine basis generation algorithm, and for supplying us with pointers to some useful information. We would also like to thank Hassan Aït-Kaci, Mike Ballantyne, Woody Bledsoe, Bob Boyer, and Roger Nasr for their comments, criticisms, and *laissez-faire* supervision. Finally, we would like to thank Mark Stickel, Hantao Zhang, and Deepak Kapur, for their insightful criticisms of an earlier draft of this paper, and for supplying benchmark times.

References

[1] Wolfram Büttner. "Unification in Datastructure Multisets". *Journal of Automated Reasoning*, 2 (1986) 75-88.

[2] Jim Christian and Pat Lincoln "Adventures in Associative-Commutative Unification" MCC Technical Report Number ACA-ST-275-87, Microelectronics and Computer Technology Corp., Austin, TX, Oct 1987.

[3] François Fages. "Associative-Commutative Unification". *Proceedings 7th International Conference on Automated Deduction*, Springer–Verlag. Lecture Notes in Computer Science, Napa Valley, (California), 1984.

[4] Albrecht Fortenbacher. "An Algebraic Approach to Unification Under Associativity and Commutativity" Rewriting Techniques and Applications, Dijon, France, May 1985, ed Jean-Pierre Jouannaud. Springer-Verlag Lecture Notes in Computer Science Vol. 202, (1985) pp. 381-397

[5] P. Gordan, "Ueber die Auflösung linearer Gleichungen mit reelen Coefficienten". *Mathematische Annalen*, VI Band, 1 Heft (1873), 23-28.

[6] Thomas Guckenbiehl and Alexander Herold. "Solving Linear Diophantine Equations". Universitat Kaiserslautern, Fachbereich Informatik, Postfach 3049, 6750 Kaiserslautern.

[7] Gérard Huet. "An Algorithm to Generate the Basis of Solutions to Homogeneous Linear Diophantine Equations". IRIA Research Report No. 274, January 1978.

[8] Gérard Huet and D.C.Oppen. "Equations and Rewrite Rules: a Survey". In Formal Languages: Perspectives and Open Problems, ed R. Book, Academic Press, 1980.

[9] J.M. Hullot. "Associative Commutative Pattern Matching". *Proceedings IJCAI-79*, Volume One, pp406-412, Tokyo, August 1979.

[10] Deepak Kapur, G. Sivakumar, H. Zhang. "RRL: A Rewrite Rule Laboratory". Proceedings of CADE-8, pp 691-692, Oxford, England, 1986.

[11] Claude Kirchner. "Methods and Tools for Equational Unification". in *Proceedings of the Colloquium on the Resolution of Equations in Algebraic Structures*, May 1987, Austin, Texas.

[12] Dallas Lankford. "New Non-negative Integer Basis Algorithms for Linear Equations with Integer Coefficients". May 1987. Unpublished. Available from the author, 903 Sherwood Drive, Ruston, LA 71270.

[13] M. Livesey and J. Siekmann. "Unification of A + C-terms (bags) and A + C + I-terms (sets)". Intern. Ber. Nr. 5/76, Institut für Informatik I, Unifersität Karsruhe, 1976.

[14] A. Martelli and U. Montanari. "An Efficient Unification Algorithm". *ACM Transactions on Programming Languages and Systems*, 4(2):258-282, 1982.

[15] Mark Stickel. "A complete unification algorithm for associative-commutative functions" *Proc. 4th IJCAI*, Tbilisi (1975), pp.71-82.

[16] Mark Stickel. "A Unification Algorithm for Associative-Commutative Functions". *JACM*, Vol.28, No.3, July 1981, pp.423-434.

[17] Mark Stickel. "A Comparison of the Variable-Abstraction and Constant-Abstraction methods for Associative-Commutative Unification" *Journal of Automated Reasoning*, Sept 1987, pp.285-289.

[18] Hantao Zhang "An Efficient Algorithm for Simple Diophantine Equations", Tech. Rep. 87-26, Dept. of Computer Science, RPI, 1987.

Unification in Finite Algebras is Unitary (?)

Wolfram Büttner

Siemens AG

Corporate Laboratories for

Information Technology

Otto-Hahn-Ring 6

D-8000 München 83

West Germany

Abstract: *Unification in algebras is widely seen as a means of improving the expressiveness and efficiency of resolution based deduction. In particular, finite algebras have recently gained considerable attention. Unification in the sense of Plotkin is based upon the notion of a set of most general unifiers, which however might either not exist or - at the other extreme - be too large. As a remedy to these conceptual drawbacks we suggest a redefinition of unification, such that with respect to this view unification in finite algebras becomes unitary. Previous work on unification in Postalgebras provides a universal unification algorithm. We will add a more application oriented approach. Applications of our methods to the switch level design of digital circuits are indicated.*

1. Introduction

The way numerous domains of interest are built up is captured adequately in the abstract notion of an algebra, say **A**, being a set A along with a set F of finitary operators. Terms over **A** are used to represent sets of objects in some domain (modeled by **A**) which share certain features.

Terms over **A** are functions defined by means of a 1-1 correspondence between F and the function symbols of a suitable termalgebra as follows:

Interpreting the function symbols occurring in a term as the corresponding operators in F, the term specifies a finitary function whose arguments and values are in A (term function).

A term t over **A** can be seen as a partially specified object in the domain of interest. Instantiating the arguments in t yields all objects satisfying this partial specification.

There is of course the problem of determining those objects which satisfy two partial specifications encoded via term functions t_1, t_2. And it is natural to ask whether the set L (t_1, t_2) of arguments (objects) on which t_1 and t_2 agree can be represented again by a term function.

This poses a functional and an algebraic question, namely to find a function with predescribed image L (t_1, t_2) and to realize this function by a term function.

Experience has shown, that this can only rarely be achieved. Separating functional and algebraic aspects provides a deeper understanding of this phenomenon and suggests a new approach to overcome these difficulties.

The paper is quite selfcontained. A certain familiarity with algebraic reasoning is assumed. As a source for "traditional" unification theory we refer to J. Siekmanns overview of the subject [Siek 86].

2. Equation solving in algebras - back to the roots

The description of point sets in $\mathbb{R}^n$ as images of mappings is well known in mathematics as an economic principle.

The unit circle, E, in $\mathbb{R}^2$ is given as image of the mapping $\delta = (\delta_1, \delta_2) : \mathbb{R} \to \mathbb{R}^2$, where $\delta_1 = \sin \phi, \delta_1 = \cos \phi$.

We say **E is parameterized** by $\mathbb{R}$. Another description of E is provided as **set of solutions of the equation** $x^2 + y^2 = 1$ or as the set of **all pairs, where the functions** $x^2 + y^2$ **and 1 agree.** The idea of parameterizing point sets which arise this way, remains meaningful in more general situations:

Let A be a set and let f, g: $A^n \to A$ be functions such that L (f, g), the set of all arguments on which f and g agree, is nonempty. A parametrization of a subset of L (f, g) is given by a mapping $\delta : A^r \to A^n$, whose image, im δ, is a subset of L (f, g). Hence $\delta = (\delta_1, \ldots, \delta_n)$ with $\delta_i : A^r \to A$.

Since $f (\delta_1 (x_1, \ldots, x_r), \ldots, \delta_n (x_1, \ldots, x_r)) = g (\delta_1 (x_1, \ldots, x_r), \ldots, \delta_n (x_1, \ldots, x_r))$, we call δ **a unifier** of f and g and abbreviate by $U\Sigma$ **(f, g)** the set of all unifiers of f and g.

Let $\delta : A^r \to A^n$ and $\tau : A^s \to A^n$ be unifiers of f and g.

We say δ is more special than τ, if there exists a mapping $\zeta : A^r \to A^s$ such that $\delta = \tau (\zeta_1, \ldots, \zeta_s)$ with $\zeta_i : A^r \to A^s$.

A necessary and sufficient condition for δ to be **more special** than τ is given by im $\delta \subseteq$ im τ. δ and δ' are **said to be equivalent** ($\delta \sim \delta'$) if im δ = im δ'. The "more special" relation induces on $\overline{U\Sigma(f, g)}$: = $U\Sigma(f, g)_{/\sim}$ a **partial order** $\leqq$ given by $\delta \leqq \tau$ iff im $\delta \subseteq$ im τ. Hence the subsets of L (f, g) serve as representatives for the equivalence classes in UΣ (f, g) and thus the lattice defined by $\leqq$ on UΣ (f, g) is isomorphic to the boolean lattice $\mathbb{P}$ (L (f, g)) of subsets of L (f, g).

A unifier δ with im δ = L (f, g) is "more general" (the dual of "more special") than every other unifier of f and g and therefore called a most general unifier. Any two most general unifiers are equivalent, induce the top element in the lattice (UΣ (f, g), $\leqq$) and provide parametrizations of L (f, g).

The unification theory outlined so far did not assume any algebraic structure on A.

Let now F be a set of finitary functions on A, turning A into an algebra **A**: = (A, F). We fix some element in A and call it 0. This algebraic structure is inherited by pointwise definition of operators to the set F_A of all finitary functions with arguments and values in A. Let $\mathbf{F_A}$ be the resulting algebra which contains T_A, the set of terms over **A**, as a subalgebra. Furthermore, $\mathbf{A^{A^n}}$ denotes the subalgebra of all (at most) n-ary functions in $\mathbf{F_A}$. Every mapping δ: A$^r \to$ A^n extends to a homomorphism $\hat{\delta}$ $\mathbf{A^{A^n}} \to \mathbf{A^{A^r}}$ by $(\hat{\delta} (f)) (x_1, \ldots, x_n)$: = $f(\delta_1, \ldots, \delta_n)$ where f $\in$ A$^{A^n}$ and $\delta = (\delta_1, \ldots, \delta_n)$ with δ_i : A$^r \to$ A .

Conversely, every homomorphism from $\mathbf{A^{A^n}}$ to $\mathbf{A^{A^r}}$ arises this way. In particular, unifiers extend to homomorphisms.

Using some standard algebraic tools, the transfer from functional to algebraic description of unifiers is formalized as follows:

Let f, g be functions from A$^{A^n}$ and let δ : A$^r \to$ A^n be a unifier of f and g.

δ induces an equivalence relation $\tilde{\delta}$ on A$^{A^n}$ by:

l, m $\in$ A$^{A^n}$, l $\tilde{\delta}$ m iff l | im δ = m | im δ. Since operators on $\mathbf{A^{A^n}}$ are defined pointwise, $\tilde{\delta}$ is a congruence relation on $\mathbf{A^{A^n}}$. In fact, it is the kernel of $\hat{\delta}$, i. e. the set of all functions in $\mathbf{A^{A^n}}$, which vanish identically on im δ. Finally, the congruence relation $\tilde{\delta}$ defined by a most general unifier δ of f and g can be characterized as the smallest congruence relation containing (f, g). As we will see later on this congruence can be computed quite well. Given the kernel of $\hat{\delta}$, the homomorphism $\hat{\delta}$ is -up to isomorphisms - an injection into $\mathbf{A^{A^r}}$. Now we may define "minimal" parametrizations of L (f, g) by r's, for which $\mathbf{A^{A^r}}$ is just big enough to contain im $\hat{\delta}$ ($\simeq \mathbf{A^{A^n}}_{/ \ker \hat{\delta}}$).

We now have a functional and an algebraic description of unifiers at our disposal. They mutually determine each other.

We will demonstrate later on how to apply algebraic machinery to finding most general unifiers.

Despite its simplicity the functional aspect of unification explains various phenomena of unification theory.

Let us review for instance unification in the sense of Plotkin [Plot 72]:

Here one requires that the functions f and g establishing an equation be term functions and unifiers of f and g be mappings whose components are term functions.

However, the less algebraic structure we have on A the fewer term functions exist. Hence, there is no reason, why surjections : $A^n \to L$ (f, g) (i. e. a most general unifiers), whose components are term functions should exist. As a remedy Plotkin introduces the set of most general unifiers, which can be thought of as a "minimal" collection of parametrizations of subsets covering L (f, g).

Unfortunately, this collection need not exist or - at the other extreme - might be too large for practical use [Baad 87]. Furthermore, an overlap of subsets covering L (f, g) often causes redundancies.

In order to overcome these conceptual drawbacks at least for finite algebras, we suggest to relax on Plotkins condition about the components of a unifier. There is a universal way to do so, i) destroying the algebraic structure on A and a more subtle one, ii) extending this structure. In any case, given term functions t_1, t_2, which agree on at least one argument, the set L (t_1, t_2) is encoded into a single unifier. However in i) the components of this unifier usually do not have any "meaning" related to the structure of A, whereas they do have in ii).

i) **Unification in Postalgebras**

Let **A**: = (A, F) be a finite algebra. Replace the structure of A by a total ordering $<$, i. e., turn A into a chain (A, $<$).

Taking maxima and minima with respect to $\langle$ we obtain two binary operators putting the structure of a distributive lattice upon A. Let 0, 1 be the bottom respectively top element of this lattice. For every singleton {a} in A \ {0} its "characteristic" function defines a unary operator C_a.

Let F': = {max, min, C_a, a | a $\in$ A } and let **A'**: = (A, F').

This algebraic structure is inherited by pointwise definition of operators to the set F_A of finitary functions with arguments and values in A and turns F_A into a Postalgebra $F_{A'}$.

The set $T_{A'}$ of terms over **A'** is of course a subset of F_A.

As Post has shown, we even have $F_A = T_A$, i. e., **A'** is functionally complete.

Therefore, for a finite set A every mapping occuring in our definition of unification and unifiers is a mapping with functions from $T_{A'}$ as components.

In particular, a surjection $\delta: A^n \to L(t_1, t_2)$ unifying t_1, t_2 $(\in T_A!)$, is a tuple of functions from $T_{A'}$. We have described in [Bütt 87] an algorithm, which computes this unifier thereby making use of the lattice structure of Postalgebras (see also [Nipk 87]).

ii) Unification in a functional by complete extension of a finite algebra

Given a finite algebra $\mathbf{A}: = (A; F)$ we will systematically add "meaningful" operators until we arrive at a set $F'' \supseteq F$ such that for $\mathbf{A}'' := (A, F'')$ we have $F_A = T_{A''}$. Since A is finite, this is guaranteed to succeed. It is more of a problem to choose F'' small and such that $T_{A''}$ is computationally "accessible".

We will make use of the following characterization of functionally complete algebras in terms of F-compatible relations due to I. Rosenberg and D. Schweigert [RoSc 84]: Recall that an m-ary relation R on A, i. e. a subset of A^m, is compatible with F, if R is a subalgebra of the product algebra $(A, F)^m$. A tolerance of $\mathbf{A}$ is a symmetric, reflexive binary relation on A which is compatible with F. A transitive tolerance of $\mathbf{A}$ is a congruence. A tolerance is called central if there exists some a in A such that $\{a\} \times A$ is a subset of R. Finally a 2/3 majority operation on $\mathbf{A}$ is a ternary function k such that $k(x, x, y) = k(y, x, x) = x$.

Now the theorem states:

Theorem: (Rosenberg, Schweigert)

Let $\mathbf{A}'': = (A, F'')$ be a finite algebra with a 2/3 majority operation and assume, that the elements of A are nullary operators.

Then $\mathbf{A}''$ is functionally complete, i. e., $T_{A''} = F_A$ iff $\mathbf{A}''$ has no nontrivial

- congruences
- central tolerances
- compatible orders with a least and a greatest element.

The theorem applies to our problems as follows:

The first task is to identify nontrivial congruences, central tolerances and compatible orders in $\mathbf{A}$. Then we decrease the number of compatible relations by adding operators which are not compatible with at least one of

these relations. Ensuring the existence of a 2/3 majority operation might be a problem. For lattices however such an operation always exists
(set $k(x, y, z) = x \cdot y + x \cdot z + y \cdot z$).
Since our primary interest is in lattices, we restrict our attention to this case.
Since **A"** is functionally complete unification in **A"** is unitary. For a given algebra A a unification algorithm still is to be devised. The next chapter shows how to do so in a special situation, which, however, is representative for a large class of lattices.

3. Unification in the 4-element boolean lattice

Unification in the 2-element boolean lattice (boolean unification) as described in [BüSi 87] has been integrated into a Prolog system and successfully applied to various tasks in circuit design [Simo 86] . Among these are verification and simplification of circuits on the gate level. While working on the gate level one always assumes that signals at the inputs are 0 or 1. Additional values are needed at switch level. A simple switch level model is based on 4 values 0, 1, U (undefined) and Z (high impedance). It is justified by physical arguments that these values are partially ordered ($\leqq$) according to the following diagram:

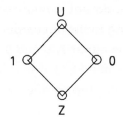

Hence A: = $\{0, 1, U, Z\}$ is a lattice, in fact it is <u>the</u> 4-element boolean lattice. If the inputs of a circuit range over A, its outputs are finitary functions with arguments and values in A. Any such function may occur at an output of a suitable circuit. Hence the arguments of the input/output relation of a circuit on switch level are elements of F_A. Now, the input/output relation of a circuit is specified in terms of the input /output relations of its components. This situation, however, can be expressed most naturally via Hornclauses over F_A. Tasks like symbolically verifying properties of a circuit on switch level can be pursued with an extended Prolog capable of performing unification in F_A.
Let now A: = $(\{0, 1, U, Z\}, \text{max}, \text{min}, - , 0, 1, U, Z)$. Here max, min denote maximum-, minimum formation respectively and - is complementation (for

brevity we write max, min as infix operators using $+$, $\cdot$). Every equation $t_1 = t_2$ with term functions t_i possesses one most general unifier (provided $L(t_1, t_2) \neq \varnothing$). However, as indicated above, we need to solve equations $f = g$ where f, g are arbitrary finitary functions over A. It is easy to find such functions, which are not term functions.

The unary function i where $i(1) = 0$, $i(0) = 1$, $i(U) = U$ and $i(Z) = Z$ is an example. This function corresponds to complementation on gate level and therefore is "meaningful".

We now apply the Rosenberg/Schweigert theorem as indicated above identifying congruences, central tolerances and compatible orders in A.

There are 2 nontrivial congruences on A given by $\{(1, Z), (0, U)\}$ and $\{(0, Z), (1, U)\}$. Now let R be a central tolerance, i. e. $\{c\} \times A \subseteq R$ for some c in A. Assume $c = Z$.

Since $(Z, 0) \in R$ we have $(0, Z) \in R$. Since also $(Z, 1) \in R$ it follows that $(0, 1) = $ max $\{ (Z,1), (Z,0) \} \in R$. With similar arguments we obtain $R = A^2$ i. e. R is a trivial congruence. The same holds for other choices of c.

Hence A does not admit nontrivial tolerances.

Now assume $<$ is an order on A, which is compatible with the operators of A. If $0 < 1$ then $1 = \bar{0} < \bar{1} = 0$ i. e. $0 = 1$ which is obviously false. Hence 0 and 1 do not compare with respect to $<$. Similarly U and Z do not compare. If neither 0 nor 1 compare with U, then $<$ is trivial. Passing to the dual order if necessary we may assume that $0 < U$. Hence $1 < Z$ and either 1, U and 0, Z do not compare or $1 < U$ and $0 < Z$. $1 < U$ implies $Z = $ min $(0, 1) < $ min $(U, U) = U$ a contradiction. Thus the only compatible order (up to duality) is given by

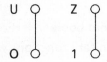

This order, however, has no least and greatest element.

Having identified the relations quoted in the theorem, we now proceed to introduce "destructive" operators. i is a good choice, since it has a meaning in the switch level model and since it is no term function. Therefore adding i to the operators of A enlarges the set of term functions. The 2 equivalence classes of a nontrivial congruence of A are <u>not</u> permuted by i. Hence they are no congruences with respect to the enlarged set of operators.

Therefore functional completenes is achieved by adding i to the operators of **A**. Let **A''** be the resulting algebra. Note, that the operators max, min and complementation induce a boolean algebra structure on F_A as well as on AA^n and i commutes wristh the boolean operators (i. e. i is an automorphism of the underlying boolean algebra). Hence every n-ary function over A is a unique sum of atoms. It is a routine matter to check that the atoms in $\mathbf{A''A''}^n$ are precisely those functions taking a value from $\{0, 1\}$ on exactly one argument and elsewhere value Z. Counting possible argument / value pairs we obtain $2 \cdot 4^n$ choices for atoms. Functions of the form $u \cdot s_1 \ldots s_n$ where $u \in \{0, 1\}$ and

$s_j \in \{x_j \cdot i(x_j), \overline{x_j} \cdot i(x_j), x_j \cdot i(\overline{x_j}), \overline{x_j} \cdot i(\overline{x_j})\}$ are obviously atoms. Since there are $2 \cdot 4^n$ of these functions we have identified the atoms in $\mathbf{A''A''}^n$.

A counting argument shows that there are 4^{4^n} different functions having the form

$$f(x_1, x_2, \ldots, x_n) = f(Z, x_2, \ldots, x_n) \cdot \overline{x_1} \cdot i(\overline{x_1})$$
$$+ f(0, x_2, \ldots, x_n) \cdot [0 \cdot x_1 \cdot i(\overline{x_1}) + 1 \cdot \overline{x_1} \cdot i(x_1)]$$
$$+ f(1, x_2, \ldots, x_n) \cdot [0 \cdot \overline{x_1} \cdot i(x_1) + 1 \cdot x_1 \cdot i(\overline{x_1})]$$
$$+ f(U, x_2, \ldots, x_n) \cdot x_1 \cdot i(x_1)$$

Since 4^{4^n} is the cardinality of AA^n (the carrier of $\mathbf{A''A''}^n$) we thus have obtained a recursive representation, which allows to express every n-ary function over A as a sum of atoms.

This representation may be used - like the Shannon expansion for boolean functions [Brya 86] - to obtain compact graph representations for elements of AA^n.

Since **A''** is functionally complete, every unification problem can be transformed into an equivalent problem of the form $f = 0$. We will now apply the algebraic machinery of the previous chapter to derive a most general unifier δ of f and 0 (Note, that 0 is the function with constant value Z). We recall, that the kernel of $\hat{\delta}$ (ker $\hat{\delta}$) consists of all n-ary functions vanishing on L (f, 0). By definition, f, $0 \in$ ker $\hat{\delta}$. Since $f = Z$ iff $i(f) = Z$, we also have $f + i$ (f) $\in$ ker $\hat{\delta}$. We claim, that ker $\hat{\delta}$ consists of all functions, which are built up by the atoms occuring in the atomic decomposition of $f + i(f)$. The functions $f + i(f) + \overline{f} + i(\overline{f})$ takes value U on every argument. Hence, it is sum of all atoms of $\mathbf{A''A''}^n$. The "zero's" of f ($= L$ (f, 0) are the arguments on which f takes value U. For every atom occuring in the decomposition of f, there must be one of these arguments, for which the atom takes a value from $\{0, 1\}$. Hence, this atom does not vanish identically on L (f, 0), which proves the claim. Now, a most general unifier $\hat{\delta}$ of f and 0 arises as follows:

Count the atoms in f + i(f), say k. Set r: = $[\log_4 k]$; since $f + i(f) + \overline{f} + i(\overline{f}) = U$, we have to have $U = \delta\,(f + i(f) + \overline{f} + i(\overline{f})) = \hat{\delta}\,(\overline{f} + i(\overline{f}))$.

Let $a_1, \ldots, a_s$ be the atoms occuring in the decomposition of $\overline{f} + i\,(\overline{f})$ and let B be the set of atoms in $A''A''^r$.

Then δ is defined by its effect on $\{a_1, \ldots, a_s\}$:

$$\delta\,(a_i) = \sum b_{ii}, \text{ where } \delta\,(i\,(a_i)) = i\,(\delta\,(a_i)), \delta\,(a_i)\cdot\delta\,(a_k) = Z \text{ for } i \neq k$$
$$b_{ij} \in B$$

and $\displaystyle\sum_{i=1}^{s} a_i = U$ (By $\sum$ we denote repeated application of the max-operator).

For $f\,(x_1, x_2) = i\,(\overline{x_1 x_2})\cdot(x_1 + \overline{x_2})$, we count that f + i (f) is built up from 28 of the 32 atoms of $A A^2$. Hence r may be chosen as 1 and, following the above recipy, a most general unifier $\hat{\delta}$ is given by $\hat{\delta}\,(x_1) = 0\cdot i\,(y) + 1\cdot y$, $\hat{\delta}\,(x_2) = 1$.

4. Conclusions

We have presented a general framework for unification in algebras, which identifies and links functional and algebraic aspects of the subject. No particular properties of the algebras, such as being equationally defined, are necessary. Within this framework it is reasonable to redefine unification such that with respect to this view, unification in finite algebras becomes unitary.

A universal unification algorithm for finite algebras is given. A refined approach dealing with the algebra of finitary functions over the 4-element boolean lattice, provides a special unification algorithm for this structure. Extending Prolog along these lines yields a logic programming language, capable to handle tasks like symbolic simulation of digital circuits on switch level.

We are currently experimenting with this Prolog extension.

5. Acknowledgements

I would like to thank Christian Hermann and Eric Tidén for helpful suggestions and fruitful discussions.

6. References

Baad 87   Baader F.,*Unification in Idempotent Semigroups is of Type Zero*,JAR2, 1986

BaDw 74   Balbes R. and Dwinger P., *Distributive Lattices,* University of Missouri Press, Columbia, U.S.A., 1974

Brya 86   Bryant R. Graph-Based *Algorithms for Boolean Functions Manipulations,* IEEE Transactions on Computers, Vol. C-35, No. 8, pg. 677 - 691, 1986

BüSi 87   Büttner W. and Simonis H., *Embedding Boolean Expressions into Logic Programming,* Journal of Symbolic Computation (to appear 1987)

Bütt 87   Büttner W, .*Application Driven Prolog Extensions,* Submitted to International Symposium on Multiple-Valued Logic, Palma de Mallorca, Spain, May 1988

Nipk 87   Nipkow T., *Unification in Functionally Complete Algebras,* Unpublished Manuscript 1987, Dept. CS, University of Manchester, England

Plot 72   Plotkin G., *Building in Equational Theories,* Machine Intelligence 7, pg. 73 - 90, 1972

Siek 86   Siekmann J.,*Unification Theory,* Proceedings, European Conference on ArtificialIntelligence, Brighton, England, July 1986

RoSc 84   Rosenberg I. and Schweigert D., *Compatible Ordesings and Tolerances of Lattices,* Preprint No. 70, Dept. of Mathematics, University of Kaiserslautern, West Germany, 1984

Simo 86   Simonis H., *Using Extended Prolog to Model Digital Circuits,* Technical Report TR-LP-2202 E.C.R.C.(European Computer Industry Research Centre), 1986

Unification in a Combination of Arbitrary Disjoint Equational Theories.

Manfred Schmidt-Schauß,

Fachbereich Informatik, Postfach 3049, Universität Kaiserslautern

West Germany

E-mail address: unido!uklirb!schauss UUCP

Abstract. The unification problem for terms in a disjoint combination $\mathcal{E}_1 + \ldots + \mathcal{E}_n$ of arbitrary theories is reduced to a combination of pure unification problems in $\mathcal{E}_j$, where free constants may occur in terms, and to constant elimination problems like: find all substitutions σ such that (the free constant) c_i no longer occurs in the term σt (modulo $\mathcal{E}_j$), where t is a term in the theory $\mathcal{E}_j$.

The algorithm consists of the following basic steps: First of all the terms to be unified are transformed via variable abstraction into pure terms belonging to one particular theory. Terms belonging to the same theory can now be unified with the algorithm for this theory. For terms in some multi-equation belonging to different theories it is sufficient to select some theory and collapse all terms not belonging to this particular theory into a common constant. Finally constant elimination must be applied in order to solve cyclic unification problems like $\langle x = f(y),\ y = g(x) \rangle$.

The algorithm shows that a combination of finitary unifying regular theories, of Boolean rings, of Abelian groups or of BNT-theories (basic narrowing terminates) is of unification-type finitary, since these theories have finitary constant-elimination problems. As a special case, unification in a combination of a free Boolean ring with free function symbols is decidable and finitary; the same holds for free Abelian groups. Remarkably, it can be shown that unification problems can be solved in the general case $\mathcal{E}_1 + \ldots + \mathcal{E}_n$ if for every i there is a method to solve unification problems in a combination of $\mathcal{E}_i$ with free function symbols. Thus, unification in a combination with free function symbols is the really hard case.

This paper presents solutions to the important open questions of combining unification algorithms in a disjoint combination of theories. As a special case it provides a solution to the unification of general terms (i.e. terms, where free function symbols are permitted) in free Abelian groups and Boolean rings. It extends the known results on unification in a combination of regular and collapse-free theories in two aspects: Arbitrary theories are admissable and we can use complete unification procedures (including universal unification procedures such as narrowing) that may produce an infinite complete set of unifiers for a special theory.

Key words: Unification, Equational theories, Decidability of Unification, Combination of equational theories, Boolean rings, Abelian groups

Acknowledgement.

This research is funded by the Deutsche Forschungsgemeinschaft.

1. Introduction.

Unification of terms with respect to an equational theory $\mathcal{E}$ [Pl72, Si75, Si86] is the problem, given a set of equations $\Gamma = \langle s_1 = t_1, \ldots, s_n = t_n \rangle_{\mathcal{E}}$ to find some or all substitutions σ such that all equations are solved i.e. $\mathcal{E}$ implies $\sigma s_i = \sigma t_i$ for all i. There are several theories $\mathcal{E}$ for which a unification algorithm or a unification procedure is known, for example commutativity (C), associtivity and commutativity or Abelian semigroups (AC), Boolean rings (BR) and free Abelian groups (AG). There are also unification procedures applicable to classes of equational theories, for example narrowing is a method which is applicable to equational theories that admit a canonical term rewriting system (TRS). For a survey on unification see [Si86, Si87].

In general the terms that are allowed as input for a unification algorithm for some theory $\mathcal{E}$ are restricted to consist of variables and function symbols that belong to $\mathcal{E}$, for example a BR-unification algorithm allows only terms built with +,*,0,1 and free constants. All useful unification algorithms known so far accept terms built with a fixed set of theory function symbols and arbitrary free constants. However, it is an open problem how to construct from an $\mathcal{E}$-unification algorithm for $\mathcal{E}$-pure terms (i.e. without free constants) a unification algorithm that also accepts terms including free constants [Bü86, Bü87]. H.-J. Bürckert and the author have given an example where unification becomes undecidable after the addition of free constants [Bü86, Sch87a, Ti86a]. We exclude this problem by assuming that free constants are permitted in terms.

The combination of unification algorithms for theories with disjoint sets of function symbols has been considered first by M. Stickel [St81] and F. Fages [Fa84] for the associative-commutative case. The algorithms accept terms built with several AC-function symbols, and free function symbols. A more general combination problem was tackled by several authors [Ye87, Ki85, Ti86a, He86]. They came up with algorithms for a combination of equational theories that obey some restrictions. C. Kirchner [Ki85] requires the theories to be simple. K.Yellick and A. Herold [Ye87, He86] require the theories to be regular and collapse-free and E. Tidén [Ti86a, Ti86b] considered the more general case of collapse-free (including nonregular) theories.

We loosen these restrictions to allow arbitrary theories in a disjoint combination. The presented algorithm can be seen as an extension of C. Kirchner's method to transform systems of multi-equations. The idea of constant-abstraction [LS78, He86] is indispensible and used heavily in our algorithm. We show that in order to solve unification problems in a combination, it is sufficient to have a unification algorithm for terms with free constants for all theories and a solution method for constant elimination problems in every theory. Alternatively we can also use a unification procedure for a combination of every theory with free function symbols. A complete solution is presented for a combination of theories, where every theory $\mathcal{E}$ in the combination satisfies one of the following cases:

i) $\mathcal{E}$ is regular
ii) $\mathcal{E}$ is a free Abelian group
iii) $\mathcal{E}$ is a free Boolean ring
iv) $\mathcal{E}$ is a BNT-theory (i.e. admits a canonical TRS and basic narrowing terminates).

Note that in this paper we will always assume that an $\mathcal{E}$-unification algorithm also accepts free constants and that all notions and definitions refer to a signature that includes infinitely many free constants.

The aim of this paper is to present the general combination algorithm and some constant elimination algorithms for special theories. We omit the completeness proofs of the algorithms as well as most of the necessary formal apparatus, since this is beyond the scope of this paper. The material can be found in [Sch87b].

In this paper we solve unification problems by transforming systems of equations rather than by giving a recursive algorithm in the Robinson-style.

2. Equational Theories.

With $\mathcal{T}(\Sigma,V)$ we denote the free term algebra over a signature Σ consisting of fixed-arity function symbols and a countably infinite set of variables V. We shall use hd(t) to denote the toplevel function symbol and $V(t)$ to denote the set of variables of a term t. A **substitution** σ is an endomorphism on $\mathcal{T}(\Sigma, V)$ such that $\{x \in V \mid \sigma x \neq x\}$ is finite. A substitutions σ can be represented by a set of variable term pairs $\sigma = \{x_1 \leftarrow t_1, \ldots, x_n \leftarrow t_n\}$ with $x_i \neq t_i$. The set $\{x_1, \ldots, x_n\}$ is the **domain** DOM(σ), the set $\{t_1, \ldots, t_n\}$ is the **codomain** COD(σ), the set of **variables introduced** by σ is denoted as $I(\sigma) := V(COD(\sigma))$. The **restriction** of a substitution σ to a set of variables W is the substitution $\sigma_{|W}$ with $\sigma_{|W} x = \sigma x$ for $x \in W$ and $\sigma_{|W} x = x$, otherwise. In order to have access to subterms of a term t, we use **occurrences** [Hu80]. The subterm of t at occurrence π is denoted by $t \backslash \pi$ and the term constructed from t by replacing the subterm at occurrence π by term s is denoted as $t[\pi \leftarrow s]$.

An **equational theory** $\mathcal{E}$ is a pair $(\Sigma_{\mathcal{E}}, E)$, where $\Sigma_{\mathcal{E}}$ is a signature and E is a set of equations. The equations in E are also called **axioms**. If a function symbol from Σ occurs in an equation in E it is called an **interpreted** function symbol, otherwise a **free** function symbol. Usually $\Sigma_{\mathcal{E}}$ only consists of interpreted function symbols and free constants. The set of all free constants in a term t is denoted by FRC(t).

An algebra A over Σ **satisfies** E or is a **model** of $\mathcal{E}$ ($A \vDash E$), if for every assignment of elements in A to variables in an equation $l = r$ in E, the corresponding equation holds in A. We say an equation $s = t$ is a **consequence** of E ($E \vDash s = t$), if every model A of E also satisfies $s = t$. We will also use $s =_{\mathcal{E}} t$ instead of $\mathcal{E} \vDash s = t$. Note that the relation $=_{\mathcal{E}}$ is a congruence relation on $\mathcal{T}(\Sigma, V)$ and that $\mathcal{T}(\Sigma, V)/=_{\mathcal{E}}$ is a free model of $\mathcal{E}$. The equivalence class of t with respect to $=_{\mathcal{E}}$ is denoted as $[t]_{\mathcal{E}}$.

A **term rewriting system** R is a set of directed equations $R = \{l_i \rightarrow r_i\}$, where $V(r_i) \subseteq V(l_i)$. The corresponding derivation relation $\xrightarrow{*}_R$ is called rewriting where $l_i \rightarrow r_i$ is used only in the given direction. A TRS is called **terminating**, if there are no infinite derivations. A TRS is called **confluent**, iff for all terms s, s_1, s_2 with $s \xrightarrow{*}_R s_1$ and $s \xrightarrow{*}_R s_2$ there exists a term s_3 with $s_1 \xrightarrow{*}_R s_3$ and $s_2 \xrightarrow{*}_R s_3$. A terminating and confluent TRS is called **canonical**. A term is in **normalform**, if no reductions are possible. In an equational theory admitting a canonical TRS every term t can be reduced to a unique normalform, denoted by $t\downarrow$.

An equational theory $\mathcal{E}$ is called **consistent**, iff there is a nontrivial model of $\mathcal{E}$, equivalently if the equation $x = y$ for different variables x and y is not a consequence of E. An equational theory $\mathcal{E}$ is called **collapse-free**, iff there is no valid equation $x =_{\mathcal{E}} t$, where t is not the variable x. An equational theory is called **regular**, iff for every valid equation $s =_{\mathcal{E}} t$ we have $V(s) = V(t)$. Equational theories are regular or collapse-free, iff the corresponding sets of axioms have this property. A theory $\mathcal{E}$ is called **simple**, iff $s =_{\mathcal{E}} t$ does never hold for a proper subterm s of some t. Note that a simple theory is regular and collapse-free, but that the converse is false [BHS87].

We extend $\mathcal{E}$-equality of terms to substitutions: Two substitutions σ and τ are equal modulo $\mathcal{E}$ over a set of variables W ($\sigma =_{\mathcal{E}} \tau \, [W]$), if $\sigma x =_{\mathcal{E}} \tau x$ for all variables $x \in W$.

We say σ is an **instance** of τ or τ is **more general** than σ over a set of variables W ($\tau \leq_{\mathcal{E}} \sigma \, [W]$), if there exists a substitution λ with ($\lambda \tau =_{\mathcal{E}} \sigma \, [W]$). Furthermore we say σ is **equivalent** to τ over W ($\tau \equiv_{\mathcal{E}} \sigma \, [W]$), iff $\tau \leq_{\mathcal{E}} \sigma \, [W]$ and $\sigma \leq_{\mathcal{E}} \tau \, [W]$.

Let $\Gamma := \langle s_i = t_i \mid i = 1, \ldots, n \rangle$ be a system of equations. A substitution σ $\mathcal{E}$-**unifies** Γ if for every equation $s_j = t_j$ in Γ we have $\sigma s_j =_{\mathcal{E}} \sigma t_j$. In this case we say σ is an $\mathcal{E}$-unifier of Γ. The set of all $\mathcal{E}$-unifiers is denoted by $U_{\mathcal{E}}(\Gamma)$.

A **complete** set $cU_{\mathcal{E}}(\Gamma)$ of unifiers of Γ is a set satisfying:

   i) $cU_{\mathcal{E}}(\Gamma) \subseteq U_{\mathcal{E}}(\Gamma)$ (correctness)

   ii) $\forall \sigma \in U_{\mathcal{E}}(\Gamma) \, \exists \tau \in cU_{\mathcal{E}}(\Gamma): \quad \tau \leq_{\mathcal{E}} \sigma \, [V(\Gamma)]$ (completeness)

A complete set is called **minimal** or a set of most general unifiers (mgus), iff additionally

   iii) $\forall \sigma, \tau \in cU_{\mathcal{E}}(\Gamma) \quad \tau \leq_{\mathcal{E}} \sigma \, [V(\Gamma)] \quad \Rightarrow \tau = \sigma$ (minimality)

Minimal sets are also designated as $\mu U_{\mathcal{E}}(\Gamma)$. Note that for fixed Γ all the sets $\mu U_{\mathcal{E}}(\Gamma)$ are equivalent [FH83].

An equational theory is called **unification based**, iff $\mu U_{\mathcal{E}}(\Gamma)$ exists for all Γ. A unification based equational theory $\mathcal{E}$ is called **unitary, finitary or infinitary** depending on the possible maximal cardinalities of $\mu U_{\mathcal{E}}(\Gamma)$ for all Γ. Theories that are not unification based are called **nullary**.

In the unification procedure described later we need constant-elimination problems to unify cyclic unification problems where several theories are involved. E. Tidén [Ti86] used a similar method for this purpose, which he called 'variable elimination'.

A **constant elimination** problem C in the theory $\mathcal{E}$ is of the form
$C := \langle c_i \notin t_{ij} \mid i = 1,...,n, j=1,...,m \rangle$, where c_i are different free constants and t_{ij} are $\mathcal{E}$-terms. The set of solutions of C is the set $U_{\mathcal{E}}(C) := \{\sigma \mid \exists t_{ij}' \ t_{ij}' =_{\mathcal{E}} \sigma t_{ij}$ and $c_i \notin FRC(t_{ij}')$ for $i = 1,...,n, j=1,...,m \}$.
A **complete set of constant eliminators** $cU_{\mathcal{E}}(C)$ is a set of substitutions, such that
  i) $cU_{\mathcal{E}}(C) \subseteq U_{\mathcal{E}}(C)$ and
  ii) For every $\theta \in U_{\mathcal{E}}(C)$ there exists a $\sigma \in cU_{\mathcal{E}}(C)$, such that $\sigma \leq_{\mathcal{E}} \theta \ [V(C)]$.
A complete set is called **minimal** or **most general,** iff additionally
  iii) $\forall \sigma, \tau \in cU_{\mathcal{E}}(C) \ \ \tau \leq_{\mathcal{E}} \sigma \ [V(C)] \ \Rightarrow \tau = \sigma$ (minimality)
Note that instances of constant eliminators of C may not be constant eliminators for C. However, in special theories like Boolean rings, for which we determine a set of constant eliminators below, it is always clear how to obtain every constant eliminator by instantiating the most general ones.

It is interesting to note that singleton constant-elimination problems can be encoded as a unification problem in a in $\mathcal{E}$ with one free function symbol. However, this is false in general as an example in [Sch87b] shows.

An example for a constant elimination problem is $\{c \notin c*x\}$ in Boolean rings, which has the most general solution $\{x \leftarrow (1+c)x'\}$ as shown in 8.1.

Constant-elimination problems that are necessary (for example all singletons) for the combination algorithm can be encoded as a unification problem in $\mathcal{E}$ together with free function symbols. However, not all such problems can be encoded this way (cf.]Sch87b]).

3. Combination of Equational Theories.

We investigate equations and unification in a combination of equational theories $\mathcal{E}_j$, $j=1,...,N$. Let $\mathcal{E}_j := (\Sigma_{\mathcal{E}j}, E_j)$ be consistent equational theories that are disjoint, i.e., the only symbols common to different $\Sigma_{\mathcal{E}j}$ are free constants. Furthermore we assume that all $\Sigma_{\mathcal{E}j}$ contain the same set of countably infinitely many free constants $\mathcal{FRC}$. We denote the disjoint combination of all $\mathcal{E}_j$ as $\mathcal{E}_1 +...+\mathcal{E}_N := (\Sigma_{\mathcal{E}1} \cup ... \cup \Sigma_{\mathcal{E}N}, E_1 \cup ... \cup E_N)$. We shall use $\mathcal{E}+$ as abbreviation for $\mathcal{E}_1 +...+\mathcal{E}_N$.

Let s be a term, then s has **syntactical theory** $\mathcal{E}_i$, if the top-level function symbol of s belongs to $\mathcal{E}_i$. A term s is called **pure** (or $\mathcal{E}$-pure) if s is a term from $\mathcal{T}(\Sigma_{\mathcal{E}i}, V)$ for some theory, otherwise a term is called a **mixed** or **general** term. We also say a term s is a **proper $\mathcal{E}_i$-term**, iff s is an $\mathcal{E}_i$-term but not a variable or free constant. A subterm s of a term t is called $\mathcal{E}_i$**-alien**, if every proper superterm of s in t is an $\mathcal{E}_i$-term, but s is not an $\mathcal{E}_i$-term. A subterm s of t is an **alien** subterm, iff s and t have a different theory and s is a maximal subterm with this property. Note that free constants or variables do not count as alien subterms. The set of free constants in a term t is denoted by $FRC(t)$ and the set of **essential free constants** in t is denoted as $ESS\text{-}FRC_{\mathcal{E}+}(t) := \{c \mid c$ is a free constant and $t =_{\mathcal{E}+} t'$ implies $c \in FRC(t')$ for all terms $t'\}$. Free constants in a term that are not essential are also called **inessential**.

In order to give an example, let $\mathcal{E}_1$ be a free Boolean ring and let $\mathcal{E}_2$ consist of free function symbols. Then the term $x+f(y)$ is a mixed $\mathcal{E}_1$-term with alien subterm $f(y)$. The term $x + a + b + b$ has a as essential free constant and b as inessential free constant, since $x + a + b + b =_{\mathcal{E}1} x + a$.

The combination $\mathcal{E}+$ has the expected properties: For $\mathcal{E}_j$-pure terms s,t we have $s =_{\mathcal{E}j} t \Leftrightarrow s =_{\mathcal{E}+} t$ and for an $\mathcal{E}_j$-pure system of equations $\Gamma = \{s_i = t_i \mid i = 1,\ldots,n\}$ the set $U_{\mathcal{E}j}(\Gamma)$ is a complete set of $\mathcal{E}+$-unifiers for $U_{\mathcal{E}+}(\Gamma)$. This result was first proved by Erik Tidén [Ti86a].

A term t in the combination has **semantical theory** $\mathcal{E}_j$, $j \in \{0,1,\ldots,N\}$ if a 'maximally collapsed' term t' with $t' =_{\mathcal{E}+} t$ has syntactical theory $\mathcal{E}_j$ (cf. [Sch87b]). For example t has semantical theory $\mathcal{E}_0$, if it is $\mathcal{E}+$-equal to a variable or a free constant. If t is $\mathcal{E}+$-equal to a pure $\mathcal{E}_j$-term, but not $\mathcal{E}+$-equal to a variable or free constant, it has semantical theory $\mathcal{E}_j$. In the example of Boolean rings and free function symbols the term $0+1*f(x)$ has semantical theory $\mathcal{E}_2$, since it is equal to $f(x)$, $x*1$ has semantical theory $\mathcal{E}_0$, since it is equal to x, and $x*f(x)$ has semantical theory $\mathcal{E}_1$, since it cannot be further simplified and the topsymbol belongs to $\mathcal{E}_1$.

4. Unification as Transformation of Systems of Equations.

We consider the process of unification as a sequence of (maybe nondeterministic) transformations that starts with a given system Γ of equations and stops with another one in solved form. This follows the ideas of [[MM82,Ki85] and is similar in spirit to solving problems in [Her30,Hu76,Co84,]. We shall also use multi-equations instead of equations since they are more appropriate. We assume that Γ is a set of multi-equations and that each multi-equation M_i is a set of terms $\{t_{i1},\ldots,t_{i,ni}\}$ also denoted as $t_{i1} = \ldots = t_{i,ni}$. Obviously every systems of equations can be considered to have this form. We use $s = t \in \Gamma$ synonymously with $s,t \in M$, where M is a multi-equation in Γ. Furthermore we assume that (generalized) merging is built-in, i.e., if $r = s$ and $s = t$ is in Γ, then also $r = t$ is in Γ, or equivalently that all multi-equations are disjoint. Note that 'merge' usually means to consider only variables common to some multi-equation and that we consider also common terms. As abbreviation we shall also use equations of the form $S = T$, where the uppercase letters denote sets of terms and $S = T$ means the conjunction of all equations $s_i = t_j$ for $s_i \in S$ and $t_j \in T$.

A **cycle** in Γ is a set $\{x_i = t_i, i = 1,\ldots,n\}$ with $x_i = t_i \in \Gamma$, where x_i is a variable and t_i is a nonvariable term, such that $x_{i+1} \in V(t_i)$ for $i = 1,\ldots,n-1$ and $x_1 \in V(t_n)$. A system of multi-equations Γ is in **sequentially solved form**, iff in every multi-equation there is at most one nonvariable term and Γ does not contain cycles. It is in **solved form**, iff no variable in Γ occurs in some proper term in Γ. Note that every sequentially solved system can be transformed into a solved one by the replacement rule defined below. It is obvious how to construct from every Γ in solved form an idempotent mgu σ_Γ for Γ.

Given an equational system Γ_0 to be solved, our goal is to have an algorithm that is able to transform Γ_0 into a solved system Γ. This algorithm is described by nondeterministic rules, such that for every given solution θ of Γ_0, there is a transformation resulting in a solved system Γ that provides a solution more general than θ. It should be kept in mind that the only relevant instantiations are those of variables in $V(\Gamma_0)$. Variables not in the original set of **significant variables** $V(\Gamma_0)$ are also called **auxiliary** variables.

First we describe some **general 'don't care'-rules** that can be applied to equational systems, i.e. these rules do not change the set of solutions. The first four rules are also referred to as **reduction rules**.

Rule: Trivial multi-equations. $\qquad M \& \Gamma \Rightarrow \Gamma$,

    if M contains only one element.

Rule: Auxiliary variables. $\qquad \Gamma \& M \Rightarrow \Gamma \& M-\{z\}$,

    if z is an auxiliary variable and does not occur elsewhere in Γ or $M - \{z\}$.

Rule: Theory-merge. $M_1 \& M_2 \Rightarrow M_1 \cup M_2$,

    if there are terms $t_1 \in M_1$ and $t_2 \in M_2$ with $t_1 =_{\mathcal{E}+} t_2$.

Rule: Equal terms. $M \Rightarrow M - \{s\}$,

    if M contains two different terms s, t with $s =_{\mathcal{E}+} t$.

Rule: Demodulation. $s = t \& \Gamma \Rightarrow s' = t \& \Gamma$

    if $s =_{\mathcal{E}+} s'$.

Rule: Replacement. $s = t \& r = l \Rightarrow s = t \& r[\pi \leftarrow t] = l$

    if $r\backslash\pi =_{\mathcal{E}} s$.

Rule: Renaming: $\Gamma \Rightarrow \{x \leftarrow x'\}\Gamma \& x = x'$,

    where $x \in V(t)$ for some t in Γ and x' is a new variable.

Rule: Unfolding. $s = t \Rightarrow s[\pi \leftarrow x] = t \& x = r$,

    if r is an alien subterm of s at occurrence π and x is a new variable.

    (this rule is also called variable-abstraction)

Note that the elimination rule $x = t \& \Gamma \Rightarrow x = t \& \{x \leftarrow t\}\Gamma$ is a special case of the replacement rule.

5. Unification of Mixed Terms.

We present in the following the basic steps, the nondeterministic rules and a strategy for unification in a combination of disjoint theories $\mathcal{E}_j$. The procedure is described such that it is (hopefully) comprehensive and termination can be seen rather directly. We do not consider all possible failure rules.

In order to design such a nondeterministic algorithm one should have in mind that a solution σ of the original system of multi-equations Γ_0 is given and that it must be possible to direct the solution process such that a solution σ_{mg} is returned that is more general than σ over $V(\Gamma_0)$. We design the steps and rules in such a way that for every nondeterministic step in this process, the number of different possibilities is finite unless an involved theory povides an infinite set of most general unifiers or constant-eliminators.

The procedure is described for a combination of N theories, since we have found no way to solve constant-elimination problems in a combination if there are algorithms for every theory, which would be required by an induction argument.

We denote the actual system of multi-equations with Γ and assume that it consists of multi-equations M_i, i.e., $\Gamma = \{M_i \mid i = 1,\ldots,M\}$.

We will use $\mathcal{E}_0$ standing for the 'theory' of free constant.

An example for this algorithm is given at the end of the paragraph.

The first step is a 'don't care'-step, i.e., no branching is necessary.

GU-Step 1.   Transform Γ into unfolded normal form.

The concept of an **unfolded normal form** (UNF) of an equation system is very important for our procedure. It has the following properties: All terms are pure, all variables contained in proper terms are auxiliary, proper terms do not contain free constants, proper terms with different syntactical theory have disjoint sets of variables, and every auxiliary variable in Γ occurs also in some proper term in Γ.

Note that it is in general not possible to remove equations $s = t$, where s and t are pure terms with different theory, due to built-in merging.

The unfolded normal form provides a partition of every multi-equation M in Γ into the following parts: a part consisting of significant variables, and for every theory $\mathcal{E}_j, j = 0,\ldots,N$ a set of terms belonging to this theory. Proper terms and free constants can be classified with respect to their outermost function symbol and auxiliary variables are classified with respect to the syntactical theory of terms they occur in.

It is easy to see that every Γ can be transformed into such an unfolded normal form by the above renaming and unfolding rules, however, this normal form may not be unique.

For example the unfolded normal form of $\langle x = f(x*y) \rangle$ is:

significant variables		Boolean ring		free functions
$\langle$ x	=	x'	=	f(z)
y	=	y'		
		x'*y'	=	z $\rangle$.

This layout of this example is arranged according to the above partitioning.

GU-Step 2.   Transform Γ into a system in separated unfolded normalform (SUNF).

This step includes branching and uses the two rules 'unification' and 'identification' described below. Furthermore it includes implicit application of reduction rules.

The SUNF has the following properties: It is an UNF and in every $\mathcal{E}_j$-part of every multi-equation there is at most one term. Furthermore we can view it as representing only unifiers σ that additionally do not unify different multi-equations, i.e. for different multi-equtions M and N we have $\sigma M \neq_{\mathcal{E}_+} \sigma N$.

A key observation is that we can use an $\mathcal{E}_j$-unification algorithm that operates only on the $\mathcal{E}_j$-part of Γ. This rule can also be seen as an extension of unification using the variable-abstraction method.

Rule: GU-Unification. $\Gamma \Rightarrow \sigma\Gamma$

   where σ is a (most general) unifier of the $\mathcal{E}_j$-related part Γ_j for some j. The mgu σ should introduce only new variables. We assume that the $\mathcal{E}_j$-part of every multi-equation in the resulting system σΓ has at most one term. This can be achieved by using the rule 'equal terms', since σ is a $\mathcal{E}_j$-unifier.

Since there may be more than one most general unifier of the $\mathcal{E}_j$-part of Γ, this rule makes branching necessary. Note that the completeness of the GU-unification rule follows by considering $\Gamma \Rightarrow \sigma\Gamma$ & $\langle\sigma\rangle$ and then by deletion of $\langle\sigma\rangle$ due to the auxiliary variables rule.

The complexity measure for showing termination is (μ_1,μ_2), where μ_1 is the number of multi-equations in the unfolded normal form and μ_2 is the number of terms (including variables) in the unfolded normal form. This measure is well-founded if we use a lexicographical ordering. Our measure is related to the Fages-measure [Fa84].

Obviously the complexity of σΓ is decreased by the GU-unification rule, since the number of multi-equations is not increased and the number of terms is properly decreased.

The following rule is a nondeterministic one and partitions the solution space into substitutions with a different identification pattern on the multi-equations in Γ.

Rule: GU-Identification. $\qquad\qquad \Gamma \Rightarrow \Gamma'$,

where in Γ' some multi-equations from Γ are joined.

more precisely, let $\sim$ be an equivalence relation on the multi-equations in Γ and then construct Γ' by joining multi-equations M_i and M_j, iff $M_i \sim M_j$.

After an identification, it is sufficient to consider only those systems of multi-equation in the search space that do not further identify multi-equations. If we write U for GU-unification and I for GU-identification, then the application sequence is like U^* or U^*IU^*. The GU-identification rule is a proviso for the application of the collapsing rule by constant-abstraction, since we can then abstract different multi-equations by different constants.

Note that the two rules unification and identification always produce systems of multi-equations in unfolded normalform. Our complexity measure shows that the application of the two rules 'Identification' and 'Unification' terminates.

The next step fixes the semantic theory of every multi-equation in Γ, i.e. it restricts the solutions for the systems such that all terms of every multi-equation have the assigned semantic theory after application of the solution.

GU-Step 3. Label all multi-equations by some $\mathcal{E}_j$, $j = 0,\ldots,N$

This step is nondeterministic, since all possible (finitely many) labelings have to be considered.

The aim of step 4 is to collapse terms with the wrong theory (according to the labeling in step 3). This step uses the method of constant-abstraction.

GU-Step 4. Transform the labeled SUNF into **constant abstracted normal form**.

This step is performed as follows: We add to every multi-equation M_i a new (auxiliary) variable y_i and then use the collapsing and replacement rule. These extra variables y_i are used as variables or constants, depending on the context. They shall play the role of constant abstractions in the collapsing rule using unification. As constituents of Γ, they are considered as variables.

For the following rules we denote the $\mathcal{E}_j$-term in the multi-equation M_i by t_{ij} (if it exists).

Let $I_{-j} := \{i \mid M_i \text{ is not labeled with } \mathcal{E}_j \text{ and } M_i \text{ contains an } \mathcal{E}_j\text{-term } t_{ij}\}$ and let

$I_{+j} := \{i \mid M_i \text{ is labeled with } \mathcal{E}_j\}$

The following rule provides the essential operation to ensure that Γ is consistent with the labeling, i.e. that a proper pure $\mathcal{E}_j$-term occurs only in multi-equations labeled $\mathcal{E}_j$.

Rule: Collapsing. $\qquad \Gamma \Rightarrow \sigma\Gamma$,

where $\mathcal{E}_j$ is a theory and σ is a (most general) unifier of the problem $\langle t_{ij} = y_i \mid i \in I_{-j}\rangle$, where y_i are the variables from above and t_{ij} is the $\mathcal{E}_j$-term of the multi-equation M_i. The y_i's are considered as constants in the unification problems. We assume that in $\sigma\Gamma$ the multi-equations σM_i with $i \in I_{-j}$ contain no $\mathcal{E}_j$-term, which can be achieved by the rule 'equal terms' and that the variables y_i remain in the multi-equation.

We can use as additional rule that $\Gamma \Rightarrow$ FAIL, if the number of multi-equations is properly decreased after application of the collapsing rule. The collapsing rule is to be applied exactly once for every theory. Note that the substitutions σ generated by this rule may have y_i's in the codomain terms. This has as consequence that the classification of terms in Γ has slightly to be changed. For proper terms and auxiliary variables ($\neq y_i$) there is no change. The variables y_i are classified according to the label of the multi-equation M_i .

After all applications of collapsing the obtained system of multi-equations has a special form. It consists of multi-equations of the form:

$X_i = y_i = t_{ij}$ for multi-equations M_i labeled by $\mathcal{E}_j$ where t_{ij} is an $\mathcal{E}_j$-term , or

$X_i = y_i$ for multi-equations M_i labeled by $\mathcal{E}_0$.

X_i denotes the part of the multi-equation consisting of significant variables. The terms t_{ij} may contain some y_k's that are labeled with a theory $\mathcal{E}$ different from $\mathcal{E}_j$. Furthermore two terms t_{ij} and $t_{ij'}$ with $j \neq j'$ have at most some variables y_i in common.

The remaining problem now is that the resulting system may have cycles, where the y_i's count as variables for determining cycles. If there are no cycles, then the system is in sequentially solved form and we are ready. Otherwise, we use constant elimination to resolve the cycles:

GU-Step 5. Select a constant-elimination problem corresponding to Γ.

This step is performed by selecting nondeterministically a constant-elimination problem C consisting of pairs $y_k \notin t_i$, where $y_k \in M_k$ plays the role of a constant, $t_i \in M_i$, and M_i and M_k are labeled with different theories. The $\mathcal{E}_j$-part C_j of C is the constant-elimination problem consisting of the pairs $y_k \notin t_i$ from C, where M_i is labeled $\mathcal{E}_j$.

GU-Step 6. Solve the constant elimination problem using the rule below.

        A solution of Γ is obtained, if afterwards the system is in sequentially solved form.

The step is performed using the following

Rule: Resolve constant-elimination problems.

    $\Gamma \Rightarrow \sigma_j \Gamma$,

        where σ_j is a (most general) solution to the $\mathcal{E}_j$-part C_j for some j. We assume that the terms appearing in $\sigma_j \Gamma$ are modified by the demodulation rule, such that the eliminated constants are in fact eliminated.

This rule is applied exactly once for every theory. The branching inherent in this rule depends solely on the number of (most general) solutions to C_j.

The described procedure has a lot of nondeterminisms. In step 2 we can choose among several most general unifiers in the unification rule, and several possibilities for identification. In step 3 there are several possibilities for labelling and in step 4 several solutions for every collapse problem. Finbally in step 5 there are several possible sensible constant-elimination problems and in addition several possible solutions to every constant-elimination problem. Nevertheless, the number of different possibilities is finite unless an involved theory povides an infinite number of unifiers or constant-eliminators.

The following theorems and corollaries hold:

5.1 Theorem. The nondeterministic algorithm described in this paragraph always terminates. Furthermore if every theory is finitary unifying and for every theory the constant-elimination problems are finitary, then the algorithm returns finitely many unifiers. ∎

5.2 Theorem. If there exist complete GU-unification procedures solving every system of equations including free constants and algorithms for the theories $\mathcal{E}_j$ that provide a complete set of constant eliminators for every constant elimination problem, then the procedure is a correct and complete procedure for solving unification problems for systems of equations in the combination $\mathcal{E}+$. ■

5.3 Corollary. If all $\mathcal{E}_j$ are of unification type finitary and there always exists a finite complete set of constant eliminators for $\mathcal{E}_j$, then $\mathcal{E}+$ is also of unification type finitary. ■

5.4 Corollary. If all $\mathcal{E}_j$ are of unification type finitary and regular, then $\mathcal{E}+$ is of unification type finitary. ■

We demonstrate the algorithm described in this paragraph by two examples.

5.5 Example. Let $\mathcal{E}_1$ be a free Boolean ring with operators $+,*,0,1$ and let $\mathcal{E}_2$ be the theory of free function symbols with operator f.

Let the original problem be $\Gamma_0 = \langle x = f(x*y)\rangle$. This problem was posed by U.Martin at the first unification workshop in Val d'Ajol as a challenge.

The significant variables are x and y.

GU-step 1 means to rename significant variables and to abstract the term x*y. The result is:

$\langle x = x' = f(z'), \quad y = y', \quad x'*y' = z'\rangle$.

The auxiliary variables are x',y' and z', where x',y' have theory $\mathcal{E}_1$ and z' has theory $\mathcal{E}_2$.

GU-step 2 now means to apply GU-unification and GU-identification.

GU-unification is not applicable.

GU-identification gives 5 possibilities: No identification, 3 pairwise identifications and to join all multi-equations.

To join the first and third means to solve $\langle x = x' = f(z') = x'*y' = z', \quad y = y'\rangle$, which has no solution, since $f(z') = z'$ is not solvable.

In order to proceed we choose the branch where no identification is made.

GU-step 3 means to label the multi-equations by some theory $\mathcal{E}_0, \mathcal{E}_1$ or $\mathcal{E}_2$.

There are 3^3 possible labelings, however, it is easy to see that for example $x = x' = f(z')$ can only be labeled with $\mathcal{E}_2$. In the branch that we shall follow we select the labeling $\mathcal{E}_2$ for the first multi-equation and $\mathcal{E}$ for the second and third.

GU-step 4 is performed by first adding the variables y_i, resulting in

$\langle (x = y_1 = x' = f(z'), \mathcal{E}_2), \quad (y = y_2 = y', \mathcal{E}_1), \quad (y_3 = x'*y' = z', \mathcal{E}_1)\rangle$.

We choose this informal notation to indicate the labeling.

The collapsing step for $\mathcal{E}_1$ is as follows:

Solve $\langle y_1 = x'\rangle$, which has the trivial result $\{x' \leftarrow y_1\}$ and as resulting system

$\langle (x = y_1 = f(z'), \mathcal{E}_2), \quad (y = y_2 = y', \mathcal{E}_1), \quad (y_3 = y_1*y' = z', \mathcal{E}_1)\rangle$.

The collapsing step for $\mathcal{E}_2$ is as follows: Solve $\langle y_3 = z'\rangle$, which has the trivial result $\{z' \leftarrow y_3\}$ and as resulting system $\langle (x = y_1 = f(y_3), \mathcal{E}_2), \quad (y = y_2 = y', \mathcal{E}_1), \quad (y_3 = y_1*y', \mathcal{E}_1)\rangle$.

These are in fact only trivial collapsing steps, which are no more than renaming variables, however, in general there are also nontrivial collapsing steps, which additionally may require branching.

GU-step 5. The system contains a cycle: $\{y_1 = f(y_3), y_3 = y_1*y'\}$, hence it is not in solved form.

The description of GU-step 5 requires that a lot of constant-elimination problems have to be considered. However, not all of them have the potential to solve the cycle. A useful heuristic is to choose only such constant elimination problems, that can solve this cycle. It is not at all clear whether it is sufficient to choose only such constant elimination problems that solve the cycle and do not contain redundant pairs. For our example we select the following $C := \{y_1 \notin y_1 {}^*y'\}$.

GU-step 6

In 8.1 it is shown, that C has $\{y' \leftarrow y''{}^*(1 + y_1)\}$ as most general solution.

Applying this to $\langle (x = y_1 = f(y_3), \mathcal{E}_2), \ (y = y_2 = y', \mathcal{E}_1), \ (y_3 = y_1{}^*y', \mathcal{E}_1) \rangle$ gives

$\langle x = y_1 = f(y_3), \ y = y_2 = y''{}^*(1 + y_1), \ y_3 = 0 \rangle$.

This system is sequentially solved and the unifier corresponding to this branch is

$\{x \leftarrow f(0), y \leftarrow y''{}^*(1+f(0))\}$.

This computation does not show that this is the only possible most general unifier, since we have not explored all possible branches. In the specialized procedure of paragraph 6 that considers a combination of an arbitrary and a simple theory we reexamine this example and show in example 6.1 that this is in fact a most general solution.

5.6 Example. This example should demonstrate that unfolding of free constants may be necessary.

Let $\mathcal{E}_1$ be the theory with function symbols f,h and axiom $\{f(h(x), x) = x\}$ and let $\mathcal{E}_2$ be the theory with function symbols g,k and axiom $\{g(k(x), x) = x\}$.

Note that the axioms of $\mathcal{E}_1$ and $\mathcal{E}_2$ can be directed and then the two theories $\mathcal{E}_1$ and $\mathcal{E}_2$ have a canonical TRS and that basic narrowing terminates, i.e. they are BNT-theories [Hul80].

The problem which we want to solve is $\langle f(x, a) = g(y, a) \rangle$, where a is a free constant.

An obvious solution to this problem is $\{x \leftarrow h(a), y \leftarrow k(a)\}$. We show how to obtain this solution with the procedure above.

The unfolded normalform is $\langle x = x', y = y', a = z' = u', f(x', z') = g(y', u') \rangle$.

Identification permits to join the last two multi-equations, which gives

$\langle x = x', y = y', a = z' = f(x', z') = u' = g(y', u') \rangle$.

Unification for $\mathcal{E}_1$ has to solve the problem $\langle z' = f(x', z') \rangle$, which results in the unifier $\{x' \leftarrow h(x''),$ $z' \leftarrow x''\}$. Application of this solution gives the system $\langle x = h(x''), y = y', a = x'' = u' = g(y', u') \rangle$.

Unification for $\mathcal{E}_2$ has to solve the problem $\langle u' = g(y', u') \rangle$, which results in the unifier

$\{y' \leftarrow k(u''), u' \leftarrow u''\}$. Application of this solution gives the system $\langle x = h(x''), y = k(u''), a = x'' = u'' \rangle$

This system is in solved form and has as most general solution $\{x \leftarrow h(a), y \leftarrow k(a)\}$.

Now suppose, that free constants are not unfolded. Then the unfolded normalform of this problem is $\langle x = x', y = y', f(x', a) = g(y', a) \rangle$. The only possibility to direct the procedure to the suggested solution is to make no identification.

In GU-step 3 we label this system as follows: $\langle (x = x', \mathcal{E}_1), (y = y', \mathcal{E}_2), (f(x', a) = g(y', a), \mathcal{E}_0) \rangle$.

GU-step 4 now adds the variables y_i: $\langle (x = y_1 = x', \mathcal{E}_1), (y = y_2 = y', \mathcal{E}_2), (y_3 = f(x', a) = g(y', a), \mathcal{E}_0) \rangle$.

The collapsing step requires to solve $\langle y_3 = f(x', a) \rangle$, where y_3 is to be considered as constant. It is easy to see by basic narrowing, that this has no solution.

The conclusion is that the procedure returns no solution that is more general than our intended solution. Hence unfolding of free constants is necessary in our procedure. ■

Remark on Efficiency. For practical purposes and for enhancing efficiency of this procedure the following ideas can be used. Completeness and termination of these improvements are very likely, but not proved in [Sch87b].

i) Do not rename variables in GU-step 1 (but make unfolding). Change the rules and steps accordingly. For example do only instantiate variables, if the resulting term is pure; Furthermore substitution components have to be added to the system, if significant variables are involved.

ii) Do not unfold free constants. Example 5.6 shows that the collapsing step has to be modified. Presumably the following mopdification is sufficient: For multi-equations labeled $\mathcal{E}_0$ we permit as alternative to adding a variable y_i also to add some free constant occurring in Γ.

iii) Replace the GU-unification step by some weaker transformation, such that for example decomposition becomes a possible step. These steps have a potential to reduce the search space if this weaker transformations do not require branching. However, these transfomation have to be chosen carefully in order to ensure termination.

iv) Do not consider all possible constant-elimination problems, but only a minimal, but sufficient subset (cf. [Sch87b]).

v) Design more special combination procedure for special cases. This idea is explored in the next paragraph if one theory is simple. ■

6. Combining an Arbitrary and a Simple Theory.

The above algorithm can be improved for the special case of a disjoint combination $\mathcal{E}+\mathcal{F}$, where $\mathcal{E}$ is arbitrary and $\mathcal{F}$ is a simple theory. The improvements over the general procedure originate in some nice properties of simple theories. A proper syntactical $\mathcal{F}$-term is also a semantical $\mathcal{F}$-term and cyclic systems of equations in $\mathcal{F}$ are not solvable. Note that this algorithm is not a specialization of the algorithm described in paragraph 5.

ASU-step 1.   Transform Γ into a system in UNF, but do not unfold free constants.
ASU-step 2.   Transform Γ into a system in separated UNF.

The rules used for step 2 are the rules ASU-unification and ASU-identification.
Rule: ASU-unification. (like GU-unification)
Rule: ASU-Identification. $\Gamma \Rightarrow \Gamma$,
    where in Γ some multi-equations containing proper $\mathcal{F}$-terms from Γ are joined. More precisely, let $\sim$ be an equivalence relation on the multi-equations in Γ such that $M_i \sim M_j$ for $i \neq j$ only if M_i as well as M_j contain a proper $\mathcal{F}$-term. Then construct Γ by joining multi-equations M_i and M_j, iff $M_i \sim M_j$.

The application sequence of these two rules can be described by $U^*(IU)^*$, which shows also that more than one identification application may be necessary.
A further difference to the general procedure is that no labeling of multi-equations is necessary. Hence the next step is a resolution of constant-abstraction:

ASU-step 3.   Abstract proper $\mathcal{F}$-terms by different constants and solve the system with respect to $\mathcal{E}$.

This step is performed as follows. We add to every multi-equation M_i that contains a proper $\mathcal{F}$-term a new variable y_i, which is used as constant-abstraction. Then we use the following collapsing rule:

Rule: ASU-Collapsing. $\qquad$ $\Gamma \Rightarrow \sigma\Gamma$,

where σ is a (most general) unifier of $\Gamma_{\mathcal{E}}$ (the $\mathcal{E}$-part of Γ including the y_i's), where the y_i are treated as constants during unification. We assume that in $\sigma\Gamma$ all multi-equations containing proper $\mathcal{F}$-terms have no $\mathcal{E}$-part, which can be achieved by the rule 'equal terms'.

ASU-step 4. Select a constant-elimination problem corresponding to Γ and $\mathcal{E}$.

This is performed by choosing nondeterministically a constant-elimination problem C consisting of pairs $y_i \notin t_j$, where $y_i \in M_i$, t_j is the $\mathcal{E}$-term in M_j, and M_i contains a proper $\mathcal{F}$-term.

ASU-Step 5. Solve the constant-elimination problem and apply the solution to Γ. A solution is obtained, if afterwards the system is in solved form.

6.1 Example: We demonstrate how to solve x $=f(x*y)$ with the ASU-procedure.

The unfolded normal form of this problem is:

$\langle x = x' = f(z'), y = y', x'*y' = z' \rangle$. ASU-unification and ASU-identification are not applicable (in contrast to GU-identification).

ASU-step 3 : $\qquad$ First we add y_1, which gives the system $\langle x = x' = y_1 = f(z'), y = y', x'*y' = z' \rangle$.

The $\mathcal{E}$-part including the y_i's is only $\langle x' = y_1 \rangle$. Application of the solution $\{x' \leftarrow y_1\}$ yields the multi-equation system: $\langle x = y_1 = f(z'), y = y', y_1*y' = z' \rangle$.

ASU-step 4. There is a cycle: $\{y_1 = f(z'), y_1*y' = z'\}$

The possible pairs for a constant-elimination problem are: $y_1 \notin y_1*y'$ and $y_1 \notin y'$.

Hence there are three possible constant-elimination problems. If $C = \{y_1 \notin y'\}$, then the solution does not resolve the cycle. The constant-elimination problem $\{y_1 \notin y_1*y', y_1 \notin y'\}$ has no solution. So we consider $C = \{y_1 \notin y_1*y'\}$. A most general constant-eliminator of this problem is $\{y' \leftarrow y''(1+y_1)\}$ (see 8.1). Application to our system gives the new system $\langle x = y_1 = f(z'), y = y''(1+y_1), 0 = z' \rangle$. This system has no cycles and hence is in sequentially solved form. A solution is $\{x \leftarrow f(0), y \leftarrow y''*(1+f(0))\}$. Since we have considered all branches, this provides a complete set of solutions.■

The completeness of the ASU procedure implies that the unification types of BR+free functions and of AG+free functions are either unitary or finitary. The example $\langle f(x)*f(y) = f(a)*f(b) \rangle$ [MN86] provides a nonunitary unification problem for the two combinations, hence the unification type is finitary in both cases.

7. Combining Collapse-free, Regular Theories.

In this paragraph we consider the case of combining two theories $\mathcal{E}_1$ and $\mathcal{E}_2$ that are collapse-free and regular. Combination algorithms for this special case have been given by K. Yellick, A. Herold, E. Tidén and C. Kirchner [Ye87, Ki85, He86, Ti86a]. Our aim is to give a very simple algorithm for this case that can be compared to theirs. The described algorithm appears to be closest to the algorithm of C. Kirchner's, which uses variable abstraction. The algorithm of A. Herold [He86] is a bit different, since he uses constant abstraction for the unification step, but I believe that this algorithm can be reformulated within our frame work.

Two important facts for a combination of two collapse-free and regular theories are (cf. [Ye87, Ki85, He86, Ti86a]:

i) if s is a proper $\mathcal{E}_1$-term and t is a proper $\mathcal{E}_2$-term, then s and t are not unifiable.

ii) if Γ has a cycle that contains proper terms from $\mathcal{E}_1$ and $\mathcal{E}_2$, then Γ is not unifiable.

The algorithm has the following basic steps:
  i) Transform Γ into UNF, but do not abstract constants by variables.
  ii) Perform (nondeterministically) unification steps on the $\mathcal{E}_1$ and $\mathcal{E}_2$-part until both the $\mathcal{E}_1$-part and $\mathcal{E}_2$-part are solved. (Note that free constants are both in the $\mathcal{E}_1$-part as well as in the $\mathcal{E}_2$-part).
  Now every multi-equation of the system that has a nontrivial $\mathcal{E}_1$-part and a nontrivial $\mathcal{E}_2$-part has the form x = t.
  iii) Check whether the resulting Γ has a cycle. If there is none, return Γ as solution.

An improvement of the algorithm can be obtained, if not the whole E_j-part is unified, but only a subsystem for which solutions are easily computable. For example if there are two variables x_1, x_2 in the $\mathcal{E}_j$-part and in the same multi-equation, then apply the unifier $\{x_1 \leftarrow x', x_2 \leftarrow x'\}$, where x' is a new variable. This is in effect the variable-canonization described in [Ki85].

This procedure is rather comprehensive and termination is obvious. Why are the termination proofs for the other procedures [Ye87, He86, Ti86a, Fa84] more complicated? In my opinion there are several reasons: All these authors do not have the notion of auxiliary variables and hence cannot get rid of the substitution components introduced by unifiers; the second reason is that using multi-equations and unfolding provides a means for structure sharing; a further reason may be that in general not the core of the problem is considered, which is the unification problem after renaming all significant variables.

8. Some Theories with Solution Procedures for Constant-Elimination Problems.

Besides regular theories, where constant-elimination problems are trivial, there are nonregular theories for which we can describe an algorithm for solving constant-elimination problems.

8.1 Boolean Rings.

The unification problem in Boolean rings is known to be decidable and unitary [MN86, BS86]. We give a method how to solve constant elimination problems C in Boolean rings.

Let $C = \{c_i \notin t_{ij} \mid i = 1,...,n, j = 1,...,m\}$ be a constant elimination problem. Let $C_0 := \{c_i \mid i = 1,...,n\}$ and let $V_0 := V\{t_{ij} \mid i = 1,...,n, j = 1,...,m\} = \{z_k \mid k = 1,...,K\}$. Let D be the set of all possible products of elements in C_0, i.e., $D := \{c_{i1}*c_{i2}*...*c_{ig} \mid \{i_1,...,i_g\} \subseteq \{1,...,n\}\}$. Note that D contains the element 1 as an empty product and hence the set D generated by C_0 has 2^n elements. We try a 'general' substitution σ with $DOM(\sigma) = V_0$. A general representation is $\sigma z_k = \Sigma\{y_{k,d}*d \mid d \in D\}$, where $y_{k,d}$ are different new variables and stand for terms not containing constants from C_0. If we apply σ to C we get the representation $\sigma t_{ij} = \Sigma\{t_{i,j,d}*d \mid d \in D\}$, where $t_{i,j,d}$ is a term not containing constants from C_0. The unification problem Γ_C corresponding to C is as follows: $\Gamma_C := \{t_{i,j,d} = 0 \mid d \in D\}$ where c_i is a factor of d, i = 1,...,n, j=1,...,m }. This unification problem does not contain constants from C_0 and is to be solved without these constants. The obtained mgu can be transformed into a solution of the constant-elimination problem C. Since Boolean rings are unitary unifying, there is at most one most general constant-eliminator necessary in Boolean rings. ❑

As an example we solve the constant elimination problem $C = \{c \notin c*y\}$: Let y := $y_a + y_b*c$, where y_a and y_b are variables that stand for terms not containing c. The problem to be solved is $c \notin c*(y_a + y_b*c)$

or equivalently $c \notin c*(y_a + y_b)$ which is in turn equivalent to the condition $y_a + y_b = 0$, since c is a free constant and y_a, y_b do not contain c. The unique solution is $y_a = y_b$, hence the most general constant-eliminator is $\{y \leftarrow y'(1+c)\}$.

8.2 Abelian Groups.

Unification in free Abelian groups is considered in [LBB84] and it is shown there that it is of type unitary and that a set of most general unifiers can be computed by solving linear Diophantine equations over the integers.

We show how to solve constant elimination problems C in Abelian groups.

Let $C = \{c_i \notin t_{ij} \mid i = 1,...,n, j = 1,...,m\}$ be a constant elimination problem. Let $C_0 := \{c_i \mid i = 1,...,n\}$ and let $V_0 := V\{t_{ij} \mid i = 1,...,n, j = 1,...,m\} = \{z_k \mid j = 1,...,K\}$.

A general solution σ of C has the form $\sigma z_k = \Sigma\{z_{k,c} \mid c \in C_0\} + z_{k,R}$, where $z_{k,c}$ is a variable standing for a term n_c*c, where n_c is an integer and $z_{k,R}$ does not contain constants from C_0. If we apply σ to t_{ij} we obtain $\sigma t_{ij} = \Sigma\{t_{i,j,c} \mid c \in C_0\} + t_{i,j,R}$, where $t_{i,j,c}$ contains all c-terms of σt_{ij}, i.e., all c's and all variables standing for a sum of c's. The condition $c_i \notin t_{i,j}$ is now equivalent to the condition $t_{i,j,ci} = 0$. Thus the solution of the whole problem C can be solved by considering the unification problem $\Gamma_C := \{t_{i,j,ci} = 0 \mid i = 1,...,n, j = 1,...,m\}$. Since unification in Abelian groups is unitary there are at most one most general constant eliminator necessary. ❑

As an example we solve the constant elimination problem $C = \{a \notin a+b+x+y, b \notin a+b+x+y\}$. We replace x by $x_a + x_b + x_R$, where x_a stands for a term $n_1 *a$, x_b stands for a term $n_2 *b$ and x_R stands for a term not containing a's nor b's. We make a similar replacement for y by $y = y_a + y_b + y_R$.

The resulting problem is $\{a \notin a+b+ x_a + x_b + x_R + y_a + y_b + y_R, b \notin a+b+ x_a + x_b + x_R + y_a + y_b + y_R\}$.

This is equivalent to the system $\langle a + x_a + y_a = 0, b + x_b + y_b = 0\rangle$. The most general solution is $\{x_a \leftarrow z_1, y_a \leftarrow (-z_1)+(-a), x_b \leftarrow z_2, y_b \leftarrow (-z_2)+(-b)\}$. This gives the most general constant eliminator $\{x \leftarrow z_1 + z_2 + x_R, y \leftarrow (-z_1)+(-a)+ (-z_2)+(-b)+y_R\}$. ∎

Note that the unification type of Boolean rings plus free function symbols

8.3 Constant-Elimination in Canonical Theories.

Let $\mathcal{E}$ be a theory with a canonical term rewriting system $R_{\mathcal{E}}$. Then the first observation is that every term t in normalform does not contain inessential free variables or constants, since the rewriting relation removes variables from terms, but does not add new variables. Hence a solution θ to a constant-elimination problem $C = \{c_i \notin t_{ij} \mid i = 1,...,n, j = 1,...,m\}$ has the property that $c_i \notin FRC((\theta t_{ij})\downarrow)$. Since we assume that an infinite number of free constants is in the signature, it is sufficient for an investigation of completeness to assume that θ is ground and normalized, eventually replacing variables by new free constants. IGiven a solution θ we consider the unification problem $\langle t_{ij} = (\theta t_{ij})\downarrow\rangle$. Let's try to solve this problem by basic narrowing [Hul80, NRS87]. Narrowing steps have to be performed only on the left hand side, (i.e. on t_{ij}) until there is a derived term t_{ij}' that is syntactically more general than $(\theta t_{ij})\downarrow$. Obviously $FRC(t_{ij}')$ does not contain c_i. Now J.-M. Hullot [Hul80] has shown that basic narrowing is a complete unification procedure for theories admitting a canonical TRS. An application of this result shows that we get a complete set of constant-eliminators, if basic narrowing is performed on the terms t_{ij} and all narrowing substitutions are returned that correspond to a set of derived terms that satisfy the elimination conditions.

For the special case of BNT-theories for which basic narrowing terminates we always obtain a finite complete sets of constant-eliminators. A criterion for termination of basic narrowing given in [Hul80] is that basic narrowing terminates on the right hand sides of the rules in a TRS.

8.4 Remark. In order to have an approximation of the solutions of a constant-elimination problem, it is possible to use an idea of E. Tidén [Ti86a]. Instead of solving $c \notin t$ solve the unification problem $t = t'$, where t' is obtained from t by replacing the constant c by a new constant c' and by renaming all variables in t. A complete set of unifiers to this problem is complete for the constant-elimination problem, but it may contain unifiers that are not eliminators, hence for the exact solution a search for the right instances is necessary. Since the application of a substitution is always correct in our procedures, such an approximation (E. Tidén called it a total complete set of eliminators) may be of practical use.

This idea gives an algorithm for solving singleton constant elimination problems in several theories. Note, however, that in general a set of constant-elimination problems cannot be encoded this way, a counterexample can be constructed from Example 11.4 in [Sch87b]. ∎

9. Decidability of Unification in $\mathcal{E}+$.

A variation of the general procedure proves the following result on decidability of unification in the combined theory $\mathcal{E}+$:

9.1 Theorem. Unification in $\mathcal{E}+$ is decidable, if for every $\mathcal{E}_i$, unification in a combination of $\mathcal{E}_i$ with free function symbols is decidable. ∎

This shows that the combination with free function symbols is in a sense already the hardest case.

The proof uses a variation of the general algorithm, where $\mathcal{E}_j$-unification steps are delayed as long as possible. We use a modified identification and semantica labelling. The last step is to solve the join of a unification problem and a constant-elimination problem. This can be performed in a combination of $\mathcal{E}_j$ with free function symbols for every j. The final argument is that the branching rate without $\mathcal{E}_j$-unification is finite.

It is an important open problem, whether there exists a theory $\mathcal{E}$, such that unification in $\mathcal{E}$ is decidable, but unification in $\mathcal{E}$ with free function symbols is undecidable. Interesting candidates for $\mathcal{E}$ are free semigroups and groups.

Conclusion.

This paper gives a unification procedure for mixed terms in a combination of arbitrary theories. This algorithm is constructed on the base of an $\mathcal{E}$-unification algorithm for every involved theory $\mathcal{E}$ and a method to solve constant-elimination problems in every theory $\mathcal{E}$. As a special case, we have shown that the unification algorithms for Abelian groups and Boolean rings can be extended to deal with free function symbols and that unification remains decidable and finitary. A unification algorithm for problems in the combination of a Boolean ring with free function symbols is going to be implemented by one of our students [CR87].

It is not clear whether there exists a general method to construct an algorithm for constant elimination from a unification algorithm as it is possible for Boolean rings and Abelian groups or whether a theory with decidable $\mathcal{E}$-unification also has decidable constant elimination problems.

Unfortunately, the described combination algorithm has a high computational complexity, for example for every significant variable in the problem we have to guess its semantical labelling. So some research is needed to recognize and eliminate redundancies in this algorithm.

The combination algorithm may also be of use to extend the Knuth-Bendix [KB70] completion procedure modulo equations as described in [LB77, PS81] such that AC1-unification instead of AC-unification can be used, since this paper provides a unification algorithm in a combination of AC-theories, AC1-theories and free function symbols. Since AC-unification problems have in general a lot of AC-unifiers, but only few AC1-unifiers, this may be useful.

The presented procedure also shows that nonregularity of a theory provides a harder problem than unification in a theory with collapse-axioms. E. Tidén [Ti86] presents results for collapse-free but nonregular theories. It should be evaluated, how these results can be adapted for our formalism.

An extension to order-sorted signatures [Sch87a] appears to be no problem. If new variables introduced by unfolding are always of sort TOP and all terms are well-sorted, then the only steps needed in addition is a weakening step.

This paper also shows that the paradigm of unfication as recursive algorithms should be replaced by the paradigm of unification as transforming systems of equations. The latter is more powerful, more intuitive and requires shorter completeness proofs.

References.

BHS87   Bürckert, H.-J., Herold A., Schmidt-Schauß, M., On Equational Theories, Unification and Decidability, LNCS 256, pp. 204-215. (Also to appear in JSC, special issue on unification)

Bü86   Bürckert, H.-J., Some relationships between Unification, Restricted Unification and Matching, in Proc. of 8th CADE, Springer, LNCS 230, pp. 514-524, (1986)

Bü87   Bürckert, H.-J., Matching - A special case of unification?, Technical report, SR-87-08

BS86   Büttner, W., Simonis, H., Embedding Boolean Expressions in Logic Programming, preprint, Siemens AG, München, (1986)

Co84   Colmerauer, A., Equations and inequations on finite and infinite trees, Proc. of the int. Conf. on FGCS, (ed. ICOT), (1984)

CL73   Chang, C., Lee, R. C., Symbolic Logic and Mechanical Theorem Proving, Academic Press, (1973)

CR87   Crone-Rawe, Bernhard, 'Unification algorithms for Boolean rings', Diplom-arbeit, Universität Kaiserslautern, (to appear)

Fa79   Fay, M., 'First Order Unification in an Equational Theory', Proc. 4th CADE, Texas, pp. 161-167, (1979)

Fa84   Fages, F., Associative-Commutative Unification, Proc. of 7th CADE (ed. Shostak, R.E.), LNCS 170, pp. 194-208, (1984)

Fa85   Fages, F., Associative-Commutative Unification, Technical report, INRIA, (1985)

FH83   Fages, F., Huet G., Complete sets of unifiers and matchers in equational theories. Proc. CAAP-83, LNCS 159, (1983).
         Also in Theoretical Computer Science 43, pp. 189-200, (1986)

Ga86   Gallier, J. H., Logic for Computer Science, Harper & Row, (1986)

Gr79   Grätzer, G. Universal Algebra, Springer-Verlag, (1979)

He86   Herold, A., 'Combination of Unification Algorithms', Proc. 8th CADE, ed. J. Siekmann, LNCS 230, pp. 450-469, (1986). Also: MEMO-SEKI 86-VIII-KL, Universität Kaiserslautern, 1985

HS87 Herold, A., Siekmann, J., Unification in Abelian Semigroups, JAR 3 (3), pp. 247-283, (1987)

HO80 Huet, G.,Oppen, D.C., Equations and Rewrite Rules, SRI Technical Report CSL-111, (1980) also in :Formal Languages: Perspectives and open problems, R. Book.(ed), Academic Press, (1982)

Hu76 Huet, G. Résolution d`Équations dans des langages d´ordre 1,2,...,ω, Thèse d`État, Univ. de Paris VII, (1976)

Hu80 Huet, G. Confluent Reductions: Abstract Properties and Applications to Term Rewriting Systems, JACM 27, 4 , pp. 797-821, (1980)

Hul80 Hullot, J.-M., Canonical Forms and Unification, Proc. 5th CADE, LNCS 87, pp.318-334, (1980)

JKK 83 Jouannaud, J.-P., Kirchner, C., Kirchner, H., 'Incremental Construction of Unification Algorithms in Equational Theories', Proc. of 10th ICALP ed J.Diaz, LNCS 154, pp. 361-373 , (1983)

Ki84 Kirchner, C., A New Equational Unification Method: A generalization of Martelli- Montanari´s Algorithm. 7th CADE, LNCS 170, pp. 224-247, (1984)

Ki 86 Kirchner,C.,: 'Computing Unification Algorithms', Conf. on Logic in Computer Science, pp. 206-216, (1986)

Ki87 Kirchner, C., Methods and Tools for Equational Unification, CNRS technical report Nr. 87-R-008, University of Nancy I, (1987)

KB70 Knuth, D.E., Bendix, P.B., 'Simple Word Problems in Universal Algebras', in: Computational Problems in Abstract Algebra, J.Leech ed. , Pergamon Press, Oxford, (1970)

LBB84 Lankford, D., Butler, D., Brady,B., Abelian group unification algorithms for elementary terms., Contemporary Math. 29, pp. 193-199, (1984)

LB77 Lankford D.S, Ballantyne A.M., Decision procedures for simple equational theories with commutative-associative axioms: complete sets of commutative-associative reductions. Report ATP-39, Dept. of Mathematics, Universoty of Texas, Austin, Texas, (1977)

LS78 Livesey, M., Siekmann, J., Unification of Sets and Multisets, SEKI technical report, Universität Karslruhe, (1978)

MM82 Martelli, A., and Montanari, U., An Efficient Unification Algorithm, ACM Trans. Programming Languages and Systems 4, 2, pp. 258-282, (1982)

Ma87 Martin, U., Unification in Boloean rings and unquantified formluae of the first order predicate calculus, (to appear in JAR)

MN86 Martin, U., Nipkov, T.,Unification in Boolean Rings,Proc. 8th CADE, LNCS 230, pp. 506-513, (1986)

NRS87 Nutt, W., Réty, P., Smolka, G., Basic Narrowing Revisited, Technical report SR-87-07, Universität Kaiserslautern, (1987)

PS81 Peterson, G.E., Stickel M.E., Complete sets of reductions for some equational theories, JACM 28,2, pp. 233-264, (1981)

Pl72 Plotkin, G., Building in equational theories, Machine Intelligence 7, pp.73-90, (1972)

Ro65 Robinson.J.A. A machine-Oriented Logic Based on the resolution principle. JACM 12,1 pp.23-41, (1965)

Sch86 Schmidt-Schauss, M., Unification under Associativity and Idempotence is of Type Nullary, JAR 2,3, pp. 277-281, (1986)

Sch87a Schmidt-Schauss, M., Computational aspects of an order-sorted logic with term declarations, thesis, (1987) , (to appear)

Sch87b   Schmidt-Schauss, M., Unification in a Combination of Arbitrary Disjoint Equational Theories, SEKI report SR-87-16, Universität Kaiserslautern, (1987)

Sh84   Shostak, R.E., Deciding Combinations of Theories, JACM 31, pp.1-12,(1984)

Si75   Siekmann, J., Stringunification, Essex university, Memo CSM-7, (1975)

Si86   Siekmann, J.H., Unification Theory, Proc. of ECAI'86, Vol II, p. vi-xxxv, Brighton, (1986)

Si87   Siekmann, J.H., Unification Theory, Journal of Symbolic Computation, (to appear)

St81   Stickel, M., 'A unification algorithm for associative-commutative functions', Journal of the ACM 28 (3), pp. 423-434 (1981)

St87   Stickel, M., 'A comparison of the variable-abstraction and constant-abstraction method for associative-commutative unification', JAR 3, pp. 285-289, (1987)

Sz82   Szabo, P., Theory of first order unification, (in German), Thesis, University of Karlsruhe, (1982)

Ti86a   Tidén, E., First-Order Unification in Combinations of Equational Theories, Thesis, Stockholm, (1986)

Ti86b   Tidén, E.: 'Unification in Combination of Collapse Free Theories with Disjoint Sets of Function Symbols', Proc. 8th CADE, LNCS 230, pp. 431-449, (1986)

Ye87   Yelick, K.A. Unification in Combinations of Collapse-free Regular Theories. J. of Symbolic Computation 3, pp. 153-181, (1987)

PARTIAL UNIFICATION FOR GRAPH BASED EQUATIONAL REASONING

Karl Hans Bläsius
IBM Deutschland GmbH
WT LILOG
Postfach 80 08 80
D-7000 Stuttgart 80
West Germany

Jörg H. Siekmann
University of Kaiserslautern
FB Informatik
Postfach 3049
D-6750 Kaiserslautern
West Germany

ABSTRACT

The problems of mechanizing equational reasoning are discussed and two prominent approaches (E-resolution and RUE-resolution) that build equality into a resolution based calculus are evaluated. Their relative strengths and weaknesses are taken as a motivation for our own approach, whose evolution is described.

The essential idea in our equational reasoning method is to store the information about partially unified terms in a graphlike structure. This explicit representation supports a goaldirected planning approach at various levels of abstraction.

Keywords: Unification, built-in equality, clause graphs with equality, planning in abstraction spaces.

This work was supported by the Sonderforschungsbereich 314, (Artificial Intelligence) of the German Research Agency (DFG).

1. Introduction

Equality is the most prominent relation to be built into an automated deduction system, as it provides an important representational basis for the formulation of first order theories [Tar 68] [Tay 79]. Unfortunately it also posed some of the most formidable obstacles against its computational treatment. In particular, the explicit and naive use of the standard equality axioms (reflexivity, symmetry, transitivity and substitution axioms) turned out to be insufficient as a basis for automated equational reasoning. Hence the last twenty years saw a proliferation of deductive techniques to incorporate the equality relation somehow directly into the inference machinery. There are currently at least four research communities with their own international conferences that support active research in equational reasoning: logic programming, unification theory, term rewriting systems and general automated deduction. Their respective aims are to develop

(i) Special purpose unification algorithms for common equational axioms such as associativity, commutativity, Boolean Rings or Abelian Groups (see [Sie 86] for a survey) .

(ii) General purpose unification algorithms based on narrowing [NRS 87] or decomposition [Kir 87] that solve equality problems in a given *class* of equational theories. They are mainly developed for logic programming with equality [SNGM 87].

(iii) Demodulation and term rewriting systems (see [HO 80] and [Buc 87] for a survey).

(iv) General inference rules to handle equality, usually in combination with a standard inference rule such as resolution [WRCS 67] [Mor 69] [Sib 69] [And 70] [Bra 75] [Sho 78] [Dig 79] [LH 85].

This paper presents a technique that integrates the second and the fourth approach with the essence of clause graph theorem proving [Kow 75]. It summarizes our experience in mechanizing equational reasoning: During a timespan of more than eight years our original plans of building equality into a connection graph based deduction system [SW 80] changed considerably in the light of experimental evidence.

Among the various methods proposed within the fourth paradigm (iv), is *paramodulation* [RW 69]: given the resolution rule and the paramodulation rule of inference, the equality axioms become superfluous except for the reflexivity axiom [Rus 87].

Although paramodulation was a substantial improvement compared to the axiomatical formalization of the equality relation, it still leads to enormous search spaces, as this inference rule can be applied almost everywhere in the clause space. For example the moderately difficult problem, that in an associative system with degree two (i.e. $x^2 = e$) we have commutativity, has an estimated search space of about 12^{10} clauses. Hence the application of paramodulation must be tightly controlled.

The problem is:

(i) how to represent the information upon which this control can be based and

ii) how to provide the equational reasoning system with a more goal oriented bahaviour.

Equational axioms - or the equality literals within a clause - have been used to pursue essentially two goals:

1. The *simplification* of a given state (i.e. a literal, a clause, a clause graph etc.)

2. The *reduction* of the *difference* between two given states

An analysis of these two goals spawned most of the above techniques and logical calculi that incorporate equality, and we like to classify them according to these two criteria.

For example if an equation can be directed such that the right hand side becomes "smaller" or "simpler" in some sense than the left hand side, this directed equation can advantageously be used in a simplification task. Demodulation or term rewriting systems are good examples for this first approach.

The second goal rejects the indiscriminate use of equality also, however it exploits the observation that no working mathematician would apply an equationally represented fact, unless used directly for some purpose, i.e. only "if needed".

Both approaches, simplification and difference reduction, have their advantages and are worth investigating. However, whereas most research is based with remarkable success on simplification mechanisms, especially on demodulation and term rewriting, difference reduction methods have found less attention, although they are at least as important for an automated deduction system.

The potential of difference reduction methods will be illustrated in this paper and a method that promises some progress is proposed.

2. Difference Reduction by E-Resolution

J. B. Morris' E-resolution [Mor 69] is the most explicit realization of the goal to use paramodulation only "if needed". The aim is to remove the differences between corresponding terms of two potentially resolvable literals (i.e. the same predicate symbol with opposite sign), such that an inference step by resolution becomes possible. E-resolution may be viewed as a sequence of paramodulation steps applied to two potentially resolvable literals until they are unifiable. It is then followed by the appropriate resolution step, and intermediate clauses are discarded.

This is a generalization of ordinary unification based resolution, and is a specialization of the more recently developed theory resolution technique of M. Stickel [Sti 85].

Similar to the "means-end" analysis in GPS [NSS 59], the role of equality in E-resolution is that of "a means to an end" i.e. to produce favorable preconditions for a resolution step. Two potentially resolvable literals could be selected by some global search strategy, followed by the search for appropriate equations, these equations are applied if the differences can be completely removed.

E-resolution defines the cooperation between resolution and equality reasoning in general terms, but the calculus does not actually support the achievement of its goal: there remains the difficult search problem to find the appropriate equality steps. Since equality of two terms with respect to a given set of equations is undecidable in general, it is impossible to continue searching for equations until the potentially resolvable literals under consideration are definitely unifiable or not. Hence the main problem is: How to find the appropriate E-resolution steps?

3. Difference Reduction Based on Partial Unification and Equality

The search problems necessitated by E-resolution are hard to solve and led to the conclusion that in most cases an E-resolution step is just too large. Realizing this, V. J. Digricoli proposed a form of partial E-resolution, which he called RUE-resolution. Essentially his calculus works similar to 'E-resolution' [Dig 79], however, a deduction step is also possible if the differences between corresponding terms cannot be removed completely. A partial unifier is applied to the literals, the remaining disagreement pairs are computed and added as negated equations to the derived clause, as in D. Brand's modification method [Bra 75].

As an example consider the following set of clauses

$$\{\{P\ q(h(f(a\ z))\ g(k(y\ c)\ f(a\ a)))\},$$
$$\{\neg\ P\ q(h(f(x\ a))\ g(h(x)\ x))\},$$
$$\{f(f(u\ v)\ w)\ =\ f(u\ f(v\ w))\},$$
$$\{b = c\},$$
$$\{h(u) = k(u\ b)\}\}.$$

The substituion $\sigma = \{x \leftarrow a,\ z \leftarrow a\}$ is a partial unifier for the literals of the first and the second clause, hence the clause

$$\{k(y\ c) \neq h(a),\ a \neq f(a\ a)\}$$

can be derived. Now the first inequation can be solved using the remaining equality literals of the clause set, but the second problem is not solvable, since the terms a and f(a a) are not equal with respect to the given equations.

But then another partial unifier could be used, for example $\tau = \{x \leftarrow f(a\ a)\}$, and the clause

$$\{k(y\ c) \neq h(f(a\ a)),\ f(a\ z) \neq f(f(a\ a)\ a)\}$$

could be derived. But now the first inequation must be solved again, performing the same operations as before. For this and other reasons Digricoli proposes to compute a "most general partial unifier" and to instantiate as late as possible. But of course the general problem remains that there is a certain indeterminism in how far a partial unifier is to be computed.

In case the standard unification algorithm of A. Robinson [Rob 65] is used, it does not matter, if partial unifiers are applied to other subterms, as there is always at most one most general unifier: If the application of this unifier leads to the unsolvability of one of the subproblems, then the whole problem is definitely unsolvable too.

However if special purpose unification algorithms are involved the situation changes significantly. Generally there may be many most general unifiers for some subproblems, and it is unknown which of these unifiers are compatible with solutions of other subproblems. The immediate application of partial unifiers to other subproblems results in an arbitrary propagation of a combination of substitutions and if there is a conflict after several steps, it is impossible to detect its origin. Only blind backtracking can help and already solved subproblems must be solved over and over again.

4. Equality Graphs

4.1. Paramodulated Clause Graphs

Our first attempt to build-in equality was based within the paradigm of connection graph proof procedures [Kow 75], where potentially resolvable literals are connected by a link. These links explicitly represent the possible resolution steps and provide a rich source of information for selection and reduction heuristics. As this representation turned out to be an indispensable tool in many current deduction systems, we tried to extend the essential idea to equational reasoning: Each side of an equation is connected by a P-link to every place in the clause graph, where it can be applied as a paramodulant [SW 80]. But alas, already the first implementation revealed: there are just far too many places and the system was drowned by the exploding number of P-links.

4.2. Constraints in Paramodulated Clause Graphs

Our second attempt to solve general equality problems was still based on paramodulated clause graphs, however they were only searched for compatible combinations of P-links. A *compatible* combination of P-links represents an executable paramodulation sequence, that modifies two literals such that they become resolvable [Blä 83]. Such paramodulation sequences can be displayed in a graph, which represents a solution of the given problem. The following example shows such a graph:

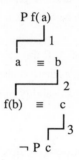

Example 1

However not every graph represents an executable, i.e. compatible, sequence of paramodulation steps as the following two examples show:

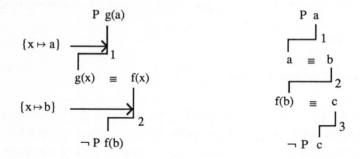

Example 2   **Example 3**

The combination of the P-links 1 and 2 in example 2 is impossible because their unifiers $\{x \to a\}$ and $\{x \to b\}$ are incompatible. In example 3 the P-links 1 and 2 are incompatible, because after paramodulation on link 1, link 2 cannot be inherited to the paramodulant Pb since the access depths do not coincide. An equality graph that cannot be executed is called *incompatible*, otherwise it is *compatible*.

The problem is to find compatible graph structures. To this end certain constraints can be stated, which are necessary but not sufficient for the compatibility of a graph and which are fast and inexpensive to test.

Some of the constraints we tested out were: (1) All unifiers of the involved P-links must merge to one most general unifier. (2) For each maximal chain of P-links in a graph the sum of all access depths must be equal to zero and each partial sum must be less than or equal to zero. (3) Each combination of P-links containing an incompatible substructure is incompatible too.

Practical experiments with a procedure based on the above constraint satisfaction method have shown however, that the set of potential graphs (i.e. combinations of P-links), which have to be created in order to test for compatibility, is still far too large, even in relatively simple examples. Especially P-links connecting variables (to everything else) make the procedure extremely inefficient. In most natural examples taken from mathematics the axioms contain many variables, and only the negated theorem contains Skolem-constants which may be effectual in constraints. Since variables can be instantiated to any terms, many of the possible combinations of operations involving axioms are compatible. In termini of graph structures: most subgraphs (combinations of P-links) are compatible, only when the subgraphs are combined or extended to the potential final solution graph, the incompatibility is detected. But for subgraphs which are in fact compatible, no constraints exist to detect an incompatibility.

The observation that such constraint satisfaction methods, albeit of great potential in artificial intelligence and logic programming, are not particularily valuable tools for equational reasoning tasks, was called the "Anti-Waltz-Effect" and discussed elsewhere [Blä 86]. The main reason for the failure of this approach is the lack of strong constraints in this application domain, which consequently do not sufficiently reduce the search space.

As a consequence the equality procedure was modified several times, and the experimental modifications finally led to an equality graph construction procedure (EGC-procedure), which constructs compatible graphs without ever creating the enormous set of incompatible ones in the first place. The essential idea is this: Starting from some initial state (some graph) a sequence of transformations is performed on each subsequent state until a compatible graph is constructed.

4.3. The Construction of Compatible Equality Graphs

As it turned out the EGC-procedure incorporates and extends the ideas underlying Digricoli's RUE-resolution and Morris´ E-resolution, however it is based on a different form of partial unification. The method will be informally described in the following, it is defined within a formal framework in [Blä 87].

We like to demonstrate the structure of equality graphs and their construction with the following example: Let E be a set of equations

$$\{g(x\ x) \equiv h(x\ b),\ h(u\ v) \equiv h(v\ u),\ h(b\ a) \equiv f(b),\ b \equiv c,\ c \equiv e\}$$

and suppose we want to resolve upon two literals, say $P(g(a\ y)\ z)$ and $\neg P(f(e)\ b)$. While the second argument of P and $\neg$P can be straightforwardly unified, the first argument presents a problem. Let us concentrate upon the first two corresponding argument terms by denoting this as our given equality problem: $< g(a\ y)\ \equiv_E\ f(e) >$.

The initial equality graph for this problem can be displayed as:

g(a y)

f(e)

where dotted lines indicate problems to be solved. The only information in this graph is that the problem $< g(a\ y)\ \equiv_E\ f(e) >$ is yet unsolved. The main discrepancies are the different toplevel symbols g and f, hence this difference must be removed by some equations. Two equations in E can be combined to form a *chain* : $g(x\ x) \equiv h(x\ b)$ - - - - $h(b\ a) \equiv f(b)$, which can be used for the removal of this discrepancy and which is therefore inserted into the graph:

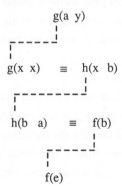

g(a y)

g(x x)   ≡   h(x b)

h(b a)   ≡   f(b)

f(e)

Now there are three subproblems to be solved: $< g(a\ y)\ \equiv_E\ g(x\ x) >$, $< h(x\ b)\ \equiv_E\ h(b\ a) >$ and $< f(b)\ \equiv_E\ f(e) >$. In all three cases the heads of both terms are equal, but now the subterms may generate new subproblems, some of which are trivially solvable. We obtain the equality graph:

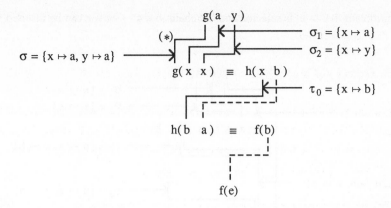

The solid links represent solved subproblems and are labeled with a substitution, empty substitutions are omitted. The substitutions σ_1 and σ_2 for the terms g(a y) and g(x x) must be checked for compatibility, i.e. they must be unifiable themselves. The result of this unification is represented as a separate link (*) labelled with the unifier $\sigma = \sigma_1 * \sigma_2$, where * is the merge operation for substitutions (essentially the most general unifier of two substitutions).

Suppose we now select the unsolved subproblem b - - - a : This subproblem is unsolvable, since we cannot build a chain of equations from the given set E to connect b and a. However instead of solving b - - - a we can create a new subproblem at a higher term level: h(x b) - - - h(b a). There exists an appropriate equation in E: h(u v) ≡ h(v u) which is now inserted into the graph with the result:

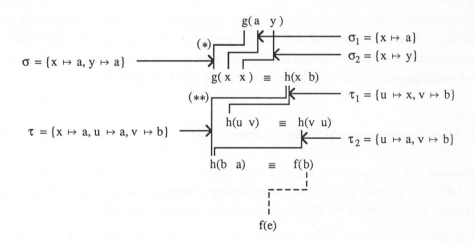

The substitutions within each chain must be checked for compatibility and the result is again represented as a separate link. In our example the unit chain h(u v) ≡ h(v u) connects the terms h(x b) and h(b a), the substitutions τ_1 and τ_2 are to be unified and the result is represented as the link (**), which is labelled with the unifier τ.

Finally the subproblem b - - - e is selected and the chain $b \equiv c$ - - - $c \equiv e$ can be inserted with the result:

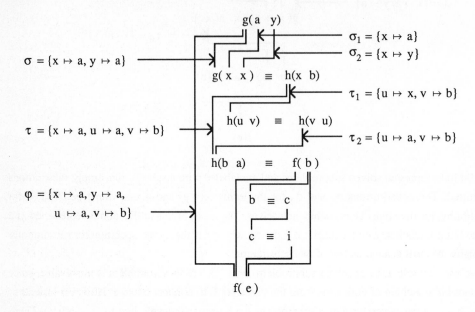

Since the substitutions σ and τ can be merged to $\varphi = \sigma * \tau$, the given equality problem is completely solved.

The transformation rules that turn a given partially solved graph (a partial plan) into a successor graph with fewer unsolved subproblems, are formally defined in [Blä 87]. They are implemented as production rules, that fire when a particular situation in the graph is encountered; they constitute the actual *metareasoning calculus*, which can be shown to be sound and complete.

4.4. Term Graphs and Substitution Graphs

Unfortunately the previous section tells only part of the story: there are still unresolved problems.
Let $< f(x\ x) \equiv_E f(g(a)\ g(b)) >$ be a given equality problem and let $E = \{a \equiv b\}$ be the set of equations. Each subproblem has a trivial solution, but they are not compatible:

In particular, we can not find a chain in E that yields an overall solution. However equations could be inserted as follows:

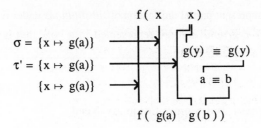

and this graph represents a final solution. Here we are using the functional reflexive axiom $g(y) = g(y)$. The notorious and well known problem, whether or not these axioms have to be used for completeness results, has been called the *paramodulation conjecture* and has been settled positively for some equational calculi [Bra 75] [Ric 78] [Rus 87].We take the view that the indiscriminate insertion of equations in situations as above (i.e. between the variable x and the term $g(b)$) is one of the major reasons for the explosion of the search space, and should not be permitted - irrespective of the theoretical completeness problem at hand: even if such permissiveness may turn a calculus into a complete one, an actual implementation would never find a solution anyway, because of its desatreous effect on the search space.

Hence we tried a different approach: equations are inserted between the unifiers themselves (i.e. the substitutions are unified under the theory) and the resulting discrepancies are removed where they occur. For the above example we obtain a graph structure as follows:

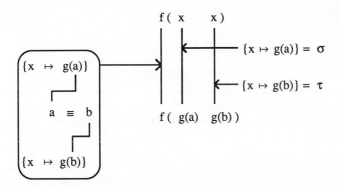

Here σ and τ are the only unifiers for the subproblems, they are not compatible and it is not permitted to insert equations between a variable x and a term t, as long as x is not a subterm of t.

Now there are two types of equality graphs that correspond to different kinds of subproblems:

- Equality graphs that represent local and partial unification of *terms* under E. The unifier of such a subproblem is never applied to other subproblems. Such graphs are called *term graphs*.
- Equality graphs that represent partial unification of *substitutions* under E. These graphs are called *substitution graphs* and are represented in a box, that is linked to the corresponding place in the term graph.

4.5. Subgraph Replacement and Computation by Need

For a given subproblem there are usually many possible equality-chains which could be inserted and which would lead to different solutions. While solving a subproblem it is impossible to know which of these alternative partial solutions is the best from a global point of view: it appears to be necessary to consider all alternatives. However, to construct a new subgraph for each possible equality-chain is far too inefficient (there would be an enormous space of graphs), hence we need an additional method to handle alternative partial solutions.

The general idea is called "computation by need" or more specifically "lazy unification" [Bür 86] and comes in the following disguise in our case: sometimes a solved subproblem has to be replaced by another subgraph, that represents a different solution of the same subproblem. For example let

$$E = \{h(f(x\ a)) \equiv g(h(x)\ x), b \equiv c, h(u) \equiv k(u\ b), f(f(u\ v)\ w) \equiv f(u\ f(v\ w))\}$$

and let $< h(f(a\ z))\ \equiv_E\ g(k(y\ c)\ f(a\ a)) >$ be the given equality problem. Suppose we already have the following equality graph:

The dotted line represents an unsolved subproblem: the unification of the substitutions σ and τ. This is unsolvable as: σ and τ are not unifiable under E. But the position that causes the incompatible assignment to the variable x can be localized within the graph, and the graph can be modified *locally* at that position with no effect on other parts of the graph, i.e. without the destruction of the hitherto existing plan for a proof. For example we can insert the equation f(u f(v w)) ≡ f(f(u v) w) between f(a z) and f(x a) and replace the subgraph

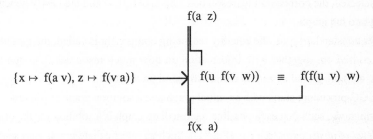

by the subgraph

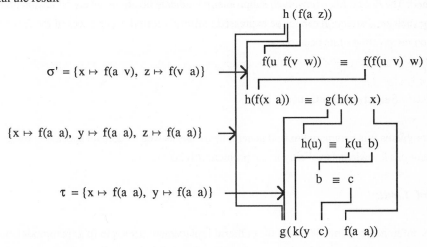

with the result

which is a solution for the given problem. This replacement of subgraphs neither destroys the solution of other subproblems nor does it destroy the global plan.

Sometimes however it may be necessary to retrieve the previous solution, hence we implemented a representation of "multiple graphs" that contains the information of the old and the new equality graph in a compact form.

4.6. Planning in Abstraction Spaces

In view of the immense search spaces an equational reasoning tool has to cope with, goal directed planning at various levels of abstraction appears to be an indispensable tool.

There are several levels of abstraction to be distinguished: the first task is to select two potentially resolvable literals as candidates. This is accomplished as follows: the clause graph representation is enhanced by socalled PER-links (potential E-resolution links) that connect literals with the same predicate symbol and opposite sign.

On the basis of domain specific knowledge, general heuristics or global planning aspects either an R-link (the standard links of a connection graph) or a PER-link is selected for execution. In case an R-link is selected, the corresponding resolution step is executed and the newly generated resolvent is connected into the graph.

In the latter nonstandard case, the equality reasoning component is called, the two literals to be made equal are passed on, together with information on how much resources (e.g. cpu-time, number of cons-cells, etc.) the global system is willing to spend on this task.

Now the EGC-procedure starts with an initial graph and a solution graph is constructed by a sequence of transformations, such that each possible intermediate graph is a solution for the original problem at a certain *level of abstraction* (see e.g. [Pla 81]). Such an intermediate graph represents a global as yet unfinished plan for the search for a solution . The abstraction is weakened by each step, and the graph is refined. The dotted lines (unsolved subproblems) indicate positions where an abstraction is used. Usually such an abstraction cannot be expressed uniformly for all occurrences of the function, but depends on the position of its occurrence.

4.7. Heuristic Selectionfunctions

For difficult problems it is paramount to find general heuristics as well as domain specific knowledge and to represent this knowledge explicitly in the problem solving system.

Weights and Limits

Heuristics for inference systems based on the subterm replacement principle (e.g. paramodulation, demodulation) are usually computed from the information implicit in a term (literal, clause, or graph) [BES 81], [WOLB 84]. For example weight, ordering, size, nesting depth etc are defined and are often based on certain symbol occurrences, like the number of different variables in a literal, etc. The main purpose of these heuristics is to control and reduce the complexity in terms of the defined measure.

Since the information of terms is available, this class of heuristics is also applicable in methods based on the difference reduction principle. Note that this is not always the case: structure sharing [BM 72], the internal representations of the "Argonne National Lab" theorem prover [LO 80], compilation of PROLOG (e.g. into the Warren Machine [War 83]) and other methods often destroy the explicitness of this information.

Difference Reduction

Equality reasoning systems based on the difference reduction principle compare two terms. A measure for the difference of two terms can be defined and heuristics can be developed to control the reduction of these differences. Just like the operators in GPS, they estimate which equations could best reduce these differences. As before the difference measure may depend on term structures and symbol occurrences. In contrast to the previous heuristics based on weights and limits, difference reduction heuristics not only restrict term sizes etc. but control the distance between terms, i.e. are directed towards the goal to make two terms equal.

Heuristics of this type are for example used by Digricoli [Dig 85] who defines a heuristic ordering by degree of unification. His ordering depends on the equality of the toplevel function symbols and on the number of unifiable subterms.

Conflict solving

In our method, the states of the inference system have a structure rich enough to incorporate both kinds of heuristics. In particular, heuristics can be incorporated that support the planning of the search for a proof from a global point of view. In the example of section 4.5 a conflict occurs in the attempt to assign two different terms f(a a) and a to the variable x. Performing a subgraph replacement, other partial solutions can be derived, possibly leading to compatible unifiers. In our example the system should create a new subgoal: extend the graph such that the variable x is assigned to a term f(. .). An appropriate operation could be activated by a heuristic prefering the insertion of an equality chain of the form

$$f(. .) \equiv f(f(. .) .)$$

between f(a z) and f(x a), such that after the insertion x is assigned to a term with topsymbol f:

$$f(a\ z)\ \text{---}\ f(. .) \equiv ... \equiv f(f(. .) .)\quad \text{---}\quad f(x\ a).$$
$$|_____|$$

The link indicates the condition which guaranties the appropriate extension and should be used by the heuristic guiding the subgraph replacement.

To sum up: at each step the graph contains the information for the correlation of the solved subproblems and also enough information to find the subproblems which still are to be solved. On the basis of this information the position can be *localized* where the graph is to be modified in order to make solutions of subproblems compatible from a global point of view.

5. Conclusion

Partial unification is an essential basis for goal directed equational reasoning as it allows to "devide and conquer": each subproblem is tackled at its time.

In general, there are many partial unifiers for a given subproblem and it is unknown which solution is compatible with solutions of other subproblems, hence the dependencies between the partial solutions (partial unifiers) have to be represented. It is claimed that equality graphs as described in this paper are a suitable data structure to represent such dependencies and dependency directed backtracking mechanisms as in truth maintenance systems can be used. Furthermore, the equality graphs represent information useful for a hierarchical planning component, which globally guides the search for a proof. Also a new class of heuristics is available that exploit the information represented by the graphs.

Although our current system solves most of the test examples known from the literature and even succeeded in finding a proof of some really hard problems, there are still problems Digricoli´s system and the deduction system at the Argonne National Laboratory can solve, but we can not. We believe that this is due to the state of development of our current implementation. In order to improve its performance currently work focusses on the development of more powerful heuristics and knowledge based planning capabilities, that exploit the rich information explicitely represented in the equality graph [Prä 87].

References

And 70   R. Anderson: *Completeness results for E-resolution,*
           Proc. Spring Joint Conf., 653-656, 1970

BES 81   K.H. Bläsius, N. Eisinger, J. Siekmann, G. Smolka, A. Herold, C. Walther:
           The Markgraf Karl Refutation Procedure,
           Proc. IJCAI, 511-518, 1981

Blä 83   K.H. Bläsius: *Equality Reasoning in Clause Graphs,*
           Proc. IJCAI, 936-939, 1983

Blä 86   K.H. Bläsius: *Against the 'Anti Waltz Effect' in Equality Reasoning,*
           Proc. German Workshop on Artificial Intelligence, Informatik-Fachberichte 124,
           Springer, 230-241, 1986

413

Blä 87   K.H. Bläsius: *Equality Reasoning Based on Graphs,*
            SEKI-REPORT SR-87-01 (Ph.D. thesis), Fachbereich Informatik, Universität
            Kaiserslautern, 1987

BM 72   R.S. Boyer, J.S. Moore: *The Sharing of Structure in Theorem-proving Programs,*
            Machine Intelligence 7, Edinburgh University Press, 101-116, 1972

Bra 75   D. Brand: *Proving Theorems with the Modification Method,*
            SIAM Journal of Comp., vol 4, No. 4, 1975

Buc 87   B. Buchberger: *History and Basic Features of the Critical-Pair/Completion Procedure,*
            Journal of Symbolic Computation, Vol. 3, Nos 1 & 2, 3-38, 1987

Bür 86   H.-J. Bürckert: *Lazy Theory Unification in Prolog: An Extension of the Warren
            Abstract Machine,*
            Proc. GWAI-86, IFB 124, Springer Verlag, 277-289, 1986

Dig 79   V.J. Digricoli: *Resolution by Unification and Equality,*
            Proc. 4th Workshop on Automated Deduction, Texas, 1979

Dig 85   V.J. Digricoli: *The Management of Heuristic Search in Boolean Experiments with
            RUE-resolution,*
            Proc. IJCAI-85, Los Angeles, 1985

HO 80   G. Huet, D. Oppen: *Equations and Rewrite Rules: A survey,*
            Technical Report CSL-111, SRI International, 1980

Kow 75   R. Kowalski: *A Proof Procedure Using Connection Graphs,*
            JACM 22, 4, 1975

Kir 87   C. Kirchner: *Methods and Tools for Equational Unification,*
            Proc. of Colloq. on Equations in Algebraic Structures, Lakeway, Texas, 1987

LH 85   Y. Lim, L.J. Henschen: *A New Hyperparamodulation Strategy for the Equality
            Relation,*
            Proc. IJCAI-85, Los Angeles, 1985

LO 80   E.L. Lusk, R.A. Overbeek: *Data Structures and Control Architecture for
            Implementation of Theorem-Proving Programs,*
            Proc. 5th CADE, Springer Lecture Notes, vol. 87, 232-249, 1980

Mor 69   J.B. Morris: E-resolution: *An Extension of Resolution to include the Equality Relation,*
            Proc. IJCAI, 1969, 287-294

NRS 87   W. Nutt, P. Rety, G. Smolka: *Basic Narrowing Revisted,*
            SEKI-Report SR-87-7, Univ. of Kaiserslautern, 1987;
            to appear in J. of Symbolic Computation, 1988

NSS 59   A. Newell, J.C. Shaw, H. Simon: *Report on a General Problem Solving Program,*
            Proc. Int. Conf. Information Processing (UNESCO). Paris, 1959

Pla 81   D. Plaisted: *Theorem Proving with Abstraction,*
            Artifical Intelligence 16, 47 - 108, 1981

Prä 87   A. Präcklein: *Equality Reasoning,*
            Internal Working Paper, Univ. of Kaiserslautern, 1987

Ric 78   M. Richter: *Logik Kalküle,*
             Teubner Verlag, 1978

Rob 65   J.A. Robinson: *A Machine-Oriented Logic Based on the Resolution Principle,*
             JACM 12, 1965

Rus 87   M. Rusinowitch: *Démonstration Automatique par des Techniques de Réécriture,*
             Thèse d'état, CRIN, Centre de Recherche en Informatique de Nancy, 1987

RW 69   G. Robinson, L. Wos: *Paramodulation and TP in first order theories with equality,*
             Machine Intelligence 4, 135-150, 1969

Sho 78   R.E. Shostak: *An Algorithm for Reasoning About Equality,*
             CACM, vol 21, no. 7, 1978

Sib 69   E.E. Sibert: *A machine-oriented Logic incorporating the Equality Axiom,*
             Machine Intelligence, vol 4, 103-133, 1969

Sie 86   J. Siekmann: *Unification Theory,*
             Proc. of European Conf. on Artificial Intelligence (ECAI), 1986
             full paper to appear in J. of Symbolic Computation, 1988

SNGM 87   G. Smolka, W. Nutt, J. Goguen, J. Meseguer: *Order-sorted equational computation,*
             in M. Nivat, H. Ait-Kaci (eds): Resolving Equational Systems, Addison-Wesley, to
             appear 1988

Sti 85   M. Stickel: *Automated Deduction by Theory Resolution,*
             Journal of Automated Reasoning Vol. 1, No. 4 (1985), 333-356

SW 80   J. Siekmann, G. Wrightson: *Paramodulated Connectiongraphs,*
             Acta Informatica 13, 67-86, 1980

Tar 68   A. Tarski: *Equational Logic and Equational Theories of Algebra,*
             in: Schmidt et.al. (eds): Contribution to Mathematical Logic, North Holland, 1986

Tay 79   W. Taylor: *Equational Logic,*
             Houston Journal of Maths, 5, 1979

War 83   D.H.D. Warren: *An Abstract Prolog Instruction Set,*
             SRI Technical Note 309, Sri International, October 1983

WOLB 84   L. Wos, R. Overbeek, E. Lusk, J. Boyle: *Automated Reasoning, Introduction and
             Applications,* Prentice Hall, 1984

WRCS 67   L. Wos, G. Robinson, D. Carson, L. Shalla: *The Concept of Demodulation in Theorem
             Proving,* J. ACM 14, 698 - 709, 1967

SATCHMO: a theorem prover implemented in Prolog

Rainer Manthey and François Bry

ECRC

Arabellastr. 17, D-8000 Muenchen 81

West Germany

Abstract

Satchmo is a theorem prover consisting of just a few short and simple Prolog programs. Prolog may be used for representing problem clauses as well. SATCHMO is based on a model-generation paradigm. It is refutation-complete if used in a level-saturation manner. The paper provides a thorough report on experiences with SATCHMO. A considerable amount of problems could be solved with surprising efficiency.

1. Introduction

In this article we would like to propose an approach to theorem proving that exploits the potential power of Prolog both as a representation language for clauses and as an implementation language for a theorem prover. SATCHMO stands for 'SATisfiability CHecking by MOdel generation'. It is a collection of fairly short and simple Prolog programs to be applied to different classes of problems. The programs are variations of two basic procedures: the one is incomplete, but allows to solve a wide range of problems with considerable efficiency; the other is based on a level-saturation organization thus achieving completeness but partly sacrificing the efficiency of the former.

Horn clause problems can be very efficiently solved in Prolog provided they are such that the Prolog-specific limitations due to missing occurs check and unbounded depth-first search are respected. As an example we mention Schubert's Steamroller [WAL 84], a problem recently discussed with some intensity: the problem consists of 27 clauses, 26 of which can be directly represented in Prolog without any reformulation and is checked for satisfiability within a couple of milliseconds by any ordinary Prolog interpreter. The idea of retaining Prolog's power for Horn clauses while extending the language in order to handle full first-order logic has been the basis of Stickel's "Prolog Technology Theorem Prover" (PTTP) [STI 84]. Stickel proposes to overcome

Prolog's limitation to non-Horn clauses by augmenting its backward reasoning mechanism by the model-elimination reduction rule. In addition he employs unification with occurs check and consecutively bounded depth-first search for achieving a complete inference system.

We propose a different way of overcoming Prolog's limitations. We introduce a second type of rules in order to be able to represent those clauses that cannot be handled by Prolog. SATCHMO can be viewed as an interpreter for this kind of rules treating them as forward rules in view of generating a model of the clause set as a whole.

A crucial point with respect to the feasibility of the approach was the observation that range-restriction of clauses may be favorably exploited when reasoning forward. A clause is called *range-restricted* if every variable occurs in at least one negative literal. (The notion 'range-restricted' was first introduced in the context of logic databases in [NIC 79]). When reasoning forward, all negative literals are resolved first. For range-restricted clauses this leads to a complete instantiation of the remaining positive literals. Ground disjunctions can be split into their component literals as has been done, e.g., in early proof procedures based on tableaux calculus [SMU 68]. The case analysis-like style of treating non-Horn clauses introduced by splitting can lead to very elegant solutions as compared with the way they are handled by PTTP. Range-restriction in combination with forward reasoning overcomes Prolog's deficencies for Horn clauses as well: infinite generation due to recursive clauses can be prevented by a subsumption test for ground atoms that is very cheap compared to 'ancestor test' and bounded search. No occurs check is needed because at least one of two literals to be unified is always ground.

A major advantage of our solution is the fact that one can afford to implement the necessary additions very easily on top of Prolog itself, whereas the additional features for the PTTP have to be implemented by extending a Prolog system. This does not prevent our approach from being implemented on a lower level as well.

A principle drawback of the approach lies in the fact that clauses which are not range-restricted may require a full instantiation of certain variables over the whole Herbrand universe. In the presence of functions this may lead to the well-known combinatorial explosion of instances. Recursive problems with functions in particular will hardly be solvable efficiently in presence of non-range-restricted clauses. As arbitrary clause sets can be transformed into range-restricted form while preserving satisfiability, one can handle instantiation naturally within the framework of range-restriction.

There are, however, much more cases than one might expect where the limitations mentioned do not harm. The efficiency SATCHMO obtains in such cases is remarkable and surprising. In order to give evidence to this claim we devote a major part of the paper to reporting about our experience with a fairly huge collection of example problems taken from recent publications. The full Steamroller, e.g., has been solved in 0.3 secs.

In an earlier paper [MB 87] we have described our model generation approach on the basis of forward rules only. In this context, model generation can be explained and justified on the basis of hyperresolution. This paper is an informal one in the sense that we don't give proofs or formal definitions. Instead we thoroughly motivate our proposal and provide full Prolog code as a specification.

Throughout the paper we employ the Prolog style of notation: Variables are represented by uppercase letters, Boolean connectives and/or by means of ',' and ';' respectively. Only a very basic knowledge about Prolog is required; as an introduction refer, e.g., to chapter 14 in [WOS 84].

In order to represent the new kind of rules inside Prolog we assume that a binary operator '--->' has been declared. Clauses that are not represented as Prolog rules may then be represented in implicational form as $(A_1, \ldots ,A_m \text{ ---> } C_1; \ldots ;C_n)$ where $\sim A_1$ to $\sim A_m$ are the negative, C_1 to C_n the positive literals in the clause. Completely positive clauses are written as $(\text{true ---> } C_1; \ldots ;C_n)$, while completely negative clauses are implicationally represented by $(A_1, \ldots ,A_m \text{ ---> false})$. Thus negation never occurs explicitly. We call the left-hand side of an implication its *antecedent* and the right-hand side its *consequent*.

2. Model Generation in Prolog

2.1 A basic procedure

It is well-known that every model of a set of clauses can be represented by a set M of positive ground atoms. The elements of M are those ground literals that are satisfied in the model, the remaining ones are - by default - assumed violated. A ground conjunction/disjunction is satisfied in M, if M contains all/some of its components. A clause (A ---> C) is satisfied in M, if Cσ is satisfied for every substitution σ such that Aσ is satisfied. Conversely, (A ---> C) is violated in M if there is a substitution σ such that Aσ is satisfied in M, but Cσ is not. 'True' is satisfied, 'false' is violated in every model.

It seems to be a very natural choice to implement model construction by asserting facts into Prolog's internal database and the test for satisfaction in a model by means of Prolog goal evalution over the "program" that consists of the facts asserted. Consider, e.g., a Prolog database containing

```
p(1).        q(1,2).
p(2).        q(2,1).
p(3).        q(2,2).
```

The clause (p(X),q(X,Y) ---> p(Y)) is satisfied in this database, because evaluation of the Prolog goal 'p(X),q(X,Y),not p(Y)' fails. On the other hand, the clause (q(X,Y),q(Y,Z) ---> q(X,Z)) is violated, as the goal 'q(X,Y),q(Y,Z),not q(X,Z)' succeeds and returns bindings X=1,Y=2,Z=1. These bindings represent a substitution such that the corresponding ground instance (q(1,2),q(2,1) ---> q(1,1)) is not satisfied in the current database.

Violated clauses can be applied as generation rules in order to further extend the database so far constructed. Initially, the database is empty and therefore clauses of the form (true ---> C) are the only ones that are violated. If there is a clause (A ---> C) and a substitution σ such that Aσ is satisfied and Cσ is violated, the clause can be satisfied by satisfying Cσ, i.e., by

- asserting Cσ, if Cσ is an atom
- creating a choice point, chosing a component atom of Cσ and adding it to the database, if Cσ is a disjunction.

Generation of 'false' indicates that the current database contradicts at least one of the completely negative clauses and thus cannot be extended into a model. One has to backtrack to a previously established choice point (if any) and to choose a different atom there (if any remains). All facts asserted between this choice point and the point where 'false' has been generated have to be retracted from the database on backtracking.

If all possible choices lead to a contradiction, the process terminates with an empty database and reports that a model could not be created. The clause set under consideration is unsatisfiable. If on the other hand a database has been constructed in which every clause is satisfied, this database represents a model of the clause set and satisfiability has been shown. In certain cases generation will never stop, because no finite database satisfies all clauses, but a contradiction does not arise either. This is due to the undecidability of satisfiability.

The model generation process outlined above can be implemented in Prolog by means of the following very simple program:

```
satisfiable :-                       satisfy(C) :-
    is_violated(C), !,                   component(X,C),
    satisfy(C),                          asserta(X),
    satisfiable.                         on_backtracking(retract(X)),
satisfiable.                             not false.

is_violated(C) :-
    (A ---> C),                      component(X,(Y;Z)) :-
    A, not C.                            !, (X=Y ; component(X,Z)).
                                     component(X,X).

on_backtracking(X).
on_backtracking(X) :-
    X, !, fail.
```

Note that 'true' is a built-in Prolog predicate that always succeeds, whereas 'false' does not succeed unless asserted.

The following example is intended to illustrate model generation in Prolog. Consider the clause set S1:

```
true ---> p(a) ; q(b)        p(X) ---> q(X) ; r(X)
q(X) ---> s(f(X))            q(X) , s(Y) ---> false
r(X) ---> s(X)               p(X) , s(X) ---> false
```

The choices and assertions made during execution of 'satisfiable' can be recorded in form of a tree:

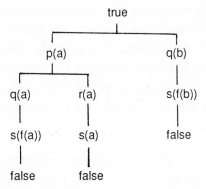

When executing 'satisfiable', this tree is traversed in a left-to-right, depth-first manner. Every possible path towards a model of S1 is closed because 'false' is finally generated. Thus, S1 has been shown unsatisfiable. If the clause (p(X),s(X) ---> false) were missing, 'false' could not be generated along the middle branch and the database constructed along this branch - consisting of p(a), r(a) and s(a) - would represent a model of the reduced clause set.

2.2   How to achieve soundness for unsatisfiability

If 'satisfiable' fails - when applied to a particular set S of clauses - this should always coincide with S being unsatisfiable. Conversely, if 'satisfiable' terminates successfully, S should in fact be satisfiable. While the latter is achieved, the former aim is not always reached: There are cases where 'satisfiable' fails although a model of S exists. This happens when a disjunction is generated that still contains uninstantiated variables shared by different components of the disjunction. Consider, e.g, the following clause set S2:

$$\text{true} ---> p(X) ; q(X)$$
$$p(a) ---> \text{false}$$
$$q(b) ---> \text{false}$$

Initially 'p(X) ; q(X)' is generated. If 'p(X)' is asserted, the disjunction is satisfied, but 'false' will be generated in the next step. The same is the case if 'q(X)' is asserted. Thus, 'satisfiable' fails. However, the set {p(b),q(a)} represents a finite model of S2 which the program was unable to find.

Soundness for unsatisfiability can be guaranteed if all disjunctions generated are completely instantiated. This is the case iff all clauses are range-restricted, i.e., if every variable in the consequent of a clause occurs in its antecedent as well. In particular, completely positive clauses - those having 'true' as their antecedent - have to be variable-free in order to be range-restricted. The example set S1 given above is range-restricted, while S2 is not. Range-restriction requires that for every variable in a clause the subset of the universe over which the variable ranges is explicitly specified inside the clause. Variables implicitly assumed to range over the whole universe are not allowed. One can expect many clauses to be range-restricted if the problem domain is somehow naturally structured. This is in particular the case if a problem is (inherently) many-sorted.

If a set S contains clauses that are not range-restricted, S nevertheless can be transformed into a set S^* that is range-restricted and that is satisfiable iff S is so. For this purpose an auxiliary predicate 'dom' is introduced and the following transformations and additions are performed:

- every clause (true ---> C) that contains variables X_1 to X_n is transformed into $(dom(X_1),...,dom(X_n)) ---> C)$
- every other clause (A ---> C) such that C contains variables Y_1 to Y_m not occuring in A is transformed into $(A,dom(Y_1),...,dom(Y_m)) ---> C)$.
- for every constant c occurring in S, a clause (true ---> dom(c)) is added; if S does not contain any constant a single clause (true ---> dom(a)) is added where 'a' is an artificial constant
- for every n-ary function symbol f occurring in S one adds a clause $(dom(X_1),...,dom(X_n)) ---> dom(f(X_1,..,X_n)))$

The 'dom' literals added to non-range-restricted clauses explicitly provide for an instantiation of the respective variables over the Herbrand universe of S. The transformation of S into its range-restricted form S^* can be compared with the transformation of a formula into its Skolemized form: although the transformed set is not equivalent to the initial set in the strict sense, a kind of weak equivalence can be observed. If the relation assigned to 'dom' (the functions assigned to the Skolem function symbols, resp.) is removed from any model of the transformed set, a model of the initial set is obtained. There is a one-to-one correspondence between the models of both sets of clauses up to the relation (functions, resp.) assigned to the additional predicate (function symbols, resp.). Therefore the transformation described preserves satisfiability. Transformation of S2 into range-restricted form yields S2*:

```
true ---> dom(a)        p(a) ---> false
true ---> dom(b)        q(b) ---> false
        dom(X) ---> p(X) ; q(X)
```

If 'satisfiable' is applied to S2* the program terminates successfully with the facts dom(a), dom(b), q(a) and p(b) in the database.

If applied to range-restricted clauses only, 'satisfiable' will be sound for unsatisfiability as well. For the rest of the paper we assume that all problems mentioned have been transformed into range-restricted form prior to checking them for satisfiability.

2.3   How to achieve refutation-completeness

As satisfiability is undecidable no theorem prover is able to successfully terminate for every satisfiable set of clauses. Unsatisfiability, however, is semi-decidable. Therefore, our program should terminate with failure for every unsatisfiable clause set. The following example set S3 is unsatisfiable:

```
true ---> p(a)                  p(X) ---> p(f(X))
p(f(X)) , p(g(X)) ---> false    p(X) ---> p(g(X))
```

When applied to S3, 'satisfiable' will generate an infinite sequence of p-atoms: p(a), p(f(a)), p(f(f(a))), ... etc. The example shows that 'satisfiable' is not complete for unsatisfiability!

This is due to the fact that the Prolog-like search strategy employed by the program is inherently unfair: at each recursive call of 'satisfiable' the database of problem clauses is searched from the top and the first violated clause found is tried to satisfy. There are problems - like S3 - where some violated clauses are never considered and thus a contradiction is never reached. Sometimes a proper ordering of clauses suffices for controlling the order in which facts are generated. In our example, however, ordering does not help!

Completeness can be achieved if clauses are generated systematically level by level. First, all atoms/disjunctions that can be generated from a given database are determined without modifying the database. Then all facts needed in order to satisfy these atoms/disjunctions are asserted altogether. If S3 would have been treated this way, the following facts would have been asserted:

```
level 1:    p(a)
level 2:    p(f(a))      q(f(a))
level 3:    p(f(f(a)))    p(f(g(a)))    p(g(f(a)))    p(g(g(a)))    false
```

A contradiction would be detected on level 3 and the attempt to generate a model for S3 would fail. The following program implements model generation on a level-saturation basis:

```
satisfiable_level :- satisfiable_level(1).

satisfiable_level(L) :-              is_violated_level(L) :-
    is_violated_level(L), !,             is_violated(C),
    on_backtracking(clean_level(L)),     not generated(L,C),
    satisfy_level(L),                    assert(generated(L,C)),
    L1 is L+1,                           fail.
    satisfiable_level(L1).           is_violated_level(L) :-
satisfiable_level(L).                    generated(L,X).

satisfy_level(L) :-                  clean_level(L) :-
    generated(L,C),                      retract(generated(L,X)),
    not C, !,                            fail.
    satisfy(C),                      clean_level(L).
    satisfy_level(L).
satisfy_level(L).
```

The program works with an intermediate relation 'generated' used for storing atoms/disjunctions violated on a level in order to be able to satisfy them after the generation process for that level has been finished. Generation of a level can be efficiently organized by means of a backtracking loop, whereas levels have to be satisfied recursively. This is because the choice points created when satisfying a disjunction have to be kept "open" to backtracking.

Although 'satisfiable' - as opposed to 'satisfiable_level' - is not complete, most of the problems to be considered in the following section will in fact turn out to be solvable by 'satisfiable'. In cases where both programs are applicable one will usually observe that the former is much more efficient than the latter.

2.4   Enhancing efficiency through Prolog derivation rules

Prolog is known to be a powerful interpreter for Horn clauses. Some design decisions made for the sake of efficieny, however, prevent Prolog from being able to handle arbitrary Horn problems.

The missing occurs check requires to avoid certain unification patterns, the unbounded depth-first strategy prevents certain recursive problems from being tractable in Prolog. For the remaining cases Prolog's power can very well be exploited if those problem clauses that are in the scope of Prolog are represented as Prolog rules, i.e., by (C :- A) instead of (A ---> C). The four Horn clauses in example set S1, e.g., can all be treated this way:

```
true ---> p(a) ; q(b)          s(X) :- p(X)
p(X) ---> q(X) ; r(X)          s(f(X)) :- q(X)
                               false :- p(X) , s(X)
                               false :- q(X) , s(Y)
```

When applied to this representation of S1, 'satisfiable' will initially work as before, i.e., p(a) and q(a) are asserted into the database in order to satisfy the two '--->'-clauses. As soon as q(a) has been asserted, however, 'false' becomes derivable. The test 'not false' performed after assertion of q(a) fails and backtracking is immediately initiated: s(f(a)) and 'false' need no more be asserted in order to run into a contradiction. Similarly the two other branches can be cut earlier due to derivability of 'false'. The tree of facts asserted has become considerably smaller than before:

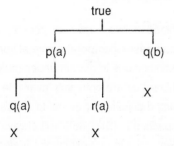

Crosses 'X' indicate that the respective branch has been closed because 'false' has become derivable.

If part of a clause set is directly represented in Prolog, 'satisfiable' can be applied without any change, provided inconsistency of the Prolog part of the problem has been tested before. This can be done by simply evaluating 'false' once before calling 'satisfiable'. The same applies to 'satisfiable_level'.

What has changed, however, is the way the model under construction is represented and satisfaction of clauses is determined. The problem clauses that have been directly represented in Prolog now serve as derivation rules. If the requirement for clauses to be range-restricted is respected, only ground literals will be derivable through these rules. Ground atoms that are required for satisfying the '--->'-clauses need not be explicitly asserted anymore, if they are derivable from the already existing facts. Thus, the model under construction is no longer represented by explicitly stored facts alone, but by all facts derivable via those problem clauses that have been represented directly in Prolog. As the Prolog goal evaluation mechanism solves goals not only over facts but through rules as well, nothing has to be changed in the programs given.

In every clause set at least the positive ground units and the completely negative clauses can always be formulated directly in Prolog. There are even problems where all clauses can be represented this way. In this case satisfiability checking reduces to a single 'false' evaluation. When using Prolog rules for clause representation one nevertheless has to be very careful in order to avoid recursion and occurs check problems. (A set of clauses is recursive, if its connection graph contains a cycle that has a unifier.) Although a complete syntactic characterization of such cases is not easy, there are relatively simple sufficient conditions, like, e.g., to avoid literals with more than one occurrence of the same variable (static occurs check) in order to avoid dynamic occurs check problems. If in doubt, the option to represent a Horn clause as a generation rule always remains. Because of range-restriction this representation will never lead to any occurs check problem.

2.5 Further variations and optimizations

The basic model generation paradigm - as outlined above - may, of course, exhibit serious inefficiencies in special situations. However, one can easily incorporate several variations into the basic procedure that may lead to considerable optimizations in certain undesirable cases. In the following we will shortly discuss three such variations that are optionally available in SATCHMO. They are intended to speed up the search for violated clauses, or permit to derive contradictions earlier. Whereas the first - called clause-set compaction in [BUT 86] - will always result in some benefit, others will pay off only if applied to problems that in fact exhibit the inefficiency they are intended to cure. Otherwise these variations may even lead to some overhead compared with the basic procedures.

1. clause-set compaction:

   Search for violated clauses may be fairly expensive in cases where the clausal formulation of a problem is highly redundant, as is the case, e.g., with

$$p(X,Y) , q(Y,Z) ---> h(X)$$
$$p(X,Y) , q(Y,Z) ---> h(Y)$$
$$p(X,Y) , s(Y,Z) ---> h(X)$$
$$p(X,Y) , s(Y,Z) ---> h(Y)$$

   The p- and h-relations have to be searched four times, the q- and s-relations twice in order to determine all instances of the clauses that are violated in a given database. If we would allow ';' to occur in the antecedent, and ',' to occur in the consequent of a rule as well, these four clauses could be compactified into the single rule

p(X,Y) , (q(Y,Z) ; s(Y,Z)) ---> h(X) , h(Y)

If this expression is tested for violation, the p-, q-, and s-relation are searched only once, the h-relation twice. In order to handle non-clausal generation rules as well, the following has to be added on top of 'satisfy':

satisfy((A,B)) :- !, satisfy(A), satisfy(B).

2. clause compilation:

When testing for contradictions by evaluating 'false', the test is performed globally over the whole database of facts. One does not take into consideration the specific fact asserted just before. As 'false' has not been derivable before this update, it can be derivable now only if the most-recently introduced literal is able to participate in a derivation of 'false'. Therefore it is possible to "focus" the contradiction test by precompiling the completely negative problem clauses into local test clauses. Consider, e.g., a set of clauses containing

false :- p(X,Y) , q(Y,Z).
q(A,B) :- s(B,A).

Precompilation would result in the following local test clauses being generated:

incompatible(p(X,Y)) :- q(Y,Z).   incompatible(q(Y,Z)) :- p(X,Y).
incompatible(p(X,Y)) :- s(Z,Y).   incompatible(s(Z,Y)) :- p(X,Y).

Once negative clauses have been compiled this way one can exploit them by slightly modifying the 'satisfy' predicate again: instead of 'assert(X),..,not false' one performs 'not incompatible(X),assert(X),...'. Thus, facts the assertion of which would directly lead to a contradiction are never asserted. Precompilation of the incompatibility rules can easily and efficiently be programmed in Prolog. A similar precompilation idea may be applied for the remaining clauses as well possibly speeding-up the search for violated clauses. We have reported about this in [BRY 87].

3. complement splitting:

When the assertion of an atom A_i chosen from a disjunction has resulted in a contradiction one may try to benefit from the information thus obtained while trying the remaining components. This can be achieved by temporarily asserting the rule (false :- A$_i$). This way any attempt to re-assert A$_i$ while trying to satisfy the respective disjunction are immediately blocked. The size of the search tree may be considerably cut down in case several occurrences of big subtrees are avoided this way. Exploiting information about attempts that have already failed for avoiding redundant work has already been suggested by Davis and Putnam in their early proof procedure. The feature can be implemented by modifying the 'component' predicate as follows:

```
component(X,(Y;Z)) :- !,
    (X=Y ;
    assert((false :- Y)),
    on_backtracking(retract((false :- Y))),
    component(X,Z)).
```

3. Experiences with SATCHMO

When reporting about experiences with a new method, one of course tends to start with "showcase" examples that are particularly well handled by the approach proposed. In our case, it turns out that several examples recently discussed in the literature are suitable for this purpose. Thus, we begin this section with discussing these examples in some detail. Then we shortly address combinatorial puzzles - a class of problems preferably taken for demonstrating the power of a theorem prover. The third and most comprehensive section will be devoted to Pelletier's "Seventy-five Problems for Testing Theorem Provers" [PEL 86]. As this collection covers a wide range of problem classes, we regard it as particularly well suited for exhibiting the potential power as well as the limits of the approach suggested.

All examples discussed have been run under interpreted CProlog Vers. 1.5 on a VAX 11/785. The solution times that will be given have been measured using the built-in predicate 'cputime'. The authors are very much aware of the fact that comparing theorem provers on the basis of cpu times is hardly ever able to do justice to the particular approaches except if all conditions are respected absolutely fairly. Reporting cpu times in this paper is not intended to compete with others, but to show which kind of examples are hard for model generation and which are easy. Moreover, we would like to demonstrate this way that theorem proving in Prolog is feasible and that the efficiency obtained when doing so may be remarkable.

3.1 Schubert's Steamroller

The steamroller problem has been presented by Len Schubert nearly a decade ago. Mark Stickel's article "Schubert's Steamroller Problem: Formulations and Solutions" [STI 86] provides an excellent overview of various attempts to solve this problem. We use the (unsorted) formulation that Stickel has proposed as a standard. As already mentioned in the introduction, this problem can be completely expressed in Prolog except for one non-Horn clause:

```
wolf(w).                              animal(X) :- wolf(X).
fox(f).                               animal(X) :- fox(X).
bird(b).                              animal(X) :- bird(X).
snail(s).                             animal(X) :- snail(X).
caterpillar(c).                       animal(X) :- caterpillar(X).
grain(g).                             plant(X) :- grain(X).

smaller(X,Y) :- fox(X) , wolf(Y).     plant(i(X)) :- snail(X).
smaller(X,Y) :- bird(X) , fox(Y).     likes(X,i(X)) :- snail(X).
smaller(X,Y) :- snail(X) , bird(Y).   plant(h(X)) :- caterpillar(X).
smaller(X,Y) :- caterpillar(X) , bird(Y).  likes(X,h(X)) :- caterpillar(X).

likes(X,Y) :- bird(X) , caterpillar(Y).

false :- wolf(X) , (fox(Y) ; grain(Y)) , likes(X,Y).
false :- bird(X) , snail(Y) , likes(X,Y).
false :- animal(X) , animal(Y) , likes(X,Y) , grain(Z) , likes(Y,Z).

animal(X) , animal(Y) , smaller(Y,X) , plant(W) , likes(Y,W) , plant(Z) ---> likes(X,Y) ; likes(X,Z).
```

Satisfiability of the Prolog part of the problem formulation can be demonstrated within 0.05 secs by evaluation of the goal 'false'. Unsatisfiability of the whole problem is proved by means of 'satisfiable' after 0.3 secs. (The best time reported in Stickel's paper was 6 secs for the unsorted version of the problem.) The choices and assertions performed during execution are as follows:

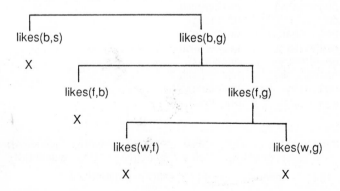

The way 'satisfiable' solves the problem corresponds pretty well to the natural language solution given by Stickel.

3.2 Lewis Carroll's "Salt-and-Mustard problem"

This problem - which stems from Lewis Carroll's "Symbolic Logic" of 1897 - has been discussed in a publication by Lusk and Overbeek in the problem corner of the Journal of Automated Reasoning [LO 85]. It is about five friends who have imposed certain rules governing which condiment - salt or mustard - to take when the five are having beef together. The problem is to check whether these rules are compatible, i.e., satisfiable. The major part of the problem consists of if-and-only-if conditions for each friend and each condiment. Each of these can be expressed by means of a non-Horn generation rule and a Horn derivation rule. Our formulation is as follows (note that some 'false'-rules have been compactified):

```
friend(barry).              salt(X)  :-  both(X).
friend(cole).               mustard(X)  :-  both(X).
friend(dix).
friend(lang).               salt(X) , mustard(X)  --->  both(X).
friend(mill).
```

```
salt(barry)      :-   oneof(cole)    ;   oneof(lang).
mustard(barry)   :-   neither(dix)   ;   both(mill).
salt(cole)       :-   oneof(barry)   ;   neither(mill).
mustard(cole)    :-   both(dix)      ;   both(lang).
salt(dix)        :-   neither(barry) ;   both(cole).
mustard(dix)     :-   neither(lang)  ;   neither(mill).
salt(lang)       :-   oneof(barry)   ;   oneof(dix).
mustard(lang)    :-   neither(cole)  ;   neither(mill).
salt(mill)       :-   both(barry)    ;   both(lang).
mustard(mill)    :-   oneof(cole)    ;   oneof(dix).
```

```
salt(barry)      --->   oneof(cole)    ;   oneof(lang).
mustard(barry)   --->   neither(dix)   ;   both(mill).
salt(cole)       --->   oneof(barry)   ;   neither(mill).
mustard(cole)    --->   both(dix)      ;   both(lang).
salt(dix)        --->   neither(barry) ;   both(cole).
mustard(dix)     --->   neither(lang)  ;   neither(mill).
salt(lang)       --->   oneof(barry)   ;   oneof(dix).
mustard(lang)    --->   neither(cole)  ;   neither(mill).
salt(mill)       --->   both(barry)    ;   both(lang).
mustard(mill)    --->   oneof(cole)    ;   oneof(dix).
```

```
oneof(X) ---> salt(X) ; mustard(X).          false :- oneof(X) ,(both(X) ; neither(X))
friend(X) ---> both(X) ; neither(X) ; oneof(X).   false :- oneof(X) , salt(X) , mustard(X).
                 false :- neither(X) , (both(X) ; salt(X) ; mustard(X)).
```

The problem has a single model:

```
salt(barry)    mustard(barry)   neither(dix)    oneof(lang)    both(barry)
salt(mill)     mustard(lang)    neither(cole)   oneof(mill)
```

'Satisfiable' finds it within 1.1 secs. Lusk and Overbeek have judged this problem to be especially hard as their theorem prover has produced 32 000 clauses for solving the problem. Although the tree to be searched before a solution is found is considerably bigger than for the Steamroller the problem is still a simple one for a model generation approach.

The two other problems discussed in the article by Lusk and Overbeek are in fact much easier: "truthtellers and liars" are dismantled within 0.1 secs, while a model for the "schoolboys problem" is found after 0.2 secs.

3.3   A non-obvious problem

Pelletier and Rudnicki [PR 86] have recently discussed a problem that is simple to state, but hard to prove. Because of its brevity we give their problem formulation as well: "Suppose there are two relations, P and Q. P is transitive, and Q is both transitive and reflexive. Suppose further the 'squareness' of P and Q: any two things are either related in the P manner or in the Q manner. Prove that either P is total or Q is total."

Our formalization of the problem - requiring a transformation into range-restricted form - is as follows:

```
dom(a).          p(X,Y) , p(Y,Z) ---> p(X,Z).
dom(b).          q(X,Y) , q(Y,Z) ---> q(X,Z).
dom(c).          q(X,Y) ---> q(Y,X).
dom(d).          dom(X),dom(Y) ---> p(X,Y) ; q(X,Y).
                 false :- p(a,b).
                 false :- q(c,d).
```

The tree to be searched by 'satisfiable' is already quite big: 348 facts are asserted and subsequently retracted again, 52 choice points are established and 53 branches are closed. The theorem is proved within 16 secs.

Pelletier/Rudnicki report about a solution time just under 2 mins, McCune reports just under 1 min [McC 86]. Just recently Walther has obtained 31.1 secs [WAL 88]. If complement splitting is applied for this example, the search tree can be considerably reduced such that the solution time goes down to 9.5 secs.

3.4   Some combinatorial puzzles

In Pelletier's collection, to be discussed below, there is another problem (number 55) posed by
Len Schubert. This problem is a simple combinatorial puzzle where the murderer of aunt Agatha
has to be found. Equality (or identity) is used in order to express that the murderer has to be
found among the three people living in the house. We conjecture that the following formulation of
the problem (not using '=') is simpler and more natural than Pelletier's:

lives(agatha).   false :- killed(X,Y) , richer(X,Y).
lives(butler).   false :- hates(agatha,X) , hates(charles,X).
lives(charles).   false :- hates(X,agatha) , hates(X,butler) , hates(X,charles)

hates(agatha,agatha).   hates(X,Y) :- killed(X,Y).
hates(agatha,charles).   hates(butler,X) :- hates(agatha,X).

true ---> killed(agatha,agatha) ; killed(butler,agatha) ; killed(charles,agatha).
lives(X) ---> richer(X,agatha) ; hates(butler,X).

After 0.05 secs a model for the problem is found indicating that aunt Agatha has killed herself.
After 0.1 secs all other possibilities have been ruled out proving that Agatha indeed must have
committed suicide.

Using this style of formulation, i.e., expressing the different choices that are existing over the finite
domain of the puzzle by means of disjunctions and expressing the remaining conditions directly in
Prolog, one can solve quite a lot of similar puzzles efficiently. We just would like to mention as
examples
  • Lewis Carroll's "lion-and-unicorn" puzzle (discussed recently in the Journal of Automated
    Reasoning [OHL 85]): 20 Prolog clauses, 1 non-Horn clause - solved in 0.1 secs
or a more substantial one:
  • the full "jobs" puzzle taken from [WOS 84]: 31 Prolog clauses, 2 non-Horn clauses -
    solved in 4.5 secs
Again the incomplete program 'satisfiable' has been sufficient.

3.5   Pelletier's seventy-five problems

All problems discussed in the following have been solved with 'satisfiable' as well, unless stated
differently.

The propositional problems 1-17 are all very easy once clausal form has been obtained. Eight of
them can be completely represented in Prolog and are all solved under 0.01 secs. Problem 12 is
the hardest among the remaining ones - a solution requires 0.15 secs.

The monadic problems 18-33 (34 has been omitted because the authors did not want to perform the "exercise" of computing 1600 clauses) are simple as well, with one exception namely problem 29. This problem - consisting of 32 clauses - requires 33 secs! (Attention: Pelletier's clausal version contains two typing mistakes.) A clause-set compaction as described above results in 23 compactified clauses and unsatisfiability of the compactified set can be shown within 4.3 secs. If in addition complement splitting is applied, the time needed goes down to 1.1 secs.

Problems 19,20,27,28, and 32 are completely expressable in Prolog and solvable in less than 0.02 secs (problem 28 is satisfiable!).

The full predicate logic problems without identity and (non-Skolem) functions 35-47 do not impose particular problems either. Problem 44 is the only one that can be completely represented in Prolog (solved under 0.01 secs). Problem 35 is the first problem for which 'satisfiable' would run forever. 'Satisfiable_level' solves it within 0.07 secs. In problem 46 the clause 'f(X) ---> f(f(X)) ; g(X)' has to be "hidden" at the end of the clause list in order to maintain applicability of 'satisfiable'. The most difficult problem in this section is problem 43, requiring 0.65 secs. Problem 47 is the Steamroller discussed earlier.

Among the remaining problems 48-69 there are nine functional problems without identity. Problems 57,59 and 60 (Pelletier's faulty clausal form corrected) can be solved by 'satisfiable' under 0.1 secs, problem 50 requires 0.55 secs (0.28 with complement splitting).

For problem 62 once again the clausal form given by Pelletier does not correspond to the non-clausal form of the theorem. If corrected the clausal form can immediately be shown satisfiable as no completely positive clauses exist.

Problems 66-69 cannot be solved by any of the two SATCHMO programs! These problems are variations of a hard recursive Horn-problem with functions. There is a single predicate ranging over the whole domain. As the problems are not range-restricted instantiation over the Herbrand universe has to be provided through the 'dom'-predicate. Consider, e.g., problem 66:

```
t(i(X),i(Y,X))  :- dom(X) , dom(Y).
t(i(i(X,i(Y,Z)),i(i(X,Y),i(X,Z)))  :- dom(X) , dom(Y) , dom(Z).
t(i(i(n(X),n(Y)),i(Y,X))  :- dom(X) , dom(Y).
t(i(X,Y)),t(X) ---> t(Y)
false  :- t(i(a,n(n(a)))).

dom(X) ---> dom(n(X)).
dom(X) , dom(Y) ---> dom(i(X,Y)).
```

Satchmo fails because the generation of new Herbrand terms via the two 'dom'-rules interferes with the generation of the necessary 't'-facts. The number of 'dom'-facts generated explodes and the comparatively few 't'-facts that can be generated on each level are "buried" by them. The only way towards possibly solving problems of this kind seems to be a careful control of Herbrand term generation: 'dom'-rules should not be applied before the other rules have not been exhaustived. As such a control feature has not yet been implemented, we do not further elaborate on this point.

Prolog's implementation of '=' cannot be used for correctly representing logical identity (except in very restricted cases). In order to represent the remaining problems with identity there are two possibilities:

1. to introduce a special equality predicate and to add the necessary equality axioms (transitivity, substitutivity etc.): this has been done for problems 48,49,51-54,56, and 58

2. to recode the problems without explicitly using identity as done in the original formulation of the three group theory problems 63-65 by Wos; problem 61 has been coded this way too

Of the problems thus augmented or recoded, 'satisfiable' was able to solve problems 48, 49, 61, 64, and 65 in less than 1 sec each. For the remaining problems 'satisfiable_level' had to be employed: of these, problem 58 was solved in 0.15 secs and problem 63 in 0.75 secs; problem 55 has been discussed above. Problems 51-54 and 56 could not be solved by either programs, due to deficiencies very similar to those responsible for failure in case of 66-69.

The last section in Pelletier's collection provides problems for studying the complexity of a proof system. The following figures are given without further comment as we have not really studied their relevance yet. Pigeonhole problems (72,73):

```
n    |    1     2     3     4     5
secs |    0.05  0.13  0.7   3.8   25.5 ...
```

Times are given for our formulation of the predicate logic version (73): the times clearly indicate exponential growth. The expository arbitrary graph problem 74 is solved in 0.18 secs.

For U-problems (71) coded as arbitrary graph problems (75) growth seems to be at most cubic:

```
n    |    1     2     3     4
secs |    0.05  0.2   0.75  2.15 ...
```

4. Conclusion

In this paper SATCHMO, a theorem prover based on model generation, is presented and experiences are described. Prolog has been used as a representation language for expressing problem clauses as well as for the implementation of SATCHMO. The approach extends Prolog while retaining its efficiency for Horn clauses as has been done by Stickel's PTTP. The additions we are proposing are, however, considerably different from Stickel's. As a consequence, SATCHMO can be implemented on top of Prolog without causing too severe inefficiencies by doing so.

As an extension of the work reported here, we would like to investigate more deeply how to benefit from further compilations of problem clauses and how to control term generation. Some considerable gain in efficiency can also be expected from investigations in more sophisticated solutions to controlling recursive Prolog-rules. Apart from this, we would like to know how SATCHMO behaves when implemented in up-to-date Prolog-systems. The simplicity of its code should make it extremely portable. Due to the splitting feature especially forthcoming parallel implementations of Prolog should be promising.

5. Acknowledgement

We would like to thank Hervé Gallaire and Jean-Marie Nicolas as well as our colleagues at ECRC for providing us with a very stimulating research ambience. The work reported in this article has benefited a lot from it. We also would like to thank the Argonne team for their interest in our work and their encouragement to present our results to the automated deduction community.

References:

[BRY 87]   Bry, F. et. al., *A Uniform Approach to Constraint Satisfaction and Constraint Satisfiability in Deductive Databases*,
ECRC Techn. Rep. KB-16, 1987 (to appear in Proc. Int. Conf. Extending Database Technology EDBT 88)

[BUT 86]   Butler, R. et. al., *Paths to High-Performance Automated Theorem Proving*, Proc. 8th CADE 1986, Oxford, 1986, 588-597

[LO 85]   Lusk, E. and Overbeek, R., *Non-Horn problems*, Journal of Automated Reasoning 1 (1985), 103-114

[MB 87]   Manthey, R. and Bry, F., *A hyperresolution-based proof procedure and its implementation in PROLOG*, Proc. of GWAI-87 (11th German Workshop on Artificial Intelligence), Geseke, 1987, 221-230

[McC 86]   McCune, B., *A Proof of a Non-Obvious Theorem*, AAR Newsletter No. 7, 1986, 5

[NIC 79]   Nicolas, J.M., *Logic for improving integrity checking in relational databases*, Tech. Rep., ONERA-CERT, Toulouse, Feb. 1979
(also in Acta Informatica 18,3, Dec. 1982)

[OHL 85]   Ohlbach, H.J. and Schmidt-Schauss, M., *The Lion and the Unicorn*, J. of Automated Reasoning 1 (1985), 327-332

[PEL 86]   Pelletier, F.J., *Seventy-five Problems for Testing Automatic Theorem*

Provers, J. of Automated Reasoning 2 (1986), 191-216

[PR 86]   Pelletier, F.J. and Rudnicki, P., *Non-Obviousness*,
AAR Newsletter No. 6, 1986, 4-5

[SMU 68]   Smullyan, R., *First-Order Logic,* Springer-Verlag, 1968

[STI 84]   Stickel, M., *A Prolog Technology Theorem Prover*,
New Generation Computing 2, (1984), 371-383

[STI 86]   Stickel, M., *Schubert's steamroller problem: formulations and solutions,*
J. of Automated Reasoning 2 (1986), 89-101

[WAL 84]   Walther, C., *A mechanical solution of Schubert's steamroller by many-sorted resolution*, Proc. of AAAI-84, Austin, Texas, 1984, 330-334
(Revised version in Artificial Intelligence, 26, 1985, 217-224)

[WAL 88]   Walther, C., *An Obvious Solution for a Non-Obvious Problem*,
AAR Newsletter No. 9, Jan. 1988, 4-5

[WOS 84]   Wos, L. et. al., *Automated Reasoning*, Prentice Hall, 1984

Term Rewriting: Some Experimental Results *

Richard C. Potter and David A. Plaisted

Department of Computer Science
University of North Carolina at Chapel Hill
Chapel Hill, North Carolina 27514

Abstract

We discuss term rewriting in conjunction with **sprfn**, a Prolog-based theorem prover. Two techniques for theorem proving that utilize term rewriting are presented. We demonstrate their effectiveness by exhibiting the results of our experiments in proving some theorems of von Neumann-Bernays-Gödel set theory. Some outstanding problems associated with term rewriting are also addressed.

Key words and phrases. Theorem proving, term rewriting, set theory.

1 Introduction

Term rewriting is one of the more powerful techniques that can be employed in mechanical theorem proving. Term rewriting allows us to prove fairly sophisticated theorems that are beyond the ability of most resolution-based theorem provers. Unlike resolution, term rewriting seems to duplicate a rule of inference that humans use in constructing proofs. In this paper, we will describe our research and results in proving theorems via term rewriting. The body of theorems we prove are set theoretic; the axiomatization of set theory employed is derived from the work of von Neumann, Bernays, and Gödel. For a list of these axioms, see [2]. The advantage of the von Neumann-Bernays-Gödel formalization is that it allows us to express set theory in first-order logic. This in turn implies that a first-order theorem prover can be used to derive set theoretic theorems. On the other hand, this formalization has a significant disadvantage in that it is very clumsy for humans to use. Second order logic is a much cleaner means for expressing the axioms of set theory.

---
* This research was supported in part by the National Science Foundation under grant DCR-8516243 and by the Office of Naval Research under grant N00014-86-K-0680.

We begin by introducing **sprfn**, the Prolog-based theorem prover we used in our research; we emphasize the formal deduction system underlying the prover. In the second section we describe the term rewriting mechanism built into **sprfn**. In the third and fourth sections we describe two theorem proving techniques utilizing term rewriting and the results of these approaches when employed in connection with **sprfn**. In each of these two sections we give examples of sample theorems that we were able to derive. We conclude by summarizing our results and addressing some problems that face term rewriting in general as well as some problems specific to term rewriting with **sprfn**.

2   The Simplified Problem Reduction Format

The theorem prover we used -- **sprfn** -- is based on a natural deduction system in first-order logic which is described in [1]. However, before we present this formal system, we would like to motivate it by describing the format on which it is based; namely, the problem reduction format. The formal deduction system implemented by **sprfn** is a refinement of the problem reduction format. Both of them embody the same goal-subgoal structure, as can be seen from what follows. The following description omits many details. For a complete discussion of the problem reduction format, see [5].

The structure of the problem reduction format is as follows. One begins with a conclusion G to be established and a collection of assertions presumed to be true. Assertions are of the form $C :- A_1, A_2, \ldots, A_n$ (implication) or P (premises) where A_i, P and C are literals or negations of literals. The implication assertion is understood to mean $A_1 \& A_2 \cdots \& A_n \rightarrow C$. The A_i's are antecedent statements, or simply antecedents, and C is the consequent of the implication. We call the conclusion G the top-goal. The process of attempting to confirm the conclusion begins with a search of the premises to see if one premise matches (is identical with or can be made identical by unification with) the goal G. If a premise P_g matches G then the conclusion is confirmed by P_g. Otherwise, the set of implications whose consequents match G is found. If the antecedent of one implication can be confirmed then one has confirmed the consequent, and hence G, which the consequent matches. Otherwise we consider the antecedents as new subgoals to be confirmed, one implication at a time. These goals are called *subgoals* because none of them is the primary goal. The process of confirming these subgoals involves repeating the method just described in connection with the top-goal.

The natural deduction system underlying **sprfn** -- the modified problem reduction format -- is based on the problem reduction format just described, although refinements are added for the sake of completeness of the deduction system. We do not have room to describe these refinements. The following description of the modified problem reduction format omits many details. For a complete discussion, see[4].

A **clause** is a disjunction of literals. A **Horn-like clause**, converted from a clause, is of the form $L :- L_1, L_2, ..., L_n$ where L and the L_i's are literals. L is called the head literal. The L_i's constitute the clause body. A clause is converted to a Horn-like clause as follows. For a given clause containing at least one positive literal, one of its positive literals is chosen as the head literal and all other literals are put in the clause body negated. For an all-negative clause, we use **false** as the head literal and form the body from positive literals corresponding to the original literals.

Now assume S is a set of Horn-like clauses. A set of inference rules, derived from S, is obtained as follows. For each clause $L :- L_1, L_2, ..., L_n$ in S, we have the following clause rule:

Clause Rules

$$\frac{\Gamma_0 \to L_1 => \Gamma_1 \to L_1, \Gamma_1 \to L_2 => \Gamma_2 \to L_2, ..., \Gamma_{n-1} \to L_n => \Gamma_n \to L_n}{\Gamma_0 \to L => \Gamma_n \to L}$$

We also have assumption axioms and a case analysis (splitting) rule. Let L be a positive literal. Then the assumption axioms and case analysis rule can be stated as follows:

Assumption Axioms

$$\Gamma \to L => \Gamma \to L \quad \text{if } L \, \varepsilon \, \Gamma$$

$$\Gamma \to \bar{L} => \Gamma, \bar{L} \to \bar{L}$$

Case Analysis (splitting) Rule

$$\frac{\Gamma_0 \to L => \Gamma_1, \bar{M} \to L, \ \Gamma_1, M \to L => \Gamma_1, M \to L}{\Gamma_0 \to L => \Gamma_1 \to L}$$

The goal-subgoal structure of this deduction system is evident. The input clause $L :- L_1, L_2, ..., L_n$ merely states that $L_1, L_2, \cdots, L_n$ have to be confirmed in order to confirm L. The corresponding clause rule for $L :- L_1, L_2, ..., L_n$ states that, if the initial subgoal is $\Gamma \to L$, then make $L_1, ..., L_n$ subgoals in succession; add to

Γ successively the literals that are needed to make each one provable; and finally, return $\Gamma_n \to L$ where Γ_n contains all the literals needed to make $L_1, \ldots, L_n$ provable.

Sprfn implements the natural deduction system just described. **Sprfn** exploits Prolog style depth-first iterative-deepening search. This search strategy involves repeatedly performing exhaustive depth-first search with increasing depth bounds. For a description of the strategy, see [6]. This search strategy is complete and can be efficiently implemented in Prolog, taking advantage of Prolog's built-in depth-first search with backtracking.

3   SPRFN and Term Rewriting

3.1   Input Format

The input to **sprfn** is formatted in Horn-like clauses. Given a set S of clauses, we convert them into Horn-like clauses as follows. For a clause containing at least one positive literal, we select one such literal to be the head, negate the remaining literals, and move them to the body of the clause. For an all-negative clause, we use **false** as the head of the clause and form the body from the positive literals corresponding to the original literals. The following example shows how to translate from clause form into the format accepted by **sprfn**. Notice the similarity of the input format syntax to Prolog program syntax.

Clause Form

$P(x) \vee Q(x)$
$\sim P(x) \vee R(x)$
$\sim Q(x) \vee R(x)$
$\sim R(a)$

Input Format for **sprfn**

$p(X) :- not(q(X))$
$r(X) :- p(X)$
$r(X) :- q(X)$
$false :- r(a)$

For input to **sprfn**, the convention is that a name starting with a capital letter is a variable name; all other names are predicate names, function names or constants. *Not* and *false* are reserved for negation and for the head of the top-level goal, respectively.

3.2   The Method of Proof

The prover attempts to prove that *false* is derivable from the input clauses. For example, given the following set of clauses:

$p(X) :- not(q(X))$
$r(X) :- p(X)$
$r(X) :- q(X)$
$false :- r(a)$

sprfn will derive the following proof:

```
false :- cases(
         (not q(a):
            (r(a) :- (p(a) :- not q(a)))),
         (q(a):
            (r(a) :- q(a)))
         )
```

Thus, *false* can be proven from the input clauses. For there are two cases to consider: (1) Suppose not q(a) is true; then we can derive *false* as follows. Since we are given that false :- r(a), we make r(a) our subgoal. Now we can derive r(a) if we can prove p(a), since we are given r(X) :- p(X). Meanwhile, we can derive p(a) if we can prove not q(a), since we are given p(X) :- not q(X). However, we are assuming not q(a), so this subgoal can be proven. (2) Suppose q(a) is true; then we can derive *false* as follows. Once again, we make r(a) our subgoal, since we are given that false :- r(a). Now we can derive r(a) if we can prove q(a), since we are given r(X) :- q(X). But we are assuming q(a), so this subgoal can be proven.

3.3   The Term Rewriting Mechanism in SPRFN

Replace. An assertion of the form **replace(<expr1>, <expr2>)** in the input signifies that all subgoals of form **<expr1>** should be replaced by subgoals of the form **<expr2>** before attempting to solve them. This is like a rewrite applied at the 'top level'. This is sound if **<expr1> :- <expr2>** is valid.

Rewrite. An assertion of the form **rewrite(<expr1>, <expr2>)** in the input signifies that all subexpressions of form **<expr1>** should be replaced by subexpressions of the form **<expr2>**. This is like a rewrite applied anywhere, not just at the top level. This is sound if the logical equivalence **<expr1> <-> <expr2>** is valid, or, in case when the expressions are terms, if the equation **<expr1> = <expr2>** is valid.

In our experiments, we translated the axioms of von Neumann-Bernays-Gödel set theory into a list of rewrite rules and then attempted to derive various theorems based on these rules. For example, consider the axiom for Subset below:

$(\forall x, y)[x \subseteq y \leftrightarrow (\forall u)[(u \in x \rightarrow u \in y)]]$

This would be translated into the following two rewrite rules, which would be given as input to the prover:

    rewrite(sub(X,Y), or(not(el(f17(X,Y),X)), el(f17(X,Y),Y))).
    rewrite(not(sub(X,Y)), and(el(U,X), not(el(U,Y)))).

Several points deserve mention. First of all, note that the single axiom gives rise to *two* rewrite rules -- a "positive" as well as a "negative" rule. This is to preserve soundness, since **sprfn** performs outermost term rewriting. The presence of the negative rewrite rule insures that whenever **sprfn** rewrites a term of the form *sub(X,Y)* with *or(not(el(f17(X,Y),X)), el(f17(X,Y),Y)))* (which implies that **sprnf** is using the positive rule) we know that this term does not appear in a negative context; for if it did, the prover would already have rewritten it using the negative rule.

We should also point out what may seem at first to be a counter-intuitive feature of these rewrite rules. Note the presence of the skolem function *f17(X,Y)* in the positive rewrite rule and the unbound variable *U* in the negative rule. One might think that the situation should be reversed. However, the correctness of this procedure can be seen by reflecting upon the following. Recall that **sprfn** performs subgoaling in attempting to prove *false*. Thus if the prover is attempting to prove *A*, let's say, and it tries to do this by trying to prove the subgoal *B*, this procedure will only be sound if it is the case that $B \rightarrow A$. Our rewrite rules must observe this fact. Hence, if we are trying to prove *A* and we attempt to do so by rewriting *A* with *B* and then trying to prove the subgoal *B*, it must be the case that $B \rightarrow A$. Or, to put the matter in Prolog symbolism, it must be the case that *A :- B*. When we skolemize the original axiom, we see that the following are logical consequences of the skolemized input clauses:

sub(X,Y) :- or(not(el(f17(X,Y),X)), el(f17(X,Y),Y))
not(sub(X,Y)) :- and(el(U,X), not(el(U,Y)))

Thus, we must express our two rewrite rules as given above.

4   Term Rewriting with a Tautology Checker

In our first experiment, we modified **sprfn** to make use of a **tautology checker**. Suppose that we wish to prove the set theoretic theorem T, which, in accordance with the procedure outline above, has been converted into the top-level goal: "false :- X".

If the flag t_test is set, then the prover will call the tautology checker tautology3(X,Y), where X is the input theorem (derived from the top-level goal "false :- X") and Y is the output consisting of the non-tautologous part (if any) of X. If X is a tautology, then the prover will halt; else, the original goal: "false :- X" is retracted and replaced in the database with the new goal: "false :- Y". The prover then proceeds to attempt to prove *false* by means of the subgoaling method described above. This method seems to work quite well. For one thing, if X is a tautology, the tautology checker allows the prover to spot this fact much sooner than if it had attempted to achieve its top-level goal by means of its subgoaling mechanism alone. For another, we have found that when X is not a tautology, by removing the tautologous portion of X and returning Y as the subgoal to be proved, we save the prover considerable time and avoid needlessly duplicated effort.

Because tautology3(X,Y) does not unify variables (it only eliminates a disjunction as a tautology if some literal L appears both negated and un-negated in the clause), as a standard practice we have included the axiom: "or(X,Y) :- prolog(tautology(or(X,Y)))" to handle cases where unifying is necessary to eliminate tautologous clauses. This allows us to invoke Prolog from within **sprfn**, and to call the Prolog predicate tautology/1 which succeeds if its input can be converted into a tautology via unification.

Thus backtracking over the elimination of a tautologous clause is still possible, but it only occurs with respect to the "or" rewrite rule. This seems more efficient than permitting backtracking into the tautology3 routine itself (which would be required if we allowed unification within tautology3).

We now exhibit two examples of the prover at work, utilizing the tautology checker.

4.1 Example 1

In this first example, we show how the tautology checker returns the non-tautologous portion of its input theorem, which is then proven by **sprfn**'s subgoaling mechanism.

Proof of Difference and Join Theorem

Our top-level goal is:

false:-eq(diff(a,b),join(a,comp(b)))

After reading in the input clauses, which contain our set theoretic rewrite rules as well as a few axioms, the prover begins by calling our tautology checker:

t_test is asserted
b_only is asserted
solution_size_mult(0.1) is asserted
proof_size_mult(0.4) is asserted

calling(tautology3(eq(diff(a,b),join(a,comp(b))),_9812))

.
.
.

After removing the tautologous portion of the theorem, tautology3 returns the following:

conjunct:
  m(f17(diff(a,b),comp(b)))
  not el(f17(diff(a,b),comp(b)),a)
  el(f17(diff(a,b),comp(b)),b)

Continue?: yes.

At this point, the tautology checker informs the user that it has a conjunction of disjunctions (in this case there is only one such disjunction) left, which it could not eliminate via tautology checking alone. It asks the user if he wishes to proceed, and in this case, we answer in the affirmative. The prover's subgoaling procedure is now invoked, and in a short time **sprfn** returns with the following:

```
proof found
false:-cases(
        (not el(f17(diff(a,b),comp(b)),a):
          (or(m(f17(diff(a,b),comp(b))),or(not el(f17(diff(a,b),comp(b)),a),
           el(f17(diff(a,b),comp(b)),b))):-(or(not el(f17(diff(a,b),comp(b)),a),
           el(f17(diff(a,b),comp(b)),b)):-not el(f17(diff(a,b),comp(b)),a)))),

        (el(f17(diff(a,b),comp(b)),a):
          (or(m(f17(diff(a,b),comp(b))),or(not el(f17(diff(a,b),comp(b)),a),
           el(f17(diff(a,b),comp(b)),b))):-(m(f17(diff(a,b),comp(b))):-
           el(f17(diff(a,b),comp(b)),a)))))

size of proof 7

8.73 cpu seconds used
5 inferences done
```

It is worth pointing out that by using the term rewriting facility *without* invoking the tautology checker, the prover was able to derive the theorem in 128.43 cpu seconds with 34 inferences. We attempted to prove the theorem using neither the tautology checker nor rewrite rules; but after letting the prover run for over two hours without finding the proof, we put it out of its misery.

4.2   Example 2

In this second example, we show the prover's term rewriting facility in action. In this particular case, the tautology checker is able to establish that the entire input theorem is a tautology; hence it is unnecessary to invoke **sprfn**'s subgoaling mechanism, since the theorem is already proven.

Proof of Power Set Theorem

Our top-level goal is:

```
false:-eq(pset(join(a,b)),join(pset(a),pset(b)))
```

After reading in the input clauses, which contain our set theoretic rewrite rules as well as a few axioms, the prover begins by calling our tautology checker:

```
t_test is asserted
b_only is asserted
solution_size_mult(0.1) is asserted
proof_size_mult(0.4) is asserted

calling(tautology3(eq(pset(join(a,b)),join(pset(a),pset(b))),_9818))
```

The rewriting mechanism displays the results of its outermost term rewriting operation:

rewrite(eq(pset(join(a,b)),join(pset(a),pset(b))),and(sub(pset(join(a,b)),
    join(pset(a),pset(b))),sub(join(pset(a),pset(b)),pset(join(a,b)))))

rewrite(sub(pset(join(a,b)),join(pset(a),pset(b))),and(sub(pset(join(a,b)),
    pset(a)),sub(pset(join(a,b)),pset(b))))

rewrite(sub(pset(join(a,b)),pset(a)),or(not el(f17(pset(join(a,b)),pset(a)),
    pset(join(a,b))),el(f17(pset(join(a,b)),pset(a)),pset(a))))

rewrite(not el(f17(pset(join(a,b)),pset(a)),pset(join(a,b))),not sub(f17(pset(
    join(a,b)),pset(a)),join(a,b)))

rewrite(not sub(f17(pset(join(a,b)),pset(a)),join(a,b)),or(not sub(f17(pset(
    join(a,b)),pset(a)),a),not sub(f17(pset(join(a,b)),pset(a)),b)))

rewrite(el(f17(pset(join(a,b)),pset(a)),pset(a)),sub(f17(pset(join(a,b)),
    pset(a)),a))

rewrite(sub(pset(join(a,b)),pset(b)),or(not el(f17(pset(join(a,b)),pset(b)),
    pset(join(a,b))),el(f17(pset(join(a,b)),pset(b)),pset(b))))

rewrite(not el(f17(pset(join(a,b)),pset(b)),pset(join(a,b))),not sub(f17(pset(
    join(a,b)),pset(b)),join(a,b)))

rewrite(not sub(f17(pset(join(a,b)),pset(b)),join(a,b)),or(not sub(f17(pset(
    join(a,b)),pset(b)),a),not sub(f17(pset(join(a,b)),pset(b)),b)))

rewrite(el(f17(pset(join(a,b)),pset(b)),pset(b)),sub(f17(pset(join(a,b)),
    pset(b)),b))

rewrite(sub(join(pset(a),pset(b)),pset(join(a,b))),or(not el(f17(join(pset(a),
    pset(b)),pset(join(a,b))),join(pset(a),pset(b))),el(f17(join(pset(a),pset(b)),
    pset(join(a,b))),pset(join(a,b)))))

rewrite(not el(f17(join(pset(a),pset(b)),pset(join(a,b))),join(pset(a),pset(b))),
    or(not el(f17(join(pset(a),pset(b)),pset(join(a,b))),pset(a)),not el(f17(
    join(pset(a),pset(b)),pset(join(a,b))),pset(b))))

rewrite(not el(f17(join(pset(a),pset(b)),pset(join(a,b))),pset(a)),not sub(f17(
    join(pset(a),pset(b)),pset(join(a,b))),a))

rewrite(not el(f17(join(pset(a),pset(b)),pset(join(a,b))),pset(b)),not sub(f17(
    join(pset(a),pset(b)),pset(join(a,b))),b))

rewrite(el(f17(join(pset(a),pset(b)),pset(join(a,b))),pset(join(a,b))),sub(f17(
    join(pset(a),pset(b)),pset(join(a,b))),join(a,b)))

rewrite(sub(f17(join(pset(a),pset(b)),pset(join(a,b))),join(a,b)),and(sub(f17(
    join(pset(a),pset(b)),pset(join(a,b))),a),sub(f17(join(pset(a),pset(b)),
    pset(join(a,b))),b)))

At this point, rewriting has been completed; the procedure **cnf_expand** is now invoked to expand the rewritten theorem into conjunctive normal form and to then eliminate all tautologous conjuncts.

```
call(0,cnf_expand(and(and(or(or(not sub(f17(pset(join(a,b)),pset(a)),a),not sub(
f17(pset(join(a,b)),pset(a)),b)),sub(f17(pset(join(a,b)),pset(a)),a)),or(or(not
sub(f17(pset(join(a,b)),pset(b)),a),not sub(f17(pset(join(a,b)),pset(b)),b)),sub(
f17(pset(join(a,b)),pset(b)),b))),or(or(not sub(f17(join(pset(a),pset(b)),pset(
join(a,b))),a),not sub(f17(join(pset(a),pset(b)),pset(join(a,b))),b)),and(sub(f17
(join(pset(a),pset(b)),pset(join(a,b))),a),sub(f17(join(pset(a),pset(b)),pset(
join(a,b))),b)))),_15815))
```

Initially, when **cnf_expand** is called, its output argument is the uninstantiated Prolog variable _15815. But when it returns, this output argument has been instantiated to the empty list, signifying that no non-tautologous portion of the theorem remains:

```
result(0,cnf_expand(and(and(or(or(not sub(f17(pset(join(a,b)),pset(a)),a),not sub(
f17(pset(join(a,b)),pset(a)),b)),sub(f17(pset(join(a,b)),pset(a)),a)),or(or(not
sub(f17(pset(join(a,b)),pset(b)),a),not sub(f17(pset(join(a,b)),pset(b)),b)),sub(
f17(pset(join(a,b)),pset(b)),b))),or(or(not sub(f17(join(pset(a),pset(b)),pset(
join(a,b))),a),not sub(f17(join(pset(a),pset(b)),pset(join(a,b))),b)),and(sub(f17
(join(pset(a),pset(b)),pset(join(a,b))),a),sub(f17(join(pset(a),pset(b)),pset(
join(a,b))),b)))),[]))
```

tautology3 returns: is_tautology

theorem_is_a_tautology

4.28 cpu seconds used
0 inferences done

We observed two very important things while running these tests. First of all, we found that including explicit rewrite rules to distribute "or" over "and" significantly slowed down the tautology checker. (Fortunately, the **cnf_expand** routine is able to test for tautologies without requiring that its input argument be in conjunctive normal form; hence employing the distribution rules is not needed.) We ran tests in which these distribution rules were used and tests in which they were not. The results are contained in the Appendix.

Secondly, we discovered that the depth to which term rewriting is allowed to take place greatly affects overall performance. For example, in the case of the Power Set theorem exhibited above, we did *not* include in our input the rewrite rules for the Subset axiom. By omitting those two rules (see the earlier section: "The Term Rewriting Mechanism in **SPRFN**") we cause the prover to regard terms of the form "sub(X,Y)" as atomic and thus it does not rewrite them. In this way, it is able to discover that the entire theorem is a tautology. On the other hand, we found that if we included the rewrite rules for the Subset axiom, then our tautology checker was no longer able to eliminate the entire theorem as a tautology; indeed, it returned a significantly long conjunction, which the subgoaling mechanism then had to prove. This took a much greater amount of time. (Cf. Table 2 of the Appendix.) For a complete summary of our test results using the tautology checker, the reader should consult the Appendix.

5 Preprocessing Input via Term Rewriting

In our second experiment, we used our term rewriting facility as a preprocessor. We discovered in our earlier experiments that, as a general rule, the more complex the theorem, the greater the number of terms that ultimately result from rewriting the theorem. In fact, we found that for certain theorems, such as the Composition of Homomorphisms theorem (see below) it was physically impossible to use the tautology checker. This was due to the fact that one term was being rewritten to a conjunction (or disjunction) of several other terms, each of which was itself subject to being rewritten into a complex of several terms and so on. Thus, nearly exponential growth of the Prolog structure occurred during the operation of the rewriting facility. This eventually caused Prolog to run out of stack long before the **cnf_expand** subroutine had a chance to eliminate any tautologous portion of the theorem.

We decided, therefore, to preprocess the theorem by reducing the size of the term that appeared as the body in the top-level goal. In general, our approach involved skolemizing the negated theorem and then using the rewriting facility to produce the initial set of input clauses. As an illustration of this technique, we present the following proof of the Composition of Homomorphisms theorem. We should point out that it was necessary to add three simple axioms in order to derive the proof; also, it was necessary once again to restrain the depth to which rewriting took place.

Proof of Composition of Homomorphisms Theorem

Our theorem is the following:

$$(\forall xh1,xh2,xs1,xs2,xs3,xf1,xf2,xf3)[(hom(xh1,xs1,xf1,xs2,xf2) \wedge$$
$$hom(xh2,xs2,xf2,xs3,xf3)) \rightarrow hom(compose(xh2,xh1),xs1,xf1,xs3,xf3)]$$

After skolemizing the negation of the theorem we have three clauses to be rewritten: hom(ah1,as1,af1,as2,af2), hom(ah2,as2,af2,as3,af3), and not(hom(compose(ah2,ah1),as1,af1,as3,af3)). Based on these clauses, the prover's term rewriting facility produced the following set of input clauses:

Clauses derived from hom(ah1,as1,af1,as2,af2):

    eq(apply(ah1,apply(af1,ord_pair(G1,G2))),
      apply(af2,ord_pair(apply(ah1,G1),apply(ah1,G2)))) :-
          el(G1,as1), el(G2,as1).
    maps(ah1,as1,as2).
    closed(as2,af2).
    closed(as1,af1).

Clauses derived from hom(ah2,as2,af2,as3,af3):

    eq(apply(ah2,apply(af2,ord_pair(G3,G4))),
      apply(af3,ord_pair(apply(ah2,G3),apply(ah2,G4)))) :-
          el(G3,as2), el(G4,as2).
    maps(ah2,as2,as3).
    closed(as3,af3).
    closed(as2,af2).

Clauses derived from not(hom(compose(ah2,ah1),as1,af1,as3,af3)):

    el(g5,as1).
    el(g6,as1).
    false :-
      eq(apply(ah2,apply(ah1,apply(af1,ord_pair(g5,g6)))),
        apply(af3,ord_pair(apply(ah2,apply(ah1,g5)),apply(ah2,apply(ah1,g6))))),
      maps(compose(ah2,ah1),as1,as3),
      closed(as3,af3),
      closed(as1,af1).

Note that our top-level goal has become:

    false :-
      eq(apply(ah2,apply(ah1,apply(af1,ord_pair(g5,g6)))),
        apply(af3,ord_pair(apply(ah2,apply(ah1,g5)),apply(ah2,apply(ah1,g6))))),
      maps(compose(ah2,ah1),as1,as3),
      closed(as3,af3),
      closed(as1,af1).

In addition to these input clauses, we added three axioms. The first two of these are trivial while the third, although non-trivial, can be derived by the prover in 24.63 cpu seconds after 15 inferences.

Axioms for proof of homomorphism theorem:

> eq(apply(XF1,S1),apply(XF2,S2)) :-
>   eq(S1,S3), eq(apply(XF1,S3),apply(XF2,S2)).
>
> el(apply(XF,X),S2) :- maps(XF,S1,S2), el(X,S1).
>
> maps(compose(X,Y),S1,S3) :- maps(Y,S1,S2),maps(X,S2,S3).

Finally, we added some extra rewrite rules which serve only to cut down on the size of data structures that result from term rewriting.

Rewrite Rules to handle large terms:

> rewrite(f32(ah1,as1,af1,as2,af2),g1).
> rewrite(f33(ah1,as1,af1,as2,af2),g2).
> rewrite(f32(ah2,as2,af2,as3,af3),g3).
> rewrite(f33(ah2,as2,af2,as3,af3),g4).
> rewrite(f32(compose(ah2,ah1),as1,af1,as3,af3),g5).
> rewrite(f33(compose(ah2,ah1),as1,af1,as3,af3),g6).
> rewrite(apply(compose(XF1,XF2),S),apply(XF1,apply(XF2,S))).

Given this preprocessed input, **sprfn** is able to derive the following proof of the theorem:

> proof found
> false:-lemma((eq(apply(ah2,apply(ah1,apply(af1,ord_pair(g5,g6))))),
>   apply(af3,ord_pair(apply(ah2,apply(ah1,g5)),
>   apply(ah2,apply(ah1,g6)))))):-[])),
>
> (maps(compose(ah2,ah1),as1,as3):-
>   maps(ah1,as1,as2),
>   maps(ah2,as2,as3)),
>
> closed(as3,af3),
> closed(as1,af1).

> size of proof 18

> 30.2333 cpu seconds used
> 14 inferences done

Note that the proof involves a lemma, which **sprfn** derived in the course of its operation. If we so desire, we can ask the prover to show us how it came up with this lemma. When we do so, it responds with the following derivation:

```
   proof of lemma:
false:-(eq(apply(ah2,apply(ah1,apply(af1,ord_pair(g5,g6))))),
      apply(af3,ord_pair(apply(ah2,apply(ah1,g5)),
      apply(ah2,apply(ah1,g6))))):-

      lemma((eq(apply(ah1,apply(af1,ord_pair(g5,g6))),
         apply(af2,ord_pair(apply(ah1,g5),apply(ah1,g6))))):-[]),

      (eq(apply(ah2,apply(af2,ord_pair(apply(ah1,g5),apply(ah1,g6))))),
         apply(af3,ord_pair(apply(ah2,apply(ah1,g5)),
         apply(ah2,apply(ah1,g6))))):-

         lemma((el(apply(ah1,g5),as2):-[])),
         (el(apply(ah1,g6),as2):-

            maps(ah1,as1,as2),el(g6,as1)))).
```

size of proof 11

26.9166 cpu seconds used
13 inferences done

6   Summary of Results

The techniques we employed allowed us to prove moderately sophisticated set theoretic theorems in rapid time with few inferences. These theorems would have been much more difficult to derive without the rewrite rules; indeed, **sprfn** was unable to derive some of them when run without the rewrite rules. Undoubtedly it would have been beyond the power of a typical resolution theorem prover to derive most of the theorems in question.

We have found that removing the tautologous portion of a theorem by means of some filter such as our tautology checker seems to speed up the derivation time, by allowing the prover to focus its attention on the non-tautologous aspects of the theorem. Furthermore, we discovered that the depth to which term rewriting is allowed greatly affects the prover's ability to arrive at a proof. Clearly, more work needs to be done in this area. At the present time, human intervention is required to adjust term rewriting depth; hopefully this can be automated to some extent in the future.

Our research leads us to conclude that preprocessing input clauses by means of rewrite rules is also highly effective in directing a theorem prover's attention towards a fast, relatively short proof. Although this kind of preprocessing is presently being done by hand, we are confident that it can be fully automated.

Finally, among the practical results that we obtained, it bears mentioning that it pays to avoid distributing "or" over "and" by means of rewrite rules.

At the same time, we discovered that there *are* limits to the power of term rewriting in connection with proving theorems from set theory. For one thing, we found that a clause's size grows almost exponentially when terms are rewritten by terms which are themselves subject to being rewritten, and so forth. Although this problem has no effect on soundness, the physical limitations of the computer itself come into play at this point, causing the prover to run out of stack before it can complete its rewriting phase.

We also realize that our procedure is not complete, if rewriting takes place at the wrong time. For example, suppose we have the rewrite rule: $B \rightarrow\!\!\!> P(x)$ and we wish to demonstrate that the following theorem is a tautology:

$$B \vee (\sim\!P(a) \wedge \sim\!P(b))$$

If we rewrite B before we distribute "or" over "and", we have:

$$P(x) \vee (\sim\!P(a) \wedge \sim\!P(b))$$

from which we can only derive:

$$(P(x) \vee \sim\!P(a)) \wedge (P(x) \vee \sim\!P(b))$$

and this is not tautologous no matter how we instantiate the variable x. Yet if we distribute "or" over "and" before rewriting B, we have:

$$(B \vee \sim\!P(a)) \wedge (B \vee \sim\!P(b))$$

from which we can derive the tautology:

$$(P(x) \vee \sim\!P(a)) \wedge (P(y) \vee \sim\!P(b))$$

since Prolog will provide a different variable each time it replaces B with $P(x)$.

This raises the following questions: Is term replacement more complete than term rewriting? How complete is term replacement for existentially quantified variables? Is replacement equivalent to delayed term rewriting? More work needs to be done before we are in a position to answer these questions.

Finally, the approaches to term rewriting that we explored are not sufficient when trying to prove theorems that require *creative insight*. For example, in one of our experiments we tried to deduce Cantor's Theorem using our rewrite rules. However, we discovered that **sprfn** was unable to find the proof without being given quite a bit of non-trivial information. Specifically, we had to provide it with axioms implying (1) that any function induces its own diagonal set and (2) that the relation which pairs a unit set with its single element is a one-one function. Once these axioms were supplied, by making use of our rewrite rules the prover was able to derive Cantor's Theorem in 33.65 cpu seconds with 12 inferences. Nevertheless, one would like the prover to be able to realize on its own that such sets and functions exist. Yet recognizing that there *is* such a thing as the diagonal of a function and that such a set might be useful in this case requires a kind of insight that goes far beyond syntactic manipulations. Unfortunately, term rewriting alone does not provide the necessary machinery for the prover to possess this kind of creative insight.

References

[1] Plaisted, D.A., 'A simplified problem reduction format', *Artificial Intelligence* **18** (1982) 227-261

[2] Boyer, Robert, Lusk, Ewing, McCune, William, Overbeek, Ross, Stickel, Mark, and Wos, Lawrence, 'Set theory in first-order logic: clauses for Gödel's axioms', *Journal of Automated Reasoning* **2** (1986) 287-327

[3] Plaisted, D.A., 'Another extension of Horn clause logic programming to non-Horn clauses', Lecture Notes 1987

[4] Plaisted, D.A., 'Non-Horn clause logic programming without contrapositives', unpublished manuscript 1987

[5] Loveland, D.W., *Automated Theorem Proving: A Logical Base*, North-Holland Publishing Co., 1978, Chapter 6.

[6] Korf, R.E., 'Depth-first iterative-deepening: an optimal admissible tree search', *Artificial Intelligence* **27** (1985) 97-109.

Appendix
Test Results Using a Tautology Checker

Table 1

	Theorems
(1)	false :- eq(union(a,b),union(b,a)).
(2)	false :- eq(join(a,b),join(b,a)).
(3)	false :- eq(union(a,a),a).
(4)	false :- eq(join(a,a),a).
(5)	false :- eq(union(a,comp(a)), universe).
(6)	false :- eq(join(a,comp(a)), 0).
(7)	false :- eq(comp(universe),0).
(8)	false :- eq(comp(0),universe).
(9)	false :- eq(comp(comp(a)),a).
(10)	false :- eq(union(a,0),a).
(11)	false :- eq(join(a,universe),a).
(12)	false :- eq(union(a,universe),universe).
(13)	false :- eq(join(a,0),0).
(14)	false :- eq(union(union(a,b),c),union(a,union(b,c))).
(15)	false :- eq(join(join(a,b),c),join(a,join(b,c))).
(16)	false :- if(sub(a,b), then(eq(join(a,b),a))).
(17)	false :- eq(comp(union(a,b)),join(comp(a),comp(b))).
(18)	false :- eq(comp(join(a,b)),union(comp(a),comp(b))).
(19)	false :- eq(join(union(a,b),union(a,comp(b))),a).
(20)	false :- eq(diff(a,b),join(a,comp(b))).
(21)	false :- eq(union(a,universe),universe).
(22)	false :- eq(join(a,union(b,c)), union(join(a,b), join(a,c))).
(23)	false :- eq(union(a,join(b,c)), join(union(a,b), union(a,c))).
(24)	false :- sub(0,a).
(25)	false :-- if(and(sub(a,b),sub(b,c)),then(sub(a,c))).
(26)	false :- if(sub(a,b),then(el(a,pset(b)))).
(27)	false :- if(disjoint(a,b),then(eq(join(a,b),0))).
(28)	false :- sub(a,union(a,b)).
(29)	false :- sub(diff(a,b),a).
(30)	false :- if(sub(a,join(b,c)), then(and(sub(a,b),sub(a,c)))).
(31)	false :- eq(pset(join(a,b)),join(pset(a),pset(b))).
(32)	false :- eq(pset(join(a,b)),join(pset(a),pset(b))).
(33)	false :- sub(prod(a,join(b,c)),join(prod(a,b),prod(a,c))).
(34)	false :- if(and(sub(a,b),sub(c,d)),then(sub(prod(a,c),prod(b,d)))).
(35)	false :- if(and(meq(a,b),meq(c,d)),then(eq(ord_pair(a,c),ord_pair(b,d)))).
(36)	false :- if(eq(a,ord_pair(b,c)),then(opp(a))).
(37)	false :- if(and(m(a),m(b)),then(sub(set(a,b),set(a,b)))).
(38)	false :- if(and(m(a),m(b)), then(eq(set(a,b),set(b,a)))).

Note that theorems (31) and (32) are the same. However, (31) was proven using a rewrite rule for the subset axiom, while (32) was proven using a replace rule for the subset axiom. Using a replace rather than a
rewrite rule prevented terms containing the "subset" predicate from being rewritten before tautology checking was performed. This allowed the prover to find the proof much faster in the case of this particular
theorem.

Table 2

Summary of Results				
	With "or-over-and" Distribution Rules		Without "or-over-and" Distribution Rules	
Theorem	Time	Inferences	Time	Inferences
(1)	3.23	0	2.5	0
(2)	4.18	0	4.14	0
(3)	1.66	0	1.61	0
(4)	1.68	0	1.76	0
(5)	5.93	4	5.31	4
(6)	5.96	6	5.85	6
(7)	3.66	4	3.5	4
(8)	2.7	2	2.68	2
(9)	5.73	4	4.86	4
(10)	2.91	2	2.53	2
(11)	4.53	4	4.46	4
(12)	4.73	4	4.33	4
(13)	2.76	2	2.73	2
(14)	9.68	0	5.51	0
(15)	10.88	0	10.88	0
(16)	7.48	4	4.91	4
(17)	10.86	0	6.64	0
(18)	18.1	0	5.94	0
(19)	9.34	0	5.33	0
(20)	10.11	5	8.73	5
(21)	4.66	4	4.53	4
(22)	20.55	0	8.44	0
(23)	19.88	0	7.73	0
(24)	1.26	2	1.18	2
(25)	12.26	8	9.63	8
(26)	3.76	4	3.21	4
(27)	18.36	14	15.85	14
(28)	0.81	0	0.78	0
(29)	0.78	0	0.84	0
(30)	40.76	16	24.04	16
(31)	217.96	32	189.38	32
(32)	4.83	0	4.28	0
(33)	3.38	0	3.11	0
(34)	63.55	32	34.63	16
(35)	15.96	0	4.93	0
(36)	69.11	23	37.78	16
(37)	67.21	0	4.25	0
(38)	109.00	0	8.84	0

These results were derived by using a tautology-checker in conjunction with rewrite/replace rules.

SUMMARY: In each case, the number of inferences required is virtually the same whether or not the "or-over-and" distribution rules are used. However, in almost every instance there is a speed-up when these rules are not used. Furthermore, as a general rule it seems that as the amount of time required to prove the theorem increases, the greater the speed-up when the "or-over-and" rules are not used.

Analogical Reasoning
and
Proof Discovery[1]

Bishop Brock, Shaun Cooper, William Pierce

Department of Computer Sciences
The University of Texas at Austin
Austin, TX 78712

Abstract

We introduce preliminary research on the problem of applying analogical reasoning to proof discovery. In our approach, the proof of one theorem is used to guide the proof of a similar theorem by suggesting analogous steps. When a step suggested by a guiding proof cannot be applied, actions are taken to bring the proofs back into correspondence, often by adding intermediate steps. Taking this approach, we have implemented a natural deduction prover which exploits analogical reasoning and has yielded some promising results in the domain of Real Analysis. We present some of these results, which include a proof of the convergence of the product of convergent sequences, using an analogous proof for the sum of convergent sequences. We also include the timing results of one experiment in which our prover's performance was compared with and without the use of analogy.

1 Motivation

Much of the work that has been done in the area of mechanical problem solving can be placed near the extremes of a spectrum. At one end, much research in Automated Theorem Proving has focused on solving difficult problems by developing efficient algorithms and fast search strategies. Many of these strategies involve an easily characterized, restricted search through a state space generated by applications of a small set of inference rules. The *Set of Support* strategy for resolution is an example. Such strategies are often justified (or even motivated) by a completeness proof. At the other end of the spectrum, much problem solving research in Artificial Intelligence has attempted to incorporate more human-like problem solving behavior. The results of these efforts tend to be knowledge-intensive. The field of Expert Systems provides many examples.

---

[1]This work was supported in part by NSF grant CCR-8613 706.

A potential limitation of the first focus is that simple search strategies may not be adequate for solving the difficult problems to which humans apply a wealth of knowledge and experience. Yet most of the research motivated by the knowledge-oriented approach is applied to domains which are less complicated than the domains to which many theorem provers are applied. It is not clear that the methods which produced many successful expert systems will yield the same success when applied to difficult domains of mathematics. Thus we are anxious to experiment with applying knowledge-intensive methods in automating proof discovery. In our current research, we have emphasized the use of analogical reasoning. For some examples of related research efforts in the application of analogical reasoning to problem solving, see [7], [9], [10], and [13].

In this paper, we will first describe how we apply analogical reasoning to theorem proving. We will then present the results of some experiments with a natural deduction implementation in the domain of Real Analysis. We will briefly discuss current and future work, and finally conclude with lessons we have learned from the research conducted so far. This paper is intended as an overview of ongoing research. Readers interested in more technical details of our implementations are referred to [4], [5], and [6].

2 Proof Discovery using a Guiding Proof

A problem solver which takes full advantage of analogical reasoning could include several facilities. Given a target problem to solve, it could retrieve a similar problem and associated solution from a large knowledge base. It could then find a mapping between the guiding problem and the target problem which summarizes how the terms of the target theorem correspond to the terms of the guiding theorem. Using the guiding solution and the mapping between the two problems, it could discover a solution to the target problem. Finally, it could store the new problem/solution pair in the knowledge base for use in solving later problems. Although we are ultimately interested in all these aspects of analogical reasoning, the research presented in this paper has been directed primarily toward the task of discovering a proof for a target theorem by following the proof for an analogous theorem. Thus our implemented system requires that the user provide the guiding theorem and proof as well as the mapping between the guiding and target theorems.

In [4], we describe a resolution based implementation which formed the basis for our first experiments. Our current implementation uses natural deduction and is described more fully in [5] and [6]. Both implementations use chaining and variable elimination (described in [2]), and both have the capability of fetching lemmas from a lemma base when necessary for completing a proof. In the following subsections, we describe how our current implementation follows a guiding proof, how it recovers when an analogy breaks down, and how it attempts to complete a proof when the

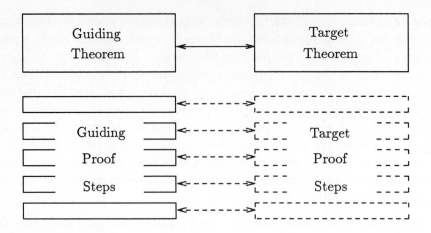

Figure 1: The analogical proof process. Arrows represent the analogy mappings. Bold portions are provided by the user.

guiding proof no longer applies.

2.1 Following the Guiding Proof

In our application of analogical reasoning to theorem proving, we treat proofs as sequences of inferences and associated intermediate steps. The mapping between theorems or between steps in a proof is characterized by pairings of subterms. Figure 1 illustrates the analogical proof process. The guiding theorem and proof are provided, as well as the mapping between the guiding and target theorems. The proof for the target theorem is constructed automatically, one step at a time, using inferences which are analogous to those used in the guiding proof.

We may think of an inference as an application of a rule to a list of arguments, which are terms occurring previously in the proof. The analogous inference is the application of the same rule to the analogous terms as determined by the analogy mapping. For the special rule of introducing a new lemma into the guiding proof, the situation is more complex. Currently, the task of fetching analogous lemmas is accomplished as follows. A special knowledge base is used to associate useful analogies with pairings of lemmas. For example, a useful analogy in real analysis pairs *plus* with *times* and certain lemmas involving the operator *plus* with certain lemmas involving the operator *times*. Another useful analogy pairs sequences with functions. The user may assign a name to such an analogy and associate this name with pairings of lemmas. When the user gives the system a target theorem to prove and a guiding proof to follow, the user may also provide the name of the analogy most appropriate to this pair. When the guiding proof uses a lemma, the system will use the appropriate lemma pairings to find the analogous lemma for use in the target proof.

As each new step is added to the target proof, the mapping between the proofs is extended to pair the terms of the new step with the terms of the corresponding guiding step. This process of updating the mapping for new steps is accomplished with the use of *labels*. Associated with each term in a proof step is a label which is a list of tags (unique identifiers). Two terms will be analogous if their labels share a common tag. A term in a new step inherits the tags from all the terms from which it was derived.

An example will clarify the way in which the system follows a guiding proof. Consider the following proofs:

	Guiding Proof[2]			Analogous Proof	
1.	*Person(X)* ⇒		1.	*Person(Y)* ⇒	
	Mortal(X)			*Paystaxes(Y)*	
2.	*Person(socrates)*		2.	*Person(franklin)*	
≫	*Mortal(socrates)*		≫	*Paystaxes(franklin)*	
≫	*Person(socrates)*	Back Chain 1	≫	*Person(franklin)*	Back Chain 1
	Proved	Match 2		Proved	Match 2.

The similarity between the two theorem statements, which is obvious to a human, will be represented in our implementation by pairings of terms. The occurrence of *person(X)* in hypothesis 1 of the guiding theorem will be paired with the occurrence of *person(Y)* in hypothesis 1 of the target theorem. Similarly, each occurrence of *socrates* in the guiding theorem will be paired with the corresponding occurrence of *franklin* in the target theorem. The mapping extends to terms at all levels in the two theorems. It should be emphasized that the mapping does not represent "global" analogies; it only pairs specific occurrences of terms. For the first proof step, the analogous inference for BACK CHAIN 1 will be the rule BACK CHAIN with the analogous argument, again 1. In the fourth step, because *Mortal(socrates)* and *Paystaxes(franklin)* were derived from analogous terms, they will be paired, as will their subterms *socrates* and *franklin*.

It is possible for a term to be mapped to several terms. Thus, one inference in the guiding proof can be associated with several possible analogous inferences. In our implementation, one of these inferences will be selected and the rest saved for backtracking. Thus different potential proofs will be examined in a depth-first manner.

---

[2]This is the natural deduction format used in this paper. The initial hypotheses are numbered. The current goal is indicated by the symbol '≫.'

2.2 Recovering When the Analogy Breaks Down

For a case where the guiding theorem is very similar to the target theorem, the method presented so far will be successful. In fact, for several of the simpler proofs we experimented with, the straightforward approach was sufficient. In many cases, however, the guiding and target theorems will differ enough that the analogy will break down: an analogous inference will simply not apply to the target proof. With the approach we have taken, the only way one proof can guide another is by suggesting analogous inference rules. We have constructed our system with the philosophy of using the guiding proof as far as possible. If the guiding inference rule has no analogue in the target proof, the system will simply ignore the inference and attempt to apply the next guiding rule. If the guiding inference does have an analogue, it will try to patch up the target proof so that the suggested inference will apply. It attempts this by adding additional steps to the target proof, maintaining the analogy mapping as it does so.

The approach the system takes in patching the target proof is determined by the way in which the analogy broke down. In the following subsections, we present three types of analogy breakdown and describe in some detail how our current implementation attempts recovery for each.

Two terms fail to unify

If a suggested inference does not apply because two terms fail to unify, the first action attempted by the recovery facility is to find an equality substitution in the lemma base which will allow one of the terms to be rewritten to something that will unify with the second. In the case of chaining[3], if the equality substitution did not apply, the two terms that failed to unify are used to find a lemma which will chain with both terms under certain restrictions. For example, if we were attempting to chain on the terms $|X + f(X)|$ and $|f(a)|$ in the following:

$$1. \quad 0 < |X + f(X)|$$
$$\gg \quad b < |f(a)|,$$

we could use the lemma $|Y + Z| \leq |Y| + |Z|$ to produce the intermediate goal $b + |Y| \leq |Y + f(a)|$ as follows:

$$1. \quad 0 < |X + f(X)|$$
$$2. \quad |Y + Z| \leq |Y| + |Z| \qquad \text{LEMMA}$$
$$\gg \quad b < |f(a)|$$
$$\gg \quad b + |Y| < |Y + f(a)| \qquad \text{CHAIN 2} \ f(a)/Z$$
$$\gg \quad b + |a| < 0 \qquad\qquad \text{CHAIN 1} \ a/X \ a/Y.$$

---

[3] *Chaining* is an inference rule for reasoning with inequalities. See [2].

The method for choosing a lemma is referred to as *double-entry fetching*. A lemma is selected so that the two terms that must be chained will unify with terms in the lemma. The advantage of this method over *single-entry fetching,* in which only one term is used to select a lemma which will eliminate the term, is that it greatly reduces the number of lemmas that are applicable. Another way in which the number of applicable lemmas is reduced is by using a more complicated algorithm for matching terms than simple unification. In the above example, the two terms used for double-entry fetching were $|X + f(X)|$ and $|f(a)|$. When the lemma fetcher searches for appropriate lemmas, it will only consider those with terms whose main functions correspond to those of the terms to be double-chained. For both of the terms in our example, the main function is the absolute value function. Thus, a lemma of the form $Y > Z \vee \lfloor Y \rfloor \leq \lfloor Z \rfloor$ would not be considered, even though it could match the terms in question. Moreover, the lemmas returned by the lemma fetcher are sorted using a heuristic determination of how well they match. Roughly speaking, they are sorted according to the number of levels at which the terms match. Thus, for our example, the lemma

$$|Y + Z| \leq |Y| + |Z|$$

would be chosen above

$$|Y| < |Z| \vee |Z| \leq |Y|$$

because the term $|X + f(X)|$ matches $|Y + Z|$ at two levels (with the absolute value function and the plus operator) whereas it matches $|Y|$ at only one.

Variable elimination prevented by a shielding term

If a variable is eliminated[4] in the guiding proof, our prover will attempt to eliminate an analogous variable in the target proof. In order for a variable to be eliminated from a clause, all occurrences of the variable in that clause must be eligible[5]. If it is not eligible, the function or predicate containing the variable and therefore preventing variable elimination is referred to as a *shielding term*. When a variable cannot be eliminated because of a shielding term, the analogy may be recovered if the shielding term can be eliminated. Suppose that in the guiding proof the variable X had been eliminated in the goal

$$0 \leq X \wedge X < a \wedge c \leq f(a)$$

---

[4] *Variable Elimination* is an inference rule for inequalities in which a formula containing a variable is replaced by an equivalent formula without the variable. See [2].

[5] A variable in a clause is *eligible* if it does not occur as an argument to any function other than $+$ or $-$, or as an argument to any predicate other than the inequality operators $<$, $\leq$, etc.

to yield

$$0 < a \land c \le f(a).$$

Suppose further that the analogous action in the attempted proof was to eliminate X from the goal

$$0 \le X \land X < a \land c \le f(X).$$

Here X cannot be eliminated, because it is shielded by the term $f(X)$. In order to recover the analogy, we need to eliminate this shielding term.

One means of accomplishing this is by chaining. For example, if the term

$$g(a) \le f(Y)$$

was a hypothesis, we could chain it with the goal, obtaining

$$0 \le X \land X < a \land c \le g(a).$$

Here the variable X is eligible, and variable elimination may be used to produce

$$0 < a \land c \le g(a).$$

This is the basic approach taken when a variable is shielded by a term in an inequality. Terms which are simliar to the shielding term with respect to the constants and the function symbols are chosen for possible chaining. To limit the number of terms which could be chained with the shielding term, the following restrictions are enforced. Only terms sharing the same main function with the shielding term are considered. All terms thus selected are sorted according to the number of levels at which they match the shielding term. This matching heuristic is similar to the one used for double-entry fetching. At this point, the term selected is chained with the shielding term if possible. If the selected term and the shielding term do not unify, a lemma is sought for chaining the two terms as described in the previous section.

In some cases, a shielding term can be eliminated by simplification. For example, if the shielding term for the variable X is $X \cdot d$ in the goal

$$X \cdot d < e \land 0 < d,$$

we may simplify this to

$$X < e/d \land 0 < d.$$

Similarly, if the shielding term is X/d in the goal

$$X/d < e \land 0 < d,$$

we may simplify this to

$$X < e \cdot d \wedge 0 < d.$$

The method is to observe that because of the second literal, d can be assumed positive (i.e., must eventually be proved positive) and therefore the first literal can be simplified by division or multiplication by d. Since this method does not introduce extra literals or result in substitutions, it is preferred to the previous approach of chaining with another clause. Thus, this method is always attempted first if the shielding term is a multiplication or division.

A variable can also be shielded by a predicate. In this case, a method for recovering the analogy is to find a hypothesis which matches the shielding predicate. The shielding predicate can then be split from the goal and eliminated in a matching step.

It should be noted that when chaining is used to eliminate a shielding term, the lemma or hypothesis used for the chaining may introduce new shielding terms. Thus the approaches presented here may have to be iterated until the variable is finally eligible.

Inference prevented by logical structure

In a natural deduction system, the goals are not placed into a standard form. Therefore, there may be impediments to following the analogy that are based strictly on the logical structure of the formulas. For example, given the proof pair:

Guiding Proof	Analogous Proof
1. $P(a)$	1. $P'(a)$
2. Q	
$\gg$ $P(X)$	$\gg$ $R \wedge (Q' \Rightarrow P'(Y))$,

the guiding proof may suggest matching $P'(Y)$ with $P'(a)$. This rule will fail in the new proof because a literal which is a conjunct cannot be matched. The prover will examine the suggested rule and determine that, in order to continue, it must split the goal. This yields:

Guiding Proof	Analogous Proof
1. $P(a)$	1. $P'(a)$
2. Q	
$\gg$ $P(x)$	$\gg$ $Q' \Rightarrow P'(y).$

Since $P'(y)$ still cannot be matched, the prover will recover again. In this case, since there is an implication, the prover will promote the antecedent to the hypothesis, allowing $P'(y)$ to be matched with $P'(a)$.

2.3 Completing a Proof Without Analogy

In experiments conducted so far, the user has always provided a guiding proof which is relevant to the attempted proof. But even a guiding proof known to be relevant is not always capable of leading an attempted proof to completion. For example, when all the steps in the guiding proof have been used to motivate steps in the attempted proof, there may still be residual subgoals to be proved. Thus, the prover must be able to discover appropriate actions to take based only upon the current state of the attempted proof. The set of methods for proving a theorem without the aid of a guiding proof is known as the *standard prover*.

The standard prover always works on the current subgoal. It selects a variable in the subgoal and tries to eliminate it. If the selected variable is shielded, the standard prover attempts to unshield it using the methods mentioned in 2.2. Whenever the subgoal contains no variables, a simple ground prover is invoked which attempts to prove the subgoal using any hypotheses which are also ground. The standard prover is similar to the one described in [4].

3 Experimental Results

Of the many areas of research in Automated Theorem Proving, the one that attempts to incorporate knowledge-intensive proof strategies into the proving process is probably the one with the greatest need for extensive experimentation. The utility of our approach cannot be established with proofs about completeness or soundness or efficiency. In experimenting with our approach to analogical proof discovery, we have focused on the domain of Real Analysis. In [4], we describe some experiments with a resolution based implementation. The major achievement of this effort involved using the proof of the theorem

$$\lim_{x \to a} f(x) = l \wedge \lim_{x \to a} g(x) = k \implies \lim_{x \to a}[f(x) + g(x)] = l + k$$

(hereafter referred to as "LIM+") to guide the discovery of the proof of the theorem

$$\lim_{x \to a} f(x) = l \wedge \lim_{x \to a} g(x) = k \implies \lim_{x \to a}[f(x) \cdot g(x)] = l \cdot k$$

(hereafter referred to as "LIM*"). In this experiment, we provided lemmas necessary for the proofs. Thus, the problem of fetching analogous lemmas from the database of lemmas was not an issue. At three points in the proof by analogy of LIM*, the system did successfully fetch a lemma (from a base of 30 lemmas) for double-entry fetching when analogous terms failed to unify. An annotated proof by analogy for LIM* is given in [4] and describes in detail how recovery is accomplished for this example when the analogy breaks down.

To find out how important a role the analogy played in our experiment, we attempted the proof of LIM* with the analogy facility turned off. The prover ran for several hours without success. We then experimented with turning the analogy facility off at different points in the proof process. Table 1 summarizes the results of this experiment.

Starting Step LIM*	Time(sec)[6]		
	With Analogy	Without Analogy	
		With Special Tautology Checker[7]	Without Special Tautology Checker
1	82.2	?	?
5	71.3	120.4	?[8]
13	45.3	67.6	94.6
16	34.2	57.8	83.8
17	35.0	51.5	41.8

Table 1: Timing Results for the LIM+/LIM* pair

We chose to base our first implementation on resolution, for reasons of simplicity. We have more recently built a natural deduction version based on the same ideas of maintaining an analogy mapping and performing analogous inferences. We felt that a natural deduction format would be preferable in the long run, as it could allow us to express more accurately the style of proof used by human mathematicians. In [5], we describe this natural deduction implementation which is the basis of our current work. Table 2 summarizes some of the experiments we have conducted with this prover.

The first entry in Table 2 is an example of the most straightforward kind of proof by analogy. The two theorems are very similar, and the analogy never broke down. The second, third, and fifth entries represent theorems which could not be proved with our first implementation. Our current implementation allows more of the hierarchical structure of the guiding proof to be expressed. This hierarchical structure better clarifies which steps in the guiding proof are relevant to the target proof. We found this to be especially important for pairs (e.g. the LIM*/LIM+ pair) where the target proof uses fewer steps than the guiding proof. We discuss this hierarchical representation of proofs further in the next section.

---

[6]Times reported are for a Symbolics 3640, running compiled LISP code.

[7]The special "tautology checker" rejected resolvents containing literals of the form $0 \leq |t|$, where t is any term.

[8]The proof had not completed in 12 hours.

Guiding Theorem	Target Theorem
A bounded, nondecreasing sequence converges.	A bounded, nonincreasing sequence converges.
$(\exists M \forall n \; \|S_n\| < M) \wedge$ $\quad (\forall n, m \; n > m \Rightarrow S_n \geq S_m) \Rightarrow$ $\exists s_0 \forall \varepsilon > 0 \exists N \forall n > N \;\; \|S_n - s_0\| < \varepsilon$	$(\exists M \forall n \; \|S_n\| < M) \wedge$ $\quad (\forall n, m \; n > m \Rightarrow S_n \leq S_m) \Rightarrow$ $\exists s_0 \forall \varepsilon > 0 \exists N \forall n > N \;\; \|S_n - s_0\| < \varepsilon$
A bounded, nondecreasing sequence converges.	An unbounded, nondecreasing sequence diverges to $+\infty$.
$(\exists M \forall n \; \|S_n\| < M) \wedge$ $\quad (\forall n, m \; n > m \Rightarrow S_n \geq S_m) \Rightarrow$ $\exists s_0 \forall \varepsilon > 0 \exists N \forall n > N \;\; \|S_n - s_0\| < \varepsilon$	$(\forall n, m \; n > m \Rightarrow S_n \geq S_m) \wedge$ $\quad (\forall M \exists n \; S_n > M) \Rightarrow$ $\forall M \exists N \forall n \geq N \;\; S_n > M$
An unbounded, nondecreasing sequence diverges to $+\infty$.	A bounded, nondecreasing sequence converges.
$(\forall n, m \; n > m \Rightarrow S_n \geq S_m) \wedge$ $\quad (\forall M \exists n \; S_n > M) \Rightarrow$ $\forall M \exists N \forall n \geq N \;\; S_n > M$	$(\exists M \forall n \; \|S_n\| < M) \wedge$ $\quad (\forall n, m \; n > m \Rightarrow S_n \geq S_m) \Rightarrow$ $\exists s_0 \forall \varepsilon > 0 \exists N \forall n > N \;\; \|S_n - s_0\| < \varepsilon$
LIM+	LIM*
$\lim_{x \to a} f(x) = l \wedge \lim_{x \to a} g(x) = k \Rightarrow$ $\lim_{x \to a}[f(x) + g(x)] = l + k$	$\lim_{x \to a} f(x) = l \wedge \lim_{x \to a} g(x) = k \Rightarrow$ $\lim_{x \to a}[f(x) \cdot g(x)] = l \cdot k$
LIM*	LIM+
$\lim_{x \to a} f(x) = l \wedge \lim_{x \to a} g(x) = k \Rightarrow$ $\lim_{x \to a}[f(x) \cdot g(x)] = l \cdot k$	$\lim_{x \to a} f(x) = l \wedge \lim_{x \to a} g(x) = k \Rightarrow$ $\lim_{x \to a}[f(x) + g(x)] = l + k$

Table 2: Example Proofs by Analogy

4   Current and Future Work

We summarize here our ongoing research and suggest directions for future work.

4.1   A Hierarchical Representation of Proofs

For theorems which are very similar, our basic strategy of applying analogous inference rules is adequate for completing proofs. For theorems which differ somewhat more, our basic strategy augmented by a facility for recovery is adequate. But for theorems which differ significantly, even the recovery facility is not enough to successfully complete a proof. The central concept here is *relevance*. A guiding proof may involve elements which could greatly aid in the discovery of a proof for a target theorem, but it will only be useful if the problem solver can determine which parts of the guiding proof are relevant. If we can solve this problem of determining relevance, we should be able to build provers which not only succeed with somewhat dissimilar theorems, but which also allow the use of *subproofs* of guiding theorems to guide the *subproofs* of target theorems. This may allow our provers to succeed with very difficult theorems by using several analogies to guide a proof.

Our current prover's success at proving LIM+ by analogy with LIM* illustrates the importance of hierarchical proof representation. Although the proof of LIM+ is simpler than that of LIM*, our first implementation was unable to succeed with this pair because it had no way of determining when to omit steps. In an experiment with our more recent implementation, we represented the dependencies between the steps of the LIM* proof. Informally, we say that one step depends on a second if the second must be executed in order for the first to be applicable. For example, the elimination of a variable may depend upon the elimination of a shielding term. These dependencies form a hierarchy. In proving LIM+ from LIM*, our system took advantage of this hierarchy by attempting higher level inferences first, resorting to lower level inferences only when necessary. In this way, the irrelevant, additional steps of the LIM* proof were ignored.

The reason for representing proofs hierarchically is to make explicit the *motivation* behind steps in the proof, so that the relevance of each step to a target proof can be determined. Our current implementation only represents dependencies between individual steps. In order to select subproofs of guiding theorems to guide subproofs of target theorems, we will also want to to represent the motivation behind entire subparts of proofs. This will involve research directed at representing more of the higher level strategies with which humans prove theorems. Future work will also involve finding ways to generate hierarchical proofs from linear proofs.

4.2 More Flexible Control Strategies

When faced with several options, our system uses heuristics to sort its alternatives. However, the overall search strategy is still depth-first. Humans, on the other hand, are very good at switching tacks as new information is gathered, always concentrating on the most promising direction. Thus, one of the improvements we plan to incorporate into our next implementation is a more sophisticated control strategy. We plan to use an agenda-based system in which alternatives can be prioritized, and in which priorities can be modified as search progresses. Much work needs to be done on learning how to assign and combine priorities. For example, we would want a branch of the search tree which follows the analogous proof closely to be assigned a higher priority than one which diverges from the analogy.

4.3 Retrieval and Mapping of Analogous Proofs

Our current implementation requires that the guiding theorem and the mapping between the guiding and target theorems be given. The focus of some of our on-going research involves automating these processes of retrieval and mapping. Some research has already been done on this problem of mapping analogues. See for example [8] and [11]. The process of retrieval requires the development of similarity metrics for determining which theorem in the database of stored theorems is most analogous to the target theorem. Mapping involves determining how best to pair subterms of the guiding and target theorems. The two problems are closely related. In order to choose which potential analogue is most similar to a target theorem, it may be necessary to examine to some depth how the potential analogues would be mapped to the target.

Developing successful facilities for retrieval and mapping will require that several difficult problems be solved. For example, when a database of hundreds or thousands of potential analogues is used, it is not feasible to carefully compare the target theorem to each stored theorem. The stored theorems must be indexed in some way which allows the system to find a target's best analogue while examining only a small portion of the database.

Mapping is also a difficult problem because of the combinatorial explosion involved in pairing subterms. One way to control this explosion is to use semantic information about the roles that subterms play. For example, the mapping facility could ignore potential pairings between terms that are not of similar types. The more sophisticated control strategies mentioned in the previous section may also be useful here. Additional experimentation will help us to learn how careful a prover has to be in determining an optimal mapping.

4.4 A Database of Theorems and Exercises

When building systems which are knowledge-intensive, a designer may tend (even if unconsciously) to tailor the knowledge to fit the problem at hand. To avoid this pitfall, we are in the process of compiling a large database of theorems and exercises for use in our experimentation. While it may be possible to tailor a system to prove one or two very difficult theorems, it is difficult to tailor it to prove hundreds of moderately difficult theorems. Our database will consist primarily of theorems and exercises taken from a standard text in Real Analysis [12]. In our experiments with proof discovery, we will attempt to prove as many of these theorems and exercises as possible. This database will also allow us to conduct realistic experiments in the retrieval of analogies. We will be more confident in a facility for analogical retrieval if it can successfully discriminate among tens or hundreds of potential analogues.

Currently, our database consists of 156 entries, including all the theorems and definitions from the first 13 sections of [12]. We may later augment this database using other texts in Real Analysis.

5 Summary and Conclusions

We have been conducting a research project aimed at using knowledge-intensive strategies in Automated Theorem Proving and have been focusing on the strategy of analogical reasoning. Our initial approach was based on a simple model of analogical proof discovery. This model involved a straightforward means of following a guiding proof and a set of methods for patching a proof when the analogy broke down. We performed some experiments with a resolution based implementation which convinced us that the model was useful.

These experiments also convinced us that several extensions were needed. To express proofs in a style closer to that of human mathematicians, we chose to use natural deduction instead of resolution. Our simple representation of a proof as a sequence of steps was inadequate to represent the motivation behind each step in the proof. This led us to incorporate a hierarchical representation for proofs. We have been able to prove theorems with this implementation where the first was unsuccessful.

But our current implementation is still not generally successful with moderately difficult theorems. We will continue to work on the hierarchical representation of proofs, look for ways to represent high level proof strategies, and begin work on an agenda based control strategy for employing these strategies. We will also work towards automating the facilities of retrieval and mapping. The compilation of a large database of theorems and exercises will aid us in conducting the experiments with which we will evaluate these efforts.

Acknowledgement

We would like to thank Woody Bledsoe for his encouragement and many helpful suggestions.

References

[1] Bledsoe, W. W., "The Use of Analogy in Automatic Proof Discovery", University of Texas Math Dept. Memo ATP-83, November 1985, MCC-AI-158-86.

[2] Bledsoe, W. W. and Hines, L. M., "Variable Elimination and Chaining in a Resolution-based Prover for Inequalities", *CADE-5*, Les Arcs, France, W. Bibel and R. Kowalski (eds.), Springer-Verlag, pp. 70-87, 1980.

[3] Bledsoe, W. W., "The UT Natural-Deduction Prover", The University of Texas at Austin, Math Department Memo ATP-17B, April 1983.

[4] Brock, B., Cooper, S. and Pierce, W., "Some Experiments with Analogy in Proof Discovery (Preliminary Report)", MCC Tech. Report AI-347-86, October 1986.

[5] Brock, B., Cooper, S. and Pierce, W., "Some Experiments with Analogy in Proof Discovery: A Natural Deduction Approach", MCC Tech. Report ACA-AI-274-87, August 1987.

[6] Brock, B., Cooper, S. and Pierce, W., *Box Users Manual*, MCC Tech. Report ACA-AI-273-87, September 1987.

[7] Carbonell, J. G., "Learning by Analogy: Formulating and Generalizing Plans from Past Experience," in *Machine Learning, An Artificial Intelligence Approach, Vol. I*, R. S. Michalski, J. G. Carbonell, and T. M. Mitchell, eds., Tioga Press, Palo Alto, CA, 1983, pp. 137-161.

[8] Gentner, D., "Structure Mapping: A Theoretical Framework for Analogy," *Cognitive Science*, vol. 7, no. 2, April 1983, pp. 155-170.

[9] Greiner, R., "Learning by Understanding Analogies", Ph.D. Thesis, Stanford University, September 1985. Technical Report STAN-CS-1071.

[10] Kling, R. E., "A Paradigm for Reasoning by Analogy", *Artifical Intelligence*, Vol. 2, pp. 147-178, 1971.

[11] Owen, S., "Heuristics for Analogy Matching", DAI Research Paper No. 280, University of Edinburgh.

[12] Ross, K. A., *Elementary Analysis: The Theory of Calculus*, Springer-Verlag, 1986.

[13] Winston, P. H., "Learning and Reasoning by Analogy", *CACM*, Vol. 23, No. 12, pp. 689-703, 1979.

Hyper-Chaining and
Knowledge-Based Theorem Proving[*]

Larry Hines

University of Texas
Austin, Texas 78759

Abstract

Many programs have built-in axioms and lemmas of various theories [2, 3,4,5,6,8,9,11]. These programs gain their efficiency by restricting and guiding the use of their axioms and lemmas; however, they do not hold their axioms or lemmas in a declarative form. Control and representation issues were intertwined with the encoding of the axioms and lemmas. As a result, the programs can not be simply combined to form more powerful programs.

This paper describes a method of building in such axioms and lemma in a declarative form while still gaining their efficiency. This is attained by utilizing a set of declaratively defined *axiom-rules* and *rule-sequences* which not only implement the axioms and lemmas but also hold strategic information concerning how and when to use them. The prover can be easily altered by adding, deleting or modifying rules; moreover, the built-in axiom-rules can also be used as components for building in further axioms and lemmas, resulting in a hierarchy of theorem proving tools.

One such tool is *hyper-chaining*. It is a rule-sequence built upon chaining (transitivity of inequality axiom) which is an axiom-rule that was, in turn, built upon an axiom-rule for the trichotomy of inequality. Using hyper-chaining, the prover has automatically proved a number of theorems including theorems on limits and the intermediate value theorem. This paper describes hyper-chaining, how it was built-in and how it was used to prove the intermediate value theorem.

1 Introduction

In order to prove non-elementary theorems, a large collection of basic axioms and lemmas must be available. However, unrestricted and unguided use of such axioms

---

[*]This work supported by National Science Foundation Grant CCR-8613 706.

explodes the search space. Human theorem provers do not suffer from this explosion; instead, humans greatly restrict the usage of each axiom. Thus, each axiom given to a mechanical theorem prover must be augmented with instructions on when and how to use it. This is the motivation for writing programs which build in the axioms along with the human's knowledge about the proper usage of the axioms. In fact, resolution [6] is a built-in version of the axiom of logic, $[(P \lor Q) \land (\neg P \lor R)] \rightarrow (Q \lor R)$, together with the restriction of unification. Other procedures which build in axioms include paramodulation [8], algebraic simplifiers, chaining [2] which builds in the transitivity of inequality, variable elimination [2] which builds in the interpolation axiom of dense fields, THM [4] which builds in induction (and more), and IMPLY [3] which builds in various axioms of logic.

The axioms and lemmas mentioned above were not encoded in a declarative form. Instead they were encoded at what may be called the program level. Control and representation issues were intertwined with the code of the axioms and lemmas. The resulting programs could not be simply combined to form more powerful programs.

The goal of this research is to raise this building in process from the program level toward the "knowledge" level and to simultaneously increase the power of the prover. This is attained by utilizing a set of declaratively defined *axiom-rules* and *rule-sequences* which not only implement the axioms or lemmas but also hold strategic information concerning how and when to use them.

Since the prover's rule are declarative defined, the prover can easily be altered either by adding or deleting rules or by modifying rules by adding, deleting or modifying their strategic components. Moreover, the built-in axiom-rules can be used not only to prove theorems, but also as components for building in further axioms or lemmas, resulting in a hierarchy of theorem proving tools.

In order to obtain the speed and efficiency of non-declarative system, we employ a principal tactic of speeding up the search in automatic theorem proving, *i.e.*, "cutting out as much as possible of the unnecessary computating in automatic deduction procedures."[7] By placing restrictions on axiom-rules and rule-sequences, the breadth of the prover's search tree can be reduced by pruning many unproductive inferences. Also, by building new axiom-rules from lower level rules, multiple smaller steps can be replaced by one larger step. This larger step is a "black box" which encapsulates the smaller steps.

However, the larger step is not simply an abbreviation for the series of smaller steps. The role of the intermediate inferences is "reduced to that of purely temporary scratchpad entities"[7] and, thus, discarded once the larger step is finished. The restrictions of the smaller steps can be strengthen with restrictions which deal with the correspondence between the smaller steps. Furthermore, since the intermediate inferences are partially derived from the axiom or lemma being built-in, some of the intermediate inferences can be executed at compile time. Thus the depth as well as the breadth of the proof tree can be collapsed.

This paper sketches the method for building in the axioms and lemmas, gives two examples of such rules (chaining and hyper-chaining),[1] and shows how they can be used to prove a number of theorems, as well as to build in new rules.

First we describe axiom-rules. To do so, it is necessary to introduce some additional terms. These include the *entry point terms* of an axiom and the *classes* and *restrictions* which are attached to these entry point terms to guide the prover in the usage of the axiom-rules.

In section 3 the axiom-rule for chaining (transitivity of inequality) axiom is presented. We demonstrate how the trichotomy axiom-rule is used to implement the chaining rule. This includes deriving the intermediate inferences between transitivity and trichotomy and the transferring to chaining of the strategy governing the use of trichotomy. That strategy is further enhanced by the user with additional classes and restrictions.

In section 4 rule-sequences are described. Next, the *hyper-chaining* rule-sequence is described. Hyper-chaining repeatedly uses the chaining axiom-rule, however the strategy governing chaining is augmented by additional requirements base upon the *dependencies between variables* as described in section 6.

Finally, we present the proof of the intermediate value theorem that the prover automatically derived using hyper-chaining.

It is important to note that the prover is only a teachable theorem prover and not a self-teaching one. Theorems, once they have been proved, may themselves be implemented as axiom-rules, and added to the rule-base for use in proving further theorems. However, the prover does not have any automatic learning process. If the user decides to build-in a theorem, he should first develop some idea concerning its proper usage. Then he may "teach" the prover the theorem and its usage. Also, as his ideas evolve the prover may likewise evolve.

Also, no attempt is made at presenting a general framework for knowledge representation. Much, but not all, of the knowledge held by the prover (and highly important to the domain studied here) is held in a declarative form (*i.e.*, axiom-rules and rule-sequences).

2 Axiom-Rules

Axiom-rules are much like ordinary clauses. However, they have two important differences. Rules are not stored in the prover's clause set; instead, rules are stored in a rule base and act on the clauses in the clause set. Also, rules, unlike other

---

[1] Axiom-rules and rule-sequences for the trichotomy of inequality, quotient cancellation, non-scalar multiplication, triangle inequality and various resolution strategies have also been built-in.

clauses, cannot be used in just any manner since they must be entered only through their entry point terms. That entry is further restricted by attached classes and restrictions.

As a brief overview, an axiom-rule may be described as a clause with certain terms identified as entry points together with a listing of their classes and restrictions which must be satisfied for the rule to be invoked. *Classes* act as filters on literals. The literals which are not filtered out become candidates to be paired with individual entry points. *Restrictions* also are filters which act on sets of literals (called matches) which have been proposed to match an axiom-rule. When a match for a rule is completed, the resolvent of the match is eligible to be placed in the prover's clause set.

2.1 Entry Point Terms

An entry point term is a term in the axiom or lemma with which some term from the clause set must be paired in order for the axiom to be invoked. For example, in the resolution axiom,

$$(\boxed{\neg P}^{\,1} \lor Q) \land (\boxed{P}^{\,2} \lor R) \rightarrow (Q \lor R),$$

the P and $\neg P$ terms are its entry point terms.

In the transitivity of inequality axiom (chaining),

$$x < \boxed{y}^{\,1} \lor \boxed{y}^{\,2} < z \lor z \leq x,$$

the two y terms are the entry point terms. These are the terms in the transitivity axiom which must be matched by two terms from the clause set. It is to these terms that we attach the classes and restrictions such as the RCF and shielding term requirements.[2]. As an example, in the following clause set the $(f\,v)$ can be paired with the first entry point, and $(f\,c)$ with the second entry point of chaining. Clause 3 is the resulting resolvent.

$$1. \quad a \leq \boxed{(f\,c)}$$

$$2. \quad \boxed{(f\,v)} \leq b \qquad P(v)$$

$$3. \quad a \leq b \qquad P(c)$$

---

[2]RCF and shielding term are defined in section 3.

2.2 Classes

In general, a class is defined by a predicate whose only input argument is a literal from the clause set. The predicate determines whether the literal is a member of the class and, thus, can match any entry point which has the class as one of its class attributes. If the predicate returns NIL then the literal is not a member of the predicate's class. If the predicate returns a nonNIL value, then the literal is a member and, if the value is numeric it is taken to be a penalty to be used in comparing literals in order to find the "best" literal.

When a new clause is added to the clause set, each of the active class predicates is run on each of the literals in the clause. Having classified the literals, each literal which satisfies all of the classes of any entry point is added to that entry point's candidates list. The order of the candidates is determined by the penalty accumulated from the classes specific to that particular entry point.

For example, the UNIT-CLAUSE class (not shown) selects literals which are contained in unit clauses. Literals which satisfy this class are allowed as candidates for any entry point which requires this class. By placing this class on an entry point of the resolution axiom, the unit resolution strategy can be employed.

2.3 Matches

Before defining restrictions, we must first define *match*. A match is a list of three things: a rule, a list of literals and a resolvent. The literals of the match have been paired with the first entry point terms of the rule. The resolvent is the clause which has been derived by the pairings. A match is *complete* and its resolvent is eligible to be added to the clause set when each entry point in the rule has been matched by a literal.[3]

For each active rule, the prover creates an *empty match* during its initialization phrase. The list of literals of an empty match is NIL and the resolvent is initialized to be the rule of the match.[4]

After initialization, the prover operates essentially as a best-first search program. The tactics and strategies of the prover are wholly determined by the classes and restrictions of the rules. The "best" node to expand next is the match which has the least penalty. The penalty of a match is based on the penalties returned by the classes and restrictions of its rule.

The prover attempts to extend the incomplete match with the lowest penalty

---

[3]This is a limitation on which axioms or lemmas that can be built-in by axiom-rules. The rule must have a set number of entry points. For example, a *hyper-chaining* axiom-rule cannot be built into the prover. However, in section 4 we define *rule sequences* with which hyper-chaining can be built into the prover.

[4]Remember that every rule is first a clause. See above.

by its first candidate as long as that penalty is less than the penalty of the best completed match. If a successful extension completes the match, it is added to the prover's list of completed matches. Otherwise, it is added to the list of incomplete matches. When no incomplete match has a penalty lower than the best completed match, the resolvent of the least penalized completed match is added to the prover's clause set and its literals are classified.

2.4 Restrictions

Restrictions are predicates which determine whether a literal can extend a given incomplete match. As with class predicates, if the restriction predicate returns NIL then the literal cannot extend the match. If each restriction of the next entry point returns a nonNIL value, then the literal successfully extends the match. The numeric values returned by the restrictions are accumulated as a penalty to be used in comparing matches in order to find the "best" match.

For example, TAUTOLOGY is a restriction which returns NIL if the resolvent of the match is a tautology.

3 Chaining and Variable Elimination

In [2], Bledsoe and Hines described routines (chaining and variable elimination) which built-in the transitivity and interpolation axiom of inequality. The strategy employed here with respect to transitivity is similar to the strategy employed in that paper. However, here it is built-in in a declarative form and can be used not only as a proof rule, but also as a subrule to build in the strategy of the hyper-chaining and rules for such axioms as the triangle inequality axiom. The earlier strategy essentially was to apply chaining to pairs of RCF terms which had opposite signs and at least one of which was a *shielding term* (see below). When all the shielding terms of a variable have been removed, *variable elimination* was immediately applied.

In the example[5] below, $(f\,x)$ of clause 2 is chained against the $(f\,c)$ of clause 1. The mgu is $\{c/x\}$. Clause 3 is the result.

$$1. \quad a \le \boxed{(f\,c)}$$

$$2. \quad \boxed{(f\,x)} \le b \qquad P(x)$$

$$\rule{4cm}{0.4pt}$$

$$3. \quad a \le b \qquad\qquad P(c)$$

---

[5]a, b and c are constants and x is a variable.

Chaining is not applied to expressions headed by plus ($+$) or times ($\cdot$). Thus in the example below the $(f\,x_\delta){+}(g\,x_\delta)$ expression of clause 1 is not chained against the $(f\,y)+(g\,y)$ expression of clause 2. In this example chaining would succeed, but in general such an "expression" orientation would require an associative-commutative unifier. Instead chaining has a "term" orientation. That is, it looks below the level of plus and times. In the example, the $(f\,x_\delta)$ term of clause 1 is chained against the $(f\,y)$ term of clause 2.[6] Thus, chaining's "term" orientation instead of an "expression" orientation avoids the problems of associative-commutative unification.

$$1. \quad (g\,c) + \varepsilon_0 < \boxed{(f\,x_\delta)} + (g\,x_\delta)$$

$$2. \quad \boxed{(f\,y)} + (g\,y) \le \varepsilon' \qquad\qquad (P\,y)$$

$$3. \quad (g\,c) + (g\,x_\delta) + \varepsilon_0 < (g\,x_\delta) + \varepsilon' \qquad (P\,x_\delta)$$

c and ε_0 are constants and δ, y and ε' are variables.

An RCF term is a non-variable term, $(f\,t_1, t_2, ..., t_n)$ where f is any function symbol other than $+$ or $\times$. If a term, t, is a RCF term and contains at least one variable, then we call t a *shielding term*. The variables contained in such shielding terms are not eligible for variable elimination.

Variable elimination is a procedure which builds in the interpolation axioms of dense linear orders without endpoints. If a variable, x, does not occur within any shielding term in a clause, C, then C can be written as the disjunction of

$$x < a_i \qquad ; i = 1, n,$$
$$b_j < x \qquad ; j = 1, m, or$$
$$P$$

where x does not occur in any a_i, b_j or any literal in P. Then C is replaced by the resolvent P or $\{b_j < a_i \; ; \; i = 1, n \; ; \; j = 1, m\}$.

Note that if n or m is zero then C is replaced by the resolvent P. Also, the rule for variable elimination is extended appropriately to include the symbol $\le$.

3.5 Building In the Rule, CHAIN

CHAIN, the rule for transitivity with full multiplication, has the clause

---

[6] Chaining would likewise be applicable to the $(g\,x_\delta)$ term of clause 1 and the $(g\,y)$ term of 2.

$$\begin{array}{cc} 1 & 2 \end{array}$$

$$(\ (0 < m_1 \cdot \boxed{B} + a_1)\ (0 < m_2 \cdot \boxed{B} + a_2)$$
$$(0 \leq m_1 \cdot a_2 - m_2 \cdot a_1)$$
$$(0 \leq m_2)$$
$$(0 \leq -m_1)).$$

It has two entry point terms, the B term in $0 < M_1 \cdot B + A_1$ and the B term in $0 < M_2 \cdot B + A_2$. To implement the strategy described above, the user attaches to the first entry point classes requiring positive RCF-terms. To the second entry point are attached classes requiring negative RCF-terms, a restriction requiring that at least one of the two RCF-terms is a shielding term and a restriction which implements variable elimination. The remainder of the strategy is inherited from the entry points' subrules.

Both entry points have as a subrule the trichotomy axiom-rule. At compile time CHAIN is expanded by matching each literal containing an entry point against the first entry point of the trichotomy rule. This is done by first creating an incomplete match which matches the first entry point of CHAIN against the first entry point of the trichotomy rule. The restrictions of the subrule's first entry point are executed which updates the match's resolvent. That resolvent, with its literals labelled by references to their parent clauses, is

$$\begin{array}{ccc} \text{TRICH 2} & \text{CHAIN 2} & \\ 0 < M_1 \cdot B + A_1 & 0 < M_2 \cdot B + A_2 & 0 \leq M_1 \cdot A_2 - M_2 \cdot A_1 \\ & 0 \leq M_2 & 0 \leq -M_1. \end{array}$$

The inherited reference to the trichotomy rule is replaced by a reference to the current entry point of CHAIN.

$$\begin{array}{ccc} \text{CHAIN 1} & \text{CHAIN 2} & \\ 0 < M_1 \cdot B + A_1 & 0 < M_2 \cdot B + A_2 & 0 \leq M_1 \cdot A_2 - M_2 \cdot A_1 \\ & 0 \leq M_2 & 0 \leq -M_1 \end{array}$$

The clause may appear much the same as the original clause for CHAIN,[7] but the classes, restrictions and other properties of the second entry point of the subrule have been propagated. These properties are augmented with the classes, restrictions and other properties of the first entry point of CHAIN.

Using the occurrence of CHAIN's second entry point within that resolvent, another match is created to match CHAIN's second entry point against the first entry point of the trichotomy rule. Again the restrictions of the first entry point of the trichotomy rule are executed; the classes, restrictions and other properties of the

---

[7]This is not always the case. The triangle inequality axiom was built-in using CHAIN, yet is significantly changed in presentation.

second entry point of trichotomy are inherited and augmented with the properties of second entry point of CHAIN; and the entry point reference to the trichotomy rule is redirected to CHAIN.

Since CHAIN has no other entry point terms, that resolvent becomes the clause for the rule. Thus, the expansion and compilation of chaining does not produce unintelligible compiled code of a LISP function. Instead it produces a clause which could be part of the prover's clause set. However, it is distinguished from other clauses by its entry points terms and their attached classes and restrictions.

3.6 An Example

Given the following clause set, the prover finds that the $F(s)$ term in clause 2 and the $F(B)$ term in clause 3 satisfy the class predicates of the first entry point of CHAIN since those terms are the only positive RCF terms. The $F(B)$ term in clause 2 and the $F(A)$ term in clause 3 satisfy the class predicates of the second entry point since they are the only negative RCF terms. Since these are the only legal candidates the search space has already been greatly reduced.

The prover successfully extends the empty match for CHAIN with $F(B)$ of clause 3, instantiating B[8] with $F(B)$, and adds it to the list of incomplete matches. The prover also successfully extends the empty match with $F(s)$ of clause 2, instantiating B with $F(s)$, and adds it to the list of incomplete matches. But the prover fails to extend the first of these incomplete matches with $F(A)$ of clause 3 since the instantiations are not consistent, and with $F(B)$ of clause 2 since neither term is a shielding term. However, the prover successfully extends the second of the incomplete matches with $F(A)$ of clause 3. The completed match's resolvent is reduced to $\square$ by the ground prover.

1. $0 \leq A$
2. $0 < -s$ $\qquad$ $0 \leq F(s) - F(B)$
3. $0 < F(B) - F(A)$

---

4. $\square$.

4 Rule Sequences

The class of axioms and lemmas which can be built-in by axiom-rules is limited. Axiom-rules have a fixed number of steps and, thus, cannot capture tactics which involve repeated application of a subrule until an arbitrary goal is satisfied. For

---

[8]The variables of the rules are in small caps.

example, axiom-rules cannot express the general loop of resolution provers (i.e., repeatedly resolve until □ is derived) and, at a more tactical level, cannot express hyper-resolution. Moreover, the strategy of *hyper-chaining* cannot be expressed by axiom-rules. So, we introduce *rule sequences*.

Rule sequences essentially provide a mechanism for building "repeat-until" into the prover. Furthermore, rule sequences extend the axiom-rule notion of "black-box" multi-step inferencing since intermediate results are wholly retained within the sequence. Only the final resolvent is returned to the prover.

A rule sequence is a sequence of *sequence steps*. A sequence step specifies which axiom-rules may be applied and under what conditions the next step of the sequence is chosen. For example, the rule sequence for hyper-resolution has one step. That step applies an axiom-rule which builds in resolution. The instructions of that axiom-rule are augmented with the additional constraints of hyper-resolution.[9] As long as there are negative literals in the resolvent, the next step of the sequence will be a repetition of the first step. When all of the negative literals have been removed, the sequence is finished.

5 Hyper-Chaining

Hyper-chaining repeatedly applies the chaining rule until a designated variable is either instantiated with a ground term or is eliminated (by so-called variable elimination). This designated variable, automatically chosen, is required to be a *top variable*[10] of the initial clause. At each step, the intermediate chaining will seek to remove the first of the remaining shielding terms of that variable.

In the example below (the proof of the continuity of the sum of two continuous functions) the objective of the first hyper-chaining was to eliminate the variable δ from clause 10, thereby obtaining clause 11. δ trivially is a top variable of clause 10 since it is the only variable in the clause. The intermediate chainings of a hyper-chaining (shown below) are discarded. Two more applications of hyper-chaining

---

[9] E.g., positive side clause. See [7].
[10] See below.

complete the proof of the theorem.

1. $0 < \delta'_{\varepsilon'}$ $\varepsilon' \leq 0$

2. $\delta'_{\varepsilon'} + x' < x_0$ $\delta'_{\varepsilon'} + x_0 < x'$ $(f\,x_0) \leq (f\,x') + \varepsilon'$ $\varepsilon' \leq 0$

3. $\delta'_{\varepsilon'} + x' < x_0$ $\delta'_{\varepsilon'} + x_0 < x'$ $(f\,x') \leq (f\,x_0) + \varepsilon'$ $\varepsilon' \leq 0$

4. $0 < \delta''_{\varepsilon''}$ $\varepsilon'' \leq 0$

5. $\delta''_{\varepsilon''} + x'' < x_0$ $\delta''_{\varepsilon''} + x_0 < x''$ $(g\,x'') \leq (g\,x_0) + \varepsilon''$ $\varepsilon'' \leq 0$

6. $\delta''_{\varepsilon''} + x'' < x_0$ $\delta''_{\varepsilon''} + x_0 < x''$ $(g\,x_0) \leq (g\,x'') + \varepsilon''$ $\varepsilon'' \leq 0$

7. $0 < \varepsilon_0$

8. $x_0 \leq x_\delta + \delta$ $\delta \leq 0$

9. $x_\delta \leq x_0 + \delta$ $\delta \leq 0$

10. $\varepsilon_0 + (f\,x_0) + (g\,x_0) < (f\,x_\delta) + (g\,x_\delta)$

     $(f\,x_\delta) + (g\,x_\delta) + \varepsilon_0 < (f\,x_0) + (g\,x_0)$ $\delta \leq 0$

    (Hyper-Chain 10 $[\delta]$: 3, 2, 6, 5, 9, 9, 8, 8)

11. $\varepsilon_0 < \varepsilon' + \varepsilon''$ $\delta'_{\varepsilon'} \leq 0$ $\varepsilon' \leq 0$ $\delta''_{\varepsilon''} \leq 0$ $\varepsilon'' \leq 0$

    (Hyper-Chain 11 $[\varepsilon']$: 1)

12. $\varepsilon_0 < \varepsilon''$ $\delta''_{\varepsilon''} \leq 0$ $\varepsilon'' \leq 0$

    (Hyper-Chain 12 $[\varepsilon'']$: 4)

13. $\square$.

$$\left\{\begin{array}{l}
\text{(Chaining 10 3)}\\
(g\,x_0)+\varepsilon_0 < (g\,x_\delta)+\varepsilon' \qquad (f\,x_\delta)+(g\,x_\delta)+\varepsilon_0 < (f\,x_0)+(g\,x_0) \quad \delta \leq 0\\
\delta'_{\varepsilon''}+x_\delta < x_0 \quad \delta'_{\varepsilon'}+x_0 < x_\delta \quad \varepsilon' \leq 0\\[4pt]
\text{(Chaining ... 2)}\\
(g\,x_0)+\varepsilon_0 < (g\,x_\delta)+\varepsilon' \qquad (g\,x_\delta)+\varepsilon_0 < (g\,x_0)+\varepsilon' \qquad\qquad \delta \leq 0\\
\delta'_{\varepsilon'}+x_\delta < x_0 \quad \delta'_{\varepsilon'}+x_0 < x_\delta \quad \varepsilon' \leq 0\\[4pt]
\text{(Chaining ... 6)}\\
(g\,x_0)+\varepsilon_0 < (g\,x_\delta)+\varepsilon' \qquad \varepsilon_0 < \varepsilon'+\varepsilon'' \qquad\qquad\qquad \delta \leq 0\\
\delta'_{\varepsilon'}+x_\delta < x_0 \quad \delta'_{\varepsilon'}+x_0 < x_\delta \quad \varepsilon' \leq 0 \quad \delta''_{\varepsilon''}+x_\delta < x_0 \quad \delta''_{\varepsilon''}+x_0 < x_\delta \quad \varepsilon'' \leq 0\\[4pt]
\text{(Chaining ... 5)}\\
\quad \varepsilon_0 < \varepsilon'+\varepsilon'' \qquad\qquad\qquad\qquad\qquad\qquad\qquad \delta \leq 0\\
\delta'_{\varepsilon'}+x_\delta < x_0 \quad \delta'_{\varepsilon'}+x_0 < x_\delta \quad \varepsilon' \leq 0 \quad \delta''_{\varepsilon''}+x_\delta < x_0 \quad \delta''_{\varepsilon''}+x_0 < x_\delta \quad \varepsilon'' \leq 0\\[4pt]
\text{(Chaining ... 9)}\\
\quad \varepsilon_0 < \varepsilon'+\varepsilon'' \qquad\qquad\qquad\qquad\qquad\qquad\qquad \delta \leq 0\\
\delta'_{\varepsilon'}+x_\delta < x_0 \quad \delta'_{\varepsilon'} < \delta \qquad\quad \varepsilon' \leq 0 \quad \delta''_{\varepsilon''}+x_\delta < x_0 \quad \delta''_{\varepsilon''}+x_0 < x_\delta \quad \varepsilon'' \leq 0\\[4pt]
\text{(Chaining ... 9)}\\
\quad \varepsilon_0 < \varepsilon'+\varepsilon'' \qquad\qquad\qquad\qquad\qquad\qquad\qquad \delta \leq 0\\
\delta'_{\varepsilon'}+x_\delta < x_0 \quad \delta'_{\varepsilon'} < \delta \qquad\quad \varepsilon' \leq 0 \quad \delta''_{\varepsilon''}+x_\delta < x_0 \quad \delta''_{\varepsilon''} < \delta \qquad\quad \varepsilon'' \leq 0\\[4pt]
\text{(Chaining ... 8)}\\
\quad \varepsilon_0 < \varepsilon'+\varepsilon'' \qquad\qquad\qquad\qquad\qquad\qquad\qquad \delta \leq 0\\
\qquad \delta'_{\varepsilon'} < \delta \qquad\qquad\qquad \varepsilon' \leq 0 \quad \delta''_{\varepsilon''}+x_\delta < x_0 \quad \delta''_{\varepsilon''} < \delta \qquad\quad \varepsilon'' \leq 0\\[4pt]
\text{(Chaining ... 8)} \quad [\text{before variable elimination of } \delta]\\
\quad \varepsilon_0 < \varepsilon'+\varepsilon'' \qquad\qquad\qquad \delta \leq 0\\
\qquad \delta'_{\varepsilon'} < \delta \qquad\qquad\qquad \varepsilon' \leq 0 \quad \delta''_{\varepsilon''} < \delta \qquad\qquad\qquad\qquad \varepsilon'' \leq 0
\end{array}\right.$$

discarded

(Hyper-Chain 10 [δ]: 3, 2, 6, 5, 9, 9, 8, 8)

11. $\quad \varepsilon_0 < \varepsilon'+\varepsilon'' \quad \delta'_{\varepsilon'} \leq 0 \quad \varepsilon' \leq 0 \quad \delta''_{\varepsilon''} \leq 0 \quad \varepsilon'' \leq 0$

6 Dependencies Between Variables

Clause 1 of the following example has two variables x and y. Removing the shielding term of y first results in a tautology,[11] but removing the shielding term of x followed by the shielding term of y derives $\square$. When the order in which the variables must be eliminated can be determined by examining the syntactic position of the variables, then dead ends such as this tautology can be avoided. For that purpose, we introduce the concept of *dependencies between variables*.

---

[11]In some examples, such a wrong first choice of variables to be eliminated forces a tautology which can only be detected after numerous chainings.

1. $F(x) \leq 0 \quad Z(y) < x \quad x \leq 1$
2. $0 < F(v)$
3. $A < Z(u)$

Remove shield of y first

4. $F(x) \leq 0 \quad 1 < x \qquad x \leq 1$

   Tautology

Remove shield of x first

5. $Z(y) < x \quad x \leq 1$

   Variable eliminatation of x rewrites 5

5. $Z(y) \leq 1$

   Now remove the shield of y

6. $\square$

A variable, y, *potentially depends* upons another variable, x in a clause C if

1. x and a shielding term, s, both occur together in a literal of C such that y is shielded by s and x is not, or

2. y potentially depends upon some variable v where v potentially depends upon x.

We will use the notation $y \geq_d x$ to signify that y potentially depends upon x.

In each of the following clauses, y potentially depends upon x. In C_1, x and the shielding term, $t(y)$, occur the literal, $t(y) < x$. y is shielded by $t(y)$; x is not. In C_2, $y \geq_d v$ and $v \geq_d x$.

C_1: $t(y) < x$

C_2: $t(y) < x \ \vee \ x < g(v)$

Two variables, x and y, are *interdependent* if

1. $x \geq_d y$ and $y \geq_d x$, or

2. s is a shielding term in C which shields both x and y.

We denote that x and y are interdependent by $x =_d y$. In each of the following clauses, $x =_d y$.

C_3: $t(y) < x \ \vee \ y < g(x)$

C_4: $t(y) < f(x)$

C_5: $t(y) < v \ \lor \ x < g(v) \ \lor \ f(x) < y$

C_6: $h(x,y) < 0$

A variable y *depends* upon another variable x (denoted $y >_d x$) if x and y are not interdependent and if

1. $y \geq_d x$,

2. $y \geq_d v$ and $x =_d v$, or

3. $v =_d y$ and $v \geq_d x$.

Two variables are *independent* if neither variable depends upon the other and if they are not interdependent. By imposing dependencies upon the independent variables of a clause, the partial ordering of dependencies can be made total. To do this, let x and y be independent variables. Choose either $y \geq_d x$ or $x \geq_d y$ and perform the closure using the second part of the definition of potential dependency and the second and third parts of the definition of dependency.

Since the number of variables in a clause is finite, there is one group of inter-dependent variables upon which all other variables depend. The variables of this group are the *top variables* of the clause. It is these variables which must be re-moved first. Referring back to the example at the beginning of this section, the only dependency in clause 1 is $y >_d x$. Thus, x is the sole top variable and must be eliminated first.

7 The Intermediate Value Theorem

We will now present the proof of the intermediate value theorem automatically found by the prover using hyper-chaining. The second-order set variable of the least upper bound principle for the reals was instantiated with the set $\{x : A \leq x \leq B \land F(x) < 0\}$.

The following is the resulting first order logic form of the theorem given to the prover.

$$\exists L \ \{A \leq L \leq B \wedge \ \forall w(A \leq w \leq B \wedge F(w) < 0 \rightarrow w \leq L)$$
$$\wedge \ \forall y(\forall z(A \leq z \leq B \wedge F(z) < 0 \wedge z \leq L \rightarrow z \leq y) \rightarrow L \leq y)\}$$
$$\wedge \ \forall u \ \{(A \leq u \leq B$$
$$\rightarrow \forall \varepsilon \ \{0 < \varepsilon$$
$$\rightarrow \exists \delta (0 < \delta \ \wedge \ \forall v \ (A \leq v \leq B \ \wedge \ u \leq v + \delta \ \wedge \ v \leq u + \delta$$
$$\rightarrow F(v) \leq F(u) + \varepsilon \wedge \ F(u) \leq F(v) + \varepsilon))\}\}$$
$$\wedge \ A < B \ \wedge \ F(A) < 0 < F(B)$$
$$\longrightarrow \exists x(A \leq x \leq B \ \wedge \ F(x) = 0)$$

The prover converted the theorem into clauses $1 - 15$ shown below. The six hyper-chainings actually leading to $\square$ are shown (only eight hyper-chainings were derived). Some ground literals are removed by a powerful ground prover based on Sup-Inf [1].[12]

The first hyper-chaining derives clause 16 by working to remove the variable y from clause 5. Clauses 6 and 7 are both used twice to remove the remaining shielding terms of y.

The second hyper-chaining centers on the variable ε in clause 16. Clause 9 is used twice to remove the shielding terms of ε. After variable elimination eliminates ε, the ground prover determines that the only way to refute the literals $L < u$ and $u < L$ is to instantiate u with L. The third hyper-chaining derives 20 from 15 and 17, instantiating variable x with the ground term L.

The central variable of the fourth hyper-chaining is the variable v of 11. The fifth hyper-chaining removes the shielding terms of ε in 21. Finally, $\square$ is derived from 20 and 22 by a hyper-chain on the variable u of 22.

---

[12]Each ground clause derived by the prover is entered into Sup-Inf's database and those literals found inconsistent are removed. Sup-Inf is also used to remove literals which are ground subsumed by the remaining literals of a clause. For example, $L < A$ would be removed from $L < A \vee L \leq A$.

(pr imv-thm hyper-chain-set)
Using rules VPCHAIN VMCHAIN within sequence HYPER-CHAIN.

1. $A < L$ 2. $L < B$

3. $A \leq Z(y)$ $L \leq y$ 4. $Z(y) \leq B$ $L \leq y$

5. $F(Z(y)) < 0$ $L \leq y$

6. $Z(y) \leq L$ $L \leq y$ 7. $y < Z(y)$ $L \leq y$

8. $0 \leq F(w)$ $w \leq L$ $w < A$ $B < w$

9. $0 < \Delta(\varepsilon\, u)$ $\varepsilon \leq 0$

10. $\Delta(\varepsilon\, u) + u < v$ $\Delta(\varepsilon\, u) + v < u$ $F(u) \leq F(v) + \varepsilon$ $\varepsilon \leq 0$
     $u < A$ $B < u$ $v < A$ $B < v$

11. $\Delta(\varepsilon\, u) + u < v$ $\Delta(\varepsilon\, u) + v < u$ $F(v) \leq F(u) + \varepsilon$ $\varepsilon \leq 0$
     $u < A$ $B < u$ $v < A$ $B < v$

12. $A < B$

13. $F(A) < 0$ 14. $0 < F(B)$

15. $0 < F(x)$ $F(x) < 0$ $x < A$ $B < x$

(HYPER-CHAIN, 5, 10, 7, 7, 6, 6)
16. $\Delta(\varepsilon\, u) + u < L$ $\Delta(\varepsilon\, u) + L \leq u$ $F(u) < \varepsilon$ $\varepsilon \leq 0$

(HYPER-CHAIN, 16, 9, 9)
17. $F(L) \leq 0$

(HYPER-CHAIN, 15, 17)
20. $F(L) < 0$

(HYPER-CHAIN, 11, 15, 8)
21. $\Delta(\varepsilon\, u) + u \leq L$ $\Delta(\varepsilon\, u) < 0$ $0 < F(u) + \varepsilon$ $\varepsilon \leq 0$
     $\Delta(\varepsilon\, u) + u < A$ $\Delta(\varepsilon\, u) + B < u$ $B \leq L$ $B < A$

(HYPER-CHAIN, 21, 9, 9, 9 ,9)
22. $u < L$ $0 \leq F(u)$
     $u < A$ $B < u$ $B \leq L$ $B < A$

(HYPER-CHAIN, 22, 20)
23. $\square$.

8 Related Work

As we mentioned earlier, many programs have built-in axioms and lemmas of various theories [2,3,4,5,6,8,9,11]. Also, Stickle's *theory resolution* [10] presents a method for building in the axioms of a theory. It provides a method to incorporate a series of inferences into a larger inference which is related in purpose to rule sequences. However, it does not use a subrule feature as the foundation for new rules.

9 Conclusion

By building in axioms and lemmas at the "knowledge" level, the use of axioms and lemmas can be controlled so as not to explode the prover's search space. Use of these axioms and lemmas transform a general purpose prover into a more powerful domain-specific one. By building in such axioms as transitivity, interpolation and triangle inequality and controlling them through classes, restrictions and rule-sequences, a real-analysis prover has been built which can automatically prove such theorems as the intermediate value theorem and the continuity of the product of two continuous functions.

Such a prover could have been written at the "program" level; however, the lemmas and axioms at the "knowledge" level can also be used as components on which to build in additional axioms and lemmas. For example, the triangle inequality axiom invokes chaining by simply stating that chaining is a subrule of its entry point terms. An equivalent effect is not easily obtained at the programming level.

The use of axioms and lemmas allows a presentation in which the main structure of a proof is emphasized, with subsidiary proofs being suppressed. Thus the proofs are easier to follow and understand. In fact, proofs need not be solely described in terms of the order in which certain inference rules were applied to certain formulas. More abstract descriptions could result from describing the activity of the rule sequences. For example, a proof which involved hyper-chaining might be described by simply noting the central variable at each step. Precisely how each of the shielding terms of the central variable is removed has been abstracted away.

Such a description might also be helpful to provers which use analogical reasoning. There such an abstract description might maintain an analogy when a more concrete description would fail.

A flexible, knowledge-based prover would also allow a learning mechanism to conjecture new restrictions. One could easily alter the rule-base to reflect the new restrictions and perform experiments in order to improve performance.

References

[1] Bledsoe, W. W., "The Sup-Inf Method in Presburger Arithmetic". The University of Texas at Austin, Math Department Memo ATP-18. December 1974. Essentially the same as: A new method for proving certain Presburger formulas. Fourth IJCAI, Tblisi, USSR, (September 1975).

[2] Bledsoe, W. W., and Hines, L. M., "Variable Elimination and Chaining in a Resolution-Base Prover for Inequalities", *Proc. 5th Conference on Automated Deduction*, Les Arcs, France, Springer-Verlag, (July 1980) 70-87.

[3] Bledsoe, W. W., and Tyson, M., "The UT Interactive Prover". University of Texas at Austin, Math Department Memo ATP 17A, (June 1978).

[4] Boyer, R. S. and Moore, J S., *A Computational Logic*, Academic Press, (1979).

[5] Plotkin, G. D. "Building Equational Theories", *Machine Intelligence*, (1972) 73-90.

[6] Robinson, J. A., "A Machine Oriented Logic Based on the Resolution Principle", *J. ACM* 12, No. 1, (1965) 23-41.

[7] Robinson, J. A., "Automatic Deduction with Hyper-Resolution", *Internat. J. Comput. Math.* I, (1969) 227-234.

[8] Robinson, J. A. and Wos, L. "Paramodulation and Theorem Proving in first order theories with equality", *Machine Intelligence*, vol. 4 (B. Meltzer and D. Michie, eds.), American Elsevier, New York, (1969) 135-150.

[9] Slagle, J. R. "ATP with Built-in Theories Including Equality, Partial Ordering and Sets", *J. ACM* 19 No. 1, (January 1972) 120-135.

[10] Stickel, M. E. "Automated Deduction by Theory Resolution", *IJCAI-85*, Los Angeles, California, (August 1985) 1181-1186.

[11] Wos, L., R. Veroff, B. Smith, and W. McCune. "The Linked Inference Principle, II: The User's Viewpoint." *CADE-7*, Napa, California, (May 1984) 316-332.

LINEAR MODAL DEDUCTIONS

Luis Fariñas del Cerro   and   *Andreas Herzig*

Langages et Systèmes Informatiques
Université Paul Sabatier
118 Route de Narbonne
31062 Toulouse Cédex

Abstract

We present a deduction method for propositional modal logics. It is based on a resolution principle for formulas written in a very simple normal form, close to the clausal form for classical logic. It allows us to extend naturally resolution and refinements of resolution to modal logics.

1   Introduction

In the last years modal logics started to be of growing interest in many domains of Artificial Intelligence, as knowledge representation, problem solving, or natural language understanding (see for example [9], [13]). This is an important reason why automated deduction techniques for modal logics must be developed. Although the traditional tableau techniques are more popular in modal logics, for some years extensions of the resolution principle to modal logics have become a new field of research (examples of it can be found in [1], [5], [7], [12], [14] or in [16] for matrix methods). After having defined the resolution principle, the next step in the resolution paradigm is the definition of refinements of resolution. This has been treated in [2], [6], where refinements have been given for formulas in a particular normal form. The latter made the resolution deductions and the completeness proof of these refinements rather tedious. In order to obtain better results we have followed the maxim "the simpler the normal form, the easier the resolution principle", and we have defined a resolution principle based on a simple normal form close to the clausal form of classical logic.

The paper is organized as follows. Section 2 introduces modal logics and a clausal normal form is presented. A linear refinement of the resolution principle based on formulas in clausal normal form is presented for an exemplary modal system in section 3, and its completeness is shown in section 4. Finally an example of deduction is developed.

2 Normal Form for Modal Logics

In this section we introduce the general family of modal logics for which a Kripe semantics can be defined. We prove that a normal form can be found for all these logics.

2.1 Modal Logics

2.1.1 Syntax

A modal language L is defined on the following pairwise disjoint sets:
- VAR: propositional variables p,q,...
- classical connectives: $\neg$, $\wedge$, $\vee$, $\rightarrow$, $\leftrightarrow$
- unary modal connective: $\square$ (necessary).

The formation rules are the classical ones together with:
- If A is a formula then $\square$A is a formula.

As usual $\lozenge$A is defined as $\neg\square\neg$A.

2.1.2 Semantics

A Kripke semantics for L is defined in terms of a model M = (W,R,m) where:
- W is a set of states,
- $R \subseteq W^2$ is a relation, and
- m:VAR $\rightarrow$ $\mathcal{P}$ (W) is a meaning function which furnishes for every propositional variable the set of states where it is true.

In order to simplify our proofs we consider only cases where the relation R is total, i.e. for every w there is a w' such that (w,w') $\in$ R.

Given a model M we say that a formula A is **satisfied** in a state w of W (noted M,w sat A) if the following conditions hold:

M,w sat p iff w $\in$ m(p) for p $\in$ VAR

M,w sat $\neg$A iff not (M,w sat A)

M,w sat A $\vee$ B iff M,w sat A or M,w sat B

M,w sat $\square$A iff for all w' $\in$ W if (w,w') $\in$ R then M,w' sat A

M,w sat $\lozenge$A iff there is a w' such that (w,w') $\in$ R and M,w' sat A.

For the other classical connectives the definitions are obtained using classical equivalences. We say that M **satisfies** A or M **is a model for** A iff there is a w in M such that

M,w sat A, and A is **valid** iff for every model M and every w in M, M,w sat A.

2.2 Clausal Normal Form

The aim of this paragraph is to define a simple normal form (in other words a given formula A is satisfiable iff the corresponding normalized formula A^I is satisfiable).

Since $\Diamond(A \wedge B)$ and $\Diamond A \wedge \Diamond B$ are not equivalent, every notion of clause must allow the $\wedge$-operator in the scope of the $\Diamond$-operator. This makes things rather complicated because (in opposition to classical logic) one clause may be unsatisfiable alone, as for example $\Diamond(p \wedge \neg p)$. Our aim is to avoid this undesirable feature. Therefore we use a technique similar to the Skolem technique for eliminating existential quantifiers. (Intuitively this is possible because $\Diamond$ is defined semantically from these quantifiers.) So we associate to each $\Diamond$-operator a Skolem constant i. The first thing to do is to define precisely the new language and its semantics.

2.3.1 Syntax

The language of LI with a set I of Skolem constants is as for L, where the unary modal connectives are □ and <i>, for every i ϵ I. Formulas are defined using classical formation rules together with : if A is a formula then □A and <i>A, for every i ϵ I, are formulas.

2.3.2 Semantics

The Kripke semantics for LI is defined using the pair MI <(W,R,m),FI>, where FI ={f_i:i ϵ I} is a set of functions from W into W, and $f_i \subseteq R$ for every i ϵ I.
The relation MI,w sat A for a formula A is defined as for L, where :
    MI,w sat <i>A   iff   MI,f_i(w) sat A.
The definitions of satisfiability and validity do not change.

2.3.3 Translation Procedure

Now we define a mapping that allows us to embed L in LI. Without loss of generality we suppose from now on that in all formulas negation appears immediately before the propositional variables.

Definition. We map a formula A of L into a formula A^I of LI by replacing each occurence of $\Diamond$ in A by <i> where i is a new constant (so in A^I every <i> occurs only once, and I is the set of new constants).

Example. If A is $\Box(\Diamond p \vee \Diamond(q \wedge \neg t))$ then A^I will be $\Box(<1>p \vee <2>(q \wedge \neg t))$, and $I = \{1,2\}$.

Lemma 1. Let A be a modal formula of L. A is satisfiable in a model for L iff A^I is satisfiable in a model for LI.

Proof. Let A^I be a modal formula of LI satisfiable in $MI = <(W,R,m),FI>$. Then A is trivially true in the model (W,R,m). In the other way the proof is by induction on the complexity of A. The base of the induction is trivial. For the induction step we present only the case where A is of the form $\Diamond B$. Suppose $\Diamond B$ has been translated to $<i>B^J$. MI,w sat $\Diamond B$ means that there is a w' such that $(w,w') \in R$ and M,w' sat B. By induction hypothesis there is a model $MJ = <(W,R,m),FJ>$, such that MJ,w' sat B^J. As $<i>$ occurs only once in A we can define the new model $MI = <(W,R,m),FI>$ where $I = J \cup \{i\}$, $FI = FJ \cup \{f_i\}$, and f_i is a new function such that $w' = f_i(w)$. This implies MI,w sat $(<i>B)^I$.

2.3.4 Clausal Normal Form

Now we define a simple clausal normal form without nested "$\wedge$", i.e. the $\vee$-operator is the only classical connective governing the literals in the clauses.

A formula of LI is said to be in **clausal normal form** if it is a conjunction in which each conjunct is in clausal form.

A formula of LI is said to be in **clausal form** if it is a disjunction (perhaps with only one disjunct) in which each disjunct is either:

    . a literal, or
    . $\Box C$, where C is a formula in clausal form, or
    . $<i>C$, where $i \in I$ and C is a formula in clausal form.

As usual we identify a conjunction $A_1 \wedge ... \wedge A_n$ with the set $\{A_1,...,A_n\}$, and each conjunct will be called **clause**. A clause with only one disjunct is called a **unit clause**. Some examples of formulas in clausal normal form are:

    $\Box(<1>p \vee \neg q \vee \Box t) \vee <2>(t \vee \Box \neg p) \vee \neg s \vee t$
    $<1>(p \vee \Box(t \vee <2>\neg t)) \vee t$

Definition. The **modal degree** d(A) of a formula A is defined as follows:

    If A is a literal, $d(A)=0$
    $d(A \Delta B)=\max(d(A),d(B))$ if Δ is $\vee, \wedge, \rightarrow$, or $\leftrightarrow$
    $d(\neg A)=d(A)$
    $d(\Delta A)=d(A)+1$ if Δ is $\Box$ or $<i>$, where $i \in I$.

Lemma 2. There is an effective procedure for constructing for any given formula A of LI an equivalent formula A' in clausal normal form.

Proof. The proof is by induction on the modal degree of A. If d(A) = 0 the proof is obtained using the classical procedure.

If A = □B then we normalize B and use the valid formula □ (C ∧ D) ↔ □C ∧ □ D. If A = <i>B then we normalize B and use the valid formulas   <i> (C ∧ D) ↔ <i>C ∧ <i>D   and <i>(C ∨ D) ↔ <i>C ∨ <i>D. Otherwise A is a classical combination of modal formulas, then we can normalize the latter and then apply the classical procedure.

2.3.5   A similar approach is proposed by G. Wrightson [17]. There, constants are appended to formulas to indicate the possible world where they are true. The semantic postulates are introduced explicitly, and tableau rules are applied to them and the given formula. In this approach constants fix worlds, whereas in our one they fix accessibility relations.

Now we shall study in an examplary way resolution for a particular modal system.

3   Modal Resolution

The rest of the paper is devoted to the definition of the resolution principle. This depends on the particular modal logic under consideration. Here we give a resolution principle for a particular system, the modal system S4, for which, as it is well known, the accessibility relation R must be reflective and transitive (see [11]). For the other classical modal system as Q and T the proofs are straightforward.

3.1   Definition of Resolution

Given a formula of S4, in a first step we eliminate all occurrences of ◊ using Lemma 1. The second step is to put the resulting formula in clausal normal form by Lemma 2. On this clausal normal form resolution will be applied.

As for classical resolution the aim of modal resolution is to produce from two given clauses a resolvent clause. To produce the resolvent clause we define a formal system and a set of simplification rules.

Formal system :

axiom:	p, ¬p ⇒ false	
∨-rule:	from A, B ⇒ C	infer A ∨ D, B ∨ E ⇒ C ∨ D ∨ E
T-rule:	from A, B ⇒ C	infer □A, B ⇒ C

□□-rule: from   □A, B ⇒ C   infer □A, □B ⇒ □C
□<i>-rule: from   □A, B ⇒ C   infer □A, <i>B ⇒ <i>C
<i><i>-rule: from   A, B ⇒ C   infer <i>A, <i>B ⇒ <i>C

Simplification rules:

The relation "A can be **simplified** into B", noted $A \approx B$, is the least congruence relation containing:

    1. $A \vee$ false $\approx A$
    2. Δ(false) $\approx$ false if Δ is □ or <i>, for $i \in I$
    3. $A \vee A \vee B \approx A \vee B$

Definition (Modal resolution rule). Given two clauses C_1 and C_2, C is a **resolvent** of C_1 and C_2 if there is a C' such that $C_1, C_2 \Rightarrow C'$ and $C' \approx C$.

We write it as $\dfrac{C_1 \quad C_2}{C}$.

Example. Given the two clauses □(<1>p ∨ q) and <1><2><1>¬p, using the formal system we obtain :

p, ¬p ⇒ false	by the axiom
<1>p, <1>¬p ⇒ <1>false	by the <i><i>-rule
<1>p ∨ q, <1>¬p ⇒ <1>false ∨ q	by T-rule and ∨-rule
□(<1>p ∨ q), <2><1>¬p ⇒ <2>(<1>false ∨ q)	by the □<i>-rule
□(<1>p ∨ q), <1><2><1>¬p ⇒ <1><2>(<1>false ∨ q)	by the □<i>-rule

and after simplification the resolvent will be <1><2>q.

Definition. Given a set of clauses S, a **linear deduction** of C_n from S with top clause C_0 is a deduction of the form

$$\frac{C_0 \qquad D_0}{C_1}$$
$$\cdots$$
$$\frac{C_k \qquad D_k}{C_{k+1}}$$
$$\cdots$$
$$C_n$$

where $C_0 \in S$, and C_{k+1} is a resolvent of C_k and D_k, and each D_k is either in S or is a C_j, for some $j < k$.

Definition. A linear deduction of the empty clause (noted "false") from S is called a **refutation** of S.

Lemma 3. If A,B $\Rightarrow$ C then A $\land$ B $\rightarrow$ C is a valid formula.
Proof. The proof is by induction on the number of rules which are necessary to establish A, B $\Rightarrow$ C. If it is an axiom the proof holds. For the induction step we have the following cases.
Case 1. The last applied rule is a $\square$<i>-rule. By induction hypothesis $\square$A $\land$ B $\rightarrow$ C is a valid formula, then $\square\square$A $\rightarrow$ $\square$(B $\rightarrow$ C) and $\square$A $\rightarrow$ (<i>B $\rightarrow$ <i>C) are valid formulas, too. Consequently $\square$A $\land$ <i>B $\rightarrow$ <i> C is a valid formula.
Case 2. The last applied rule is $\square\square$-rule. For this case we use that $\square$A $\rightarrow$ ($\square$B $\rightarrow$ $\square$C) is a valid formula if A $\rightarrow$ (B $\rightarrow$ C) is a valid formula.
For the other cases the proof is similar.

Theorem 1 (Soundness). If there is a linear refutation of S with top clause C_0 then S is unsatisfiable.
Proof. Suppose S satisfiable. When we apply ($\Rightarrow$) rules to S every resolvent will be satisfiable by lemma 3. Hence we cannot produce the empty clause.

3.2 Completeness of Linear Resolution

We use a tableau type method in order to prove the completeness of our resolution principle. So in the first part of this paragraph we define a particular tableau, called tree, for formulas in clausal normal form.

3.2.1 Tree Models for Sets of Clauses

For formulas in clausal normal form we will consider a particular type of model. A model for a set of clauses S is built up using the following **tree construction**:
1. The root of the tree is S itself.
2. On the tree we perform the following operations.

Or-operation: Choose a leaf n of the tree in which a clause C of the form A $\lor$ B or an unmarked $\square$A appears. **In the first case** append the two children $n_1=(n-\{C\}) \cup \{A\}$ and $n_2=(n-\{C\}) \cup \{B\}$ to n and **in the second case** append the child $n_1=n \cup \{A\}$ to n, and mark C in n_1. Consequently, using the or-operation as often as possible at the end in every leaf each clause is a unit clause.

And-operation: Choose a leaf n of the tree in which each clause is a unit clause. Unmark every clause in n. If for some propositional variable p ϵ VAR, p and $\neg$p appears in n, or n is included in some ancestor node, **then** do nothing, **otherwise** for each set of clauses $<i>Q_1,...,<i>Q_r$ we append a child $n_i = \{\Box A: \Box A \in n\} \cup \{Q_1,...,Q_r\}$. n_i will be called **the i-projection** of n.

This procedure stops after a finite number of steps, because each node is a set of subformulas of S, and their number is finite.

Definition. A **node** is **closed** iff
. either p and ¬p, for some p of VAR, belong to the node, or
. all (resp. some) of its children generated by an or-(resp. and-)operation are closed.

Definition. A tree is **closed** if its root is closed.

3.2.2 From Trees to Models

We build a model $MI = <(W,R,m),FI>$ from a tree T for an open set of clauses S of LI as follows.
- First if an or-operation has been applied to a node n in T, then we keep only one open child (as T is open such a child must exist), and we cut the subtrees the roots of which are the other childs.
- If a node n is included in an ancestor n' we identify n and n'.
- W is the set of nodes of T reduced by the equivalence relation on the set of nodes of the tree such that two nodes belong to the same class iff they are linked by a sequence of or-operations.
- R is the reflective and transitive closure of the relation defined by the set $\{(w,w_i):$ there is $n' \in w$ and $n_i' \in w_i$ such that n_i' is obtained from n' by an and-operation$\}$
- $f_i(w) = w_i$ if $(w,w_i) \in R$ and there are $n' \in w$ and $n_i' \in w_i$ where n_i' is the i-projection of n'. Furthermore we must make f_i total by adding arbitrary edges if necessary.
- $w \in m(p)$ iff p appears in some node of the class w.

Lemma 4. Let S be a set of clauses of LI. S is unsatisfiable implies S has a closed tree.
Proof. Suppose that S has a non-closed tree. We build the model $MI = <(W,R,m),FI>$ from this tree. Now we prove by induction on the complexity of a formula C that MI,w sat C if C belongs to a node of the equivalence class w. If C is a literal the proof holds by the definition of m, because the tree is non-closed. For the induction step we distinguish the following cases:
case 1. If $C = A \lor B$, A or B must appear in some descendent of n. Then by induction hypothesis the proof holds.
case 2. If $C = <i>Q$ then there is a node n' such that $f_i(w) = w'$ where w' is the equivalence class of n', and Q appears in n'. Then by induction hypothesis MI,w' sat Q, and consequently MI,w sat $<i>Q$.

case 3. If C= $\square$A, then A appears in n and every descendent of n. Using induction hypothesis again the proof holds.

Hence letting w_0 the equivalence class of S we have MI,w_0 sat S.

3.2.3   Completeness Proof

To prove the completeness theorem the following three lemmas are necesary:

$\lor$-**rule lemma 5.** If there are linear refutations of the sets of clauses S $\cup$ {A} with top clause C_0 and S $\cup$ {B} with top clause B then there is a linear refutation of S $\cup$ {A$\lor$B} with top clause C_0 or A$\lor$B if C_0=A.

Proof. The refutation of S $\cup$ {A} is of the form

$$
\begin{array}{cc}
C_0 & D_0 \\
\hline
& C_1 \\
& \cdots \\
& C_k \quad A \\
\hline
& C_{k+1} \\
& \cdots \\
\hline
& \text{false}
\end{array}
$$

where the first utilisation of A is against C_k. (If there is none S is refutable alone and the proof is done.)

Now we transform the inference $\dfrac{C_k \quad A}{C_{k+1}}$ into $\dfrac{C_k \quad A \lor B}{C_{k+1} \lor B}$.

In the rest of the refutation thereafter we transform for every j, j>i, the inference

$$\frac{C_j \quad D_j}{C_{j+1}} \quad \text{into} \quad \frac{C_j \lor B \quad D_j'}{C_{j+1} \lor B}, \quad \text{where}$$

$D_j' = D_j$ if $D_j = C_m$ for some m $\le$ k, or $D_j \in$ S, and $D_j' = D_j \lor B$ if $D_j = C_m$ for some m > k, or $D_j = A$.

The resulting side clauses D_j' are in S $\cup$ {A$\lor$B}, and the inferences correspond to ($\Rightarrow$) rules.

To this new deduction of false $\lor$ B $\approx$ B we append now the linear refutation of S $\cup$ {B} with top clause B and get a linear refutation of S $\cup$ {A$\lor$B} with top clause C_0 (resp. A$\lor$B if C_0=A).

T-rule lemma 6. If there is a linear refutation of the set of clauses S$\cup${A} with top clause C_0 and $\square$A $\in$ S then there is a linear refutation of S with top clause C_0 if $C_0\ne$A or $\square$A if C_0=A.

Proof. We transform each inference using A which is of the form

$$\frac{C_k \quad A}{C_{k+1}} \quad \text{into} \quad \frac{C_k \quad \Box A}{C_{k+1}} \quad \text{in virtue of the T-rule.}$$

S4-rule lemma 7. If there is a linear refutation of the set of clauses $\{\Box A_1,...,\Box A_n, Q_1,...,Q_r\}$ with top clause C_0 then there is a linear refutation of $\{\Box A_1,...,\Box A_n, <i>Q_1,...,<i>Q_r\}$ where the top clause is C_0 if $C_0 = \Box A_k$ and it is $<i>C_0$ if $C_0 = Q_k$ for some k.

Proof. We proceed by induction on the length of the refutation. If no inference has been done the proof is clear.

Else we look at the first inference $\dfrac{C_0 \quad D_0}{C_1}$.

There are three cases (except the symmetric ones).

case 1. $C_0 = Q_k$, $D_0 = Q_m$ for some k, m. Then using the $<i><i>$-rule we get the new

inference $\dfrac{<i>Q_k \quad <i>Q_m}{<i>C_1}$ (inference 1).

Now by induction hypothesis we can transform the linear refutation of the set of clauses $\{C_1,\Box A_1,...,\Box A_n, Q_1,...,Q_r\}$ with top clause C_1 into a linear refutation of the set of clauses $\{<i>C_1,\Box A_1,...,\Box A_n, <i>Q_1,...,<i>Q_r\}$ with top clause $<i>C_1$, on the top of which we put inference 1.

case 2. $C_0 = Q_k$, $D_0 = \Box A_m$. Similar to 1) we apply the $\Box<i>$-rule to get the new first

inference $\dfrac{<i>Q_k \quad \Box A_m}{<i>C_1}$, and the rest is as before.

case 3. $C_0 = \Box A_k$, $D_0 = \Box A_m$. Here there are two possible subcases (omitting again the symmetric ones) for the construction of $\Box A_k, \Box A_m \Rightarrow C_1$.

case 3.1. $\Box\Box$-rule: From $A_k, \Box A_m \Rightarrow C_1'$ infer $\Box A_k, \Box A_m \Rightarrow \Box C_1'$ and $C_1 = \Box C_1'$. By induction hypothesis we have a linear refutation of the set of clauses $\{\Box C_1',\Box A_1,...,\Box A_n, <i>Q_1,...,<i>Q_r\}$ with top clause $\Box C_1'$, and we build a new refutation as in case 1.

case 3.2. T-rule: From $A_k, \Box A_m \Rightarrow C_1$ infer $\Box A_k, \Box A_m \Rightarrow C_1$. Replacing the T-rule by a $\Box\Box$-rule: from $A_k, \Box A_m \Rightarrow C_1$ we infer $\Box A_k, \Box A_m \Rightarrow \Box C_1$. By induction hypothesis we have a linear refutation of the set of clauses $\{\Box C_1,\Box A_1,...,\Box A_n, <i>Q_1,...,<i>Q_r\}$ with top clause $\Box C_1$, and we build a new refutation as in case 1.

To prove the upward lemma we use the following definition:

Definition. Given a set of clauses S, a tree T for S, a node n of T, and a subset n' of the set of formulas in n, we define **n' is closed in n** recursively as follows.

1. For a leaf n, n' is closed in n iff n - n' is closed.

2. For an or-operation done on a clause C of n

    2a. if $C \notin n'$, n' is closed in n iff n' is closed in every son of n.

    2b. if $C \in n'$ and $C = A \vee B$, n' is closed in n iff $(n'-\{A \vee B\}) \cup \{A\}$ is closed in the son $n_1 = (n-\{A \vee B\}) \cup \{A\}$ and $(n'-\{A \vee B\}) \cup \{B\}$ is closed in the son $n_2 = (n'-\{A \vee B\}) \cup \{B\}$.

    2c. if $C \in n'$ and $C = \square A$, n' is closed in n if $n' \cup \{A\}$ is closed in the son $n_1 = n \cup \{A\}$.

3. For an and-operation applied to a node n, n' is closed in n iff there is a $<i>Q \in n$ such that the i-projection of n' is closed in the i-projection of n.

Lemma 8. Let S be a set of clauses. If S unsatisfiable and $S-\{C_0\}$ satisfiable, then S has a closed tree, such that in its root S is closed and $S-\{C_0\}$ open.

Proof. Since the set of formulas of the root is S, S will be closed in the root. The proof that $S-\{C_0\}$ is open in the root is obtained straightforward by induction on the complexity of C_0.

Upward lemma 9. Given a node n of a tree and a subset n' of n, if n' is closed in n and $n'-\{C_0\}$ is open in n, then there is a linear refutation of n' with top clause C_0.

Proof. We proceed by induction on the depth of n in the tree. If n is a leaf there are p and $\neg p$ in n', for some $p \in VAR$. As $n' - \{C_0\}$ is open, C_0 must be p or $\neg p$, and the required refutation is immediate. For the induction step suppose that by one of the operations children of n have been appended. According to the definition above we distinguish the following cases.

Case 1. An or-operation has been applied for a clause C of n.

Case 1.1. $C \notin n'$. Then n' is closed in every child of n, and by induction hypothesis there is a linear refutation of n' with top clause C_0.

Case 1.2. $C \in n'$, and $C = A \vee B$. Then there are children n_1 and n_2 of n, and $n'_1 = (n' - \{A \vee B\}) \cup \{A\}$ is closed in n_1 and $n'_2 = (n' - \{A \vee B\}) \cup \{B\}$ is closed in n_2.

Case 1.2.1. $C_0 = A \vee B$. Then $n' - \{A \vee B\}$ is open in n_1, and by hypothesis induction there are linear refutations of n'_1 with top clause A and of n'_2 with top clause B. Hence $\vee$-rule lemma 5 is applicable.

Case 1.2.2. $C_0 \neq A \vee B$. As $n' - \{C_0\}$ is open in n, $n'_1 - \{C_0\}$ is open in n_1 or $n'_2 - \{C_0\}$ is open in n_2. Let the former be true.

a) Suppose $n'_2 - \{B\}$ closed in n_2. As $n'_2 - \{B,C_0\}$ is open in n_2, by induction hypothesis there is a linear refutation of $n' - \{A \vee B\}$ with top clause C_0.

b) Suppose $n'_2 - \{B\}$ open in n_2. As n'_2 is closed in n_2, by induction hypothesis there is a linear refutation of n'_2 with top clause B. Moreover $n'_1 - \{C_0\}$ is open in n_1 and n'_1 is closed in n_1, hence by induction hypothesis there is a linear refutation of n'_1 with top clause C_0. Now $\vee$-rule lemma 5 is applicable.

Case 1.3. $C_0 \in n'$, and $C = \square A$. Then as $n' \cup \{A\}$ is closed in $n_1 = n \cup \{A\}$, by induction hypothesis there will be a linear refutation of $n' \cup \{A\}$ with top clause C_0 (resp. top clause A if $C_0 = \square A$) which will be transformed into a linear refutation of n' with top clause C_0 in virtue of T-rule lemma 6.

Case 2. n is an and-node. Then there must be a $<i>Q \in n$ such that the i-projection n'_i of n' is closed in the i-projection n_i of n. Moreover, $n'_i - \{D\}$ will be open in n_i, where D is the i-projection of C_0, i.e. $D = C_0$ for $C_0 = \Box A$ and $D = Q$ for $C_0 = {<}i{>}Q$. Now by induction hypothesis there is a linear refutation of n'_i with top clause D, and S4-rule lemma 7 warrants that there will be a linear refutation of n' with top clause C_0.

Theorem 2 (Completeness). Let S be a set of clauses. If S is unsatisfiable and $S-\{C_0\}$ satisfiable, then there is a linear refutation of S with top clause C_0.

Proof. Let S be a unsatisfiable set of clauses. Hence by Lemma 4 every tree for S is closed . Let n be the root of a tree for S. S will be closed in n and $S - \{C_0\}$ will be open in n by Lemma 8. Hence by upward Lemma 9 there will be a linear refutation of S with top clause C_0.

Example. Let us give the following unsatistiable set of clauses:

1. $\Box(p \vee {<}1{>}q \vee \Box t)$
2. ${<}1{>}\neg q$
3. ${<}1{>}\neg t \vee \Box e$
4. $\neg p$
5. ${<}2{>}\neg e$

A linear refutation of it will be:

$$
\frac{{<}1{>}\neg q \qquad \Box(p \vee {<}1{>}q \vee \Box t)}{\cfrac{\cfrac{p \vee \Box t \qquad\qquad \neg p}{\cfrac{\Box t \qquad {<}1{>}\neg t \vee \Box e}{\cfrac{\Box e \qquad {<}2{>}\neg e}{\text{false}}}}}{}}
$$

4   Conclusion

In this note we have presented a deduction method based on a simple normal form (close to the clausal form for classical logic), that allows us to extend in an easy way to modal logic the automated deduction techniques used in classical logic. The logic system on which this paper has been based is S4. For extensions of this logic as multi-S4, (i.e. S4 with an arbitrary number of modal operators) or restrictions to subsystems as T or Q, the proof of the completeness theorem is straightforward, and for S5 see [4].

The results obtained are a necessary preliminary step towards the development of resolution techniques for first order S4, which follows the same principles. For prenex S4, i.e. the fragment of first order S4 without quantifiers in the scope of modal operators, the methods and proofs are essentially the same as in this paper (see [3]). For full quantificational S4 we must extend our techniques to skolemize the □-operator [8]. H.J. Ohlbach ([15]) gives a beautiful and comprehensible work about skolem techniques for modal logics.

Acknowledgements. We thank P. Bieber for the correction of errors in a first version of the paper.

5   References

[1]   Abadi M., Manna Z. Modal theorem proving. *8CAD,* LNCS, Springer-Verlag 1986, pp 172-186.

[2]   Auffray Y., Enjalbert P., Hebrard J.J., *Strategics for modal resolution : results and problems.* Rapport du Laboratoire d'Informatique de Caen, 1987.

[3]   Balbiani Ph., Fariñas del Cerro L., Herzig A., *Declarative semantics for modal logic programs.* Rapport LSI, November 1987.

[4]   Bieber P., *Cooperating with untrusted agents.* Rapport LSI, UPS, 1988.

[5]   Cialdea M., *Une méthode de déduction automatique en logique modale.* Thèse Université Paul Sabatier, Toulouse, 1986.

[6]   Enjalbert P., Fariñas del Cerro L., Modal resolution in clausal form, *Theoretical Computer Sciences,* to appear.

[7]   Fariñas del Cerro L., A simple deduction method for modal logic. *Information Processing Letter* 14, 1982.

[8]   Fariñas del Cerro L., Herzig A., *Quantified modal logic and unification theory.* Rapport LSI, UPS, 1988.

[9]   Fariñas del Cerro L., Orlowska. E., Automated reasoning in non classical logic. *Logique et Analyse* , n° 110-111, 1985.

[10]   Fitting M.C., *Proof methods for modal and intuitionistic logics.* Methuen & Co., London 1986.

[11]   Hughes G.E., Cresswell M.J., *A companion to Modal Logic*, Methuen & Co. Ltd., London 1986.

[12]   Konolige K., Resolution and quantified epistemic logics. *8CAD*, LNCS, Springer-Verlag1986, pp199-209.

[13]   Konolige K., *A deductive model of belief and its logics.* Ph.D. Thesis, Computer Sciences Departement, Stanford University 1984.

[14]   Mints G., Resolution calculi for modal logics. *Proc. of the Academy of Estonian SSR,* 1986 (in russian).

[15]   Ohlbach H.J., *A resolution calculus for modal logics.* Report University of Kaiserslautern, 1987.

[16]   Wallen L.A., Matrix proof methods for modal logics. In *Proc. of 10th ICALP,* 1987.

[17]   Wrightson, G., Non-classical theorem proving. *Journal of automated Reasoning,* 1, 35-37, 1985.

A Resolution Calculus for Modal Logics

Hans Jürgen Ohlbach

FB Informatik, University of Kaiserslautern,W-Germany

uucp: ...!seismo!unido!uklirb!ohlbach

Abstract A syntax transformation is presented that eliminates the modal logic operators from modal logic formulae by shifting the modal context information to the term level. The formulae in the transformed syntax can be brought into conjunctive normal form such that a clause based resolution calculus without any additional inference rule, but with special modal unification algorithms, can be defined. The method works for first-order modal logics with the two operators □ and ◊ and with constant-domain Kripke semantics where the accessibility relation is serial and may have any combination of the following properties: reflexivity, symmetry, transitivity. In particular the quantified versions of the modal systems T, S4, S5, B, D, D4 and DB can be treated. Extensions to non-serial and varying-domain systems are possible, but not presented here.

Key words: modal logic, resolution principle, unification

1. Introduction

A straightforward application of the resolution principle [Ro65] to modal logic fails because two syntactically complementary literals may be in different modal contexts and are therefore not necessarily semantically contradictory. Consequently most of the proposed deduction calculi for modal logics are based on tableaux (eg. [Fit72], [Fit83]). Some attempts to adapt the ideas of the standard resolution principle to modal logics have stopped at the half-way point. The method proposed by Fariñas del Cerro [Fa85] for example cannot treat full modal logics, since quantifiers are not allowed in the scope of modalities. Abadi and Manna´s systems [AM86] restrict the application of the resolution rule to modal contexts in which it is sound. Inference across modal operators is performed by additional Hilbert-style deduction rules which generate a very redundant search space. Chan´s method [Ch87], based on an idea which is in fact very similar to the one presented in this paper, is restricted to the propositional S4 system. The most elaborated work on efficient proof systems currently seems to be L.A. Wallen´s matrix proof methods for modal logics [Wal87]. It is the first system that computes the modal context that allows an inference operation with a deterministic unification algorithm, and not by nondeterministic search. In this paper a formalism is presented that fits into the paradigm of the resolution principle for predicate logic and can therefore easily be integrated into an existing clause based resolution theorem prover. The soundness and completeness proofs as well as an extension of the method to logics with non-serial accessibility relations can be found in [Oh88].

2. An Example that Demonstrates the Basic Ideas.

Consider the formula $\quad$ ◊ ◊∀x(◊Px ∧ □Qx) ⇒ ◊(∀yPy ∧ ∀zQz) $\qquad$ (∗)

The meaning of such a formula can be described in terms of possible worlds which are connected by an accessibility relation [HC68]. A possible world is an interpretation in the predicate logic sense. It determines how the function and predicate symbols are to be interpreted in that world. The nesting of the modal operators in a formula determines which world or which interpretation respectively is actually meant. ◊$\mathcal{F}$ for instance means "there is a world b which is accessible from the current world a, such that $\mathcal{F}$ holds under the interpretation of b". □$\mathcal{F}$ means "for every world b which is accessible from the current world a, $\mathcal{F}$ holds under the interpretation of b".

This work was supported by the Sonderforschungsbereich 314 of the Deutsche Forschungsgemeinschaft and by the ESPRIT project 1033 of the European Community.

The premises $\Diamond\Diamond\forall x$ $(\Diamond Px \wedge \Box Qx)$ of the formula (❋) can therefore be expressed in words as:

$\Diamond$　　　From the initial world (say "0") there is an accessible world a,

$\Diamond$　　　from a there is an accessible world b,

$\forall x$　　such that for all x

　$(\Diamond$　　　there is a world c, accessible from b　(but depending on x, therefore c(x))

　Px　　　　such that Px holds, where P is interpreted in world c(x)

　$\wedge\Box$　　　and for all worlds u which are accessible from b

　Qx)　　　Qx holds, where Q is interpreted in world u.

The syntax transformation we are going to present in this paper records these worlds a, b, c(x) and u explicitly and attaches them as an additional "world-path" argument to the predicate and function symbols. Hence the context in which terms and atoms are to be interpreted can be recognized directly from this argument.

The above formula $\Diamond\Diamond\forall x$ $(\Diamond Px \wedge \Box Qx)$, for instance, is translated into $\forall x(P[0abc(x)]x \wedge \forall u\ Q[0abu]x)$ with the intuitive meaning:

$\forall x\ (P[0abc(x)]x$　| For all x Px holds in all worlds which are accessible via the paths 0 a b c(x)

$\wedge \forall u$　　　　| and for all admissible worlds u　(which worlds are actually admissible depends on the

　　　　　　　|　　　　　　　　　　　　world-paths in the subformulae of the quantifier.)

$Q[0abu]x)$　　| Qx holds in all worlds which are accessible via the paths 0 a b u.

In order to prove the formula (❋) by contradiction the consequence of the implication must be negated, i.e.: $\neg\Diamond(\forall yPy \vee \forall zQz)$, and moving the negation sign inside yields: $\Box(\exists y\neg Py \vee \exists z\neg Qz)$. The transformed version is $\forall v(\neg P[0v]f[0v] \vee \neg Q[0v]g[0v])$ with the intuitive meaning:

$\forall v$　　　　　| For all admissible worlds v

$(\neg P[0v]f[0v]$　| $\neg Pf[0v]$ holds in all worlds which are accessible from 0.

　　　　　　　|　　f is a skolem function that denotes the original y "that must exist",

　　　　　　　|　　　but in each world v, there may exist another y. (f has no "ordinary" argument.)

$\vee \neg Q[0v]g[0v])$　| or $-Qg[0v]$ holds in all worlds which are accessible from 0.

　　　　　　　|　　g is the second skolem function that depends also on v.

Eliminating the universal quantifiers the transformed negated formula (❋) can be written in clause form as:

　C1: $P[0abc(x)]x$　　　　　　C2: $Q[0abu]x$　　　　C3: $\neg P[0v]f[0v], \neg Q[0v]g[0v]$

There are two candidates for resolution operations, C2 with C3,2 and C1 with C3,1. Consider the first candidate, C2 with C3,2. Before generating a resolvent the atoms $Q[0abu]x$ and $Q[0v]g[0v]$ must be unified, i.e. the problem of unifying the two world-paths [0abu] and [0v] has to be solved. It is easy to see that unification is impossible unless the accessibility relation of the underlying logic is transitive. In this case $\{v \mapsto [abu]\}$ is a (most general) unifier of [0abu] and [0v]. Combining this substitution with the unifier for x and g[0v], we obtain the final unifier $\{v \mapsto [abu],$ $x \mapsto g[0abu]\}$. In case the accessibility relation is transitive and *serial* (i.e. from every world there exists an accessible world), the resolvent $\neg P[0abu]f[0abu]$ can be generated. Consider now the second candidate for resolution, C1 with C3,1, which gives rise to the unification problem $P[0abc(x)]x$ and $P[0v]f[0v]$. Again we can unify the world-paths $[0abc(x)]$ and [0v] only in logics with a transitive accessibility relation. The unifier is $\{v \mapsto [abc(x)]\}$. This substitution can be applied to the remaining terms before the unification proceeds and the second unification problem: unify x with f[0abc(x)] now fails with an occur check failure. The atoms $P[0abc(x)]x$ and $P[0v]f[0v]$ are not unifiable. Since there is no other possibility for resolution, the proof fails; and in fact, the formula (❋) is <u>not</u> a theorem.

3. Modal Logic

Now we characterize briefly the particular modal logics we shall be considering:

3.1 Syntax

The formulae are those of first order predicate logic with two additional modal operators □ (necessity) and ◊ (possibility). In order to limit the syntactic variety of modal logic (without losing expressiveness), we only consider modal logic formulae in *negation normal form* without the implication and equivalence sign, that is all negation signs are moved in front of the atoms. Arbitrary formulae can be brought into this form using the appropriate transformation rules of predicate logic and the two additional rules $\neg\Box\mathcal{F} \to \Diamond\neg\mathcal{F}$ and $\neg\Diamond\mathcal{F} \to \Box\neg\mathcal{F}$.

3.2 Semantics

A common model theory for modal logics is S. Kripke´s "possible worlds semantics" [Kr59, Kr63]. A "possible world" determines how the function and predicate symbols are to be interpreted in that world, i.e. it is an interpretation in the predicate logic sense. Different possible worlds may assign different meanings to the same symbol, but the universe is the same in all interpretations (constant-domain assumption!).There is one initial world and possibly infinitely many others which are "accessible" by an "accessibility relation" R. Two major classes of accessibility relations can be distinguished: serial and non-serial ones. We shall consider only serial accessibility relations, however in combination with the following properties or the relation R: reflexivity, symmetry and transitivity. In serial interpretations symmetric and transitive relations are also reflexive. Furthermore there is a special normal form for modal logic formulae when R is in fact an equivalence relation (modal logic S5) [HC68]. All formulae have at most one nested modal operator ("modal degree" one). In the sequel we shall assume for S5 interpretations that all formulae have been reduced to this normal form.

4. P-Logic

P-logics ("P" for Predicate logic style) as defined in this paper are syntactic variants of modal logics where the modal operators are replaced by "world-terms". A world-term represents the modal context, i.e. the sequence of nested modal operators, and is attached to the terms and atoms as an additional argument . It holds the information in which world the term or formula is to be interpreted. In order to preserve the possible worlds semantics for P-logic formulae, a world-term must denote a world and not a domain element. This suggests to formulate P-logic as a two-sorted logic with the two disjoint sorts D (for Domain) and W (for Worlds).

The world-paths that have been used in the introductory examples are syntactic variants of the world-terms and will be introduced below. World-terms are more appropriate for understanding the semantics of this structure, whereas world-paths are a suitable datastructure for the unification algorithms.

4.1 Syntax of P-Logic

As usual we begin with the definition of a signature which consists of a D-part and a W-part. The D-part, i.e. D-variable symbols, D-valued function symbols and predicate symbols, is the same as in modal logics. What is new is the W-part that is used for building world-terms. It consists of W-valued function symbols which are something like skolem functions and replace the ◊-operator, and W-variable symbols that replace the □-operator.

Definition: **(Signature of P-Logic)**

The alphabet for building P-logic terms and formulae consists of the logical connectives and the following symbols:

$\mathbb{V}_D$ is a set of *D-variable* symbols.	$\mathbb{V}_W$ is a set of *W-variable* symbols.
$\mathbb{F}_{D,n}$ is a set of (1,n)-place *D-valued function* symbols.	$\mathbb{F}_D$ is the union of all $\mathbb{F}_{D,n}$.
$\mathbb{P}_n$ is a set of (1,n)-place *predicate* symbols.	$\mathbb{P}$ is the union of all $\mathbb{P}_n$.
0 is a W-valued constant symbol (denoting the initial world).	
$\mathbb{F}_{W,n}$ is a set of (1,n)-place *W-valued function* symbols.	$\mathbb{F}_W$ is the union of all $\mathbb{F}_{W,n}$

$\Sigma_P := (\mathbb{V}_D, \mathbb{F}_D, \mathbb{P}, 0, \mathbb{V}_W, \mathbb{F}_W)$ is a *P-signature* .   ∎

Definition: **(Terms, Atoms, Literals and Formulae)**

Given a P-signature $\Sigma_P := (\mathbb{V}_D, \mathbb{F}_D, \mathbb{P}, 0, \mathbb{V}_W, \mathbb{F}_W)$,

➤ the set of *D-terms* $\mathbb{T}_D$ over Σ_P is defined as the least set such that:

     (i) $\mathbb{V}_D \subseteq \mathbb{T}_D$,   (ii) if $f \in \mathbb{F}_{D,n}$, $t_1,...,t_n \in \mathbb{T}_D$ and w is a W-term then $f(w, t_1,...,t_n) \in \mathbb{T}_D$.

➤ The set of *W-terms* $\mathbb{T}_W$ over Σ_P is defined as the least set such that:

     (i)   0 is a W-term.   (ii) if $f \in \mathbb{F}_{W,n}$, $t_1,...,t_n \in \mathbb{T}_D$ and w is a W-term then $f(w, t_1,...,t_n) \in \mathbb{T}_W$.

     (iii)   if u is a W-variable symbol and t is a W-term then u(t) is a W-term.

➤ If $P \in \mathbb{P}_n$, $t_1,...,t_n \in \mathbb{T}_D$ and w is a W-term then $P(w, t_1,...,t_n)$ is a *P-atom*.

➤ A *P-literal* is either a P-atom or a negated P-atom.

➤ P-formulae are built with P-literals, the connectives $\wedge$, $\vee$ and the universal quantifier $\forall$ as usual.

     For convenience we assume that quantified variables are standardized apart, i.e. formulae like $\forall x(\exists x\, Px) \wedge Qx$ are not allowed and should be rewritten as $\forall x(\exists y\, Py) \wedge Qx$.   ∎

Examples for P-logic terms and formulae and their modal logic counterparts:

Modal Logic	P-logic	Signature
$\square P$	$\forall w\, P(w(0))$	$P \in \mathbb{P}_0, w \in \mathbb{V}_W$
$\Diamond P$	$P(g(0))$	$P \in \mathbb{P}_0, g \in \mathbb{F}_{W,0}$
$\forall x \Diamond\, Q(x,a)$	$\forall x\, Q(h(0,x), x, a(h(0,x)))$	$Q \in \mathbb{P}_2, x \in \mathbb{V}_W, a \in \mathbb{F}_{D,0},\ h \in \mathbb{F}_{W,1}$
$\square\forall x\ (R(x) \wedge \square\ \exists y \Diamond\, R(y))$	$\forall w\, \forall x\ (R(w(0), x) \wedge \forall v\, R(k(v(w(0)), x), r(v(w(0)), x)))$	
$\uparrow$ $\uparrow$ $\uparrow\uparrow$		$R \in \mathbb{P}_1, v, w \in \mathbb{V}_W, k \in \mathbb{F}_{W,1}, r \in \mathbb{F}_{D,1}$.
w v r k		

4.2   Semantics of P-Logic

The semantics of P-logics consists of three components, the interpretation of the signature, an evaluator for terms and the satisfiability relation. First the meaning of the symbols has to be defined. The intended meaning of W-valued function symbols is to convey the meaning of the $\Diamond$-operator whose interpretation is: From a given world Σ there exists an accessible world Σ' (possibly depending on some surrounding $\forall$-quantifiers) such that .... Thus, the object that is to be assigned to a W-valued function symbol must be a function - possibly depending on some domain arguments - which maps worlds to accessible worlds.

The W-variables are intended for conveying the meaning of the $\square$-operator whose interpretation is: For all worlds Σ' which are accessible from a given world Σ .... A quantification $\forall w$... over a W-variable must therefore be restricted on some worlds; moreover the restriction has a dynamic character - it depends on the modal context. The only way to incorporate this restriction is to assign to W-variable symbols functions which map worlds to accessible worlds. On the first glance, this gives a higher order character to P-logic, but this higher order character is so weak that it can be eliminated with the standard function currying trick (see below).

Definition: **(P-Interpretation)**

Given a signature $\Sigma_P := (\mathbb{V}_D, \mathbb{F}_D, \mathbb{P}, 0, \mathbb{V}_W, \mathbb{F}_W)$:

a) By a *P-interpretation* for Σ_P we understand any tuple of the following form $(\mathbb{D}, \Sigma, \sigma, R, \Theta, \mu)$ where

> $\mathbb{D}$ is a nonempty set, the *domain* of discourse.

> Σ is a nonempty set of *D-signature interpretations*. Each D-signature interpretation is an assignment of "values" to each D-valued function symbol and predicate symbol in Σ_P as follows: To each (1,n)-place D-valued function symbol a mapping from $\mathbb{D}^n$ to $\mathbb{D}$ is assigned. To each (1,n)-place predicate symbol an n-place relation over $\mathbb{D}$ is assigned. (The elements of Σ are the "worlds".)

> σ is a *D-variable assignment*, i.e. a mapping $\mathbb{V}_D \to \mathbb{D}$.

> R is a relation on $\Sigma \times \Sigma$. (R is the accessibility relation.)

> Θ is a *W-signature interpretation* that assigns values to 0 and the W-valued function symbols: Θ assigns an element of Σ to 0 and to each (1,n)-place W-valued function symbol g an (1,n)-place world access function (see below).

> μ is a *W-variable assignment* that assigns to each W-variable a one place world access function.

When R is symmetric, two additional assumptions are necessary:

1. We assume that $\mathbb{F}_W$ contains for every (1,n)-place W-valued function symbol f a corresponding (1,n)-place "inverse" function symbol f^{-1}, and when Θ assigns an injective world access function g to f, then Θ must assign a corresponding inverse world access function (see below) g^{-1} to f^{-1}.

2. We assume that $\mathbb{V}_W$ contains for every W-variable symbol w an associated "inverse" W-variable symbol u^{-1}, and when μ assigns an injective world access function g to u then μ must also assign an associated inverse world access function g^{-1} to u^{-1}.

(Such "inverse symbols" never occur in formulae, however they may occur in substitutions! Their interpretations will be used only in situations where world access functions are injective.)

b) An (1,n)-place world access function g is a mapping $g: \Sigma \times \mathbb{D}^n \to \Sigma$ such that for every $\Sigma \in \Sigma$ and $a_1,...,a_n \in \mathbb{D}$: $R(\Sigma, g(\Sigma, a_1,...,a_n))$ holds. (g maps worlds to accessible worlds.)

c) An associated inverse world access function g^{-1} for the world access function g satisfies for every world w and domain elements $s_1,...,s_n$ the following equation: $g^{-1}(g(w, s_1,...s_n), s_1,..., s_n) = w$. ∎

The next step is to say how a term is to be evaluated in a given interpretation. The idea is to evaluate the W-term w of a D-term $f(w, t_1,...t_n)$ first, giving a signature interpretation Σ which determines the actual interpretation of f.

Definition: **(P-Evaluation of Terms)**

A P-interpretation $\mathcal{K} = (\mathbb{D}, \Sigma, \sigma, R, \Theta, \mu)$ can be turned into an interpreter for terms:

$$\mathcal{K}(t) := \begin{cases} \sigma(t) & \text{when t is a D-variable symbol.} \\ \mathcal{K}(w)(f)\,(\mathcal{K}(t_1),..., \mathcal{K}(t_n)) & \text{when } t = f(w,t_1,...,t_n) \text{ and f is a D-valued function symbol.} \\ \Theta(0) & \text{when } t = 0 \\ \mu(u)(\mathcal{K}(w)) & \text{when } t = u(w) \text{ and u is a W-variable symbol.} \\ \Theta(g)(\mathcal{K}(w), \mathcal{K}(t_1),..., \mathcal{K}(t_n)) & \text{when } t = g(w, t_1,...,t_n) \text{ and g is a W-valued function symbol.} \end{cases}$$
∎

Let $\mathcal{K}[x/a]$ denote the P-interpretation that differs from $\mathcal{K}$ only in that the variable x is assigned the value a.

Definition: **(The P-Satisfiability Relation $\models_P$)**

Let $\mathcal{K} := (\mathbb{D}, \Sigma, \sigma, R, \Theta, \mu)$ be a P-interpretation for the signature Σ_P.

$\mathcal{K} \models_P P(w, t_1,...,t_n)$	iff $\mathcal{K}(w)(P)\,(\mathcal{K}(t_1),...,\mathcal{K}(t_n))$.
$\mathcal{K} \models_P \neg P(w, t_1,...,t_n)$	iff not $\mathcal{K}(w)(P)\,(\mathcal{K}(t_1),...,\mathcal{K}(t_n))$.
$\mathcal{K} \models_P (\mathcal{F} \wedge \mathcal{G})$	iff $\mathcal{K} \models_P \mathcal{F}$ and $\mathcal{K} \models_P \mathcal{G}$.

$$\mathcal{K}\vDash_p (\mathcal{F} \vee \mathcal{G}) \qquad\qquad \text{iff} \quad \mathcal{K}\vDash_p \mathcal{F} \text{ or } \mathcal{K}\vDash_p \mathcal{G}.$$

$$\mathcal{K}\vDash_p \forall x \ \mathcal{F} \text{ where } x\in \mathbb{V}_D \qquad \text{iff} \quad \text{for every } a \in \mathbb{D}: \ \mathcal{K}[x/a]\vDash_p \mathcal{F}.$$

$$\mathcal{K}\vDash_p \forall u \ \mathcal{F} \text{ where } u\in \mathbb{V}_W \qquad \text{iff} \quad \text{for every world access function g: } \mathcal{K}[u/g]\vDash_p \mathcal{F}. \qquad\qquad \blacksquare$$

Definition: **(P-Models)**

A P-interpretation $\mathcal{K}$ is a *P-model* for a P-formula $\mathcal{F}$ iff $\mathcal{K}\vDash_p \mathcal{F}$ ($\mathcal{K}$ *P-satisfies* $\mathcal{F}$).

A P-formula $\mathcal{F}$ is *P-satisfiable* if a P-model for $\mathcal{F}$ exists. It is *P-unsatisfiable* if no P-model for $\mathcal{F}$ exists. $\qquad\qquad \blacksquare$

5. Translation from Modal Logic Syntax to P-Logic Syntax.

We must define how to translate the modal logic signature into P-logic signature and the modal logic formulae into P-logic fromulae.

1. Translation of the signature: We construct an initial P-signature $\Sigma_p := (\mathbb{V}_D, \mathbb{F}_D, \mathbb{P}, \{0\}, \varnothing)$ where $\mathbb{V}_D, \mathbb{F}_D, \mathbb{P}$ are the same variable, function and predicate symbols as in the modal logic formula.

2. Translation of terms and formulae: We define a translation function Π that takes a modal logic formula $\mathcal{F}$, translates it into a P-logic formula $\Pi(\mathcal{F})$ and updates the signature Σ_p with the generated W-variables that replace the $\square$-operator and the skolem functions for the $\exists$-quantifier and the $\Diamond$-operator. Π needs an auxiliary function π that makes the recursive descent into the modal formulae and terms. π records as a second argument the modal context in form of a W-term w and as a third argument the universally quantified variables D-vars. Take Σ_p to be a "global variable" that is updated during the recursive descent.

D-vars + x $\qquad$ means the concatenation of a list D-vars $= (x_1 \ldots x_n)$ with x, the result is $(x_1 \ldots x_n x)$.

f(w+ D-vars) $\qquad$ denotes the term $f(w, x_1, \ldots, x_n)$ where D-vars $= (x_1 \ldots x_n)$

The transformation rules are:

1. The toplevel call is: $\Pi(\mathcal{F}) := \pi(\mathcal{F}, 0, ())$ $\qquad\qquad$ where () is the empty list.

2. $\quad \pi(\ \mathcal{F}\wedge \mathcal{G}, w, \text{D-vars}) := \pi(\mathcal{F}, w, \text{D-vars}) \wedge \pi(\mathcal{G}, w, \text{D-vars})$

3. $\quad \pi(\ \mathcal{F}\vee \mathcal{G}, w, \text{D-vars}) := \pi(\mathcal{F}, w, \text{D-vars}) \vee \pi(\mathcal{G}, w, \text{D-vars})$

4. $\quad \pi(\ \forall x\mathcal{F}, w, \text{D-vars}) \quad := \forall x \ \pi(\mathcal{F}, w, \text{D-vars} + x)$

5. $\quad \pi(\ \square\mathcal{F}, w, \text{D-vars}) \quad := \forall u \ \pi(\mathcal{F}, u(w), \text{D-vars})$

$\qquad\qquad\qquad$ u is added as a new W-variable to Σ_p.

6. $\quad \pi(\ \exists x\mathcal{F}, w, \text{D-vars}) \quad := \pi(\mathcal{F}, w, \text{D-vars})[x\leftarrow f(w+\text{D-vars})]$ $\qquad$ (i.e. replace x by f(w+D-vars))

$\qquad\qquad\qquad$ f is added as a new (1,|D-vars|)-place D-valued function symbol to Σ_p.

7. $\quad \pi(\ \Diamond\mathcal{F}, w, \text{D-vars}) \quad := \pi(\mathcal{F}, g(w +\text{D-vars}), \text{D-vars})$

$\qquad\qquad\qquad$ g is added as a new (1,|D-vars|)-place W-valued function symbol to Σ_p.

Let P be an n-place predicate symbol and let f be an n-place function symbol.

8. $\quad \pi(\ \neg P(t_1,\ldots,t_n), w, \text{D-vars}) \qquad := \neg P(w, \pi(t_1, w, \text{D-vars}),\ldots, \pi(t_n, w, \text{D-vars}))$

9. $\quad \pi(\ P(t_1,\ldots,t_n), w, \text{D-vars}) \qquad := P(w, \pi(t_1, w, \text{D-vars}),\ldots, \pi(t_n, w, \text{D-vars}))$

10. $\quad \pi(\ f(t_1,\ldots,t_n), w, \text{D-vars}) \qquad := f(w, \pi(t_1, w, \text{D-vars}),\ldots, \pi(t_n, w, \text{D-vars}))$

11. $\quad \pi(\ x, w, \text{D-vars}) \qquad\qquad\qquad := x \qquad\qquad\qquad$ where x is a D-variable $\qquad\qquad \blacksquare$

Theorem: **(Soundness and Completeness of the Translation Algorithm)**

$\mathcal{H}$ is a satisfiable modal formula if and only if $\Pi(\mathcal{H})$ is a P-satisfiable P-formula.

The proof follows the recursion of π and assures that the information about the modal context is correctly shifted from the nesting of the modal operators to the W-terms. $\qquad\qquad \blacksquare$

These results are the basis for a complete proof procedure: In order to prove that a modal logic formula is unsatisfiable, it is sufficient to prove that the translated P-logic formula is P-unsatisfiable.

6. Conjunctive Normal Form

Since the P-logic syntax contains no existential quantifier and no modal operators, a transformation of an arbitrary formula to an equivalent set of clauses is essentially the same as in predicate logic, but without the need to skolemize.

7. World-Paths - An Alternative Syntax for W-terms.

W-terms contain W-variable symbols in a functional position and are therefore higher order terms. For many purposes, especially for the definition of the unification algorithms, the purely first order "world-path" syntax that has been used in the introductory examples is much more convenient. The transition from W-terms to world-paths can be easily explained on the semantic level, where the function symbols and W-variable symbols are interpreted as functions, such that the currying operation is applicable. The currying operation transforms an n-place function f into an n-1-place function f^c that produces a one-place function which, when applied to the remaining argument returns the same value as f when applied to all n arguments at once, i.e. $f(s_1,...,s_n) = s_1 f^c(s_2,...,s_n)$. Currying can be used to remove the "world argument" from the interpretation of an (1,n)-place W-valued function symbol, leaving a function that is applicable to domain elements only. For a nested function call like $f(g(w, s_1,...,s_n), t_1,...,t_m)$ the equivalent curried function call looks like $g(w, s_1,...,s_n) f^c(t_1,...,t_m) = (w\ g^c(s_1,...,s_n)) f^c(t_1,...,t_m) = w\ (g^c(s_1,...,s_n) \circ f^c(t_1,...,t_m))$ where $\circ$ is the composition of functions. A corresponding term that can be interpreted in such a way is (apply $(\circ\ g^c(s_1,...,s_n)\ f^c(t_1,...,t_m))$ w) or still simpler a list $[w\ g^c(s_1,...,s_n)\ f^c(t_1,...,t_m)]$, where the associativity of $\circ$ can be exploited to remove the parentheses of nested terms.

A term $f^c(t_1,...,t_m)$ without a W-term as a first argument will be called a *CW-term* (curried world term) in the sequel. The interpretation of a CW-term $f(t_1,...,t_n)$ occurring in a world-path is that of a curried function mapping the corresponding domain elements $\mathfrak{K}(t_1),...,\mathfrak{K}(t_n)$ to a function which accepts a world and returns another world. The interpretation of W-variable symbols remains unchanged. A world-path evaluates under·an interpretation $\mathfrak{K}$ to the same world as the corresponding W-term. Since there is a unique correspondence between a W-term and a world-path, we can use freely these two syntactic versions for denoting worlds.

Examples of the different syntactic versions:

Modal Logic	P-logic, W-term Syntax	World-path Syntax
$\square P$	$\forall w\ P(w(0))$	$\forall w\ P[0\ w]$
$\Diamond P$	$P(g(0))$	$P[0\ g]$
$\forall x\ \Diamond\ Q(x,a)$	$\forall x\ Q(h(0,x), x, a(h(0,x)))$	$Q([0\ h(x)], x, a[0\ h(x)])$
$\square \forall x\ (R(x) \wedge$	$\forall w\ \forall x\ (R(w(0), x) \wedge$	$\forall w\ \forall x\ (R([0\ w], x) \wedge$
$\square\ \exists y\ \Diamond\ R(y))$	$\forall v\ R(k(v(w(0)), x), r(v(w(0)), x)))$	$\forall v\ R([0\ w\ v\ k(x)], r[0\ w\ v]))$
$\Diamond\square\Diamond P$	$\forall w\ P(h(w(g(0))))$	$\forall w\ P[0\ g\ w\ h]$ ∎

8. Substitutions

One of the most important notions is that of a substitution as a mapping from terms to terms. Since we deal with functional variables, we must use the notions of the λ-calculus for correlating syntactic objects with functions.

Definition: **(Substitutions)**

➤ A *substitution* (in W-term syntax) is a finite set $\{x_1 \mapsto t_1,...,x_k \mapsto t_k\}$ of assignments of D-terms to D-variables and λ-expressions $\lambda(w)\ g_n(g_{n-1}(...\ g_1(w, s_{1,1},...,s_{1,k1}), s_{2,1},..., s_{2,k2}),...)$ to W-variables where the g_i are $(1,k_i)$-place W-valued function symbols and the $s_{i,j}$ are D-terms which do not contain w as a subterm. Here n may be 0 in which case $\lambda(w)$ w is just the identity mapping.

➤ A substitution component $v \mapsto \lambda(w)\ g_n(g_{n-1}(...\ g_1(w, s_{1,1},...,s_{1,k1}), s_{2,1},..., s_{2,k2}),...)$ for a W-variable has in world-path syntax the following structure: $v \mapsto [g_1(s_{1,1},...,s_{1,k1}) ... g_n(s_{n,1},...,s_{n,kn})]$. $v \mapsto []$ is the representation of the identity $\lambda(w)$ w.

➤ Substitutions can be turned into mappings from terms to terms, literals to literals and clauses to clauses using the inductive definition of terms and the β-reduction rule of the λ-calculus such that the following equation for all $(1,n)$-place function and predicate symbols f and terms w and t_i hold: $f(w,t_1,...,t_n)\sigma = f(w\sigma,t_1\sigma,...,t_n\sigma)$. An application of a substitution $\sigma = \{x_1 \mapsto t_1,..., x_n \mapsto t_n\}$ to a term t has the intuitive meaning that all occurrences of D-variables x_i are simultaneously replaced by t_i and all occurrences of W-variables x_j in a subterm $x_j(s)$ (in W-term syntax) are simultaneously replaced by $(\lambda(w) g(...(w, ...)...)(s))$ and immediately reduced with the β-reduction rule to $g(...(s,...)...)$. Furthermore, an application of a substitution with an inverse function in symmetric interpretations like $\lambda(w) g^{-1}(w)$ to a term $g(s)$ immediately reduces the term to $g^{-1}(g(s)) = s$.

Thus, the rewrite rule $g^{-1}(g(v, a_1,..., a_n), a_1,..., a_n)) \to v$ will implicitly be applied whenever it is possible.

➤ The application of a substitution $\{x_i \mapsto [g_1...g_n]\}$ in world-path syntax splices the partial world-path $[g_1...g_n]$ into each world-path at the place of an occurrence of x_i, i.e. $[0...a\ x_i b\ ...]\{x_i \mapsto [g_1...g_n]\} = [0...a\ g_1...g_n\ b\ ...]$. The corresponding rewrite rule that reduces world-paths with inverse functions in symmetric interpretations is: $[g(a_1,..., a_n)\ g^{-1}(a_1,..., a_n)] \to []$.

➤ The composition $\sigma\tau$ of two substitutions σ and τ is denoted as: $t(\sigma\tau) = (t\sigma)\tau$ for a term t.

➤ $\sigma_{|V}$ is the restriction of a substitution σ to a set V of variables.

➤ A substitution σ is *idempotent* iff $\sigma\sigma = \sigma$. ∎

Examples for substitutions:

W-term syntax:	$\sigma = \{x \mapsto f(y), v \mapsto \lambda(w)\ g(u(h(w, a(0))))\}$	(u is a W-variable)
World-path syntax:	$\sigma = \{x \mapsto f(y), v \mapsto [h(a[0])\ u\ g]\}$.	
$f(k(v(l(0))), x)\sigma$	$= f(k(\lambda(w)\ g(u(h(w, a(0))))\ (l(0)), f(y))$	(W-term syntax)
	$= f(k(g(u(h(l(0), a(0))))\ (l(0)), f(y))$	(β-reduction)
$f([0\ l\ v\ k], x)\sigma$	$= f([0\ l\ h(a[0])\ u\ g\ k], f(y))$	(World-path syntax) ∎

Substitutions are usually the syntactic equivalents to the variable assignments that are used in the definition for the semantics of variables. Therefore we must find syntactic conditions for substitutions that reflect the restrictions imposed on the semantic objects assigned to variables. The definition of W-variable assignments in P-interpretations has the restriction, that only functions which map worlds to accessible worlds can be assigned to a W-variable. A substitution $\sigma = \{v \mapsto \lambda(w)\ h(g(w))\}$, for instance, has no meaningful interpretation as a W-variable assignment unless the accessibility relation is transitive. Due to the interpretation of the W-valued function symbols f and g, the corresponding semantic function $\lambda(\Sigma)h'(g'(\Sigma))$ would map a world Σ to a world Σ' which is accessible in two steps, and only transitivity guarantees that Σ' is directly accessible from Σ. In non-transitive interpretations, substitution components like $v \mapsto \lambda(w)\ g(w)$ with only one nested function in the codomain can be interpreted as W-variable assignments. Therefore we introduce the notion of *R-compatible* substitutions with a syntactic condition that guarantees that R-compatible substitutions can be interpreted as W-variable assignments.

Defintion: Given an accessibility relation R, a substitution σ is called *R-compatible*

iff either R is transitive (there is no restriction for transitive interpretations) or

every substitution component for W-variables has the form:

$u \mapsto \lambda(w)\ g(w, s_1,...,s_k)$		or
$u \mapsto \lambda(w)\ g^{-1}(w, s_1,...,s_k)$	in case R is symmetric	or
$u \mapsto \lambda(w)\ w$	in case R is reflexive.	∎

9. Prefix-Stability - An Invariant on the Structure of Terms.

Terms of a translated modal logic formula are "prefix-stable", i.e. they have the property that all occurrences of a W-variable have identical subterms. A term like $f(w(0), g(w(v(0))))$ for example never occurs. This property prevents the unification algorithms for transitive interpretations from becoming infinitary, and should be preserved also for resolvents. Unifiers must therefore be "prefix-preserving", i.e. the unified instances must again be prefix-stable.

An important property of prefix-stable terms is that they have no double occurrences of a W-variable at the toplevel of a world-path (toplevel linearity). Furthermore a W-variable in a prefix-stable term cannot occur in its own prefix (prefix-linearity). Nevertheless it is possible, that a W-variable occurs a second time behind its first occurrence in a prefix-stable world-path. An example is: $[0 \ w \ g(a[0 \ w \ v])]$.

10. Unification

The unification algorithms for P-logic terms, we are going to define work on the world-path syntax. The unification of world-paths that must produce R-compatible substitutions is therefore the only difference to unification of first order predicate logic terms.

Unification when the accessibility relation has no special properties.

R-compatible substitutions are allowed to substitute a partial world-path with *exactly* one CW-term for a W-variable. World-paths like $[0 \ v \ a]$ and $[0 \ b \ w]$ are therefore unifiable with a unifier $\{v \mapsto b, w \mapsto a\}$, whereas the two terms $[0 \ v \ a]$ and $[0 \ b \ u \ w]$ would require a non-R-compatible substitution $\{v \mapsto [b \ c], w \mapsto a\}$. They are not unifiable.

In general two world-paths are unifiable when they have equal length and the CW-terms are pairwise unifiable with compatible unifiers. Thus, the world-paths can be treated like ordinary terms and with the exception that the argument lists may be of different length, there is no difference from unification of first-order predicate logic terms. *There is at most one most general unifier* that is unique up to renaming for every unification problem.

Unification when the accessibility relation is reflexive only.

The substitution component $w \mapsto []$ represents the assignment of the identity mapping to a W-variable. It is R-compatible because in reflexive interpretations a world is accessible from itself. Therefore R-compatible substitutions are allowed to substitute a partial world-path with *at most* one CW-term for a W-variable. The substitution components $w \mapsto []$ remove a variable completely from a world-path such that the world-paths $[0 \ v \ a]$ and $[0 \ b \ u \ w]$ are unifiable with the two independent unifiers $\{v \mapsto b, u \mapsto [], w \mapsto a\}$ and $\{v \mapsto b, u \mapsto a, w \mapsto []\}$. The unification algorithm must consider all possibilities to remove W-variables w by the substitution component $w \mapsto []$ and to unify the CW-terms in the reduced world-paths pairwise. Since there are only finitely many variables to be removed, there are *at most finitely many most general unifiers* for each unification problem.

Unification when the accessibility relation is symmetric only.

In symmetric interpretations, each W-valued function symbol has an associated inverse function. A substitution component $w \mapsto a^{-1}$ is therefore suitable for collapsing the partial world-path $[a \ w]$ into $[a \ a^{-1}] = []$. R-compatible substitutions are allowed to substitute a partial world-path with *exactly* one CW-term or an "inverse" CW-term for a W-variable. The "inverse" v^{-1} of a W-variable is also allowed, because the interpretation of a W-variable is also a function whose inverse exists in symmetric interpretations. For example the two world-paths $[0 \ v \ w]$ and $[0]$ are unifiable with a unifier $\{w \mapsto v^{-1}\}$. The unification algorithm must consider all possibilities to collapse a W-variable w and its predecessor t in the world-path by the substitution component $w \mapsto t^{-1}$ to the empty path $[]$ and to unify the CW-terms in the reduced world-paths pairwise. Since there are only finitely many variables to be collapsed, there are *at most finitely many most general unifiers* for each unification problem.

The symmetric case is the first case where the prefix-stability of terms can be exploited: When a W-variable w and its predecessor t in a world-path have been collapsed with the substitution component $w \mapsto t^{-1}$, we know that in all other terms occurring in the current clause set, t is the predecessor of w. The application of $w \mapsto t^{-1}$ to an arbitrary term containing w in the clause set will therefore collapse [t w] to []. No inverse CW-term will ever occur in an instantiated term.

Unification when the accessibility relation is reflexive and symmetric only.
The two basic ideas for reflexivity and symmetry can simply be joined. The unification algorithm must consider all possibilities to remove W-variables w by the substitution component $w \mapsto []$ and to collapse a W-variable w and its predecessor t in the world-path by the substitution component $w \mapsto t^{-1}$ into the empty path [] and to unify the CW-terms in the reduced world-paths pairwise. Since there are only finitely many variables to be collapsed or removed, there are *at most finitely many most general unifiers* for each unification problem.
For example the two world-paths [0 a u v] and [0 w] are unifiable with two independent unifiers $\{u \mapsto a^{-1}, v \mapsto w\}$ and $\{v \mapsto u^{-1}, w \mapsto a\}$.

Unification when the accessibility relation is transitive only.
R-compatible substitutions may substitute arbitrary partial world-paths for a W-variable. A unifier for the two world-paths [0 v c d] and [0 a b w d] is $\{v \mapsto [a b], w \mapsto c\}$, but the substitution $\{v \mapsto [a b w'], w \mapsto [w' c]\}$ with a new W-variable w' is also a unifier. Thus, we must consider variable splitting like in the unification algorithm for associative function symbols, which is in general infinitary. Fortunately it turns out that the toplevel linearity of the world-paths is sufficient to *keep the unification finitary*.
The unification algorithm for two world-paths $s = [s_1 \ldots s_n]$ and $t = [t_1 \ldots t_m]$ works from left to right.
Roughly speaking it consists of the three main steps:
1. Unify the term s_1 with t_1 and call the world-path unification algorithm recursively for $[s_2 \ldots s_n]$ and $[t_2 \ldots t_m]$.
2.a When s_1 is a W-variable then for $i = 2, \ldots, m$ create the substitution component $s_1 \mapsto [t_1 \ldots t_i]$ and call the world-path unification algorithm recursively for $[s_2 \ldots s_n]$ and $[t_{i+1} \ldots t_m]$.
2.b When t_i is a W-variable, then split t_i into the two new W-variables [u v], create the substitution component $\{s_1 \mapsto [t_1 \ldots t_{i-1} u], t_i \mapsto [u v]\}$ and call the world-path unification algorithm again for $[s_2 \ldots s_n]$ and $[v \ t_{i+1} \ldots t_m]$.

Unification when the accessibility relation is reflexive and transitive only.
The ideas for the reflexive case and transitive case can be joined without further problems. The algorithm for the transitive case must simply be augmented with a step that removes W-variables w with a substitution component $w \mapsto []$. There are still at most *finitely many most general unifiers* for each unification problem.

Unification when the accessibility relation is an equivalence relation (modal logic S5)
World-paths for S5 interpretations in normal form of modal degree one consist of at most two CW-terms, i.e. they look like [0] or [0 t]. Two world-paths [0] and [0 t] can only be unified when t is a variable, the unifier is $t \mapsto []$. Two world-paths [0 s] and [0 t] can be unified when s and t are unifiable. Therefore there is *at most one most general unifier* for each unification problem.

The Unification Algorithm
Only one algorithm is defined that takes the accessibility relation *type* R as an additional parameter and branches internally to the R-depending algorithm for world-paths. (R may be just a list of key words like ´reflexive´, ´symmetric´ and ´transitive´.) The main control loop of the algorithm is similar to the Robinson algorithm for first-order terms .

($s \in t$ is true if s is a subterm of t. $Var(s_1 \ldots s_n)$ denotes the set of variables occurring in $s_1, \ldots, s_n$.)

Function   Unify-terms (s, t, R)

Input: s and t are either empty lists or two prefix-stable terms or atoms.

Output: A complete set of idempotent and prefix-preserving unifiers for s and t.

If s = t   then Return {ø}   (ø is the empty substitution, i.e. the identity.)

If s is a variable   then Return If s ∈ t then ø else {{s ↦ t}}

If t is a variable   then Return If t ∈ s then ø else {{t ↦ s}}

If Var(s, t) = ø or s = () or t = () or topsymbol(s) ≠ topsymbol(t)   then Return ø.

If s and t are CW-terms   then Return Unify-termlists(arguments(s), arguments(t), R)

Let s =: f(v,s_1,...,s_n) and t =: f(w,t_1,...,t_n)

　　Let Ξ := Unify-world-paths (v, w, R)

　　　Return $\bigcup_{\xi \in \Xi} \{\xi\theta \mid \theta \in$ Unify-termlists $((s_1,...,s_n)\xi, (t_1,...,t_n)\xi, R)\}$

Function   Unify-termlists (s, t, R)

Input: Two prefix-stable termlists s =: $(s_1...s_n)$ and t =: $(t_1...t_m)$.

Output: A complete set of prefix-preserving idempotent unifiers for s and t.

If s = t then Return {ø}.

Let Ξ := Unify-terms (s_1, t_1, R)

　　Return $\bigcup_{\xi \in \Xi} \{(\xi\theta)_{|Var(s,t)} \mid \theta \in$ Unify-termlists $((s_2...s_n)\xi, (t_2...t_n)\xi, R)\}$.

Function   Unify-world-paths (s, t, R)

Input: Two world-paths.

Output: A complete set unifiers for s and t.

If s = t   then Return {ø}.

If s = () or t = ()   then Return ø.

Return Case R denotes an accessibility relation which is

　　　without special properties   then Unify-termlists (s, t, R)

　　　reflexive and (or) symmetric   then Unify-world-paths-reflexive-or-symmetric (s, t, R)

　　　transitive   then Unify-world-paths-transitive (s, t, R)

　　　reflexive and transitive   then Unify-world-paths-transitive (s, t, R)

　　　reflexive, symmetric and transitive then Unify-world-paths-equivalence (s, t, R).

Function   Unify-world-paths-reflexive-or-symmetric (s, t, R)

Input: Two world-paths s =: $[s_1 ... s_n]$ and t =: $[t_1 ... t_m]$.

　　　　　　R ∈ {{reflexive}, {symmetric}, {reflexive, symmetric}}.

Output: A complete set of unifiers for s and t.

Let Λ := Unify-instantiated-wps (Unify-terms (s_1, t_1, R), $[s_2...s_n]$, $[t_2...t_m]$, R)

If R is reflexive only then n' := 1, m' := 1 else n' := n, m' := m.

For i = 1, ..., n'   Λ:= Λ ∪ Unify-instantiated-wps (Unify-collapse $([s_1...s_i], R)$, $[s_{i+1}...s_n]$, t, R).

For i = 1, ..., m'   Λ:= Λ ∪ Unify-instantiated-wps (Unify-collapse $([t_1...t_i], R)$, s, $[t_{i+1}...t_m]$, R).

Return Λ.

Function Unify-world-paths-transitive (s, t, R)

Input: Two world-paths $s =: [s_1 \ldots s_n]$ and $t =: [t_1 \ldots t_m]$. $R \in \{transitive\}, \{reflexive, transitive\}\}$.

Output: A complete set of unifiers for s and t.

Let $\Lambda := \emptyset$

  For $i = 0, \ldots, m$ ($i = 0$ is the collapsing case when R is reflexive.)

    Let $\Xi := $ Unify-prefix $(s_1, [t_1 \ldots t_i], R)$

        $\Lambda := \Lambda \cup$ Unify-instantiated-wps $(\Xi, [s_2 \ldots s_n], [t_{i+1} \ldots t_m], R)$.

    $\Lambda := \Lambda \cup$ Unify-split (s, t, i, R).

  Repeat the For loop with s and t exchanged.

  Return $\{\lambda_{|Var(s,t)} \mid \lambda \in \Lambda\}$.

Function Unify-world-paths-equivalence (s, t, R)

Input: Two world-paths s and t. $R = \{reflexive, symmetric, transitive\}$.

Output: A complete set of unifiers for s and t.

Return Case $(s, t) = ([0], [0])$ then $\{\emptyset\}$

                $([0], [0w])$ or $([0w], [0])$ where w is a variable then $\{\{w \mapsto []\}\}$

                $([0], [0r])$ or $([0r], [0])$ then $\emptyset$

                $([0r], [0q])$ then Unify-terms (r, q, R).

Function Unify-instantiated-wps (Ξ, s, t, R)

Input: Ξ is a set of substitutions, s and t are two world-paths.

Output: A complete set of unifiers for s and t which are smaller than some element of Ξ.

Return $\bigcup_{\xi \in \Xi} \{(\xi\theta)_{|Var(s,t)} \mid \theta \in$ Unify-world-paths $(s\xi, t\xi, R)\}$.

Function Unify-collapse(t, R)

Input: $t =: [t_1, \ldots, t_n]$ is a world-path. $R \in \{\{reflexive\}, \{symmetric\}, \{reflexive, symmetric\}\}$.

Output: A complete set of unifiers which collapse t into [].

Case $n = 0$ Return $\{\emptyset\}$.

  $n > 0$ Let $\Lambda := \emptyset$

      If R is reflexive and t_1 is a variable then

             $\tau := \{t_1 \mapsto []\}$; $\Lambda := \Lambda \cup \{\tau\theta \mid \theta \in$ Unify-collapse$([t_2, \ldots, t_n]\tau, R))$

      If R is symmetric and $t_1 \neq 0$ and $n > 1$ then

        For $i = 2, \ldots, n$

          If t_i is a variable then

            $\tau := \{t_i \mapsto [t_1^{-1}]\}$

            $\Xi := \{\tau\theta \mid \theta \in$ Unify-collapse$([t_2, \ldots, t_{i-1}], R))$

            $\Lambda := \Lambda \cup \bigcup_{\xi \in \Xi} \{\xi\theta \mid \theta \in$ Unify-collapse$([t_{i+1}, \ldots, t_n]\xi, R))$

        Return Λ.

Function Unify-prefix (s_1, t, R)

Input: A CW-term s_1, a world-path $t =: [t_1 \ldots t_n]$. R is transitive (and reflexive).

Output: If either s_1 is a variable or $n = 1$: A complete set of unifiers for $[s_1]$ and $[t_1 \ldots t_n]$.

Return   If $n = 1$ then   Unify-terms (s_1, t_1, R).

       elseif s_1 is a variable and $s_1 \notin t$ and either $n > 0$ or R is reflexive

               then $\{\{s_1 \mapsto t\}\}$.

       otherwise $\emptyset$.

Function   Unify-split (s, t, i, R)

Input: Two world-paths s =: $[s_1...s_n]$ and t =: $[t_1...t_m]$, a positive integer
                and a transitive (and possibly reflexive) accessibility relation R.

Output: A complete set of unifiers for s and t.

If i > 1 and t_i is a variable and $s_1 \notin [t_1...t_i]$ <u>then</u>

    <u>Let</u> $\xi := \{s_1 \mapsto [t_1...t_{i-1}u], t_i \mapsto [u\ v]\}$ where u and v are new variables

        <u>Return</u> $\{\theta_{|Var(s,t)} \mid \theta \in$ Unify-instantiated-wps $(\{\xi\}, [s_2...s_n], [v\ t_{i+1}...t_m], R)\}$

<u>else</u> <u>Return</u> $\emptyset$.

Theorem The unification algorithm terminates and it is sound and complete, i.e. it computes a complete set of prefix-preserving idempotent most general unifiers.   ∎

11. The Resolution Rule

There is no significant difference to the resolution rule for predicate logic.

Definition: Let $C = Ptl_1 \vee ... \vee Ptl_n \vee C'$ and $D = \neg Psl_1 \vee ... \vee \neg Psl_m \vee D'$ be two clauses with no variables in common and let σ be a most general prefix-preserving and R-compatible unifier for the termlists $tl_1,...,tl_n$ and $sl_1,...,sl_m$, i.e. $tl_1\sigma = ... = tl_n\sigma = sl_1\sigma = ... = sl_m\sigma$. Then the clause $(C' \vee D')\sigma$ is called a *resolvent* of C and D.   ∎

Theorem The resolution rule is sound and complete.

The completeness proof follows the ideas of the completeness proof of the resolution rule for predicate logic using term interpretations (an analogy to Herbrand interpretations) and semantic trees.   ∎

A **final example** shall illustrate the whole procedure. The example proves that in the modal system D (serial accessibility relation with no special properties) Löb´s Axioms $\square(\square G \Rightarrow G) \Rightarrow \square G$ imply the formula $\square P \Rightarrow \square\square P$ that characterizes transitive accessibility relations. Let $G := P \wedge \square P$. The theorem to be proved is

    $\mathcal{F} := (\square(\square(P \wedge \square P) \Rightarrow (P \wedge \square P)) \Rightarrow \square(P \wedge \square P)) \Rightarrow (\square P \Rightarrow \square\square P)$.

Step 1: Negation of $\mathcal{F}$ yields:

    $\neg((\square(\square(P \wedge \square P) \Rightarrow (P \wedge \square P)) \Rightarrow \square(P \wedge \square P)) \Rightarrow (\square P \Rightarrow \square\square P))$.

Step 2: The negation normal form of $\neg\mathcal{F}$ is:

    $(\Diamond(\square(P \wedge \square P) \wedge (\neg P \vee \Diamond\neg P)) \vee \square(P \wedge \square P)) \wedge (\square P \wedge \Diamond\ \Diamond\neg P)$

Step 3: Translation into P-logic syntax:

    $((\forall u\ (P[0au] \wedge \forall v\ P[0auv]) \wedge (\neg P[0a] \vee \neg P[0ab])) \vee \forall w(P[0w] \wedge \forall x\ P[0wx]))$
    $\wedge (\forall y\ P[0y] \wedge P[0cd])$

Step 4: Conjunctive normal form:

    $\forall u\ (P[0au] \wedge \forall v\ P[0auv]) \wedge$
    $(\neg P[0a] \vee \neg P[0ab] \vee \forall w(P[0w]) \wedge$
    $(\neg P[0a] \vee \neg P[0ab] \vee \forall x\ P[0wx]) \wedge$
    $\wedge \forall y\ P[0y] \wedge P[0cd]$

Eliminating the universal quantifiers gives the usual clause notation:

C1:	P[0au]	C2:	P[0auv]
C3:	¬P[0a], ¬P[0ab], P[0w]	C4:	¬P[0a], ¬P[0ab], P[0wx]
C5:	P[0y]	C6:	P[0cd]

Step 5: A resolution refutation:

C4,1 & C5	→	R1: ¬P[0ab], P[0w′x′]	(unifier = {y ↦ a})
R1,1 & C6	→	R2: ¬P[0ab]	(unifier = {w′ ↦ c, x′ ↦ d})
R2 & C1	→	R3: empty	(unifier = {u ↦ b}) ■

Conclusion

A clause based resolution calculus has been developed for several first order modal logics. The two most significant advantages of this calculus are:

➤ Instantiation and inference across modal operators can be controlled by a uniform and deterministic unification algorithm. Thus the large search space generated by the usual instantiation rules and operator shifting rules in tableau based systems is eliminated.

➤ The method is absolutely compatible with the paradigm of the standard resolution principle, i.e. there is no need to develop specialized theorem provers for modal logics. Only a slight modification of an existing predicate logic resolution based theorem prover is sufficient, in other words most of the sophisticated implementation and search control techniques, for instance the connection graph idea [Ko75, Ei85], which have been developed for predicate logic can immediately be applied to modal logics as well. This is an indirect advantage which, however, should not be underestimated: almost 25 years of experience with the resolution principle are now available to modal logic theorem proving as well.

What else can be done?

My hope is that the basic ideas presented in this work are powerful enough to open the door to efficient theorem proving in a much larger class of non-standard logics than the relatively simple modal systems I have examined so far. Let me therefore sketch some ideas for further work in this area.

Resolution in Modal Logics with Non-Serial Accessibility Relations.

A serious complication arises in non-serial models where a world may have no accessible world at all. A formula □P for example may become true in such an interpretation, not because P is true, but because the quantification "for every accessible world" is empty. This phenomenon has two consequences for formulae in the P-logic translation:

1. A formula ∀u P[0u] must not be straightforwardly instantiated with a non-variable term t because a world-path [0t] in a formula P[0t] requires the existence of an accessible world.

2. The two literals P[0u] and ¬P[0u] in a formula ∀u P[0u]∧ ¬P[0u] are not necessarily complementary because the quantification may be empty.

The introduction of an artificial predicate "End(w)" which is true in an interpretation $\mathcal{K}$ iff $\mathcal{K}$(w) is the last world, i.e. $\mathcal{K}$(w) exists and there is no further world accessible from $\mathcal{K}$(w), provides a solution to both problems:

1. The instantiation operation can now introduce a condition expressing its admissibility. For example P[0u] instantiated with {u ↦ a} yields End[0] ∨ P[0a], i.e. either 0 is the last world or P holds in the world denoted by [0a]. The resolution operation therefore inserts such "End-literals" as a residue into the resolvent.

2. Further End-literals are to be inserted into the resolvent in order to handle the case that the resolution literals are not

contradictory. For example the resolvent of P[0u]∧ ¬P[0u] would be End[0]. This literal can for instance be resolved with a literal Q[0a] where a is not a variable because End[0] expresses that 0 is the last world whereas [0a] postulates the existence of an accessible world a.

A sound and complete calculus based on this idea is given in [Oh88]. The subjects mentioned below, however, are still open research problems.

Equality Reasoning in Modal Logics

Consider the formula $\mathcal{F}$: a = b ∧ □P(a). Since the second occurrence of a is in the scope of the □-operator and may therefore be interpreted differently to the first occurrence, it is not possible to replace a by b and to deduce □P(b). Thus, an unrestricted application of a replacement operation like the paramodulation rule [RW69] is not sound in the modal logic syntax. In P-logic syntax, the modal context is available at each term and can be used to influence a deduction operation. The translated formula $\Pi(\mathcal{F})$: a[0] = b[0] ∧ ∀ u P([0u] a[0u]) for instance can safely be paramodulated when the accessibility relation is reflexive, the unifier for a[0] and a[0u] is {u ↦ []}, and the paramodulant is P([0] b[0]). (We assume the equality predicate to be rigid!) Thus, equality reasoning by paramodulation should be no problem in P-logic. The paramodulation rule need not be changed, just the accessibility relation dependent unification algorithms must be applied for unifying one side of an equation, which is always a D-term, with the subterm of the literal to be paramodulated. More work, however, needs to be done for adapting term rewriting and Knuth-Bendix completion techniques to P-logic. They must be able to handle more than one most general unifier and the term orderings must take the world-path strings into account.

Many-Sorted Modal Logics

Resolution and paramodulation calculi for sorted first order predicate logic have been published for instance in [Wa87] and [SS85]. It has been shown that only two slight modifications of the unification algorithm and one modification of the paramodulation rule are necessary for handling hierarchical sort structures: A variable x of sort S1 and a variable y of sort S2 can only be unified when there is a common subsort of S1 and S2. A variable x of sort S1 can only be unified with a term t of sort S2 if S2 = S1 or S2 is a subsort of S1. The paramodulation rule must take care that a paramodulation operation with an equation whose two sides have different sorts does not increase the sort of the paramodulated term. Adapting these ideas to P-logic should be no problem when the sort structure and the sort declarations for the function symbols do not depend on the modal context. This is the case in most applications where only fixed sorts like "Integer", "Real" etc. occur.

Varying-Domain Modal Logics

In varying-domain interpretations there is no universal domain, but each world has its own domain which may or may not intersect with the domain of other worlds. That means universally quantified domain variables depend on the modal context. The idea is now to modify the translation function Π such that the W-term that characterizes the modal context is attached to the domain variables as well. Unification of such a world-depending D-variable x[p] with a D-term $f([q],t_1,...,t_n)$ is only possible when the world-paths p and q are unifiable. To demonstrate this, let us try to prove the Barcan formula ∀x□Px ⇒ □∀xPx which does not hold in varying-domain models. If the proof fails, we have some evidence that the idea is sufficient. The P-logic clause form of the negated Barcan formula ∀x□Px ∧ ◊∃x¬Px is: C1: P([0u] x[0])   C2: ¬P([0a] f[0a])

In fact, the two world-paths [0] and [0a] of the variable x and the symbol f are not unifiable and no refutation is possible. In constant-domain interpretations on the other hand, where x has no world-path, there is the unifier {u ↦ a, x ↦ f[0a]}.

Multiple Operator Modal Logics

In the most simple multiple operator modal logics the two operators □ and ◊ are just multiplied and each copy is associated with its own accessibility relation. In such a logic we can index □ and ◊ with an integer and transfer this index to the P-logic syntax to obtain indexed W-variables and W-valued function symbols. The unification of a W-variable with index i and a partial world-path $[p_1...p_n]$ with indices $i_1,...,i_n$ is then possible only when all indices equal i. Even operators indexed with a set of integers should be easy to handle. If the semantics of an operator $\square_{\{i,...,k\}}$ is disjunctive, i.e. $\square_{\{i,...,k\}} \Leftrightarrow \square_i \vee ... \vee \square_k$ then the unification of the corresponding CW-terms with index sets s_I and t_K must compute just the intersection $I \cap K$. If the semantics is conjunctive, i.e. $\square_{\{i,...,k\}} \Leftrightarrow \square_i \wedge ... \wedge \square_k$ then two non-variable CW-terms are unifiable only when their index sets are equal and a W-variable is unifiable with a CW-term only when the index set of the variable is a subset of the index set of the CW-term.

Other Properties of the Accessibility Relation

Let us come back to modal logics with one pair of operators $\square$ and $\lozenge$ and consider some accessibility relations with properties other than reflexivity, symmetry and transitivity.

Linear Accessibility Relations.

Linear means that there is just one sequence of worlds. The interesting case, where the interpretation of the two modal operators is not identical is when the accessibility relation R is transitive, i.e. a total ordering. In this case for two given worlds it can always be determined which one is farther away from the initial world. The consequence is that for example a formula like $\lozenge\square P \wedge \lozenge\square\neg P$ is unsatisfiable when in addition R is serial. The reason is that for the two worlds denoted by the two $\lozenge$-operators, all worlds "behind" that one which is farthest away from the initial world, are also accessible form the other world. In other words there is no linear and serial interpretation where the intersection of the worlds denoted by the two $\square$-operators is empty, and the formula requires P and $\neg$P to hold in these worlds. In order to find this contradiction in the P-logic version of the formula: $\forall u\ P[0au] \wedge \forall v\ \neg P[0bv]$, we must be able to unify [au] and [bv]. A plausible unifier could be $\{[au] \mapsto [max(a,b)w], [bv] \mapsto [max(a,b)w]\}$ where $max(a,b) = a$ if R(b,a) holds, otherwise $max(a,b)=b$. The figure below shows an interpretation falsifying the formula.

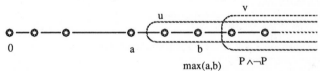

Euclidian Accessibility Relations.

The accessibility relation of the modal system S 4.3 is reflexive, transitive and has the additional property that, if from a world a two worlds b and c are accessible, then either b is accessible from c or vice versa:

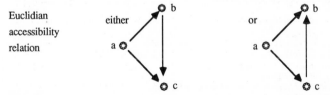

The unification of the two world-paths [0abv] and [0ac] for instance is possible, provided c is accessible from b. The unifier must therefore be a tuple consisting of a substitution and the negation of the condition, i.e. the relations that must hold if the unification was wrong. In our example it could look like $(\{[bv] \mapsto c\}, R(c,b))$. The resolution operation must insert these literals as a residue into the resolvent, similar to the residue in non-serial interpretations. Furthermore resolution between such literals R(s,t) and literals containing partial world-paths [t s] must be allowed.

Accessibility Relations with finite R-chains.

One of the models for the modal system G that is characterized by Löb´s axioms $\square(\square\mathcal{F} \Rightarrow \mathcal{F}) \Rightarrow \square\mathcal{F}$ has a transitive accessibility relation where all chains $R(a_1,a_2)$, $R(a_2,a_3)$,... terminate. In order to demonstrate where the methods developed in this work fail for the system G, let us try to prove Löb´s axiom for the predicate P assuming only transitivity of R. The P-logic clause form for the negated theorem $\neg \square(\square P \Rightarrow P) \Rightarrow \square P$ is:

C1: ¬P[0va], P[0v]　　　　　　　　　C2: ¬P[0b]

Resolution between C2 and C1,2 produces ¬P[0ba] which is not very helpful. The second possibility for resolution i the self resolution of C1 with a unifier $\{v \mapsto [v´a]\}$ giving the resolvent ¬P[0v´aa], P[0v´]. Now there is another self resolution possibility giving ¬P[0v"aaaa], P[0v"] and a third time self resolution becomes possible etc: C1 carries tl potential to generate a world-path with a sequence of infinitely many non-variable CW-terms, and that contradicts the assumption that there are no infinite R-chains. Simply eliminating the first literal and generating P[0v] as a new clause would be a sound deduction in this example. The new clause could then be resolved with C2 giving the desired refutation. The potential to generate infinitely long world-paths with C1 has been indicated by a weak unifier (without renaming) $\{v \mapsto [va]\}$ that puts a non-variable term behind the substituted variable itself. Adding a rule for detecting these situations and removing literals from clauses that can produce infinitely long world-paths might lead to a feasible calculus for G.

Acknowledgements

I would like to thank Andreas Nonnengart for many useful discussions and contributions, and Lincoln A. Wallen whose talk about his modal Matrix method, given in spring 1987 in Munich, inspired me to this work. Norbert Eisinger and Jörg Siekmann read drafts of this paper and gave many useful hints.

References

AM86　M. Abadi, Z. Manna. *Modal Theorem Proving.*
　　　　In Proc. of CADE 86, pp. 172-189, 1986.

Ch87　Chan, M. *The Recursive Resolution Method for Modal Logic.*
　　　　New Generation Computing, 5, 1987, pp 155-183.

Ei85　N. Eisinger. What you always wanted to know about connection graphs.
　　　　Proc. of 8th Conference on Automated Deduction, Oxford, 1985.

Fa85　L. Fariñas del Cerro. *Resolution modal logics.*
　　　　In Logics and Models of Concurrent Systems, (K.R. Apt, ed.), Springer 1985, pp 27-55.

Fit72　M.C. Fitting. *Tableau methods of proof for modal logics.*
　　　　Notre Dame Journal of Formal Logic, XIII:237-247,1972.

Fit83　M.C. Fitting. *Proof methods for modal and intuitionistic logics.*
　　　　Vol. 169 of Synthese Library, D. Reidel Publishing Company, Dordrecht, 1983.

HC68　G.E.Hughes, M.J.Cresswell, *An Introduction to Modal Logics.*
　　　　Methuen &Co., London, 1968.

Ko75　R. Kowalski: *A Proof Procedure Using Connection Graphs.*
　　　　J.ACM Vol. 22, No.4, 1975.

Kr 59　S. Kripke. *A Completeness Theorem in Modal Logic.*
　　　　J. of Symbolic Logic, Vol 24, 1959, pp 1-14.

Kr 63　S. Kripke. *Semantical analysis of modal logic I, normal propositional calculi.*
　　　　Zeitschrift für mathematische Logik und Grundlagen der Mathematik,Vol. 9, 1963, pp 67-96.

Oh88　H.J. Ohlbach, *A Resolution Calculus for Modal Logics.*
　　　　Thesis, FB Informatik, Univ. of Kaiserslautern (forthcoming 1988).

Ro65　Robinson, J.A. *A Machine Oriented Logic Based on the Resolution Principle.*
　　　　J.ACM Vol. 12, No. 1, pp 23-41, 1965.

RW69　Robinson, G., Wos,L., *Paramodulation and theorem provcing in first order theories with equality.*
　　　　Machine Intelligence 4, American Elsevier, New York, pp. 135-150, 1969.

SS 85　M. Schmidt-Schauß. *A Many-Sorted Calculus with Polymorphic Functions Based on Resolution and Paramodulation.* Proc. of 9th IJCAI, Los Angeles, 1985, 1162-1168.

Wal87　L.A.Wallen *Matrix proof methods for modal logics.*
　　　　In Proc. of 10th IJCAI, 1987.

Wa87　C. Walther: *A Many-sorted Calculus Based on Resolution and Paramodulation.*
　　　　Research Notes in Artifical Intelligence, Pitman Ltd., London, M. Kaufmann Inc., Los Altos, 1987.

Solving Disequations in Equational Theories

Hans-Jürgen Bürckert, FB Informatik, Universität Kaiserslautern,

Postfach 3049, D-6750 Kaiserslautern, FR Germany

E-mail: uklirb!buerckert@unido

Abstract: Disunification is the problem to solve a system $\langle s_i = t_i: 1 \leq i \leq n, p_j \neq q_j: 1 \leq j \leq m \rangle$ of equations and disequations. Solutions are substitutions for the variables of the problem that make the two terms of each equation equal, but let those of the disequations different. We investigate this in the case, where equality is defined by an equational theory E. We show how E-disunification can be reduced to E-unification, that is solving equations only, and give a disunification algorithm for equational theories provided there is a unification algorithm. In fact this result shows that for theories, where the solutions of all unification problems can be represented by finitely many substitutions, there is also a finite representation of the solutions of all disunification problems. As an application we discuss how AC1-disunification can be used as a method to represent many AC-unifiers by a few AC1-disunifiers.

Keywords: Equational theories, E-unification, E-disunification, solving equations and disequations.

1. Introduction

The problem of solving equations in a given equational theory, also known as E-unification (G. Plotkin 1972, J. Siekmann 1978, G. Huet & D.C. Oppen 1980, F. Fages & G. Huet 1983, 1986), is well established now and has many applications in artificial intelligence and computer science (J. Siekmann 1987). Closely related is the problem of solving disequations, that is, finding solutions for problems of the form $\langle p \neq q \rangle$, or more generally solving a combination of equations and disequations $\langle s_i = t_i: 1 \leq i \leq n, p_j \neq q_j: 1 \leq j \leq m \rangle$. Solutions to these problems are substitutions of the variables in the terms s_i, t_i, p_j, q_j, such that the instantiated terms of the equations become equal, while the instantiated terms of the disequations have to be different – both with respect to a given equational theory. In the case of the empty theory equality and difference are defined purely syntactical (J.A. Robinson 1965, A. Colmerauer 1984).

Originally A. Colmerauer (1984) discusses the problem of solving equations *and* disequations in the framework of logic programming, and he gives an algorithm to solve these disunification problems. H. Comon (1986) investigates the problem in the more general framework of a set of alternative systems of equations and disequations allowing also certain variables to be parameters. Solutions are substitutions of the (non-parameter) variables solving at least some of the alternatives for arbitrary substitutions for the parameter variables. The reason for this generalizations are certain applications in algebraic specification. In fact he proposes disunification as a tool for proving sufficient completeness of an algebraic specification defined by a set of rewrite rules. J.-L. Lassez et al. (1987) prove some theoretical properties of disunification problems, especially with respect to solvability equations and disequations and redundancy of disequations. C. Kirchner & P. Lescanne (1987) again take up the problem and give a set of rules for transforming disunification problems into solved forms. However, all these papers deal with uninterpreted signatures, that is, they investigate solving equations and disequations with respect to the empty equational theory, i.e., syntactical equality.

Recently several papers are tackling the general problem of disunification in an arbitrary equational theory. A first investigation is due to W. Buntine (1986). He generalizes (A. Colmerauer 1984) for logic programming with equality (cf. J. Jaffar et al. 1986). H. Comon & P. Lescanne (1988) take up the approach of H. Comon (1986) and of C. Kirchner & P. Lescanne (1987) and generalize it for the theory of associative and commutative function; H. Comon (1988) has some general results in disunification with respect to arbitrary equational theories.

In the following sections we present a formal framework for solving equation and disequations in an arbitrary equational theory E. We introduce the necessary algebraic notations in order to define E-disunification problems and their solutions. Since there is some trouble with the common notion of solutions defined by substitutions we generalize this notion and give some representation theorem that leads to an algorithm for solving E-disunification problems by solving suitable E-unification problems. This implies that, when the solutions of all E-unification problems can be represented by finite sets of substitutions, we also have a finite representation for the solutions of all E-disunification problems. Finally we discuss how we can represent large sets of most general AC-unifiers by a few solutions to certain AC1-disunification problems and we give some further applications for disunification.

2. Terms, Substitutions, Equations and Disequations

We assume the reader to be familiar with the notions of universal algebra and equational logic as needed for unification theory; we are consistent with the notations of the common literature (G. Grätzer 1979, S. Burris & H.P. Sankappanavar 1979, W. Taylor 1979, J.H. Siekmann 1987, G. Huet & D.C. Oppen 1980, F. Fages & G. Huet 1986, H.-J. Bürckert, A. Herold & M. Schmidt-Schauß 1987). We require the notion of algebras, homomorphisms, congruences, and quotient algebras as can be found for example in G. Grätzer (1979) and just recall some of the important notations of unification theory.

Given signature Σ of finitely many *function symbols* coming with an arity and a possibly infinite set $\mathcal{V}$ of *variable symbols* the Σ-*term algebra over* $\mathcal{V}$ is denoted by $T_\Sigma(\mathcal{V})$; its carrier is the set of terms constructed with $\mathcal{V}$ and Σ in the common way, and its operations f^T map terms $t_1,\ldots,t_n$ to the term $f(t_1,\ldots,t_n)$. If $\mathcal{V}$ is the empty set, $T_\Sigma(\emptyset)$ is called the *ground* Σ-*term algebra*, abbreviated by T_0. The set of Σ-endomorphisms on the term algebra $T_\Sigma(\mathcal{V})$ that move at most a finite set of variables - the Σ-*substitutions* - is denoted by SUB_Σ. Σ-*equations* and Σ-*disequations over* $\mathcal{V}$ are pairs of terms denoted by $s = t$ and $s \neq t$, respectively. We drop Σ if it is fixed and clear form the context, and we assume $\mathcal{V}$ to be infinite and fixed and abbreviate $T_\Sigma(\mathcal{V})$ by T.

For any syntactical object O like terms, substitutions, sets or tupels of terms we write $\mathcal{V}(O)$ for the set of variables occurring in this object. For a substitution σ we call the finite set of variables $DOM\sigma = \{x \in \mathcal{V}: \sigma x \neq x\}$ the *domain*, the finite set of terms $COD\sigma = \{\sigma x: x \in DOM\sigma\}$ the *codomain*, and $VCOD\sigma = \mathcal{V}(COD\sigma)$ the set of variables *introduced by* σ. We represent a substitution σ by the finite set of its variable assignments or *substitution components* $\{x \leftarrow \sigma x: x \in DOM\sigma\}$. If $VCOD\sigma$ is empty, we call σ a ground substitution. The *restriction* $\sigma/_V$ of a substitution σ to a set of variables V is defined by $\sigma/_V x = \sigma x$ for all $x \in V$ and $\sigma/_V x = x$, otherwise.

3. Unification and Disunification

Usually a system of equations and disequations $\langle s_i = t_i: 1 \leq i \leq n, p_j \neq q_j: 1 \leq j \leq m \rangle$ is to be solved with respect to a given algebra $\mathcal{A}$, that is, we are looking for assignments to the variables in the system with elements of $\mathcal{A}$ – or equivalently we are searching for homomorphisms from the term algebra T to the algebra $\mathcal{A}$–, such that the s_i and t_i are mapped to the same element, while the p_j and q_j are mapped to different elements in $\mathcal{A}$.

Now, in unification theory we solve equations in the quotient $T/=_E$ of the term algebra modulo a given equational theory. Here an *equational theory* is defined as the least congruence $=_E$ on the term algebra induced by a set of term pairs (equations) E, that is closed with respect to application of substitutions. For our applications we also want to restrict ourselves to solving equations and disequation with respect to this quotient $T/=_E$. By the definition of the quotient this is equivalent to find substitutions σ, such that σs_i and σt_i are in the same equivalence class (i.e., $\sigma s_i =_E \sigma t_i$), while the σp_j and σq_j are in different equivalence classes (i.e., $\sigma p_j \neq_E \sigma q_j$). Obviously, when σ solves an equation, also every instance $\lambda\sigma$ solves it, while this need not be the case for disequations. An immediate consequence is, that, when an equation is solvable in $T/=_E$, it is also solvable in every other algebra, in contrast again to disequations.

Let us denote the problem to resolve a system $\Gamma \cup \Delta$ of equations and disequations in the algebra $\mathcal{T}/_E$ as an *E-disunification problem*, written

$$\langle \Gamma, \Delta \rangle_E := \langle s_i = t_i : 1 \leq i \leq n, p_j \neq q_j : 1 \leq j \leq m \rangle_E.$$

An *E-solution* of $\langle \Gamma, \Delta \rangle_E$ is every substitution $\sigma \in \mathcal{SUB}$, such that $\sigma s_i =_E \sigma t_i$ $(1 \leq i \leq n)$ and $\sigma p_j \neq_E \sigma q_j$ $(1 \leq j \leq m)$, where $DOM\sigma = \mathcal{V}(\Gamma, \Delta)$. The set of these solutions is denoted $U_E(\Gamma, \Delta)$. If the system is only an equation system Γ, it is called an *E-unification problem*, and its E-solutions are called *E-unifiers*. The set of all E-unifiers is denoted by $U_E(\Gamma)$. For unification problems one is usually only interested in a subset of $U_E(\Gamma)$ that represents all these E-solutions. Representation is defined in terms of instantiation on some set of variables W:

δ is an *E-instance on W* of σ $(\delta \geq_E \sigma [W])$ iff there is some λ with $\delta x =_E \lambda \sigma x$ for all $x \in W$.

The E-solution sets of unification problems are closed under E-instantiation on $\mathcal{V}(\Gamma)$ or on some superset V of $\mathcal{V}(\Gamma)$. Every E-instance of an E-unifier is again an E-unifier. Hence we define representative sets of E-unifiers or *complete sets of E-unifiers* $cU_E(\Gamma)$ by the property that the union of all E-instances (on V) of the elements of $cU_E(\Gamma)$ is exactly the set of all E-solutions $U_E(\Gamma)$.

Unfortunately the solutions of systems with disequations cannot be represented by the above instantiation method. The reason for this is that now the solution sets are no longer closed under instantiation. For example the disunification problem $\langle x = y, x \neq a \rangle_\emptyset$ in the empty theory has a solution $\{x \leftarrow v, y \leftarrow v\}$ by our definition, however, the instance $\{x \leftarrow a, y \leftarrow a\}$ is of course not a solution.

4. Substitutions with Exceptions

The idea to overcome this problem is to define a more general notion of solutions (cf. W. Buntine 1986, H. Comon 1986, J.-L. Lassez et al. 1987, P. Lescanne & C. Kirchner 1987, G. Smolka et al. 1987). As we have seen, a substitution represents the set of all its instances, hence in order to describe all instances except certain ones, we propose *differences* of substitutions mirroring the difference of the corresponding instance sets. For example all solutions of $\langle x = y, x \neq a \rangle_\emptyset$ can be represented by the difference of $\{x \leftarrow v, y \leftarrow v\}$ and $\{v \leftarrow a\}$. Thus we define a *substitution with exceptions* as a pair $\sigma - \Psi$ of a substitution σ and a family of substitutions $\Psi = \{\psi_\iota : \iota \in I\}$, the exceptions.

In general there is no chance to have finitely many exceptions for every solution of a disunification problem:
Let $\langle x = y, a.x \neq x.a \rangle_A$ be a disunification problem with respect to an associative function ".". Solutions to this problem are all substitutions of x and y with the same arbitrary string except the infinitely many strings a^n, $n \geq 1$. They are represented by the following substitution with exceptions:

$$\{x \leftarrow v, y \leftarrow v\} - \{\{x \leftarrow a^n\} : n \geq 1\} \text{ or equivalently } \{x \leftarrow v, y \leftarrow v\} - \{\{v \leftarrow a^n\} : n \geq 1\}.$$

The example also demonstrates two interesting ways to choose substitutions with exceptions to represent the solutions of a disunification problem $\langle \Gamma, \Delta \rangle_E$. Depending on the applications we represent the solutions by substitutions with exceptions $\sigma - \Psi$, such that for all $\psi \in \Psi$ either $DOM\psi \subseteq DOM\sigma$ or $DOM\psi \subseteq VCOD\sigma$. The first form is more suitable for testing, whether or not a given substitution is an instance of the substitution with exception, while the second form supports generation of instances represented by the substitution with exceptions (of course provided there are only finitely many exceptions, cf. definition below). In order to distinguish the two forms, we denote the second one by $\sigma - \Psi_\sigma$ since the exceptions here depend on the (codomain of the) substitution, while in the first form we can restrict ourselves to substitutions with exceptions $\sigma - \Psi$ that are independent in the sense that the substitution and its exceptions have the same distinguished domain $V = DOM\sigma = DOM\psi$ (for all $\psi \in \Psi$) and variable disjoint codomains.

Definition: Let W be any set of variables, and let us say a substitution λ is an *E-instance on W* of a set of substitutions Ψ (abbreviated $\lambda \geq_E \Psi [W]$) iff it is an instance of some $\psi \in \Psi$. Then:

1. We call a substitution λ an *E-instance on W* of a substitution with exceptions $\sigma-\Psi$ (first form), iff λ is an E-instance on W of σ, but not an E-instance on W of Ψ. We abbreviate this by $\lambda \geq_E \sigma-\Psi \, [W]$.

2. We call λ an *E-instance on W* of a substitution with exceptions $\sigma-\Psi_\sigma$ (second form), iff there is some γ, such that $\lambda =_E \gamma\sigma \, [W]$, but γ is not an E-instance on $\mathcal{V}(\sigma W)$ of Ψ. We abbreviate this by $\lambda \geq_E \sigma-\Psi_\sigma \, [W]$.

A substitution with exceptions is *E-consistent on W*, iff it has at least one E-instance on W, otherwise it is *E-inconsistent on W*.

Obviously the two notions are consistent with the common instance notion for substitutions as defined above, i.e. $\lambda \geq_E \sigma \, [W]$ iff λ is an instance of σ, where σ is considered as a substitution with (empty) exceptions of any of the two forms. However, while a substitution always has instances, this may no longer be the case for substitutions with exceptions: $\{x \leftarrow f(v)\} - \{\{x \leftarrow f(w)\}\}$ or equivalently $\{x \leftarrow f(v)\} - \{\{v \leftarrow w\}\}$ have no instances. The reason for this is that in the first form the substitution can be an instance of one of its exceptions and in the second form the exceptions span the whole set of substitutions.

Inconsistency Lemma: *A substitution with exceptions $\sigma-\Psi$ (first form) has no E-instances on W, iff σ is an E-instance on W of Ψ and a substitution with exceptions $\sigma-\Psi_\sigma$ (second form) has no E-instances on W, iff ε (the identity) is an E-instance on $\mathcal{V}(\sigma W)$ of Ψ_σ*

From the computational point of view substitutions with exceptions will only make sense, if we can decide, whether or not a substitution with exceptions is consistent. Sufficient conditions are that there are at most finitely many exceptions and that we can decide the problem, if a substitution σ is an E-instance on W of another substitution ψ. The latter decision can be done by solving the E-unification problem $\langle \sigma x = \psi x : x \in W \rangle_E$, where the variables of ψx are treated as constants. This is known as *E-matching* (cf. H.-J. Bürckert 1987 for a discussion of the relationships between E-matching and E-unification).

For certain applications (cf. H. Comon 1986) we are only interested in ground solutions of the disunification problems. In this case, we have to test whether or not a substitution with exceptions has ground instances ("ground consistency"). Hence we need – for example to test a substitution with exceptions of the second form for E-instances – a decision procedure for testing, if all ground substitutions are E-instances of a (finite) set of substitutions. An example will demonstrate, what may happen:

The substitutions $\{x \leftarrow a\}$ and $\{x \leftarrow f(v)\}$ span all ground substitutions (w.r.t. the variable set $\{x\}$), if the signature contains only the constant a and the unary function f. That is, $\sigma-\{\{x \leftarrow a\}, \{x \leftarrow f(v)\}\}$ has no ground instances.

5. Representation of Solutions

As for standard substitutions we are interested in representative sets of substitutions with exceptions. Therefore we define a set of substitutions with exceptions SE to be a complete representation for the solutions of a disunification problem $\langle \Gamma, \Delta \rangle_E$, iff the instances of the elements of SE are exactly the solutions of the disunification problem:

(i) $\lambda \geq_E \sigma-\Psi \, [V]$ for some $\sigma-\Psi \in SE$ implies λ solves $\langle \Gamma, \Delta \rangle_E$ (correctness)

(ii) λ solves $\langle \Gamma, \Delta \rangle_E$ implies $\lambda \geq_E \sigma-\Psi \, [V]$ for some $\sigma-\Psi \in SE$ (completeness)

These definitions are consistent with the corresponding definitions for E-unification (cf. J. Siekmann 1987). Correctness and completeness for a set of E-unifiers is defined as correctness and completeness for the set of substitutions considered as set of substitutions with (empty) exceptions. In this case, if they exist, one is usually interested in *minimal* representative sets of E-unifiers, that are complete sets $\mu U_E(\Gamma)$ with the additional property:

$$\forall \sigma, \tau \in \mu U_E(\Gamma) \text{ with } \delta \geq_E \sigma \, [V] : \sigma = \tau \quad \text{(minimality)}.$$

Let us call a theory *finitary* (w.r.t. disunification), if every disunification problem has a finite complete set of substitutions with exceptions representing all its solutions; and we call it *unitary* (w.r.t. disunification), if the solutions of every disunification problem can be represented by a single substitution with exceptions. Analoguously finitary and unitary theories w.r.t. unification are defined (cf. J. Siekmann 1987).

The following **Representation Theorem** shows that the solutions of a disunification problem can be obtained as instances of certain sets of substitutions with exceptions that are generated by solving unification problems only. It depends on some obvious set transformations (cf. J.-L. Lassez et al. 1987):

$$U_E(\Gamma,\Delta) = U_E(\Gamma) - \bigcup\{U_E(p = q): p \neq q \in \Delta\} = U_E(\Gamma) - \bigcup\{U_E(\Gamma, p = q): p \neq q \in \Delta\}.$$

Part 1 of the theorem corresponds to the first transformation, part 2 and 3 to the second one. It essentially says that in this transformation the sets of solutions can be replaced by complete solution sets. Notice, that when they exist, we can also use minimal ones.

Representation Theorem: *Let* $\langle \Gamma, \Delta \rangle_E$ *be a disunification problem, let* $cU(\Gamma)$ *be a complete solution set for the unification problem* $\langle \Gamma \rangle_E$, *and let* $V \supseteq \mathcal{V}(\Gamma, \Delta)$.
1. *Let* $\Psi := \bigcup \{cU(p = q): p \neq q \in \Delta\}$, *where* $cU(p=q)$ *are complete solution sets (on V) for the unification problems* $\langle p = q \rangle_E$ *for all* $p \neq q$ *in* Δ. *Then:*
$$\lambda \text{ solves } \langle \Gamma, \Delta \rangle_E \text{ iff } \lambda \geq_E \sigma - \Psi \, [V] \text{ for some } \sigma \in cU(\Gamma).$$
2. *Let* $\Phi_\sigma := \bigcup \{cU(\sigma p = \sigma q): p \neq q \in \Delta\}$, *where* $cU(\sigma p = \sigma q)$ *are complete solution sets (on* $\mathcal{V}(\sigma V)$) *for the problems* $\langle \sigma p = \sigma q \rangle_E$ *for all* $p \neq q$ *in* Δ. *Then:*
$$\lambda \text{ solves } \langle \Gamma, \Delta \rangle_E \text{ iff } \lambda \geq_E \sigma - \Phi_\sigma [V] \text{ for some } \sigma \in cU(\Gamma).$$
3. *Let* $\Theta_\sigma := \bigcup \{cU(\sigma V = \psi V): \psi \in \Psi\}$ *for all* $\sigma \in cU(\Gamma)$, *where* $cU(\sigma V = \psi V)$ *are complete solution sets (on* $\mathcal{V}(\sigma V \cup \psi V)$) *for the problems* $\langle \sigma v = \psi v: v \in V \rangle_E$ *for all* $\psi \in \Psi$. *Then:*
$$\lambda \text{ solves } \langle \Gamma, \Delta \rangle_E \text{ iff } \lambda \geq_E \sigma - \Theta_\sigma [V] \text{ for some } \sigma \in cU(\Gamma).$$

Proof: 1. λ solves $\langle \Gamma, \Delta \rangle_E$ then $\lambda s =_E \lambda t$ for all $s = t \in \Gamma$ and $\lambda p \neq_E \lambda q$ for all $p \neq q \in \Delta$. Hence there is some $\sigma \in cU(\Gamma)$ with $\lambda \geq_E \sigma[V]$. Assume $\lambda \geq_E \Psi[V]$, then there is some $p \neq q \in \Delta$ and some $\psi \in cU(p = q)$ with $\lambda \geq_E \psi[V]$. Hence $\lambda p =_E \lambda q$, a contradiction. Conversely let $\lambda \geq_E \sigma[V]$ for some $\sigma \in cU(\Gamma)$, but not $\lambda \geq_E \Psi[V]$. Then obviously $\lambda s =_E \lambda t$ for all $s = t \in \Gamma$. Assume $\lambda p =_E \lambda q$ for some $p \neq q \in \Delta$, then λ is an instance of some $\psi \in cU(p = q)$. That is $\lambda \geq_E \Psi[V]$, a contradiction.
Part 2 and 3 are proved analoguously (cf. H.-J. Bürckert 1987b). ■

Obviously the theorem holds also, if we are interested in ground solutions only. The above substitutions with exceptions also represent all ground solutions of the given disunification problem.

As an immediate corollary of the Representation Theorem we have that the type unitary or finitary for a given theory is preserved with respect to both unification and disunification, and the solutions can be represented by substitutions with finitely many exceptions.

Corollary: *Let E be an equational theory.*
a) E is unitary with respect to unification iff E is unitary with respect to disunification. In this case the solutions of disunification problems can be represented by one substitution with finitely many exceptions.
b) E is finitary with respect to unification iff E is finitary with respect to disunification. In this case the solutions of disunification problems can be represented by a finite set of substitution with finitely many exceptions.

The above results show that, provided we have a unification algorithm $E\text{-}UNIFY(\Gamma)$ for a theory E, that always computes a finite, complete set of E-unifiers for a system of equations Γ and an algorithmus $IS\text{-}CONSISTENT(\sigma - \Psi)$ for E-consistency tests of substitutions with exceptions, we also get a disunification algorithm $E\text{-}DISUNIFY(\Gamma,\Delta)$ for E, that computes a finite and complete set of substitutions with (finitely many) exceptions solving the disunification problem $\langle \Gamma, \Delta \rangle_E$. The consistency test can be done by E-unification.

We only give the algorithm corresponding to part 1 of the Representation Theorem; it can easily be extended for part 2 or 3 of the theorem.

Algorithm $E\text{-}DISUNIFY(\Gamma, p_1 \neq q_1, \dots, p_n \neq q_n)$

Input: A system of equations Γ and disequations $p_1 \neq q_1, \dots, p_n \neq q_n$

    1. $cU := E\text{-}UNIFY(\Gamma)$, $cU_1 := E\text{-}UNIFY(p_1 = q_1),\dots, cU_n := E\text{-}UNIFY(p_n = q_n)$

    2. $\Psi := cU_1 \cup \dots \cup cU_n$

    3. $SE := \{\sigma - \Psi : \sigma \in cU \text{ and } IS\text{-}CONSISTENT(\sigma - \Psi)\}$

Output: A finite, complete set SE of substitutions with (finitely many) exceptions solving the input system or the empty set, if the input system has no solutions.

Notice, that the system is not solvable, iff the equation part is unsolvable or none of the computed substitutions with exceptions has an instance. If some of the unification problems corresponding to the disequations have no solution, then these disequations are redundant.

6. Propagation and Merging of Solutions

In order to solve systems of equations one usually solves them sequentially: The equations are solved one by one and the solutions of former equations are substituted into the later ones (*propagation*). An alternative method is to solve all equations quasi-parallel and then to unify the solutions (*merging*). It is well-known that the two approaches lead to complete sets of solutions for a system, provided the intermediate steps produce complete solution sets (the proofs are very similar to those of the Representation Theorem and can be found for example in H.-J. Bürckert 1987a; cf. also H.J. Ohlbach 1986 and A. Herold 1987). We want to generalize these results for substitutions with exceptions as solutions for systems of equations and disequations.

Propagation Theorem: *Let $\langle \Gamma_1, \Delta_1 \rangle_E$ and $\langle \Gamma_2, \Delta_2 \rangle_E$ be disunification problems, let $V \supseteq \mathcal{V}(\Gamma_1, \Delta_1, \Gamma_2, \Delta_2)$. Let cU_1 be a complete solution set (on V) for $\langle \Gamma_1 \rangle_E$, let cU_σ be complete solution sets (on $\mathcal{V}(\sigma V)$) for the problems $\langle \sigma\Gamma_2 \rangle_E$ for all $\sigma \in cU_1$, and let $\Psi := \bigcup \{cU(p = q): p \neq q \in \Delta_1 \cup \Delta_2\}$, where $cU(p=q)$ are complete solution sets (on V) for $\langle p = q \rangle_E$ for all $p \neq q$ in $\Delta_1 \cup \Delta_2$. Then the set $cU := \{(\tau\sigma)/_V - \Psi: \tau \in cU_\sigma, \sigma \in cU_1\}$ is a complete solution set for $\langle \Gamma_1, \Gamma_2, \Delta_1, \Delta_2 \rangle_E$.*

Proof: It is well-known that $\{(\tau\sigma)/_V: \tau \in cU_\sigma, \sigma \in cU_1\}$ is a complete solution set for $\langle \Gamma_1, \Gamma_2 \rangle_E$. Hence the theorem follows with the Representation Theorem. ∎

Merging Theorem: *Let $\langle \Gamma_1, \Delta_1 \rangle_E$ and $\langle \Gamma_2, \Delta_2 \rangle_E$ be disunification problems, let $V \supseteq \mathcal{V}(\Gamma_1, \Delta_1, \Gamma_2, \Delta_2)$. Let cU_i be complete solution sets (on V) for $\langle \Gamma_i \rangle_E$ $(i = 1, 2)$, and let $cU(\sigma V = \tau V)$ be complete solution sets (on $\mathcal{V}(\sigma V \cup \tau V)$) for $\langle \sigma v = \tau v: v \in V \rangle_E$ for all $\sigma \in cU_1$ and $\tau \in cU_2$. Let $\Psi := \bigcup \{cU(p = q): p \neq q \in \Delta_1 \cup \Delta_2\}$, where $cU(p=q)$ are complete solution sets (on V) for the problems $\langle p = q \rangle_E$ for all $p \neq q$ in $\Delta_1 \cup \Delta_2$, and finally letet $\Theta_{\sigma\tau} := \bigcup \{cU(\sigma V = \psi V = \tau V): \psi \in \Psi\}$ for all $\sigma \in cU_1$ and $\tau \in cU_2$, where $cU(\sigma V = \tau V = \psi V)$ are complete solution sets for $\langle \sigma v = \psi v, \psi v = \tau v: v \in V \rangle_E$. Then the set $cU := \{(\delta\sigma)/_V - \Theta_{\sigma\tau}: \delta \in cU(\sigma V = \tau V), \sigma \in cU_1, \tau \in cU_2\}$ is a complete solution set for the system $\langle \Gamma_1, \Gamma_2, \Delta_1, \Delta_2 \rangle_E$. (Notice, that $\delta\sigma =_E \delta\tau [V]$).*

Proof: Again the proof follows with the Representation Theorem and the fact that $cU := \{(\delta\sigma)/_V: \delta \in cU(\sigma V = \tau V), \sigma \in cU_1, \tau \in cU_2\}$ is a complete solution set for $\langle \Gamma_1, \Gamma_2 \rangle_E$. ∎

Notice, that the symmetry of the Merging Theorem implies the independence of the solutions from the ordering of equations and disequations in the system.

We could also formulate a propagation theorem for the second form of substitutions with exceptions and a merging theorem for the first form. The first and second part of the Representation Theorem are special cases of propagation and merging, where Δ_1 and Γ_2 are empty. Additionally we could have some versions of merging or propagation analoguously to part 3 of the Representation Theorem. However, these cases are similar to the above two versions, but still more technical.

7. AC-Unification is AC1-Disunification

We want to demonstrate by an example how disunification can be used to prevent extensive splitting of solutions. Instead of solving equations in the theory AC of an associative and commutative function f, we can solve them in the theory $AC1$, that is AC with some unit 1, but with the constraints that the variables must not become equal to 1, that is, we solve them together with the disequations $x \neq 1$ for all the variables. When we avoid evaluation of the constraints $x \neq 1$ in these AC1-disunification problems as long as possible, we can avoid the exponential growth of the number of solutions. For example in theorem proving applications of E-unification (G. Plotkin 1972) the idea is to propagate these constraints into resolvents that are created with the substitutions solving the equation part only, but only when these substitutions are not instances of the collected exceptions (i.e. the substitution with collected(!) exceptions has still some instances).

Both AC- and AC1-unification problems can be solved by computing the minimal non-negative integer solutions of suitable corresponding linear Diophantine equations (M.Livesey & J. Siekmann 1976, M.E. Stickel 1976, 1987, J.M. Hullot 1980, F. Fages 1985, A. Fortenbacher 1985, C. Kirchner 1985, W. Büttner 1986, A. Herold & J. Siekmann 1986, A. Herold 1986). In the case of AC1-unification there is a one-to-one correspondence between the Diophantine solutions and the most general AC1-unifiers, while in the case of AC-unification we must in addition instantiate all subsets of the variables of these most general AC1-solutions with the unit 1 to obtain the most general AC-unifiers. By this post-process in general the number of AC-solutions is growing exponentially in the number of variables introduced by the AC1-solutions: The problem $\langle f(x, f(y, z)) = f(v, f(v, f(v, v)))\rangle_{AC1}$ (i.e. $\langle xyz = v^4\rangle_{AC1}$ in a more readable string notation) has one most general solution introducing 15 new variables, while the corresponding AC-problem has about 2^{15} (i.e. more than 32 000) most general AC-unifiers (H.-J. Bürckert et al. 1988). Hence it is convenient to avoid an explicit generation of all these most general AC-unifiers by representing them by AC1-unifiers. How this representation works is stated by the following theorem on the relationship between AC- and AC1-unification (A. Herold & J.H. Siekmann 1986, A. Herold 1987).

__Theorem:__ 1. A substitution σ solves the AC-unification problem $\langle s_i = t_i : 1 \leq i \leq n \rangle_{AC}$ iff σ solves the AC1-problem $\langle s_i = t_i : 1 \leq i \leq n \rangle_{AC1}$ and $\sigma x \neq_{AC1} 1$ for all variables $x \in V = \mathcal{V}(s_i, t_i : 1 \leq i \leq n)$.
2. Let μU_{AC1} be a minimal set of AC1-unifiers for $\langle s_i = t_i : 1 \leq i \leq n \rangle_{AC1}$ then the set
$$\mu U_{AC} := \{(v\sigma)/_V : v \in N_\sigma, \sigma \in \mu U_{AC1}, v\sigma x \neq 1 \; \forall x \in V\}$$
with $N_\sigma := \{\{w \leftarrow 1 : w \in W\} : W \subseteq VCOD\sigma\}$ is a minimal set of AC-unifiers for $\langle s_i = t_i : 1 \leq i \leq n \rangle_{AC}$.

Hence the AC-unification problem $\langle s_i = t_i : 1 \leq i \leq n \rangle_{AC}$ can equivalently be considered as an AC1-disunification problem $\langle s_i = t_i : 1 \leq i \leq n, x \neq 1 : x \in V \rangle_{AC1}$ with $V = \mathcal{V}(s_i, t_i : 1 \leq i \leq n)$. The solutions of a system of AC-equations now can be solved step by step, inheriting the solutions of already solved equations (by applying the substitutions) to the still unsolved ones and propagating the constraints that substituted terms must not be equal to 1.

8. Further Applications

There are several further applications of E-disunification: For example, logic programming (J.A. Goguen & J. Meseguer 1986, J. Jaffar et al. 1986, J.H. Gallier & S. Raatz 1986, H.-J. Bürckert 1986) as well as term rewriting (D.S. Lankford & R.M. Ballantyne 1977, G.E. Peterson & M.E. Stickel 1981, M.E. Stickel 1984, J.P. Jouannaud &

H. Kirchner 1984) are extended to unification or matching with respect to equational theories. The approach of A. Colmerauer - disunification as a constraint solving process for logic programming - is naturally generalized to the above extension. Also H. Comon's application for disunification to show sufficient completeness for algebraic specifications given by term rewriting systems might be extended this way (equational theories may be used to specify non-free datatypes).

There are also applications for resolution based theorem proving (C.-L. Chang & R.C.-T. Lee 1973, L. Wos et al. 1984), where disunification can be used to avoid redundancies in the search space. Consider for example the clause $\{P(x), \neg P(y), Q(x,y)\}$, where any resolution step with the third literal $Q(x,y)$ of this clause that identifies x and y leads to a resolvent that is a tautology. Hence this is an unnecessary step and we should only look for resolution candidates that do not identify these arguments: Unification of $Q(x,y)$ with some literal $\neg Q(s,t)$ of another clause can be done under the constraint $x \neq y$, i.e., we have the disunification problems $\langle x = s, y = t, x \neq y \rangle_{\emptyset}$. For theorem proving systems with built-in E-unification procedures this will result in E-disunification problems. Obviously one is not interested in substitutions as solutions for these problems, but in substitutions with exceptions for easy generation of instances of the substitutions that fulfill the constraints, i.e., the exceptions. Similar constraints can be formulated for other redundancy tests in theorem proving procedures, as for example the subsumption or the purity tests.

A rather similar application can be found in unification theory itself, namely for certain combination procedures. Some work on the combination of equational theories use the constant abstraction method for subterms (A. Herold 1987, M. Schmidt-Schauß 1987). Here alien subterms are replaced by new free constants, such that the unification algorithm for pure terms of the top theory can be applied. The abstraction constants have again to be replaced by the corresponding subterms, whereby some post-unification has to be done, since the subterms also might contain variables that are already used in the pure unification step. But since in further steps certain identifications of these abstracted subterms are necessary to retain completeness, one can do these post-unification steps under the constraints that the subterms need not to be identified; we again have E-disunification problems.

9. Conclusion

We gave the formal framework for solving equations and disequations under arbitrary equational theories. As solutions we considered substitutions with exceptions $\sigma - \Psi$ representing all substitutions for the variables of the disunification problems by being instances of σ, but not of the exceptions $\psi \in \Psi$. We reduced the computation of these substitutions with exceptions completely to solving equations and gave an algorithm based on this reduction.

We already mentioned some papers dealing with disunification in equational theories. Closely related to ours is the approach of W. Buntine (1986), who independently had a very similar result as the Representation Theorem and an analoguous E-unification algorithm based on this result. Generalizing the work of A. Colmerauer (1984) he starts with logic programs containing equality and he deals with closed world semantics obtained by *equality completion* of the programs, in order to handle disequations (cf. J. Jaffar et al. 1986). A slight disadvantage of this approach may be that for equality completion infinitely many axioms have to be introduced. Our algebraic approach of solving equations and disequations over the E-quotient of the free term algebra seems to be easier to grasp. However, he has a very elegant revised notion of instantiation for unifiers leading to a general notion of representation of solutions (cf. J.-L. Lassez et al. 1987, P. Lescanne & C. Kirchner 1987, G. Smolka et al. 1987).

H. Comon & P. Lescanne 1988 and H. Comon 1988 deal with parametrized disunification problems consisting of disjunctions of systems of equations and disequations. H. Comon & P. Lescanne give a set of rules defining a disunification algorithm for the above kind of problems in the case of the syntactical theory and the theory AC of associativity and commutativity. H. Comon has a complete disunification procedure for any theory with an existing unification algorithm, and a disunification algorithm for certain special theories he calls "quasi-free". He also has a disunification algorithm for parameterless problems in theories that are finitary with respect to unification. He shows that E-disunification is not even semi-decidable, when unification is undecidable in a theory E (H. Comon 1987/88). In both papers solved forms are used that differ from ours, and it is not quite clear how they compare.

Since the number of exceptions in the solutions of a disunification problem may be infinite and hence it cannot be tested whether such a substitution with exceptions represents an empty solution set in general, the question will arise, whether E-disunification can become undecidable, although E-unification is decidable. The following reduction, due to M. Schmidt-Schauß, shows that this may happen. Let

$$DA = \{f(x, f(y, z)) = f(f(x, y), z), f(g(x, y), z) = g(f(x, z), f(y, z)), f(x, g(y, z)) = g(f(x, y), f(x, z))\}$$

be the theory of distributivity and associativity of two binary functions f, g and some free constants, where unification is known to be undecidable (P. Szabo 1982, J. Siekmann & P. Szabo 1986). If we add a constant 0 with the axioms $f(x, 0) = f(0, x) = 0$, $g(x, 0) = g(0, x) = 0$ then unification in the resulting theory $DA0$ will become decidable again (notice, that DA-unification of two terms, where one of them is a ground term, is decidable). However, disunification is undecidable, since the DA0-disunification problems $\langle \Gamma, x \neq 0 : x \in \mathcal{V}(\Gamma) \rangle_{DA0}$, where no term of Γ contains the constant 0, are equivalent to the DA-unification problems $\langle \Gamma \rangle_{DA}$.

Here are some open questions to be considered in the future:
Can the substitutions with exceptions be resolved to standard substitutions, if we are interested in ground solutions only? That is, are there complete sets of substitutions representing all ground solutions of a given E-disunification problem?
This problem is only solved for the empty theory (H. Comon 1986, C. Kirchner & P. Lescanne 1987, J.-L. Lassez et al. 1987).
Are there decision procedures for testing "ground consistency" of a substitution with exceptions, i.e., is it possible to decide, if a substitution with exceptions has some ground instances?
These questions make only sense in the case of finite complete representation sets. Then they may be useful for example to prove sufficient completeness of algebraic specifications with non-free datatypes specified by E.
Can the extended completion procedures for term rewriting systems modulo equational theories be adapted to use disunification to avoid explosion of candidates for critical pairs?
This should especially be investigated for the case of term rewriting systems modulo AC.
Is there at least a semi-decision procedure for E-disunification, when E-unification is decidable?
This problem was mentionned by H. Comon (1987/88). He also has some results to our last question:
What happens, if we allow alternatives of systems of equations and disequations and/or parameters?

Acknowledgements

I gratefully acknowledge many useful discussions with W. Nutt, M. Schmidt-Schauß, and G. Smolka on the topics of this paper. The discussion with H. Comon helped to correct some mistakes in a former draft of this paper. I also want to emphasize once more the contribution of M. Schmidt-Schauß, who had the above idea for the reduction of undecidability of disunification.

The research in this paper was partially supported by the German Bundesministerium für Forschung und Technologie and the Nixdorf Computer AG under contract of the Joint Research Project ITR8501A, and by the Deutsche Forschungsgemeinschaft in the frame of the Special Research Project SFB 314.

References

W. Buntine: *A Theory of Equations, Inequations, and Solutions for Logic Programming.* New South Wales Institute of Technology, 1986.
H.-J. Bürckert: *Lazy Theory Unification in PROLOG: An Extension of the Warren Abstract Machine.* Proc. of 10th German Workshop on Art. Intelligence, Springer, 1986, p. 277-288.
H.-J. Bürckert: *Matching - A Special Case of Unification?* SEKI-Report, Universität Kaiserslautern, 1987.
H.-J. Bürckert: *Solving Disequations in Equational Theories.* SEKI-Report, Universität Kaiserslautern, 1987.
H.-J. Bürckert, A. Herold & M. Schmidt-Schauß: *On Equational Theories, Unification, and Decidability.* Proc. of 2nd Conf. on Rewriting Techniques and Applications, Springer, LNCS 256, 1987, p. 204-215; to appear also in J. of Symb. Comp., Special Issue on Unification (ed. C. Kirchner), 1987.

H.-J. Bürckert, A. Herold, J. Siekmann, M. Stickel & M. Tepp: *Opening the AC-Unification Race*. In preparation, 1988

W. Büttner: *Unification in the Datastructure Multisets*. J. of Automated Reasoning, Vol. 2, No. 1, 1986, p. 75-88.

S. Burris, H.P. Sankappanavar: *A Course in Universal Algebra*. Springer, 1979.

C.-L. Chang & R.C.-T. Lee: *Symbolic Logic and Theorem Proving*. Academic Press, 1973.

A. Colmerauer: *Equations and Inequations on Finite and Infinite Trees*. Proc. of Conf. on Fifth Gen. Comp. Syst., ICOT, 1984, p. 85-99.

H. Comon: *Sufficient Completeness, Term Rewriting Systems, and Anti-unification*. Proc. of Conf. on Automated Deduction, Springer LNCS 230, 1986, p. 128-140.

H. Comon: *Private communications*. 1987/88.

H. Comon: *Unification et Disunification. Théorie et Applications*. Thesis (in French), Université de Grenoble, 1988.

H. Comon & P. Lescanne: *Equational Problems and Disunification*. Université de Grenoble and Centre de Rech.en Inform. de Nancy, 1988.

F. Fages: *Associative-Commutative Unification*. Proc. of 7th Conf. on Automated Deduction, Springer, LNCS 170, 1984, p. 194-208.

F. Fages & G. Huet: *Complete Sets of Unifiers and Matchers in Equational Theories*. Proc. of CAAP'83, Springer, LNCS 159, 1983, p. 205-220; see also J. of Theoret. Comp. Sci. 43, 1986, p. 189-200.

A. Fortenbacher: *An Algebraic Approach to Unification under Associativity and Commutativity*. Proc. of Conf. on Rewriting Techniques and Applications, Springer, LNCS 202, 1985, p. 381-397.

J. Gallier & S. Raatz: *SLD-Resolution Methods for Horn Clauses with Equality Based on E-Unification*. Proc. of Int. Symp. on Logic Programming, 1986

J.A. Goguen & J. Meseguer: *EQLOG - Equality, Types, and Generic Modules for Logic Programming*. In: *Logic Programming: Functions, Relations, and Equations*. Prentice Hall, 1986, p. 295-363.

G. Grätzer: *Universal Algebra*. Springer, 1979.

A. Herold: *Combination of Unification Algorithms*. Proc. of 8th Conf. on Automated Deduction, Springer, LNCS 230, 1986.

A. Herold: *Combination of Unification Algorithms in Equational Theories*. Dissertation, Universität Kaiserslautern, 1987.

A. Herold & J.H. Siekmann: *Unification in Abelian Semigroups*. MEMO-SEKI, Universität Kaiserslautern, 1986.

A. Herold, J.H. Siekmann & M.E. Stickel: *Benchmarks for AC-Unification*. Private communication, 1987.

G. Huet & D.C. Oppen: *Equations and Rewrite Rules: A Survey*. In: *Formal Languages: Perspectives and Open Problems*.(ed. R. Book), Academic Press, 1980.

J.M. Hullot: *Compilation des Formes Canoniques dans des Théories Equationelles*. Thèse (in French), Université de Paris-Sud, 1980

J. Jaffar, J.-L. Lassez & M. Maher: *Logic Programming Language Scheme*. In: *Logic Programming: Functions, Relations, Equations*. (eds. D. DeGroot, G. Lindstrom), Prentice Hall, 1986.

J.P. Jouannaud & H. Kirchner: *Completion of a Set of Rules Modulo a Set of Equations*. Proc. of 11th ACM Conf. on Principles of Programming Languages, 1984.

C. Kirchner: *Methodes et Outils de Conception Systematique d'Algorithmes d'Unification dans les Théories Equationelles*. Thèse de Doctorat d'Etat (in French), Université de Nancy, 1985.

C. Kirchner & H. Kirchenr. *Implementation of a General Completion Procedure Parametrized by Built-in Theories and Strategies*. Proc. of EUROCAL Conf., 1985.

C. Kirchner & P. Lescanne: *Solving Disequations*. Proc. IEEE 2nd Symp. on Logic in Comp. Sci., 1987.

D. Lankford & R.M. Ballantyne: *Decision Procedures for Simple Equational Theories with Commutative-Associative Axioms: Complete Sets of Commutative-Associative Reductions*. Internal Report, University of Texas, Austin, 1977.

J.-L. Lassez, M.J. Maher & K. Marriot: *Unification Revisited*. Technical Report, IBM Yorktown Heights, 1987.

M. Livesey & J.H. Siekmann: *Unification of AC-Terms (Bags) and ACI-Terms (Sets)*. Int. Report, Essex University, 1975, and Universität Karlsruhe, 1976.

H.J. Ohlbach: *Link Inheritance in Abstract Clause Graphs*. J. of Automated Reasoning, Vol 3, No 1, 1987, p. 1-34.

G.E. Peterson & M.E. Stickel: *Complete Sets of Reductions for Equational Theories with Complete Unification Algorithms*. JACM, Vol 28, No 2, 1981, p. 322-364.

G. Plotkin: *Building in Equational Theories*. Machine Intelligence 7, 1972, p. 73-90.

J.A. Robinson: *A Machine Oriented Logic Based on the Resolution Principle*. JACM, Vol 12, No 1, 1965, p. 23-41.

M. Schmidt-Schauß: *Combination of Unification Algorithms in Arbitrary Disjoint Equational Theories*. SEKI-Report, Universität Kaiserslautern, 1987, also in this proceedings.

J.H. Siekmann: *Unification and Matching Problems*. PH.D. Thesis, Essex University, 1978.

J.H. Siekmann: *Unification Theory. A Survey*. J. of Symb. Comp., Special Issue on Unification (ed. C. Kirchner), 1987.

J.H. Siekmann & P. Szabo: *The Undecidability of the DA-unification Problem*. SEKI-Report SR-86-19, Universität Kaiserslautern, 1986.

G. Smolka, W. Nutt, J.A. Goguen & J. Meseguer: *Order-Sorted Equational Computation*. SEKI-Report, Universität Kaiserslautern, 1987.

M.E. Stickel: *A Complete Unification Algorithm for Associative-Commutative Functions*. Proc. of 4th Int. Joint Conf. on Art. Intelligence, Tblisi, 1975, p. 71-82.

M.E. Stickel: *Unification Algorithms for Artificial Intelligence*. Ph. D. Thesis, Carnegie-Mellon University, 1976.

M.E. Stickel: *A Unification Algorithm for Associative-Commutative Functions*. JACM, Vol 28, No 3, 1981, p. 423-434.

M.E. Stickel: *A Case Study of Theorem Proving by the Knuth-Bendix Method Discovering that $X^3 = X$ implies Ring Commutativity*. Proc. of 7th Conf. on Automated Deduction, Springer, LNCS 170, 1984, p. 248-258.

M.E. Stickel: *A Comparison of the Variable-Abstraction and Constant-Abstraction Methods for Associative- Commutative Unification*. J. of Automated Reasoning, Vol 3, No 3, 1987, p. 285-289.

P. Szabo: *Unifikationstheorie erster Ordnung*. (In German), Dissertation, Universität Karlsruhe, 1982.

W. Taylor: *Equational Logic*. Houston Journal of Mathematics 5, 1979.

L. Wos, R. Overbeek, E. Lusk, J. Boyle: *Automated Reasoning - Introduction and Applications*. Prentice Hall, 1984.

ON WORD PROBLEMS IN HORN THEORIES

Emmanuel Kounalis   *Michael Rusinowitch*

LRI   CRIN
Universite de Paris-Sud   Campus Scientifique BP 239
91405 Orsay Cedex   54506 Vandoeuvre les Nancy
FRANCE   FRANCE
kounalis@lri.lri.fr   rusi@crin.crin.fr

ABSTRACT

We interpret Horn clauses as conditional rewrite rules. Then we give sufficient conditions so that the word problem can be decided by conditional normalization in some Horn theory. We also show how to prove theorems in the initial models of Horn theories.

keywords: Horn clause, resolution, term-rewriting system, word problems, initial model, inductionless induction.

INTRODUCTION

Horn Logic, a restriction of first order logic, has provided a most useful logical basis for many applications in Computer Science: expert systems, data-bases, algebraic specifications, logic programming, formal calculus... The Prolog language has been designed in this powerful framework. Some attempts has been made to build in a Prolog-like language the important relation of equality: EQLOG (Goguen Meseguer 84) where equations can be added to the Horn Clauses, SLOG (Fribourg 85), both of them are not complete without serious restrictions on the input programs. These approaches make use of equations as rewrite rules in order to simplify goals. Our method is based on a complete refutational strategy, which handle an equation as a rewrite-rule when it is orientable, and when its pre-conditions of application are smaller than the equation for some well chosen ordering, which ensures that it can be recursively checked. This technique is more flexible than the conditional-term rewriting systems and avoids its drawbacks, since

1. it does not fail when an equation is not orientable,

2. it ignores rules with preconditions bigger than the conclusion.

Our procedure has been implemented, and we have obtained some complete sets of conditional rules that no other system could previously derive. (see example 2 and appendix).

Moreover Horn clauses set can be provided with an initial model where a fact is true only if it can be proved with the first-order logic classical deduction rules. This model is the one of interest in data-bases (closed-world assumption (see Reiter 78)), and abstract data-types (initial algebra (Remy 82)(Padawitz 85)). In the following we propose to extend methods of (Jouannaud Kounalis 86) for proving theorems by induction in equational theories to Horn theories.

1. PRELIMINARIES AND NOTATIONS

In this section we review standard concepts and notation. Let F be a set of function symbols graded by an arity function. Let X be a set of variables. The algebra of terms on F and X is denoted by $T(F,X)$. We call $T(F)$ the set of ground terms on F, which is the set of terms where no variable occurs. Let P be a set of predicate (or relation) symbols. The equality symbol "=" is a particular element of P whose arity is 2. The set of atoms is denoted by $A(P,F,X)$, and the set of ground atoms (i.e. the Herbrand Base) by $A(P,F)$. A set of formulae is E-satisfiable if it admits a model where "=" is interpreted as equality. Otherwise it is E-unsatisfiable. The entailment relation will be denoted by $\not\models$. Hence, $S \not\models C$ means that every equality model of S is a model of C. A substitution is a mapping σ from X to $T(F,X)$ with $\sigma(x)=x$ but for a finite set of variables. The subset of elements x where $\sigma(x) \neq x$ is the domain of σ and is denoted by $Dom(\sigma)$. The result of applying a substitution σ to an object t is denoted by $t\sigma$. The set of variables occurring within an object t is represented by $V(t)$.

2. A COMPLETE SET OF INFERENCE RULES FOR HORN LOGIC WITH EQUALITY

A long-standing problem in theorem proving is to efficiently handle the equality relation. Attempting to solve it, Wos & Robinson proposed to build equality within new inference rules such as "paramodulation". In the mean time, Knuth & Bendix designed for the purely equational case, a completion procedure which, given a set of equations, orient them, simplify them by each other, solves ambiguities and, if it eventually stops, returns a canonical system. In this case, the word problem (w.r.t. the input theory) has a straightforward solution: terms are equal iff they reduce to the same irreducible form when fully simplified by the final set of oriented equations.

It has been proved in (Hsiang Rusinowitch 87) that Knuth & Bendix algorithm (Knuth Bendix 70) uses a refutationally complete set of inference rules (for equational logic) to generate all the consequences of the initial set of equations. The completion procedure can be interpreted as an attempt to saturate a set of equations by inference rules, in hope to derive a finite set which has no non-trivial consequence. The nice feature of this approach is that it can be easily generalized to Horn Clauses. Starting from a set of Horn Clauses, and from a refutationally complete system of inference rules, we iterate these rules on the initial set. If the procedure halts, it provides a very efficient way to solve word problems by what can be called "conditional normalization", which is a combination of backward chaining (trying to check pre-conditions before using a rule for simplification) and forward chaining (releasing a rule for simplification). Horn Clause heads are used as simplifiers, whose pre-conditions are gathered within the negative literals. Results of ours naturally meet Conditional Term Rewriting Systems (Remy 82) (Kaplan 87) (Jouannaud Waldmann 86) (Ganzinger 87). However, as UKB compared to KB, our procedure offers more flexibility: non-orientable equations do not cause the procedure to fail necessarily; moreover, the pre-conditions of a rule need neither to be smaller than the rule itself, nor to belong to another theory, as it is the case for some aforementionned works.

Our set of inference rules, denoted by INF, is merely a restriction to Horn clauses of the inference system given in (Lankford 75). The main feature of this system is that any inference step always involves the maximal literals of the parent clauses, where the maximality notion is defined relatively to a *complete simplification ordering* on Herbrand Universe (see (Peterson 83) (Hsiang Rusinowitch 87)) which will be denoted by >. We recall the definition of such an ordering:

2.1. COMPLETE SIMPLIFICATION ORDERINGS

A complete simplification ordering < is an ordering on $A(P,F,X) \cup T(F,X)$ havinwith the following properties:

O1. *< is well founded*

O2. *< is total on $A(P,F) \cup T(F)$*

O3. *for every $w,v \in A(P,F,X) \cup T(F,X)$ and every substitution $\theta : w < v$ implies $w\theta < v\theta$*

O4. *for every $t,s \in T(F)$, $t<s$ implies $w[o \leftarrow -t] < w[o \leftarrow -s]$*

O5. *for every $t,s,a,b \in T(F)$, with $t \leq s$, $b \leq a$ and $w \in A(P,F)$*

      1. if s is a subterm of w and w is not an equality then $(s=t) < w$.

      2. if s is a strict subterm of a or b then $(s=t) < (a=b)$

O6. *if $(u=w) < A < (u=v)$, $u>w$ and $u>v$, where u,v and w are ground terms, and A a ground atom then there is a ground term t such that A is equal to the atom $(u=t)$.*

We write $s \rightarrow t$ instead of $s=t$ whence $s>t$. Examples of complete simplification orderings are given in (Peterson 83)(Hsiang Rusinowitch 86).

FACTORING

    If $L_1, L_2, ..., L_k$ are literals of a clause C which are unifiable with mgu θ, and for every atom $A \in C-\{L_1,...,L_k\}$, $L_1\theta \nleq A\theta$, then the clause $C\theta- (L_2\theta \ V...V \ L_k\theta)$ is a factor of C.

RESOLUTION

    If $C_1 = L_1 \ V \ C_1'$ and $C_2 = L_2 \ V \ C_2'$ are clauses such that L_1 and $\neg L_2$ are unifiable with mgu θ and

1. $\forall\ A \in C_1'$, $L_1\theta \not\leq A\theta$ and $\forall\ A \in C_2'$, $L_2\theta \not\leq A\theta$

2. if L_1 is an equality literal then C_2 is x=x

then $C_1'\theta\ V\ C_2'\theta$ is a resolvent of C_1 and C_2

PARAMODULATION

Let C_1 be a clause (s=t) V C_1' . Let C_2 be another clause which has a non-variable subterm s' at occurrence n in a literal L_2, which is unifiable with s with mgu θ. We suppose that $s\theta \not\leq t\theta$, and

1.$\forall\ A \in C_1'$, $(s=t)\theta \not\leq A\theta$ and $\forall\ A \in C_2\text{-}\{L_2\}$, $L_2\theta \not\leq A\theta$

2.if L_2 is an equation, s' occurs in the largest member of L_2.

then, ($C_2[n\leftarrow t]\ V\ C_1'$)$\theta$ is a paramodulant of C_1 into C_2 at n.

2.2. REMARKS ON THE RESTRICTIONS OF THE RULES

Let us emphasize that the reflexive functional axioms are not needed, and that paramodulation is never performed into variables: these two conditions ensures efficiency of the inference system. Let us consider the clause $\neg$ s(a)=s(b) V a=b. We are not allowed to use it as an active clause for paramodulation, since the head (positive literal) is smaller than the conditional part(negative literals). The intuitive reason of such a restriction is that we want to bring back the problem of checking an equality to less complex subproblems. On the contrary our paramodulation rule favours use of $\neg$ a=b V s(a)=s(b) since it decreases the "size" of problems.

Now we present a set of deletion rules, which are fundamental for efficiency. Moreover, for many E-satisfiable systems, the deletion inference rules prevent the generation of infinitely many new clauses.

PROPER_SUBSUMPTION: The clause C_1 properly subsumes C_2 if C_1 subsumes C_2 and C_2 does not subsume C_1. The subsumption rule states that one may delete from S any clause which is properly subsumed by another clause in S.

SIMPLIFICATION: If the unit equation s=t is in S and $C_2[s\theta]$ is a clause in S which contains an instance $s\theta$ of s, and $s\theta > t\theta$, and there is an atom A in $C_2[s\theta]$ such that A> ($s\theta$=$t\theta$), then the clause $C_2[t\theta]$ is a simplification of $C_2[s\theta]$ by s=t. The simplification rule states that one may replace a clause in S by the same one after it has been simplified.

TAUTOLOGY_DELETION: The tautology deletion rule states that we can delete the tautologies.

CLAUSAL_SIMPLIFICATION: If the unit literal L is in S, then we can replace any clause in S which contains a negated instance of L, by the same clause where this instance has been deleted.

2.3. FAIRNESS HYPOTHESIS

As in Knuth-Bendix algorithm, in order to get a (refutationally) complete strategy, we need some fairness assumption to ensure that no crucial inference will be postponed forever. Given an initial set of clauses S_0, the derivation $S_0 \to S_1 \to, ..., \to S_i \to$, where S_i is obtained from S_{i-1} by one inference step, is **fair** if : $\forall j$ { $R \in \cap_{i \geq j} RP(S_i)$ *implies that R is subsumed by a clause in* $\cup_{i \geq 0} S_i$}, where RP(T) denotes the whole set of resolvents, factors and paramodulants which can be infered (in one step) from a set of clauses T. From now on we suppose that INF is submitted to a scheduler which let only fair derivations be generated. We say that a clause is normalized by S when it is fully simplified by the unit oriented equations available in S. We also suppose that before adding a new clause, it is always immediately normalized and checked whether it is subsumed by a clause in S.

2.4. THEOREM

The set INF is refutationally complete for Horn logic (with equality). To be more precise: any fair derivation of an E-unsatisfiable set of clauses containing x=x, yields the empty clause.

The proof uses an extension of the semantic tree methods. It is fully detailed in (Rusinowitch 87).

3. SATURATED HORN SETS

3.1. DEFINITION: SATURATED SET

A set of Horn Clauses S is **saturated** for INF if S is E-satisfiable and if no new clause can be infered from S. To be more precise, every clause in RP(S) either is a tautology or a clause which is subsumed by some element of S. Saturation can be interpreted as a generalization of the notion of local confluence (Huet 80) for term-rewriting systems. The **saturation algorithm** consists in applying, iteratively, the inference rules of INF to some input set S.

EXAMPLE 1. The following set of clauses is saturated:

  1. $P(0)$ 2. $\neg P(x)$ or $P(s(x))$

If we apply resolution without any restriction, an infinite set of clauses is generated: $P(s(0))$, $P(s(s(0)))$,...,$P(s^i(0))$,... Now, if we apply the INF system of rules, there is no resolvent between clauses 1. and 2., because $P(x) < P(s(x))$ in any complete simplification ordering.

EXAMPLE 2. Let us consider a clausal specification of the inf_or_equal predicate (i) in the set of integers (s and p are the successor and predecessor functions) (Kaplan 87):

0. $x = x$.   5. $\neg i(o,x) = t$ or $i(o,s(x)) = t$.

1. $s(p(x)) = x$.   6. $\neg i(o,x) = f$ or $i(o,p(x)) = f$.

2. $p(s(x)) = x$.   7. $i(s(x),y) = i(x, p(y))$.

3. $i(o,o) = t$.   8. $i(p(x),y) = i(x, s(y))$.

4. $i(o,p(o)) = f.$ 9. $\neg t = f.$

When the saturation algorithm is applied to 0..9 with the lexicographic RPO (Dershowitz 87) and the following precedence :$i>p>s>o>t>f$, it stops after deriving two new clauses:

10. $\neg i(o,s(x)) = f$ or $i(o,x) = f.$ by o-paramodulation of 2. and 6.

11. $\neg i(o,p(x)) = t$ or $i(o,x) = t.$ by o-paramodulation of 1. and 5.

Let us notice that the conditional term rewriting approach cannot handle rules like 10. and 11. since the pre-conditions are bigger than the conclusions. Another important feature of our method is the use of negative assumptions such as 9. to avoid divergence of the saturation process. Without 9., infinitely many clauses would be generated; however, in presence of 9., these clauses are immediately subsumed. The clause $\neg t = f$ is an inductive theorem of the specification 1,...,8. Adding inductive theorems to a specification, can be viewed as a way to approximate the initial model by first-order formulas. This remark has also been exploited in (Ganzinger 86).

Any canonical term rewriting system whose equations are oriented by a complete simplification ordering is saturated by INF.

3.2. DEFINITION: INPUT DERIVATION

An **input derivation** D, from a set of clauses S, is a sequence of clauses $(C_1,...,C_n)$ such that $C_1 \in S$ and each C_{i+1} is either a factor of C_i or a resolvent of C_i and a clause of S, or a paramodulant of a clause of S **into** C_i. C_1 is called the top clause. If C_n is the empty clause then D is an input refutation of S.

3.3. THEOREM

Let C be a negative clause and S be a saturated Horn set (containing x=x). Then $S \cup \{C\}$ is E-unsatisfiable iff it admits an input refutation with top clause C.

4. OPERATIONAL SEMANTICS: HORN CLAUSES AS REWRITE RULES

From a formal point of view, the restrictive format of paramodulation and resolution that we are using can be interpreted in terms of conditional term rewriting operations. However, we apply a clause to a ground term as a conditional rule only if the clause has a condition which is smaller than the conclusion. For instance, the clauses 10. and 11. in Example 2 will never be used as rewrite rules. In order to get decidable rewriting relations, we shall restrict our study to "ground preserving" sets of Horn Clauses:

4.1. DEFINITION:

A Horn set S is **ground preserving** if (i): every positive equational literal s=t which belongs to some clause of S, either is orientable or satisfies V(s)=V(t). (ii): for every clause C of S, the variables appearing in the negative literals also appear in the positive literal.

4.2. DEFINITION OF A CONDITIONAL TERM REWRITING RELATION

Let S be a Horn set, A and B two ground terms or literals. Let $\neg C \vee s=t$ be an element of S, where C is a conjunction of positive literals. Then

A → B with $\neg C \vee$ s=t in the context S

if there is a substitution θ such that: $A = A[s\theta]$, $B = A[t\theta]$, $s\theta > t\theta$, $S \not\models C\theta$ and $C\theta < (s\theta = t\theta)$.

Let A be a literal and $\neg C \vee A'$ be an element of S, where C is a conjunction of positive literals and A' is positive. Then

A → TRUE with $\neg C \vee$ A' in the context S

if there is a substitution θ such that $A = A'\theta$, $S \not\models C\theta$ and $C\theta < A'\theta$.

We write **A -S→B** to express that A→B with some clause of S (in the context of the theory S). The reflexive transitive closure of -S→ is denoted by -S→*. Notice that, due to the noetherian property of >, there is no infinite chain of -S→ rewriting. Our next result states that the truth of a fact can be decided by "-S→ rewriting":

4.3. PROPOSITION: Let A be a ground positive literal and S a ground preserving saturated Horn set (containing x=x). Then S $\not\models$ A iff A -S→* TRUE.

The proof uses noetherian induction on A w.r.t. <.

Since the only way to rewrite an equation to TRUE is by using the clause (x=x) of S, we have immediately, under the same hypothesis, a Church-Rosser property:

4.4. COROLLARY: If S $\not\models$ s=t, then there is a term β such that s -S→* β *←S- t.

4.5. COROLLARY: The relation -S→ has the ground confluence property: if s *←S- t -S→* u then there exists a term v such that s -S→* v *←S- u.

It is decidable whether a ground term is reducible by -S→ when S is a ground-preserving saturated Horn set. Moreover, every ground term admits a unique normal form w.r.t. the rewriting relation -S→.

EXAMPLE 2 (continued): the system 0,...,11 is saturated and ground preserving. Moreover, the rules

10 and 11 (and 0 !) are useless for the rewriting relation we defined. Therefore we have proved that 1,..,8 are ground confluent. This is the first (automatized) proof of this fact. This final system of rules can now be used to compute by normalization the inf_or_equal relation. For instance:

i(s(o),p(s(p(o))))→i(s(o),p(o))→i(o,p(p(o)))→f by applying successively rules 1. 7. and 6. For the last step we have to check recursively that i(o,p(o))=f.

5. WORD PROBLEMS

We shall suppose now that, when we are extending the set of symbols F, it is possible to extend "<" as a complete simplification ordering on the new Herbrand universe (it works with most of the simplification orderings (Hsiang Rusinowitch 87)). Let us notice that S $\not\models$ s=t is equivalent to S $\not\models$ (s=t)ψ where ψ is a substitution applying every variable of s=t to a new constant. This remark shows that when a theory is presented by a Horn set S which is ground preserving and saturated, the equality of two terms s and t can be checked by testing the identity of the -S→ normal forms of the ground terms sψ and tψ. Therefore:

5.1. THEOREM: The word problem is decidable in theories axiomatized by ground-preserving saturated Horn sets.

We give in (Kounalis Rusinowitch 87) two examples of ground convergent systems, which have been obtained with our implementation of the saturation algorithm.

6. PROOFS IN THE INITIAL MODEL

The nice feature of Horn sets is that they admit an initial model which is the most interesting one in many applications: applicative programming languages, abstract interpreter definitions, algebraic data-type specifications, data-bases etc.. However, a fact cannot, in general, be proved in the initial model by mere inference reasoning via INF: some kind of induction is necessary. We show in this section how to make proofs in the initial model by a very simple extension of the our inference system and how the method applies to proofs of simple properties of arithmetic theory. We feel that these methods, in a long run, will challenge succesfully the alternative method of Boyer and Moore. We first consider the case where the equality predicate does not occur in S, then consider the general case.

6.1. HORN SETS WITHOUT EQUALITY

6.1.1. Definition: Given a horn set S (without equality), a clause C is inductively reducible w.r.t. S iff for any ground substitution θ: either there exists a positive literal P in C such that Pθ is reducible w.r.t. S, or there exists a negative literal ¬N in C such that Nθ is irreducible w.r.t. S.

Example: Let S be the set {P(0), ⌐P(x) or P(s(x))}. Then the atom P(x) is inductively reducible w.r.t. S. The key theorem to the method is the following:

6.1.2. Theorem: Let S be a ground-preserving saturated Horn set (without equality). Then a clause C is valid in the initial model of S iff C is inductively reducible w.r.t. S.

Example: Let S be the saturated set { Eq(x,x), i(0,s(x)), ⌐i(x,y) or i(s(x),s(y))} where i is intended to be the inf relation on natural numbers. Then the following statements are valid in the initial model of S, since they are inductively reducible: i(x,s(x)), ⌐i(x,x), i(x,y) or i(y,x) or Eq(x,y) (trichotomy law). Note that the last statement is not a Horn clause.

6.2. HORN SETS WITH EQUALITY

We shall now extend the previous method to deal with equality. The method is based on the following remark: the ground normal forms related to the relation $-S\to$ are not modified when we add to S a clause which is valid in its initial model.

6.2.1. Definition: Given a Horn set S, a positive non equational literal A (resp. an equation s = t with s > t) is inductively reducible w.r.t. S iff all ground instances of A (resp of s) are reducible w.r.t. S.

6.2.2. Lemma (Jouannaud Kounalis 1986): Let S be a ground-preserving Horn set and let us assume that the positive literal A is inductively reducible w.r.t. S. Then a ground term or a ground atom is in normal form for S iff it is in normal form for S $\cup$ {A}.

In words, inductive reducibility of an atom A ensures that no normal form becomes reducible, once this atom has been added to the set of clauses. But, this is not enough for A to be valid in the initial model of S, because two different normal forms could get equal in the new set of clauses. This is avoided by considering sets of clauses which have a unique normal form property: we have seen in section 4 that saturated sets have this property:

6.2.3. Theorem: Let S be a ground-preserving saturated Horn set and let A be a positive literal. Let us assume that S $\cup$ {A} is saturated, too. Then A is valid in the initial model of S iff A is inductively reducible w.r.t. S.

So in order to prove (or disprove) theorems in the standard model defined by a Horn set we start from a saturated Horn set; then we add the atom to be proved and apply the set of inference rules INF to the augmented Horn set. Before any normalized atom is added, it is checked for inductive reducibility. This last result also shows a kind of semi-completeness of our system: Provided we have a suitable ordering >, and a Horn set in which inductive reducibility is decidable, we shall disprove in a finite time every atom which is not valid in the initial model.

6.3. CHECKING FOR INDUCTIVE REDUCIBILITY

Inductive reducibility is in general undecidable. We have obtained some interesting decidable sub-cases by enforcing some restrictions on the structure of the horn sets. The algorithm we propose for horn sets without equality uses the notion of "test set" as it was defined in the equational case (Jouan-naud Kounalis 86): checking for the inductive reducibility of an object amounts to test the reducibility of a finite number of ground instances of this object. The ground instances to be considered have a depth which is determined by the depth of the literals of S. This is detailed in (Kounalis Rusinowitch 88).

Example: Let S be the set { P(0) , ¬P(x) or P(s(x)) }. Then for the atom P(x) is we take as test set {P(0), P(s(0)). P(s(s(0))), P(s(s(s(0)))) }. We can easily verify that every atom in the test set is reducible and thus P(x) is inductively reducible w.r.t. S.

7. CONCLUSION

We have presented a very natural and effective interpretation of Horn clauses as rewrite rules. We have described a completion algorithm to build convergent sets which allows to decide word problems by conditional normalization. Moreover, we have shown how to prove theorems in the initial model of a Horn set, by extending the equational notion of inductive reducibility. Our approach can be improved in several ways: first we think that the saturation algorithm would succeed more often if we could incorporate the axioms for commutativity and associativity in the unification algorithm, as suggested by Plotkin. Moreover, the generation of many clauses could be avoided by using a notion of critical pair criteria as in Kuchlin. Last, the test for inductive reducibility should be improved to apply to a larger class of Horn sets.

8. REFERENCES

FRIBOURG L. SLOG: A logic programming language interpreter based on clausal superposition and rewriting. Proc. of the 1985 Symposium on Logic Programming. Boston, MA pp. 172-184, july 1985.

DERSHOWITZ. N Termination. In : Rewriting Techniques and Applications, J.P. Jouannaud, ed., Lect. Notes in Comp. Sci., vol.202, Springer, 180-224, 1985

GANZINGER. H Ground Term Confluence in Parametric Conditional Equational Specifications, Proceeding of 4th Symposium on Theoretical Aspects of Computer Science Passau, RFA, February 1987

GOGUEN J. , MESEGUER J. Eqlog: Equality, Types, and Generic Modules for Logic Programming. J. of Logic Programming, Vol.1, Number 2, pages 179-210, 1984.

HSIANG. J., RUSINOWITCH M. On word problems in equational theories. ICALP 1987

JOUANNAUD J. P. , KOUNALIS E. Automatic proofs by induction in equational theories without constructors. Proc. of the Symposim on Logic in Computer Science, Cambridge, MA, pp 358-366, June 1986.

JOUANNAUD. J.P, WALDMANN. B Reductive Conditional Term Rewriting Systems, Proc. 3rd IFIP Conf. on Formal Description of Programming Concepts Lyngby, Danemark, 1986

HUET G., Confluent reduction: abstract properties and applications to term-rewriting systems. JACM 27, 797-821. (1980).

KAPLAN S. Simplifying Conditional Term Rewriting Systems: Unification, Termination Confluence, to appear in J. of Symb Comp.

KNUTH. D, BENDIX. P Simple Word Problems in Abstract Algebra, Leech J.ed, Pergamon Press, pp. 263-297, 1970

KOUNALIS E., RUSINOWITCH M., On words problems in Horn theories, to appear as rapport INRIA. Nancy (1988).

KUCHLIN , W. A confluence criterion based on the generalised Newman lemma. proc. of EUROCAL, B.Caviness, LNCS 204, (1985) pp. 390-399.

LANKFORD D. S. Canonical inference. Memo ATP-32, Dept. of Math. and Comp. Sc., University of Texas, Austin, Texas, 1975.

PADAWITZ P. Horn Clause specifications: a uniform framework for abstract data types and logic programming. Universitat Passau, MIP-8516 December 1985.

PLOTKIN, G. Building-in equational theories, Machine Intelligence 7, B.Meltzer and D.Mitchie,eds, American Elsevier, New-York (1972) pp. 73-90.

REITER R. On closed world databases, in "Logic and databases. eds Gallaire H. , Minker J. Plenum Press, New York 55-76(1978).

REMY. J.L Etude des syste'mes de re'e'criture conditionnelle et applications aux types abstraits alge'briques, The'se d'Etat, I.N.P.L, Nancy, 1982

RUSINOWITCH. M, Demonstration automatique par des techniques de Re'e'criture. Thesis. Universite' de Nancy 1, 1987.

Canonical Conditional Rewrite Systems[*]

Nachum Dershowitz[1], *Mitsuhiro Okada*[2], *and G. Sivakumar*[1]

[1]Department of Computer Science
University of Illinois at Urbana–Champaign
Urbana, Illinois 61801, U.S.A.

[2]Department of Computer Science
Concordia University
Montreal, Quebec H3G 1M8, Canada

Abstract

Conditional equations have been studied for their use in the specification of abstract data types and as a computational paradigm that combines logic and function programming in a clean way. In this paper we examine different formulations of conditional equations as rewrite systems, compare their expressive power and give sufficient conditions for rewrite systems to have the "confluence" property. We then examine a restriction of these systems using a "decreasing" ordering. With this restriction, most of the basic notions (like rewriting and computing normal forms) are decidable, the "critical pair" lemma holds, and some formulations preserve canonicity.

1. Introduction

Conditional rewriting systems arise naturally in the algebraic specification of data types and have been studied largely from this perspective [Remy–82, Kaplan–84, Bergstra–Klop–82]. See also [Brand–Darringer–Joyner–78]. With differing restrictions on left–hand sides and conditions, useful results have been obtained about the confluence of such systems. More recently, conditional rewriting systems have been shown to provide a natural computational paradigm combining logic and functional programming [Dershowitz–Plaisted–85, Fribourg–85, Goguen–Meseguer–86]. A program is a set of conditional rules and a computation is the process of finding a substitution that makes two terms equal in the underlying equational theory.

---

[*] This research was supported in part by the National Science Foundation under Grant DCR 85–13417. The second author is also partly supported by the Grant of the Committee on Aid to Research Activity of Faculty of Engineering and Computer Science (Concordia University), Fonds pour la Formation de Chercheurs et l'Aide a la Recherche (Quebec) and the Natural Science and Engineering Research Council (Canada).

In this paper we study the theory of conditional rewrite systems. In Section 2, we present various formulations of conditional equations as rewrite systems and compare their expressive power. We also examine the differing restrictions under which we can prove confluence of the various systems and the equivalence of these formulations to the underlying equational theory if they are canonical. In Section 3, we identify a class of systems which are "decreasing". For these systems, the basic notions of rewriting are all decidable and the critical pair lemma holds as for unconditional systems. We show that decreasing systems extend the classes of "simplifying" and "reductive" systems that had been proposed earlier.

2. Conditional Equations

A *positive–conditional* equation is of the form

$$s_1{=}t_1 \wedge \cdots \wedge s_n{=}t_n : s = t$$

where $n \geq 0$ and the $s_i = t_i$ are equations, containing universally quantified variables. The ":" may be thought of as implication with $s = t$ as the conclusion and $t_i = s_i$ as the premises. In this paper, we consider only *equational* consequences and proofs. Note that equational logic lacks a "law of excluded middle". The relation between equational proofs and first–order ones is studied in [Dershowitz–Plaisted–87].

We define the one–step replacement relation $\longleftrightarrow$ and its reflexive–transitive closure $\overset{*}{\longleftrightarrow}$ as follows: If $s_1{=}t_1 \wedge \cdots \wedge s_n{=}t_n : s = t$ is a conditional equation, σ is a substitution, u is a term, π is a position in u and $t_i\sigma \overset{*}{\longleftrightarrow} s_i\sigma$ for $i = 1, \cdots, n$, then $u[s\sigma]_\pi \longleftrightarrow u[t\sigma]_\pi$ where $u[s\sigma]_\pi$ denotes the term u with $s\sigma$ as a subterm at position π and $u[t\sigma]_\pi$ is the term obtained by replacing $s\sigma$ by $t\sigma$. We write $E \vdash s = t$ if $s \overset{*}{\longleftrightarrow} t$ for a set E of positive–conditional equations.

For example, using equations

$$
\begin{aligned}
0 + y &= y \\
x + y = z : \quad s(x) + y &= s(z)
\end{aligned}
$$

we have $s(0) + s(0) \longleftrightarrow s(s(0))$ since $0 + s(0) \longleftrightarrow s(0)$.

2.1. Conditional Rewrite Systems

Conditional rules are conditional equations with the equation in the conclusion oriented from left to right. A conditional rule is used to rewrite terms by replacing an instance of the left–hand side with the corresponding instance of the right–hand side (but not in the opposite direction) provided the conditions hold. A set of conditional rules is called a conditional rewrite system. Depending on what criterion is used to check conditions different rewrite relations are obtained for any given system R (see below). Once a criterion is chosen, we can define the one–step rewrite relation $\rightarrow$ and its reflexive–transitive closure $\overset{*}{\rightarrow}$ as follows:

$u[l\sigma]_\pi \longrightarrow u[r\sigma]_\pi$ if $c: l \longrightarrow r$ is a rule, σ is a substitution, u is a term, π is a position in u and $c\sigma$ satisfies the criterion.

A term t is *irreducible* (or in *normal* form) if there is no term s such that $t \longrightarrow s$. We say that two terms s and t are *joinable*, denoted $s \downarrow t$, if $s \stackrel{*}{\longrightarrow} v$ and $t \stackrel{*}{\longrightarrow} v$ for some term v. A rewrite relation $\longrightarrow$ is said to be *noetherian* if there is no infinite chain of terms $t_1, t_2, \cdots, t_k, \cdots$ such that $t_i \longrightarrow t_{i+1}$ for all i. A rewrite relation $\longrightarrow$ is said to be *confluent* if the terms u and v are joinable whenever $t \stackrel{*}{\longrightarrow} u$ and $t \stackrel{*}{\longrightarrow} v$. It is *locally confluent* if the terms u and v are joinable whenever $t \longrightarrow u$ and $t \longrightarrow v$ (in one step). A rewrite system R is *canonical* if its rewrite relation is both noetherian and confluent.

There are a fair number of different ways of formulating conditional equations as rewrite rules:

Semi–Equational systems
Here we formulate rules as $s_1 = t_1 \wedge \cdots \wedge s_n = t_n : l \longrightarrow r$, where the conditions are still expressed as equations. To check if a condition holds we use the rules bidirectionally, as identities, and check if $s_i\sigma \stackrel{*}{\longleftrightarrow} t_i\sigma$.

Join systems
Here we express rules as $s_1 \downarrow t_1 \wedge \cdots \wedge s_n \downarrow t_n : l \longrightarrow r$. The conditions are now checked in the rewrite system itself by checking if $s_i\sigma$ and $t_i\sigma$ are joinable. Note the circularity in the definition of $\longrightarrow$. The base case, of course, is when unconditional rules are used or the conditions unify syntactically. This definition is the one most often used; see [Kaplan–84, Jouannaud–Waldmann–86, Dershowitz–Okada–Sivakumar–87].

Normal–Join systems
Here rules are written $s_1 \downarrow' t_1 \wedge \cdots \wedge s_n \downarrow' t_n : l \longrightarrow r$. This is similar to join systems except that $s_i\sigma$ and $t_i\sigma$ are not only joinable, but also have a common reduct that is irreducible. (A sufficient condition for this is that the common reduct not contain any instance of a left–hand side.)

Normal systems
A special form of normal–join systems has all conditions of the form $s_i \stackrel{!}{\longrightarrow} t_i$ (meaning that $s_i \stackrel{*}{\longrightarrow} t_i$ and t_i is an irreducible ground term).

Inner–Join systems
Here rules are written $s_1 \downarrow^i t_1 \wedge \cdots \wedge s_n \downarrow^i t_n : l \longrightarrow r$. We require that $s_i\sigma$ and $t_i\sigma$ are joinable by *innermost rewriting*. That is, in rewriting these terms, one applies a rule at some position only if all proper subterms are already in normal form.

Outer–Join systems

Here rules are written $s_1\downarrow^o t_1 \wedge \cdots \wedge s_n\downarrow^o t_n: l \rightarrow r$. We require that $s_i\sigma$ and $t_i\sigma$ are joinable by *outermost rewriting*. That is, in rewriting these terms, one applies a rule at some position only if no rule can be used above (at a superterm).

Meta–Conditional systems

Here we allow any (not necessarily recursively enumerable) predicate p in the conditions. For example, we may have conditions like $s \in S$ (for some term s and set S), $x \xrightarrow{!} x$ (x is already in normal form), or $l > r$ (for some ordering $>$). We write $p:\ l \rightarrow r$.

Most of the formulations above have been considered by different authors with slight variations. For example, Bergstra and Klop in [Bergstra–Klop–86] restrict their attention to systems which are *left–linear* (no left–hand side has more than one occurrence of any variable) and non–overlapping (no left–hand side unifies with a renamed non–variable subterm of another left–hand side or with a renamed proper subterm of itself). With these restrictions on left–hand sides, they refer to semi–equational systems as of *Type I*, join systems as of *Type II* and normal systems as of *Type III$_n$*. They also prove that, with these restrictions on left–hand sides, *Type I* and *Type III$_n$* systems are confluent. Meta-conditional systems with *membership* conditions were proposed in [Toyama–87].

2.2. Sufficient Conditions for Confluence

An interesting question to address is what are the criteria under which each formulation is confluent. For syntactic criteria we have to consider overlaps between left–hand sides of rules called "critical pairs" defined below.

Let R be a rewrite system with rules $c: l \rightarrow r$ and $p: g \rightarrow d$ renamed to share no variables.

Definition 1. If g unifies with a non–variable subterm of l at position λ via a substitution σ, then the conditional equation $(c \wedge p)\sigma: r\sigma = l\sigma[d\sigma]_\lambda$, is a *critical pair* of the two rules.

A system is *non–overlapping* (or unambiguous) if it has no critical pairs. A critical pair $c: s = t$ is *feasible* if there is a substitution σ for which $c\sigma \xrightarrow{*} true$. A critical pair $c: s = t$ is *joinable* if for all feasible substitutions σ, there exists a term v such that $s\sigma \xrightarrow{*} v$ and $t\sigma \xrightarrow{*} v$. A critical pair is an *overlay* if the two left–hand sides unify at the root.

The *depth* of a rewrite is the depth of recursive evaluations of conditions needed to determine that the matching substitution is feasible. We define it formally for join systems.

Definition 2. The depth of an unconditional rewrite is 0; the depth of a rewrite using a conditional rule $u{\downarrow}v: l \rightarrow r$ and substitution σ is one more than the maximum of the depths of the two derivations used to show $u\sigma{\downarrow}v\sigma$. The depth of a n–step derivation $t \overset{*}{\rightarrow} s$ is the maximum of the depths of each of the n steps.

We write $t \underset{k}{\rightarrow} s$ if $t \rightarrow s$ and the depth of the rewrite step is no more than k. Similarly $t \underset{k}{\overset{*}{\rightarrow}} s$ will mean that the maximum depth in that derivation is at most k.

A critical pair between rules $c: l \rightarrow r$ and $p: g \rightarrow d$ is *shallow–joinable* if for each feasible substitution σ, $l\sigma \underset{m}{\rightarrow} r\sigma$, $l\sigma \underset{n}{\rightarrow} l\sigma[d\sigma]$ then there exists a term v, $r\sigma \underset{n}{\rightarrow} v$ and $l\sigma[d\sigma] \underset{n}{\rightarrow} v$. That is, the critical pair is joinable with the corresponding depths preserved. In particular, critical pairs between

unconditional rules should be joinable unconditionally.

For example, the rules $f(x) \rightarrow g_1(x)$ and $h(x){\downarrow}c: f(x) \rightarrow g_2(x)$ overlap to yield a critical pair $h(x){\downarrow}c: g_1(x) = g_2(x)$ (which is an also an overlay). If we also had a rule $h(0) \rightarrow c$, then we would have a feasible instance of this critical pair for the substitution $x \mapsto 0$. This instance would be shallow joinable, if for some term t, $g_1(0) \underset{1}{\overset{*}{\rightarrow}} t \overset{*}{\leftarrow} g_2(0)$ since $f(0) \rightarrow g_1(0)$ and $f(0) \underset{1}{\rightarrow} g_2(0)$.

For unconditional systems, the Critical Pair Lemma (see [Knuth–Bendix–70]) states that a system is locally confluent iff all its critical pairs are joinable. For conditional systems this is true only for some of the formulations above; in general, stronger restrictions are needed. By Newman's Lemma, a noetherian system is confluent iff it is locally confluent. We list below some sufficient conditions for the confluence of the various formulations.

Semi–Equational systems
Noetherian and critical pairs are joinable [proof straightforward].

Join systems
Decreasing (see next section for definition) and critical pairs are joinable [Section 3].

Join systems
Noetherian and all critical pairs are overlays and joinable [Dershowitz–Okada–Sivakumar–87].

Normal systems
Noetherian, left–linear and critical pairs are shallow–joinable [Dershowitz–Okada–Sivakumar–87].

Inner systems
Noetherian and critical pairs are joinable.

2.3. Strength of Rewrite systems

Let E be a set of conditional equations. By $E \vdash s = t$, we mean that $s \overset{*}{\longleftrightarrow} t$ is provable in E. Similarly, if R is a rewrite system (in any of the formulations), we use $R \vdash s \downarrow t$, to mean that s and t are joinable using the rules in R.

Definition 3. R and E have the same *logical strength* if $E \vdash s = t$ *iff* $R \vdash s \downarrow t$. Similarly, two rewrite systems R and R' have the same logical strength if $R \vdash s \downarrow t$ *iff* $R' \vdash s \downarrow t$. We say that R is *stronger* than R' if any two terms joinable using R' are joinable using R, but not the converse.

Figure 1 depicts the relative strength of the various formulations. In the figure, $A \rightarrow B$, means that A is stronger than B in general. That is, if we take a system of type B and just change the connective in conditions to convert to a system of type A (for example, $s \downarrow^i t$ to $s \downarrow t$ to covert an inner–join to a join system), then we have that what is provable in B is also provable in A. In particular, if B is canonical then so is A. The converse is, of course, not true in general.

We now state and prove some of the equivalences and relationships between the various systems.

Proposition 1: If a join system is noetherian, then it is equivalent to the corresponding normal–join system (obtained by changing conditions of the form $s \downarrow t$ to $s \downarrow^1 t$).

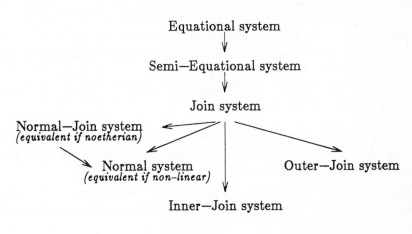

Figure 1

Proposition 2: We can convert a join system R to an equivalent normal system R' by a conservative extension (using new function symbols) provided that we allow the normal system to be non–left–linear (have repeated variables in left–hand sides).

Proposition 3: Let R (with conditions of the form $s{\downarrow}t$) be a canonical (confluent and noetherian) join system R' the corresponding semi–equational systems (change conditions to $s = t$) and E the underlying equational system (change conditions to $s = t$ and $l \rightarrow r$ to $l = r$). The following are equivalent:

(1) $u = v$ is provable in E, that is, $E \vdash u = v$

(2) u and v have a common reduct in R, that is, $R' \vdash u{\downarrow}v$

(3) u and v have a common reduct in R', that is, $R \vdash u{\downarrow}v$

Proof. Proposition 1 is easy to see, for the noetherian property implies that if two terms are joinable, then they have a common reduct that is irreducible. The translation mechanism for Proposition 2 uses two new function symbols eq and $true$. We add a new rule $eq(x,x) \rightarrow true$ and change conditions to the form $eq(s,t){\downarrow}true$. With this translation, it is easy to prove that for any two terms s and t not having the new function symbols eq and $true$, we have $R \vdash s{\downarrow}t$ iff $R' \vdash s{\downarrow}t$. The argument for Proposition 3 is by induction on the depth of a proof. The interesting case is when $u = v$ is provable in E and we wish to show $u{\downarrow}v$ in R. By induction on the depth we first show that the sub-proofs in E can be replaced by rewrite proofs and then using the confluence of R we can show that $u{\downarrow}v$. □

Under the assumption of canonicity the various weaker formulations of join systems are also equivalent to the corresponding join system and, hence, to the underlying equational system. We prove this below for inner systems.

Theorem 1. *Let R be a canonical inner–join (or outer–join, normal–join) system and R' the corresponding join system obtained by replacing conditions of the form $s \downarrow^i t$ ($s \downarrow^o t$, $s \downarrow^! t$, respectively) by $s{\downarrow}t$. R is equivalent to R'. That is for any terms s and t, $R \vdash s{\downarrow}t$ iff $R' \vdash s{\downarrow}t$.*

It is easy to see that if s and t have a common reduct in an inner–join system, then they have a common reduct in the join system (the same proof holds). To show the other direction we prove an even stronger version.

Lemma 1. *For any terms s and t, if $s{\downarrow}t$ in the inner–join system R' then $s \downarrow^i t$ in R. That is, s and t are joinable by innermost rewriting in the inner join system.*

Proof. The proof of this lemma is carried out by induction on the depth of $s{\downarrow}t$ in R'. If s joins t at depth 0 in R', the same rewrite sequence is also valid in the inner–join system since we do not use any conditional rules. So, $s{\downarrow}t$ in R too. Since R is canonical, this also means that $s \downarrow^i t$ in R. Otherwise, if s and t are joinable at depth $n+1$, every condition, say $u{\downarrow}v$, used for rewriting is provable in depth at most n in R'. By the induction hypothesis, the corresponding conditions $(u \downarrow^i v)$ are all provable in R. Therefore, the rewrite steps also apply in R, hence $s{\downarrow}t$ in R. Again, since R is confluent and noetherian, $s{\downarrow}t$ in R implies $s \downarrow^i t$ in R. $\qquad\square$

We can sum up by saying that if any formulation as a rewrite system is canonical, then it is equivalent to the corresponding equational system.

3. Decreasing Systems

By the *reduction ordering* $>_R$ of a rewrite system, we mean the irreflexive–transitive closure $\overset{+}{\to}$ of the reduction relation. That is, $t_1 >_R t_2$ if $t_1 \overset{+}{\to} t_2$. The reduction ordering is *monotonic*. That is, if $t >_R s$ then $u[t] >_R u[s]$ for any context $u[\cdot]$. By the *proper subterm ordering* $>_s$ we mean the well–founded ordering $u[t]>_s t$ for any term t and non–empty context $u[\cdot]$. In this section, we will use the join system formulation of conditional rules to illustrate definitions and results.

Definition 4. A conditional rewrite system is *decreasing* if there exists a well–founded extension $>$ of the proper subterm ordering such that $>$ contains $>_R$ and $l\sigma > s_1\sigma, \cdots, t_n\sigma$ for each rule $s_1{\downarrow}t_1 \wedge \cdots \wedge s_n{\downarrow}t_n\colon l \to r$ $(n \geq 0)$ and substitution σ.

Note that the second condition restricts all variables in the condition to also appear on the left–hand side. In general, a decreasing ordering need not be monotonic.

Proposition 4: If a rewrite system is decreasing then it has the following properties:

(1) The system is terminating.

(2) The basic notions are decidable. That is, for any terms s, t
    i) one–step reduction ("does $s \to t$?")
    ii) finite reduction ("does $s \overset{*}{\to} t$?")
    iii) joinability ("does $s \downarrow t$?")
    iv) normal form or reducibility ("is s irreducible?")
    are all decidable.

Proof. That the system is terminating is obvious from the well–foundedness of $>$. The decidability of basic notions is proved by transfinite induction on $>$, as follows. We first consider the following property: "Given a term t we can find the set of normal forms of t." If t has no instance of a left–hand side of any rule as a subterm then t is irreducible and it is its only normal form. Otherwise, let $t = u[l\sigma]$ for some rule $u_1{\downarrow}v_1 \wedge \cdots \wedge u_n{\downarrow}v_n\colon l \to r$ By our two conditions on decreasingness we have that $t=u[l\sigma] > l\sigma$ and $l\sigma > u_i\sigma, v_i\sigma$. By induction, since $t>u_i, v_i$, we can compute the set of normal forms of u_i, v_i for each i and check if the rule applies. If it does then $t \longrightarrow u[r\sigma]$. Similarly (using each matching rule) we can compute all the terms, say $s_1, \ldots, s_n$, that t rewrites to in one–step. By induction hypothesis one can enumerate the normal forms for each s_i. Then the union of these is the set of normal forms for t.

Other basic properties can be shown decidable likewise. □

The following are some sufficient conditions for decreasingness:

Simplifying systems [Kaplan–84, Kaplan–87]

A conditional rewrite system R is *simplifying* if there exists a simplification ordering $>$ (in the sense of [Dershowitz–82]) such that $l\sigma > r\sigma, s_1\sigma \cdots, t_n\sigma$, for each rule $s_1 \downarrow t_1 \wedge \cdots \wedge s_n \downarrow t_n: l \longrightarrow r$ $(n \geq 0)$.

Reductive systems [Jouannaud–Waldmann–86]

A conditional rewrite system is *reductive* if there is a well–founded mono-tonic ordering $>$ such that $>$ contains the reduction ordering $>_R$ and $l\sigma > s_1\sigma, \cdots, t_n\sigma$ for each rule $s_1 \downarrow t_1 \wedge \cdots \wedge s_n \downarrow t_n: l \longrightarrow r$ $(n \geq 0)$.

Both simplifying systems and reductive systems are special cases of decreas-ing ones. To see this for simplifying systems, note that simplification orderings contain the subterm ordering, by definition. For reductive systems, note that no monotonic well–founded ordering can have $s > t$ for a proper subterm s of t. So we can extend the monotonic ordering with the subterm property and get a well–founded ordering as in [Jouannaud–Waldmann–86]. The following is an example of a system that is decreasing but neither simplifying nor general reductive:

$$
\begin{array}{rcl}
b & \longrightarrow & c \\
f(b) & \longrightarrow & f(a) \\
b \downarrow c \quad : \qquad a & \longrightarrow & c
\end{array}
$$

This is not reductive because there is no monotonic extension of the reduction ordering (which has $f(b) >_R f(a)$) that can have $a > b$.

Decreasing systems also satisfy the critical pair lemma.

Theorem 2. *For any decreasing system, if every critical pair is joinable, then the system is confluent, hence canonical.*

Proof. The proof is implicit in [Jouannaud–Waldmann–86] where they impose stronger conditions for their definition of a reductive system (as explained earlier) but use essentially the same conditions that we have for decreasingness in their proof. □

Were we to omit "the subterm property" in our definition of a decreasing ordering, then the critical pair lemma no longer holds. We illustrate this with the counter–example shown in Table 1. All critical pairs are joinable, yet the term $h(f(a))$ has two normal forms d and $k(f(b))$. Although we have $h(f(a)) \longrightarrow h(f(b))$ and $h(g(b)) \longrightarrow k(g(b)) \longrightarrow d$, to determine if $h(f(b)) \longrightarrow h(g(b))$ using the last rule we have to check the condition $h(f(b)) \downarrow d$

		c	$\rightarrow$	$k(f(a))$
		c	$\rightarrow$	d
		a	$\rightarrow$	b
		$h(x)$	$\rightarrow$	$k(x)$
		$h(f(a))$	$\rightarrow$	c
		$k(g(b))$	$\rightarrow$	d
$h(f(x)) \downarrow d$	:	$f(x)$	$\rightarrow$	$g(x)$

Table 1

which leads to a cycle. Note that if we converted this to a semi–equational system we would have that $h(f(b)) \longleftrightarrow^* d$ and the last rule can be applied and the system is confluent.

This example satisfies the conditions for decreasing systems except the subterm property. The reduction ordering of the above rewrite system is embeddable into the well–founded ordering $<_\infty$ (which, however, does not have the subterm property) of Takeuti's system $O(2,1)$ of ordinal diagrams (one of the two major systems of proof theoretic ordinals). See [Okada–Takeuti–87] for definitions. Also, $<_\infty$ satisfies the additional condition for decreasingness (each term in the condition– d and $h(f(x))$– is smaller than the left–hand side—$f(x)$—of that rule). So it is clear that well–foundedness alone is not sufficient. For more details see [Okada–87, Dershowitz–Okada–88].

We saw in the previous section that, while the confluence of a join system implies the confluence of the corresponding semi–equational system (without any other restriction), the converse is not true in general. We now show that if we restrict our attention to decreasing systems the converse does hold. That is, under the assumption of decreasingness the two formulations, semi–equational systems and join systems make no difference with respect to confluence.

Theorem 3. *If a decreasing semi–equational system (conditions of the form $s = t$) is confluent, then the corresponding join system (with conditions changed to $s \downarrow t$) is also confluent.*

It is convenient to introduce the following notations. By a *direct proof* of $s = t$, we mean a rewrite proof of the form $s \downarrow t$. That is, s and t are joinable. By a *completely direct* proof of $s = t$, we mean a direct proof of $s = t$ (i.e, of $s \downarrow t$) in which every subproof of the conditions (during application of conditional rules) is also direct. For instance, if a substitution instance of the form $s_1\sigma = t_1\sigma \wedge \cdots \wedge s_n\sigma = t_n\sigma$: $l\sigma \rightarrow r\sigma$ of a conditional rule is used in the proof with subproofs of $s_i\sigma = t_i\sigma$, then each of these subproofs is also direct. If for a given proof P of $s = t$ there is a completely direct proof P' (of $s \downarrow t$), then we say that the proof P is completely normalizable.

Lemma 2 (Complete Normalizability Lemma). *For any confluent and decreasing semi–equational system, every proof is completely normalizable.*

Proof. This is proved using transfinite induction on the well–founded decreasing ordering. It is easily seen that every proof in a decreasing system can be made direct if the system is confluent. By using the properties of decreasing-ness, we can show that every top–level subproof is smaller in the decreasing ordering and, hence, can be made completely direct by the induction hypothesis. Then this lemma follows. □

Theorem 3 is a direct consequence of the Complete Normalizability Lemma. Thus, if a decreasing system is confluent in the semi–equational formulation, then it is confluent as a join system.

4. Conclusion

We have studied different formulations of conditional rewrite rules and their expressive power and identified a class of decreasing systems for which most of the interesting notions of rewriting are decidable and which satisfy the critical pair lemma. Decreasing systems have weaker restrictions than the simplification systems or reductive systems studied previously. We have shown that straightfor-ward attempts at weakening these restriction further (by dropping the subterm property) do not work. For this class of systems we have also shown that two of the formulations as rewrite systems are equivalent with respect to confluence.

REFERENCES

[Bergstra–Klop–82] Bergstra, J. A., and Klop, J. W. "Conditional rewrite rules: Confluence and termination", Report IW 198/82 MEI, Mathematische Centrum, Amsterdam, 1982.

[Bergstra–Klop–86] Bergstra, J. A., and Klop, J. W. "Conditional rewrite rules: confluency and termination", *JCSS* **32** pp. 323–362 , 1986.

[Brand–Darringer–Joyner–78] Brand, D., Darringer, J. A., Joyner, W. J. "Completeness of conditional reductions". Report RC 7404, IBM Thomas J. Watson Research Center, December 1978.

[Dershowitz–82] Dershowitz, N. "Orderings for term–rewriting systems". *Theoretical Computer Science* **17**(3), pp. 279–301 (March 1982).

[Dershowitz–Okada–88] Dershowitz, N., and Okada, M. "Proof–theoretic techniques for term rewriting theory". *Proceedings of the third annual Symposium on Logic in Computer Science,* Edinburgh, July 1988 (to appear).

[Dershowitz–Okada–Sivakumar–87] Dershowitz, N., Okada, M., and Sivakumar, G. "Confluence of Conditional Rewrite Systems". *First International Workshop on Conditional Rewriting Systems*, Orsay, France (July 1987) (to appear in Lecture Notes in Computer Science, Springer, Berlin).

[Dershowitz–Plaisted–85] Dershowitz, N., and Plaisted, D. A. "Logic programming *cum* applicative programming". *Proceedings of the 1985 Symposium on Logic Programming*, Boston, MA, pp. 54–66 (July 1985).

[Dershowitz–Plaisted–87] Dershowitz, N., and Plaisted, D. A. "Equational programming". In: *Machine Intelligence 11*, J. E. Hayes, D. Michie and J. Richards, eds., 1987 (to appear).

[Fribourg–85] Fribourg, L. "SLOG: A logic programming language interpreter based on clausal superposition and rewriting". *Proceedings of the 1985 Symposium on Logic Programming*, Boston, MA (July 1985), pp. 172–184.

[Goguen–Meseguer–86] Goguen, J. A., and Meseguer, J. "EQLOG: Equality, types and generic modules for logic programming". In *Logic Programming: Functions, relations and equations* (D. DeGroot and G. Lindstrom, eds.), Prentice–Hall, Englewood Cliffs, NJ, pp. 295–363, 1986.

[Jouannaud–Waldmann–86] Jouannaud, J. P., and Waldmann, B. "Reductive Conditional term rewriting systems". *Proceedings of the Third IFIP Working Conference on Formal Description of Programming Concepts*, Ebberup, Denmark.

[Kaplan–84] Kaplan, S. "Fair conditional term rewriting systems: Unification, termination and confluency", Laboratoire de Recherche en Informatique, Université de Paris–Sud, Orsay, France, November 1984.

[Kaplan–87] Kaplan, S. "Simplifying conditional term rewriting systems: Unification, termination and confluence", *Journal of Symbolic Computation* (to appear).

[Knuth–Bendix–70] Knuth, D. E., and Bendix, P. B. "Simple word problems in universal algebras". In: *Computational Problems in Abstract Algebra*, J. Leech, ed. Pergamon Press, Oxford, U. K., 1970, pp. 263–297.

[Okada–87] Okada, M. "A logical analysis for theory of conditional rewriting". *First International Workshop on Conditional Rewriting Systems*, Orsay, France (July 1987) (to appear in Lecture Notes in Computer Science, Springer, Berlin).

[Okada–Takeuti–87] Okada, M. and Takeuti, G. "On the theory of quasi–ordinal diagrams", in *Logic and Combinatorics*, ed. S.Simpson, Contemporary Mathematics 68 (1987), American Mathematical Society.

[Remy–82] Rémy J.-L., "Etude des systèmes de réécriture conditionnels et applications aux types abstraits algébriques" Thèse, Institut National Polytechnique de Lorraine, July 1982.

[Toyama–87] Toyama, Y. "Term rewriting systems with membership conditions". *First International Workshop on Conditional Rewriting Systems*, Orsay, France (July 1987) (to appear in Lecture Notes in Computer Science, Springer, Berlin).

Program Synthesis by Completion with Dependent Subtypes*

Paul Jacquet

LIFIA -IMAG Institut National Polytechnique
46, Avenue Felix Viallet
38031 Grenoble cedex
France

Abstract

In this paper we explore the possibility of specifying a restricted form of conditional equations within, what we call, a dependent subtypes discipline. Applying recent results on order sorted computation to the lattice structure of dependent subtypes we show how the Knuth-Bendix completion procedure can be used to synthesize a restricted class of programs, a subject which brought about this work.

Key words and phrases. Program Synthesis, Conditional and Order Sorted Rewriting.

1 Introduction

Rewriting and the Knuth-Bendix completion procedure are both paradigms for computation and proof. The former works by simplifying terms. Dershowitz [1] has shown how the latter can be used to interpret logic programs, written as a set of equivalence preserving rules, as well as to synthesize such rewrite programs from specifications. In this framework the analogy between proofs and programs, already elaborated in other contexts [2], resides in the way the completion procedure can be used. As with other deductive methods, e.g. [3], [4], for program synthesis we start with a specification of the predicate (i.e. boolean function) p, to be synthesized, expressed in a suitable logic. In the context of this paper it will be equational logic.

Let us assume that we are given rewrite programs in the sense of Dershowitz [1] for the

---

* This work has been partially supported by the CNRS (PRC Programmation et Outils pour l'IA).

theory E of our problem (except obviously for *p*), that is to say a set R of rules such that R is sound with respect to E, terminates for all true (in E) ground terms, and reduces all true ground input terms to the constant *true*. Then we can add the specifying equation to these rules and run the completion procedure to prove the specification. If the procedure does not abort, and executes fairly, the specification is proved and, as a "side effect" we are left with a set of rules which allows to reduce non-deterministically all true ground terms to the irreducible constant *true*. Thus we get a rewrite program for *p*.

Despite the numerous reasons which may prevent the completion procedure from succeeding reasonably (e.g. failure in the choice of a "good" ordering, or in the choice of strategy to guide the completion, absence of the necessary auxiliary definitions,...) this approach seems promising from the unified view of programs and specifications which it offers.

This work is an attempt to deal with conditional programs in a restricted case, where conditions must be expressed only on variables, by coding conditional information inside the rules with, what we call, a dependent subtyping. The ordered structure of dependent subtypes, determined statically , and connected with an appropriate order sorted unification algorithm, are used to propagate conditions during the completion process. A similar idea can be found in [5] to propagate features in a hierarchy of feature types. Due to lack of place, all proofs are omitted; they will appear in a forthcoming report.

2 Notations and Basic Notions

Let us suppose that our general framework allows to specify hierarchical rewriting systems. Following [6], [7] we recall some basic notions, assuming familiarity with the many-sorted algebras and term rewriting systems "folklore" [8], [9].

Let $\Sigma = (S, \Omega)$ be a *signature* with *sorts set* S and finite family Ω of S-sorted *operators*, and X be a S-indexed family of denumerable sets of *variables* . $T_{\Sigma(X)}$ denotes the family of sets of *terms* constructed using Ω and X (it also denotes the term algebra). For a term t, Var(t) denotes the set of all variables that occur in t, and O(t) the set of all occurrences of t. t/u denotes the subterm of t at occurrence u and t[u ← t'] the term obtained by replacing t/u by t' in t. If X is empty, we write T_Σ to denote the family of sets of *ground terms*.

A *specification* SP = (S, Ω, E) consists of a signature Σ = (S, Ω) and a set E of Σ-equations. We denote $\equiv_E$ the smallest congruence relation generated by E over $T_{\Sigma(X)}$.

SP is considered as a *term rewriting system* RT if each equation in E is seen as a rewrite rule.

A conditional rule, denoted c :: l → r, is a rule where l and r are of the same sort and c is of sort boolean; moreover we have Var(c) ⊆ Var(l) and Var(r) ⊆ Var(l).

Definition 2.a:

A term t is *conditionally rewritten* to t', and we note t →$_{RT}$ t', if there exists a rule c :: l → r in RT, a sub-term t/u of t and a substitution σ such that:

- t/u = σ(l)
- σ(c) ≡$_E$ *true*   and   t' = t[u ← σ(r)].

Let $\rightarrow_{RT}$ (or simply $\twoheadrightarrow$) denote the reflexive and transitive closure of $\rightarrow_{RT}$, and $\leftrightarrow$ the reflexive, transitive and symmetric closure of $\rightarrow_{RT}$. If RT is confluent and *true* is irreducible by R then $t \equiv_E t'$ iff $t \twoheadrightarrow t_1$ and $t' \twoheadrightarrow t_1$ for some t_1 [10].

A *hierarchical* specification is a finite sequence of specifications $(SP_0, SP_1, \ldots, SP_n)$ such that:
$$\forall i \in [1, n], \ SP_{i-1} \subseteq SP_i$$
That is to say:

    (i) $S_{i-1} \subseteq S_i, \Omega_{i-1} \subseteq \Omega_i, E_{i-1} \subseteq E_i$

    (ii)   For every equation $l = r$ in $E_i - E_{i-1}, l \in T_{\Sigma i}(X) - T_{\Sigma_{i-1}}(X)$ and $Var(r) \subseteq Var(l)$.

A hierarchical specification is *sufficiently complete* iff for each term t in $T_{\Sigma i}$, if the sort of t is in S_{i-1}, then there exists a term t' in $T_{\Sigma_{i-1}}$ such that $t \equiv_{Ei} t'$, where $i \in [1, n]$.

A hierarchical rewriting system is a sequence of rewriting systems $(RT_0, RT_1, \ldots, RT_n)$ such that the corresponding specification $(SP_0, SP_1, \ldots, SPn)$ is a hierarchical specification.

Definition 2.b:

Let $RT = (RT_0, RT_1, \ldots, RT_n)$ be a hierarchical rewriting system. A term t is *hierarchically rewritten* to t', and we note $t \rightarrow_{H,RT} t'$, if there exists a rule $c :: l \rightarrow r$ in R_i, a subterm t/u of t and a substitution σ such that:

- $t/u = \sigma(l)$
- $\sigma(c)$ is a term in $T_{\Sigma_{i-1}}(X)$.
- $\sigma(c) \twoheadrightarrow_{H,RT} true$ and $t' = t[u \leftarrow \sigma(r)]$.

If $RT = (RT_0, RT_1, \ldots, RT_n)$ is sufficiently complete, then $\leftrightarrow_{H,RT}$ is equal to $\leftrightarrow$ on ground terms [10].

In the following, the symbol set Ω_i is assumed to be implicitly associated with RT_i, and the terms in $T_{\Sigma_0}(X)$ are called primitive terms. By definition RT_0 includes a boolean system where the constant *true* is irreducible. In the sequel we use $RT = (RT_0, RT_1)$ to denote a hierarchical rewriting system with two levels. Following [7] we introduce the notion of *contextual rewriting*.

A contextual term is a pair of terms denoted $c :: t$ where c is of sort boolean.

Definition 2.c:

Given $RT = (RT_0, RT_1)$, a contextual term $c :: t$ is contextually rewritten to $c' :: t'$, and we note $c :: t \rightarrow_{RT,C} c' :: t'$, if there exists a rule $c'' :: l \rightarrow r$ in R, a subterm t/u of t and a substitution σ such that:

- $t/u = \sigma(l)$.
- $\sigma(c)$ is a term in $T_{\Sigma_0}(X)$ such that $t' = t[u \leftarrow \sigma(r)]$ and $c' = c \wedge \sigma(c'')$.

We now define, using this framework, a notion of rewrite program specification analogous to that of Dershowitz [1]. Let $RT = (RT_0, RT_1)$ be a specification where:

- $S_0 = \{bool, \tau\}$ To simplify the definition we assume only one sort τ different from bool. The extension to more than one sort is straightforward.

- $\Omega_0 = \Omega_{bool} \cup \Omega_{C\tau} \cup \Omega_{R\tau}$ Where $\Omega_{C\tau}$ is the set of constructors for objects of sort τ and $\Omega_{R\tau}$ a set of predicates defined on objects of sort τ.

- $R_0 = R_{bool} \cup R_{R\tau}$ Where $R_{R\tau}$ contains defining rules for predicates in $\Omega_{R\tau}$.

such that R_0 is confluent, terminating and contains only unconditional rules. Moreover *true* and *false* are irreducible in R_0.

- $S_1 = S_0$
- $\Omega_1 = \Omega_0 \cup \Omega_{D\tau}$ Where $\Omega_{D\tau}$ are input/output relations simulating defined operators on objects of sort τ.

- $R_1 = R_0 \cup R_{D\tau}$ Where $R_{D\tau}$ contains defining rules for relations in $\Omega_{D\tau}$.

such that RT_1 is sufficiently complete and consistent with respect to RT_0.

A specification RT verifying these conditions is called a **rewrite program specification** (RPS for short).

Examples: A RP Specification of natural numbers.
- $S_0 = \{bool, nat\}$
- $\Omega_0 = \Omega_{bool} \cup \{0 : \to nat, s : nat \to nat, eq : nat \times nat \to nat\}$
- $R_0 = R_{bool} \cup \{ eq(0, 0) \to true, eq(0, s(y)) \to false, eq(s(x), 0) \to false,$
       $eq(s(x), s(y)) \to eq(x,y)\}$
- $S_1 = S_0$
- $\Omega_1 = \Omega_0 \cup \{add : nat \times nat \times nat \to bool\}$
- $R_1 = R_0 \cup \{ eq(x, y) :: add(x, 0, y) \to true, not(eq(x, y)) :: add(x, 0, y) \to false,$
       $true :: add(x, s(y), s(z)) \to add(x, y, z)\}$

Note that to "compute" with add, using narrowing for instance, we only need the first and the last rules which reduce true ground terms to the constant *true*.

3 Dependent Subtypes

Let $RT = (RT_0, RT_1)$ be a rewrite program specification. In the following we use τ to denote the carrier of sort τ of the initial algebra specified by RT. We will use τ both to denote a sort or a type.

Let x be a variable of type τ (we assume that there exist denumerably many such variables) and r a binary relation specified in RT_0, we define the following subset of τ:

$$r[x] = \{ y \in \tau \mid r(y, x) \twoheadrightarrow_{R0} true\}$$

Let us suppose that ξ stands for a variable or a ground term of type τ, then $r[\xi]$ denotes a **dependent subtype** (DST) of τ with ξ as its *argument*.

If $\wedge$ denotes the intersection of relations. We obviously have:

$$(r \wedge r')[x] = r[x] \cap r'[x]$$

The complement of $r[\xi]$ in τ is denoted $\overline{r}[\xi]$, and we have $r[\xi] \cup \overline{r}[\xi] = \tau$ and $r[\xi] \cap \overline{r}[\xi] = \varnothing$ where $\varnothing$ denotes the void sub-type.

For a given ξ of sort τ and a binary relation r we can associate the following Boolean lattice of subtypes:

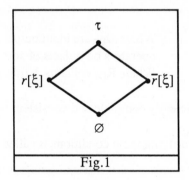

Fig.1

This construction can be easily extended to the case of more than one relation.

For example, the order relation *leq* can be decomposed as the union of two relations *less* $\vee$ *eq* where *eq* denotes the equality relation on objects of sort τ and *less* the quasi-ordering associated to *leq*. We have:

$$less\ [\xi] \cup eq\ [\xi] = leq[\xi] \text{ and } less\ [\xi] \cap eq\ [\xi] = \varnothing .$$

In this case we can associate with the order relation the following lattice:

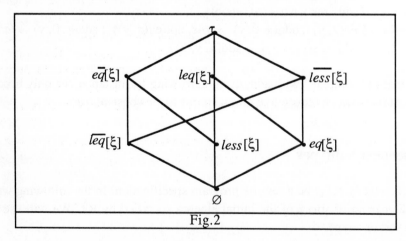

Fig.2

In the sequel we assume that the lattice associated to an order relation is that of Fig. 2, notice that if *leq* is total $\overline{less}$ is also a total order relation. These examples show that subtypes are not necessarily disjoint, a value of the underlying type τ may be of more than one subtype. Furthermore, note that if the set of relations being considered is finite then the associated Boolean lattice (denoted L in the following) is also finite and therefore complete.

4 Typed Terms, Dependent Subtypes Calculus

Typed terms are inductively defined by the following rules :

Definition 4.a

(i)	x:t	where x is a variable and t a type.
(ii)	x:$\{t_1,t_2,\ldots,t_n\}$	where x is a variable and t_i's are DST.
(iii)	f($e_1,e_2, \ldots,e_n$)	where f is a function symbol and the e_i's are typed terms.

Examples:		
	(1) in(x:int, y:$\{eq[x],leq[10]\}$.l:list(int))	is a typed term.
	(2) f(x:*less*[0],y:$\{leq[x], eq[z]\}$)	is a typed term.
	(3) f(x:int,h(y):*less*[x])	is not a typed term.

Note: $\xi:\{t_1\}$ may be shortened to $\xi:t_1$.

Below, we give contextual conditions under which a typed term is well-typed. These conditions are routine when we use only types, the introduction of DST gives rise to new conditions.

Definition 4.b

(i)	x:t	is a well typed term.
(ii)	x:$\{t_1,t_2,\ldots,t_n\}$	assume that the t_i's are of the form $r_i[\xi_i]$, if x:$\{t_1,t_2,\ldots,t_n\}$ occurs in a term M, then M is well typed if and only if for all ξ_i's we have:

       - if ξ_i is a variable then $\xi_i \in$ Var(M)
(note that Var(x:$\{t_1,t_2,\ldots,t_n\}$) = $\{x\}$).
       - if ξ_i is a ground term, no particular condition.

Note that a sub-term of a well-typed term may be not well-typed. From the point of view of classical type analysis all the ξ_i's have the same type which is also the type of x.

(iii)	f($e_1,e_2,\ldots,e_m$)	if e_i is an expression of type t_i and t_1 x t_2 x ... x $t_m \to t$ is the type of f then f($e_1,e_2,\ldots,e_m$) is a well typed term of type t.

Example: Only term (1) in the previous example is well typed.

For every term M we denote $V_{ST}(M)$ the subset of $Var(M)$ containing variables x for which there exists a DST $r[x]$ occurring in M.

4.1 Subtypes Inversion

If r is a relation belonging to the lattice, it is interesting to know if r^{-1} belongs to this lattice as well. If this is the case, then we can deduce from x:r[y] the new type r^{-1} [x] for y.
1. If r is an equivalence relation:
r and $\overline{r}$ are symmetric, i.e. $r^{-1} = r$ and $\overline{r}^{-1} = \overline{r}$.
2. If r is a total ordering:
We have the following propositions (recall that r is the join of a quasi-ordering r' and of the equality relation):

Proposition 4.a : $r^{-1} = \overline{r'}$
Proposition 4.b : $r'^{-1} = \overline{r}$.
Proposition 4.c: $\overline{r^{-1}} = r'$.
Proposition 4.d: $r'^{-1} = r$.

4.2 Subtypes Composition

In order to understand what the two declarations x:r[y] and y:q[z] mean, when they occur in the same term, we have to study how these relations compose to know if $r \circ q$ belongs to the lattice, thus allowing to infer a new type for x.
Proposition 4.e: If r and q are both equivalence or order relations such that $q \subseteq r$ then:
$$r \circ q = q \circ r = r$$
Proposition 4.f: If r is an equivalence relation then:
$$r \circ \overline{r} = \overline{r} \circ r = \overline{r}$$
Proposition 4.g: If r is an order relation and q the associated quasi-ordering then:
$$r \circ q = q \circ r = q$$

4.3 DST-Calculus

In a given term a variable may have, directly or indirectly, different DST .Below, we shall offer rules which allow to compute the set E_M of all the DST for the variables of a given term. Note that, although we allow only variables to be typed with DST, due to rule (4), E_M may contain declarations like ξ:r[ξ'] where ξ is a ground term.

Let M be a well typed term, and E_M the following associated set:
$$\{\xi_1: r_1[\xi_1'], \xi_2: r_2[\xi_2'],\dots, \xi_n: r_n[\xi_n']\}$$

initialized with all the explicit declarations occurring in M.

$$(1) \quad \frac{E_M \cup \{\ \xi : r[\xi'] \ , \ \xi : r'[\xi'] \}}{E_M \cup \{\ \xi : (\ r \wedge r\ ')[\xi'] \}}$$

$$(2) \quad \frac{E_M \cup \{\ \xi : \varnothing \}}{\bot}$$

$$(3) \quad \frac{E_M \cup \{\ \xi : r[\xi'] \ , \ \xi' : r'[\xi''] \}}{E_M \cup \{\ \xi : r[\xi'] \ , \ \xi' : r'[\xi''], \ \xi : (r \circ r')[\xi''] \}} \qquad \text{if } (r \circ r')[\xi] \in L$$

$$(4) \quad \frac{E_M \cup \{\ \xi : r[\xi'] \ \}}{E_M \cup \{\ \xi : r[\xi'] \ , \ \xi' : r^{-1}[\xi] \ \}} \qquad \text{if } r^{-1}[\xi] \in L$$

$$(5) \quad \frac{E_M \cup \{\ g : r[g'] \ \}}{E_M} \qquad \text{if } r(g, g') \text{ reduces to } true$$

$$(6) \quad \frac{E_M \cup \{\ g : r[g'] \ \}}{\bot} \qquad \text{if } r(g,g') \text{ reduces to } false$$

Note: If m is the cardinal of L, and n the number of different variables or ground terms occurring, as the arguments of DST, in a term, the number of different declarations is bounded by $n^2 \times m$. So the process of type calculus, for a given term, always terminates.

M is said *correctly typed* if and only if $E_M \neq \bot$, otherwise M may be discarded.

Note that, due to rule (1), for every ξ', a given ξ has at most one dependent subtype with ξ' as its argument.

5 Computing with Dependent Subtypes

A dependent subtyped rule is a rule $l \to r$ where l and r are correctly typed terms and $Var(r) \subseteq Var(l)$. Note that such a rule can be trivially transformed into a contextual rule by bringing back DST information into the context, because conditions of well-typing are consistent with those of well-formed contextual rules.

Moreover, dependent subtyped rules can also be considered as order sorted rules owing to the lattice structure of DST. We briefly recall some results about order sorted rewriting which are necessary to understand the following example, while we refer the reader to [12], [13], [14] for a complete presentation.

Note that, because of the lattice structure of DST, signatures are regular, that is to say every correctly typed term has a least type. To avoid semantical problems, we assume that the void subtype $\varnothing$ cannot occur in a correctly typed term (the rules of DST-calculus are consistent with this hypothesis).

In the following, by typed variable we understand a variable typed with DST.

Definition 5.a

In this framework *substitutions* are *type-preserving*. That is to say, they are homomorphic extensions of variable assignments, which are different from identity on a finite domain (denoted $Dom(\sigma)$), with the additional property that for every $x \in Dom(\sigma)$ and for every $\xi \in V_{ST}(x)$ we have:

$$type(\sigma(x), \sigma(\xi)) \leq type(x, \xi)$$

where $type(x, \xi)$ denotes the DST of x depending on ξ, and $\leq$ the ordering on types.

We obviously extend the domain of application of σ to DST by considering them as terms.

5.1 DST-Unification

Quoting [13] we recall that, at the heart of each unification algorithm, a variable symbol x has to be unified with a term M. Hence we found a sequence of statements like the following:

(1) If $x = M$ then return({}).
(2) If $x \in Var(M)$ then stop with failure.
(3) return({<x, M>}).

The extension proposed by Walther[13], to obtain a many-sorted unification algorithm, fails with DST essentially for two reasons:

- When we unify two typed variables $x:r[\xi]$ and $y:r'[\xi']$, then if ξ and ξ' are variables we do not know which variable(s) will be substituted for ξ and ξ'.

- We cannot unify a typed variable with a term which is not a variable nor a ground term because some term(s) in the resulting substitution will not be well-typed.

The idea, like in [Sch 86], lies in delaying the introduction of DST's constraints. Hence we replace the previous sequence by the following:

(a) If x is a non typed variable then: execute steps (1), (2) and (3) as previously.
(b) If x is a typed variable:
(b1) If M is a typed variable then: return({<x, z>, <M, z>}) where z is a new variable.
(b2) If M is not a ground term nor a variable then: stop with failure.
(b3) return({<x, M>}).

If the correctly typed terms P and Q are unifiable with the previous algorithm, we are left with a substitution σ_s without type information. The next step is to introduce type constraints.

Let $\xi:T_\xi$ stand for $\{\xi:r_1[\xi_{i1}]\ \xi:r2[\xi_{i2}],\ldots,\xi:r_n[\xi_{in}]\}$, and σ_s a substitution.

- We obviously extend the domain of application of σ_s to sets like $\xi:T_\xi$.
- We denote $gstc(\xi:T_\xi)$ (greatest subtype calculus) the set obtained from $\xi:T_\xi$ by repeated application of rules (1), (2), (5) and (6) of DST-calculus.

Let x be a typed variable of type T_x and T_M the type of M. We construct a new substitution σ equal to σ_S where:

- Each pair of pairs $<x, z>$, $<M, z>$ in σ_S, where M is a typed variable, is replaced by:

$$<x:T_x, z:T_z>, <M:T_M, z:T_z>$$

where $z:T_z = gstc(\sigma_S(z:(T_x \cup T_M)))$

If $z:T_z = \perp$ then stop with failure.

- Each pair $<x, M>$ in σ_S, where M is a non typed variable , is replaced by:

$$<x:T_x, M:T'_M>$$

where $M:T'_M = gstc(\sigma_S(x:T_x))$

- Each pair $<x, M>$ in σ_S, where M is a ground term, is replaced by:

$$<x:T_x, M>$$

- Each pair $<y, M>$ in σ_S, where y is a non typed variable and M is a typed variable, is replaced by:

$$<y, M:T'_M>$$

where $M:T'_M = gstc(\sigma_S(M:T_M))$

Theorem 5.1: If P and Q are two unifiable correctly typed terms, the previous algorithm will compute a substitution σ such that $\sigma(P)$ and $\sigma(Q)$ are two, correctly typed, equal terms.

6 Example

In this section we illustrate how the completion procedure (modified by the introduction of the previous unification algorithm) may be used to synthesize a rewrite program from an equational specification with dependent subtypes. We start with a hierarchical specification of lists of natural numbers, where RT_0 specifies constructors, of sort nat and list, and some predicates on nat.

$S_0 = \{$bool, nat, list$\}$

$\Omega_0 = \Omega_{bool} \cup \{0 : \rightarrow$ nat, s : nat $\rightarrow$ nat, nil : $\rightarrow$ list[nat], . : nat x list[nat] $\rightarrow$ list[nat], $less$: nat x nat $\rightarrow$ bool, eq : nat x nat $\rightarrow$ bool$\}$

$R_0 = R_{bool} \cup \{eq(0,0) \rightarrow true, eq(0, s(y)) \rightarrow false, eq(s(x), 0) \rightarrow false,$ $eq(s(x), s(y)) \rightarrow eq(x, y), less(x, 0) \rightarrow false, less(0, s(y)) \rightarrow true,$ $less(s(x), s(y)) \rightarrow less(x, y) \}$

We assume that the lattice of Fig.2 is constructed with eq and $less$.

RT_1 specifies rewrite programs used in the specification of max, a rewrite program which computes the maximum element of a given list.

$S_1 = S_0$

$\Omega_1 = \Omega_0 \cup \{$in : nat x list[nat] $\rightarrow$ bool, sup : nat x list[nat] $\rightarrow$ bool$\}$

$R_1 = R_0 \cup \{(1) \quad$ in(x,nil) $\rightarrow false \qquad (4) \quad$ sup(x, nil) $\rightarrow true$

$(2) \quad$ in(x, x.l) $\rightarrow true \qquad (5) \quad$ sup(x, y:leq[x].l) $\rightarrow$ sup(x, l)

(3)   in(x,y: $\overline{eq}$ [x].l) → in(x, l)   (6)   sup(x, y: $\overline{leq}$ [x].l) → *false* }

Notice that, to avoid the use of extra variables, rule (2) is not written in(x, y:eq[x].l) → *true*. Assume that for max we have the following specification:

(S)   in(x, l) & sup(x, l) = max(x, l)

Which asserts that x is the maximum element of l if and only if x belongs to l and x is greater or equal to all the elements of l. In order to find a new definition for max, the ordering supplied to the completion procedure should make 'in' and 'sup' greater than max. Therefore (S) is oriented from left to right.

a) Forming a critical pair with (S) and (1) we get, after normalization, the following rule:

(i)   max(x, nil) → *false*

b) With (S) and (2) we get:
$$sup(x, x.l_1) = max(x, x.l_1)$$
Notice that rule (5) is applicable to the left part because *eq* is smaller than *leq* in the lattice. We get now the following subproblem:

(S') sup(x, l_1) = max(x, x.l_1)

b1) With (S') and (4) we get:

(ii)   max(x, x.nil) → *true*

b2) With (S') and (5) we get, after normalization:
$$max(x, x.(z:leq[x].l_2)) = sup(x, l_2)$$
(S') is used to rewrite the right part (we assume for (S') the same orientation as for (S)), introducing a recursive call, we get then the following rule:

(iii)   max(x, x.(z:*leq*[x].l)) → max(x, x.l)

b3) With (S') and (6) we get:

(iv)   max(x, x.(z: $\overline{leq}$ [x].l)) → *false*

c) Coming back to (S), with (6) we get:

(v)   max(x, y: $\overline{leq}$ [x].l) → *false*

d) With (S) and (3) we get:

$$in(x, l_1) \,\&\, sup(x, y: \overline{eq}\,[x].l_1) = max(x, y: \overline{eq}\,[x].l_1)$$

Rule (5) is applicable on the right part ,we have $\overline{eq}\,[x] \cap leq[x] = less[x]$, so we get:

$$in(x, l_1) \,\&\, sup(x, y{:}less[x].l_1) = max(x, y{:}less[x].l_1)$$

rewritten to: $in(x, l_1) \,\&\, sup(x, l_1) = max(x, y{:}less[x].l_1)$

Using (S) to rewrite the left part we get the following rule:

$$(vi) \quad max(x, y{:}less[x].l) \rightarrow max(x, l)$$

Application of other rules does not produce new rules, hence we are left with a rewrite program for max (rules *(i)* to *(vi)*).

Notice that specially to obtain subproblem (S') and rule *(vi)*, this method prunes the futile paths that a classical contextual rewriting system has to explore. Moreover, the new typing for y in *(vi)* is obtained directly, but the price to pay is to compute the closure of all declarations within a term

7 Conclusion

We have shown that computing with dependent subtypes is a simple mechanism which efficiently combines the effects of conditional and order sorted rewriting in a restricted case. The main restrictions concerns the kind of relations allowed to express conditions (essentially equivalence and total order relations) as well as the fact that only variables and ground terms may be typed. From the point of view of program synthesis, we are left with a system allowing to synthesize a restricted class of rewrite programs, typically those for which one can find a specification respecting the previous restrictions. However a lot of problems remain and we will list, below, some possible further research directions :

- Adapting the approach to allow the synthesis of functional rewrite programs. This will probably lead to a modification of the completion procedure and an extensive use of narrowing.

- Finding a suitable way to include the definition of the lattice within the system. It seems that the concept of genericity [15] is a good way to deal with this problem.

- Extending the unification algorithm to many-sorted equational theories [16]. This will allow to type any term.

Acknowledgement. I would like to thank the members of the SADA group at LIFIA, and also Bertram Fronhöfer and Jean-Luc Rémy for valuable discussions that contributed to this work.

References

[1]   N. Dershowitz. *Computing with Rewrite Systems.*
Information and Control Vol.65, N° 2/3, May/June 1985.

[2] J.L. Bates and R.L. Constable. *Proofs as programs*.
TR 82-530, Dept. of Computer Science, Cornell University, March 1983.

[3] W.Bibel. *Syntax-directed, semantics-supported program synthesis*.
Artificial Intelligence 14, 1980, pp. 234-261.

[4] Z. Manna and R.J. Waldinger. *A deductive approach to program synthesis*.
ACM Transactions on Programming languages and Systems Vol. 2, No. 1,
January 1980, pp 90-121.

[5] G. Smolka and H. Aït-Kaci. *Inheritance Hierarchies: Semantics and Unification*.
MCC Technical report AI-057-87.

[6] J-L. Remy and H. Zhang. REVEUR4: *A system for validating conditional algebraic
specifications of abstract data types*. Proc. of 6th ECAI, Pisa, pp. 563-572 , 1984.

[7] H. Zhang and J-L. Remy. *Contextual Rewriting*. Proc. 1st. Conference on
Rewriting Techniques and Applications.LNCS 202, pp. 46-62, 1985.

[8] J. Meseguer and J.A. Goguen. *Initiality, induction, and computability*.
in Algebraic methods in semantics, M. Nivat and J.C. Reynolds, eds.
Cambridge University Press, 1985, pp 459-541.

[9] G. Huet and D.C. Oppen. *Equations and rewrite rules: A survey*.
in Formal Language Theory: Perspectives and Open Problems.R. Book, ed.
Academic Press, New York, 1980, pp 349-405.

[10] J-L. Remy. *Etudes des systèmes de réécriture conditionnelles et applications
aux types abstraits algébriques*. Thèse d'état, Université de Nancy 1, 1982.

[11] M. Navarro and F. Orejas. *On the equivalence of hierarchical and non-
hierarchical rewriting on conditional term rewriting systems*.Eurosam 84, Oxford.

[12] R.J. Cunningham and A.J.J. Dick. *Rewrite Systems on a lattice of types*.
Acta Informatica 22, 1985, pp. 149-169.

[13] C. Walther. *A classification of many-sorted unification problems*. Proc. 8nth
International Conference on Automated deduction,LNCS 230, 1986, pp. 525-537.

[14] G. Smolka, W. Nutt, J. Meseguer and J.A. Goguen. *Order-sorted
equational computation*. Presented at the Colloquium on the Resolution of
Equations in Algebraic Structures. Lakeway, Texas, May 4-6 1987.

[15] D.Bert and R.Echaed. *Design and implementation of a generic, logic and
functional programming anguage*. Proc. ESOP'86, LNCS 213, pp. 119-132.

[16] M. Schmidt-Schauβ. *Unification in many-sorted equational theories*Proc. 8nth
International Conference on Automated deduction, LNCS 230, 1986, pp. 538-552.

Reasoning about Systems of Linear Inequalities

Thomas Käufl

Institut für Logik, Komplexität und Deduktionssysteme
University of Karlsruhe
Postfach 6980
7500 Karlsruhe 1
FR of Germany

1. Introduction

The procedures described in this paper reduce systems of linear inequalities over the rational and integral numbers. If such a system turns out to be satisfiable, subsumed comparisons are eliminated and the entailed equations are determined. This may give rise to further simplification.

The reduction procedures presented use the Sup-Inf-Method developed by Bledsoe, (see [BLE] or [SHO],) a complete decision procedure for systems of linear inequalities over the rational numbers. A system of linear inequalities over the integers given it can find out its unsatisfiability only, if it is unsatisfiable if considered as system over the rationals. But there are other reduction procedures outlined in chapter 3, which make possible to increase the class of systems over the integers recognized to be satisfiable by the Sup-Inf-Method. These procedures and some results taken from the theory of polyhedral sets allow to construct a procedure deciding most systems of linear inequalities over the integers without use of quantifier elimination. This algorithm is presented in chapter 5 together with its foundations. Additionally the class of systems of inequalities decided by it is characterized.

The procedures outlined in this paper are part of a simplifier being itself part of the program verifier "Tatzelwurm". (A sketch of the simplifier is presented in [KÄU85].) Theorems obtained in program verification frequently contain systems of inequalities containing variables for the integers and for the rationals. The design of the procedures will observe this property.

The paper is organized in the following way: After the presentation of some terminology in the subsequent chapter auxiliary reduction procedures are outlined in the third. In the following chapter we define the notions of subsumed comparisons and entailed equations. A reduction algorithm for systems of linear inequalities over the rational numbers will conclude this chapter. The next one contains a reduction procedure for systems of linear inequalities over the integers together with its foundations. A brief discussion why we have developed a new procedure will conclude the paper.

2. Notation and Terminology

Vectors and matrices are italicized. A vector is assumed to be a column of numbers. a^t is the transposed vector, θ the zero-matrix or zero-vector. $a^t b$ denotes the inner product of the vectors a and b.

Abbreviations and Symbols

t possibly indexed denotes an arithmetical term, c, d denote numbers and q denotes an integer number. $\mathbf{Z}$ is the set of the integral numbers and $\mathbf{Q}$ the set of the rational numbers.

Let t be a term and ρ one of the relation symbols $=, \neq, <, \leq, >$ or $\geq$. $t \, \rho \, 0$ is said to be *canonical*, if
1. the coefficients and the constant of t are integers with 1 as their greatest common divisor;
2. ρ is different from $<$ and $>$, if $t \, \rho \, 0$ a comparison over the integers.

It is possible to transform an arbitrary linear comparison into one of the format t ρ 0. As only linear comparisons are considered in this paper, the term linear is omitted frequently.

By a *system of inequalities* we mean a set S of non-negated canonical and linear comparisons. It may contain equations. If it consists of equations only, it is said to be a *system of equations*. We shall write systems of comparisons as conjunctions or as sets whatever notation is more appropiate. t ρ 0 is said to be a *comparison over* Z, if each variable occuring in t stands for an integer, else it is a *comparison over* Q.

For ease in presentation we shall also use the format t ρ c for a canonical comparison. In this case c is a constant and the constant of t is assumed to be zero. Also $a^t x$ ρ c with *a* as vector of the coefficients and *x* as vector of the variables of t will be used.

A comparison 0 ρc is *inconsistent*, if the number c does not fulfill the relation denoted by ρ, *consistent* otherwise.

Let S be a system of inequalities. $[S]_x t$ ($[t_1]_x t$) is the system (term) obtained by replacing each occurence of x by the term t. x ⊆ S (x ⊆ t) means that x occurs in S, t resp. and x ⊄ S (x ⊄ t) that S, t resp. does not contain an occurence of x. Each variable is assumed to range over one of the sorts integer or rational.

3. Auxiliary Reduction Procedures

3.1. The Diophantine Reduction

The procedure reduces canonical comparisons over Z. It is due to King [KIN69].

Let ρ be one of the relation symbols ≤ and ≥, t ρ c a canonical comparison over Z, d the greatest common divisor of the coefficients of t and q be the result of the integer division of c by d. The reduction is performed using the following rules:

(D1) $t \geq c \Rightarrow t/d \geq q + 1$

(D2) $t \leq c \Rightarrow t/d \leq q$

3.2. Elimination of Variables in Systems of Inequalities

A variable x is eliminated by use of the rule

(E) $(x = t \wedge S) \Rightarrow (x = t \wedge [S]_x t)$, provided x ⊄ t

After application of (E) the comparisons occuring in $[S]_x t$ are transformed into canonical ones if necessary and comparisons becoming consistent are removed. x is said to be an *eliminable variable*.

An equation t = 0 can always be transformed into the equivalent x = t' with x ⊄ t', provided x is variable for rational numbers. If x is variable for integers, t = 0 is equivalent to c x = t', x ⊄ t', where c ≠ ±1 in general. Nevertheless one can replace c x = t' ∧ S by c x = t' ∧ $[S]_x t'/c$, since c x = t' is satisfied only for those assignments of numbers to the variables occuring in t' for which t'/c ∈ Z. The subsequent transformation of the comparisons of $[S]_x t'/c$ into canonical ones will remove the fractions occuring as coefficients.

Eliminate (S)
S is a system of inequalities containing variables for integers and rationals.

Value: A system S' simplified by repeated application of rule (E) or *false* if S turns out to be unsatisfiable.

The algorithm may be found in [KÄU87]. Before each application of (E) it is checked whether there is a variable for the integers with coefficient ±1. If this is the case, these variables are eliminated else it is checked whether there is a variable for the rationals. If there is such a variable, one of these variables will be eliminated else a variable for the integers with arbitrary coefficient will be eliminated. After each application of (E) the comparisons over the integers are reduced using the diophantine reduction. If S contains a satisfiable system of linear diophantine equations with the variables $x_1, \ldots, x_n$, its solution will be determined. It is well known that the solution can be written in the format $x_i = f_i(p)$, $1 \le i \le n$, where the f_i are multilinear functions. The x_i will be eliminated and the elements of p considered as new variables.

4. The Sup-Inf-Reduction

In order to simplify a system of linear inequalities over **Q** the Sup-Inf-Reduction performs the tasks:
  1. Decision of the satisfiability
  2. Elimination of subsumed inequalities
  3. Deduction of entailed equations
The second and third task is executed only, if the system is satisfiable.

The kernel of the Sup-Inf-Reduction is the Sup-Inf-Method developed by Bledsoe [BLE] and refined by Shostak [SHO]. The Sup-Inf-Method decides the unsatisfiability of systems of inequalities over the rational numbers. As it does not use the diophantine properties of the integers, the method is not complete for systems of inequalities over the integers. Nevertheless the procedure may be applied to systems of inequalities over **Z**. In this case the system is considered as system over **Q**. Thus its satisfiability is not found out in general, but as discussed in section 4.3 and 4.4 each subsumed inequality is eliminated and each entailed equation is deduced. Hence systems of inequalities over **Z** are treated in this chapter too.

A system S of inequalities can be transformed into an equivalent disjunction $S_1 \lor \ldots \lor S_k$ ($k \ge 1$) of systems of inequalities where each comparison has the format $t \le c$. (See [BUN] or [KÄU87].) An equation $t = c$ could be replaced by $t \le 0 \land t \ge 0$. But it is more convenient to use the equation for the elimination of a variable and to apply the Sup-Inf-Reduction to the new system obtained by cancelling the equation. We shall discuss this simplification in section 4.2.

4.1. Systems with Variables over Z and Q

We exclude the case where the system S can be partitioned into a system S_1 over **Z** and a system S_2 over **Q**. In order to eliminate the subsumed inequalities (section 4.3) and to determine the entailed equations (section 4.4) one considers S as system over the rationals. Then by use of methods adopted from quantifier elimination the variables over **Q** are eliminated. By this a disjunction of systems of inequalities over **Z** is obtained the satisfiability of which can be checked by methods presented in chapter 5. (For the elimination of the variables over **Q** see [KÄU87].)

Therefore it is sufficient to consider only systems of inequalities containing either variables over **Q** or variables over **Z**.

4.2. Equations in Systems of Inequalities

Let $S' = S \setminus \{t = 0\}$ and $t = 0$ be equivalent to $cx = t'$ ($x \nmid t'$). We have $c = 1$, if x is a variable for a rational number. If x is a variable for an integer, $c \ne \pm 1$ in general. x can be eliminated in S'. Let S_1 be the new system.

Proposition: Let x be a variable for a rational number or $c = 1$, if x is variable for an integer.

1. S is not satisfiable iff S_1 is not satisfiable.
2. Assume $x \notin g$: $S \Vdash g = 0$ iff $S_1 \Vdash g = 0$
3. Assume $x \lessdot g$ and let g_1 be the term obtained after elimination of x in g: $S \Vdash g = 0$ iff $S_1 \Vdash g_1 = 0$

Proof: See [KÄU87]. ∎

Let S be a system of inequalities and S' be obtained after application of the procedure eliminate (section 3.2.2) to S. Then the equations occuring in S' may be cancelled before submitting this system to the Sup-Inf-Reduction. The proposition above guarantees the correctness of this simplification.

4.3. Elimination of Subsumed Inequalities

Definition 1: Let S be a system of inequalities. An inequality $t \leq c$ is termed *subsumed*, if $t \leq c \in S$ and $S \setminus \{t \leq c\} \Vdash t \leq c$. (\ is the set theoretical difference.)

Because of $S \setminus \{t \leq c\} \Vdash t \leq c$ iff $(S \setminus \{t \leq c\}) \wedge \neg t \leq c$ is unsatisfiable, subsumed inequalities can be discovered using the Sup-Inf-Method.

S subsumes $t \leq c$ over **Z**, iff S subsumes $t \leq c$ over **Q**. Therefore a system of comparisons over **Z** can be treated as system over **Q** when the subsumed inequalities are eliminated.

4.4. Deduction of Entailed Equations

Definition 2: Let S be a satisfiable system of inequalities. An equation $t = c \notin S$ is termed *entailed*, if $S \Vdash t = c$.

If S entails an equation, then in general it entails an infinite set of them. For example the system

$$y \leq z+1 \wedge x+1 \leq y \wedge x \geq z$$

entails the equations

$$[*] \quad x = z \text{ and } y = z+1$$

but also $y + x = 2 z + 1$ and in general any linear combination

$$\lambda(x - z) + \mu(y - z) = \mu$$

of [*].

Therefore we must show: If S entails equations, a set $\{t_1 = c_1: 1 \leq i \leq n\}$ is entailed by S such that $S \Vdash t = c$ iff $\{t_1 = c_1: 1 \leq i \leq n\} \Vdash t = c$. The if-part of this equivalence holds because of the transitivity of the implication. The proof of the only-if part may be found in [KÄU86] or in [KÄU87].

Definition 3: $t = c$ is *linear combination* of the equations $\{t_i = c_i: 1 \leq i \leq n\}$, if there are rational numbers $\lambda_1,..., \lambda_n$ with $\lambda_1 t_1 + ... + \lambda_n t_n = t$ and $\lambda_1 c_1 + ... + \lambda_n c_n = c$.

Theorem 4: Suppose $Sx \leq c$ entails an equation. Then there is a subsystem $S_0 x \leq c_0$ such that any equation entailed by $Sx \leq c$ is a linear combination of the equations of $S_0 x = c_0$ and $S_0 x = c_0$ is entailed by $Sx \leq c$ too.

Proof: See [KÄU86] or [KÄU87]. ∎

The theorem allows to prove the equivalence stated above. $S_0 x = c$ is the system of equations looked for and as any equation $g^t x = c$ entailed by $Sx \leq c$ is a linear combination of the equations of $S_0 x = c$, $g^t x = c$ is entailed by $Sx \leq c$ too.

Now we must show how $S_0 x = c_0$ is determined. In order to do this, we need an important property the Sup-Inf-Method has. Suppose x is a variable occuring in the system S. If the set

$\{c: [S]_x{}^c$ is satisfiable$\}$

has an upper bound, the Sup-Inf-Method computes the least upper bound. (See Shostak [SHO].) The least upper bound of x will be denoted by $\sup_S(x, \varnothing)$.

Theorem 5: Suppose S is a system of inequalities $t_i \leq c_i$ $(1 \leq i \leq n)$. Let z be a fresh variable, $1 \leq j \leq n$, T the system consisting of $t_i \leq c_i$ $(1 \leq i \leq n, i \neq j)$ and $t_j + z \leq c_j$.
Then $S \Vdash t_j = c_j$ iff $\sup_T(z, \varnothing) = 0$.

Proof: See [KÄU87]. ■

Theorem 5 allows to sketch the procedure determining the entailed equations.

Entailed Equations
Let $S = \{t_i \leq c_i : 1 \leq i \leq n\}$ be satisfiable and z a fresh variable. For each i evaluate

$\sup_{T_i}(z, \varnothing)$, with $T_i = (S \setminus \{t_i \leq c_i\}) \wedge t_i + z \leq c_i$.

If $\sup_{T_i}(z, \varnothing) = 0$, S entails $t_i = c_i$.

Because of $S \Vdash t = c$ over $\mathbf{Z}$ iff $S \Vdash t = c$ over $\mathbf{Q}$ a system of inequalities over $\mathbf{Z}$ may be treated as system over $\mathbf{Q}$.

4.5. The Sup-Inf-Reduction

The procedure simplifies systems of inequalities over $\mathbf{Q}$ but can be applied to systems of inequalities over $\mathbf{Z}$ too. These systems are treated as if they were systems over $\mathbf{Q}$. Therefore the procedure does not find out the unsatisfiability of every system over $\mathbf{Z}$, since it does not exploit the diophantine properties of the integers.

Sup-Inf-Reduction (S)
S is a system of inequalities of the format $t \leq 0$. It is assumed that the procedure Eliminate, see section 3.2, is applied to S.

1. Determine whether S is unsatisfiable. If this is the case return "unsatisfiable S" else continue with step 2.
2. Eliminate the subsumed inequalities. Let S' be the system of inequalities obtained. Apply step 3 to it.
3. Determine the entailed equations $s_i^t x = c$ and replace the inequalities $s_i^t x \leq c$ by these equations. The system obtained is returned.

Remark
The behaviour of the algorithm is improved by some modifications. If S entails an equation $x = c$, where c is a number, this equation can be determined during execution of step 1. After computation of an entailed equation this one is used to eliminate a variable before step 3 is begun resp. continued.

5. A Reduction Procedure for Systems of Linear Inequalities over Z

In this chapter a reduction procedure for systems of inequalities over Z is presented. The procedure is incomplete but theorem 11 allows to characterize the systems of inequalities the satisfiability of which can be decided. To this end we must observe that a system of inequalities over Z defines a polyhedral subset of Q^n, if the system contains n disjoint variables and if it is considered as system over Q Therefore we shall adopt some terminology and some notions used in convex linear algebra.

Definition 6: Let S be a system of inequalities.
1. The set of all x fulfilling S is the *admissible domain* of S.
2. The *admissible domain* D of x is the set of all values $c \in Q$ for which $[S]_x{}^c$ is satisfiable. D and x
   are said to be *bounded*, if there is a $b \geq 0$ such that $|c| < b$ for all $c \in D$.
3. The admissible domain of S is *bounded*, if each variable occuring in S is bounded.

Admissible domains of systems of inequalities, which are polyhedral sets, are a subclass of the convex sets.

5,1. The Main Theorem for Polyhedral Sets

Polyhedral sets are the admissible domains of systems of inequalities. The main theorem stated in this section tells that they can also be defined by an appropriate sum of vectors.

Theorem 7: The admissible domain of a system S of inequalities is defined by

$$\sum_{i=1}^{M} \alpha_i\, v_i + \sum_{j=1}^{N} \beta_j\, w_j + \sum_{k=1}^{L} \gamma_k\, g_k$$

In this linear combination the

v_i constitute a convex combination, $\alpha_i \geq 0$ and $\sum \alpha_i = 1$,

$w_j \neq \theta$ span a cone and $\beta_j \geq 0$,

g_k are a basis of a subspace.

Proof: See for example Judin/Golstein [JG]. The theorem is due to Motzkin [MOT]. ∎

Assume we are given a system S of inequalities with an unbounded admissible domain B. Then it is easily seen that in the characterization of B according to theorem 7 $N \geq 1$ or $L \geq 1$ must hold. The subsequent lemma is an immediate consequence of this theorem.

Lemma 8: Let $a_i{}^t x \leq c_i$, $1 \leq i \leq m$ be a system of inequalities and its admissible domain B as in theorem 7. Then for all j and i $a_i{}^t w_j \leq 0$ and $a_i{}^t g_k = 0$.

Proof: Suppose $a_1{}^t w_1 > 0$. We have $a_1{}^t (\sum \alpha_i v_i + \beta_1 w_1) \leq c_1$ since $\sum \alpha_i v_i + \beta_1 w_1 \in B$. Define $d_1 = \sum \alpha_i a_1{}^t v_i$ and $d_2 = a_1{}^t w_1$. (Note $d_2 > 0$.) Then β_1 is bounded by $(c_1 - d_1)/d_2$ which contradicts theorem 7.
In the same way the assumption $a_1{}^t g_1 > 0$ yields a contradiction. ∎

5.2. The Integral Satisfiability of Unbounded Polyhedral Sets

x is said to be an **integral point**, if each component of x is an integer.

Definition 9: Let S be a system of inequalities. If S does not contain or imply an equation, S is said to be *strict*.

Lemma 10: Let $a_i{}^tx \leq c_i$, $1 \leq i \leq m$ be a strict system S of inequalities and its admissible domain B as in theorem 7. Then there exists a y such that

$$y = \sum_{i=1}^{M} \alpha_i \, v_i + \sum_{j=1}^{N} \beta_j \, w_j + \sum_{k=1}^{L} \gamma_k \, g_k$$

and each $\beta_j > 0$, $\gamma_k \neq 0$, $\alpha_i \geq 0$, $\sum \alpha_i = 1$ and for all i $a_i{}^t \, y < c_i$.

Proof: See [KÄU87]. ∎

Now we are ready to present a sufficient criterion for the satisfiability of a subclass of unbounded polyhedral sets by integral points.

Theorem 11: Let $a_i{}^tx \leq c_i$, $1 \leq i \leq m$ be a strict system S of inequalities with n variables and the admissible domain

$$B = \{ \sum_{i=1}^{M} \alpha_i \, v_i + \sum_{j=1}^{N} \beta_j \, w_j + \sum_{k=1}^{L} \gamma_k \, g_k \colon \beta_j \geq 0, \, \alpha_i \geq 0, \, \sum \alpha_i = 1 \}$$

Whenever an a_i is selected, if there is a w_j such that $a_i{}^t w_j < 0$, then S is fulfilled by an integral point.

Proof: Without loss of generality we may assume the coefficient vector a_i of each inequality to fulfill

$$\| a \| = \sqrt{a \, a} \; = 1$$

Because of lemma 10 there exists a $y \in B$ such that for all i $a_i{}^t y < c_i$. Furthermore if

$$y = \sum \alpha_i v_i + \sum \beta_j w_j + \sum \gamma_k g_k$$

each of the β_j and γ_k is different from zero. Now we consider the cone

$$y \, (\delta) = \sum \alpha_i \, v_i + \sum (\beta_j + \delta_j) w_j + \sum \gamma_k \, g_k, \text{ with } \delta = (\delta_1, \dots, \delta_M) \text{ and } \delta_j \geq 0 \text{ for all j}$$

for which

$$y \, (\delta) \in B \text{ for all } \delta \geq \theta$$

holds. If $\delta \, ' < \delta \, ''$, then $a_i{}^t y \, (\delta \, '') < a_i{}^t y \, (\delta \, ')$, because to each a_i there is a w_j such that $a_i{}^t w_j < 0$. As the coefficient vector of each inequality is a unit-vector $c_i - a_i{}^t y \, (\delta \, ')$ is the distance of $y(\delta)$ from $a_i{}^t x = c_i$. δ can be increased in such a way that for all i $c_i - a_i{}^t y(\delta') > n$. Then the sphere $K = \{x \colon \| x - y \, (\delta) \| \leq n \}$ is a subset of B. K contains an integral point.

Let e_i be the i-th unit-vector and $| \mu_i | < 1$ such that the i-th component of $y \, (\delta) + \mu_i e_i \in \mathbf{Z}$. The point

$$p = y \, (\delta) + \sum_{i=1}^{n} \mu_i e_i$$

has integral components only and because of

$$\| p - y \, (\delta) \| = \sqrt{\sum_{i=1}^{n} \mu_i^2} < n$$

p belongs to K. Hence p is an integral point fulfilling S. ∎

Remark

In the subsequent section we shall show that every system of inequalities can be transformed into a strict one.

5.3. A Reduction Procedure for Systems of Inequalities over Z

Reduce (S)

Let $S = \{a_i^t x \le c_i : 1 \le i \le m\}$. S is assumed to be submitted to the procedure Eliminate already and to contain no equation. Additionally each inequality of S is supposed to be simplified by the diophatine reduction.

1. The Sup-Inf-Reduction is applied to S. If S turns out to be unsatisfiable or if the admissible domain of a variable does not contain an integer, return "not satisfiable S". If entailed equations are deduced apply step 2 to S, else step 3.

2. Suppose $S = T \wedge E$ where E consists of equations only and T does not contain an equation.

   2.1. If E is not satisfiable, return: not satisfiable S

   2.2. E is satisfiable: Let $x_i = f_i(p)$, $1 \le i \le k$, be the solution of E and eliminate the x_i occuring in T.
   T becomes empty: return E. If an inconsistent comparison is obtained, return "not satisfiable S". If the diophantine reduction was applied during elimination, apply Reduce to the system obtained and return the result. In all other cases continue with step 3.

3. Let S_1 be the system obtained in the previous steps.

   3.1. There is a bounded variable x, l the lower and u the upper bound of its admissible domain.

   For each $j \in Z$ with $l \le j \le u$ evaluate the result of Reduce($[S_1]_x^j$).

   If one of the results is "satisfiable", return "satisfiable S" otherwise "not satisfiable S".

   3.2. Check the satisfiability of S according to theorem 11.

Remarks

1. The determination of the admissible domain is a by-product of the application of the Sup-Inf-Method. This result is used in step 3.1.

2. The procedure used for the elimination of variables applies the diophantine reduction to the new inequalities obtained. An inequality changed by this reduction it is possible that new equations are entailed. Therefore in step 2.2 Reduce is called in order to deduce these equations.

3. If a variable is bounded, a case analysis is performed. Thus step 3.2 executed the admissible domain of S is unbounded. The procedure is incomplete. If the admissible domain does not fulfil the criterion of theorem 11 nothing can be said about the satisfiability of S. We shall discuss this topic in more detail in the subsequent section.

4. An algorithm determining the w_j being necessary for the application of theorem 11 is contained in [KÄU87].

5. S_1 (step 3) is a strict system of inequalities, since after execution of step 2 E may be cancelled.

Examples

1. In order to find out the integral unsatisfiability of the system

$$2x + 5z + 1 \le 0 \wedge 2y + 3z + 2 \le 0 \wedge 2x + 4y + 11z + 5 \ge 0$$

one must determine its entailed equations

$$2x + 5z + 1 = 0 \wedge 2y + 3z + 2 = 0.$$

If the left one e.g. is used in order to eliminate x, the new system

$$2x + 5z + 1 = 0 \wedge 2y + 3z + 2 = 0 \wedge 2y + 3z + 2 \geq 0$$

is obtained. The second equation subsumes the inequality and permits to eliminate z in the first one. By this the equation is replaced by 6x - 10y - 7 = 0, which is found out to be unsatisfiable by the diophantine reduction.

2. Assume we are given the system $2x + 3y \geq 1 \wedge 5x + 3y \leq 1 \wedge 2x + y \leq 1$. The admissible values for x lie between 0 and 1/2 and those for y between 0 and 7/11 but assigning 0 to x and y does not satisfy S. By this the unsatisfiability of the system is established executing step 3.1. of the procedure.

3. In order to establish the satisfiability of

$$S = \{-x \ -y - 2z \leq 0, -x + y \leq 0, -y \ -z \leq 2, -x + 2y + z \leq 2\}$$

one must execute step 3.2. of the procedure, i.e. apply theorem 11. One has

$$w_1 = (1, 0, 0)^t, \quad w_2 = (2, 1, 0)^t.$$

It is easily seen, that theorem 11 implies the integral satisfiability of S.

5.4. The Incompleteness of the Procedure

Let $a_i{}^t x \leq c_i$, $1 \leq i \leq m$ be a strict system S of inequalities and

$$B = \{ \sum_{i=1}^{M} \alpha_i v_i + \sum_{j=1}^{N} \beta_j w_j + \sum_{k=1}^{L} \gamma_k g_k \colon \beta_j \geq 0, \alpha_i \geq 0, \sum \alpha_i = 1 \}$$

its unbounded admissible domain. The integral satisfiability cannot be decided using the procedure of section 5.3, if N = 0 or $a_i{}^t w_j = 0$ for all i and j. (In the threedimensional space B will be a prism, if N = 0. In the plane the problem will be trivial: S consists of at least two inequalities $a^t x \leq c_1$ and $a^t x \leq c_2$ with $c_1 > c_2$. If the diophantine reduction is applied to these inequalities already, $a^t x = c_1$ is satisfied by integral points and by this we obtain the integral satisfiability of S.)

As systems of linear inequalities over **Z** are a subclass of Presburger-Arithmetic, their satisfiability is decidable. Thus if the procedure of section 5.3 fails, one of the procedures deciding this theory could be applied. (See for these procedures Kreisel/Krivine [KR] or Cooper [COO72].) If one of the general decision procedures must be applied to S, and if $N \geq 1$ for the admissible domain B, it is possible to transform B into a polyhedral set for which N = 0 and which is integral satisfiable iff B is. The system of inequalities belonging to this new polyhedreal set will be smaller than S and thus the satisfiability is checked with less expense. (For further details see [KÄU87].)

5.5. Some Improvements of the Procedure

If in the system $a_i{}^t x \leq c_i$ each $c_i \geq 0$, the system is integral satisfiable.

If a variable occurs in each inequality not negative (not positive), the system the satisfiability of which is to be decided, may be replaced by a simpler one. To this end one cancels each inequality in which this variable occurs. An empty system obtained, the given system is satisfiable. If this is not the case, the reduction procedure must be applied to the new system. (See [KÄU87] for further details.) If in step 3 of the procedure of section 5.3 the system contains two inequalities equivalent to $c_1 \leq a^t x$ and $a^t x \leq c_2$ a case split is possible. For each j with $c_1 \leq j \leq c_2$ one replaces the two inequalities by $a^t x = j$ and applies the procedure to these systems.

6. Why a New Reduction Procedure for Systems of Linear Inequalities over the Integers?

1. One could ask whether an integer linear programming algorithm might be applicable. But the algorithm of Gomory cannot be trusted to terminate on unbounded polyhedral sets. (See Dantzig [DAN].)

2. The procedure shall be used for the simplification of systems of linear inequalities occuring in big theorems. Frequently these systems are satisfiable. In this case it is convenient to deduce the entailed equations and to remove subsumed comparisons, since then a simplification of the entire theorem will be possible. (See [KÄU86] for an example.)

3. It is intended to apply the procedure to theorems obtained in program verification. These theorems frequently contain comparisons with variables for integers and rationals. This must be observed by a reduction procedure.

4. The factorization of the coefficients of comparisons is necessary in order to transform them into canonical ones. Thus the diophantine reduction is not expensive. In many cases the application of this reduction transforms systems of inequalities over Z being satisfiable if considered as systems over Q into new systems which are not satisfiable over Q. Together with the deduction of entailed equations the number of systems having this property is increased.

5. In nearly all applications of a simpler version of the procedure to theorems obtained in program verification it was observed that an unsatisfiable system of inequalities over Z was not satisfiable over Q too. Thus a reduction procedure seems to be reasonable which applies algorithms with little expense first before using algorithms with higher costs, viz. proof by cases or application of theorem 11. Also the deduction of entailed equations makes possible to dispense with these procedures in most cases.

Acknowledgement: I am greatly indebted to Martin Tins who read an earlier version of this paper and implemented one of the reduction procedures.

This work was partially supported by the Deutsche Forschungsgemeinschaft SFB 314, "Künstliche Intelligenz - Wissensbasierte Systeme".

References

[BLE]   W.W. Bledsoe: A New Method for Proving Certain Presburger Formulas
          4th Int. Joint Conference on Artificial Intelligence,
          Tiblisi, pp.15-21: 1975
[BUN]   A. Bundy: The Computer Modelling of Mathematical Reasoning
          London: 1983; Academic Press
[COO72]   D.C. Cooper: Theorem Proving in Arithmetic without Multiplication
          Machine Intelligence 8, B. Meltzer and D. Michie, eds., New York: 1972
[DAN]   G.B. Dantzig: Lineare Programmierung und Erweiterungen
          Berlin, Heidelberg, New York: 1966
[JG]   D.B. Judin, E.G. Golstein: Lineare Optimierung I
          Berlin: 1986
[KÄU85]   Th. Käufl: The Simplifier of the Program Verifier "Tatzelwurm"
          Österreichische Artificial Intelligence-Tagung 1985
          Informatik-Fachberichte; Berlin, Heidelberg, New York: 1985; Springer
[KÄU86]   Th. Käufl: Program Verifier "Tatzelwurm": Reasoning about Systems of Linear Inequalities
          8th International Conference on Automated Deduction 1986
          Berlin, Heidelberg, New York: 1986; Springer
[KÄU87]   Th. Käufl: Reasoning about Systems of Linear Inequalities
          Technical Report 16/87, University of Karlsruhe, Institut für Logik, Komplexität und
          Deduktionssysteme: 1987
[KIN69]   J.C. King: A Program Verifier
          Ph.D. Thesis, Carnegie Mellon University, Pittsburgh: 1969
[KR]   G. Kreisel, J.-L. Krivine: Modelltheorie
          Berlin, Heidelberg, New York: 1972
[MOT]   T.S. Motzkin: Beiträge zur Theorie der linearen Ungleichungen
          Dissertation
          Zürich: 1936
[SHO]   R. Shostak: On the SUP-INF Method for Proving Presburger Formulas
          JACM 24/4, pp. 529 - 543: 1977

A Subsumption Algorithm
Based on Characteristic Matrices

Rolf Socher

FB Informatik, Universität Kaiserslautern

Postfach 3049, D-6750 Kaiserslautern, W.-Germany

Abstract: The elimination of subsumed clauses is a powerful rule in automated theorem provers. Since subsumption tests in general, however, must be repeated very often, the efficiency of the subsumption test is crucial for its use. Hence several algorithms to test whether one given clause subsumes another have been developed. One of them is the S-link test based on the connection graph procedure. The S-link test essentially tries to compute a merge of matching substitutions between two clauses. This paper presents an improvement of the S-link test based on the reduction of possible matches.

1. Introduction

Rules that eliminate redundant information play a prominent role among the methods to reduce the search space of proof procedures. One of these rules is the deletion of subsumed clauses. Since this is a very powerful rule which seems to be essential for solving more complicated problems [KM84], some efforts have been made to develop procedures that test clauses for subsumption. On the other hand the subsumption test must be repeated very often and hence its efficiency is decisive for its use. There are two basic ideas to test, whether a clause C subsumes another clause D: The first idea amounts to derive a contradiction from $C \wedge \neg D$ and the second to find a matching substitution σ tha maps C to D. The first method [CL73] is very expensive (in fact, the stronger notion of implication instead of the subsumption is tested), whereas the efficiency of the second method depends on the strategy of searching for the appropriate matching substitution σ. A strategy based on the second method and on the connection graph proof procedure is provided in [Ei81]. In a connection graph the information whether two literals are unifiable (or complementary unifiable, respectively) is explicitly given by the links between the literals. This provides an efficient preselection which singles out those clauses D that do not possess the appropriate links to the clause C. Having preselected the candidates one tries to compose the matching substitutions from literals in C to literals in D to find the matcher σ from C to D. In some cases a lot of such compositions are possible and hence the search for σ may become expensive. Especially this may occur if theory unification is involved, because there may be more than one most general theory unifier between two literals.

The method presented in this paper tries to improve this search by imposing restrictions on the possible matching substitutions. It is based on the idea of giving the variables and literals of a clause a characteristic property, which in fact denotes the information about the occurrences of variables in the various argument positions of a literal. An order for these characteristics is defined and it is shown that this order is compatible with the matching substitution σ from C to D. Thus all matchers that do not respect this order can be singled out.

2. Basic Notions

We use the standard notions of first order logic: Given pairwise disjoint alphabets, let $\mathbb{V}$ be the set of **variable** symbols; $\mathbb{F}_n$ the set of n-ary **function** symbols and $\mathbb{F} = \bigcup \mathbb{F}_n$. Let $\mathbb{P}_n$ be the set of n-ary **predicate** symbols and $\mathbb{P} = \bigcup \mathbb{P}_n$. Furthermore let $\mathbb{T}$ be the set of **terms**, i.e. $\mathbb{T}$ is the smallest set with $\mathbb{V} \subseteq \mathbb{T}$ and $f(t_1,...,t_n) \in \mathbb{T}$ for all $f \in \mathbb{F}_n$ and $t_i \in \mathbb{T}$. If $P \in \mathbb{P}_n$ and $t_i \in \mathbb{T}$ we call $P(t_1,...,t_n)$ an **atom**. **Literals** are signed atoms (+A or -A) and $\mathbb{L}$ is the set of all literals. **Clauses** are disjunctions of literals; in general we write them as sets. If C is a clause we write |C| for the number of literals of C.

For any object o containing variables we define $\mathbb{V}(o)$ as the set of all **variables occuring in** o.

A **substitution** is a mapping from variables to terms identical almost everywhere. It is extended to mappings on terms, atoms, literals and clauses by the usual homomorphism. In general we will represent a substitution σ with $x_i\sigma=t_i$ for $1 \leq i \leq n$ by the set of pairs $\{x_1 \rightarrow t_1,...,x_n \rightarrow t_n\}$. The **domain** of a substitution σ is the set $dom(\sigma):=\{x \in \mathbb{V} \mid x\sigma \neq x\}$.

Two substitutions σ and τ are called **strongly compatible**, if $\sigma\tau = \tau\sigma$.

A **matching substitution** from a term or literal s to a term or literal t is a substitution μ such that $s\mu=t$.

A **unifier** for two terms or literals s and t is a substitution σ such that $s\sigma = t\sigma$. The set of all unifiers of two terms s and t is denoted by U(s,t). If L is a literal and C is a clause, then $U(L,C):= \bigcup_{c \in C} U(L,c)$.

3. Subsumption

Let C and D be clauses. Then C σ-**subsumes** D, if $|C| \leq |D|$ and there is a substitution σ such that $C\sigma \subseteq D$. In this paper we deal only with σ-subsumption and simply use the term "subsumption" (Loveland [Lo78] uses "subsumption" for the implication $C \Rightarrow D$). We remark that the condition $|C| \leq |D|$ is not necessary for the following results. In the following we assume that D is a ground clause (i.e. contains no variables) when C subsumes D is tested. This is no loss of generality [GL85].

There are some well known algorithms to test whether a given clause subsumes another (e.g. [CL73], [St73], [GL85] and [Ei81]). We shortly describe an algorithm that is based on connection graphs. The property of connection graphs that is important in this context is the existence of links between unifiable literals, the so called S-links. The following definition and theorem can be found in [Ei81].

3.1 Definition:

Let C and D be clauses and L a literal from C. We define uni(C,L,D) as the set of all matching substitutions mapping L to some literal in D.

3.2 Theorem:

Let $C = \{L_1,...,L_n\}$ and D be clauses. Then C subsumes D iff $|C| \leq |D|$ and there is an n-tuple $(\sigma_1,...,\sigma_n) \in \times_{i=1..n} uni(C,L_i,D)$ such that the σ_i are pairwise strongly compatible.

3.3 Example:

Given the set $\{C,D_1,D_2,D_3\}$ of clauses with $C=\{Pxy,Qyc\}$, $D_1=\{Pac,Rbc\}$, $D_2=\{Puv,Qvw\}$ and $D_3=\{Pab,Pba,Qac\}$ one wants to find out, which clauses are subsumed by C. D_1 can be excluded, since the literal Qyz from C is not unifiable with any literal in D_1. D_2 cannot be a candidate either, since uni(C,Qyc,D_2)=$\emptyset$. For D_3 we obtain the two pairs (σ_1,τ) and (σ_2,τ), where $\sigma_1 = \{x \to a, y \to b\}$, $\sigma_2 = \{x \to b, y \to a\}$ and $\tau = \{y \to a\}$. From these two pairs only (σ_2,τ) is strongly compatible and thus C subsumes D_3.

This example shows that in order to find the clauses that are subsumed by a given clause $C=\{L_1,...,L_n\}$ first there is a preselection of those clauses that are connected to every literal in C by the S-links of the connection graph. If D is such a clause, then each literal in C is unifiable with some literal in D. For such a candidate clause D the subsumption algorithm is accomplished by a test of all elements of $\times_{i=1..n} uni(C,L_i,D)$ on strong compatibility.

Subsumption tests involving long clauses with more than one matching substitution for each literal may require an expensive search of all elements of the cartesian product. Especially this can arise when there is more than one most general unifier for two literals due to theory unification.

The aim of this paper is to reduce the effort for this test by a further preselection of the possible matching substitutions.

4. Characteristic Matrices

In this section we define a characteristic property of literals that provides a restriction for matching substitutions σ for the test C subsumes D. It rests upon the observation, that to each occurrence of a variable v at some argument position in C there must correspond an occurrence of $v\sigma$ at the same argument position in D.

First we define the characteristic only for clauses C that contain neither function symbols nor constants. It is easy to see that in this case we can further assume without loss of generality that D does not contain function symbols either.

4.1 Example:

Let $C=\{L_1,L_2\}$ and $D=\{K_1,K_2,K_3\}$ with $L_1=Pxy$, $L_2=Pyz$, $K_1=Pab$, $K_2=Pac$, $K_3=Pdb$. We have $|uni(C,L_1,D)| = |uni(C,L_2,D)| = 3$ and hence there are 9 possible combinations which are all incompatible. The reason is that the variable y occurs at first argument position in L_1 and at second position in L_2 and there is no argument in D that occurs in different literals at first and at second position. One could denote this by attaching to each argument v a pair (m,n) with $m,n \in \{0,1\}$ where m=1 iff v occurs at first argument position in some literal and n=1 iff v occurs at second argument position in some literal. Then y is attached the pair (1,1), whereas all arguments of D are attached either (1,0) or (0,1).

In this example all literals have the same predicate symbol. In general we do not just simply encode the occurrence of an argument v at position i by 0 or 1, we take the set of all predicate symbols where v occurs in position i instead. Furthermore we define the characteristic not for variables but for literals. This leads to the following definition:

4.2 Definition:

(i) For each literal or term L of the form $P(t_1,...,t_n)$ we define a **topsymbol selector** π_0 by $\pi_0(L):=P$ and **argument selectors** π_j by $\pi_j(L) := t_j$ for $1 \leq j \leq n$.

(ii) Let C be a clause and $L=P(t_1,...,t_n) \in C$. Let $m=\max\{k \mid \exists L \in C: L=Q(s_1,...,s_k)\}$, i.e. M is the maximal arity of predicate symbols occurring in literals in C. .
Then we define the **characteristic matrix** of L with respect to C, $\chi:=\chi_C(L)$ by
$\chi_{ij}:= \{\pi_0(K) \mid K \in C$ and $\pi_i(K)$ is defined and $\pi_j(L)=\pi_i(K)\}$ for $1 \leq j \leq n$ and $1 \leq i \leq m$.
In the following when writing characteristic matrices we omit the set braces and we write 0 for the empty set.

(iii) We define a relation $\leq$ on characteristic matrices χ and ψ of the same dimension by $\chi \leq \psi$ iff $\chi_{ij} \subseteq \psi_{ij}$ for all appropriate i,j.

4.3 Example:

Let $C = \{L,K\}$ with $L=Pxy$, $K=Qyx$. Then

$$\chi(L) = \begin{bmatrix} P\ Q \\ Q\ P \end{bmatrix} \text{ and } \chi(K) = \begin{bmatrix} Q\ P \\ P\ Q \end{bmatrix}$$

4.4 Lemma:

Let C be a clause whith no function and constant symbols and D a clause with no function symbols. If C σ-subsumes D, then $\chi_C(L) \leq \chi_D(L\sigma)$ holds for each $L \in C$.

Proof:

Suppose $C\sigma \subseteq D$ and $L \in C$. Let $A := \chi_C(L)$ and $B := \chi_D(L\sigma)$. It follows that $L = Px_1...x_n$ for some P and x_i and that the dimensions of A and B are equal. Let $P \in A_{ij}$ for some i,j. This implies that there is some $K \in C$ with $\pi_0(K) = P$ and $\pi_i(K) = \pi_j(L) = x_j$. Then $K\sigma \in D$ and $\pi_i(K\sigma) = \pi_j(L\sigma) = x_j\sigma$. Furthermore we have $\pi_0(K\sigma) = P$, from which follows $P \in B_{ij}$. Hence we have shown $A_{ij} \subseteq B_{ij}$ for all appropriate i,j and therefore $\chi_C(L) \leq \chi_D(L\sigma)$. ∎

Lemma 4.4 provides a restriction on the possible matching substitutions that is sufficient in many cases to exclude subsumption at all: If there is some literal L in C such that there exists no literal K in D with $\chi_C(L) \leq \chi_D(K)$ and $\pi_0(L) = \pi_0(K)$, then C cannot subsume D. The following example is taken from [GL85] and shows that the complexity of the subsumption test can be reduced.

4.5 Example:

Let $C_m = \{Pxy_1z_1, Pz_1y_2z_2,...,Pz_{m-2}y_{m-1}z_{m-1}, Pz_{m-1}xz_m\}$ and $D_k = \{Pab_1a,...,Pab_ka\}$. Each pair $(L_i,K_j) \in C_m \times D_k$ is unifiable. Let σ_{ij} be the unifier of (L_i,K_j).

a) The S-link test for subsumption needs k^m steps in the worst case and k^2 steps in the best case to detect that C does not subsume D:

For each literal $L_i \in C_m$ we have $uni(L_i,D) = \{\sigma_{ij} \mid 1 \leq j \leq k\}$ whence $|uni(L_i,D)| = k$. Therefore $|\times_{i=1..m}uni(L_i,D)| = k^m$. The number of steps needed for the search of a strongly consistent m-tuple $(\sigma_1,..,\sigma_m) \in \times_{i=1..m}uni(L_i,D)$ depends on the ordering of the literals in C_m and on the search strategy. Depth-first search (with uncontrolled backtracking) will always yield the worst case with complexity k^m. Breadth-first search will yield the best case when one starts with the literals L_1 and L_m and the worst case when one ends with one of the literals L_1 or L_m.

b) The subsumption test described in lemma 4.4 needs at most 2 steps:

We get the following characteristic matrices for the literals in C:

$$\chi(L_1) = \begin{bmatrix} P & 0 & P \\ P & P & 0 \\ 0 & 0 & P \end{bmatrix} \quad \chi(L_i) = \begin{bmatrix} P & 0 & P \\ 0 & P & 0 \\ P & 0 & P \end{bmatrix} \text{ for } 2 \leq i \leq m-1 \text{ and } \chi(L_m) = \begin{bmatrix} P & P & P \\ 0 & P & 0 \\ P & 0 & P \end{bmatrix}$$

and all the literals in D have the same characteristic matrix:

$$\chi(K) = \begin{bmatrix} P & 0 & P \\ 0 & P & 0 \\ P & 0 & P \end{bmatrix}$$

for each $K \in D$. We have to test if $\chi(L_i) \leq \chi(K)$ for the three matrices above. At least after two steps one can recognize that $\chi(L_1) \nleq \chi(K)$ (respectively $\chi(L_m) \nleq \chi(K)$).

In automated theorem proving subsumption tests are usually repeated very often. Therefore the computation of the characteristic matrices for a clause C can be made once for many subsumption tests.

Of course one has to realize that the computation of the set S of characteristic matrices for the clause C is an additional effort that has to be put into the subsumption algorithm. Thus one has to weigh the costs of the method against its possible gains. One can even put more effort into the computation of the set S: It is sufficient to test only those literals L_i for which $\chi(L_i)$ is maximal with respect to $\leq$, whether there is $K \in D$ with $\chi(L_i) \leq \chi(K)$. Since in the example above $\chi(L_i) \leq \chi(L_m)$ for $2 \leq i \leq m-1$ we need only consider the set $S' = \{\chi(L_1), \chi(L_m)\}$.

In order to extend the characteristic to arbitrary literals, we introduce some notation. The construction of the characteristic of a literal L depends only upon the occurrences of variables in L. Therefore we consider a literal of the form $P(t_1,...,t_n)$ as a function $P'(x_1,...,x_m)$ of its variables:

4.6 Definition:

Let L be a literal or term of the form $P(t_1,...,t_n)$ and let $(x_1,...,x_m)$ be the multiset of variables occurring in L. Then L can be written as a function P' of $x_1,...,x_m$. We write $L' := P'(x_1,...,x_m)$. If C is a clause, then we write $C' := \{L' \mid L \in C\}$.

4.7 Lemma:

Let L and K be literals or terms and σ, τ substitutions. Then $L\sigma = K\tau$ iff $L'\sigma = K'\tau$. ∎

The computation of the characteristic of the clause D requires only the consideration of those literals in D that are connected by S-links to some literal in C, i.e. those literals, that are instantiations of some literal in C. We write those literals with the predicate symbols occuring in C'. Therefore we have the following

4.8 Definition:

Let C and D be clauses. Then
$$M(D,C) := \{L'\mu \mid L \in C, \mu \in uni(C,L,D)\}.$$

4.9 Example:

Let $C = \{L_1, L_2, L_3\}$ with $L_1 = P(x,y)$, $L_2 = P(f(y),d)$ and $L_3 = R(a,g(x))$ and $D = \{P(a,b), P(f(b),d), R(a,f(b))\}$. For the construction of C' we introduce the new predicate symbols Q_1, Q_2, Q_3. The meanings of the Q_i can be illustrated by
$$Q_1 = P, Q_2 = P(f(.),d), Q_3 = R(a,g(.))$$
Then $C' = \{Q_1(x,y), Q_2(y), Q_3(x)\}$. We have the following unifiers:
$$uni(C,L_1,D) = \{\{x \rightarrow a, y \rightarrow b\}, \{x \rightarrow f(b), y \rightarrow d\}\}$$
$$uni(C,L_2,D) = \{\{y \rightarrow b\}\}$$
$$uni(C,L_3,D) = \emptyset$$

This yields

$$M(D,C) = \{ Q_1(a,b), \ Q_1(f(b),d), \ Q_2(b)\}$$

4.10 Lemma:

Let C and D be clauses. C subsumes D iff C' subsumes M(D,C).

Proof:

Assume there is a substitution σ with $C\sigma\subseteq D$. Let $K\in C'$. Then there is an $L\in C$ with $L'=K$. We have $K\sigma=L'\sigma\in M(D,C)$, since $\sigma\in$ uni(C,L,D). Hence $C'\sigma\subseteq M(D,C)$.

Now suppose there is a substitution σ with $C'\sigma\subseteq M(D,C)$. This implies that for each $L\in C$ there are K' with $K\in C$ and $\tau\in$ uni(C,K,D) with $L'\sigma=K'\tau$. From Lemma 4.4 we obtain $L\sigma=K\tau\in D$. This proves $C\sigma\subseteq D$. ∎

4.11 Corollary:

Let C and D be arbitrary clauses. If C σ-subsumes D, then $\chi_{C'}(L) \le \chi_{M(D,C)}(L\sigma)$ holds for each $L\in C'$.

Lemma 4.4 only gives a necessary condition for subsumption. The next lemma characterizes a situation, where it is sufficient to find to each literal L in the clause C a unifiable literal K in the clause D with $\chi(L)\le\chi(K)$.

4.12 Lemma:

Let C and D satisfy the assumptions of 4.4. Suppose that no two literals in C have the same predicate symbol and that there exists at most one literal in C that possesses more than one S-link to D. Then C subsumes D iff for each literal $L\in C$ there is a $K\in D$ with $\chi_C(L)\le\chi_D(K)$ and $\pi_0(L)=\pi_0(K)$.

Proof:

Suppose that for each literal $L\in C$ there is a $K\in D$ with $\chi_C(L)\le\chi_D(K)$ and $\pi_0(L)=\pi_0(K)$.

Let $C = \{L_0,L_1,...,L_n\}$ and let L_0 be the only literal in C that possesses eventually more than one S-link to D. Let uni(C,L_0,D)=$\{\sigma_1,...,\sigma_k\}$ and uni(C,L_j,D)=$\{\tau_j\}$ for $1\le j\le n$. We show that $(\sigma_i,\tau_1,...,\tau_n)$ is strongly compatible for all $1\le i\le k$. Let L and L' be different literals from C. Let $\varphi\in$ uni(C,L,D) and $\psi\in$ uni(C,L',D). Then $\varphi\ne\psi$ and at least one of L and L' is not equal to L_0, say $L\ne L_0$. Then uni(C,L,D)=$\{\varphi\}$ holds. We show that φ and ψ are strongly compatible:

Let $x\in$ dom(φ)∩dom(ψ). Then there are i,j such that $\pi_i(L)=x$ and $\pi_j(L')=x$.

Let $A:=\chi_C(L)$, $B:=\chi_D(L')$, $A':=\chi_C(L\varphi)$ and $B':=\chi_D(L'\psi)$. Then $\pi_0(L')\in A_{ij}$. Now there is a $M\in D$ with $\chi_C(L)\le\chi_D(M)$ and $\pi_0(L)=\pi_0(M)$. Since uni(L,D)=$\{\varphi\}$ we have $M=L\varphi$. This implies $A\le A'$ and from this follows $\pi_0(L'\psi) = \pi_0(L')\in A'_{ij}$. According to the definition of A this yields $\pi_i(L\varphi)=\pi_j(L'\psi)$. Hence $x\varphi = (\pi_i(L))\varphi= \pi_i(L\varphi) = \pi_j(L'\psi) = (\pi_j(L'))\psi= x\psi$.

We have shown that the elements of $(\sigma_i,\tau_1,...,\tau_n)$ are pairwise strongly compatible, and this implies that C subsumes D. ∎

The next example shows that the condition, that at most one literal of C is connected with more than one S-link to D, cannot be dropped:

4.13 Example:

Let $C=\{L_1,L_2\}$ with $L_1=Pxy$, $L_2=Qyx$ and $D=\{K_1, K_2, K_3, K_4\}$ with $K_1=Pab$, $K_2=Pcd$, $K_3=Qbc$ and $K_4=Qda$. Then $|uni(C,L_1,D)| = |uni(C,L_2,D)| = 2$ and the two characteristic matrices for C are those of example 4.3. We have $\chi_D(K_1) = \chi_D(K_2) = \chi_C(L_1)$ and $\chi_D(K_3) = \chi_D(K_4) = \chi_C(L_2)$ but C does not subsume D.

Now we can formulate the algorithm that improves the S-link test for C subsumes D:

Input: Clauses $C=\{L_1,...,L_n\}$ and D.

Output: True, if C subsumes D and false otherwise.

1. If there is a literal L in C that does not possess an S-link to a literal in D, then return false.
2. Compute $uni(C,L_i,D)$ for each literal L_i of C. If $uni(C,L_i,D)=\emptyset$ for some i, then return false.
3. Compute L' and M(D,C) and the set of all characteristic matrices of all literals according to 4.2.
4. If there is any matrix χ of L' such that for all matrices ψ of M(D,C) $\chi\not\ddagger\psi$ then return false.
5. If all predicate symbols of literals in L' are different and all literals except one possess exactly one S-link to D, then return true.
6. For all literals L_i in C: Remove from $uni(C,L_i,D)$ each σ with $\chi(L_i')\not\ddagger\chi(L_i\sigma)$.
7. If there is some $(\sigma_1,...,\sigma_n) \in \times_{i=1..n}uni(C,L_i,D)$ such that the σ_i are pairwise strongly compatible then return true, otherwise return false.

5. Summary

We have given an improvement of the S-link test for subsumption. Furthermore it was shown by some examples how the remaining complexity of the S-link test can be reduced with this method. On the other hand some cost has to be invested for the computation of the characteristic matrices of literals. This cost can be higher than the advantage of the method if the clause set is small. But for more complex examples the costs of the method surely will amortize.

The subsumption test based on characteristic matrices is currently being implemented in the Markgraph Karl Theorem Prover [KM84].

Acknowledgement

My thanks are due to Jörg Siekmann who read an earlier draft of the paper. Above all, I would like to thank my colleague Norbert Eisinger for his support and thorough reading of the paper which resulted in improvements both in style and substance.

References

[CL73]   Chang, C.L.; Lee, R.C.: Symbolic Logic and Mechanical Theorem Proving. Academic Press, New York (1973).

[Ei81]   Eisinger, N.: Subsumption and Connection Graphs. *Proc. of the 7th IJCAI*, Vancouver, 480-486 (1981).

[GL85]   Gottlob, G.; Leitsch, A.: On the Efficiency of Subsumption Algorithms. *Journal of the ACM*, 32/2, 280-295 (1985).

[KM84]   Karl, Mark G. Raph: The Markgraph Karl Refutation Procedure. Interner Bericht, SEKI-Memo MK-84-01, Universität Kaiserslautern (1984).

[Lo78]   Loveland, D.W.: Automated Theorem Proving : A Logical Basis. Fundamental Studies in Computer Science, North Holland (1978).

[St73]   Stillman, R.B.: The Concept of Weak Substitution in Theorem Proving. *Journal of the ACM*, 20/4, 648-667 (1973).

A Restriction of Factoring in Binary Resolution

Arkady Rabinov

Department of Computer Science
Stanford University
Stanford, California 94305

Abstract

Since the unconstrained use of factoring can lead to a growing number of irrelevant and redundant clauses, it is important to consider the strategies restricting the factoring in binary resolution. We propose in this paper FE-resolution – a strategy restricting factoring in binary resolution to the literals that are equivalent in the sense that each of them is the result of renaming the variables in any other. We show that FE-resolution is complete.

Key words and phrases. Factoring, binary resolution, theorem proving.

1 Introduction

We are going to study some restriction of binary resolution where only factoring satisfying a special equivalence principle is permitted. We call this resolution **FE-resolution** and show that this restricted principle is complete.

Factoring occurs when the set of literals resolved upon in one of the premises of resolution is not a singleton. Since the situations in which factors are needed are rare in practice[1] and since the unconstrained use of factoring can lead to a growing number of irrelevant and redundant clauses, resolution is often used without factoring. On the other hand, it is well known that the resolution principle is not complete unless factoring is permitted. For example, it is impossible to infer the empty clause from the clauses $\{P(u), P(v)\}$ and $\{\neg P(u), \neg P(v)\}$ without using factoring.

Maslov[2] noticed that it is not necessary to permit arbitrary factoring in binary resolution. He considered the effect of factoring on a particular literal and defined it to be *substantial* (he called it "essential") for this literal if at least one variable is replaced by a term which is not a variable, or if a variable which has a single occurrence in the clause is replaced by a variable which has two or more occurrences in the resulting factor. He announced the following theorem:

Factors that are substantial for each literal in the set resolved upon can be prohibited without loss of completeness of binary resolution.

A proof of Maslov's theorem was published by Sharonov[4]. He also noted that this restriction of factoring is incompatible with the deletion strategies.

FE-resolution proposed in this paper is a stronger restriction of factoring. We claim that resolution is still complete if factoring is restricted only to positive literals and moreover, the literals in the set resolved upon are equivalent in the sense that each of them is the result of renaming the variables in any other.

The following example shows that FE-resolution is incompatible with the deletion strategies. FE-resolution prohibits any factoring in the following two clauses $\{P(x, y, z, b), P(y, x, b, z)\}$ and $\{\neg P(x, y, z, b), \neg P(y, x, b, z)\}$ (here a and b are constants). On the other hand, any of the four resolvents obtained from these two clauses without using factoring is a tautology. Furthermore, the result of any resolution (obtained without prohibited factoring) among the two original clauses and the four additional resolvents is subsumed by one of them.

We do not know whether FE-resolution is compatible with other strategies.

2 FE-Resolution

We refer to Robinson[3] for the definitions of a substitution, a substitution component, a variable of a component, a term of a component, the x-standardization, the y-standardization, and the most general unifier.

We follow Robinson's definition of resolution:

A *resolvent* of clauses C_1 and C_2 is any clause of the form

$$(C_1 \setminus D_1)\xi_x\sigma \cup (C_2 \setminus D_2)\xi_y\sigma,$$

where

$D_1 \subseteq C_1$, $D_2 \subseteq C_2$, D_1 and D_2 are not empty;

ξ_x is the x-standardization of C_1, ξ_y is the y-standardization of C_2;

σ is the most general unifier for all atoms of $D_1\xi_x$ and $D_2\xi_y$;

$D_1\xi_x\sigma = \{A\}$ and $D_2\xi_y\sigma = \{\neg A\}$, where A is an atom.

We call D_1 and D_2 the *key sets* (or *sets resolved upon*). We also call D_1 the *positive* key set and D_2 the *negative* key set.

Recall that a clause C_1 is an *instance* of clause C_2 if there exists a substitution σ such that $C_1 = C_2\sigma$. We say that clauses C_1 and C_2 are *equivalent* and write $C_1 \approx C_2$ if each of them is an instance of the other. In other words, $C_1 \approx C_2$ iff C_1 and C_2 can be obtained from each other by renaming variables. It is obvious that relation $\approx$ is reflexive, symmetric and transitive.

Example: The literals $p(y, x)$ and $p(x, y)$ are equivalent, also the literals $p(f(w), y, y)$ and $p(f(u), v, v)$ are equivalent. But the literals $p(x, y)$, $p(f(w), y)$, $p(f_1(w), y)$, $p(x, a)$ (here a is a constant), $p(x, x)$ and $\neg p(x, y)$ are **not** equivalent to each other.

Definition An *FE-resolvent* of clauses C_1 and C_2 is any resolvent of C_1 and C_2 such that negative key set is a singleton and all literals in its positive key set are equivalent.

Example: Consider the clause

$$\{p(x,y),p(z,z),p(f(w),y),p(f_1(w),y),p(x,a),p(y,x),...\}$$

(where a is a constant). All six literals, explicitly shown above can participate in factoring in twenty nine various combinations of two or more literals, but only one combination $\{p(x,y),p(y,x)\}$ is allowed for FE-resolvent.

Definition *FE-resolution* is the restriction of binary resolution in which only FE-resolvents are generated.

The main result is that FE-resolution is complete.

Theorem 1 (Completeness of FE-resolution). *If there is a refutation of a set of clauses S by binary resolution, then there is a refutation of S by FE-resolution.*

3 Proof of the Theorem

We start with some lemmas and definitions.

Lemma 1. *If σ is the most general unifier then the variable of any component of σ does not occur in the term of any component.*

This property of the most general unifier follows immediately from Robinson's defining algorithm.

Lemma 2. *If σ is the most general unifier then*

$$\sigma = \sigma\sigma.$$

This property of most general unifiers follows immediately from Lemma 1.

An arbitrary substitution σ defines a function on the set of all literals. Any clause C is a set of literals, and by $C\sigma$ we denote the image of this set under σ. The following simple properties of the relationship between clauses and substitutions are obvious set-theoretical facts.

Lemma 3. *If C_1 and C_2 are clauses and σ is a substitution then*

$$(C_1 \cup C_2)\sigma = C_1\sigma \cup C_2\sigma,$$

$$C_1\sigma \setminus C_2\sigma \subseteq (C_1 \setminus C_2)\sigma,$$

$$(C_1 \cap C_2)\sigma \subseteq C_1\sigma \cap C_2\sigma,$$

and

$$\text{if } C_1 \subseteq C_2 \text{ then } C_1\sigma \subseteq C_2\sigma.$$

A substitution is *invertible* if the terms of all components are different variables.

The *inverse* of an invertible substitution $\sigma = \{x_1/v_1, ..., x_i/v_i\}$ is the substitution $\sigma^{-1} = \{v_1/x_1, ..., v_i/x_i\}$.

By definition, a substitution, which is the x-standardization or the y-standardization of some clause, is invertible. It is also obvious that if all variables of a literal L are among the variables of the components of an invertible substitution σ then $L\sigma\sigma^{-1} = L$. Therefore we have the following lemma:

Lemma 4. *If σ is the x-standardization or the y-standardization of a clause C and $C_1 \subseteq C$ then $C_1\sigma\sigma^{-1} = C_1$.*

Recall that a clause C *subsumes* a clause D if and only if there is a substitution σ such that $C\sigma \subseteq D$.

We will need the following version of the well known Lifting Lemma.

Lifting Lemma. *If C_1 subsumes K_1, C_2 subsumes K_2, and if K is a resolvent of K_1 and K_2, then either*

- C_1 *subsumes K or*

- C_2 *subsumes K or*

- *there exists a resolvent C of C_1 and C_2 such that C subsumes K.*

The ultimate goal of the following lemmas is to show that an arbitrary resolution can be replaced with a sequence of FE-resolutions with the result which subsumes original resolvent.

We consider the atom A and clause C to be fixed for the Lemmas 5-10.

The lemmas and the proofs do not depend on the meaning of A and C, however this meaning could provide some useful insight. Therefore, we present the following informal explanation: The atom A is the atom to which the key sets of the original clauses are unified in the original resolution, and the clause C is the result of the original resolution. The *key part* defined below is the result of various transformations and mixes of the original key sets, whereas the rest of the clause represents various transformations and mixes of the differences between original clauses and their key sets. We will prove that key part can be made empty, and at the same time the rest of the clause will subsume the result of original resolution.

Lemma 5. *There exists a constant M (with respect to literal A) with the following property:*

for any set of literals X if

- A *is an instance of any of the literals in X and*

- *any two literals in X are not equivalent to one another*

then the number of elements of X is less then M.

Proof: Let M be a set of different symbols in A which are not variables, and let the number of those symbols be m. Let n be the total number of symbols in A, N be the set of variables $x_1, ..., x_n$, S be a set of all the different string of elements of $N \cup M$ of the length less or equal to n. Obviously S is finite. And so let the cardinality of S be $M - 1$.

For arbitrary $L \in X$ consider its x-standardization L'. It is obvious that $L' \approx L$. Since A is an instance of L, A is also an instance of L'. Therefore the length of L' is less than or equal to the length of A and the only symbols which could be in L' are the symbols from M and the variables x_i. Since the total number of symbols in L' is less than or equal to n, not more than n different variables can belong to L' and therefore all these variables are in N. Therefore, it is clear that L' coincides with some $s \in S$. Since no two elements of X are equivalent, the x-standardization of another element of X must coincide with another element of S, consequently the x-standardization is an injection of X into S. This proves that the cardinality of X less than M.□

Let K be a clause and $L \in K$. We say that L is *minimal* in K if every literal $L' \in K$, such that L is an instance of L', is in turn an instance of L.

Lemma 6. *In any clause there exists a minimal literal.*

Proof: By induction on the number of literals in the clause.□

Recall that a clause is *positive (negative)* if all its literals are positive (negative).

Let B be a clause and let B_1 be a subset of B (possibly empty). We say that B_1 is a *key part* of B if there exists a substitution σ such that $B_1\sigma \subseteq \{A, \neg A\}$ and $(B \setminus B_1)\sigma \subseteq C$.

Lemma 7. *If*

(i) a nonempty set P_2 is a subset of the positive set P_1, which in turn is a key part of the clause P, and if

(ii) a nonempty negative set N_2 is a subset of N_1, which in turn is a key part of the clause N

then

A: *P and N can be resolved upon P_2 and N_2,*

B: *There exists Q_1 such that Q_1 is a key part of Q, where Q is a resolvent of P and N resolved upon P_2 and N_2, and*

C: *The number of negative literals in Q_1 is less than the number of negative literals in N_1.*

Proof:

A: Let ξ_x be the x-standardization of P and let ξ_x^{-1} be its inverse as defined earlier. Similarly, let ξ_y be the y-standardization of N, and let ξ_y^{-1} be its inverse. Since P_1 is a key part of P, we have for some σ_P

$$P_1\sigma_P \subseteq \{A, \neg A\}, \tag{1}$$

and

$$(P \setminus P_1)\sigma_P \subseteq C. \tag{2}$$

Similarly, we have for some σ_N

$$N_1\sigma_N \subseteq \{A, \neg A\}, \tag{3}$$

and

$$(N \setminus N_1)\sigma_N \subseteq C. \tag{4}$$

Let ζ_x be a substitution derived from the substitution $\xi_x^{-1}\sigma_P$ by deleting all components whose variables are not among the terms of ξ_x. Similarly, let ζ_y be a substitution derived from the substitution $\xi_y^{-1}\sigma_N$ by deleting all components whose variables are not among the terms of ξ_y.

All variables of the substitution ζ_x are x_i and all variables of the substitution ζ_y are y_i. Therefore the set $\zeta_x \cup \zeta_y$ is also a substitution. Denote this substitution by σ. Since none of the variables of $P_1\xi_x$ and $P\xi_x$ are among the variables of the deleted components of $\xi_x^{-1}\sigma_P$, it is obvious that $P_1\xi_x\zeta_x = P_1\xi_x\xi_x^{-1}\sigma_P$ and $P\xi_x\zeta_x = P\xi_x\xi_x^{-1}\sigma_P$. Similarly, $N_1\xi_y\zeta_y = N_1\xi_y\xi_y^{-1}\sigma_N$ and $N\xi_y\zeta_y = N\xi_y\xi_y^{-1}\sigma_N$. From these equations, the definition of σ and Lemma 4 it follows that

$$P_1\xi_x\sigma = P_1\xi_x\zeta_x = P_1\sigma_P, \tag{5}$$

$$N_1\xi_y\sigma = N_1\xi_y\zeta_y = N_1\sigma_N, \tag{6}$$

$$P \setminus P_1\xi_x\sigma = P \setminus P_1\xi_x\zeta_x = P \setminus P_1\sigma_P \tag{7}$$

and

$$N \setminus N_1\xi_y\sigma = N \setminus N_1\xi_y\zeta_y = N \setminus N_1\sigma_N. \tag{8}$$

From (3), (6) it follows that

$$N_1\xi_y\sigma \subseteq \{A, \neg A\} \tag{9}$$

and from (1), (5) it follows that

$$P_1\xi_x\sigma \subseteq \{A, \neg A\}. \tag{10}$$

This means that all atoms of $P_2\xi_x$ and $N_2\xi_y$ are unifiable. Since P_2 is positive and N_2 is negative, P_2 and N_2 can be used as key sets.

B: Let σ_1 be the most general unifier for $P_2\xi_x$ and $N_2\xi_y$. The resolvent of P and N resolved upon P_2 and N_2 is

$$Q = (P \setminus P_2)\xi_x\sigma_1 \cup (N \setminus N_2)\xi_y\sigma_1. \tag{11}$$

Since by (9) and (10) σ is unifier for $P_2\xi_x$ and $N_2\xi_y$, we have for some σ_z

$$\sigma = \sigma_1\sigma_z.$$

Together with Lemma 2 this implies

$$\sigma_1\sigma = \sigma_1\sigma_1\sigma_z = \sigma_1\sigma_z = \sigma. \tag{12}$$

We are going to prove that

$$Q_1 = (P_1 \setminus P_2)\xi_x\sigma_1 \cup (N_1 \setminus N_2)\xi_y\sigma_1 \tag{13}$$

is a key part of Q.

Since $P_1 \subseteq P$ and $N_1 \subseteq N$, Lemma 3 together with (11) implies

$$Q_1 \subseteq Q. \tag{14}$$

From (13) using Lemma 3 together with (12) we can infer

$$
\begin{aligned}
Q_1\sigma &= ((P_1 \setminus P_2)\xi_x\sigma_1 \cup (N_1 \setminus N_2)\xi_y\sigma_1)\sigma \\
&= (P_1 \setminus P_2)\xi_x\sigma \cup (N_1 \setminus N_2)\xi_y\sigma \\
&\subseteq P_1\xi_x\sigma \cup N_1\xi_y\sigma
\end{aligned}
$$

From this and (9) and (10) we have

$$Q_1\sigma \subseteq \{A, \neg A\}. \tag{15}$$

From (11) and (13) using Lemma 3 and (12) we can infer

$$
\begin{aligned}
(Q \setminus Q_1)\sigma &= (((P \setminus P_2)\xi_x\sigma_1 \cup (N \setminus N_2)\xi_y\sigma_1) \setminus ((P_1 \setminus P_2)\xi_x\sigma_1 \cup (N_1 \setminus N_2)\xi_y\sigma_1))\sigma \\
&\subseteq (((P \setminus P_2)\xi_x \cup (N \setminus N_2)\xi_y) \setminus ((P_1 \setminus P_2)\xi_x \cup (N_1 \setminus N_2)\xi_y))\sigma \\
&\subseteq (((P \setminus P_2)\xi_x \setminus (P_1 \setminus P_2)\xi_x) \cup ((N \setminus N_2)\xi_y \setminus (N_1 \setminus N_2)\xi_y))\sigma \\
&\subseteq (((P \setminus P_2) \setminus (P_1 \setminus P_2))\xi_x \cup ((N \setminus N_2) \setminus (N_1 \setminus N_2))\xi_y)\sigma \\
&\subseteq ((P \setminus P_1)\xi_x \cup (N \setminus N_1)\xi_y)\sigma
\end{aligned}
$$

It follows from (2), (4), (7) and (8) that

$$((P \setminus P_1)\xi_x \cup (N \setminus N_1)\xi_y)\sigma \subseteq C,$$

and therefore

$$(Q \setminus Q_1)\sigma \subseteq C.$$

Together with (14) and (15) this proves that Q_1 is key part of Q.

C: Since P_1 is positive $(P_1 \setminus P_2)\xi_x\sigma_1$ has no negative literals, and since N_2 is negative and not empty, $(N_1 \setminus N_2)\xi_y\sigma_1$ has fewer negative literals than N_1.

This concludes our proof of Lemma 7. $\square$

Let B, X be sets of literals. If no literal in B is an instance of any literal in X, then we say that B is *separated* from X.

Lemma 8. *Let* P, P_1, P_2, N, N_1, N_2 *be as in the statement of Lemma 7. If in addition*

(iii) N_1 *is separated from the set* X *and*

(iv) $P_1 \setminus P_2$ *is separated from the set* X

then Q_1 constructed in Lemma 7 is separated from X.

Proof: Assume that Q_1 is not separated from X. This means that there exists a literal $q \in Q_1$ and a literal $x \in X$, such that x is an instance of q. By definition (13) of Q_1, this means that either a) there exists $n \in N_1 \setminus N_2$ such that $n\xi_y\sigma_1 = q$ and consequently x is an instance of n, which contradicts (iii), or b) there exists $p \in (P_1 \setminus P_2)$ such that $p\xi_x\sigma_1 = q$ and consequently x is an instance of p, which contradicts (iv). These contradictions prove Lemma 8.□

Lemma 9. *If*

- *a nonempty set P_2 is a subset of a positive set P_1, which in turn is a key part of a clause P, such that $P_1 \setminus P_2$ is separated from a set X, and if*

- *a set N_1 is a key part of the clause N, such that N_1 is separated from the set X, and if*

- *all elements of P_2 are equivalent with respect to relation $\approx$*

then there exists a clause Q which is deducible by FE-resolution from N and P, and which has a positive key part Q_1 which is separated from X.

Proof: Consider the set M of all triples $\langle C, C_1, c \rangle$ where a) C is a clause , for which there exists an FE-resolution from the clauses P and N; b) C_1 is a key part of C such that C_1 is separated from X; and c) c is a number of negative literals in C_1. Let n be the number of negative literals in N_1. Since the triple $\langle N, N_1, n \rangle$ obviously belongs to M, it is clear that M is not empty.

Let the triple $\langle N', N_1', n' \rangle$ be a triple with minimal third component. We are going to prove that $n' = 0$. Assume that $n' > 0$. This means that N_1' has at least one negative literal L. Let $N_2' = \{L\}$. Since P, P_1, P_2, N', N_1', N_2' and X satisfy the conditions of Lemma 7 and Lemma 8, there exists a resolvent Q of P and N' mentioned in the conclusions of Lemma 7 and Lemma 8. Since all literals of P_2 are equivalent, and since N_2' is a singleton, the resolution mentioned in Lemma 7 and Lemma 8 is an FE-resolution, and therefore there exists an FE-resolution of Q from the clauses P and N'. Let Q_1 be a key part of Q mentioned in Lemma 7 and Lemma 8. Let q be the number of negative literals in Q_1. The choice of Q and Q_1 implies that $\langle Q, Q_1, q \rangle \in M$ and $q < n$. This contradicts the definition of n and proves that $n = 0$. This means that N_1' has no negative literals and proves that N' and N_1' are required Q and Q_1. This concludes our proof of the Lemma 9.□

Lemma 10. *If P has a positive key part P_1 and N has a negative key part N_1, then there exists a clause with empty key part and this clause is deducible by FE-resolution from the clauses P and N.*

Proof: Consider the set M of quadruples $\langle X, c, C, C_1 \rangle$ where

a) X is a set of positive literals such that A is an instance of any $x \in X$, and any two literals in X are not equivalent with respect to the relation $\approx$;

b) c is the number of literals in X;

c) C is a clause deducible by an FE-resolution from the clauses P and N;

d) C_1 is a key part of C, such that C_1 is positive and C_1 is separated from X.

Since $\langle \emptyset, 0, P, P_1 \rangle \in M$, M is not empty. From Lemma 5 it follows that there exists a constant K such that all $\langle X_C, c, C, C_1 \rangle$ in M satisfy $c \leq K$. Let $\langle X', p', P', P_1' \rangle$ be a quadruple with the maximal second component. We want to prove that P_1' is empty and therefore P' is required.

Assume that P_1' is not empty. By Lemma 6, there exist a literal L minimal in P_1'. Since P_1' is a key part, it follows that A is an instance of L. Consider the set $X' \cup \{L\}$, and let q be a number of literals in it. Since P_1' separated from X', L is not equivalent to any literal in X. Consequently $L \notin X'$ and therefore

$$q > p'. \tag{16}$$

Let $P_2' = \{p : p \in P_1' \wedge L \approx p\}$. Since L is minimal in P_1', our definition of P_2' implies that there is no literal in $P_1' \setminus P_2'$ of which L is an instance. Alternatively, $P_1' \setminus P_2'$ is separated from $\{L\}$. Since P_1' is separated from X' and consequently $P_1' \setminus P_2'$ is separated from X', this means that $P_1' \setminus P_2'$ is separated from $X' \cup \{L\}$. It is also obvious that since all literals of N_1 are negative and all literals of $X' \cup \{L\}$ are positive, N_1 is separated from $X' \cup \{L\}$. We see that P_2', P_1', P', $X' \cup \{L\}$, N_1 and N satisfy the conditions of Lemma 9. Therefore there exists a Q which is deducible by FE-resolution from N and P' and which has a positive key part Q_1 which is separated from $X' \cup \{L\}$. Since P' is deducible by FE-resolution from N and P, Q is also deducible by FE-resolution from N and P. Therefore $\langle X' \cup \{L\}, q, Q, Q_1 \rangle \in M$.

But this is impossible, because our selection of the quadruple $\langle X', p', P', P_1' \rangle$ as a quadruple in M with the maximal second component contradicts (16). This shows that our assumption that P_1 is not empty cannot hold. This in turn proves that P has empty key part.

This concludes our proof of Lemma 10.$\square$

Main Lemma. *If C is a resolvent of P and N then there exists a clause C' which subsumes C and which is deducible from P and N by FE-resolution .*

Proof: Let $P_1 \subseteq P$ and $N_1 \subseteq N$ be key sets. Let σ be the most general unifier of P_1 and N_1. Let $A = P_1 \sigma$. It is obvious that P_1 is a key part of P and N_1 is a key part of N. It follows from Lemma 10 that there exists a clause with empty key part and which is deducible by FE-resolution from P and N. By definition of key part this means that this clause subsumes C.

Theorem 1 (Completeness of FE-resolution). *If there is a refutation of the set of clauses S by resolution, then there is a refutation of S by FE-resolution.*

Proof: Let $K_0, K_1, \ldots, K_n$ be a refutation of S, i.e. a sequence of clauses such that each K_i is either a clause in S or a resolvent of clauses preceding K_i, and $K_n = \square$. We will prove by induction on k that for each k, $k \leq n$, there exists a clause C_k such that C_k subsumes K_k and C_k is deducible from S by FE-resolution.

Consider the case $k = 0$. It is obvious that $K_0 \in S$ and thus K_0 is also deducible by FE-resolution. Let $C_0 = K_0$. Obviously C_0 subsumes K_0.

Assume now that the inductive hypothesis holds for all $i < k$. If $K_k \in S$ then let $C_k = K_k$. It is obvious that C_k is deducible from S by FE-resolution and that C_k subsumes K_k.

If $K_k \notin S$ then K_k is a resolvent of K_i and K_j, where $i < k$ and $j < k$. By the inductive hypothesis C_i subsumes K_i and C_j subsumes K_j. Now the Lifting Lemma implies that either

a) C_i subsumes K_k, in which case we just let $C_k = C_i$, or

b) C_j subsumes K_k, in which case we just let $C_k = C_j$, or

c) there exists a resolvent C of C_i and C_j such that C subsumes K_k. In this case the Main Lemma implies that there exists a C_k such that C_k is deducible from C_i and C_j by FE-resolution and such that C_k subsumes C. And so C_k subsumes K_k. Since by the inductive hypothesis C_i and C_j deducible from S by FE-resolution, this means that C_k is deducible from S by FE-resolution. This concludes the inductive step.

Considering $k = n$ we can conclude that there exists C_n such that C_n is deducible by FE-resolution from S, and C_n subsumes K_n. Since K_n is $\square$, this implies that C_n is $\square$ and so there exists a refutation of S by FE-resolution. This concludes our proof of Theorem 1.$\square$

4   Acknowledgements

Vladimir Lifschitz introduced me to the problems of factoring and to the results of Maslov and Sharonov. I am also indebted to him for his time and patience.

References

[1] M.Genesereth and N. Nilsson, *Logical Foundations of Artificial Intelligence*, Los Altos, CA, Morgan Kaufmann, 1987.

[2] S. Maslov, Proof-search Strategies for Methods of the Resolution Type, *Machine Intelligence* **6**, 1971.

[3] J. Robinson, A Machine-Oriented Logic Based on the Resolution Principle, *Journal of the Association for Computing Machinery* **12**, 1965.

[4] V. Sharonov, Strategy of defactorization in resolution method (Russian) *Issledovania po Prikladnoi Matematike*, Kazan, USSR, 1974.

SUPPOSITION-BASED LOGIC
FOR AUTOMATED NONMONOTONIC REASONING

Philippe Besnard [1] — *Pierre Siegel* [2]

[1] IRISA
　Campus de Beaulieu
　35042 Rennes Cédex
　FRANCE

[2] GIA
　Université d'Aix-Marseille II
　13288 Marseille Cédex
　FRANCE

Abstract: We present a first-order logical system based on a standard first-order language which is enriched with new predicate symbols - supposition predicates. The logic is standard. The interesting non-standard features of a nonmonotonic logic are captured by considering whether a given consequence depends on the special predicates. A supposition theory contains formulas of the original language and special axioms governing the behaviour of the suppositions. Our system, when applied to suppositions theories, has many of the desired features of a nonmonotonic logic. We also discuss proof procedures issues.

1 INTRODUCTION

Nonmonotonic reasoning deals with implicit information, particularly information that cannot be extensively stated. A well-known example of a piece of knowledge from which arises a case of reasoning in face of implicit information is
"Typically, birds can fly".
This sentence means that a given bird should be ascribed the property of being able to fly insofar as doing so **does not yield a contradiction**. For instance, if we only know that "Tweety is a bird", we conclude that "Tweety can fly"; but if we additionally know that "Tweety can't fly", then we no longer draw our previous conclusion "Tweety can fly". The word "nonmonotonic" is intended to reflect the fact that adding further information ("Tweety can't fly") to one's knowledge K may cause one to abandon conclusions ("Tweety can fly") drawn in light of K alone.

Most existing nonmonotonic logics rely on condition "**does not yield a contradiction**" being given the meaning "**is consistent**" or "**is possible**". This appears to be quite a direct and natural thing to do: for instance, the above statement about birds reads
"If it is possible (consistent) that a given bird can fly then that bird can fly".
Expectedly enough, in those nonmonotonic logics, theories are then, roughly speaking, added information of the kind "it is possible (consistent) that...". As a first consequence, such an account of the constraint "does not yield a contradiction" makes it necessary for the nonmonotonic logics of the kind discussed to enjoy the following property:
If p then it is not possible (consistent) that not p
Further consequences concern relationships between the connectives such that negation and disjunction for instance, and the formal account of "possible" developed in such a nonmonotonic logic. In fact, one gets logical systems of excessive complexity (Boolos 1979) without even coming close to anything satisfactory for our purposes. In view of the rather pragmatic nature of nonmonotonic reasoning as performed by human beings, such an approach seems to be unreasonable. In contrast, it presumably makes more sense to

This research has been sponsored by CNRS (GRECO PRC IA)

follow human practice by taking nonmonotonic reasoning to rely on supposition but not on possibility, the latter notion being much stronger than the former. Indeed, it may very well happen that what is assumed is not possible. At a general level, making suppositions is free of constraints. On the other hand, specifying particular conditions under which certain suppositions can be made turns out to be of great interest. For instance, one surely wants not to suppose that a given bird can fly if it can be deduced that that bird cannot fly. So, no matter how weak a formalization of a notion of supposition might be, it is a good candidate for underlying a logical system for nonmonotonic reasoning. In this paper, we explore this issue by presenting a supposition-based nonmonotonic logic.

From a formal point of view, the supposition-based logic described in this paper is restricted to a first order language without modality. Theories we are dealing with are then actually first order theories. This enables us to apply standard properties of first order logic so as to get various results about our logic, especially compactness results from which we can define proof procedures for theories we call supposition theories.

2 DEFINITIONS AND PROPERTIES

2.1 Supposition language and supposition extensions of theories

Definition: Let L be a first order language with equality. For each formula p of L, a new relational symbol $suppose_{[p]}$ is introduced, whose arity is the number of free variables of p (if p is closed then $suppose_{[p]}$ is a propositional symbol). These new relational symbols are called **supposition symbols**.

Definition: The **supposition language** LS associated with L is the first order language obtained from L by introducing the set RS of supposition symbols of L.

Definition: If rs is a supposition symbol of arity n, and $t_1, \ldots, t_n$ are closed terms of LS then the positive literal $rs(t_1,\ldots,t_n)$ of LS is a **supposition**. If $n=0$, rs is a propositional symbol.

If p is the formula $\forall x\,(\,r(x) \rightarrow s(y)\,)$ of L, the corresponding new relational symbol of LS is $suppose_{[\forall x\,(\,r(x)\,\rightarrow\,s(y)\,)]}$ that has arity 1 because p has exactly one free variable y. Using this supposition symbol, we can construct literals such as

$$suppose_{[\forall x\,(\,r(x)\,\rightarrow\,s(y)\,)]}(z)$$
$$suppose_{[\forall x\,(\,r(x)\,\rightarrow\,s(y)\,)]}(y).$$

Notice that in the latter, the symbol y occurring in the supposition symbol has nothing to do with the variable y that is the argument in the literal. The reader should keep in mind that $suppose_{[\forall x\,(\,r(x)\,\rightarrow\,s(y)\,)]}$ is **only a notation** for the relational symbol of LS associated with $\forall x\,(\,r(x) \rightarrow s(y)\,)$. The two literals given above are not suppositions, as opposed to their closed instances like

$$suppose_{[\forall x\,(\,r(x)\,\rightarrow\,s(y)\,)]}(a)$$
$$suppose_{[\forall x\,(\,r(x)\,\rightarrow\,s(y)\,)]}(f(g(a),b)).$$

We insist that LS be a first order language. This enables us to apply properties and employ proof procedures from first order logic. We briefly discuss at the end of the paper other ways to give a logical account of a concept of supposition (modality, higher order).

Definition: If F is a consistent set of closed formulas of LS, and S a set of suppositions such that $F \cup S$ is consistent, then $E = F \cup S$ is an **extension** of F. Furthermore, if E is maximally consistent then E is a **maximal extension** of F.

It is important that an extension of a consistent set of closed formulas of LS be defined as that set supplemented with suppositions, that is, **closed literals**. This enables us to prove the following property about the existence of maximal extensions and to consider resolution methods to devise proof procedures for our logic.

2.2 Properties

Existence property: Let F be a set of closed formulas of LS.
1) Every extension of F is included in a maximal extension of F.
2) If F is consistent, then F has a maximal extension.

Proof.
The proof of 2) follows from that of 1) that we now give. Let $E_0 = F \cup S$ be an extension of F, and let $\{E_i\}$ be the set of extensions of F that contain E_0. We must show that $\{E_i\}$ has a greatest element for set inclusion. By Zorn's lemma, this holds if every totally ordered (with respect to set inclusion) subset of $\{E_i\}$ has an upper bound in $\{E_i\}$.
Let then $\{O_i\}$ be a totally ordered subset of $\{E_i\}$. Every O_i is consistent and is of the form $F \cup S \cup S_i$ where S_i is a set of suppositions.
Take $O = \cup_i O_i = F \cup S \cup_i S_i$. Since O is an upper bound for every O_i, it remains to prove that O is an extension of F. But O results from adding $S \cup_i S_i$ to F. To conclude the proof, it is then sufficient to show that O is consistent. Since O is a set of closed formulas, the compactness theorem tells us that O is consistent if every finite subset of O is consistent. Now, every finite subset X of O is contained in some extension E_i of F. It is immediate (definition of extensions) that this extension is consistent and so is X.

Notation: In order to improve readability, we write $F \wedge f$ instead of $F \cup \{f\}$ where f is a formula and F a set of formulas.

Compactness property: Let p be a closed formula.
1) If $E = F \cup S$ is a maximal extension of F, then $E \models p$ iff there exists a finite subset $\{s_1, ..., s_k\}$ of S such that $F \wedge s_1 \wedge ... \wedge s_k \models p$
2) The formula p is valid in some maximal extension of F iff there exists a finite number $s_1, ..., s_k$ of suppositions such that

$$F \wedge s_1 \wedge ... \wedge s_k \models p \qquad \text{and} \qquad F \wedge s_1 \wedge ... \wedge s_k \text{ is consistent}$$

Proof.

1) By the compactness theorem, $E = F \cup S \models p$ iff there exists a finite subset of E in which p is valid. For such a subset $\{s_1,...,s_k\}$ of S, evidently, $F \wedge s_1 \wedge ... \wedge s_k \models p$.

2) Let $E = F \cup S$ be a maximal extension in which p is valid. By 1), there exists a finite subset $\{s_1,...,s_k\}$ of S such that $F \wedge s_1 \wedge ... \wedge s_k \models p$. Since E is an extension, E is consistent and so is $F \wedge s_1 \wedge ... \wedge s_k$. Conversely, if $F \wedge s_1 \wedge ... \wedge s_k$ is consistent then $F \cup \{s_1,...,s_k\}$ is an extension of F, which, by the existence property, is contained in some maximal extension E of F. That p is valid in E follows from $F \wedge s_1 \wedge .. \wedge s_k \models p$

Proof-theoretic property: Let p be a closed formula.

1) If $E = F \cup S$ is a maximal extension of F, then $E \models p$ iff there exists a finite subset $\{s_1, ..., s_k\}$ of S such that

$$F \wedge s_1 \wedge ... \wedge s_k \models p$$
$$\Leftrightarrow \quad F \wedge \neg p \models \neg s_1 \vee ... \vee \neg s_k$$
$$\Leftrightarrow \quad Skolem(F \wedge \neg p) \models \neg s_1 \vee ... \vee \neg s_k$$

2) The formula p is valid in some maximal extension of F iff there exists a finite number $s_1, ..., s_k$ of suppositions such that

	$F \wedge s_1 \wedge ... \wedge s_k \models p$	and $F \wedge s_1 \wedge ... \wedge s_k$ is consistent
$\Leftrightarrow$	$F \wedge \neg p \models \neg s_1 \vee ... \vee \neg s_k$	and $F \not\models \neg s_1 \vee ... \vee \neg s_k$
$\Leftrightarrow$	$Skolem(F \wedge \neg p) \models \neg s_1 \vee ... \vee \neg s_k$	and $Skolem(F) \not\models \neg s_1 \vee ... \vee \neg s_k$

$Skolem(F \wedge \neg p)$ is the set of clauses obtained from skolemizing $F \wedge \neg p$, that is, replacing variables that are existentially quantified by functional terms with Skolem symbols. Be careful that a closed literal in which Skolem symbols occur and whose relational symbol is a supposition symbol, is not a supposition. The reason for this is that Skolem symbols are not in the language LS. Hence, in order to prove that p is valid in some extension, it doesn't help to exhibit a negative clause with an occurrence of such a literal.

Proof.

By the compactness property, 2) follows from 1) that we now prove. If $E = F \cup S \models p$ then $F \wedge s_1 \wedge .. \wedge s_k \models p$ by the compactness property. So, $F \wedge \neg p \models \neg s_1 \vee .. \vee \neg s_k$. The last equivalence (it would not hold if some formula $\neg s_i$ were not closed) is proved by:

$$F \wedge \neg p \models \neg s_1 \vee ... \vee \neg s_k$$
$$\Leftrightarrow \quad Skolem(F \wedge \neg p \wedge s_1 \wedge ... \wedge s_k) \quad \text{is inconsistent (Herbrand theorem)}$$
$$\Leftrightarrow \quad Skolem(F \wedge \neg p) \cup Skolem(s_1 \wedge ... \wedge s_k) \quad \text{is inconsistent}$$

because $Skolem(u \wedge v)$ is $Skolem(u) \cup Skolem(v)$ whenever u and v are closed formulas

$$\Leftrightarrow \quad Skolem(F \wedge \neg p) \wedge s_1 \wedge ... \wedge s_k \quad \text{is inconsistent}$$

because $Skolem(u)$ is u itself whenever u is a variable-free clause

$$\Leftrightarrow \quad Skolem(F \wedge \neg p) \models \neg s_1 \vee ... \vee \neg s_k$$

2.3 About proof procedures

The proof-theoretic property yields proof procedures, not complete in the general case, that can be based on resolution methods ($Skolem(F \wedge \neg p)$ being actually a set of clauses). Indeed, in order to prove that a closed formula p is valid in at least one extension of F it is enough to exhibit a variable-free clause, all of whose literals are denials of suppositions, which is deducible from $Skolem(F \wedge \neg p)$ but not from $Skolem(F)$. Since we are dealing with nonmonotonic reasoning, it is not surprising that the proof procedures just outlined are incomplete: the best illustration stems from the condition "...not deducible from $Skolem(F)$" by which the undecidability of satisfiability comes into play.

The last equivalence stated in the proof-theoretic property suggests that a promising approach to devise proof procedures for supposition-based logic centers around particular resolution methods especially incremental ones or even better, the production field method.

A resolution method is said to be **incremental** if, given a set of clauses C, it first focusses on the search for a deduction of a clause c from C prior to enlarging the search to a deduction of c from a superset C' of C. Here this means that if a clause c all of whose literals are denials of suppositions were valid in $Skolem(F)$ then that clause would be found before any clause of $Skolem(F \wedge \neg p)$ not in $Skolem(F)$ is resolved upon. So, a single run of an incremental resolution method can jointly handle both steps (establishing the deducibility of an appropriate clause c from $Skolem(F \wedge \neg p)$ and the nondeducibility of c from $Skolem(F)$). Previous applications of incremental resolution methods to nonmonotonic reasoning are due to Bossu & Siegel (1982), Besnard & al. (1983),...

If CS is the set of clauses all of whose literals are denials of suppositions then an incremental resolution method should give all clauses of CS deducible from $Skolem(F)$ prior to giving the clauses of CS deducible from $Skolem(F \wedge \neg p)$. This is a variant of:
        "Given a set of clauses P,
        find those clauses of P deducible from a particular set C of clauses"
where P is called a **production field** (Siegel 1987). In practice, a production field is most usefully specified by means of characteristic properties enjoyed by the clauses it consists of. In order to find those clauses of P valid in C, it is theoretically sufficient to test, for each clause c of P, whether $C \models c$, by aiming at proving the inconsistency of $C \cup \{\neg c\}$. Such an approach is of course unrealistic in the general case because the production field is just too large. A means to solve this general problem is by using a resolution method based on literals being ordered inside clauses: A-ordering (Kowalski & Hayes 1969), saturation (Bossu & Siegel 1985),... These methods generate all resolvants for resolution on head literals (recall that literals are ordered in a clause) and the resulting set of clauses (free of tautologies and not allowing for the subsumption of clauses) is said to be saturated. If the ordering used forces literals that are denials of suppositions to occur in the last part of clauses then a saturated set contains all clauses of CS deducible from $Skolem(F \wedge \neg p)$. This kind of method is decidable for the class of groundable clauses (Bossu & Siegel 1985). From SL-resolution (Kowalski & Kuehner 1971), Siegel (1987) devised a far more efficient method which does not effectively construct saturated sets.

3 SUPPOSITION THEORIES

3.1 Definition and consistency property

Of course, the first order language LS defined above offers a rather poor account of the concept of supposition from a semantical point of view. This is deliberate because we are looking for a convenient and rather general framework in which one can express theories displaying various aspects of the notion of supposition. There are two ways a logic may reflect, at least partially, semantical features of a notion of supposition. First, this can be done through dedicated axiom schemata or inference rules. Second, properties reminiscent of the idea of supposition can be described within specific theories. The former way yields a more uniform account of the idea of supposition but the latter allows for more flexibility as various aspects of the concept of supposition can be captured. We now describe a case in the latter way: a special class of theories of our supposition-based logic is defined in which constraints are imposed upon occurrences of supposition symbols. These constraints are rather weak, as we limit ourselves to the description of a minimal framework. Stronger constraints may be imposed, these are investigated in a forthcoming paper.

Definition: A **supposition theory** of LS is the union $G = W \cup D$, of a set W of closed formulas of L and a set D of **supposition rules**, each of which corresponds to
— a closed formula of the form:
$$\forall X \, \forall Y \, (\ suppose_{[p]}(X) \wedge q \ \rightarrow \ (\, r \wedge p \,) \,)$$
    where p, q and r are formulas of L and X consists of all free variables of p
— and to a closed axiom schema of the form:
$$suppose_{[\forall X \forall Z \, (\alpha \rightarrow p)]} \ \rightarrow \ \forall X \, \forall Z \, (\alpha \rightarrow suppose_{[p]}(X) \,)$$
    where Z consists of all free variables of α (which is any formula of L).

A supposition theory $G = W \cup D$ is then a first order theory of LS. Also, keep in mind that $suppose_{[\forall X \forall Z \, (\alpha \rightarrow p)]}$ is only a notation for a propositional symbol in which symbols X and Y have nothing to do with the sequences of variables X and Y.

Consistency property:
Let W be a closed formula of L and D a set of supposition rules.
1) If W is satisfiable then $G = W \cup D$ is satisfiable.
2) If W is satisfiable then $G = W \cup D$ has an extension.

Proof.
 1) If M is a model of W, the interpretation identical to M over L in which $suppose_{[p]}$ is false for all formulas p of L is a model of G since D is equivalent to a set of implicative formulas in all of which at least one supposition symbol occurs left to the $\rightarrow$ connective.
2) By 1), if W is consistent then so is G. By the existence property, G has an extension.

3.2 Discussion

The closed formula attached to a supposition rule can be seen as meaning the following:

"p being supposed, if q is true then r is true" $\qquad$ (1)

"p being supposed, if q is true then p is true" $\qquad$ (2)

The first sentence expresses a conditional relationship between q and r ("if q then r") subordinate to the supposition p. The second sentence simply means that whenever that supposition is made and the requirements of the conditional relation are met ("q is true"), then the conditions of the supposition should be verified ("p is true").

The axiom schema attached to a supposition rule provides a minimal restriction to the use of suppositions. The problem of semantics for supposition symbols is ignored but this at least discards uses of suppositions that are in flagrant contradiction with any intuitively uncontroversial features of the notion of supposition. Moreover, the framework provided here allows for a number of interpretations of the notion of supposition. For instance, it allows for an interpretation requiring the general rule "if $\vdash p$ then $\vdash suppose_{[p]}$", so that the notion of supposition involved gets closer to the notion of possibility. Through the general rule "if $\vdash suppose_{[p]}$ then $\vdash p$" the notion of supposition involved gets closer to the notion of belief developed in autoepistemic logic (Moore 1985). However, first order logic is in general not sufficient if such rules are to be taken into account and this is a fundamental difference between most nonmonotonic logics and our supposition-based logic.

A significant instance of the axiom schema is the following one:

$$suppose_{[\forall X(1 \to p)]} \to \forall X\, (\, 1 \to suppose_{[p]}(X)\,)$$

where 1 denotes any tautology. This instance means that if we suppose the universal closure of a formula then we suppose all instances of that formula. Such instances play a central role in the derivation of universally quantified formulas valid in maximal extensions

Another significant instance of the schema is the next one:

$$suppose_{[\forall X(p \to p)]} \to \forall X\, (\, p \to suppose_{[p]}(X)\,)$$

where $\forall X\, (p \to p)$ is a valid formula whose supposition is $suppose_{[\forall X(p \to p)]}$ from which it is possible to deduce $\forall X\, (p \to suppose_{[p]}(X)\,)$.

4 EXAMPLES

That "typically, students are young people" can be expressed by a supposition rule with formula e and axiom schema s given next:

e. $\qquad \forall x\, (\, Stu(x) \wedge suppose_{[Young(x)]}(x) \to Young(x)\,)$

s. $\qquad \forall x\, \forall Y\, (\, suppose_{[\forall x\, \forall Y\, (\alpha \to Young(x))]} \to (\alpha \to suppose_{[Young(x)]}(x)\,)\,)$

The relational symbol $suppose_{[Young(x)]}$ of LS has arity 1 as it is attached to the open formula $Young(x)$ of L. Similarly, $suppose_{[\forall x\, Young(x)]}$ is the supposition symbol attached to the closed formula $\forall x\, Young(x)$. As often happens, e is indeed of the form $\forall X\, \forall Y\, (suppose_{[p]}(X) \wedge q \to (r \wedge p))$ where p and r are the same.

Example 1: Dealing with general laws

Let $G = W \cup D = \{e\} \cup \{s\}$, a supposition theory in which W is empty and D consists of the unique supposition rule given above. We can then deduce the universally quantified formula

a1. $\forall x \ (Stu(x) \rightarrow Young(x))$ " All students are young people"

G has only one maximal extension, resulting from adding to G all the suppositions of the language LS, in particular $suppose_{[\forall x \ Young(x)]}$, $suppose_{[\neg \forall x \ Young(x)]}$, $suppose_{[Young(baby)]}$, $suppose_{[\neg Young(baby)]}$, $suppose_{[\forall x \ Stu(x)]}$ ...

In this extension, $suppose_{[\forall x \ 1 \rightarrow Young(x)]}$ yields the formula a1 using e **and** s. If the axiom schema s were to be missing, it would not be possible to derive that quantified formula but only instances of it, as happens in default logic (Reiter 1980). Precisely, the instance of s to be used is

s1. $suppose_{[\forall x \ 1 \rightarrow Young(x)]} \rightarrow \forall x \ (1 \rightarrow suppose_{[Young(x)]}(x))$

and the result follows immediately since the maximal extension contains

$suppose_{[\forall x \ 1 \rightarrow Young(x)]}$.

As regards a resolution-based proof procedure, we show that a1 is valid in some extension of G by computing $Skolem(G \wedge \neg a1)$. We get four clauses:

   $\neg Stu(x) \ \vee \ \neg suppose_{[Young(x)]}(x) \ \vee Young(x)$

   $\neg suppose_{[\forall x \ 1 \rightarrow Young(x)]} \vee \ suppose_{[Young(x)]}(x)$

   $Stu(sk)$

   $\neg Young(sk)$

where sk is a Skolem symbol. After elementary resolution steps, we get the clause

   $\neg suppose_{[\forall x \ 1 \rightarrow Young(x)]}$

which is not valid in $Skolem(G)$, so that it shows that a1 is valid in an extension of G. The variable-free clause

   $\neg suppose_{[Young(x)]}(sk)$

is also a resolvent for the initial four clauses but cannot be taken as a solution because sk is not a symbol of LS so that $suppose_{[Young(x)]}(sk)$ is not a supposition.

Example 2: Instantiations and exceptions

We now take $G = W \cup D$ where D is as above and W is:

   $F \ = \{ Stu(baby), Stu(paul), \neg Young(paul) \}$

Again, G has only one maximal extension E very much like above but in which of course there is none of the two suppositions $suppose_{[Young(paul)]}$ and $suppose_{[\forall x \ Young(x)]}$. By using F, e and s1 it can be proved that the denials of these two suppositions are valid in G. Also, E contains $suppose_{[Young(baby)]}$ from which $Young(baby)$ can be deduced.

Of course, it is no longer possible to derive

a1. $\forall x \ (Stu(x) \rightarrow Young(x))$

but it is possible to derive

a2.　$\forall x$　$(Stu(x) \wedge (x{\neq}paul) \rightarrow Young(x))$.

This can be established from the following formulas of G :

F.　　$Stu(baby) \wedge Stu(paul) \wedge \neg Young(paul)$

e.　　$\forall x$　$(Stu(x) \wedge suppose_{[Young(x)]}(x) \rightarrow Young(x))$

s2.　$\forall x$　$(suppose_{[\forall x\ x{\neq}paul\ \rightarrow\ Young(x)]} \rightarrow (x{\neq}paul \rightarrow suppose_{[Young(x)]}(x)))$

The result follows from e and s2 as the maximal extension of G contains the supposition

$suppose_{[\forall x\ x{\neq}paul\ \rightarrow\ Young(x)]}$.

Turning to a resolution-based proof procedure, we get from $Skolem(G{\wedge}\neg a2)$ the clauses:

$\neg Stu(x) \vee \neg suppose_{[Young(x)]}(x) \vee Young(x)$

$\neg suppose_{[\forall x\ x{\neq}paul\ \rightarrow\ Young(x)]} \vee x{=}paul \vee suppose_{[Young(x)]}(x)$

$Stu(sk)$

$\neg Young(sk)$

$sk \neq paul$

where sk is a Skolem symbol. From these clauses resolution yields the negative clause

$\neg suppose_{[\forall x\ x{\neq}paul\ \rightarrow\ Young(x)]}$.

5 COMPARISONS AND CONCLUSION

5.1 Comparison with circumscription

Among all existing nonmonotonic logics, the one our supposition-based logic is closest to turns out to be circumscription (McCarthy 1980). Both stick to a pure first order language. In this way, there is no difficulty about having equivalent semantical and syntactical approaches and it is possible for methods of automated deduction devised for first order logic to be applied. As another consequence, both logics behave satisfactorily with respect to connectives and quantifiers.

Roughly speaking, circumscription (of a predicate $p(X)$) can be approximated via the supposition rule

$\forall X$　$(suppose_{[\neg p(X)]}(X) \rightarrow \neg p(X))$

provided that the notion of inference is taken to be validity in all maximal extensions instead of validity in at least one maximal extension.

Problems with circumscription have been pointed out by Etherington et al. (1985). As opposed to our supposition-based logic, circumscription need not preserve the consistency of theories. Also, circumscription cannot affect the equality relation: circumscribing the predicate $p(x)$ in the theory consisting of the unique formula $p(a)$ does not yield the conclusion $\neg p(b)$ because $a{\neq}b$ is not valid in the initial theory. In fact, it is very much as though all theories contained the supposition rules

$\forall x\ \forall y$　$(suppose_{[x=y]}(x,y) \rightarrow x{=}y)$

$\forall x\ \forall y$　$(suppose_{[x{\neq}y]}(x,y) \rightarrow x{\neq}y)$.

5.2 Comparison with default logic and autoepistemic logic

Problems related to the way connectives and quantifiers behave surface in logics such as default logic because their languages are not pure first order. Basically, they are propositional logics as shown by their quantified versions that are rather limited generalizations involving quantification. An illustration is furnished by example 1 which can be stated in those logics with the result that it is possible, for any given student, to conclude that that student is young whereas there is no way to conclude that all students are young. As regards connectives, here is an example about disjunction: in default logic, given typically if a then c and typically if b then c, it happens that c cannot be concluded from a or b.

5.3 Conclusion

The language of our supposition-based logic is not as appealing as it could be. Instead of working with a set of supposition symbols, we could define one non-extensive predicate "suppose" of higher order or a modal operator. This complexifies problems about semantics and proof procedures to such an extent that we are led to define fragments that do not clearly seem stronger than the logic developed here. It would be more interesting to supplement LS with other supposition symbols so as to get a language LSS. By closure under this operation we would get a language LS* enjoying the properties given in the paper.

The logic presented has many advantages from the standpoint of the formalization of nonmonotonic reasoning. It also has advantages in view of the fact that it allows for pure first order proof procedures to be applied. But it introduces the equality predicate and this appears to be for the time being, the main burden of our logic due to computational reasons

6. REFERENCES

Besnard, Quiniou & Quinton (1983) A theorem-prover for a decidable subset of default logic, *Proc. AAAI-83*, 27-30.

Boolos (1979) *The unprovability of consistency*, Cambridge University Press.

Bossu & Siegel (1982) Nonmonotonic reasoning and databases, in: *Advances in database theory* (ed. Gallaire, Minker & Nicolas), Plenum Press, 239-284.

Bossu & Siegel (1985) Saturation, nonmonotonic reasoning and closed-world assumption, *A.I. 25*, 13-63.

Etherington, Mercer & Reiter (1985) On the adequacy of predicate circumscription for closed world reasoning, *Comp. Int. 1*, 11-15.

Kowalski & Hayes (1969) Semantic trees in automatic theorem-proving, in: *Machine Intelligence 4* (ed. Meltzer & Michie), American Elsevier, 87-101.

Kowalski & Kuehner (1971) Linear resolution with selection function, *A.I. 2*, 227-260.

McCarthy (1980) Circumscription - A form of nonmonotonic reasoning, *A.I. 13*, 27-39.

Moore (1985) Semantical considerations on nonmonotonic logic, *A.I. 25*, 75-94.

Reiter (1980) A logic for default reasoning, *A.I. 13*, 81-132.

Siegel (1987) *Représentation de connaissances en calcul propositionnel*, thèse, Marseille.

Argument-Bounded Algorithms as a Basis for Automated Termination Proofs

Christoph Walther
Universität Karlsruhe [1]

Abstract We present a method to automate termination proofs for recursively defined algorithms. We introduce the concept of an argument-bounded algorithm, show the important role of these algorithms for termination proofs, and present a procedure how to recognize argument-bounded algorithms. Using argument-bounded algorithms, hypotheses (in form of first-order formulas) how a given algorithm may terminate, can be synthesized by machine. The proposed procedure for termination proofs works automatically, i.e. without any direct or indirect human support. We give several examples to illustrate the power of our technique. Our proof procedure has been implemented and proved successful on several example runs, including the database from [Boyer and Moore, 1979].

1 Introduction

A central problem in the development of correct software is to verify that algorithms always terminate, provided the intended operations have decidable domains. Non-terminating algorithms compute partial operations, hence machine resources are wasted if a given input is not in the domain of the implemented operation. Also, manpower is wasted with the debugging of those algorithms, and the frustration caused by non-terminating programs is a common experience of programmers and computer scientists. Therefore, techniques to verify termination are of considerable interest in computer science, but, of course, as the halting problem is the "classical" undecidable problem, we cannot find a procedure which proves or falsifies the termination of all algorithms.

The termination of an algorithm is proved by invention of a _well-founded order relation_ for the algorithm and then verification that the arguments of each recursive call are less - in the sense of the invented order - than the input initially given to the algorithm. This paper is concerned with the automatization of termination proofs for a certain class of algorithms. We will consider here only algorithms, which terminate according to well-founded orders based on the so-called _size order_. This order compares data objects by their _size_, e.g. stacks are compared by their depth, lists by their length, trees by the number of their nodes etc. In this paper we shall present a technique to prove termination automatically for this class of algorithms, where throughout this paper, "automatic" is used as a synonym for "without any direct or indirect human support".

Let us illustrate our approach to termination proofs by an introductory example: Suppose we want to design a sorting algorithm for lists of natural numbers. This data structure can be defined in our notation by

      structure empty add(head:number tail:list):list

---

[1] Author's address: Institut für Logik, Komplexität und Deduktionssysteme, Universität Karlsruhe, Postfach 6980, D 7500 Karlsruhe 1, Federal Republic of Germany. This work was supported in part by the _Sonderforschungsbereich 314_ "Künstliche Intelligenz und Wissensbasierte Systeme" of the _Deutsche Forschungsgemeinschaft._

where number (standing for the natural numbers) denotes another (data) structure defined elsewhere. The symbols empty and add are the *constructors* of list, i.e. each list equals empty or else can be constructed by applying add to elements of the structures number and list. The symbols head and tail are the *selectors* of add and serve as kinds of inverse operations to the constructor, yielding the first element of a list and the list with the first element removed respectively.

As an example of an algorithm in our notation, we define an algorithm remove, which removes all occurrences of a number from a list:

> *function* remove(n:number x:list):list $\Leftarrow$
>> *if* x$\equiv$empty *then* x
>> *if* x$\equiv$add(head(x) tail(x)) $\wedge$ head(x)$\equiv$n *then* remove(n tail(x))
>> *if* x$\equiv$add(head(x) tail(x)) $\wedge$ $\neg$head(x)$\equiv$n *then* add(head(x) remove(n tail(x)))

Given an algorithm *function* minimum(x:list):number $\Leftarrow$ ... , which returns a minimal element of a non-empty list x (in the sense of some order relation for number defined elsewhere), we can define a sorting algorithm[2] for list by:

> *function* sort(x:list):list $\Leftarrow$
>> *if* x$\equiv$empty *then* x
>> *if* x$\equiv$add(head(x) tail(x)) *then* add(minimum(x) sort(remove(minimum(x) x))).

Now suppose we have to verify the termination of sort. Then a typical argumentation would read as: "To prove the termination of sort, we have to find a well-founded order relation $<_{\mathcal{R}}$ such that remove(minimum(x) x) $<_{\mathcal{R}}$ x , whenever x$\equiv$add(head(x) tail(x)) holds true. Our first observation is, that remove(n x) always returns x or a list strictly shorter than x, in symbols

$$(1) \qquad \text{remove}(n\ x) \leq_{\#} x\ ,$$

hence $<_{\#}$ could be the well-founded order we are looking for. Now searching for a condition, which guarantees, that remove(n x) $<_{\#}$ x, we find out that this requirement is satisfied iff n is a *member* of x. So, if we define list-membership as

> (2) *function* member(n:number x:list):bool $\Leftarrow$
>> *if* x$\equiv$empty *then* false
>> *if* x$\equiv$add(head(x) tail(x)) $\wedge$ head(x)$\equiv$n *then* true
>> *if* x$\equiv$add(head(x) tail(x)) $\wedge$ $\neg$head(x)$\equiv$n *then* member(n tail(x))

we can prove that

$$(3) \qquad \forall n\text{:number}\ \forall x\text{:list}\quad \text{member}(n\ x)\equiv\text{true} \leftrightarrow \text{remove}(n\ x) <_{\#} x\ .$$

---

[2] This algorithm also "purges" a list, i.e. multiple occurences of list elements are eliminated. We could have defined a "real" sorting algorithm using *delete* [Boyer and Moore, 1979] instead of *remove*. However, we prefer this version of *sort*, because we need *remove* for subsequent illustrations. The reader may verify, that our technique works for *delete* in the same way as it does for *remove*.

Our next observation is, that each non-empty list contains its minimum, i.e. we can also prove

(4) $\forall$x:list   x$\equiv$add(head(x) tail(x)) $\rightarrow$ member(minimum(x) x)$\equiv$true .

From (4) and (3) we obtain finally

$\forall$x:list   x$\equiv$add(head(x) tail(x)) $\rightarrow$ remove(minimum(x) x) $<_{\#}$ x

and the termination of sort is proved".

Our termination proof procedure mechanizes this chain of reasoning as an instance of the *generate-and-test* paradigm, cf. [Winston, 1984]: Analyzing algorithms, such as remove above, we are able to *recognize* properties as (1) and *synthesize* algorithms as (2) such that properties as (3) invariantly hold true, thus generating hypotheses, as e.g. (4), how a given algorithm, as e.g. sort, may terminate. Then we use an *induction theorem prover* as a "tester" to verify the generated hypotheses and this completes the termination proof for the algorithm under consideration.

One may observe, that in our example the definition of minimum, of course, is relevant for the semantics of sort to compute an ordered permutation of a purged list, but is *completely irrelevant* for the *termination* of sort, as long as formula (4) holds true. Hence sort also terminates if we let minimum return the last element of a non-empty list (where sort reverses a purged list), or even if minimum(x) returns head(x) (where sort computes the purged list). Therefore, the premises generated by our technique are *necessary* and *sufficient* for the termination of the algorithm under consideration.

Since our approach does not require special theorem proving abilities, we will concentrate on the "generate"-mode of our technique. We will assume instead that we have access to a powerful induction theorem prover. For instance, the system of Boyer and Moore [Boyer and Moore, 1979] would satisfy our requirements.

Note, that induction theorem proving systems have to solve their own termination problems (and in fact, the work presented here was inspired by the development of such a system [Biundo et al., 1986]). Our approach to termination proofs is almost the same as the approach implemented in the system of Boyer and Moore. But an essential difference is that the "generate"-part is left to a *human user* of the Boyer-Moore system, who has to invent *auxiliary algorithms*, as e.g. member, and to discover so-called *induction lemmata*, as e.g. formula (3) above, to *guide the system* for a termination proof. Hence our proposal can be viewed as an attempt to mechanize the Boyer-Moore user in this respect.

The remainder of this paper is organized in the following way: After giving some formal preliminaries in Section 2, we settle the environment, in which we define and use data structures and algorithms in Section 3. Section 4 proposes a *semi*-automatic technique for proving termination of algorithms. There we assume that certain knowledge, which is relevant for a termination proof (as e.g. (1), (2) and (3) in the example above), is already given.

The remaining sections are concerned with an *automatic acquisition* of this knowledge: In Section 5 we present a procedure, which recognizes, whether an algorithm is *argument-bounded*, i.e. whether the result of an algorithm always is less than or equal to one of its arguments (in the sense of the size order). With this technique, for instance, remove will be recognized as argument-bounded automatically, and a fact as (1) above will be found by machine. Section 6 shows, how a *difference-algorithm* can be synthesized automatically for each argument-bounded algorithm. This algorithm returns true for a list of arguments, if and only if the given argument-bounded algorithm (applied to the same arguments) returns a result *strictly less* than one of its arguments. Finally, we show in Section 7, how each difference-algorithm is *optimized*. In the example above, an optimized algorithm like member will be synthesized for remove, such that a property like (3) invariantly holds true. This completes the presentation of our approach to automated termination proofs, and we conclude with some remarks in Section 8.

The use of an induction theorem prover is crucial for our approach, not only for proving the termination hypotheses generated, but also for the recognition of argument-bounded algorithms and the optimization of the difference-algorithms. We also give the formulas, which have to be proved for these reasons, in order to convince a reader that we do not shift the problem of automating termination proofs to a theorem proving problem, which is hard to solve for a machine. However, we do not claim, that our proposal is a panacea for the automatization of termination proofs. We illustrate in Section 7, where our method fails (and how to recover) - either because a non argument-bounded algorithm is defined by a user, or because an algorithm does not compute an argument-bounded operation.

We also investigate the soundness of our approach, but with the halting problem being undecidable, we cannot display any completeness proof for our technique. Hence we must content ourselves with *pragmatic* aspects to judge the usefulness of our proposal. We present a digest of algorithms in [Walther, 1987], and show which termination hypotheses are generated for them. The reader may verify, that we do not use definitions, which are especially tailored for our method. There we also show the difference-algorithms synthesized for each argument-bounded algorithm in the digest. This paper is a shortened version of the technical report [Walther, 1987], to which the reader is referred for any omitted proofs.

2 Formal Preleminaries

Subsequently, we shall assume familiarity with the basic notions of first-order logic and of equality reasoning, as given, for instance, in [Goguen et al., 1978; Huet and Oppen, 1980].

2.1 Syntax

We assume we are given a non-empty set S of *sort symbols* which are names for the various domains under consideration. Given any non-empty set M, we let M* denote the set of all finite sequences (or strings) of elements of M, including the empty sequence λ, and we use M^+ as an abbreviation for $M^* \setminus \{\lambda\}$. For an S-indexed family of sets $M = (M_s)_{s \in S}$ and any $w = s_1 \ldots s_k \in S^*$, M_w denotes the set of all finite sequences $m_1 \ldots m_k \in M^*$ with $m_i \in M_{s_i}$ for $1 \leq i \leq k$, where $M_\lambda = \{\lambda\}$.

An S-*sorted signature* $\Sigma = (\Sigma_{w,s})_{w \in S^*, s \in S}$ is an S^+-indexed family of pairwise disjoint sets. Any $f \in \Sigma_{w,s}$ is a *function symbol* of *rank* ws, *arity* w and of *sort* s. $f \in \Sigma_{w,s}$ is called *reflexive* iff s occurs in w and *irreflexive* otherwise. Each S-sorted signature is separated into a *constructor part* and a *defined part*. The constructor part of Σ, denoted Σ^c, is a signature such that (1) $\Sigma^c_{w,s} \subseteq \Sigma_{w,s}$ and (2) $\Sigma_{\lambda,s} \subseteq \Sigma^c_{\lambda,s}$ for all $w \in S^*$ and all $s \in S$. The elements of Σ^c are called *constructor function symbols* or *constructors* for short. The *defined part* Σ^d of Σ is a signature such that $\Sigma^d_{w,s} = \Sigma_{w,s} \setminus \Sigma^c_{w,s}$ for all $w \in S^*$ and all $s \in S$. An element of Σ^d is called a *defined function symbol* or a *defined symbol* for short.

$V = (V_s)_{s \in S}$ is an S-indexed family of non-empty, pairwise disjoint, and infinite sets. Elements of V_s are called *variables* of sort s. We assume that $V \cap \Sigma = \emptyset$. For any $s \in S$, $T(\Sigma, V)_s$ denotes the set of all (*well formed*) Σ-*terms* (over Σ and V) of sort s, which is mutually recursively defined as the smallest subset of $(\Sigma \cup V)^*$ satisfying (1) $V_s \subseteq T(\Sigma, V)_s$ and (2) $ft^* \in T(\Sigma, V)_s$, if $f \in \Sigma_{w,s}$ and $t^* \in T(\Sigma, V)_w$ for some $w \in S^*$. To ease readability, we shall use also parentheses, i.e. f(t*), when we write terms in *examples.* $\mathcal{T}(\Sigma,\mathcal{V})=(\mathcal{T}(\Sigma,\mathcal{V})_s)_{s\in S}$ is the set of all Σ-terms (over Σ and $\mathcal{V}$). A Σ-term t is called an *f-term* iff t=ft* for some f$\in\Sigma_{w,s}$ and some t*$\in\mathcal{T}(\Sigma,\mathcal{V})_w$. For a Σ-term t, $\mathcal{V}$(t) is the set of all variable symbols in t and we define $\mathcal{V}(T)=\cup_{t\in T}\mathcal{V}$(t) for a set of Σ-terms T.

$\mathcal{F}(\Sigma,\mathcal{V})$ is the set of all (*well formed*) Σ-*formulas* (over Σ and $\mathcal{V}$), built with the junctors $\wedge, \vee, \rightarrow,$ $\leftrightarrow$, and $\neg$, and the quantifier symbol $\forall$ from the *atomar formulas* TRUE, FALSE (denoting *truth* and *falsity*) and $t_1\equiv t_2$ (for all $t_1,t_2\in\mathcal{T}(\Sigma,\mathcal{V})_s$). A Σ-formula $t_1\equiv t_2$ is called a Σ-*equation*, i.e. $\equiv$ denotes the *syntactic* equality. A Σ-*literal* is an atomar formula or a Σ-formula of the form $\neg\varphi$, where φ is an atomar formula. $Lit(\Sigma,\mathcal{V})$ denotes the set of all Σ-literals. For a quantifier-free Σ-formula φ, $\mathcal{V}(\varphi)$ is the set of all variable symbols in φ and we define $\mathcal{V}(\Phi)=\cup_{\varphi\in\Phi}\mathcal{V}(\varphi)$ for a set of quantifier-free Σ-formulas Φ. For a finite and non-empty subset $\Phi=\{\varphi_1, ... ,\varphi_n\}$ of $\mathcal{F}(\Sigma,\mathcal{V})$, $\vee\Phi$ is an abbreviation for the Σ-formula $[\varphi_1\vee ... \vee \varphi_n]$, and $\vee\emptyset$ is defined as the Σ-formula FALSE. A mapping $\sigma:\mathcal{V}_s\rightarrow\mathcal{T}(\Sigma,\mathcal{V})_s$ with σ(x)=x almost everywhere is called a Σ-*substitution*. SUB$(\Sigma,\mathcal{V})$ is the set of all Σ-substitutions. Σ-substitutions are extended as *endomorphisms* to mappings from $\mathcal{T}(\Sigma,\mathcal{V})_s$ to $\mathcal{T}(\Sigma,\mathcal{V})_s$. We shall apply Σ-substitutions also to quantifier-free Σ-formulas. Given a sequence $x^* = x_1...x_{|w|} \in \mathcal{V}_w$, σ(x*) denotes the corresponding sequence $\sigma(x^*) = \sigma(x_1)...\sigma(x_{|w|}) \in \mathcal{T}(\Sigma,\mathcal{V})_w$.

2.2 Semantics

Given an S-sorted signature Σ, a Σ-algebra A is a pair $(\mathcal{A},\alpha)$, where $\mathcal{A}=(\mathcal{A}_s)_{s\in S}$ is an S-indexed family of sets and $\alpha=(\alpha_f)_{f\in\Sigma}$ is a Σ-indexed family of mappings, such that $\alpha_f:\mathcal{A}_w\rightarrow\mathcal{A}_s$ for each f$\in\Sigma_{w,s}$. $\mathcal{A}_s$ is called the *carrier* for s and α_f is the *operation* of A named by f. When A is known from the context, we will write, for instance, *foo* instead of α_{foo} in subsequent examples. A Σ-algebra A is a *standard Σ-algebra* iff A, viewed as a Σ^c-algebra, is an *initial* algebra, cf. [Goguen et al., 1978].

A mapping $a:\mathcal{V}_s\rightarrow\mathcal{A}_s$ is an A-*assignment* for the Σ-algebra A. Given an A-assignment a, some $x^*\in\mathcal{V}_w$, and some $a^*\in\mathcal{A}_w$, $a[x^*/a^*]$ denotes the A-assignment, which assigns the variables in x* to the corresponding members of a*, and otherwise agrees with a. A standard Σ-*interpretation* I is a pair (A,a_I) such that a_I is an A-assignment and A is a standard Σ-algebra. $I[x^*/a^*]$ denotes the standard Σ-interpretation $(A,a_I[x^*/a^*])$. I is a *standard model* of a Σ-formula φ (or a set of Σ-formulas Φ) iff I is a standard Σ-interpretation, which satisfies φ (or Φ respectively), in symbols $I \models \varphi$ (or $I \models \Phi$). Since we do not need assignments for *closed* Σ-formulas (or sets of *closed* Σ-formulas), we use standard Σ-algebras also as standard Σ-interpretations and standard models for closed formulas. Given an S-indexed family $\mathcal{A}=(\mathcal{A}_s)_{s\in S}$ and w$\in S^*$, a relation $<_\mathcal{R}\subseteq\mathcal{A}_w\times\mathcal{A}_w$ is an *order* (relation) of $\mathcal{A}_w$ iff $<_\mathcal{R}$ is an *irreflexive* and *transitive* relation. $<_\mathcal{R}$ is a *well-founded order* of $\mathcal{A}_w$, if in addition $<_\mathcal{R}$ satisfies the *minimum condition*, i.e. each non-empty subset $\mathcal{B}_w$ of $\mathcal{A}_w$ has (at least) one $<_\mathcal{R}$-minimal element, cf. [Cohn, 1981].

3 Specifications

To have a framework to deal with data structures and algorithms formally, we introduce the notion of a *specification*: A specification S is a triple (S,Σ,Φ), where S is a set of sort symbols, Σ is an S-sorted signature and Φ is a finite set of closed Σ-formulas, called the *axioms* of S. A Σ-algebra M is a *standard model* of the specification S iff M is a standard model of Φ. S is *admissible* iff S has a standard model, which is *unique* up to Σ-*isomorphism*. For an admissible specification S, the *theory* Th(S) of S is defined as Th(S) = $\{\varphi \in \mathcal{F}(\Sigma,\mathcal{V}) \mid \varphi$ is closed and M $\vDash \varphi\}$, where M is a standard model of S. Hence each admissible specification[3] *specifies a theory* and we may ask, whether a Σ-formula φ holds in the theory, i.e. whether $\varphi \in$ Th(S).

3.1 Theorem Proving by Induction

Since we are interested in automated deduction, we have to formulate *syntactical* requirements, which are sufficient for $\varphi \in$ Th(S). Suppose, we have a sound and complete first-order calculus $\mathcal{K}$, where $\Psi \vdash \psi$ denotes that ψ can be deduced in $\mathcal{K}$ from the set of hypotheses Ψ. Then $\Phi \vdash \varphi$ implies $\varphi \in$ Th(S), but having only the axioms of S as hypotheses, only Σ-formulas can be deduced, which are true in *all* models of Φ. Hence we may miss Σ-formulas, which are true in a standard model, but false in a nonstandard model of Φ, and, of course, these are the formulas in which we are mainly interested.

To overcome this problem, one uses formula schemas to formulate *induction axioms* for an admissible specification S (called S-*induction axioms* for short), cf. [Aubin, 1979; Boyer and Moore, 1979] for proposals of such formula schemas. We shall not go into formal details here, but we demand only, that each S-induction axiom of an admissible specification S is *true* in the standard model of S, i.e. $I(S) \subset$ Th(S), where $I(S)$ denotes the set of all S-induction axioms obtained as instances of *particular* formula schemas for S-induction axioms. Hence $\Phi \cup I(S) \vdash \varphi$ implies $\varphi \in$ Th(S), and this is the fundamental idea (conventional[4]) automated theorem proving by induction is based on.

Subsequently we shall use $\vdash_S \varphi$ as an abbreviation for "$\Phi \cup I(S) \vdash \varphi$", where S denotes an admissible specification. Hence one can read "$\vdash_S \varphi$" as "φ is provable from Φ by an induction theorem proving system", as e.g. by one of the systems described in [Aubin, 1979; Boyer and Moore; 1979; Biundo et al., 1986].

3.2 Data Structures and Algorithms

But how do we obtain an admissible specification? Given an admissible specification S, we *extend* S by a new data structure or by a new algorithm, defined on the data structures already in S. But we demand, that these extensions satisfy certain requirements to guarantee that the resulting specification S' is admissible. We start with the *initial* specification $(\mathcal{S},\Sigma,\Phi)$, which is defined by $\mathcal{S}=\{\text{bool}\}$, $\Sigma_{w,s}=\varnothing$, except $\Sigma^c_{\lambda,\text{bool}}=\{\text{true, false}\}$ and $\Phi=\{\neg\text{true}\equiv\text{false}, \forall\text{b:bool } \text{b}\equiv\text{true} \vee \text{b}\equiv\text{false}\}$.

An admissible specification S can be extended by an S'-*structure* D for a new sort symbol *s*, by defining the *constructor* symbols, which are used to build the elements of the new data structure. Additionally, all *selector* symbols, which belong to the constructors, and the signature for all these function symbols are specified. As an example, we have seen already the data structure for list in Section 1. Another example is the data structure for number, given as

structure 0 succ(pred:number):number

Here 0 and succ are constructors and pred is the (only) selector of succ. The S'-structure D for *s* is *admissible* for a specification S=$(\mathcal{S},\Sigma,\Phi)$ iff all function symbols in D and the new sort *s* are not mentioned in S and all function symbols in D are different. Then S'=$(\mathcal{S}\cup\{s\},\Sigma',\Phi')$ is an *admissible exten-*

---

[3] We have borrowed the term "specification" from [Goguen et al., 1978], where, however, Φ is a set containing only Σ-*equations*.

[4] Here we regard induction proof procedures, which evolve from modifications of the *Knuth-Bendix completion procedure* as non-"convential" - see [Kapur and Musser, 1987] for a discussion of this approach.

sion of S, if Σ'^c is Σ^c extended by the new constructors, Σ'^d is Σ^d extended by the new selectors and Φ' is obtained by extending Φ with the set of *representation formulas* REP_s for s, i.e. a set of closed Σ'-formulas axiomatizing properties of the new data structure. For the structure list, for instance, we would obtain the following set of representation formulas REP_{list}

$\{ \quad \forall k\text{:list } \forall n\text{:number } \neg\text{empty}\equiv\text{add}(n\ k), \qquad\qquad \forall k\text{:list } k\equiv\text{empty} \vee k\equiv\text{add}(\text{head}(k)\ \text{tail}(k)),$

$\qquad \forall k\text{:list } \forall n\text{:number } \text{head}(\text{add}(n\ k))\equiv n, \qquad \forall k\text{:list } \forall n\text{:number } \text{tail}(\text{add}(n\ k))\equiv k,$

$\qquad \text{head}(\text{empty})\equiv 0, \qquad\qquad\qquad\qquad\qquad\qquad \text{tail}(\text{empty})\equiv\text{empty},$

$\qquad \forall k_1,k_2\text{:list } \forall n_1,n_2\text{:number } \text{add}(n_1\ k_1)\equiv\text{add}(n_2\ k_2) \rightarrow n_1\equiv n_2 \wedge k_1\equiv k_2 \ \}.$

We may also extend a specification S by a new algorithm: For a specification $S'=(S',\Sigma',\Phi')$, an expression F given as

$$\textit{function } f(x^*\text{:w})\text{:s} \Leftarrow \textit{if } \varphi_1 \textit{ then } r_1 \ ... \ \textit{if } \varphi_k \textit{ then } r_k$$

is an S'-*algorithm* for f iff each φ_i is a quantifier-free Σ'-formula and each r_i is a Σ'-term of sort s, where $x^*\in V_w$ and x^*:w abbreviates $x_1\text{:}s_1...x_n\text{:}s_n$ such that $x^*=x_1...x_n$ and $w=s_1...s_n$. The expressions "*if* φ_i *then* r_i" are the *cases* of F, φ_i is the *condition* and r_i is the *result* of case i. F is a *recursive* S'-algorithm iff at least one case i of F is *recursive*, i.e. φ_i or r_i contains an f-term. The set of *definition formulas* DEF_F associated with F is defined as

$$\text{DEF}_F=\{ \ \forall x^*\text{:w } \varphi_1 \rightarrow fx^*\equiv r_1 \ , \ ... \ , \ \forall x^*\text{:w } \varphi_k \rightarrow fx^*\equiv r_k \ \}.$$

An S'-algorithm F is *case-unique* in a specification $S=(S,\Sigma,\Phi)$ with $S\subset S'$ and $\Sigma\subset\Sigma'$ iff $\vdash_S [\forall x^*\text{:w } \varphi_i \rightarrow \neg\varphi_j]$ for all cases i and j with $i\neq j$, i.e. there is *at most* one condition, which is satisfied for a given input. F is *case-complete* in S iff $\vdash_S [\forall x^*\text{:w } \varphi_1 \vee ... \vee \varphi_k]$, i.e. there is *at least* one condition, which is satisfied for a given input. F *terminates* in S iff for each Σ-algebra $M=(A,\alpha)$, which (as a Σ-algebra) is a standard model of Φ, there is some well-founded order relation $<_R$ of A_w such that for each $a^*\in A_w$, for each recursive case i of F and for each f-term $f\delta_{i,h}(x^*)$ in case i (where $\delta_{i,h}$ is a substitution)

$$M[x^*/a^*] \models \varphi_i \quad \text{implies} \quad M[x^*/a^*](\delta_{i,h}(x^*)) <_R a^* \ .$$

An S'-algorithm F for f is *admissible* for a specification $S=(S,\Sigma,\Phi)$ iff $f\notin\Sigma$, F is case-unique, case-complete and terminating in S and $S'=(S',\Sigma',\Phi')$ is an *admissible extension* of S by F iff $S'=S$, $\Sigma'=\Sigma\cup\{f\}$, $\Phi'=\Phi\cup\{\text{DEF}_F\}$ and F is admissible for S.

We say, that a specification S' *contains* a structure D or an algorithm F iff either S' is an admissible extension of a specification S by D or by F or S' is an admissible extension of a specification S containing D or F. Using admissible extensions, we can now formulate a sufficient criterion for obtaining admissible specifications with the following theorem:

Theorem 3.1 If S is the initial specification or S is an admissible extension of an admissible specification by an S-structure or by an S-algorithm, then S is admissible. ∎

Based on this theorem we can think of a software development system, which is implemented as a system maintaining admissible specifications: After booting, the system comes up with the initial specification. Then a user of the system may define the data structures and the algorithms she is concerned with in form of S-structures and S-algorithms. The system checks each input for admissibility and extends its current specification on success. Otherwise, the input is rejected with a failure indication.

To implement the admissibility test for S-algorithms, the system has access to an *induction theorem proving system*, which performs the tests for case-uniqueness and for case-completeness. The remaining requirements for admissibility (except *termination*) are only of trivial syntactic nature. Hence for the remainder of this paper, we are concerned with the problem to supply our system with a facility to recognize the *termination* of certain S-algorithms automatically.

4 Argument-Bounded Operations

4.1 The Size Order

Well-founded orders for termination proofs are frequently obtained by comparing the size of the data objects under consideration. For instance, *stacks* are compared with their *depth*, *lists* with their *length*, *trees* with their *number of nodes* etc. We obtain an *abstract* notion of *size* by counting (using the so-called s-*size mapping*) how many applications of *reflexive* constructors are necessary to construct a data object, where the substructures involved are ignored. Now comparing the sizes of a pair of data objects of sort s with the usual $<_N$ relation on the natural numbers $\mathbb{N}$, we obtain the *size order*, i.e. a *well-founded order* for the data structure s. For the remainder of the paper $(\mathcal{A},\alpha)$ stands for an arbitrary standard Σ-algebra A, and we define:

Definition 4.1 For each $s \in S$, the s-*size mapping* $\#_s:\mathcal{A} \to \mathbb{N}$ in A is defined as:

(1) $\quad \#_s(a) = 0$ $\qquad\qquad\qquad\qquad\qquad$, if $a \in \mathcal{A}_{s'}$ and $s \neq s'$,

(2) $\quad \#_s(\alpha_c(a^*)) = 0$ $\qquad\qquad\qquad\quad$, if $c \in \Sigma^c_{w,s}$, $a^* \in \mathcal{A}_w$ and c is irreflexive, and

(3) $\quad \#_s(\alpha_c(a^*)) = 1 + \#_s(a^1) + ... + \#_s(a^{|w|})$ $\quad$, if $c \in \Sigma^c_{w,s}$, $a^* \in \mathcal{A}_w$ and c is reflexive.

The *size order* in A, denoted $<_\#$, is a subset of $\mathcal{A} \times \mathcal{A}$ defined as : $a_1 <_\# a_2$ iff $\#_s(a_1) <_N \#_s(a_2)$ and a_1, $a_2 \in \mathcal{A}_s$ for some $s \in S$. ∎

Condition (1) ensures that substructures of the structure under consideration are ignored. Hence, for instance, we do not count the occurences of succ if we compare elements of the structure list. Therefore add(succ(succ(0)) empty) $<_\#$ add(0 add(0 empty)).

With condition (2), all data objects build with irreflexive constructors are minimal elements of the size order. Consequently, for the s-*expressions* of LISP [McCarthy et al., 1962], defined in our notation by

structure atom(index:number) nil cons(car:sexpr cdr:sexpr):sexpr ,

we obtain *nil* and all data objects of form *atom*(n) as minimal elements in $\mathcal{A}_{sexpr}$. Based on the size order we shall now introduce the concept of an *argument-bounded* operation:

Definition 4.2 Let $f \in \Sigma^d_{w,s}$ and $p \in \{1,...,|w|\}$. Then α_f is a *p-bounded* operation iff $\alpha_f(a^*) \leq_{\#} a^p$ for all $a^* \in \mathcal{A}_w$. α_f is an *argument-bounded* operation iff α_f is a *p-bounded* operation for some $p \in \{1,...,|w|\}$. Subsequently we let $\Gamma(A) = (\Gamma_p(A))_{p \in \mathbb{N}}$ be a family of sets such that $\Gamma_p(A) \subset \Sigma^d$ for all $p \in \mathbb{N}$. $\Gamma_p(A)$ is a set of *p-bounded function symbols in* A iff α_f is a p-bounded operation for all $f \in \Gamma_p(A)$. $\Gamma(A)$ is a family of *argument-bounded function symbols in* A iff $\Gamma_p(A)$ is a set of p-bounded function symbols in A for all $p \in \mathbb{N}$. For the remainder of this paper we will write Γ_p and Γ if A is known from the context. ■ 5

Argument-bounded operations are frequently used in computer science to define algorithms recursively. For instance, all *reflexive* selectors are 1-bounded, provided they return their argument if applied to a constructor they do not belong to. Hence car, cdr, pred and tail denote 1-bounded operations, if we assume that $car(nil)=cdr(nil)=nil$, $car(atom(a))=cdr(atom(a))=atom(a)$, $tail(empty)=empty$ and $pred(0)=0$.

In arithmetic, for instance, we have the difference, the (truncated) quotient and the remainder operation as 1-bounded operations, because $n-m \leq_{\mathbb{N}} n$, $n/m \leq_{\mathbb{N}} n$, and $(n \bmod m) \leq_{\mathbb{N}} n$, if we assume that $0-m=0$, $n/0 \leq_{\mathbb{N}} n$, and $(n \bmod 0) \leq_{\mathbb{N}} n$. The remainder is also 2-bounded, i.e. $(n \bmod m) \leq_{\mathbb{N}} m$, if $(n \bmod 0) = 0$ is satisfied.[6]

In the programming language COMMONLISP [Steele, 1984], for instance, nthcdr, member and remove denote 2-bounded operations, and intersection would denote a 1-bounded one, provided the order of elements in the first argument would reflect the order of elements in the result. Also our algorithm for remove, cf. Section 1, computes a 2-bounded operation, and one can observe that all these operations are frequently used for recursive definitions of algorithms.

4.2 The Γ-Bound

Hence we have to benefit from argument-bounded operations, when we are concerned with the automatization of termination proofs. But since we have only defined *semantical* notions so far, we also have to develop *syntactical* concepts for our purpose. The starting idea is quite simple: we use the technique of *estimation* when we prove inequalities. Suppose, for instance, we have to verify that $(n-m)/(p+1) \leq_{\#} n$ for all natural numbers n,m and p (where in this case $\leq_{\#}$ agrees with $\leq_{\mathbb{N}}$). Since the (truncated) quotient is 1-bounded, we know that $(n-m)/(p+1) \leq_{\#} (n-m)$ and with the difference being 1-bounded we find that $(n-m) \leq_{\#} n$. Hence the inequality is proved. This technique for proving inequalities is formalized with the so-called Γ-*bound*, i.e. a relation on terms, which (for a given family Γ of argument-bounded function symbols) mirrors the semantical $\leq_{\#}$ relation on the *syntactical level* .

Definition 4.3 Let Γ be a family of argument-bounded function symbols in A. Then $r \in \mathcal{T}(\Sigma,\mathcal{V})_s$ is a Γ-*bound* of $q \in \mathcal{T}(\Sigma,\mathcal{V})_s$ in A, denoted $q \leq_{\Gamma} r$, iff (i) $q=r$, or (ii) $q=fq^*$ and $q^p \leq_{\Gamma} r$ for some $f \in \Gamma_p$. ■

Let us illustrate this definition with an example:

---

5 Do not confuse $a_1 \leq_{\#} a_2$ with $\#_s(a_1) \leq_{\mathbb{N}} \#_s(a_2)$! For instance, $\#_{list}(sort(a)) \leq_{\mathbb{N}} \#_{list}(a)$ for *all* $a \in \mathcal{A}_{list}$, but $\#_{list}(sort(a)) = \#_{list}(a)$ and $sort(a) \neq a$ for some $a \in \mathcal{A}_{list}$, hence $sort(a) \, \$_{\#} \, a$ and for some $a \in \mathcal{A}_{list}$.

6 Note, that we need such additional requirements only for exceptional arguments, as e.g. car(nil) or division by zero, which would normally result in an error indication. We would not need these requirements, if we could use *partial* operations. But since the semantics of first-order logic demand *total* operations, we feel free for these cases to stipulate the values, which are convenient for us.

Example 4.1 Given tail$\in \Sigma^d_{\text{list,list}} \cap \Gamma_1$ and remove$\in \Sigma^d_{\text{number,list,list}} \cap \Gamma_2$, we obtain for all $n \in \mathcal{T}(\Sigma, \mathcal{V})_{\text{number}}$ and all $x \in \mathcal{T}(\Sigma, \mathcal{V})_{\text{list}}$

(1) $x \leq_\Gamma x$, by condition (i), hence

(2) tail(x)$\leq_\Gamma x$, by condition (ii), hence

(3) remove(n tail(x))$\leq_\Gamma x$, by condition (ii), hence

(4) tail(remove(n tail(x)))$\leq_\Gamma x$, by condition (ii).   ∎

One can observe, that in these examples $q \leq_\Gamma r$ is a sufficient condition for the objects denoted by q and r to be in the $\leq_\#$ relation, no matter which values are assigned to the variables in the terms. The following theorem proves that this observation generally holds true:

Theorem 4.1 Let $q, r \in \mathcal{T}(\Sigma, \mathcal{V})$ and $x^* \in \mathcal{V}_w$ such that $q \leq_\Gamma r$ and $\mathcal{V}(\{q, r\}) \subset \mathcal{V}(x^*)$. Then $A[x^*/a^*](q) \leq_\# A[x^*/a^*](r)$ for all $a^* \in \mathcal{A}_w$.   ∎

4.3 The Γ-Difference Equivalent

Having obtained a syntactical requirement for $\leq_\#$, we shall now develop a syntactical requirement for $<_\#$: Given q and r such that $q \leq_\Gamma r$, we are looking for a condition which is equivalent with $A[x^*/a^*](q) <_\# A[x^*/a^*](r)$. To this effect, we obviously now have to take the assignments for the variables into account. So let us assume, that a literal $D_{p,f}(x^*)$ is associated with each p-bounded function symbol f, expressing that $f(x^*)$ returns a value stricly less than x^p. We call such a literal a *p-difference literal* of f. For instance, member(n x)≡true is a 2-difference literal of remove(n x) and x≡add(head(x) tail(x)) is a 1-difference literal of tail(x). Hence we define:

Definition 4.4 For each $f \in \Gamma_p \cap \Sigma^d_{w,s}$, we let $D_{p,f}(x^*) \in Lit(\Sigma, \mathcal{V}(x^*))$ where $x^* \in \mathcal{V}_w$. $D_{p,f}(x^*)$ is called a *p-difference literal of f in A* iff

$$A[x^*/a^*] \models D_{p,f}(x^*) \quad \text{iff} \quad \alpha_f(a^*) <_\# a^p \quad \text{for all } a^* \in \mathcal{A}_w. \quad ∎$$

Now given, for instance, remove(n tail(x))$\leq_\Gamma$tail(x)$\leq_\Gamma x$, we find that remove(n tail(x)) is *strictly less* than x iff at least one of these inequalities is strict. But this can be expressed by our difference literals, i.e. we obtain member(n tail(x))≡true ∨ x≡add(head(x) tail(x)). Hence given q and r, we inspect the inequalities constituting $q \leq_\Gamma r$ and collect the associated difference literals in a set. Now the object denoted by q is *strictly less* than the object denoted by r iff at least one of the literals in the set is true. We formalize this idea with the Γ-*difference equivalent*:

Definition 4.5 Given $q, r \in \mathcal{T}(\Sigma, \mathcal{V})$ such that $q \leq_\Gamma r$, the Γ-*difference equivalent* of q and r, denoted $\Delta_\Gamma(q, r)$, is a subset of $Lit(\Sigma, \mathcal{V})$ recursively defined as:

(i) $\Delta_\Gamma(q, r) = \varnothing$, if q=r, else

(ii) $\Delta_\Gamma(q, r) = \{D_{p,f}(q^*) \mid p \in P\} \cup \bigcup_{p \in P} \Delta_\Gamma(q^p, r)$,
     if q=fq* and $q^p \leq_\Gamma r$ for some $f \in \Gamma_p \cap \Sigma^d_{w,s}$, where P={p'∈{1,...,|w|} | f∈$\Gamma_{p'}$ and $q^{p'} \leq_\Gamma r$}.   ∎

We illustrate this definition with an example:

Example 4.2 We obtain for the terms of Example 4.1:

$\Delta_\Gamma(x, x) = \emptyset$, by condition (i), hence

$\Delta_\Gamma(\text{tail}(x), x) = \{x \equiv \text{add}(\text{head}(x)\ \text{tail}(x))\ \}$, by condition (ii), hence

$\Delta_\Gamma(\text{remove}(n\ \text{tail}(x)), x) = \{x \equiv \text{add}(\text{head}(x)\ \text{tail}(x)), \text{member}(n\ \text{tail}(x)) \equiv \text{true}\ \}$, with condition (ii). ■

From the examples above we may observe, that if we assign the variables in a given pair of terms q and r (satisfying $q \leq_\Gamma r$) such that the evaluation of q is *strictly* less than the evaluation of r, then some of the literals in $\Delta_\Gamma(q,r)$ evaluate to *true* under the same assignment. Hence the difference equivalent provides us with a *syntactical* characterization of the $<_\#$ relation and we prove with the following theorem that this observation generally holds true:

Theorem 4.2 Let $q, r \in T(\Sigma, V)$ and $x^* \in V_w$ such that $q \leq_\Gamma r$ and $V(\{q, r\}) \subset V(x^*)$. Then $A[x^*/a^*] \models \vee \Delta_\Gamma(q,r)$ iff $A[x^*/a^*](q) <_\# A[x^*/a^*](r)$ for all $a^* \in \mathcal{A}_w$. ■

4.4 A Syntactical Termination Criterion for Algorithms

Using the above definitions and results we are now able to formulate a sufficient criterion for the termination of an S-algorithm, where here, and for the remainder of this paper, $S = (S, \Sigma, \Phi)$ is an arbitrary admissible specification with standard model $M = (\mathcal{A}, \alpha)$ and Γ is a family of argument-bounded function symbols in M:

Theorem 4.3 Let S' be a specification and let F be an S'-algorithm for f given as in Section 3.2. Then F terminates in S, if there exists some non-empty $P \subset \{1, ..., |w|\}$ such that for each f-term $f\delta_{i,h}(x^*)$ in a recursive case i of F (with condition φ_i)

(1) $\delta_{i,h}(x^p) \leq_\Gamma x^p$ for all $p \in P$,   and   (2) $\vdash_S [\ \forall x^*:w\ \varphi_i \to \vee \cup_{p \in P} \Delta_\Gamma\ (\delta_{i,h}(x^p), x^p)\ ]$.

Sketch of Proof For the sake of simplicity, let us assume that $P = \{1, ..., |w|\}$. Then with condition (1) and Theorem 4.1, all arguments of f in a recursive call are less than or equal to the corresponding arguments initially given to f. With condition (2) and Theorem 4.2, at least one of these inequalities is strict. Hence F must terminate. Obviously, this argumentation works also for $P \neq \{1, ..., |w|\}$, as long as $P \neq \emptyset$. ■

Note, that all algorithms, which terminate with this theorem must terminate according to the *lexicographical* size order for *all* permutations of (a sublist of) the argument list[7]. The formulas of condition (2) in Theorem 4.3 are called the *termination hypotheses* of F, and, obviously, these are the formulas we are looking for. Based on Theorem 4.3, we can supply our software development system with a *semi*-automatic facility to recognize the termination of (some) S'-algorithms: We let a (trustworthy) hu-

---

[7] We can weaken this requirement in Theorem 4.3, such that algorithms terminating lexicographically with *some* permutation of (a sublist of) the argument list will be recognized, cf. [Walther, 1987]. Since this necessitates some formal clutter, we dispense with this extension here. Note, that the termination of all algorithms given in [Boyer and Moore, 1979] (except the one defined for *Ackermann's function*) can be proved with Theorem 4.3.

man user declare a family $\Gamma(S) \subset \Sigma^d$ of function symbols to denote argument-bounded operations and let her also state the difference literals for them. Then as a test for the termination of a new S'-algorithm F, the system checks whether conditions (1) and (2) of Theorem 4.3 are satisfied:

Firstly, the system has to compute P as a subset of all argument positions in $\{1,...,|w|\}$ such that $\delta_{i,h}(x^p) \leq_{\Gamma(S)} x^p$ for all $p \in P$ and all recursive cases i.[8] Then it computes $\Delta_{i,h} = \cup_{p \in P} \Delta_{\Gamma(S)}(\delta_{i,h}(x^p), x^p)$, provided $P \neq \varnothing$. Note that the computation of $\leq_{\Gamma(S)}$ and $\Delta_{\Gamma(S)}$ is *trivial* and an implementation is obtained directly from the Definitions 4.3 and 4.5. Then each termination hypothesis $[\forall x^*:w \; \varphi_i \rightarrow \vee \Delta_{i,h}]$ is given to the induction theorem prover. On success, the termination of the new algorithm is proved (cf. Theorem 4.3), provided $\Gamma(S)$ is a family of argument-bounded function symbols in the standard model of the current specification S, and the literals given are in fact difference literals for the function symbols in $\Gamma(S)$.

Let us resume our example from the introduction: Suppose that we ask the system to extend its current specification by the algorithm R for remove. Then the system recognizes

$$\delta_{2,1}(x) = \delta_{3,1}(x) = tail(x) \leq_{\Gamma(S)} x , \qquad\qquad\qquad\qquad \text{, computes}$$

$$\Delta_{\Gamma(S)}(\delta_{2,1}(x), x) = \Delta_{\Gamma(S)}(\delta_{2,1}(x), x) = \{x \equiv add(head(x) \; tail(x))\} \qquad\qquad \text{, and proves}$$

$$\vdash_S [\; \forall n:number \; \forall x:list \; x \equiv add(head(x) \; tail(x)) \wedge (\neg)head(x) \equiv n \rightarrow x \equiv add(head(x) \; tail(x)) \;] .$$

With a proof of the generated termination hypotheses, the termination of remove is verified. In the same way, the algorithm for member is accepted and the system's current specification is extended with both algorithms. Now assume that remove is declared as 2-bounded and member(n x)≡true is given as the 2-difference literal for remove by the user. Then confronted with the algorithm for sort, the system recognizes

$$\delta_{2,1}(x) = remove(minimum(x) \; x) \leq_{\Gamma(S)} x , \qquad\qquad \text{and computes}$$

$$\Delta_{\Gamma(S)}(\delta_{2,1}(x), x) = \{member(minimum(x) \; x) \equiv true\}.$$

Finally, with an (induction) proof of the termination hypothesis, i.e.

$$\vdash_S [\; \forall x:list \; x \equiv add(head(x) \; tail(x)) \rightarrow member(minimum(x) \; x) \equiv true \;] \; ,$$

which, of course, should not be a challenge for an induction prover, the termination of sort is verified.

However, the success of this procedure depends on a *trustworthy* human user, because the system's derivations would be unsound, if function symbols were declared argument-bounded by mistake or incorrect difference literals were given for them. Also, we need an *experienced* human, because the difference literals have to be chosen skillfully: Of course, with $\neg f(x^*) \equiv x^p$ we always have a p-difference literal for a p-bounded function symbol f. But this simple solution causes proof technical problems - we defer a discussion of this topic to Section 7.

Hence for the remainder of this paper we are concerned with the problem to *recognize* algorithms, which compute argument-bounded operations, and to *synthesize* (useful) difference literals for them automatically, in order to get rid of any human support.

---

[8] Note, that this set corresponds to the *measured subset* of [Boyer and Moore, 1979]. Therefore - as pointed out by *Boyer* and *Moore* - it is advantageous for subsequent induction proofs (viz. for the recognition of adequate induction axioms) to compute *all minimal* subsets P of $\{1,...,|w|\}$ which satisfy the requirements in Theorem 4.3.

5 Argument-Bounded Algorithms

To recognize argument-bounded operations, we make a destinction between operations denoted by *selector functions* and operations, which are computed by *algorithms*. As a starting point, let us consider the selectors given for the structures in a specification S: Since each standard model of S satisfies the representation formulas for the structures contained in S, we can prove, that each *reflexive* selector is 1-bounded.

Theorem 5.1 Let S contain a structure for some $s \in S$, let c be a constructor of s, and let $b^1,...,b^m$ be the selectors of c. Then for each *reflexive* selector $b \in \Sigma^d_{s,s}$ of c, α_b is a 1-bounded operation and $x \equiv cb^1x ... b^mx$ is a 1-difference literal of b in M. ∎

Consequently, for a specification containing the structure for list as given in Section 1, we know that tail denotes a 1-bounded operation and $x \equiv add(head(x)\ tail(x))$ is a 1-difference literal of tail. Also, for the structure number as given in Section 3, pred denotes a 1-bounded operation and $x \equiv succ(pred(x))$ is a 1-difference literal for pred.

For an operation computed by an algorithm, we have to inspect the algorithm looking for conditions which imply that the computed operation is argument-bounded. But what do these conditions look like? Consider, for instance, the algorithm R for remove from the introduction, and let us prove by induction that R computes a 2-bounded operation:

For the base case we have *remove(n x)* = χ, hence *remove(n x)* $\leq_\#$ χ. Now assume as the induction hypothesis that *remove(n tail(χ))* $\leq_\#$ *tail(χ)*. For *remove(n x)* = *remove(n tail(χ))* we have *remove(x)* $\leq_\#$ *tail(χ)* by the induction hypothesis and *remove(n x)* $\leq_\#$ χ because tail is 1-bounded, i.e. *tail(χ)* $\leq_\#$ χ. If *remove(n x)* = *add(head(χ) remove(n tail(χ)))*, then χ = *add(head(χ) tail(χ))* and we obtain with the induction hypothesis *remove(n x)* $\leq_\#$ *add(head(χ) tail(χ))* = χ. Therefore R computes a 2-bounded operation.

Analyzing the above proof, we observe, that we do not use very intrinsic properties of R to obtain it. All we need is induction and a frequent usage of our knowledge about $\leq_\#$. But since we have the Γ-bound as a syntactical means to describe $\leq_\#$, we can formulate *syntactical* requirements for an algorithm (as, e.g. $x \leq_\Gamma x$ and tail(x) $\leq_\Gamma x$ in the example above), which ensure that the computed operation is argument-bounded. Hence we are able to define a *schema* for algorithms, such that each instance of this schema, called a *p-bounded algorithm*, computes a p-bounded operation.

Definition 5.1 Let S contain an algorithm G for some $g \in \Sigma_{w,s}$ given as:

$$function\ g(x^*:w):s \Leftarrow if\ \varphi_1\ then\ r_1\ ...\ if\ \varphi_k\ then\ r_k.$$

Then G is a *p-bounded algorithm* for some $p \in \{1,...,|w|\}$ iff for all cases i of G

(1) r_i contains no g-term and $r_i \leq_\Gamma x^p$, or

(2) r_i contains exactly one g-term $g\delta_i(x^*)$ such that (i) $r_i \leq_\Gamma g\delta_i(x^*)$ and (ii) $\delta_i(x^p) \leq_\Gamma x^p$, or

(3) r_i contains exactly one g-term $g\delta_i(x^*)$ such that for some $r \in \mathcal{T}(\Sigma, \mathcal{V})_s$ and some constructor c of

s with selectors $b^1,...,b^m$:   (i) $\vdash_S [\forall x^*\!:\!w\ \varphi_i \rightarrow x^P \equiv cb^1 x^P ... b^m x^P]$

(ii) $r_i = cb^1 x^P ... b^{n-1} x^P\ r\ b^{n+1} x^P ... b^m x^P$, (iii) $r \leq_\Gamma g\delta_i(x^*)$, and (iv) $\delta_i(x^P) \leq_\Gamma b^n x^P$.   ∎

The above conditions (1) and (2) are *independent* of the definition given for the data structure of s. Condition (3), however, *depends* on the definition of a particular data structure for s. For $s=number$ or $s=list$, for instance, we instantiate condition (3) to the following conditions (3') and (3") respectively:

(3') r_i contains exactly one g-term $g\delta_i(x^*)$ such that for some $r \in \mathcal{T}(\Sigma, \mathcal{V})_{number}$

(i) $\vdash_S [\forall x^*\!:\!w\ \varphi_i \rightarrow x^P \equiv succ(pred(x^P))]$,

(ii) $r_i = succ(r)$, (iii) $r \leq_\Gamma g\delta_i(x^*)$, and (iv) $\delta_i(x^P) \leq_\Gamma pred(x^P)$

(3") r_i contains exactly one g-term $g\delta_i(x^*)$ such that for some $r \in \mathcal{T}(\Sigma, \mathcal{V})_{list}$

(i) $\vdash_S [\forall x^*\!:\!w\ \varphi_i \rightarrow x^P \equiv add(head(x^P)\ tail(x^P))]$,

(ii) $r_i = add(head(x^P)\ r)$, (iii) $r \leq_\Gamma g\delta_i(x^*)$, and (iv) $\delta_i(x^P) \leq_\Gamma tail(x^P)$.

For instance, the algorithm R for remove is a 2-bounded algorithm, because with $x \leq_\Gamma x$, condition (1) is satisfied in the 1st case, with remove(n tail(x)) $\leq_\Gamma$ remove(n tail(x)) and tail(x) $\leq_\Gamma$ x condition (2) is satisfied in the 2nd case, and with $\vdash_S [$ $\forall$n:number $\forall$x:list x$\equiv$add(head(x) tail(x)) $\wedge$ $\neg$head(x)$\equiv$n $\rightarrow$ x$\equiv$add(head(x) tail(x)) $]$, r = remove(n tail(x)), remove(n tail(x)) $\leq_\Gamma$ remove(n tail(x)) and tail(x) $\leq_\Gamma$ tail(x) condition (3) is satisfied in the 3rd case.

The algorithm given for sort in Section 1 is not a 1-bounded one, because neither condition (2,i) nor condition (3,ii) are satisfied in the 2nd case. However, we do not worry about this, because sort does not denote a 1-bounded operation.[5]

Note, that we have *pure syntactical requirements* in our definition of a p-bounded algorithm. Most of them are trivial features, imposing restrictions on the syntactical form of a term or comparing a pair of terms with the Γ-bound. All we need is the computation of the Γ-bound and, consequently, these features are decidable and easy to compute.

As an exception, however, it may be necessary to prove that certain Σ-formulas belong to the theory of a specification, cf. condition (3,i), in order to recognize an algorithm as a p-bounded one. Hence it is *undecidable*, whether an algorithm is p-bounded or not. But the theorem proving problems involved with this test (cf. the formula generated for remove) usually do not constitute a challenge for an induction theorem prover, as experiments with our system indicate.

However, since we may generate Σ-formulas ψ such that $\nvdash_S [\forall x^*\!:\!w\ \varphi_i \rightarrow \psi]$ when we test for condition (3,i), we have to supply our induction prover with a *halting criterion*. We can think of a sophisticated procedure looking for counterexamples, cf. [Aubin, 1979; Boyer and Moore, 1979], but any stupid facility, such as the limitation of CPU-time, memory or kilowatt-hours vasted, may help if a "intelligent" equipment is not at hand.

The syntactical restrictions given for a p-bounded algorithm allow us to prove (in the same way as we did for *remove*), that the computed operation is a p-bounded one:

Theorem 5.2   If S contains a p-bounded algorithm G for some g$\in \Sigma$, then α_g is a p-bounded operation.   ∎

6 Synthesis of Difference Algorithms

Having a means to recognize argument-bounded operations, we can think of a system which extends its family of argument-bounded function symbols automatically whenever a specification is extended by an argument-bounded algorithm. However, we cannot take advantage of this facility, as long as we do not have the difference literals of the argument-bounded function symbols obtained thereby. But how do we find these difference literals ?

Given a p-bounded algorithm G for some g∈ $\Sigma_{w,s}$, we *synthesize* an algorithm Δ^pG for some new function symbol Δ^pg∈ $\Sigma'_{w,bool}$ (where Σ' is Σ extended by Δ^pg), such that Δ^pG(t*) returns true if and only if G(t*) returns something size-smaller than t^p. Hence with Δ^pg(x*)≡true we have found a p-difference literal $D_{p,g}$ of g, provided we succeeded in the synthesis of Δ^pG.

We construct Δ^pG, called a *p-difference algorithm* of G, inductively by using the cases of G. Since G is p-bounded, we know that G(t*) returns something size-smaller than or equal to t^p. Hence we let Δ^pG(t*) return false, if we can prove that G(t*) returns t^p, and we let Δ^pG(t*) return true, if we can prove that G(t*) returns something size-smaller than t^p. We let Δ^pG(t*) perform the same recursion as G(t*) does, in all other cases. For instance, for an 1-bounded algorithm M for minus given as

function minus(n,m:number):number ⟸
   if m≡0 *then* n
   if ¬m≡0 *then* minus(pred(n) pred(m))

we have to construct an algorithm Δ^1M for Δ^1minus: Since minus(n m) = n for the 1st case of M, we obtain "*if* m≡0 *then* false" as the 1st case of Δ^1M. For the 2nd case of M, we have minus(n m) = minus(pred(n) pred(m)) and with minus (and pred) being 1-bounded, we know that minus(pred(n) pred(m)) $\leq_\Gamma$ pred(n) $\leq_\Gamma$ n. Obviously, minus(n m) returns something size-smaller than n iff at least one of these inequalities is strict. For the rightmost inequality this can be expressed by the Γ-difference equivalent Δ_Γ(pred(n), n) and we obtain as the 2nd case of Δ^1M "*if* ¬m≡0 ∧ n≡succ(pred(n)) *then* true ". But what to return for " ¬m≡0 ∧ n≡0 "? Then minus(pred(n) pred(m)) $\leq_\Gamma$ pred(n) = n and the result depends only on the strictness of the remaining inequality. But since we are in an *inductive* construction of Δ^1M , we can express the strictness of this inequality as Δ^1minus(pred(n) pred(m)). Hence we obtain "*if* m≡0 ∧ n≡0 *then* Δ^1minus(pred(n) pred(m)) " as the 3rd case of Δ^1M. Summing up, we have constructed Δ^1M as:

function Δ^1minus(n,m:number):bool ⟸
   if m≡0 *then* false
   if ¬m≡0 ∧ n≡succ(pred(n)) *then* true
   if ¬m≡0 ∧ n≡0 *then* Δ^1minus(pred(n) pred(m))

We may observe from this example, that all we need to construct a p-difference algorithm Δ^pG from G, is a frequent usage of our knowledge about $<_\#$. But since we have the Γ-difference equivalent as a syntactical means to describe the $<_\#$ relation, we can formulate syntactical requirements for a p-bounded algorithm, which allow us to define the cases for a p-difference algorithm as we did in the example above. Hence we are able to give a *uniform construction*, which for any p-bounded algorithm G for some g yields a *p-difference algorithm* $\Delta^P G$ for some new function symbol $\Delta^P g$, such that $D_{p,g} = \Delta^P g(x^*) \equiv$ true is a p-difference literal of g:

Definition 6.1 Let S contain a p-bounded algorithm G for some $g \in \Sigma_{w,s}$ given as

$$function \ g(x^*{:}w){:}s \Leftarrow if \ \psi_1 \ then \ r_1 \ ... \ if \ \psi_k \ then \ r_k$$

and let Σ' be Σ extended by a new function symbol $\Delta^P g \in \Sigma'_{w,bool}$. Then we define the *difference terms* $b_i \in \mathcal{T}(\Sigma', \mathcal{V})_{bool}$ and the *difference sets* $\Delta_i \subset \mathcal{F}(\Sigma, \mathcal{V})$ for the cases i of G by

(1) $b_i =$ false $\quad$ and $\quad \Delta_i = \Delta_\Gamma(r_i, x^p)$, $\qquad\qquad\qquad$ if r_i satisfies Definition 5.1 (1),

(2) $b_i = \Delta^P g \delta_i(x^*)$ and $\Delta_i = \Delta_\Gamma(r_i, g\delta_i(x^*)) \cup \Delta_\Gamma(\delta_i(x^p), x^p)$, $\quad$ if r_i satisfies Definition 5.1 (2), and

(3) $b_i = \Delta^P g \delta_i(x^*)$ and $\Delta_i = \Delta_\Gamma(r, g\delta_i(x^*)) \cup \Delta_\Gamma(\delta_i(x^p), b^n x^p)$, $\quad$ if r_i satisfies Definition 5.1 (3).

The *p-difference algorithm* $\Delta^P G$ of G is an algorithm for $\Delta^P g \in \Sigma'_{w,bool}$ defined as:

$$function \ \Delta^P g(x^*{:}w){:}bool \Leftarrow if \ \varphi_{1,1} \ then \ true \ if \ \varphi_{1,2} \ then \ b_1 \ ... \ if \ \varphi_{k,1} \ then \ true \ if \ \varphi_{k,2} \ then \ b_k,$$

where we define $\varphi_{i,1} = \psi_i \wedge \vee \Delta_i$ and $\varphi_{i,2} = \psi_i \wedge \neg \vee \Delta_i$ for each case i of G. ∎

Given the 2-bounded algorithm R for remove , for instance, we have for case 1 of R $b_1 =$ false and $\Delta_1 = \varnothing$, hence we obtain cases 1.1 and 1.2 of $\Delta^2 R$ as:

$$if \ x \equiv empty \wedge FALSE \ then \ true \quad and \quad if \ x \equiv empty \wedge \neg FALSE \ then \ false$$

For case 2 of R, we compute $b_2 = \Delta^2$remove(n tail(x)) and $\Delta_2 = \{ \ x \equiv add(head(x) \ tail(x)) \ \}$, hence we obtain cases 2.1 and 2.2 of $\Delta^2 R$ as:

$$if \ x \equiv add(head(x) \ tail(x)) \wedge head(x) \equiv n \wedge x \equiv add(head(x) \ tail(x)) \ then \ true$$
$$if \ x \equiv add(head(x) \ tail(x)) \wedge head(x) \equiv n \wedge \neg x \equiv add(head(x) \ tail(x)) \ then \ \Delta^2 remove(n \ tail(x))$$

Finally, we compute for case 3 of R $b_3 = \Delta^2$remove(n tail(x)) and $\Delta_3 = \varnothing$, hence we obtain cases 3.1 and 3.2 of $\Delta^2 R$ as:

$$if \ x \equiv add(head(x) \ tail(x)) \wedge \neg head(x) \equiv n \wedge FALSE \ then \ true$$
$$if \ x \equiv add(head(x) \ tail(x)) \wedge \neg head(x) \equiv n \wedge \neg FALSE \ then \ \Delta^2 remove(n \ tail(x))$$

Obviously, we use *pure syntactical features* in our construction of a p-difference algorithm. All we need is the computation of the Γ-difference equivalent, and, consequently, the synthesis of a p-difference algorithm is a trivial task. We can prove with the following theorem, that each p-difference algorithm is admissible for a specification, and, of course, that a p-difference literal is obtained with it:

Theorem 6.1 If S contains a p-bounded algorithm G for some g∈ $\Sigma_{w,s}$, then the p-difference algorithm Δ^pG of G is admissible for S. If in addition S contains the p-difference algorithm Δ^pG of G, then for any x*∈ $\mathcal{V}_w$, Δ^pgx*≡true is a p-difference literal of g in M.

7   Optimizations

Since we use the definition formulas of a p-difference algorithm in subsequent termination proofs, we are interested in obtaining a p-difference algorithm as simple as possible. This is of proof-technical relevance, because termination proofs become thereby less complicated. Hence we employ the induction theorem prover to *optimize* the synthesized difference algorithms. We use several optimization techniques, cf. [Walther, 1987], the most important of which are called case subsumption and recursion elimination:

7.1   Case Subsumption

Case subsumption replaces a case "*if* ψ ∧ φ *then* b" in an algorithm with "*if* ψ *then* b", provided $\vdash_S$ [∀x*:w ψ → φ] and eliminates this case from the algorithm, if $\vdash_S$ [∀x*:w ψ → ¬φ]. Since the cases of our difference algorithms are given as " *if* ψ_i∧∨Δ_i *then* true " and " *if* ψ_i∧¬∨Δ_i *then* b_i ", we let our system check for $\vdash_S$ [∀x*:w ψ_i → ∨Δ_i] and, if it fails, for $\vdash_S$ [∀x*:w ψ_i → ¬∨Δ_i]. For the difference algorithm Δ^2R from Section 6, the system has to prove

$\vdash_S$ [∀x:list x≡empty → ¬FALSE] ,

$\vdash_S$ [∀n:number ∀x:list x≡add(head(x) tail(x)) ∧ head(x)≡n → x≡add(head(x) tail(x))] , and

$\vdash_S$ [∀n:number ∀x:list x≡add(head(x) tail(x)) ∧ ¬head(x)≡n → ¬FALSE].

Of course, the proofs are trivial, hence the system comes up with the optimized version of Δ^2R as :

function Δ^2remove(n:number x:list):bool ⇐
   if x≡empty *then* false
   if x≡add(head(x) tail(x)) ∧ head(x)≡n *then* true
   if x≡add(head(x) tail(x)) ∧ ¬head(x)≡n *then* Δ^2remove(n tail(x))

Obviously, we have found the algorithm, which we called member in the introduction. After this optimization, the algorithm for Δ^2remove has only one recursive case, and therefore is simpler than the algorithm for remove with two recursions in it. This is the reason, why we prefer difference literals obtained with our difference algorithms, instead of difference literals of the form ¬f(x*)≡x^p. Termination proofs involving a difference algorithm allow simpler induction schemas than proofs involving the argument-bounded algorithm, because usually difference algorithms have less recursions.[9]

---

[9] We can see from Definition 6.1, that no difference algorithm has more recursions than the p-bounded algorithm it belongs to. Another advantage in the usage of difference literals is, that the result terms of a difference algorithm generally are simpler than the result terms of the corresponding p-bounded algorithm. For instance, the algorithm Δ^2R returns Δ^2remove(n tail(x)) in the 3rd case, whereas R returns add(head(x) remove(n tail(x))).

7.2 Recursion Elimination

Sometimes an induction proof is not necessary at all, when proving a termination hypothesis, provided the difference algorithm synthesized is non-recursive or, at least, represents a *recursive* formulation of a *non-recursive* property. In the latter case we attempt to eliminate the recursions in the difference algorithm $\Delta^P G$ by replacing a case "*if* φ *then* $\Delta^P g \delta(x^*)$" with a non-recursive case "*if* φ *then* b" , provided $\vdash_S [\forall x^*:w \ \varphi \rightarrow \delta(\varphi) \vee \delta(\varphi_{n_1}) \vee \dots \vee \delta(\varphi_{n_j})]$, where "*if* φ_{n_1} *then* b", ... , "*if* φ_{n_j} *then* b" are also non-recursive cases of $\Delta^P G$. For the algorithm $\Delta^1 M$ from Section 6, for instance, our system replaces the recursion with false and obtains

function Δ^1minus(n,m:number):bool $\Leftarrow$
if m$\equiv$0 *then* false
if $\neg$m$\equiv$0 $\wedge$ n$\equiv$succ(pred(n)) *then* true
if $\neg$m$\equiv$0 $\wedge$ n$\equiv$0 *then* false

because it verified $\vdash_S [\forall$ n,m:number $\neg$m$\equiv$0 $\wedge$ n$\equiv$0 $\rightarrow$ ($\neg$pred(m)$\equiv$0 $\wedge$ pred(n)$\equiv$0) $\vee$ pred(m)$\equiv$0)] .

7.3 Algorithm Transformation and Further Improvements

We improve our method by extending the definitions of the Γ-bound and the Γ-difference equivalent with additional cases, such that, for instance, empty $\leq_\Gamma$ x and n $\leq_\Gamma$ succ(n) hold true. Without these extensions, remove would not be a 2-bounded algorithm, if we let it return empty instead of x in the non-recursive case. We also define a *normal form* for algorithms and allow *more* than one recursion in a p-bounded algorithm, cf. Definition 5.1. All extensions enlarge the class of algorithms, for which our technique is successful in the automated generation of termination hypotheses, cf. [Walther, 1987].

However, our method can fail if the user of our system is not cooperative. For instance, she may define an algorithm, which is not p-bounded, but computes a p-bounded operation. This can happen, if the algorithm is non-optimal in some sense. Using the method of *algorithm transformation*, cf. [Walther, 1987], our system attempts to transform the algorithm into a p-bounded algorithm, which computes the same operation. For instance, the recursive case of the algorithm Q for quotient as defined in [Boyer and Moore, 1979] reads in our notation: "*if* $\neg$lessp(i j) $\wedge$ $\neg$j$\equiv$0 *then* succ(quotient(minus(i j) j))". Our system proves the termination of Q automatically. But Q is not a 1-bounded algorithm, only because minus(i j) $\leq_\Gamma$ pred(i) does not hold. However, after an inspection of the definition of minus, our system decides to replace minus(i j) in the above case by minus(pred(i) pred(j)), because it proved with $\vdash_S$ [$\forall$ i,j:number $\neg$lessp(i j) $\wedge$ $\neg$j$\equiv$0 $\rightarrow$ $\neg$j$\equiv$0], that this replacement does not alter the computed operation. Now the system verifies minus(pred(i) pred(j)) $\leq_\Gamma$ pred(i) (since minus is 1-bounded) and consequently accepts the modified algorithm for quotient as a 1-bounded one.

One can see from the above examples, that the theorem proving problems involved with these optimizations (cf. the formulas generated for Δ^2remove, for Δ^1minus and for quotient) do not constitute a challenge for an induction theorem prover. Fortunately, this observation usually holds true as experiments with our system [Biundo et al., 1986] indicate. However, since we may generate optimization formulas φ such that $\nvdash_S \varphi$ (for instance, if we attempt to eliminate the recursion in Δ^2remove), we have a second reason to supply our induction prover with a *halting criterion* (cf. Section 5).

7.4 The Role of the User

Our method must fail, however, if the user defines an algorithm, which does not compute a p-bounded operation because the results for some "do'nt care"-arguments (as e.g. on division by zero) are ill-defined. For instance, the algorithm G for greatest.factor(x y) as defined in [Boyer and Moore, 1979] contains the case: "*if* $\neg y \equiv 0 \wedge \neg y \equiv succ(0) \wedge$ remainder(x y)$\equiv 0$ *then* y". G is not 1-bounded, only because y $\leq_\Gamma$ x does not hold. We modify G yielding G' by replacing the result y in the above case with min(x y) (which computes the minimum of a pair of natural numbers). Since min is 1-bounded, we have min(x y) $\leq_\Gamma$ x and G' is a 1-bounded algorithm. Obviously, the results of G and G' differ only if the algorithms are applied to (0, n+2). The reader may verify, that this modification has no consequences for the theorems and algorithms in the database, which refer to greatest.factor, including the algorithm for prime.factors, cf. [Boyer and Moore, 1979]. With this modification, our system is able to prove the termination of prime.-factors. It is interesting to observe, that both systems have to prove almost the same (difficult) theorems for this example - our system when proving the termination hypotheses synthesized for prime.factors, cf. [Walther, 1987], and the system of *Boyer* and *Moore* when proving the induction lemmata proposed by the user, cf. [Boyer and Moore, 1979]. The reason is, that the premises of the induction lemmata used by *Boyer* and *Moore* are much stronger than neccessary to prove the termination of prime.factors.

8 Conclusion

With the results of the preceeding sections we are now able to design a system, which recognizes the termination of S-algorithms *without any guidance of a human user*: We resume our implementation from the end of Section 4, and all that remains is to supply our system with an automatic facility to recognize argument-bounded operations and to synthesize the difference literals for them.

For each admissible specification $S_n=(S,\Sigma,\Phi)$ with standard model M_n, we associate S_n with a family $\Gamma(S_n) \subset \Sigma^d$ of argument-bounded function symbols in M_n. We also define a literal $D_{p,f}$ for each $f \in \Gamma_p(S_n)$, such that $D_{p,f}$ is a p-difference literal of f in M_n :

For the *initial specification* S_0, we define $\Gamma_p(S_0):=\varnothing$ for all $p \in \mathbb{N}$, and, of course, $\Gamma(S_0)$ is a family of argument-bounded function symbols in M_0. Now assume that S_n with $n>0$ is an admissible extension of an admissible specification S_{n-1} with standard model M_{n-1}, and assume further, that $\Gamma(S_{n-1})$ is a family of argument-bounded function symbols in M_{n-1}.

If S_n is an admissible extension of S_{n-1} by an S_n-*structure* for some $s \in S$, we let $\Gamma_1(S_n):= \Gamma_1(S_{n-1}) \cup \{b \in \Sigma^d_{s,s}|b$ is a reflexive selector of some constructor $c \in \Sigma^c_{w,s}\}$, $D_{1,b} := x \equiv cb^1x \dots b^{|w|}x$ (where $x \in \mathcal{V}_s$), and we define $\Gamma_p(S_n):=\Gamma_p(S_{n-1})$ for all $p \in \mathbb{N}\backslash\{1\}$. With Theorem 5.1, $\Gamma(S_n)$ is a family of argument-bounded function symbols in M_n, and $D_{1,b}$ is a 1-difference literal of b in M_n.

Now suppose, that S_n is an admissible extension of S_{n-1} by an S_n-*algorithm* F for some $f \in \Sigma^d_{w,s}$. Using Definition 5.1, we let the system test for each $p \in \{1,\dots,|w|\}$, whether F is a p-bounded algorithm (possibly after an algorithm transformation, cf. Section 7.3), and we define P as the set of all indices p, for which this test succeeded. Then for all $p \in P$, we define $\Gamma_p(S_n):= \Gamma_p(S_{n-1}) \cup \{f\}$, $D_{p,f} := \Delta^P fx^* \equiv true$ (where $x^* \in \mathcal{V}_w$), and we let $\Gamma_{p'}(S_n):=\Gamma_{p'}(S_{n-1})$ for all $p' \in \mathbb{N}\backslash P$.

Now, using Definition 6.1, we let the system compute the p-difference algorithms $\Delta^P F$ for all $p \in P$. Then $\Delta^P F$ is optimized by case subsumption and recursion elimination as discussed in Sections 7.1 and 7.2. Finally, the system extends its current specification S_n successively by the p-difference algorithms just computed and optimized, yielding a specification $S_{n+|P|}$.

With Theorems 3.1 and 6.1, $S_{n+|P|}$ is an admissible specification, with Theorem 5.2, $\Gamma(S_{n+|P|}):=\Gamma(S_n)$ is a family of argument-bounded function symbols in each standard model $M_{n+|P|}$ of $S_{n+|P|}$, and with Theorem 6.1 each literal $D_{p,f}$ is a p-difference literal of f in $M_{n+|P|}$.

Let us summarize our example from the introduction: We saw at the end of Section 4, how the termination of remove and sort was recognized automatically, however under the given assumption, that remove is 2-bounded and member(n x)≡true is a 2-difference literal for remove. We saw in Section 5, how to recognize by machine, that remove is 2-bounded. We gave an automated synthetization of Δ^2remove in Section 6 yielding the 2-difference literal Δ^2remove(n x)≡true for remove. Finally we showed how this algorithm is optimized by machine, and, obviously, this resulted in the same algorithm as we gave for member in the introduction. Hence we obtained a proof of sort's termination, without any user guidance.

Our technique has been implemented in the INKA-system [Mill, 1987], an induction theorem prover under development at the University of Karlsruhe [Biundo et al., 1986; SFB, 1987]. As a benchmark for the system, we have also used the example database from [Boyer and Moore, 1979]: There 73 algorithms terminate with selector functions and 6 terminate with argument-bounded algorithms. Our system is able to prove the termination of all algorithms automatically. Hence our system is able to synthesize all the *induction lemmata* [Boyer and Moore, 1979], which have to be submitted to the Boyer-Moore system by a human user. Also, 3 algorithms in the database terminate with a well-founded order different from the size order, viz. normalize, gopher and samefringe, but, of course, termination proofs necessitated for algorithms terminating not with the size order cannot be solved by our method.

Acknowledgements I am indebted to Norbert Eisinger, Maritta Heisel and to my colleagues from the INKA-group, Susanne Biundo, Birgit Hummel and Dieter Hutter, for their comments on an earlier draft of this paper.

References

Aubin, R. *Mechanizing Structural Induction*. Theoretical Computer Science, vol 9, 1979.

Biundo, S., Hummel, B., Hutter, D. and Walther, C. *The Karlsruhe Induction Theorem Proving System*. Proceedings 8th CADE, Springer Lecture Notes Comp. Sc., vol. 230, 1986.

Boyer, R.S. and J S. Moore *A computational Logic*. Academic Press, 1979.

Cohn, P.M. *Universal Algebra*. D. Reidel, 1981.

Goguen, J.A., Thatcher, J.W. and Wagner, E.G. *An Initial Algebra Approach to the Specification, Correctness, and Implementation of Abstract Data Types*. In *Current Trends in Programming Methodology*, R.T. Yeh (Ed.), Prentice Hall, 1978.

Huet, G. and Oppen, D.C. *Equations and Rewrite Rules: A Survey*. In *Formal Language Theory: Perspectives and Open Problems* , R. Book (Ed.), Academic Press, 1980.

Kapur, D. and Musser, D.R. *Proof by Consistency*. Artificial Intelligence, vol 31, no 2, 1987.

McCarthy, J., Abrahams, P.W., Edwards, D.J., Hart, T.P. and Levin, M.I. *LISP 1.5 Programmers's Manual*. The MIT Press, 1962.

Mill, M. *Implementierung eines Verfahrens für Terminierungsbeweise*. Diplomarbeit, Fakultät für Informatik, Universität Karlsruhe, 1987.

Sonderforschungsbereich (SFB) 314 *Arbeits- und Ergebnisbericht für die Jahre 1985-1986-1987*. Universität Karlsruhe, 1987.

Steele, G.L. *Common Lisp - The Language*. Digital Press, 1984.

Walther, C. *Argument-Bounded Algorithms as a Basis for Automated Termination Proofs*. Interner Bericht 17/87, Fakultät für Informatik, Universität Karlsruhe, 1987.

Winston, P.H. *Artificial Intelligence*. Addison-Wesly, 1984.

Two Automated Methods in Implementation Proofs *

Leo Marcus and Timothy Redmond
Computer Science Laboratory
The Aerospace Corporation
Box 92957
Los Angeles, CA 90009

Abstract

We sketch the theory of mapping between levels and composition of state transitions, two important components in proofs of implementation. We show how they are automated in the State Delta Verification System and give examples of their use.

Key words and phrases. Program correctness, program verification, microcode verification, implementation.

1   Introduction

The State Delta Verification System, SDVS, is a system for checking proofs about the course of a computation, i.e., proofs of "program correctness."

In this paper we give a brief overview of state deltas (in Section 2), and then describe two automated capabilities in SDVS version 6: mapping between levels in proofs of implementation (Section 3) and composition of state changes (Section 4), and show how they are used in the system. For more details see the papers [12] and [11]. This introduction consists of three sections: a short description of SDVS and informal treatments of mapping and composition.

*This work was supported by the Aerospace Sponsored Research program

1.1 SDVS

We begin with a short description of SDVS. SDVS can trace its origins to [5] as a pragmatic approach to certain verification problems. Since then it has undergone extensions ([9]) and formalizations ([12], [11], [10].) The current version of SDVS is written in Common Lisp and runs on the Symbolics Lisp Machine. Its main use has been to check proofs of microprogram correctness; for example, SDVS has recently been used in the verification of a subset of the instruction set of the Bolt, Beranek, and Newman C/30 computer ([4] and [3].) Thus, there are facilities for taking as input formal descriptions of the lower (or host, or micro-) machine and the upper (or target, or macro-) machine, the microprogram, and the correspondence between the two machines. These components are combined into a claim that the lower machine together with the microprogram implements the upper machine. The user's proof of correctness is checked by SDVS for consistency with this claim.

The user communicates to SDVS through several languages. The *user interface language* allows interactive proof building and querying. The *proof language* is used to write the proof for the system to check. The *state delta language* is used to write the theorems to be proved and to describe the relevant programs and specifications. Finally, the hardware description language ISPS ([1]) is translated by the module *TR* into the state delta language.

The proof language can be divided into dynamic and static parts. The dynamic part controls the state transitions made by the system. It includes constructs for proof by symbolic execution (corresponding to sequential execution), proof by cases (corresponding to branching), and proof by induction (corresponding to loops). The static part deals with proving that certain assumptions imply certain conclusions about a given state.

For simple theories where efficient decision procedures exist and are implemented, SDVS will derive all conclusions without any user-input proof. Examples of such theories are equality over uninterpreted function symbols and some fragments of naive set theory. For more complicated domains, the current philosophy (and implementation) allow the user to write proofs by having the system notice progressively more difficult conclusions, storing the newly verified conclusions and using them as lemmas on which to base the next conclusion. The derivation from a given set of lemmas to the next conclusion may be automatic in some cases or may require the user to designate that an axiom or a previously proved lemma is to be invoked.

The system may be run in interactive mode, batch mode, or (as in most real applications) as a combination of the two. In the interactive mode, the user writes the proof in SDVS with help from system prompts, the system executing each proof command as it is written. Expressions are written in standard infix notation (e.g., x+y). A batch proof can either be written using the Symbolics Emacs editor or can be saved from an interactive session to be re-executed later.

1.2 Mapping

Program verification frequently makes use of the notion of one program implementing a specification. Thus, the phrase "verification of a computer program" often refers to a proof that the program implements some high level specification of what is expected of the program. This concept is well known; see for example [2].

Essentially, then, we have two specifications given to us. One specification, the upper specification, gives a high level description of how one expects a machine to behave. The second specification, the lower specification, gives a description of how the low level machine (executing the program) will behave. For example, in the context of microcode verification (see [3], or [7], [6] and other papers in that special journal issue) the upper specification would be a specification of how the machine responds to machine language instructions. The lower specification would describe how the machine is implemented by the microcode. The problem is to show that a machine satisfying the lower specification will satisfy all the requirements of the upper specification.

First, we shall formally define the notion of mapping, a description of the correspondence between the high level specification and the low level specification. We will focus attention on some particular types of mappings (Example 4 and Example 5.) These mappings include the following features

- the values of the upper variables are functionally dependent on the values of the lower variables,

- the upper specification only corresponds to a certain time interval in the lower specification. This allows one to use the mapping to specify that the upper machine is implemented by the lower machine as long as certain variables in the lower machine remain constant. This is useful in microcode verification when the upper machine is implemented by a micromachine which is executing microcode from a read-only memory.

Then for these specific types of mappings we prove theorems which allow one to "express" an upper specification in the lower specification language (Theorem 4 and Theorem 5.) This allows one to prove that a lower specification correctly implements an upper specification by proving that the translation of the upper specification into the lower language follows from the lower specification (Theorem 3.)

Other verification systems also deal with the issue of correspondence between levels. For example, in FDM, Ina Jo produces a theorem of implementation based on a mapping in a manner similar to ours; see [13] for an informal description. On the other hand Gypsy ([8]) does not have an explicit concept of level.

1.3 Composition

Some of the work which went into the C/30 verification involved summarizing the action of a section of microcode into a single statement. In the course of the proof, these statements (summaries) were first proved and then used to prove more complex statements. Since at the time of the C/30 verification, no facility existed for automatically determining the net effect of several state transitions done in succession, the statements summarizing the action of sections of the microcode were determined by hand. This made the C/30 verification substantially more difficult. This experience showed the need for an automatic facility for "composing" several state changes into a single state change.

Another reason for needing a composition capability stems from the fact that the markpoint to markpoint (MPISPS[9]) semantics for ISPS[1] is defined in terms of composition of state changes.

The specification of the composition algorithm developed to deal with these issues was defined by the nature of the verification problem that the composer is supposed to solve. First, since the composer is used to compute the MPISPS semantics of ISPS, the composer must give exact results (neither too weak nor too strong). The composer must be fast because it has to calculate the composition of hundreds of state changes in a reasonable time. The composer must also be optimized to handle an especially simple type of state change; a large portion of the state changes generated from a computer program take the form of assignments. Finally, it must deliver a readable composition for most of the input that it receives.

We give a formal definition of state transitions and the composition of state transitions. An algorithm for calculating compositions of state transitions is described in a high level manner. The description of the algorithm given in this paper loosely corresponds to the algorithm now used by SDVS to calculate compositions.

2 Computational Models

We begin by laying a semantic base for computation and for the implementation of one computation by another computation. We will start by giving a formal definition of a computational domain. Informally, during a computation, the program variables take on values from the computational domain.

Definition 1 *A computational domain is a triple $\langle \mathcal{A}, \mathcal{L}, \mathcal{D} \rangle$ such that*

- *$\mathcal{A}$ is a model for the typed first order language $\mathcal{L}$.*

- *the language $\mathcal{L}$ has types* place, domain *and* architecture. *Furthermore the type* architecture *represents a boolean algebra.*

- *the language $\mathcal{L}$ does not contain the symbols .*, #, $\rightsquigarrow$, δ, *or* $\| * \|$

- *$\mathcal{D}$ is a finite set of statements from the language $\mathcal{L}(.)$ where "." is a function symbol from type place to type domain. All occurrences of "." in sentences from $\mathcal{D}$ are applied to constant place symbols from the language $\mathcal{L}$. The sentences in $\mathcal{D}$ are called declarations.*

Intuitively the notion is that the objects of type place are in a one to one correspondence with the variables of the computation. We think of them as being actual places or registers in a machine. The variables take on values which are the objects of type domain.

Definition 2 *A computational model over a computational domain $(\mathcal{A}, \mathcal{L}, \mathcal{D})$ is a pair, $\langle T, (\sigma_t)_{t \in T} \rangle$, where*

- *T is a linearly ordered set with a minimal element,*

- *for each $t \in T$, σ_t is a function from the elements of $\mathcal{A}$ of type place to the elements of $\mathcal{A}$ of type domain.*

T is called the timeline of the computational model. For any $t \in T$, σ_t is called a state in the computation.

For any $t \in T$ and any object p of type place in $\mathcal{L}$, the "contents of p at time t" refers to the value of $\sigma_t(p)$.

The symbols "." and "#" will be function symbols from type place to type domain. Suppose that ϕ is a first-order sentence in the language $\mathcal{L}(.)$ and $t \in T$. $\langle T, (\sigma_t)_{t \in T} \rangle \models_t \phi$ is defined to mean that the model for $\mathcal{L}(.)$ which extends $\mathcal{A}$ by interpreting $.(p)$ as $\sigma_t(p)$ satisfies the sentence ϕ. This definition can then be extended to the language of state deltas containing dots and pounds $\mathcal{L}_{SD}$ based on $\mathcal{L}$ as explained in Definition 4.

3 State Deltas

In this section we give the definition of state deltas (Definition 4) preceded by two examples. Lack of space prevents a more leisurely presentation, as in [9].

Technically, SDVS checks proofs of state deltas. For example, SDVS can handle proofs of claims of the form

"*if P is true now, then Q will become true in the future.*"

If P is a program (perhaps with some initial conditions) and Q is an output condition, then the above claim is an input-output assertion about P. SDVS can also prove claims of the form

"if P is true now, then Q is true now."

In this case, if P is a program and Q is a specification, then the claim asserts the correctness of P with respect to Q.

The first "if-then" statement above is the basic building block of state deltas. The view of the world captured by state deltas is that there are places (to be thought of as abstract machine registers, also called program variables in other contexts) that can "hold" contents. A state of a computation or a machine is an association of contents with places. Two or more places may be interdependent, or actually overlap because the same bit locations belong to both of them.

Underlying our approach is the equivalence between programs and certain kinds of theories, in the sense that a program specifies a set of computations, a computation can be thought of as a relational structure associating a state to each point in a linearly ordered set (the timeline), and a set of these computational models can be specified by a theory. In the following discussion. state deltas are used to describe sets of computational models.

Example 1 *Consider the following state delta.*

```
[SD     pre:  .A > 0
        comod: (A)
        mod:  (A)
        post: (#A = .A+1)]
```

A state delta has four fields: pre *(precondition),* comod *(comodification list),* mod *(modification list), and* post *(postcondition.) The above state delta is true at a given time (call it "now") in a given computational model if at any time in the future when the contents of* A *are greater than 0 (*pre*) and the contents of* A *have not changed between now and then (*comod*), there is a still later time when the contents of* A *are one more than they were when we checked the precondition (*post*), and only the contents of* A *have changed between precondition time and postcondition time (*mod*).*

Much added expressive power comes from allowing the precondition and postcondition to contain state deltas in addition to first order sentences. This is well-defined since all one must do is evaluate the truth of the precondition and postcondition at certain times, and this evaluation can be done for state deltas as well as for "static" sentences.

Example 2 *The following state delta is true: (note that commas indicate conjunction)*

```
[SD     pre:  (.A = 1,
               [SD    pre:  (.A > 0)
```

```
                        comod: NIL
                        mod: (A)
                        post: (#A = .A+1)])
            comod: NIL
            mod: (A)
            post: (#A = 1000)]
```

It can be interpreted as a claim about the computation that is represented by the state delta embedded in the precondition. Call the embedded state delta S; if the contents of A are 1 and S is constantly active, then the claim is that definitely at some later time the contents of A will be 1000. This claim does not provide any more information about the values of A. For example, one cannot infer that they increase monotonically; between the times when A takes on the values 1, 2, ..., 1000, ..., A can take on arbitrary values. Also, nothing is specified about the length of the time interval between these increasing values, nor about how long A has a particular value.

Definition 3 *A state delta is a formula*

```
    [SD     pre: P
            comod: C
            mod: M
            post: Q]
```

where P *is in* $\mathcal{L}(.)$ *or is in* $\mathcal{L}_{SD}$, Q *is in* $\mathcal{L}(.,\#)$ *or is in* $\mathcal{L}_{SD}$ *, and* C *and* M *are lists of places.* $\mathcal{L}_{SD}$ *is the set of all first-order combinations of state deltas.*

Definition 4 *The state delta*

```
    [SD     pre: P
            comod: C
            mod: M
            post: Q]
```

is true in a computational model at time t_0 *if for any later time* $t_1 \geq t_0$ *at which* P *is true and the contents of the places in* C *have not changed between* t_0 *and* t_1, *there is some still later time* $t_2 \geq t_1$ *at which* Q *is true and the contents of the places not in* M *are unchanged between* t_1 *and* t_2.

4   Mappings

4.1   A definition for Mapping

Now we define the semantic basis for the notion of implementation. Informally, a mapping describes how a lower machine implements an upper machine. We want the definition to allow for a general correspondence, in fact any correspondence which will describe how the lower level mimics the upper level. A mapping will thus describe a relationship between computations of the lower machine and computations of the upper machine. The following makes formal what we mean by an instance of a lower machine implementing an upper machine.

Definition 5 *A* mapping model *from the upper domain* $(\mathcal{A}^u, \mathcal{L}^u)$ *to the lower domain* $(\mathcal{A}^l, \mathcal{L}^l)$ *is a pair,* $(\mathcal{M}^l, \mathcal{M}^u)$, *where*

- $\mathcal{M}^l$ *and* $\mathcal{M}^u$ *are computational models over the lower and upper domains respectively and*

- *the timeline of* $\mathcal{M}^u$ *is a subset of the timeline of* $\mathcal{M}^l$.

A mapping *from an upper domain* $(\mathcal{A}^u, \mathcal{L}^u)$ *to a lower domain* $(\mathcal{A}^l, \mathcal{L}^l)$ *is a set of mapping models between those same domains.*

Example 3 *Suppose that the upper domain and the lower domain differ only in that the places in the upper model,* $\mathcal{A}^u$, *are a subset of the places in the lower model,* $\mathcal{A}^l$. *Then there is a unique mapping consisting of mapping models*

$$(\langle T^l, (\sigma_t^l)_{t \in T^l}\rangle, \langle T^u, (\sigma_t^u)_{t \in T^u}\rangle)$$

which satisfy

$$T^u = T^l$$

$$\forall t \in T^u \sigma_t^u = \sigma_t^l | \mathcal{P}^u$$

where $\mathcal{P}^u$ *is the set of objects in* $\mathcal{A}^u$ *of type* **place***. Roughly speaking, in this mapping, the upper computation can be obtained from the lower computation by forgetting the contents of the places not in the upper domain.*

Whenever we are discussing computational models over the upper and lower domains, we will use $\mathcal{M}^u = \langle T^u, (\sigma_t^u)_{t \in T^u}\rangle$ and $\mathcal{M}^l = \langle T^l, (\sigma_t^l)_{t \in T^l}\rangle$ as the defaults respectively. We will also use the state delta logic to describe the lower and upper computational models. We will use π^u (π^l) to represent sets of sentences from the upper (lower) state delta logic.

Now we are ready to define formally the notion of implementation.

Definition 6 π^l *implements* π^u *with respect to the mapping* μ, *denoted*

$$\mu \vdash \pi^l \hookrightarrow \pi^u.$$

iff for any $(\mathcal{M}^l, \mathcal{M}^u)$ *in* μ *one has*

$$[\mathcal{M}^l \models \pi^l] \longrightarrow [\mathcal{M}^u \models \pi^u]$$

Theorem 1 *Implementation satisifes reflexivity (with respect to the identity mapping, any sentence implements itself) and transitivity (if $\mu_1 \vdash \pi_1 \hookrightarrow \pi_2$ and $\mu_2 \vdash \pi_2 \hookrightarrow \pi_3$ then $\mu_3 \vdash \pi_1 \hookrightarrow \pi_3$ where μ_3 is the relational composition of μ_1 and μ_2.)*

Now that implementation has been defined, we can define an important property of mapping. Note that with the general definition above, with respect to the empty mapping any sentence implements any other sentence. It is up to the "user" to define a suitable class of mappings to match his intended notion of implementation. The suitability here refers to the ease (or possibility) of recovering the target from the host. Intuitively, it seems desirable that if π^l implements π^u with respect to a mapping μ and if the theory π^l is consistent, then the theory π^u should be consistent. For example, we want every π^l to implement $\pi^u = \text{TRUE}$, but only $\pi^l = \text{FALSE}$ to implement $\pi^u = \text{FALSE}$, because that is the "hardest" specification to achieve.

A property of the mapping μ which will ensure this is called the *lifting property* (defined below).

Definition 7 *μ satisfies the lifting property iff for any computational model $\mathcal{M}^l$ over the lower domain there is a computational model $\mathcal{M}^u$ over the upper domain such that $(\mathcal{M}^l, \mathcal{M}^u) \in \mu$.*

Theorem 2 *If*

- *π^l implements π^u with respect to a mapping μ,*

- *the theory π^l is consistent, and*

- *μ satisfies the lifting property*

then the theory π^u is consistent.

We feel that some version of the lifting property is essential for any meaningful mapping. All mappings considered in this paper will satisfy the lifting property.

The following example shows how the mapping can be used to specify that the upper machine is implemented by the lower machine as long as certain variables in the lower machine remain constant. A typical example of a variable in the lower machine which one would wish to fix would be the read-only memory ROM of the microcode that a lower machine executes.

Example 4 *Suppose that the set of upper places is a subset of the set of lower places. Let ROM be a set of places in the lower that are not present in the upper. There is a unique mapping consisting of mapping models $(\mathcal{M}^l, \mathcal{M}^u)$ satisfying*

$$T^u = \{t \in T^l : \forall p \in ROM \forall t_1 \in T^l \, (t_1 \leq t \rightarrow \sigma_t(p) = \sigma_{t_1}(p))\}$$

$$\forall t \in T^u \sigma_t^u = \sigma_t^l | \mathcal{P}^u$$

where $\mathcal{P}^u$ is the set of places in the upper model. Informally, in this mapping lower states are only relevant to the upper as long as the places in ROM have not been modified. If ROM is empty then this example is equivalent to Example 3. This mapping is lifting since T^l has a minimal element.

Example 5 Informally this example describes a mapping where each lower state maps up to an upper state in such a way that the upper places can be calculated as a function of the values of the lower places. To define this mapping we need the following:

- a set of objects, $\{q_1, \ldots, q_m\}$, from $\mathcal{A}^l$ of type place

- for each object, $p \in \mathcal{A}^u$ of type place, a function

$$\nu_p : \overbrace{\mathcal{D}^l \times \ldots \times \mathcal{D}^l}^{m \text{ products}} \to \mathcal{D}^u$$

where $\mathcal{D}^u$ and $\mathcal{D}^l$ are the set of elements of $\mathcal{A}^u$ and $\mathcal{A}^l$ of type domain.

The mapping consists of mapping models which satisfy

- for all $p \in \mathcal{P}^u$ and all $t \in T^u$

$$\sigma_t(p) = \nu_p(\sigma_t(q_1), \ldots, \sigma_t(q_m))$$

- $T^u = T^l$

This mapping is lifting.

We could have defined this mapping in several ways. First of all, if the set of lower level places is finite (a reasonable assumption), we could have just taken the q_i's to be all those places. If we wanted to attach some significance to the particular choice of which q_i's correspond to which p's (e.g., minimal set), then it would be hard to parametrize the functions ν_p as in Theorem 4.

4.2   Expressibility of Upper Statements in the Lower Theory

In this section we will give a method by which one can "express" the statement that π^l implements π^u in terms of a statement from the lower theory (see Theorem 3). The reason that we are interested in this phenomenon is that it is useful for verification of implementation theorems. If one has a verification system which can prove statements in a temporal logic then this theorem gives a method for using the verification system to prove that a lower theory implements an upper theory. Thus the object of this section is to show that, under certain circumstances, upper statements are expressible in the lower theory. The following makes formal what is meant by "expressing" a statement from the upper theory in the lower theory.

Definition 8 *A sentence ϕ^u in $\mathcal{L}_{SD}^u$ is said to be expressible in $\mathcal{L}_{SD}^l$ with respect to the mapping μ iff there is a sentence ϕ^l in $\mathcal{L}_{SD}^l$ such that for all $(\mathcal{M}^l, \mathcal{M}^u) \in \mu$, we have*

$$[\mathcal{M}^l \models \phi^l] \longleftrightarrow [\mathcal{M}^u \models \phi^u]$$

Under these circumstances ϕ^l will be denoted by $\mu\{\phi^u\}$. (This is not meant to be taken to mean that ϕ^l is unique - $\mu\{\phi^u\}$ may not be well-defined.)

Theorem 3 *If π^u is expressible in the lower logic with respect to μ, then for all π^l*

$$[\mu \vdash \pi^l \hookrightarrow \pi^u] \longleftrightarrow [\pi^l \vdash \mu\{\pi^u\}].$$

This is the statement of the claim that, under the right conditions, a proof of implementation can be done completely at the lower level.

Proof: 1) $\rightarrow$ Assume $\mathcal{M}^l \models \pi^l$ and let $\phi^l \in \mu\{\pi^u\}$. We must show $\mathcal{M}^l \models \phi^l$. By lifting, there is $\mathcal{M}^u$ such that $(\mathcal{M}^l, \mathcal{M}^u) \in \mu$. By implementation $\mathcal{M}^u \models \pi^u$, so by expressibility $\mathcal{M}^l \models \phi^l$.

2) $\leftarrow$ Let $(\mathcal{M}^l, \mathcal{M}^u) \in \mu$, $\mathcal{M}^l \models \pi^l$. We must show $\mathcal{M}^u \models \pi^u$. By hypothesis, $\mathcal{M}^l \models \mu\{\pi^u\}$. So by expressibility, $\mathcal{M}^u \models \pi^u$.

QED

Now we shall reconsider Example 5 . The next theorem is the basis for the implementation command in SDVS (see Example 6.)

Theorem 4 *If μ is a mapping defined as in Example 5 and the function*

$$p, x_1, \ldots, x_m \mapsto \nu_p(x_1, \ldots, x_m)$$

is expressible in $\mathcal{L}_{SD}^l$, as are the places q_1, ..., q_n, then every upper level statement is expressible in the lower level.

Proof: Let the term $\psi(p, x_1, \ldots, x_m) \in \mathcal{L}^l$ interpret $\nu_p(x_1, \ldots, x_m)$. Suppose that ϕ^u is any statement from $\mathcal{L}^u$. If ϕ^l is obtained from ϕ^u by replacing all terms of the form $.p$ in ϕ^u with $\psi(p, .q_1, \ldots, .q_m)$ then ϕ^u is expressible in the lower theory as ϕ^l.

QED

To illustrate the flexibility of this technique we will give another theorem without proof.

Theorem 5 *If*

- *μ is as in Example 4*

- *there is a predicate, ψ on places from $\mathcal{L}^l$ such that p is a place from the upper theory iff ψ(p)*

then any statement in the upper theory can be expressed in the lower theory.

4.3   Mappings in SDVS

In this section we give the details of the method that mappings are incorporated into SDVS using Theorem 4.

The proof language must allow the user to specify the mapping (correspondence) between the places of one state delta and another that implements it, or more generally, between the states of one computation and another that implements it. In SDVS, a mapping is an assignment for each target (upper level) place of a tuple of host (lower level) places such that the value of the target place is a function of the values of the associated host places.

The command **implementation** fills the role of "theorem-constructor". It takes (prompts the user for) a theorem name, the upper-level specification (as the markpoint-to-markpoint translation of an isps program), the lower-level specification (as the translation of an isps program), the formula containing the mapping functions, the places in the host which must be constant for the implementation to be valid, and the invariants for host state changes which must hold for the implementation to be valid.

The result is a theorem (state delta) which denotes the implementation of the upper level by the lower level. The precondition of the theorem contains the lower level specification, the constant formulas, and some equalities which provide names for certain sets of places in the lower level, specifically,

- names for the set of all lower level places,

- the set of all mapped lower level places,

- the set of all unmapped lower level places,

- and the set of all constant lower lever places.

The comod and mod of the theorem are empty. The postcondition of the theorem contains n+2 items, where there are n state deltas in the upper level specification. The first item is an alldisjoint predicate stating the disjointness of the sets of mapped onto lower level places. The second item is a state delta representing the validity of the upper level declarations and the one-to-oneness of certain mapping functions. The next n items are upper level state deltas which have been transformed in to lower level theorems.

The mapping construct can either take the form

- mapping(.tplace, $f(.hplace_1, \ldots, .hplace_n)$) where f is some explicit function, for example, mapping(.tplace, .hplace), or

- mapping(.tplace, $f(.hplace_1, \ldots, .hplace_n$,values(tval$_1$, $f(hval_1^1), \ldots, hval_n^1), \ldots,$ $tval_k, f(hval_1^k, \ldots, hval_m^k)))$, where the tvals are possible values of tplace and the hvals are possible values of the the hplaces.

The constant field is used to state which places of the host are constant (for example, the ROM in the micromachine specification) The invariants field, if needed, is used to specify the significant states in the lower level machine. We do not consider these two fields in more detail here.

In the following simple example of the implementation command, italic font is for user type-in and ordinary small font is used for code or machine type-back (including quantifiers and Greek.) Here are the lower machine host.isp, and upper machine target.isp.

Example 6

```
! Sets 16-bit X to 0.          !Sets the 30-bit Z to 1.
! Sets 16-bit Y to 1.
HOST { US} := (              TARGET {US} := (
**Registers**                **Registers**
X<15:0>, Y<15:0>             Z<29:0>
**Process**                  **Process**
TOP { MAIN} := BEGIN         TOP { MAIN} := BEGIN
X _ 0 NEXT Y _ 1            Z _ 1
END                          END
)                            )
```

Here is the mapping specification from target places to host places:

```
formulas mappings:mapping(.z,(.x @ .y)<29:0>)
                  mapping(.target\upc,map\upc(.host\upc),
values(target\started,map\upc(host\started),
target\halted,map\upc(host\halted)))
```

Next, we invoke the implementation command:

```
<sdvs.1> implementation
theorem name: implementation.thm
upper-level spec: mpisps file name: TARGET.ISP
starting mark point[]: CR
ending mark points[]: CR
preconditions[]: CR
lower-level spec: isps file name: HOST.ISP
mappings: formulas(mappings)
constants[]:CR
invariants[]:CR
```

Implementation theorem 'implementation.thm' created.

Here is the theorem "implementation.thm" (state delta) that was created:

[sd pre: (isps(host.isp),
       implementation.thm.places = union(x,y,host\upc),
       implementation.thm.mapped.places = union(x,y,host\upc),
       implementation.thm.unmapped.places = diff(implementation.thm.places, implementation.thm.mapped.places))
comod:
mod:
post: (alldisjoint(union(x,y),host\upc),
       [sd pre:(true)
       comod: (all)
       mod:
       post: $(\forall\alpha 1\forall\alpha 2(lh(\alpha 1) = 16$ & $lh(\alpha 2) = 16 \rightarrow lh(\alpha 1@\alpha 2) < 29 : 0 >) = 30))]$,
       [sd pre: (.host\upc = host\started)
       env:
       mod: (x,y,host\upc,implementation.thm.unmapped.places)
       post: (#host\upc = host\halted, (#x @ #y)<29:0> = 0(28) @ 1(2))])]

Note that the three clauses in the postcondition correspond to the architecture of the target, the declarations of the target, and and the correctness of the functionality of the host. The proof is simply:

proof implementation.proof: (prove impl.thm
          proof:
             (prove g(2)
                  proof: NIL,
            prove g(3)
                  proof: execute))

The symbols g(2) and g(3) refer to the second and third goals of the postcondition. The first goal g(1) was proved automatically without even telling the system to attempt it. The goal g(2) was also proved automatically, but since it is a state delta, the user must explicitly direct SDVS to begin the proof. The goal g(3) was proved by execution.

5 Composition

5.1 Preliminaries

The purpose of this section is to define composition of state transitions, prove a theorem about composability of state transition, and give some indication how composition is implemented in SDVS. We will define a specification language for describing state transitions in a slightly more abstract way than state deltas.

Returning to Definition 1 the symbol δ will be a predicate symbol taking one argument of type **architecture** and the function symbol $\| * \|$ is from type **place** to type **architecture**. We will assume for the purposes of this paper that $A \models \|p\| \neq 0$ for all constant place symbols p. Informally the type **architecture** is inserted into the language $\mathcal{L}$ to keep track of dependencies between places. Thus if the value of the place p is somehow "dependent" on the value of the place q then one would expect that $\|p\| \leq \|q\|$ would hold. The purpose of the objects of type **architecture** becomes clearer when one starts thinking about state transitions. The type **architecture** gives one a language for describing which places are modified. Informally state transition instances should consist of a starting state, a finishing state (if achieved), and a description of what had to be modified to achieve the finishing state. The description of the modified places will be expressed in terms of an element of the type architecture.

The following definition makes this formal.

Definition 9 *A state transition model is either a singleton* $\langle \sigma_1 \rangle$ *where* σ_1 *is a state or a triple* $\langle \sigma_1, \sigma_2, m \rangle$ *where*

- σ_1 *and* σ_2 *are a states,*

- m *is an element of* $\mathcal{A}$ *of type* architecture, *and*

- $\|p\| \wedge m = 0$ *implies that* $\sigma_1(p) = \sigma_2(p)$.

If a state transition model has the form $\langle \sigma_1, \sigma_2, m \rangle$ *then one says that the final state is achieved.*

σ_1 *is called the initial state of either* $\langle \sigma_1 \rangle$ *or* $\langle \sigma_1, \sigma_2, m \rangle$.

σ_2 *is called the final state of* $\langle \sigma_1, \sigma_2, m \rangle$.

m *is called the modlist of the state transition* $\langle \sigma_1, \sigma_2, m \rangle$.

The word modlist is taken from SDVS where (currently) the only possible type of value for m has the form $\|p_1\| \vee \ldots \vee \|p_n\|$. This element is represented as the list $(p_1, \ldots, p_n)$. See Definition 4.

The transition formula saying that the contents of p will increase by one is expressed by the statement

$$(\#p = .p + 1) \,\&\, \delta(\|p\|)$$

The $\delta(\|p\|)$ signifies that the final state is reached and that only the variable p is modified in the process. If one wishes to express the state transition which says "if the value of p is greater than zero then the value of p will increase by one" one uses the formula,

$$.p > 0 \rightsquigarrow (\#p = .p + 1 \,\&\, \delta(\|p\|))$$

Now we will give a recursive definition for the syntax and semantics of transition formulas.

Definition 10 *A transition formula is defined recursively by the following statements:*

- *if P is in the language $\mathcal{L}(.)$ and ϕ is a transition formula then $P \rightsquigarrow \phi$, $P \,\&\, \phi$, and $\exists g \phi$ are transition formulas*

- $P \& \delta(m)$ *(or equivalently $\delta(m) \& P$) is a transition formula if P is in the language $\mathcal{L}(., \#)$ and m represents an element of type* architecture *in $\mathcal{L}$.*

Suppose that ρ is a state transition model, P is from the language $\mathcal{L}(.)$, Q is from the language $\mathcal{L}(., \#)$, and m is an element of type architecture *in the language $\mathcal{L}$. The semantics of transition formulas is defined recursively by the following statements:*

- *$\rho \models P \leadsto \phi$ iff $\rho \models P$ implies that $\rho \models \phi$ and $\rho \not\models P$ implies that ρ does not reach its final state*

- *$\rho \models P \& \phi$ iff the initial state of ρ satisfies P and $\rho \models \phi$*

- *$\rho \models \exists g \phi$ iff there is an interpretation of g such that $\rho \models \phi$*

- *$\rho \models Q \& \delta(m)$ iff ρ achieves its final state with a modlist less than the interpretation of m and $\rho \models Q$.*

We are interested in executing transition formulas in succession. The transition formula obtained by 'executing' two state transitions in succession is called the composition of the two state transitions. To make this more precise consider the following definition.

Definition 11 *Suppose that ρ_1 and ρ_2 are state transition models. The notions of the composability and the composition of ρ_1 and ρ_2 are defined through three cases.*

- Case 1: The state transition model ρ_1 does not achieve its final state.
  The state transition models ρ_1 and ρ_2 are said to be composable and their composition is ρ_1.

- Case 2: The state transition model ρ_1 achieves its final state but ρ_2 does not achieve its final state.
  The transition formulas are said to be composable iff the final state of ρ_1 coincides with the initial state of ρ_2. Under these circumstances their composition is the transition formula which does not achieve its final state and whose initial state coincides that of ρ_1

- Case 3: Both ρ_1 and ρ_2 achieve their final state.
  The state transition models ρ_1 and ρ_2 are said to be composable iff the final state of ρ_1 coincides with the initial state of ρ_2. In this case the composition is the state transition model whose

  - *initial state is the initial state of ρ_1*
  - *final state is the final state of ρ_2*
  - *the modlist is equal to the boolean sum of the modlist of ρ_1 and ρ_2*

In each case the composition is denoted $\rho_1 \circ \rho_2$.

Definition 12 *If ϕ_1 and ϕ_2 are transition formulas then the composition of ϕ_1 and ϕ_2 (if it exists) is a transition formula ϕ such that for any state transition model ρ, $\rho \models \phi$ iff there exists state transition models, ρ_1 and ρ_2, such that*

- *$\rho_1 \models \phi_1$ and $\rho_2 \models \phi_2$*

- *ρ_1 and ρ_2 are composable and their composition is ρ*

In this situation the transition formula ϕ is called the composition of the transition formulas ϕ_1 and ϕ_2 and is denoted $\phi_1 \circ \phi_2$.

Example 7 *Suppose that there are no declarations and there is exactly one object, p, of type place. We will assume this for each of the examples of this section. If we first increment the value of the variable p by one and then increment the value of p by two, then the net result is to increment the value of p by three. This concept is roughly equivalent to saying that the composition of the transition formulas,*

$$\#p = .p + 1 \ \& \ \delta(\|p\|)$$
$$\#p = .p + 2 \ \& \ \delta(\|p\|)$$

is the transition formula

$$\#p = .p + 3 \ \& \ \delta(\|p\|)$$

Example 8 *This example involves state transitions in which the state transition is conditional, i.e. one may or may not reach the next state. Consider the transition formula:*

$$.p > 1 \rightsquigarrow \#p = .p + 1 \ \& \ \delta(\|p\|).$$

In this example, one does not know that one will reach the next state unless $.p > 1$. Thus when one composes this transition formula with another transition formula, the second transition formula may not be applicable if $.p \leq 1$. To illustrate this, consider the following transition formulas:

$$.p > 1 \rightsquigarrow (\#p = .p + 1) \ \& \ \delta(\|p\|)$$
$$\#p = .p + 2 \ \& \ \delta(\|p\|)$$

Their composition is

$$.p > 1 \rightsquigarrow (\#p = .p + 3) \ \& \ \delta(\|p\|)$$

Example 9 *This final example shows how a condition on a transition formula will move back through a composition. Note that the composition of the two transition formulas:*

$$(\#p = .p + 1) \ \& \ \delta(\|p\|)$$
$$.p > 1 \rightsquigarrow (\#p = .p + 2) \ \& \ \delta(\|p\|)$$

is the transition formula:

$$.p > 0 \rightsquigarrow (\#p = p + 3) \ \& \ \delta(\|p\|)$$

5.2 Theorem and Algorithm

The following theorem gives a large class of transition formulas which are composable.

Theorem 6 *Suppose that ϕ_1 and ϕ_2 are transition formulas and*

- *for all occurrences of terms of the form $\#x$ in ϕ_1 and all occurrences of terms of the form $.x$ or $\#x$ in ϕ_2, x is a constant symbol of type* place.

Then ϕ_1 and ϕ_2 are composable.

The proof of this theorem will now be briefly motivated with a simple example. Consider the transition formulas

$$\#p = .p + 1 \ \& \ \delta(\|p\|)$$

$$\#p = .p + 2 \ \& \ \delta(\|p\|)$$

The composition of these two transition formulas could have been represented as

$$\exists g(g = .p + 1 \ \& \ \#p = g + 2) \ \& \ \delta(\|p\|)$$

Here the variable g is being used to represent the contents of p after the first state transition has occurred and before the second state transition has occurred. This transition formula could then be simplified to the more readable form

$$\#p = .p + 3 \ \& \ \delta(\|p\|)$$

Proof: Let $\{p_1, \ldots p_n\}$ be the union of all the places occurring

- as an argument to a pound in ϕ_1

- as an argument to a dot or pound in ϕ_2

- as an argument to a dot in one of the declarations

Choose variable symbols $\{g_1 \ldots g_n\}$ which are not used in the statement of the declarations or in the statements of ϕ_1 or ϕ_2. Suppose that the declarations are $\{\psi_1, \ldots \psi_k\}$. Then it is easy to see that the composition of ϕ_1 and ϕ_2 is ϕ_1 with its outermost expression of the form $\rho \ \& \ \delta(m_1)$ replaced by

$$\exists g_1 \ldots \exists g_n$$
$$\rho[\#p_i/g_i] \ \& \ (\psi_1 \ \& \ \ldots \ \& \ \psi_k)[.p_i/g_i]$$
$$\& (\|p_1\| \wedge m_1 = 0 \to g_1 = .p_1) \ \& \ \ldots \ \& \ (\|p_n\| \wedge m_1 = 0 \to g_n = .p_n)$$
$$\& \phi_2[.p_i/g_i, \delta(x)/\delta']$$

where δ' is

$$((\|p_1\| \wedge x = 0 \rightarrow \#p_1 = g_1) \ \& \ \ldots) \ \& \ \delta(x \vee m_1)$$

QED.

Note that the proof of Theorem 6 is constructive. This means that we can use the technique given in the proof of Theorem 6 as a basic composition algorithm. However, there are some significant refinements that must be made in order to get a usable algorithm. The general formulas that the basic algorithm gives must be optimized in certain common cases, such as assignment and declarations. For lack of space, we cannot go into the details here. The interested reader may refer to [11].

5.3   Composition in SDVS

Composition is used internally in processing mpisps, and it can also be called explicitly by the user in interactive mode via the command **compose**.

The following example illustrates using the composer in a proof of the state delta. Let s5 be the state delta

```
[sd pre: (covering(all, x,y,upc,tmp),formulas(machine), .upc=1)
      comod:
      mod: (all)
      post: (#x=.x+1, #y=.y)]
```

where **machine** is the set { s1, s2, s3, s4 } , equaling respectively:

```
[sd pre: (.upc = 1)
   comod:
   mod: (upc,tmp)
   post: (#tmp = .x,#upc = .upc + 1)]
[sd pre: (.upc = 2)
   comod:
   mod: (x,upc)
   post: (#x = .y,#upc = .upc + 1)]
[sd pre: (.upc = 3)
   comod:
   mod: (y,upc)
   post: (#y = .tmp,#upc = .upc + 1)]
[sd pre: (.upc=4)
   comod:
   mod: (y,upc)
   post: (#y = .y + 1, #upc=1)]
```

Of course s5 could be proved by direct execution (indicated by the command ∗):
`<sdvs.1>prove s5 *`

```
open -- [sd pre: (covering(all,x,y,upc,tmp),formulas(machine),.upc = 1)
            comod:
            mod: (all)
            post: (#x = .x + 1,#y = .y)]
apply -- [sd pre: (.upc = 1) mod: (upc,tmp) post: (#tmp = .x, #upc = .upc + 1)]
apply -- [sd pre: (.upc = 2) mod: (x,upc) post: (#x = .y, #upc = .upc + 1)]
apply -- [sd pre: (.upc = 3) mod: (y,upc) post: (#y = .tmp, #upc = .upc + 1)]
apply -- [sd pre: (.upc = 4) mod: (y,upc) post: (#y = .y + 1, #upc = 1)]
```

```
apply -- [sd pre: (.upc = 1) mod: (upc,tmp) post: (#tmp = .x, #upc = .upc + 1)]
apply -- [sd pre: (.upc = 2) mod: (x,upc) post: (#x = .y, #upc = .upc + 1)]
apply -- [sd pre: (.upc = 3) mod: (y,upc) post: (#y = .tmp, #upc = .upc + 1)]
close -- 7 steps/applications
```

But we are really just interested in applying s1, s2, and s3 in succession. So let us make a state delta which will embody the same effect as that successive application.

```
<sdvs.2> compose
composed sd name: composedsd
Do you wish to compose sds from the proof stack? (y or n) [n]:n
sd []: s1
sd []: s2
sd []: s3
sd []: CR
declarations[]: covering(all, x, y, tmp, upc)


Experimental Composer
Composed
[sd pre: (.upc = 1)
   comod:
   mod: (y,x,tmp,upc)
   post: (#upc = 3 + 1,#y = .x,#x = .y,#tmp = .x)]
```

Now we can use the following as a proof:

```
(prove s5
      proof:(prove composedsd
                  proof: (apply s1,
                          apply s2,
                          apply s3,
                          close),
             apply composedsd,
             apply s4,
             apply composedsd,
             close))
```

References

[1] Barbacci, M. R., Barnes, G.E., Cattell, R. G., and Sieiworek, D.P., "The ISPS Computer Description Language", Tech. report CMU-CS-79-137, Carnegie-Mellon University, Computer Science Department, August 1979

[2] Birman, A. and Joyner, W. H., "A Problem-Reduction Approach to Proving Simulation Between Programs", *IEEE Transactions on Software Engineering,* 2 (1976) 87-96

[3] Cook, J. V.,"Final Report for the C/30 Microcode Verification Project", Tech Report ATR-86(6771)-3, The Aerospace Corporation, 1986

[4] Cook, J. V., "C/30 Proof", Tech. report ATR-86(6771)-4, The Aerospace Corporation, 1986.

[5] Crocker, S. D., "State Deltas: A Formalism for Representing Segments of Computation, PhD Thesis, University of California, Los Angeles, 1977.

[6] Damm, W., Doehmen, G., Merkel, K., and Sichelschmidt, M., "The AADL/S* Approach to Firmware Design Verification", *IEEE Software,* 3 (1986) 27-37

[7] Dasgupta, S., Wilsey, P., and Heinanen, J., "Axiomatic Specifications in Firmware Development Systems", *IEEE Software,* 3 (1986) 49-58

[8] Good, D., Akers, R., Smith, L. " Report on Gypsy 2.05", Tech Report ICSCA-CMP-48, Institute for Computing Science, University of Texas, Austin, February, 1986

[9] Marcus, L., "SDVS 6 Users' Manual", Tech Report ATR-86A(2778)-4, The Aerospace Corporation, 1987

[10] Marcus, L., Redmond, T., and Shelah, S."Completeness of State Deltas", Tech Report ATR-85(8354)-5, The Aerospace Corporation, 1985

[11] Redmond, T., "Composition of State Changes and Program Verification", Tech Report ATR-86A(2778)-3, The Aerospace Corporation, 1987

[12] Redmond, T. and Marcus, L., "Mapping between Levels and Proofs of Implementation", Tech Report ATR-86A(8554)-5, The Aerospace Corporation, 1987

[13] Scheid, J., Martin, R., Anderson, S., and Holtsberg, S., "INA-JO Specification Language Reference Manual, Release 1" TM 6021/001/02, System Development Corporation, January 1986

A New Approach to Universal Unification and Its Application to AC-unification

Mark Franzen,
Department of Mathematics, Northwestern University

Lawrence J. Henschen,
Department of Computer Science, Northwestern University

Abstract

A complete unification algorithm for simple theories is described. This algorithm is an extension of the variable abstraction approach. Since the associative-commutative (AC) theory is simple, an immediate consequence of this algorithm is a new AC-unification procedure. A partial correctness proof is given and some preliminary termination results for the AC case are presented.

1. Introduction

Given two terms built from function symbols, variables and constants, the classical unification problem is to find a substitution for the variables in the terms which makes the two terms equal. Equational unification extends this problem to that of finding a set of substitutions which makes the two terms equal under a set of equational axioms. From an algebraic point of view, this is equivalent to solving equations in a variety.

Equational unification has a large number of applications in Computer Science, particularly in Automated Deduction [Siekmann 1984]. However, every equational theory requires that a special unification algorithm be discovered and implemented. Hence, much study has been given to the problem of automatically determining a unification procedure for a given equational theory. There are two basic approaches: a universal unification algorithm and the combination of known unification procedures.

A universal unification algorithm is a procedure which takes as input an equational theory E and a pair of terms ⟨s,t⟩ and generates a complete set of E-unifiers for s and t. A universal algorithm, based on narrowing, is known for equational theories which admit a canonical term rewriting system [Fay, 1979],[Hullot, 1980]. Recently, several procedures extending Martelli and Montanari's method [1982] of transformations have been presented which solve the unifcation problem for a wide class of equational theories ([Kirchner, 1985],[Martelli, Moiso & Rossi, 1986] and [Gallier & Snyder, 1987]).

A second approach is that of combining known special purpose unification algorithms. Yelick [1985], Kirchner [1985], Herold [1985], and Tiden [1986] have all presented different procedures for combining known unification algorithms for theories which have disjoint sets of function symbols. These procedures all require that the theories be collapse-free (a theory is collapse-free if it contains no axioms of the form x = t, where x is a variable and t is a non-variable term).

In this paper, a universal unification procedure for *simple* equational theories is presented. (A theory is simple iff it contains no equations of the form s = t, where s is a proper subterm of t. This is a somewhat stronger restriction than collapse-freeness. See Section 4.1.) Our procedure is a generalization of the variable abstraction approach of Stickel [1981] and Yelick [1985]. We require that the axioms for the theory contain only one function symbol and that a complete set of unifiers be known for a special pair of terms, $\langle s_{ab}, t_{ab} \rangle$, the *basic abstracted terms*. In many cases, these restrictions can be relaxed or eliminated as discussed in Section 6.

The procedure is defined in Section 3, a partial correctness proof is given in Section 4 and some termination results are stated for the AC theory in Section 5.

2. DEFINITIONS AND NOTATIONS

We will use the standard concepts and notations of equational theories, unification, etc. (see e.g. Huet & Oppen, [1980]). However, for the sake of reference and clarity, some definitions and basic facts are repeated here.

Let $\mathbb{F}$ be a set of function symbols with associated arity, and $\mathcal{V}$ a set of variables. Let $\mathbb{T}$ be the set of all terms constructed in the usual way using the symbols in $\mathbb{F}$ and the variables in $\mathcal{V}$. Given a term t, let $\mathcal{V}(t)$ be the set of variables in t, and F(t) the function symbols in t. A *compound term* is a term which is not a variable or a constant.

Substitutions will be denoted by a set of variable-term pairs, $\{v_1 \leftarrow t_1, v_2 \leftarrow t_2, \ldots \}$ and $ will denote the set of all substitutions. The *composition* of two substitutions σ and τ is written $\sigma \circ \tau$ and is defined by $(\sigma \circ \tau)t = \sigma(\tau t)$. The *empty substitution*, i.e. the mapping which sends every variable to itself, will be written ε. Also, for a substitution σ, define:

$$D(\sigma) = \{v \in \mathcal{V} \mid \sigma v \neq v\} \qquad \text{(domain)}$$
$$R(\sigma) = \{\sigma v \mid v \in D(\sigma)\} \qquad \text{(range)}$$
$$I(\sigma) = \{v \in \mathcal{V}(t) \mid t \in R(\sigma)\} \qquad \text{(range variables)}.$$

An *equation* is a pair of terms, written s = t. For a set of equations E, the *equational theory presented by E* is the set of equations, E^*, formed by taking the finest congruence on $\mathbb{T}$ containing all pairs $\sigma s = \sigma t$ for all s = t in E and σ in $\$$. Thus E^* is the set of all equations derivable by a finite application of the rules: reflexivity, symmetry, transititivy, replacement and substitution (see e.g. Burris & Sankappanavar, [1981]). We will denote this congruence relation on terms by $s =_E t$, i.e. $s =_E t$ iff $s = t \in E^*$.

We extend this congruence relation to substitutions by:

$$\sigma =_E \alpha \;\text{iff}\; \forall v \in \mathcal{V}, \; \sigma v =_E \alpha v.$$

If this relation holds only over a set of variables W we write:

$$\sigma =_E^W \alpha \;\text{iff}\; \forall v \in W, \; \sigma v =_E \alpha v.$$

A substitution α is *more general* than σ on W, written $\alpha \leq_E^W \sigma$ iff:

$$\exists \theta \in \$, \; \theta \circ \alpha =_E^W \sigma.$$

Define an equivalence relation on substitutions by:

$$\sigma \equiv_E^V \alpha \;\text{iff}\; \sigma \leq_E^V \alpha \;\&\; \alpha \leq_E^V \sigma.$$

Given a set of equations, E, and two terms s,t we say a substitution σ is an *E-unifier* of s and t iff $\sigma s =_E \sigma t$. Let $U_E(s,t)$ be the set of all such unifiers (We will write U_E when the terms are clear from context). In general, this set is infinite; we represent it by a *complete set of unifiers*, CSU_E, from which U_E can be generated by taking all instances of substitutions in the CSU_E. If every element of a CSU_E is necessary for completeness, the set is called a *minimal complete set of unifiers for s and t*, written $\mu CSU_E(s,t)$. Formally, we have:

DEFINITION. Let s and t be terms, and $V = \mathcal{V}(s) \cup \mathcal{V}(t)$. A set of unifiers, Σ, is a $\mu CSU_E(s,t)$ iff:

1. $\Sigma \subseteq U_E(s,t)$ (Soundness)

2. $\forall \sigma \in U_E(s,t), \; \exists \alpha \in \Sigma, \; \alpha \leq_E^V \sigma$ (Completeness)

3. $\forall \sigma_1, \sigma_2 \in \Sigma, \; \sigma_1 \leq_E^V \sigma_2 \Rightarrow \sigma_1 = \sigma_2$ (minimality)

4. $\forall \sigma \in \Sigma, \; D(\sigma) = V \;\&\; D(\sigma) \cap R(\sigma) = \emptyset$ (protectiveness)

When it exists, a μCSU_E is unique up to $\equiv_E^V$ [Fages & Huet, 1983].
Depending on E, the size of the μCSU_E may be bounded. If E = $\emptyset$, then
a μCSU_E has at most one element. If E contains only the associative
and commutative axioms (the AC theory), then a μCSU_E is always finite.
If E contains only the associative axiom, then there are pairs of
terms for which the μCSU_E is infinite. (For a recent classification
of equational theories in the unification hierarchy, see [Siekmann,
1986].) Equational theories for which the μCSU_E is always finite are
finitary unifying or just *finitary*. For any theory, if there is a
finite CSU_E, then a μCSU_E can always be found by filtering out all
non-minimal unifiers through matching.

We say an E-unification procedure is *partially correct* if,
whenever it terminates, it generates a complete set of E-unifiers.
If, in addition, the procedure always terminates, then it is *totally
correct*. Furthermore, we say a procedure is an E-unification
procedure for the *variable-only case*, if it generates a CSU_E for input
terms composed only of variables and function symbols occuring in E.

We also need to define unifiers of substitutions. A substitution σ
E-unifies the substitutions α and β iff $\sigma \circ \alpha =_E \sigma \circ \beta$. We then extend the
definitions of $U_E(\alpha,\beta)$ and $CSU_E(\alpha,\beta)$ to substitutions in the natural
way.

3. THE ALGORITHM

We now present a partially correct E-unification algorithm for the
variable-only case, where E is a set of equations which contains only
one function symbol and which presents a simple theory. First, we
need some definitions.

If $t = f(r_1,\ldots,r_n)$ is a compound term, define the *abstraction* of
t, written $\hat{t}$, by $\hat{t} = f(x_1,\ldots,x_n)$, where the x_i are fresh variables.
(This corresponds to the notion of a *variable abstraction* in
[Stickel, 1981]). Define the *preserving substitution*, denoted $P(t,\hat{t})$,
by $P(t,\hat{t}) = \{x_1 \leftarrow r_1,\ldots,x_n \leftarrow r_n\}$. Note that the preserving
substitution takes an abstraction back to its original term, i.e.
$P(t,\hat{t})(\hat{t}) = t$.

For two substitutions α and σ, the term *corresponding pair of
terms for* v, refers to a pair of terms $\langle t_1,t_2 \rangle$, where $t_1 = \alpha v$,
$t_2 = \sigma v$, for some $v \in \mathcal{V}$.

Let $\{x_1,\ldots,x_n,y_1,\ldots,y_n\}$ be a fixed set of variables. Then define
two special terms: $s_{ab} = f(x_1,\ldots,x_n)$ and $t_{ab} = f(y_1,\ldots,y_n)$, called
the *basic abstracted terms*. Before we begin the algorithm, we assume

that we have available a μCSU_E for the basic abstracted terms. We call this set U^{AB}. (In practice, this set must be finite. However, if U^{AB} is infinite, our algorithm will enumerate an infinite complete set of unifiers, see [Franzen, 1988]).

The basic idea of the algorithm can now be stated:

Suppose that we are to unify two terms s and t. If one of the terms (say s) is a variable, and $s \notin \mathcal{V}(t)$, we return a set containing the single substitution $\{s \leftarrow t\}$. If $s \in \mathcal{V}(t)$, return $\emptyset$. If both terms are compound terms with head symbol f we proceed as follows:

1. Form the abstractions $\hat{s}, \hat{t}$ and the preserving substitution, $\beta = P(s, \hat{s}) \cup P(t, \hat{t})$, for s and t.

2. Let $U^{AB}(\hat{s}, \hat{t}) = \{\alpha \circ \sigma \circ \tau \mid \sigma \in U^{AB}\}$, where τ is the substitution which renames the variables in $\hat{s}$ and $\hat{t}$ to the corresponding variables in the basic abstracted terms (i.e. $\tau\hat{s} = s_{ab}$, $\tau\hat{t} = t_{ab}$) and α renames the variables in $I(\sigma)$ to fresh variables. Note that $U^{AB}(\hat{s}, \hat{t})$ is a μCSU_E for $\hat{s}$ and $\hat{t}$.

3. For each $\theta \in U^{AB}(\hat{s}, \hat{t})$, obtain a complete set of E-unifiers for the substitutions θ and β. Let Σ be the union of these sets. Return Σ as it is a $CSU_E(s, t)$.

Step 3 is accomplished by using the procedure MAP-UNIFY$_E$, which E-unifies the substitutions θ and β by recursively calling the main unification algorithm for the corresponding pair of terms $\langle t_1, t_2 \rangle$, for each variable $v \in D(\theta) \cup D(\beta)$. This procedure is based on K.Yelicks's MAP-UNIFY algorithm [Yelick, 1985].

The algorithm is shown in Figure 1. It is given in the form of a procedure UNIFY$_E$, which takes as input two terms and returns a complete set of unifiers for these terms. UNIFY$_E$ is mutually recursive with MAP-UNIFY$_E$.

4. Proof of Partial Correctness

We present in this section a partial correctness proof for UNIFY$_E$. For simplicity, we have stated the algorithm in such a way that non-protective unifiers are generated. Hence, we do not prove that the set generated by the algorithm is protective. However, it is obvious that any unifier can be renamed so that it is protective.

4.1 Restrictions

For UNIFY$_E$, we assume that E is a set of equations containing only variables and a single function symbol, f, of arity greater than 0 and

```
UNIFY_E = proc(s,t :term) returns(set of substitution)
    case
        s ∈ 𝒱 & t ∈ 𝒱:                                    ;Case 1
            return {{s ← t}}
        s ∈ 𝒱 & t ∉ 𝒱 & s ∈ 𝒱(t):                         ;Case 2
            return ∅
        s ∈ 𝒱 & t ∉ 𝒱 & s ∉ 𝒱(t):                         ;Case 3
            return {{s ← t}}
        t ∈ 𝒱 & s ∉ 𝒱 & t ∈ 𝒱(s):                         ;Case 4
            return ∅
        t ∈ 𝒱 & s ∉ 𝒱 & t ∉ 𝒱(s):                         ;Case 5
            return {{t ← s}}
        else                                              ;Case 6
            β := P(s,ŝ) ∪ P(t,t̂)
            𝒜 := U^AB(ŝ,t̂)
            Σ := U_{θ∈𝒜} MAP-UNIFY_E(θ,β)
            return Σ
    end
end UNIFY_E


MAP-UNIFY_E = proc(β,θ:substitution) returns(set of substitution)
    Σ_0 = {ε}
    j := 0
    for v ∈ D(β) ∪ D(θ) do
        j := j + 1
        Σ_j := { α∘σ | σ ∈ Σ_{j-1} & α ∈ UNIFY_E(σ(βv),σ(θv)) }
    end
    return Σ_j
end MAP-UNIFY_E
```

FIGURE 1.   A procedure for unifying two variable-only
terms in a simple theory with only one function symbol.

that E presents a simple theory. We also assume that the input terms,
s and t, are made up of only variables and the function symbol f.
(Relaxation of these restrictions is discussed in Section 6.2). We
give the neccessary definitions and some basic facts.

An equation, s = t, where either t or s is a variable and the
other term is a non-variable, is called a *collapse equation*. An
equation is *subterm collapsing* iff one side of the equation is a
proper subterm of the other. A set of equations is *simple* iff it
contains no subterm collapsing equations and it is *collapse-free* iff
it contains no collapse equations.

Let #(X,s) denote the number of occurences of a symbol X in a term
s, where X ∈ 𝒱 ∪ 𝔽. An equation, s = t, is a *permutation equation* iff
for every symbol X ∈ 𝒱 ∪ 𝔽, #(X,s) = #(X,t). A set of equations is
permutative if it contains only permutation equations. Associativity
(the theory presented by {f(x,f(y,z)) = f(f(x,y),z)}) is an example of

a permutative theory.

We need the following three propositions: (for proofs of the first two, see e.g. Bürckert, Herold & Schmidt-Schauß [1987])

Proposition 1. A set of equations, E, is permutative (collapse-free) iff E^* is permutative (collapse-free).

Proposition 2. (i) Every permutative theory is simple.
(ii) Every simple theory is collapse free.

Proposition 3. Let E be a set of equations which presents a simple theory. If v is a variable and t is a non-variable term with $v \in \mathcal{V}(t)$, then $U_E(v,t) = \emptyset$.

    Proof. Let $\sigma \in \$$. Since v occurs in t, the term σv is a proper subterm of σt. Therefore $\sigma v = \sigma t$ is a subterm collapsing equation and thus must not be in E^*.

4.2 THE PARTIAL CORRECTNESS THEOREM

Lemma 1. The procedure MAP-UNIFY$_E(\theta,\beta)$ terminates returning a CSU$_E(\theta,\beta)$ if all its calls to UNIFY$_E(t_1,t_2)$ terminate returning a CSU$_E(t_1,t_2)$.

    Proof. By induction on $k = \#(D(\theta) \cup D(\beta))$. For a detailed proof see [Yelick, 1987].

Theorem 1. If UNIFY$_E(s,t)$ terminates returning Σ, then Σ is a CSU$_E(s,t)$.

    Proof. By induction on the depth of recursion in UNIFY$_E$. The base cases are Cases 1-5 of UNIFY$_E$. Case 6 is an inductive case.

Case 1: Both terms are variables, hence all unifiers are instances of the unifier $\{s \leftarrow t\}$.

Cases 2,4: By Proposition 3, s and t have no E-unifiers.

Cases 3,5: One term is a variable (v) not contained in the other term (t_1). The substitution $\{v \leftarrow t_1\}$ is a unifier, of which all others are instances.

Case 6:

Soundness. Assume $\sigma \in \Sigma$. Thus, $\sigma \in$ MAP-UNIFY$_E(\theta,\beta)$ for some $\theta \in U^{AB}(\hat{s},\hat{t})$, and $\beta\hat{t}=t$, $\beta\hat{s}=s$. Then, using Lemma 1 and the induction hypothesis, σ is an E-unifier of θ and β. Hence,

$$
\begin{aligned}
\sigma s &= \sigma(\beta\hat{s}) &&\text{since } \beta \text{ is a preserving substitution}\\
&=_E \sigma(\theta\hat{s}) &&\text{as } \sigma \text{ unifies } \beta \text{ and } \theta\\
&=_E \sigma(\theta\hat{t}) &&\text{as } \theta \text{ unifies } \hat{s} \text{ and } \hat{t}\\
&=_E \sigma(\beta\hat{t}) &&\text{as } \sigma \text{ unifies } \beta \text{ and } \theta\\
&= \sigma t &&\text{since } \beta \text{ is a preserving substitution.}
\end{aligned}
$$

Completeness. Let α be an arbitrary E-unifier for s and t and let $D = \mathcal{V}(s) \cup \mathcal{V}(t)$. W.l.o.g. we assume that $D(\alpha) \subseteq D$. We must show that $\tau \leq_E^D \alpha$ for some $\tau \in \Sigma$.

Let $\hat{s}$, $\hat{t}$ be the abstractions of s and t, respectively. Let $U^{AB}(\hat{s},\hat{t})$ be a μCSU for $\hat{s}$, $\hat{t}$ and let $\beta = P(t,\hat{t}) \cup P(s,\hat{s})$ be the preserving substitution. Define:

$$V = \mathcal{V}(\hat{s}) \cup \mathcal{V}(\hat{t}).$$

Notice that $D(\beta) = V$ and $I(\beta) = D$. Also, by renaming $U^{AB}(\hat{s},\hat{t})$ we may assume that

(eq 1) $\qquad D \cap I(\theta) = \emptyset \qquad\qquad$ for each $\theta \in U^{AB}(\hat{s},\hat{t})$.

Now $\qquad\quad \alpha s =_E \alpha t \qquad\qquad$ since α is an E-unifier of s,t,

hence $\qquad \alpha(\beta\hat{s}) =_E \alpha(\beta\hat{t})$

thus $\qquad\;\; \alpha\circ\beta \in U_E(\hat{s},\hat{t})$.

Therefore, because $U^{AB}(\hat{s},\hat{t})$ is a CSU for $\hat{s}$ and $\hat{t}$, we must have that

(eq 2) $\qquad \alpha\circ\beta =_E^V \sigma\circ\theta \qquad\qquad$ for some $\theta \in U^{AB}(\hat{s},\hat{t})$, $\sigma \in \$$.

Notice that since θ is protective,

$\qquad\qquad D(\theta) = V, \qquad\qquad$ hence we can assume that

$\qquad\qquad D(\sigma) \subseteq I(\theta).$

Therefore, from eq. 1, and the fact that $D(\alpha) \subseteq D$, we have

$\qquad\qquad D(\sigma) \cap D(\alpha) = \emptyset.$

Thus we may define

$$\tilde{\sigma} = \sigma \cup \alpha.$$

Note that this implies that $\tilde{\sigma} =^D \alpha$.

Lemma 2. $\tilde{\sigma}\circ\theta =_E \tilde{\sigma}\circ\beta$.

$\qquad$ **Proof.** Since $D(\theta) = D(\beta) = V$, the equality need only be proved for each $v \in V$. Hence for $v \in V$:

$\qquad\qquad \tilde{\sigma}(\beta v) \quad = \quad \alpha(\beta v) \qquad\qquad$ since $\mathcal{V}(\beta v) \subseteq I(\beta) = D$ and $\tilde{\sigma} =^D \alpha$

$\qquad\qquad\qquad\quad =_E \sigma(\theta v) \qquad\quad$ by eq 2

$\qquad\qquad\qquad\quad = \tilde{\sigma}(\theta v).$

$\qquad$ The last equality follows from the fact that $\mathcal{V}(\theta v) \cap D(\alpha) = \emptyset$ (from eq. 1), and that $\tilde{\sigma} = \sigma \cup \alpha$. $\qquad\square$

Now, completing the proof of the Theorem, from Lemma 2, we see that $\tilde{\sigma}$ is an E-unifier of θ and β. Therefore, from Lemma 1 and the induction hypothesis, we must have that

$$\tau \leq_E \tilde{\sigma}, \text{ for some } \tau \in \text{MAP-UNIFY}_E(\theta,\beta).$$

Thus, since $\tilde{\sigma} =^D \alpha$, we have

$$\tau \leq^D_E \alpha, \text{ for some } \tau \in \Sigma. \quad \blacksquare$$

5. Applying the Algorithm to the AC Theory

To demonstrate our procedure, we apply it to the theory presented by the associative and commutative axioms (the AC theory). While efficient, complete and terminating AC-unification algorithms are known [Stickel, 1981][Livesey & Siekmann, 1976], they depend on solutions to a system of linear diophantine equations. However, our approach yields a complete AC unification procedure not involving solutions to diophantine equations.

Another reason for choosing the AC theory to apply our method is that the AC theory raises an interesting theoretical question: why does the combination of an infinitary theory (A) with a finitary theory (C) result in a finitary theory (AC)? Since our method applies to all three of these theories, it is hoped that it will be useful in investigations into this kind of question.

Since the associative and commutative axioms are permutation equations, the AC theory is permutative (Proposition 1) and hence simple (Proposition 2). Therefore, our UNIFY$_{AC}$ algorithm is a partially correct AC-unification procedure. Unfortunately, there are pairs of terms for which UNIFY$_{AC}$ doesn't terminate. However, UNIFY$_{AC}$ will terminate for certain classes of terms, and experimental results indicate that the algorithm can be modified so that it will terminate for all pairs of terms and still retain completeness.

A set of 7 unifiers comprises a μCSU_{AC} for the basic abstracted terms $f(x_1, x_2)$ and $f(y_1, y_2)$ in the AC theory. The 7 elements in this set, U^{AB}, are:

$$\theta_1 = \{x_1 \leftarrow r_0, \; x_2 \leftarrow r_1, \; y_1 \leftarrow r_1, \; y_2 \leftarrow r_0\}$$
$$\theta_2 = \{x_1 \leftarrow r_0, \; x_2 \leftarrow f(r_2, r_1), \; y_1 \leftarrow r_1, \; y_2 \leftarrow f(r_0, r_2)\}$$
$$\theta_3 = \{x_1 \leftarrow f(r_2, r_0), \; x_2 \leftarrow r_1, \; y_1 \leftarrow f(r_2, r_1), \; y_2 \leftarrow r_0\}$$
$$\theta_4 = \{x_1 \leftarrow r_0, \; x_2 \leftarrow r_1, \; y_1 \leftarrow r_0, \; y_2 \leftarrow r_1\}$$
$$\theta_5 = \{x_1 \leftarrow r_0, \; x_2 \leftarrow f(r_2, r_1), \; y_1 \leftarrow f(r_0, r_2), \; y_2 \leftarrow r_1\}$$
$$\theta_6 = \{x_1 \leftarrow f(r_2, r_0), \; x_2 \leftarrow r_1, \; y_1 \leftarrow r_0, \; y_2 \leftarrow f(r_2, r_1)\}$$
$$\theta_7 = \{x_1 \leftarrow f(r_1, r_3), \; x_2 \leftarrow f(r_0, r_2), \; y_1 \leftarrow f(r_2, r_3), \; y_2 \leftarrow f(r_0, r_1),\}$$

UNIFY$_{AC}$ terminates under certain conditions:

Theorem 2. Let s be a term in which there are no repeated variables. Then UNIFY$_{AC}(s,t)$ terminates and returns a CSU$_{AC}$ for s and t.

Proof. The proof of termination is by Noetherian induction. We use a complexity measure on pairs of terms defined by

$$weight(s,t) = (m,n)$$

where $m = \max \{ \#(v,r) \mid r \in \{s,t\}, v \in \mathcal{V}(r)\}$

$$n = \sum_{v \in V(s)} \#(v,s) + \sum_{v \in V(t)} \#(v,t)$$

(For example, $weight(\ f(x,f(y,y))\ ,\ f(u,f(u,u))\ ) = (3,6)\ )$.

The proof of this theorem suggests using the complexity measure $weight$ to prune the branches of the tree to force the algorithm to terminate. This is the basis for the algorithm $UNIFY2_{AC}$, obtained from $UNIFY_{AC}$ by pruning according to weight (See figure 2). Experimental results suggest the following:

Conjecture. $\forall s,t \in \mathbb{T}$, $UNIFY2_{AC}(s,t)$ terminates returning a $CSU_{AC}(s,t)$.

```
UNIFY2_AC = proc(s,t :term) returns (set of substitution)
    case
        s ∈ 𝒱 & t ∈ 𝒱:                                        ;Case 1
            return {{s ← t}}
        s ∈ 𝒱 & t ∉ 𝒱 & s ∈ 𝒱(t):                            ;Case 2
            return ∅
        s ∈ 𝒱 & t ∉ 𝒱 & s ∉ 𝒱(t):                            ;Case 3
            return {{s ← t}}
        t ∈ 𝒱 & s ∉ 𝒱 & t ∈ 𝒱(s):                            ;Case 4
            return ∅
        t ∈ 𝒱 & s ∉ 𝒱 & t ∉ 𝒱(s):                            ;Case 5
            return {{t ← s}}
        else                                                  ;Case 6
            β := P(s,ŝ) ∪ P(t,t̂)
            𝒜 := U^AB(ŝ,t̂)
            (m,n) := weight(s,t)
            Σ := ⋃_{θ∈𝒜} MAP-UNIFY2_AC(θ,β,(m,n))
            return Σ
    end
end UNIFY2_AC

MAP-UNIFY2_AC = proc(β,θ:substitution,(m,n):pair of integer)
                        returns(set of substitution)
    Σ_0 := {ε}
    j := 0
    for v ∈ D(β) ∪ D(θ) do
        j := j + 1
        Σ_j := { α∘σ | σ ∈ Σ_{j-1} & α ∈ UNIFY2_AC(σβv,σθv)
                      & weight(σβv,σθv) < (m,n)  }
    end
    return Σ_j
end MAP-UNIFY2_AC
```

FIGURE 2. A modified $UNIFY_{AC}$ procedure which forces termination using the $weight(s,t)$ complexity measure.

5.2 An Example

We illustrate the $UNIFY2_{AC}$ algorithm using an example which would cause the $UNIFY_{AC}$ algorithm to cycle. From Theorem 2, we see that the $UNIFY_{AC}$ algorithm terminates if one term has no repeated variables. Hence, we choose two terms with repeated variables for our example.

Let $s = f(w,w)$, $t = f(f(u,u),z)$ be the input terms. Case 6 of $UNIFY2_{AC}$ is the appropriate case. The abstracted terms $\hat{s} = f(x_1,x_2)$, $\hat{t} = f(y_1,y_2)$ are formed along with the preserving substitution $\beta = \{x_1 \leftarrow w , x_2 \leftarrow w , y_1 \leftarrow f(u,u) , y_2 \leftarrow z\}$. We set $\mathcal{A} = U^{AB} = \{\theta_1, \ldots ,\theta_7\}$, the μCSU_{AC} for $\hat{s}$ and $\hat{t}$ as listed in Section 5. We see that $weight(s,t) = (2,5)$, so we set $m = 2$, $n = 5$. $MAP\text{-}UNIFY2_{AC}(\beta,\theta_i)$ is then called for $i = 1,\ldots,7$. In our example, we'll trace only $MAP\text{-}UNIFY2_{AC}(\beta,\theta_7)$.

The loop in $MAP\text{-}UNIFY2_{AC}(\beta,\theta_7)$ is executed 4 times as $D(\beta) \cup D(\theta_7)$ contains 4 variables. In tracing $MAP\text{-}UNIFY2_{AC}(\beta,\theta_7)$, we'll consider these variables in the order x_1,x_2,y_1,y_2:

$j = 1$: Here $v = x_1$ and the corresponding pair of terms is $\langle\beta v,\theta_7 v\rangle = \langle w,f(r_1,r_3)\rangle$. The weight of this pair is $(1,3)$, thus, $\Sigma_1 = UNIFY2_{AC}(w,f(r_1,r_3)) = \{\sigma\}$, where $\sigma = \{w \leftarrow f(r_1,r_3)\}$.

$j = 2$: Now $v = x_2$ and applying σ to the corresponding pair of terms $\langle\beta v,\theta_7 v\rangle$ yields $\langle\sigma(\beta v),\sigma(\theta_7 v)\rangle = \langle f(r_1,r_3),f(r_0,r_2)\rangle$. The weight of this pair is $(1,4)$, thus we send it to $UNIFY2_{AC}$, which returns a CSU_{AC} for this pair. This set includes the seven AC-unifiers:

$\tau_1 = \{r_0 \leftarrow q_1, r_1 \leftarrow q_0, r_2 \leftarrow q_0, r_3 \leftarrow q_1\}$
$\tau_2 = \{r_0 \leftarrow q_1, r_1 \leftarrow q_0, r_2 \leftarrow f(q_0,q_2), r_3 \leftarrow f(q_2,q_1)\}$
$\tau_3 = \{r_0 \leftarrow f(q_2,q_1), r_1 \leftarrow f(q_2,q_0), r_2 \leftarrow q_0, r_3 \leftarrow q_1\}$
$\tau_4 = \{r_0 \leftarrow q_0, r_1 \leftarrow q_0, r_2 \leftarrow q_1, r_3 \leftarrow q_1\}$
$\tau_5 = \{r_0 \leftarrow f(q_0,q_2), r_1 \leftarrow q_0, r_2 \leftarrow q_1, r_3 \leftarrow f(q_2,q_1)\}$
$\tau_6 = \{r_0 \leftarrow q_0, r_1 \leftarrow f(q_2,q_0), r_2 \leftarrow f(q_2,q_1), r_3 \leftarrow q_1\}$
$\tau_7 = \{r_0 \leftarrow f(q_2,q_3), r_1 \leftarrow f(q_1,q_3), r_2 \leftarrow f(q_0,q_1), r_3 \leftarrow f(q_0,q_2)\}$

Hence, $\Sigma_2 = \{\tau_i \circ \sigma: i = 1,..,7\}$.

$j = 3$: $v = y_1$. Applying each substitution in Σ_2 to the corresponding pair of terms for v yields seven pairs of terms. However, all but three of these pairs have weight at least as great as (m,n):

substitution	pair	weight
$\tau_1 \circ \sigma$	$\langle f(u,u), f(q_0,q_1) \rangle$	$(2,4)$
$\tau_2 \circ \sigma$	$\langle f(u,u), f(f(q_0,q_2), f(q_2,q_1)) \rangle$	$(2,6)$
$\tau_3 \circ \sigma$	$\langle f(u,u), f(q_0,q_1) \rangle$	$(2,4)$
$\tau_4 \circ \sigma$	$\langle f(u,u), f(q_1,q_1) \rangle$	$(2,4)$
$\tau_5 \circ \sigma$	$\langle f(u,u), f(q_1, f(q_2,q_1)) \rangle$	$(2,5)$
$\tau_6 \circ \sigma$	$\langle f(u,u), f(f(q_2,q_1), q_1) \rangle$	$(2,5)$
$\tau_7 \circ \sigma$	$\langle f(u,u), f(f(q_0,q_1), f(q_0,q_2)) \rangle$	$(2,6)$

Hence, only the three pairs with weight $(2,4)$ are sent to UNIFY2_{AC}. (Notice that the four pairs with weight $(2,5)$ or $(2,6)$ would cause UNIFY_{AC} to cycle. For example, the pair corresponding to $\tau_5 \circ \sigma$ is just a renaming of the original pair $\langle s,t \rangle$.) Thus,

$$\begin{aligned}
\Sigma_3 = \ & \{ \alpha \circ \tau_1 \circ \sigma \mid \alpha \in \text{UNIFY2}_{AC}(f(u,u), f(q_0,q_1)) \} \\
& \cup \{ \alpha \circ \tau_3 \circ \sigma \mid \alpha \in \text{UNIFY2}_{AC}(f(u,u), f(q_0,q_1)) \} \\
& \cup \{ \alpha \circ \tau_4 \circ \sigma \mid \alpha \in \text{UNIFY2}_{AC}(f(u,u), f(q_1,q_1)) \} .
\end{aligned}$$

$j = 4$: $v = y_2$. Every substitution in Σ_3 is applied to the corresponding pair of terms for v, $\langle \beta v, \theta_7 v \rangle = \langle z, f(r_0,r_1) \rangle$. We'll consider only one such pair.

Notice that the substitution $\alpha = \{u \leftarrow q_1\}$ is in the set $\text{UNIFY2}_{AC}(f(u,u), f(q_1,q_1))$, thus, $\alpha \circ \tau_4 \circ \sigma \in \Sigma_3$. Applying this substitution to the corresponding pair, $\langle z, f(r_0,r_1) \rangle$ yields the pair $\langle z, f(q_0,q_0) \rangle$, which has weight $(1,3)$. A CSU_{AC} for this pair contains only the substitution $\mu = \{z \leftarrow f(q_0,q_0)\}$. Hence,

$$\begin{aligned}
\mu \circ \alpha \circ \tau_4 \circ \sigma = \{ & w \leftarrow f(q_0,q_1),\ r_0 \leftarrow q_0,\ r_1 \leftarrow q_0,\ r_2 \leftarrow q_1, \\
& r_3 \leftarrow q_1,\ u \leftarrow q_1,\ z \leftarrow f(q_0,q_0) \}
\end{aligned}$$

is an element of Σ_4. Restricting this substitution to the variables in s and t yields:

$$\{ w \leftarrow f(q_0,q_1),\ u \leftarrow q_1,\ z \leftarrow f(q_0,q_0) \}.$$

We can check that this substitution is indeed an AC-unifier of s and t. In fact a μCSU_{AC} for s and t contains only this substitution (up to $\equiv_{AC}^{\{w,u,z\}}$). Hence, UNIFY2_{AC} has generated a CSU_{AC} for the pair $\langle s,t \rangle$.

6. Concluding Remarks

We presented a new universal unification algorithm for simple theories on one function symbol and used it to derive a new

AC-unification procedure. A requirment of our method is that a CSU_E be known for the basic abstracted terms. Thus, our algorithm reduced the E-unification problem to the problem of unifying the basic abstracted terms. A proof of partial correctness was given, along with some preliminary termination results.

6.1 Comparisons to Related Work

Recall that a universal unification procedure based on narrowing is known [Hullot, 1980]. This procedure requires that the theory, E, be canonical. A class of theories is known for which this narrowing algorithm will generate a μCSU_E [Siekmann & Szabo, 1981], [Herold 1982]. Our algorithm does not require a canonical theory but works for a different class of theories - the simple theories. However, our algorithm will not, in general, generate a minimal set of unifiers and there are finitary theories (e.g. AC) for which the basic algorithm, $UNIFY_E$, will not terminate.

Universal unification procedures based on the Martelli & Montanari method view the problem as a set of equations to be unified (see e.g. Kirchner [1985] or Gallier & Snyder [1987]). Our procedure treats the problem as a unification of substitutions. While the two views seem to have equivalent power, one may be preferred to the other for efficiency reasons. (See [Kirchner, 1985] for some comments on the relative efficiency of the two approaches.)

Also, specialized unification procedures are known for many simple theories. For example, Stickel [1981] and Livesey & Siekmann [1976] have independently presented complete AC-unification algorithms. Both of these algorithms depend on solutions to a system of linear diophantine equations. Preliminary experimentation indicates that our $UNIFY2_{AC}$ algorithm is less efficient than either of these known algorithms. It is unclear whether or not an algorithm based on our universal approach (or indeed any universal approach) can be tailored in specific cases to match the efficiency of a specially designed algorithm.

6.2 Loosening the Restrictions

The variable-only restriction does not limit the algorithm. By Proposition 2, E must be collapse-free. Hence, we may use one of the combination procedures (e.g. [Tiden 1986]) to automatically extend the algorithm to the general case. This allows the algorithm to work for terms containing uninterpreted function symbols or more than one instance of operators satisfying the axioms of E. Also, the restriction that f be of arity 1 or greater is not very severe, as the unification problem for a theory on one nullary function symbol is not very difficult.

6.3 Future Work

An important problem left open by this work is that of finding a termination condition for finitary theories. The UNIFY2$_{AC}$ procedure is an attempt to do this for the AC theory. However, the termination proof of Theorem 2 depends on specific properties of the AC theory and it appears that a termination proof for the UNIFY2$_{AC}$ procedure would also depend on properties of the AC theory. Hence, a significant problem is to extend the termination result to the wider class of finitary, simple theories. A related problem is that of modifying the algorithm so that it generates a μCSU_E (i.e. no redundant unifiers are generated).

Another open problem is that of finding the set U^{AB} for a given theory. It is possible to find this set automatically for a certain class of theories [Franzen, 1988]. This is similar to Kirchner's method of automatically generating the mutation transformation for syntactic theories [Kirchner, 1986].

It may also be possible to extend the algorithm to theories containing more than one function symbol. One way to do this may be to require that a CSU_E be known for every possible pair of basic abstracted terms. For example, if E contained the function symbols f and g, one would need a CSU_E for each of the following pairs of basic abstracted terms: $\langle f(..),f(..)\rangle$, $\langle f(..),g(..)\rangle$, $\langle g(..),g(..)\rangle$. Then, depending on the head symbols of the input terms s and t, one of these CSU_Es could be used in the main procedure.

References

Burris, S., Sankappanavar, H.P. (1981). *A Course in Universal Algebra*, Springer-Verlag.

Bürckert, H.-J., Herold, A., and Schmidt-Schauß, M. (1987). *On Equational Theories, Unification and Decidability*, Proc. of RTA'87, Pierre Lescanne, ed., Springer-Verlag, LNCS 256, 204-215.

Fages, F. and Huet, G. (1983). *Unification and Matching in Equational Theories*, Proc. of CAAP'83, Springer-Verlag, LNCS 159, 205-220.

Fay, M. (1979). *First Order Unification in an Equational Theory*, Proc. of 4th Workshop on Automated Deduction, 161-167.

Franzen, M. (1988). *A Unification Algorithm for Simple Equational Theories*, Doctoral Thesis, Northwestern University, forthcoming.

Gallier, J.H. and Snyder, W. (1987). *A Genereal Complete E-Unification Procedure,* Proc. of RTA'87, Pierre Lescanne, ed., Springer-Verlag, LNCS 256, 216-227.

Herold, A. (1982). *Universal Unification and a Class of Equational Theories,* Proc. of GWAI-82 (ed. W. Wahlster), Springer-Verlag, IFB-82, 177-190.

Herold, A. (1985). *A Combination of Unification Algorithms,* in Proc. of 8th CADE, Springer-Verlag, LNCS 230, 450-469.

Huet, G., Oppen, D.C. (1980). *Equations and Rewrite Rules: A Survey,* in *Formal Languages: Perspectives and Open Problems,* R. Book, ed., Academic Press, 349-405.

Hullot, J.M. (1980). *Canonical Forms and Unification,* Proc. of 5th CADE, Springer-Verlag, LNCS 87, 318-334.

Kirchner, C. (1985). *Méthodes et outils de conception systématique d'algorithmes d'unification dans les théories equationelles,* Thèse de doctorat d'état, Université de Nancy I.

Kirchner, C. (1986). *Computing Unification Algorithms,* 1st IEEE Symposium on Logic in Computer Science, Cambridge, Massachusetts, June 1986, 206-216.

Livesey, M., Siekmann, J. (1976). *Unification of A + C Terms (Bags) and A + C + I Terms (Sets),* Technical Report Interner Bericht Nr. 3/76, Institut für Informatik I, Universität Karlsruhe.

Martelli, A., Moiso, C., Rossi, G.F. (1986). *An Algorithm for Unification in Equational Theories,* Proc. 1986 Symposium on Logic Programming, Salt Lake City, 180-186.

Martelli, A., Montanari, U. (1982). *An Efficient Unification Algorithm,* ACM TOPLAS, Vol. 4, No. 2, 258-282.

Siekmann, J., Szabo, P. (1981). *Universal Unification and Regular ACFM Theories,* Proc. of IJCAI-81, Vancouver.

Siekmann, J. (1984). *Universal Unification,* Proc. of 7th CADE, Springer-Verlag, LNCS 170, 1-42.

Siekmann, J. (1986). *Unification Theory,* Proc. of ECAI'86, Vol. II, vi-xxxv.

Stickel, M.E. (1981). *A Unification Algorithm for Associative-Commutative Functions,* JACM, vol. 28, 423-434.

Tiden, E. (1986). *Unification in Combinations of Collapse-Free Theories with Disjoint Sets of Function Symbols,* Proc. of 8th CADE, Springer-Verlag, LNCS 230, 431-449.

Yelick, K. (1985). *Combining Unification Algorithms for Confined Regular Equational Theories,* Proc. of RTA'85, Springer-Verlag, LNCS 202, 365-380.

Yelick, K. (1987). *Unification in Combinations of Collapse-Free Regular Theories,* J. Symbolic Computation, vol. 3, 153-181.

An Implementation of a Dissolution-Based System Employing Theory Links [†]

Neil V. Murray
SUNY at Albany

Erik Rosenthal
Wellesley College

Abstract. *We have been developing an automated deduction system based on path dissolution, an operation that was first introduced in [7]. Preliminary experimental results are promising. The next major phase in the development of that system will be the inclusion of a "theory-link processor." In this paper, we describe those experimental results and some of the meta theory required for that next phase.*

1. Introduction

We introduced path resolution in [6]. Much of that paper was devoted to analyzing the structure of logical formulas in negation normal form (NNF), something we found necessary in order to define path resolution precisely. That work raised many questions and led us in several directions. Early experiments indicate that the most important of these may be path dissolution, a path-based inference mechanism that operates on quantifier-free predicate calculus formulas in NNF.

Dissolution is somewhat unusual as compared with other inference technologies. First, it replaces the formula that is dissolved upon, producing an equivalent formula, and it is strongly complete at the ground level: *Any* sequence of dissolution steps will eventually produce a linkless formula. Secondly, one cannot restrict attention to conjunctive normal form (CNF) when employing dissolution: A single application generally produces a formula that is not in CNF *even if the original formula is.* Thirdly, dissolution does not lift directly to the general level. In [8] we discussed methods for employing dissolution as a resolution chain recognizer; the result was a hybrid rule of inference that we call *partial path resolution.* Dissolution may also be used at the first order level if one treats the dissolvent as an inference (i.e., conjoins it to the original formula) rather than as a

---

[†] This research was supported in part under grant CCR-8600848 awarded by the National Science Foundation.

replacement.

Dissolution operates on certain unsatisfiable subformulas by killing all c-paths that pass through the subformula; that is, it creates a new formula whose c-paths consist of exactly those from the original formula that do not pass through the unsatisfiable subformula. The strong completeness at the ground level stems from the fact that the number of c-paths is strictly reduced by each dissolution step.

In [9] we introduced *theory links*, a notion closely related to Stickel's theory resolution [12]. The idea is to treat a collection of clauses as a theory. For example, if $\sim E(x) \lor A(x)$ ("elephants are animals") is a clause, then $E(x) \land \sim A(x)$ is a theory link. The idea is that a theory link (or any conjunction including it) denotes FALSE in the theory; hence any c-path containing a theory link can be removed. In [9], we showed that, in most situations, theory links may be treated as if they are ordinary links. In this paper, we consider dissolving on chains that include theory links.

Most proofs are omitted for lack of space.

2. Preliminaries

We present here a brief description of path dissolution and of our analysis of formulas in NNF— greater detail can be found in [7] and in [6]. The overlap is intended to make this paper reasonably self-contained.

We believe this material to be quite intuitive and straightforward *when viewed from the perspective of paths.* Our somewhat unusual notation was designed precisely to make this view clear. Since embracing any new notation extracts a penalty in the form of extra effort from the reader, the inevitable temptation will be to follow these developments by mentally converting semantic graphs and manipulations of them back to a more familiar notation. If, instead, the reader spends the extra up-front time to familiarize himself with the 'path semantics' view and then proceeds on that basis, the intuition behind subsequent developments will be considerably more transparent.

We assume the reader to be familiar with the notions of *atom, literal, formula, resolution,* and *unification.* We consider only quantifier-free formulas in which all negations are at the atomic level.

A *semantic graph* is empty, a single node, or a triple *(N,C,D) of nodes, c-arcs,* and *d-arcs,* respectively, where a node is a literal occurrence, a c-arc is a conjunction of two non-empty semantic graphs, and a d-arc is a disjunction of two non-empty semantic graphs. Each semantic graph used in the construction of a semantic graph is be called an *explicit subgraph.* We use the notation $(G,H)_c$ for the c-arc from G to H and similarly use $(G,H)_d$ for a d-arc. In either case, all other arcs are either in G or in H, and we call (say) $(G,H)_c$ the *final arc* of the entire graph. We consider an empty graph to be an empty disjunction, which is a contradiction.

As an example, the formula

$$((A \wedge B) \vee C) \wedge (\sim A \vee (D \wedge C))$$

is the graph

$$
\begin{array}{ccc}
A \rightarrow B & & \overline{A} \\
\downarrow & \rightarrow & \downarrow \\
C & & D \rightarrow C
\end{array}
$$

Note that horizontal arrows are c-arcs, and vertical arrows are d-arcs.

The formulas we are considering are in *negation normal form* (NNF) in that all negations are at the atomic level, and the only connectives used are AND and OR.

One of the keys to our analysis is the notion of *path*. Let G be a semantic graph. A *partial c-path through G* is a set c of nodes such that any two are connected by a c-arc, and a *c-path* is a partial c-path that is not properly contained in any partial c-path. We similarly define d-path using d-arcs instead of c-arcs.

Lemma 1. Let G be a semantic graph. Then an interpretation I satisfies (falsifies) G iff I satisfies (falsifies) every literal on some c-path (d-path) through G.

Informally, a semantic graph is equivalent to the conjunction of its d-paths (one of its DNF equivalents) and to the disjunction of its c-paths (one of its CNF equivalents).

We will frequently find it useful to consider subgraphs that are not explicit; that is, given any set of nodes, we would like to examine that part of the graph consisting of exactly that set of nodes. The previous example is shown below on the left. The subgraph relative to the set $\{A, \overline{A}, D\}$ is the graph on the right.

$$
\begin{array}{ccccc}
A \rightarrow B & & \overline{A} & & \overline{A} \\
\downarrow & \rightarrow & \downarrow & & \downarrow \\
C & & D \rightarrow C & A \rightarrow & D
\end{array}
$$

The subgraph of G with respect to a set of nodes N is denoted G_N; see [9] for a precise definition.

The most important subgraphs are the *blocks*. A *c-block C* is a subgraph of a semantic graph with the property that any c-path p that includes at least one node from C *must pass through C*; that is, the subset of p consisting of the nodes that are in C must be a c-path through C. In the example above, the c-path $\{A, B, D, C\}$ passes through the subgraph depicted on the right since it includes the nodes of $\{A, D\}$, which is a c-path of the subgraph. The c-path $\{C\,\overline{A}\}$ does not pass through the subgraph since it contains only the partial c-path $\{\overline{A}\}$ of the subgraph, and hence this subgraph is not a c-block. A *d-block* is similarly defined with d-paths, and a *full block* is a subgraph that is both a c-block and a d-block. One way to envision a full block is to imagine the formula stored as an n-ary tree in which each internal node is a conjunction or a disjunction, and each leaf is a literal. Assume the tree is managed so that no child of a conjunction is a conjunction, and no child of a disjunction is a disjunction. Each node in such a tree then represents an explicit subgraph; its children are its *fundamental subgraphs*. The isomorphism theorem from [6] states that a full block is the union of some fundamental subgraphs of a single explicit subgraph.

3. Path Dissolution

Both path resolution and path dissolution can be understood as generalizations of well-known rules that operate on single links in formulas in CNF. Binary resolution and the Prawitz matrix reduction rule are, respectively, the two aforementioned well-known rules. See [6] for a detailed discussion of path resolution. To help motivate path dissolution, we begin with an informal description of the 'path semantics' view of both binary resolution and path dissolution.

The example below shows a resolution inference on the left, and a dissolution replacement on the right.

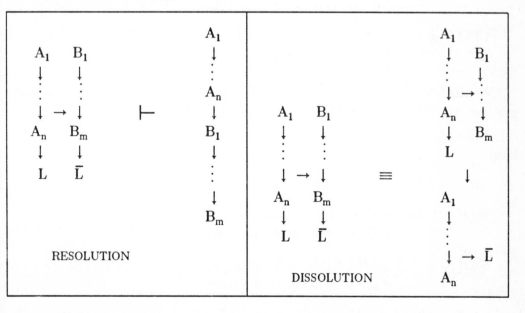

The resolution rule is justified by noting that a satisfying c-path cannot go through the link and hence must contain one of the A_i's or one of the B_i's. The resolvent may be thought of as the union of the subclauses through which the link "forces" a satisfying c-path.

In contrast, the dissolution rule *kills every c-path in the entire graph that passes through the link.* The dissolvent graph has exactly the c-paths of the original *except the ones through the link.* No interpretation satisfies such paths, so their removal leaves the meaning of the graph unchanged.

When the graph is not in CNF, full blocks X and Y take the place of the clauses. If the chain dissolved upon is H $(H \subset (X,Y)_c)$, and if $H_X = H \cap X$, then we must compute new structures whose c-paths are precisely those of X that *go through H_X* and those that *miss H_X* (similarly for Y and H_Y). The former structure is called the *c-path extension* of H_X and is denoted $CPE(H_X, X)$; the latter is (under the restrictions of Lemma 3) the *strong split graph of H_X in X* and is denoted $SS(H_X, X)$.

To define these objects precisely, let H be an arbitrary subgraph of G, let $F_1, \ldots, F_k$ be the fundamental subgraphs of G that meet H, and let $F_{k+1}, \ldots, F_n$ be those that do not. We use H_{F_i} for the subgraph of G relative to nodes in $H \cap F_i$. Then

$$CPE(\phi, G) = \phi \quad \text{and} \quad CPE(G,G) = G.$$

$$CPE(H,G) = CPE(H_{F_1}, F_1) \vee \ldots, \vee CPE(H_{F_n}, F_n)$$
$$\text{if the final arc of } G \text{ is a d-arc}$$

$$CPE(H,G) = CPE(H_{F_1}, F_1) \wedge \ldots, \wedge CPE(H_{F_k}, F_k) \wedge F_{k+1} \wedge \ldots, \wedge F_n$$
$$\text{if the final arc of } G \text{ is a c-arc}$$

Lemma 2. The c-paths of $CPE(H, G)$ are precisely the c-paths of G that pass through H.

Using the same notation we define the strong split graph of H in G as follows:

$$SS(\phi, G) = G \quad \text{and} \quad SS(G,G) = \phi.$$

$$SS(H,G) = SS(H_{F_1}, F_1) \vee \ldots, \vee SS(H_{F_n}, F_n)$$
$$\text{if the final arc of } G \text{ is a d-arc}$$

$$SS(H,G) = (SS(H_{F_1}, F_1) \vee \ldots, \vee SS(H_{F_k}, F_k)) \wedge F_{k+1} \wedge \ldots, \wedge F_n$$
$$\text{if the final arc of } G \text{ is a c-arc}$$

Lemma 3. If H is a c-block in G, then SS(H, G) is isomorphic to the subgraph of G relative to the nodes that lie on c-paths that miss H.

We define a *chain* in a graph to be a set of pairs of c-connected nodes such that each pair can simultaneously be made complementary by an appropriate substitution. A *link* is an element of a chain, and a chain is *full* if it is not properly contained in any other chain. A graph G is *spanned* by the chain K if every c-path through G contains a link from K; in that case, we call K a *resolution chain* for G.

Path dissolution operates on a resolution chain by constructing a semantic graph whose c-paths are exactly those that do not pass through the chain. This is not practical for all resolution chains: A special type of chain that we call a *dissolution chain* is required. Since single links always form dissolution chains, the class is not too specialized. The construction of the dissolvent from such a chain is straightforward.

A resolution chain H is a *dissolution chain* if it is a single c-block or if it has the following form: If M is the smallest full block containing H, then $M = (X, Y)_c$, where $H_X = H \cap X$ and $H_Y = H \cap Y$ are each c-blocks.

Given a dissolution chain H, define DV(H, M), the *dissolvent of H in M*, as follows: If H is a single c-block, then DV(H, M) = SS(H, M). Otherwise (i.e., if H consists of two c-blocks), then

$$
\text{DV}(H, M) \quad = \quad
\begin{matrix}
\text{CPE}(H_X, X) & \rightarrow & \text{SS}(H_Y, Y) \\
& \downarrow & \\
\text{SS}(H_X, X) & \rightarrow & \text{CPE}(H_Y, Y) \\
& \downarrow & \\
\text{SS}(H_X, X) & \rightarrow & \text{SS}(H_Y, Y)
\end{matrix}
$$

Intuitively, DV(H, M) is a semantic graph whose c-paths miss at least one of the c-blocks of the dissolution chain. The only paths left out are those that go through the dissolution chain and hence are unsatisfiable. Notice that we may express DV(H, M) in either of the two more compact forms shown in Figure 1 (since CPE(H_X, X) $\cup$ SS(H_X, X) = X and CPE(H_Y, Y) $\cup$ SS(H_Y, Y) = Y):

$$
\begin{matrix}
X & \rightarrow & \text{SS}(H_Y, Y) \\
\downarrow & & \\
\text{SS}(H_X, X) & \rightarrow & \text{CPE}(H_Y, Y)
\end{matrix}
\qquad \text{or} \qquad
\begin{matrix}
\text{SS}(H_X, X) & \rightarrow & Y \\
& & \downarrow \\
\text{CPE}(H_X, X) & \rightarrow & \text{SS}(H_Y, Y)
\end{matrix}
$$

Figure 1.

Note that the three representations are semantically equivalent but are not in general isomorphic; in particular their d-paths need not be the same. The c-paths of all three representations, however, are identical; they consist of exactly those c-paths in M that do not pass through H.

Theorem 1. Let H be a ground dissolution chain in a graph G, and let M be the smallest full block containing H. Then M and $DV(H, M)$ are equivalent.

Proof: We show that any interpretation I satisfies M iff I satisfies $DV(H, M)$. The proof refers specifically to the compact version of $DV(H, M)$ shown on the right in Figure 1.

Suppose first that I satisfies M and let p be a c-path through M satisfied by I. Then $p = p_x p_y$, where $p_x = p \cap X$ is a c-path through X and $p_y = p \cap Y$ is a c-path through Y.

Case 1: p_x misses H_X. Since H_X is a c-block (by the definition of dissolution chain), p_x is a c-path through $SS(H_X, X)$ by Lemma 3. Then $p_x p_y$ is a c-path through $SS(H_X, X) \rightarrow Y$ and hence through $DV(H, M)$.

Case 2: p_x hits H_X. Since H_X is a c-block, p_x passes through H_X and is a c-path in $CPE(H_X, X)$ by Lemma 2.

Case 2a: p_y misses H_Y. Then, by an argument similar to that of Case 1, p_y is a c-path through $SS(H_Y, Y)$. As a result, p is a c-path through $CPE(H_X, X) \rightarrow SS(H_Y, Y)$ and hence through $DV(H, M)$.

Case 2b: p_y hits H_Y. Then since H_Y is a c-block, p_y passes through H_Y, p passes through H, and hence p contains a link. Thus I does not satisfy p.

The converse is straightforward and left to the reader. □

We may therefore select an arbitrary dissolution chain H in G and replace the smallest full block containing H by its dissolvent, producing (in the ground case) an equivalent graph. We call the resulting graph the *dissolvent of G with respect to H* and denote it $Diss(G, H)$.

The graph formed by dissolution has strictly fewer c-paths than the old one: The proof of Theorem 1 shows that all c-paths are preserved *except those that pass through H*, and they are removed. Therefore, the two graphs are semantically equivalent, and, as a result, finitely many dissolutions (bounded above by the number of c-paths in the original graph) will yield a graph without links. If this graph is empty, then the original graph was spanned; if not, then every (necessarily linkless) c-path characterizes a model of the original graph.

As an example, consider the graph in Figure 2 below.

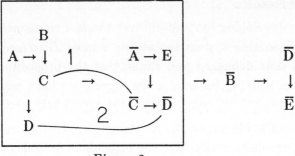

Figure 2.

Links 1 and 2 form a dissolution chain; the graph in Figure 3 below is the result of dissolving on this chain using the second of the two more compact forms of the dissolvent. The structure in the upper box of Figure 3 consists of the part of the graph in which c-paths miss the left half of the chain (C ∨ D) and are therefore unrestricted on the right; the structure in the lower box consists of that part of the graph containing c-paths that go through the left half of the chain and miss the right half of the chain (∼C ∨ ∼D).

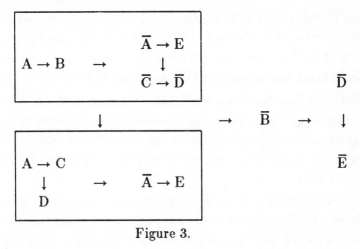

Figure 3.

Observe that there are eighteen c-paths in the first graph but only fourteen in the second. Notice also that the second graph *replaces* the first. The two are equivalent: The fourteen c-paths in Figure 3 appear in Figure 2, and each of the remaining four in Figure 2 passed through the dissolution chain and was therefore unsatisfiable.

4. Experimental Results

We have been developing (on and off over the last six months) a dissolution-based automated deduction system tentatively named *Dissolver*. It operates on single links and is now debugged and running at the propositional level. While the system has been running for only a few weeks, we have tested it on several problems known to be difficult, most notably the *pigeonhole formulas.*

In a rather remarkable paper [4], A. Haken proved that resolution requires exponentially long proofs to verify that the pigeonhole formulas are unsatisfiable. Prior experience with them had already indicated that they were difficult to deal with. Except for Tseitin's extended resolution [15], an extraordinarily non-deterministic technique, there is no system known to admit polynomial length proofs of these formulas. (If they are treated together as a single first order problem, theorem provers with induction can prove the result fairly easily.)

For this discussion, we use P_n to denote a set of clauses that state that given n pigeonholes and n+1 pigeons, each pigeon can be assigned a hole without a roommate. Preliminary comparisons with ITP, a theorem prover developed at Argonne National Laboratories [5], indicate that Dissolver is quite efficient.

We should mention that ITP was designed to be portable and accessible to other users; Dissolver is still in diapers. The point is, the comparison should be regarded as a ball park comparison between hyper-resolution and path dissolution, and not as a comparison between ITP and Dissolver*.

We ran ITP and Dissolver on a VAX 750. Although both are written in Pascal, the similarity ends there: They are fundamentally different in concept and design, so the comparison is based solely on cpu time. Since Dissolver works with formulas in negation normal form, we ran it with both NNF and CNF versions of the pigeonhole problems.

Using hyper-resolution, ITP took more than ten minutes to determine that P_3 is unsatisfiable. (It took almost two-and-a-half hours using binary resolution.) Dissolver solved the CNF version of P_3 in fifteen seconds and the NNF version in less than three seconds! P_4 is a large problem: forty-five clauses with one

---

* Implementation quality of course plays a large role in the efficiency of a system. The earliest running version of Dissolver had two major design flaws. (There may be others yet to be discovered.) We first ran it on a VAX 8550, a machine some seven times faster than the VAX 750 on which we made the comparisons with ITP. It took twelve hours to handle P_3 in CNF. We eliminated one design flaw, and Dissolver then dispensed with P_3 in less than three seconds! We next ran it on P_4, and the run took eleven hours. The second design flaw was eliminated, and Dissolver then proved that P_4 was unsatisfiable in twenty-one minutes. (The running time for P_3 was lowered to two seconds.) Our comparisons were made under the assumption that when the good folks at Argonne developed ITP, they did not make the kinds of blunders we made.

hundred literals. Dissolver solved the NNF version of P_4 in one minute, twenty-two seconds, and the CNF version in one hour, forty-one minutes. ITP was unable to handle P_4 after twenty-four hours.

We must emphasize that these results are preliminary, but at least on the propositional pigeonhole formulas, dissolution appears to be more effective than hyper-resolution. We of course are unable to make predictions of efficiency at the general level until the first order system is developed.

5. Negation Normal Form

The difference in the running times of the CNF and NNF versions of the pigeonhole formulas is striking. The primary reason is that there are many fewer c-paths in NNF. With P_4, for example, the difference is a *factor* of 2^{24} (2^{26} c-paths versus 2^{50} c-paths). Dissolution is intrinsically NNF-based: Even if a formula is in CNF, one application of dissolution will typically create a formula that is not in CNF.

It can be argued that the NNF versions are more natural than the CNF versions: The CNF formulas have a collection of two clauses that state, "If pigeon i is in hole j, then pigeon k is not in hole j." That clause is repeated for each pigeon. The statement in NNF is, "If pigeon i is in hole j, then none of the other pigeons is in hole j."

Nevertheless, it is unreasonable to assume that a theorem prover will never have to deal with formulas in clause form, even if there is a more compact representation. Since NNF is so much to the advantage of Dissolver, a preprocessor that factors out literals that are common to several clauses (if such exist) should lead to an improvement in running time. (Indeed, in view of the cpu times used by Dissolver on the two versions of P_4 and considering the fact that CNF can be exponentially larger than NNF, the improvement may be quite substantial.) One of the next stages in the development of Dissolver will be the addition of such a preprocessor.

We ran one small experiment that indicates the potential value of such preprocessing. In [10] Pelletier lists several propositional problems for theorem provers. Problem 12 is:

$$(p \Leftrightarrow q \Leftrightarrow r) \Leftrightarrow \sim (p \Leftrightarrow q \Leftrightarrow r)$$

The clause form of this problem consists of eight three-literal clauses. ITP refuted this clause set in seven seconds; Dissolver required almost three minutes. Yet when given the NNF version of this problem, Dissolver used only one second. The reason is straightforward: the clause form has $3^8 = 6561$ c-paths whereas the

NNF version has 32. *

By factoring common literals in different orders, different formulas (decidedly non-isomorphic as graphs) can result. But factoring the above example in any reasonable manner will yield an impressive reduction in c-paths. In fact, a factoring algorithm that can recognize common *non-atomic* subformulas would reduce this problem to a graph with three c-paths! As a result, we suspect that such preprocessing will play an important role, not only for Dissolver, but for other non-clausal theorem proving techniques. Furthermore, it may be useful to apply such factoring techniques *during* the deduction.

We should point out that a simple version of such preprocessing that handles only common literals can be done in polynomial time. While explosive growth can occur when putting a formula into *clause* form, our goal is to go in the opposite direction, which of course shrinks the formula: The input is possibly exponentially larger than the desired output.

6. Dissolution and Theory Links

Any clause in a formula can be removed from the formula and used as a theory link (for a detailed discussion, see [9]), but not all theory links work well with dissolution because of the restrictions on a dissolution chain. That restricted form— two c-blocks— means that the best choice for theory links are two-literal clauses. For the purposes of this discussion, by a theory we mean a set of such two-clauses, and by a theory link, we mean the negation of one clause that is in the theory or that is a consequence of the theory.

As an example, consider P_2 (two pigeonholes and three pigeons). Let Pij be the predicate that means, "Pigeon i is in hole j." In clause form, P_2 may be written

$$
\begin{array}{ccccccccc}
\overline{P11} & \overline{P11} & \overline{P12} & & \overline{P12} & \overline{P21} & \overline{P22} & & P11 & P21 & P31 \\
\downarrow & \rightarrow & \downarrow & \rightarrow & \downarrow & \rightarrow & & \downarrow & \rightarrow & \downarrow & \rightarrow & \downarrow & \rightarrow & & \downarrow & \rightarrow & \downarrow & \rightarrow & \downarrow \\
\overline{P21} & \overline{P31} & \overline{P22} & & \overline{P32} & \overline{P31} & \overline{P32} & & P12 & P22 & P32
\end{array}
$$

---

* Between the time of the original submission of this paper to the program committee and the preparation of the camera-ready copy, a preliminary factoring algorithm and an improved link-selection strategy were added to Dissolver. The factoring routine runs in a fraction of a second on any reasonably sized formula. Dissolver now handles the CNF version of P4 in less than two minutes, and the CNF version of the Pelletier example requires only one-fifth of a second.

Each negative clause states that if pigeon i is in hole j, then pigeon k is not in hole j, and each positive clause states that each pigeon is in some hole. Observe that for P_n, $n > 2$, there are more negative clauses, but each is still a two-clause; the positive clauses, however, contain n literals. As a result, for any of the pigeonhole formulas, the collection of all negative clauses may be removed and treated as a theory. The remaining formula has no ordinary links— all literals are positive. There are of course many theory links: For any i, j, and k, {Pij, Pkj} is a theory link. This makes semantic sense: The clause $\sim$Pij $\vee$ $\sim$Pkj says that if pigeon i is in hole j, then pigeon k is not in hole j, and the corresponding theory link says that pigeon i and pigeon k are both in hole j.

The above example illustrates the nature of a theory link: a set of literals with the property that all cannot be true under an interpretation satisfying the theory. Of course, an ordinary link automatically has this property. Since it is this property that makes Theorem 1 hold, Theorem 2 is intuitively obvious. The proof is straightforward, but we omit it here.

Theorem 2. Let H be a ground dissolution chain comprised of any combination of ordinary links and theory links in a graph G, and let M be the smallest full block containing H. Then M and DV(H, M) are equivalent, provided that we interpret equivalent to mean *in the presence of the theory.*

The key to using theory links with dissolution then is a preprocessor that removes two-clauses and creates a list of theory links. Since the theory consists of two-clauses, all consequences of the theory are clauses of length at most two (by the completeness of resolution, and the fact that the resolvent of a pair of two-clauses is a clause of length one or two). Forming the full dissolvent (see [8]) of the theory and surveying d-paths with at most two distinct literals will produce the complete set of theory links*. This set is at most quadratic in the size of the original theory because any set of two-clauses is at most quadratic in the number of its distinct atoms.

The issue of computing all consequences does not arise for the pigeonhole problems, because all literals in the two-clauses are negative; the theory contains no links and its clauses already include all consequences. For other problems, we may wonder if computing the consequences of the theory has benefits that outweigh the cost. This is a good question for future exploration.

One strategy that would avoid the consequence issue is to install as theory links only negations of clauses in the theory. This can be done in linear time.

---

* The full dissolvent of *any* semantic graph (defined to be the formula obtained by repeatedly dissolving until no links remain) contains explicitly every prime implicant as a d-path. This is easily seen from the consequence completeness of resolution and the fact that the full dissolvent has no links.

But, since there may be consequences of the theory required for a proof, the formula could be non-empty and linkless after several dissolution steps *even if it is unsatisfiable.*

This difficulty arises in other situations as well. One is the case when all clauses in the original formula are two-clauses. In that case, the theory is the entire formula. Observe that the remaining formula is an empty *conjunction*, not the empty clause.

A related problem occurs when a formula is satisfiable. One of the nice properties of dissolution with ordinary links is that it produces all models of a formula; that is, after dissolving away all links, the remaining c-paths may be interpreted as a list of all possible ways of satisfying the original formula. With a (not fully dissolved) theory present, this is not the case since some c-paths in the theory may contain ordinary links.

All of these difficulties can be handled in the same manner: If the formula becomes linkless but is not the empty graph, bring back one clause at a time from the theory. This is not expensive. The original preprocessing took linear time. Moreover, dissolving on a single theory link is equivalent to dissolving on a chain consisting of two ordinary links. As a result, even if the entire theory has to be brought back, we expect the theory links to improve processing time.

Note also that dissolving with theory links is not difficult. Pointers from theory links to literal occurrences in the formula will make finding them virtually the same process as the one for finding ordinary links. The dissolution operation is purely structural, depending only on the location of the links.

The potential value of using theory links can be seen by considering the pigeonhole problems again. After removing the two-clauses, P_3 becomes

$$
\begin{array}{cccccccc}
\text{P11} & & \text{P21} & & \text{P31} & & \text{P41} \\
\downarrow & \rightarrow & \downarrow & \rightarrow & \downarrow & \rightarrow & \downarrow \\
\text{P12} & & \text{P22} & & \text{P32} & & \text{P42} \\
\downarrow & & \downarrow & & \downarrow & & \downarrow \\
\text{P13} & & \text{P23} & & \text{P33} & & \text{P43}
\end{array}
$$

This formula has 3^4, or 81, c-paths; the CNF version of P_3 has $(2^{18})(3^4)$—over twenty million c-paths! More generally, P_n has $2^{n^2(n+1)/2}$ c-paths through the negative two-clauses and n^{n+1} through the positive clauses, so the number of c-paths in the theory links version is reduced by a factor of $2^{n^2(n+1)/2}$. Needless to say, the use of theory links will dramatically improve the running time of Dissolver on these formulas; a significant speedup can be expected on any formula with many two-clauses.

While theory links have not yet been added to Dissolver's arsenal, we have run Dissolver on problems that simulate the theory links versions of the pigeonhole formulas. The simulated P_4 was dispensed with in 34 seconds. In CNF, P_5 has eighty-one clauses and 180 literals— a large problem indeed— too large for Dissolver or for ITP. Dissolver did handle the NNF version of P_5 after almost 34 hours. When we simulated the theory links version, Dissolver required 2 hours 24 minutes. *

There are a number of questions that are likely to be in the reader's mind. One is, if NNF preprocessing and theory links are both added to Dissolver, which should take preference? The answer may be to first use the technique that most reduces the number of c-paths, and theory links wins that contest hands down because the clauses are *removed* from the formula. Note that if the two-clauses become theory links in the pigeonhole formulas, the remaining clauses cannot be factored, so the NNF preprocessor will have no effect.

One might also ask, will Dissolver with theory links do as well as it did with the simulations? The answer is: better! The simulated versions of the pigeonhole formulas have the identical path structure but n times as many literals. As a result, most steps, which consist primarily of creating nodes in and deleting nodes from an n-ary tree, require n times the processing time in the simulated run.

The addition of theory links seems only to allow Dissolver to solve P_{n+1} in about as much time as required to solve P_n without theory links. One might ask, is this much of an improvement? The answer is yes. The size of P_n is cubic in n, so P_{n+1} is a much larger problem than P_n.

7. First Order Considerations

In view of its strong completeness at the ground level, dissolution cannot lift directly to the first order level. As we indicated in the introduction, dissolution can be used as a resolution chain recognizer (see [8]), and it can also be employed at the general level if it is used as an inference rather than as a replacement. This is the approach we intend to take in the system currently under development, although a long range goal is the implementation of partial path resolution.

The extent to which dissolution will prove effective at the first order level will ultimately be determined experimentally. Predicting such performance is hazardous at best, but some aspects of dissolution can be analyzed with respect to lifting in advance of experimental results. Strong completeness will, of course, not lift, and if that were its only important characteristic, dissolution's

---

* The new factoring routine and the link deletion strategy allow solution of this problem in about 25 minutes.

effectiveness for the first order case would be quite doubtful. But this is not so: As we pointed out in Section 5, dissolving away all links in a ground formula results in a formula that *explicitly* contains all of the original formula's consequences. In the first order case we also get a complete set of consequences *with respect to instances of the original formula that correspond to the selected links.* (The exact correspondence is not obvious, as the example below will show.) As a result, the dissolution process *in any one copy* of a general formula requires no backtracking.

We believe that this process of 'carrying along all consequences' reduces the efficiency of individual inferences, but it permits totally deterministic proof discovery in the ground case. It may help limit the usual explosion of the search space in the general case. ITP using hyper-resolution is, in contrast, searching for all proofs in parallel. At the other end of the spectrum, Stickel's Prolog Technology Theorem Prover (PTTP) can take advantage of propositional problems by employing unbounded depth-first search [13,14]. This fact, along with the remarkable efficiency of single PTTP steps, allows it to handle P_4 and P_5 in time amounts that are embarrassingly small. (P_5 in less than three seconds! Host machine unknown.) But our hope for dissolution is rekindled when we remember that PTTP is the product (directly and indirectly) of twenty-five years of active research into issues of both implementation and theory.

The first order dissolution technique we initially plan to implement will be as follows. Create a copy of the original graph and then dissolve *with replacement* on the copy as much as possible. The idea is to drive the copy toward some instantiated linkless consequence, which is then conjoined to the original graph. (If we are lucky, the consequence will be empty!) The process can then be repeated, with preference given to those links not used in previous iterations. Note that in general, as dissolution is applied, some links not yet used in the copy will simply vanish, their literals having become instantiated in ways inconsistent with their original unifiers.

We demonstrate this approach on the example in Figure 4. Links 1 through 6 are compatible. Regardless of the order in which they are activated, the resulting graph contains all c-paths except those through any of the six links.

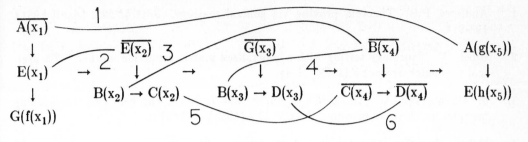

Figure 4.

We omit the individual dissolution steps; successively activating the four dissolution chains comprised of the sets of links $\{5,3\}$, $\{4,6\}$, $\{2\}$, and $\{1\}$ results in the graph of Figure 5.

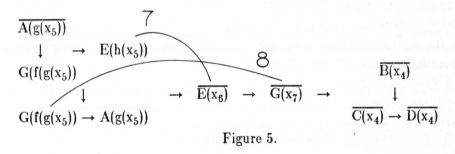

Figure 5.

The semantic graph above has 6 c-paths, whereas the original one has 48. Dissolving on links 7 and 8 yields the empty graph.

The example above has the interesting property that although one instance of the formula is insufficient to demonstrate a contradiction, dissolution produces the empty graph using only one copy. We have not yet investigated this phenomenon and thus cannot satisfactorily characterize it. It does indicate that the relationship between "iterations of forming full dissolvents" and "instances required for a proof set" is not necessarily straightforward and is worthy of future study.

Acknowledgements

We are grateful to Mark Stickel for taking the time to discuss with us the many issues related to propositional versus first order deduction, and for running PTTP on P_4 and P_5. Kiralee McCauley and Cynthia Shettle are Wellesley College students who coded a substantial part of Dissolver; without their efforts, it would not yet be running.

References

1. Andrews, P.B. Theorem proving via general matings. *J.ACM* 28,2 (April 1981), 193-214.

2. Bibel, W. Tautology testing with a generalized matrix reduction method. *Theoretical Computer Science* 8 (1979) 31-44.

3. de Champeaux, D. Sub-problem finder and instance checker, two cooperating modules for theorem provers. *J.ACM* 33,4 (1986), 633-657.

4. Haken, A. The intractability of resolution. *Theoretical Computer Science*, 39 (1985), 297-308.

5. Lusk, E., and Overbeek, R. A portable environment for research in automated reasoning. *Proceedings of CADE-7*, Napa, CA, May 14-16, 1984. In *Lecture Notes in Computer Science*, Springer-Verlag, Vol. 170, 43-52.

6. Murray, N.V., and Rosenthal, E. Inference with Path Resolution and Semantic Graphs. *J.ACM* 34,2 (April 1987), 225-254.

7. Murray, N.V., and Rosenthal, E. Path dissolution: A strongly complete rule of inference. *Proceedings of the 6th National Conference on Artificial Intelligence*, Seattle, WA, July 12-17, 1987, 161-166.

8. Murray, N.V., and Rosenthal, E. Inferencing on an arbitrary set of links. *Proceedings of the 2nd International Symposium on Methodologies for Intelligent Systems*, Charlotte, NC, October 1987. In *Methodologies for Intelligent Systems*, (Ras, Z. and Zemankova, M., eds.) North-Holland, 1987, 416-423.

9. Murray, N.V., Rosenthal, E. Theory links in semantic graphs. *Proceedings* of the 8th International Conference on Automated Deduction, Oxford, England, July 1986. In *Lecture Notes in Computer Science*, Springer-Verlag, Vol. 230, 353-364.

10. Pelletier, F.J. Seventy-Five Problems for Testing Automatic Theorem-Provers. *Journal of Automated Reasoning* 2 (1986) 191-216.

11. Robinson, J.A. Automatic deduction with hyper-resolution. *International Journal of Computer Mathematics*, 1 (1965), 227-234.

12. Stickel, M.E. Automated deduction by theory resolution. *J. Automated Reasoning*, 1,4 (1985), 333-355.

13. Stickel, M.E. A Prolog technology theorem prover: implementation by an extended Prolog compiler. *Proceedings of the 8th International Conference on Automated Deduction*, Oxford, England, July 1986. In *Lecture Notes in Computer Science*, Springer-Verlag, Vol. 230, 573-587.

14. Stickel, M.E. Personal Communication. November 1987.

15. Tseitin, G. S. On the complexity of derivations in propositional calculus. *Structures in Constructive Mathematics and Mathematical Logic*, Part II, A. O. Sliosenko, ed. (1968), 115-125.

Decision Procedure for Autoepistemic Logic

Ilkka Niemelä

Digital Systems Laboratory
Department of Computer Science
Helsinki University of Technology
Otakaari 5 A, 02150 Espoo, Finland

Abstract

Autoepistemic logic is a nonmonotonic logic for modeling the beliefs of an ideally rational agent who reflects on his own beliefs. Autoepistemic logic has been lacking a decision procedure to answer the question: given a set of premises describing the beliefs of an agent does the agent believe a given formula (is the given formula derivable from the given set of premises). Two derivability relations are defined for propositional autoepistemic logic corresponding to the agent's own point of view and the point of view of an external observer of the beliefs of the agent. Decision procedures based on analytic tableaux are then given and proved correct for both of the relations.

Key words and phrases. Nonmonotonic logic, theorem proving, analytic tableaux.

1 Introduction

Commonsense reasoning is nonmonotonic, i.e. on the basis of incomplete information we often draw conclusions that we later retract because we learn more about the situation. In the past few years many attempts have been made to formalize nonmonotonic reasoning. Some of the most interesting formalization attempts are the logics of McDermott and Doyle [1,2]. These logics, however, have peculiarities that suggest they do not quite capture the intuitions given as the basis of their development. Autoepistemic logic suggested by Moore [3] avoids these peculiarities by reconstructing nonmonotonic logic as a model of an ideally rational agent's reasoning about his own beliefs.

The following example clarifies the concept of autoepistemic reasoning. We are given that Tweety is a bird and that most birds can fly. We formalize this in autoepistemic logic as follows:

$$\text{BIRD(TWEETY)}, \tag{1.1}$$

$$\forall x((\mathrm{BIRD}(x) \land \neg L \neg \mathrm{CAN\text{-}FLY}(x)) \to \mathrm{CAN\text{-}FLY}(x)) \qquad (1.2)$$

In autoepistemic logic the intended reading of a formula Lp is that the agent believes p. Given this set as premises describing the beliefs of an agent we can deduce in autoepistemic logic that the agent believes that Tweety can fly because he does not believe that it cannot. On the other hand, if we add $\neg \mathrm{CAN\text{-}FLY}(\mathrm{TWEETY})$ to the premises, the agent no longer believes that Tweety can fly. This shows also that the meaning of a formula in autoepistemic logic is context-sensitive. It depends on the set of premises in which the formula is embedded. As Moore [3] points out, autoepistemic logic is nonmonotonic because it is indexical; the operator L changes its meaning with context as do the indexical words in natural language.

Autoepistemic logic has been lacking a decision procedure to answer the question: given a set of premises describing the beliefs of an agent does the agent believe a given formula. In this paper we give a decision procedure for propositional autoepistemic logic based on analytic tableaux [5]. We proceed as follows. First we give a formal definition of the total set of beliefs of an agent which is called an autoepistemic theory. Then we define two derivability relations for autoepistemic logic corresponding to the agent's own point of view and the point of view of an external observer of the beliefs of the agent. Decision procedures based on analytic tableaux are then given and proved correct for both of the relations.

2 Autoepistemic Logic

The language $\mathcal{L}$ of autoepistemic logic is the language of propositional logic augmented by a monadic operator L ($\neg L \neg$ is abbreviated M).

In autoepistemic logic the primary objects of interest are the sets of formulas which are interpreted as the total beliefs of an agent. These sets are called autoepistemic theories. A formula of the form Lp will be true with respect to an agent iff (if and only if) p is in his set of beliefs. This is formalized as follows.

An *autoepistemic interpretation* of an autoepistemic theory T is an assignment of truth values to formulas of the language of the theory T that satisfies the following conditions:

1. It follows the usual truth recursion of propositional logic.
2. A formula Lp is true in it iff $p \in T$.

An *autoepistemic model* of an autoepistemic theory T is an autoepistemic interpretation of T in which all the formulas of T are true.

An autoepistemic theory T is *sound* with respect to an initial set of premises A iff every autoepistemic interpretation of T in which all the formulas of A are true is an autoepistemic model of T. An autoepistemic theory T is *semantically complete* iff T contains every formula that is true in every autoepistemic model of T.

An autoepistemic theory T is *stable* if it satisfies the following conditions:

1. if $p_1, \ldots, p_n \in T$ and $p_1, \ldots, p_n \vdash q$, then $q \in T$ ($\vdash$ is the derivability relation of propositional logic).
2. If $p \in T$, then $Lp \in T$.
3. If $p \notin T$, then $\neg Lp \in T$.

Autoepistemic theory T is *grounded* in a set of premises A iff

$$T \subseteq Th(A \cup \{Lp \mid p \in T\} \cup \{\neg Lp \mid p \notin T\})$$

where $Th(A) = \{p \mid A \vdash p\}$. Moore [3] proves that T is grounded in A iff T is sound with respect to A and that T is stable iff T is semantically complete.

Moore [3] argues that the possible sets of beliefs that an ideally rational agent might hold, given A as his premises, ought to be just the extensions of A that are semantically complete and sound with respect to A. We call such sets the *stable expansions* of A. Given A as premises we give the necessary and sufficient condition for an autoepistemic theory to be a stable expansion of A.

Theorem 2.1 *Let A be a given set of premises. An autoepistemic theory T, for which $A \subseteq T$ holds, is semantically complete and sound with respect to premises A iff it satisfies the condition*

$$T = Th(A \cup \{Lp \mid p \in T\} \cup \{\neg Lp \mid p \notin T\}) \qquad (2.1)$$

Proof. ($\leftarrow$) Let T satisfy the condition (2.1). Then T is grounded in A and Moore [3] proves that T is grounded in A iff T is sound with respect to A. The condition (2.1) implies also that T is stable and Moore [3] shows that T is stable iff T is semantically complete. $A \subseteq T$ is implied also by (2.1).

($\rightarrow$) Let $A \subseteq T$ and T be semantically complete and sound with respect to premises A. Because it is sound, it is grounded in A and $T \subseteq Th(A \cup \{Lp \mid p \in T\} \cup \{\neg Lp \mid p \notin T\})$. Because it is semantically complete, it is stable and therefore $Th(T) \subseteq T$ holds. We still have to show that $Th(A \cup \{Lp \mid p \in T\} \cup \{\neg Lp \mid p \notin T\}) \subseteq T$. We accomplish this by showing that $A \cup \{Lp \mid p \in T\} \cup \{\neg Lp \mid p \notin T\} \subseteq T$ because this implies $Th(A \cup \{Lp \mid p \in T\} \cup \{\neg Lp \mid p \notin T\}) \subseteq Th(T) \subseteq T$.

Assume that there exists $q \in A \cup \{Lp \mid p \in T\} \cup \{\neg Lp \mid p \notin T\}$ such that $q \notin T$. Because $A \subseteq T$ and $q \notin T$, $q \notin A$. So $q \in \{Lp \mid p \in T\} \cup \{\neg Lp \mid p \notin T\}$. If $q \in \{Lp \mid p \in T\}$, then q is some Lr such that $r \in T$. But T is stable so $Lr \in T$ and $q \in T$, which is a contradiction. So $q \in \{\neg Lp \mid p \notin T\}$. Then q is some $\neg Lr$ such that $r \notin T$. Because T is stable, $q \in T$, which is also a contradiction and so $T = Th(A \cup \{Lp \mid p \in T\} \cup \{\neg Lp \mid p \notin T\})$. $\square$

Moore [3] does not give any formal definition for derivability in autoepistemic logic. Here we propose two possible definitions. We would like to define the derivability relation for autoepistemic logic in such a way that a formula is derivable from a set of premises exactly when the agent believes in the formula, i.e. the formula belongs to the stable expansion of the premises. There are difficulties in this approach. A set of premises may have several stable expansions as for example the

α	α_1	α_2
$T(p \wedge q)$	Tp	Tq
$F(p \vee q)$	Fp	Fq
$F(p \rightarrow q)$	Tp	Fq
$T \neg p$	Fp	Fp
$F \neg p$	Tp	Tp

β	β_1	β_2
$F(p \wedge q)$	Fp	Fq
$T(p \vee q)$	Tp	Tq
$T(p \rightarrow q)$	Fp	Tq

Table 1: α- and β-formulas.

set $\{\neg Lp \rightarrow q, \neg Lq \rightarrow p\}$ or it can happen that there are no stable expansions as e.g. for the set $\{\neg Lp \rightarrow p\}$.

From the agent's point of view there can be alternative sets of beliefs or even no set of beliefs at all. So we define the derivability relation $\mid\sim_i$ to characterize the agent's point of view of his own beliefs:

Definition 2.2 $A \mid\sim_i p$ *iff there is a stable expansion T of A such that $p \in T$.*

An external observer cannot know which one of the possible alternative sets of beliefs the agent has chosen, so an external observer can safely conclude that the agent believes a formula only if the formula is in all the possible sets of beliefs of the agent. We define the derivability relation $\mid\sim_e$ to capture this notion.

Definition 2.3 $A \mid\sim_e p$ *iff $p \in T$ for all stable expansions T of A.*

The derivability relation $\mid\sim_i$ corresponds the notion of arguability defined by Mc-Dermott and Doyle for their non-monotonic logic [1] and the derivability relation $\mid\sim_e$ equals to McDermott and Doyle's notion of provability [1].

3 Analytic Tableaux

The decision procedure for propositional autoepistemic logic is based on analytic tableaux for propositional logic [5], which we introduce in this section.

A *signed formula* is an expression of the form Tp or Fp where p is an (unsigned) formula. By the *conjugate* of a signed formula we mean the result of changing "T" to "F" or "F" to "T". Thus the conjugate of Tp is Fp and the conjugate of Fp is Tp. We use the letter "α" to stand for any signed formula of the form given in the column α in Table 1. For any such formula α, we define the two formulas α_1 and α_2 as given in Table 1. Similarly we define β, β_1, β_2 according to Table 1.

An analytic tableau is a dyadic tree whose points are signed formulas. The points thus have at most two successors. The tree has a unique point called the root which is not a successor of any other point. A point is an end point if it has no successors. A path is a finite or denumerable sequence of points, beginning with the root, which is such that each term of the sequence (except the last, if there is one) is the predecessor of the next. For every point x in the tree there exists a unique path P_x whose last term is x. By a branch we shall mean a path whose last term is an end point of the tree, or a path that is infinite.

Given two dyadic trees T_1 and T_2, whose points are occurrences of signed formulas, we call T_2 a direct extension of T_1 if T_2 can be obtained from T_1 by one application of the operation (1) or (2):

1. If some α occurs on the path P_q, where q is an end point of T_1, then we may adjoin either α_1 or α_2 as the sole successor of q. (In practice, we usually successively adjoin α_1 and then α_2.)

2. If some β occurs on the path P_q, where q is an end point of T_1, then we·may simultaneously adjoin β_1 as the left and β_2 as the right successor of q.

T is an *analytic tableau* for a signed formula p if there exists a finite sequence $(T_1, T_2, \ldots, T_n = T)$ such that T_1 is a 1-point tree whose origin is p and, for each $i < n$, T_{i+1} is a direct extension of T_i.

A branch of a tableau is *closed* if it contains some signed formula and its conjugate, otherwise it is *open*. A tableau is closed if all of its branches are closed. We call a branch of the tableau complete if for every α which occurs in the branch, both α_1 and α_2 occur in the branch and for every β which occurs in the branch, at least one of β_1, β_2 occurs in the branch. A tableau is *completed* if all of its branches are either closed or complete. $\Gamma(p)$ denotes the completed analytic tableau for the signed formula Fp.

4 Decision Procedure

The decision procedure for autoepistemic logic must find all the stable expansions of a given set of premises A and for derivability relation $\mathrel{\mid\!\sim}_i$ check whether one of them contains the given formula p and for derivability relation $\mathrel{\mid\!\sim}_e$ check whether all of them contain the given formula p. A stable expansion T of a set of premises A is a set of formulas which satisfies the (fixed point) condition $T = Th(A \cup \{Lp \mid p \in T\} \cup \{\neg Lp \mid p \notin T\})$. In the decision procedure all the relevant stable expansions are found using analytic tableaux.

First the tableau $\Gamma(A \to p)$ is constructed to find all the formulas of the form Lq which can be involved in closing the tableau $\Gamma(A \to p)$ and thus showing that p is contained in a stable expansion of A. For each of these formulas the decision procedure must decide whether the formula or its negation is to be included in the stable expansion. The choices have to be made in such a way that the expansion of A induced by the choices satisfies the fixed point condition given above.

The fixed point condition requires that if a formula Lq is chosen to be in the stable expansion, q must belong to the expansion and if $\neg Lq$ is in the expansion, q cannot belong to the expansion. But here the problem is exactly the same as for the first tableau, i.e. does q belong to the expansion. So for each such Lq and $\neg Lq$ a tableau $\Gamma(A \to q)$ is constructed and the procedure is repeated until no new Lq formulas can be found.

As the choice between the inclusion of an Lq formula or its negation to a stable expansion may depend on which other such formulas are included, the decisions have to be made consistently. The consistency can be verified using the constructed analytic tableaux by adding to every branch of every tableaux the formula TLq if Lq is chosen to be included in the expansion and the formula FLq if $\neg Lq$ is to be included. The decisions are consistent if for each Lq chosen tableau $\Gamma(A \rightarrow q)$ is now closed and for each $\neg Lq$ chosen tableau $\Gamma(A \rightarrow q)$ is still open. Every relevant stable expansion can be found by testing every possible combination of choices for consistency.

To present the detailed algorithm for the decision procedure the following definitions are used. Given a finite set of formulas A and a formula p, we define the *tableau structure* $\langle A, p, \Gamma^*, X \rangle$ as follows. Γ^* is the tableau $\Gamma(A \rightarrow p)$[1] and X is the smallest set such that $\Gamma^* \in X$ and if $\Gamma' \in X$ and TLq or FLq appears in some open branch of Γ', then $\Gamma(A \rightarrow q) \in X$. A *labeling* of a tableau structure is an assignment of one label, either OPEN or CLOSED, to each of the tableau in X.

The following algorithm AEL decides whether $A \mathrel{\vdash_i} p$ $(A \mathrel{\vdash_e} p)$ given a finite set of formulas $A \subseteq \mathcal{L}$ and a formula $p \in \mathcal{L}$.

Algorithm AEL:

1. The tableau structure $\langle A, p, \Gamma^*, X \rangle$ is generated.
2. All the labelings of the tableau structure are tested for acceptability. This test consists of first adding to all the branches in all the tableaux in X the formula TLq if the tableau $\Gamma(A \rightarrow q) \in X$ is labeled CLOSED and the formula FLq if the tableau $\Gamma(A \rightarrow q) \in X$ is labeled OPEN. The labeling is acceptable if all the tableaux labeled OPEN have now some branch open and all the tableaux labeled CLOSED have all the branches closed. $A \mathrel{\vdash_i} p$ $(A \mathrel{\vdash_e} p)$ holds iff there exists an acceptable labeling where Γ^* is labeled CLOSED (Γ^* is labeled CLOSED in all the acceptable labelings).

We give an example to clarify the algorithm. Let

$$A = \{\neg Lp \rightarrow q, \neg Lq \rightarrow p\} \tag{4.1}$$

We examine whether $A \mathrel{\vdash_i} p$ or $A \mathrel{\vdash_e} p$ holds. Figure 1 shows the tableau structure for A, p. Closed branches are marked with $\times$ symbols. Because TLq appears in an open branch of $\Gamma^* = \Gamma(A \rightarrow p)$, tableau $\Gamma' = \Gamma(A \rightarrow q)$ must be built. No other tableaux are needed as TLp and TLq are the only formulas of the form TLr or FLr which appear in Γ^* or Γ'. The tableau structure has two acceptable labelings:

$$\{\langle \Gamma^*, OPEN \rangle, \langle \Gamma', CLOSED \rangle\} \text{ and } \{\langle \Gamma^*, CLOSED \rangle, \langle \Gamma', OPEN \rangle\}$$

For example the first one is acceptable because after adding FLp and TLq to every branch in both of the tableaux, Γ^* has a branch open but Γ' is closed. Because

---

[1] If A is a finite set of formulas and it appears in a formula, it stands for the conjunction of its elements. If it is empty, it stands for the formula $p \vee \neg p$.

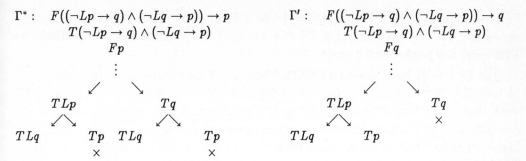

Figure 1: Tableau structure $\langle A, p, \Gamma^*, \{\Gamma^*, \Gamma'\}\rangle$.

there is an acceptable labeling, where Γ^* is labeled OPEN and another one, where it is labeled CLOSED, $A \mathrel{\vert\!\sim}_i p$ holds but $A \mathrel{\not\vert\!\sim}_e p$ does not.

Theorem 4.1 *The algorithm AEL always halts and finds all the acceptable labelings of the tableau structure.*

Proof. As the set A is finite, only a finite number of tableaux can be constructed. Once this is done, there are only a finite number of labelings for the tableau structure. These can all be checked for acceptability. □

To prove the correctness of the decision procedure we require the analytic tableau method to have the following two properties.

1. It is complete, i.e. p is provable ($\vdash p$) iff $\Gamma(p)$ is closed.
2. It is exhaustive, by which we mean the following. Let A be a finite set of formulas and B be a finite set of atomic formulas or their negations. If $A \cup B \vdash p$ but $A \not\vdash p$ then there appears some Fq such that $q \in B$ or Tq such that $\neg q \in B$ in every open branch of $\Gamma(A \to p)$.

Smullyan [5] proves the completeness of the analytic tableau method for propositional logic. Exhaustiveness can be shown in the following way. Assume that $\Gamma(A \to p)$ has some open branch θ that contains no Fq, $q \in B$ or Tq, $\neg q \in B$. If we add to every branch of $\Gamma(A \to p)$ Tq for every $q \in B$ and Fq for every $\neg q \in B$, we have the tableau $\Gamma(A \cup B \to p)$, which is open because θ is an open branch. By the deduction theorem $A \cup B \vdash p$ iff $\vdash A \cup B \to p$ and $\vdash A \cup B \to p$ iff $\Gamma(A \cup B \to p)$ is closed. But this is a contradiction as $\Gamma(A \cup B \to p)$ is supposed to be open.

Lemma 4.2 *If T is a stable expansion of A, there is an acceptable labeling of the tableau structure $\langle A, p, \Gamma^*, X\rangle$ such that $p \in T$ iff Γ^* is labeled CLOSED in the labeling.*

Proof. Let T be a stable expansion of A. By theorem 2.1 $T = Th(A \cup \{Lp \mid p \in T\} \cup \{\neg Lp \mid p \notin T\})$ holds for T. We shall construct the acceptable labeling from T. In the tableau structure a tableau $\Gamma(A \to q)$ is labeled CLOSED if $q \in T$ and OPEN if $q \notin T$. We show that this labeling is acceptable.

Let $\Gamma(A \to q)$ be labeled OPEN. Assume that the tableau is closed. Then there exists sets A^+ and A^- such that $A \cup A^+ \cup A^- \vdash q$, where $A^+ = \{Lr \mid$

$\Gamma(A \rightarrow r) \in X, r \in T\}$ and $A^- = \{\neg Lr \mid \Gamma(A \rightarrow r) \in X, r \notin T\}$. But then $A \cup \{Lp \mid p \in T\} \cup \{\neg Lp \mid p \notin T\} \vdash q$ and $q \in T$, which is a contradiction. So $\Gamma(A \rightarrow q)$ has some branch open.

Let $\Gamma(A \rightarrow q)$ be labeled CLOSED. Then $q \in T$ and there are two minimal sets of formulas $X^+ = \{Lr_1, \ldots, Lr_n \mid r_i \in T\}$ and $X^- = \{\neg Ls_1, \ldots, \neg Ls_m \mid s_j \notin T\}$ such that $A \cup X^+ \cup X^- \vdash q$. If $X^+ = X^- = \emptyset$, $\Gamma(A \rightarrow q)$ is closed because of the completeness of the tableau method. Otherwise, by exhaustiveness there has to be some FLr_i or TLs_j in every open branch of $\Gamma(A \rightarrow q)$. Then $\Gamma(A \rightarrow r_i) \in X$ or $\Gamma(A \rightarrow s_j) \in X$. But these tableaux are labeled CLOSED and OPEN respectively and then the branch in $\Gamma(A \rightarrow q)$ is closed because it contains also TLr_i or FLs_j respectively. So all the tableaux labeled OPEN have some branch open and all the tableaux labeled CLOSED have all the branches closed. $\square$

Lemma 4.3 *If there is an acceptable labeling of the tableau structure $\langle A, p, \Gamma^*, X \rangle$, there is a stable expansion T of A such that, for every tableau $\Gamma(A \rightarrow q) \in X$, the tableau is labeled CLOSED iff $q \in T$.*

Proof. We construct the stable expansion T of A from the acceptable labeling. Let R_0 be the set of formulas Lq such that $\Gamma(A \rightarrow q) \in X$ is labeled CLOSED and $\neg Lq$ such that $\Gamma(A \rightarrow q) \in X$ is labeled OPEN in the labeling. Let $Lq_1, Lq_2, \ldots$ be an enumeration of all the formulas of the form $Lq \in \mathcal{L} - \{Lr \mid Lr \in R_0 \text{ or } \neg Lr \in R_0\}$ such that if Lq_i is a subformula of Lq_j, then $i < j$. Let $T_0 = Th(A \cup R_0)$.

Define R_{i+1} and T_{i+1} for $i = 0, 1, \ldots$ as follows:

$$R_{i+1} = \begin{cases} R_i \cup \{Lr_{i+1}\} & \text{if } q_{i+1} \in T_i \\ R_i \cup \{\neg Lr_{i+1}\} & \text{otherwise} \end{cases}$$

$$T_{i+1} = Th(A \cup R_{i+1})$$

Let $T = \bigcup_{i=0}^{\infty} T_i$ and $R = \bigcup_{i=0}^{\infty} R_i$. Clearly $T_i \subseteq T_{i+1}$ and $T = Th(A \cup R)$. By theorem 2.1 the set of formulas T is a stable expansion of A iff $T = Th(A \cup \{Lp \mid p \in T\} \cup \{\neg Lp \mid p \notin T\})$ holds. So we can prove that T is a stable expansion of A by showing that $R = \{Lp \mid p \in T\} \cup \{\neg Lp \mid p \notin T\}$. We use the notation $L^+ = \{Lp \mid p \in T\}$ and $L^- = \{\neg Lp \mid p \notin T\}$.

First we show that $L^+ \cup L^- \subseteq R$. Let $Lq \in L^+$ and thus $q \in T$. We shall show that $Lq \in R$. If $Lq \in R_0$, then $Lq \in R$. If $Lq \notin R_0$, then either $\neg Lq \in R_0$ or Lq is some Lq_i. Assume that $\neg Lq \in R_0$. Then tableau $\Gamma(A \rightarrow q)$ is OPEN and $A \cup R_0 \nvdash q$. So $q \notin T_0$ but $q \in T$ and there exists $k \geq 1$ such that $q \notin T_{k-1}$ but $q \in T_k$, i.e. $A \cup R_{k-1} \nvdash q$ but $A \cup R_{k-1} \cup \{Lq_k\} \vdash q$ (or $A \cup R_{k-1} \cup \{\neg Lq_k\} \vdash q$). By exhaustiveness there would be FLq_k (or TLq_k) in every open branch of the tableau $\Gamma(A \cup R_{k-1} \rightarrow q)$. As there would be only subformulas of $A \cup R_{k-1} \rightarrow q$ in the tableau and as Lq_k is not a subformula of R_{k-1} because for every $Lq_j \in R_{k-1}$, $j < k$, it must be a subformula of $A \rightarrow q$. But in this case FLq_k (or TLq_k) appears also in the tableau $\Gamma(A \rightarrow q) \in X$ and therefore there is a tableau $\Gamma(A \rightarrow q_k) \in X$. So $Lq_k \in R_0$ or $\neg Lq_k \in R_0$. But this is a contradiction because Lq_k is in the enumeration. Thus

q is some q_i. Assume that $q_i \notin T_{i-1}$. As $q_i \in T$, there is $k \geq i$ such that $q_i \in T_k$ and $q_i \notin T_{k-1}$. So $A \cup R_{k-1} \not\vdash q_i$ but $A \cup R_{k-1} \cup \{Lq_k\} \vdash q_i$ (or $A \cup R_{k-1} \cup \{\neg Lq_k\} \vdash q_i$). As $k \geq i$, Lq_k is not a subformula of Lq_i. Then it has to be a subformula of A and therefore $\Gamma(A \to q_k) \in X$. But this is a contradiction, because either $Lq_k \in R_0$ or $\neg Lq_k \in R_0$ and then Lq_k cannot belong to the enumeration. So $L^+ \subseteq R$.

Let $\neg Lq \in L^-$ and thus $q \notin T$. If $\neg Lq \in R_0$, then $\neg Lq \in R$. If $\neg Lq \notin R_0$, then either $Lq \in R_0$ or Lq is some Lq_i. Assume that $Lq \in R_0$. Then $\Gamma(A \to q)$ is CLOSED and $q \in T_0 \subseteq T$, which is a contradiction. So q is some q_i. As $q_i \notin T$, $q_i \notin T_{i-1}$ and so $\neg Lq_i \in R_i \subseteq R$. Thus $L^- \subseteq R$ and $L^+ \cup L^- \subseteq R$.

Now we shall show that $R \subseteq L^+ \cup L^-$. Let $Lq \in R$. We shall show that $q \in T$, which implies that $Lq \in L^+$. If $Lq \in R_0$, there is a CLOSED tableau for $A \to q$ and $A \cup R_0 \vdash q$. So $q \in T_0$ and $q \in T$. If $Lq \notin R_0$, then as $Lq \in R$, q is some q_i. So $Lq \notin R_{i-1}$ but $Lq \in R_i$. So $q \in T_{i-1}$ and $q \in T$.

Let $\neg Lq \in R$. We shall show that $q \notin T$, which implies that $\neg Lq \in L^-$. If $\neg Lq \in R_0$, there is a OPEN tableau for $A \to q$. Assume that $q \in T$. Then there is some $k \geq 1$ such that $q \notin T_{k-1}$ but $q \in T_k$, i.e. $A \cup R_{k-1} \not\vdash q$ but $A \cup R_{k-1} \cup \{Lq_k\} \vdash q$ (or $A \cup R_{k-1} \cup \{\neg Lq_k\} \vdash q$). By exhaustiveness there would be FLq_k (or TLq_k) in every open branch of the tableau $\Gamma(A \cup R_{k-1} \to q)$. As above Lq_k must be a subformula of $A \to q$ and therefore there is a tableau $\Gamma(A \to q_k) \in X$, which is a contradiction because Lq_k is in the enumeration. Thus $q \notin T$.

It remains to be shown that if $\neg Lq \notin R_0$ but $\neg Lq \in R$, then also $q \notin T$. If $\neg Lq \notin R_0$ but $\neg Lq \in R$, then q is some q_i. Assume again that $q_i \in T$. Then there is some $k \geq i$ such that $A \cup R_{k-1} \not\vdash q_i$ but $A \cup R_{k-1} \cup \{Lq_k\} \vdash q_i$ (or $A \cup R_{k-1} \cup \{\neg Lq_k\} \vdash q_i$). By exhaustiveness there would be FLq_k (or TLq_k) in every open branch of the tableau $\Gamma(A \cup R_{k-1} \to q_i)$. As Lq_k is not a subformula of R_{k-1} or q_i because of the enumeration, it must be the subformula of A. But then FLq_k (or TLq_k) appears also in every tableaux in X, so there is a tableau $\Gamma(A \to q_k) \in X$. This is again contradiction and so $q_i \notin T$. Thus $R \subseteq L^+ \cup L^-$ and $R = L^+ \cup L^-$.

We still have to show that for every tableau $\Gamma(A \to q) \in X$, the tableau is labeled CLOSED iff $q \in T$. If $\Gamma(A \to q)$ is CLOSED, then $A \cup R_0 \vdash q$ and $q \in T_0 \subseteq T$. If $\Gamma(A \to q)$ is OPEN, then $\neg Lq \in R_0 \subseteq R$ and $\neg Lq \in L^-$. So $q \notin T$. $\qquad\square$

Theorem 4.4 *If A is finite, then it is decidable whether $A \mathrel{|\!\sim_i} p$.*

Proof. By theorem 4.1 AEL always halts and finds all the acceptable labelings for the tableau structure $\langle A, p, \Gamma^*, X \rangle$. Now we have to show that $A \mathrel{|\!\sim_i} p$ iff there is acceptable labeling such that Γ^* is labeled CLOSED in it. Assume that there is an acceptable labeling such that Γ^* is labeled CLOSED in it. Then by lemma 4.3 there is stable expansion T of A such that $p \in T$ and thus $A \mathrel{|\!\sim_i} p$. On the other hand, let Γ^* be labeled OPEN in all the acceptable labelings. Assume that $A \mathrel{|\!\sim_i} p$ holds. Then by lemma 4.2 there is an acceptable labeling such that Γ^* is labeled CLOSED, which is a contradiction. So $A \mathrel{|\!\sim_i} p$ iff there is acceptable labeling such that Γ^* is labeled CLOSED in it. $\qquad\square$

Theorem 4.5 *If A is finite, then it is decidable whether $A \mathrel{|\!\sim_e} p$.*

Proof. We have to show that $A \mathrel{|\!\sim_e} p$ iff Γ^* is labeled CLOSED in all the acceptable labelings. Let Γ^* be labeled CLOSED in all the acceptable labelings. Assume that $A \mathrel{|\!\not\sim_e} p$ holds. Then there is a stable expansion T of A such that $p \notin T$. By lemma 4.2 there is an acceptable labeling such that Γ^* is labeled OPEN in it but this is a contradiction and $A \mathrel{|\!\sim_e} p$ holds. On the other hand, if Γ^* is labeled OPEN in one of the acceptable labelings then by lemma 4.3 there is a stable expansion T of A such that $p \notin T$ and therefore $A \mathrel{|\!\not\sim_e} p$ holds. $\qquad\square$

5 Conclusions

We have defined two possible derivability relations for autoepistemic logic to capture the following intuition: given a set of premises describing the beliefs of an agent does the agent believe a given formula (is the given formula derivable from the given set of premises). One of the relations corresponds to the agent's own point of view of his beliefs and the other one corresponds to the point of view of an external observer of the beliefs of the agent. We have given and proved correct decision procedures based on analytic tableaux for both of the relations.

Experimental versions of the decision procedures have been implemented in Prolog and integrated into the HLM (Helsinki Logic Machine) theorem proving system [4] developed for theorem proving and model checking in intensional logics.

Acknowledgements

The financial support for this work was provided by the Academy of Finland.

References

[1] McDermott, D., and Doyle, J. Non-monotonic logic I. *Artificial Intelligence 13* (1980), 41–72.

[2] McDermott, D. Non-monotonic logic II. *Journal of the Association for Computing Machinery 29* (1982) 1, 33–57.

[3] Moore, R.C. Semantical considerations on nonmonotonic logic. *Artificial Intelligence 25* (1985), 75–94.

[4] Niemelä I., and Tuominen, H. A system for logical expertise. In *STEP-86 Symposium Papers: Methodology, Volume 2*, M. Karjalainen, J. Seppänen, and M. Tamminen, Eds. (Espoo, Finland, Aug 19–22). Finnish Society of Information Processing Science, Helsinki, 1986, pp. 44–53.

[5] Smullyan, R.M. *First-Order Logic.* Springer-Verlag, Berlin, 1968.

Logical Matrix Generation and Testing

Peter K. Malkin and Errol P. Martin
Automated Reasoning Project,
RSSS, Australian National University,
GPO Box 4, Canberra, ACT 2600 Australia.

Abstract

Logical matrices are generalisations of truth-tables. They provide power-ful computational tools for dealing with inference systems. A matrix in which all of the axioms and rules of a theory are true (or designated) but in which some query statement is false (or undesignated) shows that the query state-ment cannot be derived from the given theory. In this paper we discuss the use of matrices in logical research and in automated theorem-proving, and we conjecture further uses in the area of general constraint satisfaction. The com-putational ease of use of matrices is partly overshadowed by the computational difficulty of selecting the best or even a suitable matrix for a given task. Little is known about the relative efficiency of various matrices. We discuss our im-plementation of a general matrix generate and test algorithm which is designed to assist with research into these questions.

1 Introduction

This article describes logical matrices in the context of our implementation of a matrix generation and testing program, which we call MGT. The purpose of this program is to generate sets of matrices satisfying a specified set of axioms and rules; or alternatively, to test existing sets of such matrices against new formulas.

Logical matrices are generalisations of truth-tables, as are found, for example, in many-valued logics. In §2 we set down the main ideas and terminology used in connection with matrices. The key property is that a matrix which satisfies all the rules and axioms of a theory provides a truth-like model of the theory. Because the matrix is a model of the theory, though in general not an exact model, it can be used to check inferences from the theory. This procedure is not limited to many-valued logic, and can be applied quite generally to model any nonstandard logic.

In this paper we will discuss the use of matrices in two areas, research in pure logical theory (§3) and in limiting the search space by node-pruning in automated theorem-provers (§4). We also conjecture further uses of matrices in the area of constraint satisfaction (§5). We conclude (§6) with a discussion of our implementation, which is a prototype designed to assist research in the problems mentioned throughout the paper.

2   Definitions

In this paper we will restrict the logical vocabulary entirely to pure implication, hence we will only be concerned with matrix models for the implication connective. This is not as severe a restriction as it may appear, because implication is the main connective studied in many non-standard logical systems.

The Boolean implication $A \rightarrow B$ is counted as true when either the antecedent A is false, or the conclusion B is true, or both. This leaves only the situation where the antecedent is true and the conclusion is false, and here $A \rightarrow B$ is false. We can represent this information in the form of a diagram or truth-table, with 'is a true proposition' represented just by 'T' (or some other specific symbol), and 'is a false proposition' represented by 'F' (or some other symbol, distinct from that chosen for 'is true').

The table for the Boolean implication connective is thus:

$\rightarrow$	F	T
F	T	T
* T	F	T

It is tables of this sort that we call matrices. The sentences of a formal language can be interpreted by means of a matrix: Assign truth values to the atomic formulas of the language, and then use the matrix to compute the value of complex sentences.

An assignment which yields the value T for a sentence provides a *model* for the sentence. If a sentence evaluates to T for *all* assignments of values from the matrix to the atomic terms, then the sentence is a *tautology*. For example: $a \rightarrow a$, $a \rightarrow (b \rightarrow a)$ and $(a \rightarrow b) \rightarrow ((b \rightarrow c) \rightarrow (a \rightarrow c))$ are tautologies, while $(a \rightarrow b) \rightarrow b$ is not.

As it happens, the two-valued truth-table matrices are completely sufficient to model classical logic. That is, all of the theorems of Boolean logic are tautologies according to the truth-tables, and all of the tautologies are theorems of logic. This is a special feature of the situation with some logical systems, that they have matrices which not only model the logic, but do so *characteristically*. We are concerned with theorem-proving in less restrictive logics and it is here that matrices with more than two values are of use. Non-standard logics have models in these generalized matrices but they are not characteristic models.

The final property we will note in the truth-tables is the concept of an order among the truth values, which represents the consequence relation of any logic modelled by the matrix. In the boolean case this ordering is F ≤ T. In the general case we can expect the induced consequence ordering of a matrix to be a partial order.

In a generalised logical matrix the situation is similar to the case of a many-valued logic, that is, there are a number of truth-values. The notions of truth and falsity are retained in a matrix by dividing the values into non-empty subsets of *designated* and *undesignated* values.

The requirement for validity (or tautologousness) now translates to: a formula is *valid* in a matrix if it has a *designated* value regardless of how the atomic terms are instantiated. The difference is that instead of just evaluating to a single true value, the formula may evaluate to any one of the several possible designated values.

In a matrix of the general sort we are considering, it can no longer be assumed that all of the rules and axioms of classical logic are valid, or even that the matrix has a simple axiomatisation of its tautologies. All that is required is that the matrix satisfy all the axioms and rules of some logic in which we are interested, without being characteristic for that logic. (Indeed, many interesting nonstandard logics such as modal and relevant logics are not many-valued at all, in the sense that no finite matrix is characteristic for the logic.)

The consequence ordering of truth-values in a general matrix is got by defining $a \leq b$ just when $a \rightarrow b$ is designated. Thus each possible matrix of a certain size can be viewed as a decision to *order* the truth-values in a certain way.

3 Matrices and logic research

Matrices as a research tool in logic are especially useful when very little is known about the algebraic structure of the logic, so that the partial models given by a set of matrices for the logic are the only available models.

This was once the case with *modal logics*, for instance. The possible worlds models of the last 25 years now give more information. Currently research in *relevant* logics [1] is greatly assisted by matrix considerations. No doubt the study of default reasoning logics and nonmonotonic logics will soon be making use of matrix techniques.

The chief uses of matrices in logic research are:

Theoremhood Checking:

Matrices provide a strong tool for indicating theoremhood. Suppose there exists a matrix satisfying a logic which also refutes some formula; then the formula cannot be a theorem of the logic in question. On the other hand, suppose that all of the matrices (of specific dimension and of specific designated and undesignated values) which validate all of the rules and axioms of logic **L** are available, and suppose that all of these matrices do validate a formula *A*. This situation provides very strong *heuristic evidence* of the theoremhood of *A* in **L**.

Independence Checking:

A rule or axiom of a theory is *independent* in the theory if it cannot be derived from the other rules and axioms of that theory. Matrices are used constantly in logical research to investigate various axiomatizations of logical systems.

Empirical Data:

Matrices often allow general characteristics of a theory to be identified. For example, if two purported representations of a theory produce the same set of validating matrices, then there is good reason to believe that the two representations are equivalent.

Heuristic Evidence of Structure

The algebraic structure of a logic is ordered by the consequence relation of the logic. In general the implication or entailment ordering will not be linear. (Despite the theorem $(A \to B) \lor (B \to A)$ of Boolean logic, it is clear that we do not regard it as true, that of any two propositions A and B, either A implies B, or B implies A.)

Investigation of the algebraic consequence ordering is one of the deepest problems in logic, and it is one which can be assisted by the heuristic evidence provided by matrices. Finite matrices essentially give a finite *snap-shot* of the logical systems they validate, and the induced consequence relations of a set of matrices for a logic can give evidence about the algebraic structure of a logic

Investigations of the implication ordering of a matrix or set of matrices for a logic can thus be used to reveal general characteristics and properties of the actual ordering imposed by the logic.

4   Matrices and Theorem Proving

Matrix techniques have proved to be extremely valuable in implementing automated theorem-provers for various non-standard logics. Since logical matrices are excellent for theoremhood checks, judicious application of known valid matrices for the logic can reduce the search space.

The idea of the method is essentially the same as in the classic theorem proving research of Gelernter [2], where geometric models were used to control the size of the search space in a theorem prover for axiomatic geometry.

The Kripke Matrix Filter:

The Kripke Automated Theorem Prover (Thistlewaite, McRobbie and Meyer [10]) employs logical matrices to cut its search space.

Before explaining Kripke's current matrix pruning method, the overall theorem proving technique must first be understood. The Kripke system proves theoremhood in a particular system of relevant logic **LR**. In relevant logic, standard resolution techniques do not apply, for two reasons. In the first place, the vocabulary of relevant logics includes all the Boolean connectives, and a set of relevant connectives which carry most of the inferential weight. However, these connectives do not allow normal-isation of formulas to conjunctive normal form, so resolution cannot be applied to all formulas of the system.

Secondly, resolution amounts to the suppression of necessary truths such as (in this case) excluded middles. The relevant logic position is that this kind of move is the cause of the various counterintuitive facts about standard logic, such as that all inconsistent theories are equivalent and trivial.

As a result, implementing a theorem prover in **LR** is difficult. The search space in the proof search tree grows exponentially, each node creating its own subtree unless the node is itself an axiom. The matrix filter prunes the proof search tree by checking potential proof nodes against several matrices known to validate *LR*.

The time reduction resulting from this matrix filtering method has been dramatic in actual use. In combination with other techniques, the theorem prover is viable and has contributed to the understanding of relevant logic. The problem is that just as in theorem and independence checks, formulas can get through a matrix test even though they are not theorems.

It has been conjectured by Thistlewaite, McRobbie and Meyer[10] that using larger or more carefully chosen matrices will result in more formulas being refuted, which should yield speedups. Larger matrices are presumably more accurate models of logical systems on the average than smaller ones. What remains an open question is whether the cost of using larger matrices is paid for by the reduction in the overall search space. Larger matrices are more costly to check than smaller ones simply because the number of instantiations required is greater. The check of a formula with 9 propositional variables with a 10×10 matrix requires 9^{10} different instantiations while the same formula checked with a 3×3 matrix requires only 9^3, an exponentially large difference. The trade off, then, is between possibly reducing a search space which grows exponentially and a filtering procedure with an exponentially increased cost associated with it. What is not known is how to compute the optimal matrix size for a given check.

Relative Effectiveness of Matrices

Thistlewaite, McRobbie and Meyer [10] also consider whether a matrix filter could be constructed in which the matrix applied at a node is dynamically chosen with a view to being a likely refuter of the node, if it can be refuted. It has proved extraordinarily hard to define the right notion here of a good matrix. Clearly some matrices are more effective than others, and in fact some three-valued matrices are very effective. An open problem, then, is the construction of particular matrices that are especially good at refuting formulas, and the development of measures of strength of matrices. If matrices could be custom-made to refute **LR** non-theorems, then the cost of using larger matrices might be offset by the greater number of nodes that could be pruned in a given proof-search.

General Automated Theorem Prover Applications:

The utility of especially powerful refuting matrices is not limited to the Kripke matrix filtering system. If custom matrices do refute enough formulas to pay for the cost of generating and checking them, then such matrices could be used in other automated theorem provers as well.

5 Matrices and Constraint Satisfaction

It is conjectured that matrix techniques will be found useful in constraint logic programming (CLP). (See, e.g. Jaffar, et.al.[3].)

The main features of CLP implementations are: an extra domain of values for the constraints; special constraint predicates or relations over that domain, e.g., typically equalities or inequalities would be among the constraint relations. Goals written in the new language are called constraints. The intention is that the theorem proving program will collect constraint goals as a special case, and take special action to determine if the constraints can be jointly satisfied.

The usual case considered by Jaffar, et.al.[3], is where the extra domain is that of the real numbers. However it seems plausible that the constraints could also be propositional constraints, expressing implications of various kinds. This is considered in Morishita et.al.[4], but the constraints are all Boolean conditions.

If arbitrary propositional constraints are to be introduced, say in order to model hypotheses and theory change, then matrix techniques would appear to provide a promising method of showing that the constraints are consistent , while simultaneously providing some interpretation of the ground terms as a solution.

6 Implementation Issues – Generation Methods

A matrix is a model of specific dimension with both designated and undesignated values which validates all of the rules and axioms of a given logical system. Given this definition, there are two presumptively computable tasks:

1. Generate **all** matrices of user-specified dimension and with user-specified sets of designated and undesignated values which validate all of the rules and axioms of a logical system defined by the user, and

2. Identify those matrices contained in a file indicated by the user, which validate a formula or set of formulas chosen by the user.

The MGT system is a general matrix generate and test system which can perform these tasks. It is intended to provide a research environment to assist enquiry into the open problems mentioned above. MGT allows the user to specify the matrix dimension, the designated and undesignated values, and a set of rules and axioms, and then produces all of the matrices which validate the rule/axiom set. It also allows the user to specify a rule/axiom set and a file which contains matrices; in this case the output is all of the matrices in this file which validate the rule/axiom set.

The goal is to develop a research tool which would be used to deliver useful data for current and future research in the fields of automated theorem-proving and logic. To this end, a goal of the program is to allow a wide range of logical systems to be researched while maintaining efficient execution. To some extent this has been achieved. The user can choose between many predefined rules and axioms which afford the fastest generation and test times since code has been specially written to deal with them. If another rule or axiom is required, the user can specify it directly in propositional form and the system will attempt to validate it, but with time penalty.

Despite this flexibility, the MGT system cannot handle certain logical systems because of logical assumptions integrated into its design. Some of these assumptions and the limitations they cause will be discussed further below.

The Problem:

The main problem is to produce *all* matrices of a given size and designation which validate some chosen set of rules and axioms. Computationally this goal quickly becomes unmanageable as the size of the matrices increases because of the enormous resulting search space. Without the aid of specialized generation techniques, even matrices as small as 4×4 cannot be processed in a reasonable period of time.

We now briefly discuss the various means by which this problem has become computationally tractable.

Possible Values Table:

The search space of possible matrices can be represented by a **possible values table** (*pvt*). The *pvt* contains the values that the cells of the implication matrices being generated can have. By taking all possible combinations of these cell values, this table represents all of the possible matrices that have to be checked.

The total number of possible matrices for a logic becomes unmanageably large beyond 3×3. Also, the naive search space contains redundancies of various sorts, for instance, superfluous elements a and b for which both $(a \to b)$ and $(b \to a)$ are designated. Such values should be collapsed into a single value. There will also be isomorphisms among the set of valid matrices.

There are thus two complementary strategies for producing the required set of valid matrices: reduction of the search space by preliminary analysis of the *pvt*, and the use of efficient algorithms in traversing the space. The greatest efficiencies in the present research have been realized with preliminary analysis and pruning of the *pvt*.

Total-ordering:

The first reduction of the *pvt* is got from an observation of R.K.Meyer, that an equivalent search space can be constructed in which the elements of the matrices are integers, *and* the matrices always have $(d \rightarrow u)$ undesignated. The theory of this simplifying move is too detailed to explain here; for general reference the work of Slaney [9] is recommended.

The use of integers as truth-values allows the use of the natural numerical ordering. The imposition of this total ordering gives us the following constraint: If $(a \rightarrow b)$ is designated, then a is numerically less than or equal to b. As a result of this constraint all of the cells below the main diagonal must be undesignated, which generally results in a large reduction of the naive search space size.

Total-ordering of the truth-values has two other direct consequences. The first is that all matrices produced will validate the rule of *modus ponens:* that when A and $A \rightarrow B$ are both theorems, then B must also be a theorem. The other consequence is that all matrices will be antisymmetric, i.e., that both $(a \rightarrow b)$ and $(b \rightarrow a)$ are designated only if $a = b$. This removes superfluous elements. The range of logics which can be studied with MGT is, therefore, limited to theories with these properties. Although this precludes the study of certain somewhat obscure logics, the time savings realized from the imposition of the total-ordering constraint enables much larger matrices to be managed.

Axiomatic Constraints

The most effective means of cutting the search space has proved to be by a preliminary analysis of the logic, designed to remove obviously bad combinations of cells. There is a whole range of constraints of this sort, all discovered on an *ad hoc* piecemeal basis. We give one extremely simple example: In most logics, though not all, the law of *Identity*, viz. $A \rightarrow A$ is required to be valid. This imposes an immediate constraint on the *pvt*, in that the main diagonal must now contain *only* designated values.

In Slaney [9] and the work of Pritchard and Meyer [7], constraints of this sort are used to great effect. However, in general MGT cannot take advantage of axiomatic constraints because the design goal of the system is that it should be free of assumptions about the target logic.

Generate and Test

The main algorithm for finding all validating matrices for a logic is the generate and test paradigm. Searching by these means is well-known to be highly inefficient, because of bad backtracking and thrashing. The algorithm we use employs a dependency-directed backtracking method to reduce the effect of the first problem.

The method, which we adapted from [5] and [6], depends on analysing the refutations obtained at a given stage of the search, and using this information to choose an appropriate backtrack point.

This method of non-chronological backtracking still does not help with thrashing and in fact our system still exhibits thrashing behaviour. The behaviour is reduced however by further modifications to the basic generate and test paradigm, in which the search space is partly re-organised dynamically to reduce the possibility that a known refutation will re-occur at a later stage.

Results

The system is now able to routinely handle several of the relevant logics, and can produce matrices in a reasonable time for up to 8×8. Research into the *data* produced by the program, for the purpose of investigating the problems mentioned above, has just begun.

References

[1] Anderson A.R. and Belnap N.D. Jr. (1975), *Entailment: The Logic of Relevance and Necessity*, Princeton, Princeton University Press.

[2] Gelernter, H. (1959), Realization of a geometry-theorem proving machine, Proceedings of the International Conference on Information Processing, UNESCO House, pp. 273-282. Reprinted in Siekmann and Wrightson (1983), pp. 99-117.

[3] Jaffar, J. and Lassez, J-L. (1987), Constraint Logic Programming, Proceedings of the Fourteenth ACM Symposium on Principles of Programming Languages , pp.111-119.

[4] Morishita, S., Numao, M. and Hirose, S. (1987), Symbolical construction of truth value domain for logic program, Proceedings of the 4th International Conference on Logic Programming (ed. J-L. Lassez), Cambridge, Mass., MIT Press.

[5] Pritchard, P.A. (1978), And now for Something Completely Different , ms., Australian National University.

[6] Pritchard, P.A. (1979), Son of Something Completely Different , ms., University of Queensland.

[7] Pritchard, P.A. and Meyer, R.K., (1977), On computing matrix models of propositional calculi , ms. ANU.

[8] Siekmann, J. and Wrightson, G. (1983), Automation of Reasoning: Classical Papers on Computational Logic , 2 vols. Springer-Verlag.

[9] Slaney, J.K. (1980), Computers and Relevant Logic: A project in computing matrix model structures for propositional logics , Ph.D. thesis, Australian National University.

[10] Thistlewaite, P.B., McRobbie, M.A., and Meyer, R.K. (1988), Automated Theorem-Proving in Non-Classical Logics , London, Pitman.

Optimal Time Bounds for Parallel Term Matching

Rakesh M. Verma   I.V. Ramakrishnan

Department of Computer Science
State University of New York at Stony Brook
Stony Brook, NY 11794

Abstract

Term Matching is a fundamental operation in term rewriting, functional programming and logic programming. Parallel algorithms for this operation have attracted much attention recently. However nontrivial lower bounds for term matching are as yet unknown. In this paper, we obtain lower bounds on parallel time for this problem. We also establish the tightness of our lower bounds for some representations and several models, by giving matching upper bounds with as few processors as possible.

Key words and phrases. complexity, optimal bounds, parallel term matching.

1 Introduction

Term matching is an important problem that arises very often in functional/equational programming and symbol manipulation systems like term rewriting. Informally, given two terms, which are expressions over function symbols $f_1, f_2, \ldots$ and variables $v_1, v_2, \ldots$, the *term matching* problem is to determine an assignment of values to the variables in one of the terms – which is called the pattern and the other is referred to as the subject – so as to render the two terms equivalent. For example, the terms $f_1(f_2(v_1, v_2), f_3(v_1), v_2)$ and $f_1(f_2(f_5, f_4(f_6)), f_3(f_5), f_4(f_6)))$ become equivalent on assigning $v_1 = \cdot f_5$ and $v_2 = f_4(f_6)$. If all variables have at most one occurrence in a term, then the term is said to be linear otherwise it is said to be non-linear. In case of non-linear terms the substitutions for different occurrences of the same variable must be identical. Note that term matching is a special case of unification, in which one of the two terms (for all practical purposes) may be considered to be variable free.

Parallel algorithms for term matching have been proposed in [5], [6], [10] and [12] on the PRAM (Parallel Random Access Machine) model [7]. This model consists of a collection of processors, each with a unique integer ID which it knows, that share a global memory and execute a single program in lock step. There are variants of the PRAM model which handle conflicting reads and writes to the same global memory location differently. The strongest variant is the CRCW PRAM (Concurrent Read Concurrent Write) model that allows simultaneous reading and writing to the same location by more than one processor. The CREW PRAM (Concurrent Read Exclusive Write) allows simultaneous reading by more than one processor, but disallows simultaneous writes. The weakest variant is the EREW PRAM (Exclusive Read Exclusive Write) in which simultaneous reading and simultaneous writing are both prohibited. Three different variants of the CRCW PRAM have been proposed in the literature. The first and the weakest is the Common PRAM in which all processors attempting to write in the same location simultaneously must write the same value. The second is the Arbitrary PRAM in which one of the processors attempting writes to the same location simultaneously succeeds but no assumption can be made as to which processor succeeds. The third and the strongest variant is the Priority PRAM in which processors have a fixed priority according to their id's and the processor with the highest priority, among the competing processors, succeeds.

Although several parallel algorithms for term matching have been proposed on the various PRAM's, lower bounds on time for parallel term matching are as yet unknown. In this paper we settle several open problems regarding the complexity of term matching. Specifically we show that on any CRCW PRAM term matching requires $\Omega(\log n/\log\log n)$ time independent of representation of terms, as long as processors are restricted to a polynomial in the input size. This means that the lower bounds hold for all three representations of terms: strings stored in arrays and trees (hence DAG's also). We also prove that this lower bound is tight for the string representation, by giving corresponding optimal time upper bounds on the Arbitrary PRAM. Note that on the Arbitrary PRAM string matching can be done in $O(1)$ time. Thus our lower bounds for term matching imply that the presence of variables makes term matching provably more difficult than string matching. We prove $\Omega(\log n)$ time lower bounds for term matching on the CREW PRAM for all three representations. We also show that the lower bounds are tight for the string and tree representations. All our lower bounds hold for both linear as well as non-linear terms, whereas our upper bounds are for the general case of non-linear terms. Thus non-linearity is not the inherent source of difficulty in uninterpreted term matching.

The PRAM model is a very convenient model for studying the inherent parallel complexity of problems because one does not have to worry about the messy details of interprocessor communication and various distributions of data (different distributions may affect parallelism by avoiding hot spots etc.). However, with present day technology it is impractical. Thus our lower bound results establish that even under ideal conditions (such as a global memory accessible by all processors in unit time), there is a limit to the parallelism achievable for term matching. Our upper bounds primarily show that (in some cases) the lower bounds cannot be improved. For our upper bounds on the PRAM to have far reaching practical implications it is necessary to establish the additional cost at which one can realize the global memory assumed by the PRAM. Some work is being done in this area. More importantly, we believe that the techniques developed here to obtain parallelism will not lose relevance on the more practical models.

2 Preliminaries

Because of space limitations we shall assume that the reader is familiar with the definitions in [10]. We include here only those which are absolutely necessary. We shall use superscripts (I, $I\!I$ etc.) whenever we want to indicate the arities of function symbols with non-zero arity. A term can be represented in three different ways: as a labeled Directed Acyclic Graph, as a fully parenthesized string stored in an array or by a *labeled directed tree* defined in the obvious way. The DAG's and the trees are ordered.

We will denote the directed edge from n_1 to n_2 by (n_1, n_2) and its label by label((n_1, n_2)). The label of a node n will be denoted by label(n). We shall assume that the labeled directed tree is available in the form of an inverted tree. Such a tree is specified by three arrays Parent, Label and Edge_label. For each node i, Parent[i] contains the node number of the parent of i, Label[i] contains label(i) and Edge_label[i] contains label((i, Parent[i])).

Let us assume that the two terms s and t are represented as *labeled directed trees*. Without loss of generality let t denote the subject. Informally, the sequence of edge-labels on the path from a node to its root is referred to as its *e-path*. Term matching of s and t involves: identification of the *corresponding* nodes of the two trees representing the terms s and t (these are the nodes that have identical e-paths to their respective roots); checking whether the labels are same in case the corresponding nodes are labeled with function symbols, called the *homogeneity* check and checking whether the substitutions for different instances of the same variable are identical. We call this the *consistency* check.

We need the majority problem in order to obtain the lower bounds. The majority problem is defined as follows: Given n bits $x_1, \ldots, x_n$ output a 1 iff at least $n/2$ bits are 1. For convenience, throughout this paper we shall assume that n is even. We assume that the reader is familiar with the notion of a constant depth many-one reduction denoted $\leq_m^{cd}$ and constant depth truth table reduction denoted by $\leq_{tt}^{cd}$ [2]. All our reductions shall assume that each processor knows the size of the input (a standard assumption made in parallel processing).

3 Lower Bounds

Proposition 1 *The Boolean OR, AND of $n^{O(1)}$ bits stored in an array can be found in $O(1)$ time using $n^{O(1)}$ processors on the Common CRCW PRAM.*

Proof: Straightforward.

Proposition 2 *The maximum of n distinct numbers stored in an array can be found in $O(1)$ time using n^2 processors on the Common CRCW PRAM.*

Proof: Omitted.

We shall use the following results.

Theorem 1 ([4]) *The OR of n bits requires $\log_{(5+\sqrt{21})/2} n$ time on the CREW PRAM (independent of the number of processors and memory cells used).*

Theorem 2 ([1]) *Any CRCW PRAM requires $\Omega(\log n/\log\log n)$ time to compute the majority function of n bits, if processors are bounded by a polynomial in n (even if the number of memory cells is infinite !).*

Theorem 3 *If processors are bounded by a polynomial in n,* **any** *CRCW PRAM requires $\Omega(\log n/\log\log n)$ time for term matching, where n is the number of symbols in the input terms and the input is in the form of two labeled ordered inverted trees.*

Proof: The lower bound is obtained by giving a constant depth reduction (using processors bounded by a polynomial in the input size) from majority to term matching, i.e. majority $\leq_m^{cd}$ term matching with trees. Let $x_3, \ldots, x_{n+2}$ be any instance of the majority problem, where the x_i's are boolean values. We construct an instance of the term matching problem, over the variables $V = \{x, y\}$ and the function symbols $F = \{f^{I\!I}, 0^I, 1^I, a\}$, as follows (the constant a is needed to satisfy arity constraints). First pad the input with exactly one extra 0 as x_1 and one extra 1 as x_2. Assign to each index i, $1 \leq i \leq n+2$, $(i-1)^2$ processors. These processors find in constant time the maximum index $1 \leq j < i$ for which $x_j = x_i$ (proposition 2). Once such an index is found, a representative processor p_i for each index i, stores j in Parent[i] for $1 \leq i \leq n+2$ (if there is no such index then Parent[i] $= 0$). Parent$[n+3] = max\{j | x_j = 1\}$, Parent$[n+4] = max\{j | x_j = 0\}$ and Parent[0] $= 0$. Now in another global array Label$[0 : n+4]$ the value x_i is stored for each $1 \leq i \leq n+2$, Label$[n+3]$ = Label$[n+4] = a$. A third array called Edge_label$[1 : n+4]$ is set to one for each i such that Parent[i] $\neq 0$. Edge_label[i] is set to 1 and 2 respectively for $i = 1, 2$. Label[0] is set to f, i.e. the node labeled f is the root of the term constructed. The three arrays constitute the subject term in the form of a labeled inverted ordered tree. The pattern is always a fixed term consisting of a tree with a vertex labeled f as the root, a vertex labeled variable x as its first child (edge_label 1) and a chain of $n/2+1$, 1's with the first 1 as the second child of the root and the chain ends with a variable y. The label of each edge from a 1 to its parent, if the parent is not the root, is one and similarly the label of the edge from the node labeled y to the last node labeled 1 in the chain is one. We note that both the terms can be constructed by $O(n^3)$ processors in $O(1)$ time, on the Common CRCW PRAM. It is now easy to see that the two terms so constructed, match iff the output for the majority problem is a one for this input instance. The lower bound now follows from theorem 2. ∎

Several interesting features of this proof deserve further mention. It is now clear that term matching with linear terms also has the same lower bound. In fact term matching with *just one variable* is at least as hard as majority (we can get rid of the 0's part of the construction, thus eliminating one variable). Notice that the proof does not in any way exploit the fact that time must be spent for the consistency check etc. In fact the lower bound result holds for any parallel algorithm which takes as input two ordered inverted trees and finds all pairs of corresponding nodes in the two trees. The theorem holds for all CRCW PRAM's because the reduction can be carried out on the weakest CRCW PRAM.

Theorem 4 *Majority $\leq_{tt}^{cd}$ labeled tree isomorphism. Hence the lower bound on time from theorem 3 holds for labeled tree isomorphism also.*

Proof sketch: The subject is constructed as in theorem 3 above. We construct $n/2 + 1$ patterns $P_{n/2+1}, \ldots, P_{n+1}$, where pattern P_i consists of the root labeled by f with the left subtree identical to the left subtree of the subject and a chain of i $1's$ terminated by the constant symbol a, as its right subtree. The entire construction can be done in constant time using polynomially many processors. Now by proposition 1, we can OR the results of the isomorphism tests for these $n/2 + 1$ patterns against the subject in constant time. It is easy to see that the output is a 1 iff there are at least $n/2$ $1's$ in the corresponding instance of majority. ∎

This must be contrasted with the fact that string matching can be done in $O(1)$ time on the CRCW PRAM using processors equal to the sum of the lengths of the two strings. Hence, in general, problems with inverted trees as the input representation may require more parallel time compared to the string representation. We have given one specific example where this is actually the case. We close our discussion of the tree representation with an easy corollary of the above theorems.

Corollary 1 *The lower bound on parallel time for majority also holds for (1) computing the length of a list whose elements are stored in an array (2) list ranking (3) finding the height of an inverted tree and (4) computing the level of a node in an inverted tree.*

Proof: This result comes from an examination of the proofs of the previous theorems. For a definition of list ranking see [3]. For the first problem, note that the list does not necessarily fill all the locations in the array. ∎

Theorem 5 *Term matching with trees on the EREW or CREW PRAM, requires $\Omega(\log n)$ time.*

Proof: We give a constant depth reduction from the boolean And problem to term matching with trees. The subject term consists of a root labeled f_n with n children corresponding to the n input values, with the input values as labels of these vertices and the indices as edge labels. The second term consists of a root labeled f_n and n children all labeled with 1's and again the edge labels range from 1 to n. It is easy to see that the two terms can be constructed by n processors in constant time on the EREW PRAM. The two terms constructed match iff the output is a 1 for the boolean And problem. The lower bound now follows from theorem 1. Note that by a simple fanout tree construction we can get rid of the assumption that we have an infinite set of function symbols f_n, one for each n. Specifically, the construction can be modified so as to use only one function symbol f of arity 2 and the constants 0, 1.   ∎

Corollary 2 *The lower bound of theorem 2 also applies to (1) the equivalence problem for two ground terms (2) pattern matching on strings and (3) labeled ordered tree isomorphism.*

Proof: Observe that both trees constructed in the proof of theorem 2 are ground, i.e. there are no variables.   ∎

3.1   Term Matching with String Representation

When terms are represented as strings stored in arrays the lower bound is relatively more difficult to obtain. The difficulty lies in computing, in constant time the location where a symbol must be stored in the array. It turns out that this is not very difficult if we allow a "reasonably" small number of spaces in the strings representing the two terms or at most n redundant trailing right parenthesis. However, we shall not make such assumptions. Although, the lower bound for term matching with strings can be obtained by using the weaker notion of constant-depth truth table reducibility we shall prove a much stronger result viz., majority constant-depth many-one reduces to term matching with strings. In the following theorem, the model is a Common CRCW PRAM.

Theorem 6 *Majority $\leq_m^{cd}$ term matching, when terms are represented as strings stored in arrays.*

Proof: Let $x_1, \ldots, x_n$ be an instance of majority. Without loss of generality assume that $x_1 = 1$ and $x_n = 0$ (we can always pad the input with a 0 and a 1 so that this is true). We construct the terms t_s stored in array $S[6 : 13n + 2]$ and t_p stored in array $P[1 : 7n/2 - 1]$ as follows. The set of variables is $V = \{y_1, \ldots, y_{n/2+1}\}$ and the set of function symbols is $F = \{h^{I\!\!I}, f^I, k^I, a\}$. We shall also use an array Temp[1:2n]. All arrays are initialized to the character $\#$ in $O(1)$ time using $O(n)$ processors. Now processor p_i for $1 \leq i \leq n$ reads input x_i and stores it in Temp[i] if $x_i = 1$ and Temp[$n + i$] otherwise. Thus we have separated the 1's from the 0's. Processor p_i for $1 \leq i \leq 2n$ reads Temp[i] and does the following. If Temp[i] $= \#$ do nothing. If Temp[i] $\neq \#$ and $i < 2n$ then store h in $S[6i]$, a left-parenthesis in $S[6i + 1]$ and f (k) in $S[6i + 2]$ if $x_i = 1$ ($x_i = 0$). Processor p_{2n} stores the term $k(a)$ in locations $S[12n]$ to $S[12n+3]$. Locations $S[12n + 4]$ through $S[13n + 2]$ are set to right-parentheses, as there are exactly $n - 1$ occurrences of h. Now, using $O(n^3)$ processors, overall, the distance between each occurrence of the symbols f or k and the nearest h (if any) is found (from proposition 2, this can be done in constant time). Note that all these distances are multiples of 3. With $O(n^2)$ processors in constant time all these distances are padded with terms, $f(\ldots f(a) \ldots)$, of the appropriate size (e.g., for a distance of 3 use (a) for 6 use (f(a)) etc.). This completes the construction of the subject, which has as many occurrences of f (as children of h) as there are 1's in the instance of majority and as many k's as 0's. The pattern is always a fixed term of the form $h(f(y_1), \ldots, h(f(y_{n/2}), y_{n/2+1}) \ldots)$, with $n/2$ occurrences of f. It is easy to see that it can also be constructed in constant time with $O(n)$ processors. Now we claim that t_p matches t_s iff there are at least $n/2$ 1's in the corresponding instance of majority. ∎

The $\Omega(\log n / \log \log n)$ time lower bound for all CRCW PRAM's for term matching with strings follows from this theorem. It is easy to see that the reduction on the CREW model, given in the previous section, can be carried out so as to output terms as strings stored in arrays, whence the $\Omega(\log n)$ time lower bound for term matching with strings follows on the CREW model.

4 Optimal Time Upper Bounds

In this section we show how to improve the upper bounds of [12] and [10] on the CREW model, when terms are represented as trees. We also design optimal time algorithms for the Arbitrary (and hence Priority also) CRCW PRAM and the CREW PRAM, when terms are represented as strings stored in arrays. To the best of our knowledge these are the first algorithms for term matching with the string representation. We shall first present optimal algorithms for the string representation. To derive the optimal algorithm (on the CREW model) for the tree representation we get rid of the relatively complicated processor allocation problem by linearizing the two trees into strings and then invoking our optimal algorithm for strings. We note that a direct algorithm is also possible but for lack of space we shall omit it here. Our algorithms shall use the sublogarithmic time algorithms in [3] for computing prefix sums in parallel. In [3] it is shown that the prefix sums of n numbers, each of $\log n$ bits, can be computed in time $O(\log n / \log \log n)$ using $n \log \log n / \log n$ processors.

Our algorithm for term matching with strings stored in arrays consists of the following steps:

1. {For each left (right) parenthesis find its level and the address of the corresponding right (left) parenthesis} Assign to each symbol a value 0, +1 or −1 in a separate array as described in Table 1. Now the prefix algorithm of [3] is used to compute the nesting level of each symbol in S (the subject). Sort all parentheses on the pair (level, location in S). The corresponding parentheses are now in adjacent locations. For each parenthesis store the location of its corresponding parenthesis in an array.

2. {Find the order of the non-parenthesis symbols, in S, within each corresponding parenthesis pair Q, ignoring all symbols inside a pair of parentheses nested within Q} Find, for each non-parenthesis symbol, the first left parenthesis to its left. For each pair of corresponding parentheses copy all symbols immediately nested in this pair into a separate array. Each symbol is assigned the value 1 and the parallel prefix algorithm is used to determine the order of each symbol within a parentheses pair.

⋆ refers to any non-parenthesis entry

Symbol in $A[i-1]$	Symbol in $A[i]$	$B[i].level$
⋆	⋆	0
⋆	(	+1
⋆	)	0
(	⋆	0
(	(	not possible
(	)	not possible
)	⋆	−1
)	)	−1
)	(	not possible

Table 1: Table used in step 1

3. {Obtain the e-paths, to the root, of all non-parenthesis symbols in S} Find for each non-parenthesis symbol s_i all the left parentheses in which it is nested and get the order of the function symbols preceding these parentheses. Observe that s_i is nested inside left parenthesis l_i if location(l_i) < location(s_i) < location(corresponding right parenthesis of l_i). Store this information in a separate array for each s_i.

4. Step 4 consists of repeating the above three steps for the pattern.

5. {Find corresponding nodes} For each non-parenthesis symbol p_i in the pattern find the symbol s_i in the subject such that e-path$(r_1, s_i) =$ e-path(r_2, p_i) . To each symbol in the pattern there corresponds exactly one symbol in the subject with the same e-path. If for some symbol in the pattern there is no such symbol in the subject, then halt and output No.

6. The homogeneity check is trivial once the corresponding nodes have been determined.

7. {Consistency check} Determine all occurrences of the same variable in the pattern. Since the corresponding symbols in the subject for these occurrences have been determined, we just have to compare the terms beginning at these symbols in the subject for syntactic equality.

We note that some of the ideas here have been used by [10] for converting terms represented as strings into trees. However their ideas do not yield tight upper bounds immediately. The proof of correctness is straightforward and therefore omitted.

Complexity on the Arbitrary CRCW PRAM: There is an obvious implementation of this algorithm which requires $O(n^3)$ processors and $O(\log n / \log \log n)$ time, but we shall show that $O(n^2 \log n)$ processors are enough to meet the lower bound on time. For

Step 1, we do not explicitly sort the pairs but rather, in an array Temp[1:n] processor p_i stores in location i the level of the symbol $S[i]$ if $S[i]$ is a parenthesis and 0 otherwise. Now for each j such that $Temp[j] \neq 0$ and S[j] is a left parenthesis we can find the min $\{i | Temp[i] = Temp[j] \wedge S[i] =')' \wedge i > j\}$ with $O(n)$ processors in $O(\log \log n)$ time using the algorithm of Shiloach and Vishkin [11]. Thus step 1 requires $O(n^2)$ processors and $O(\log n / \log \log n)$ time. It is not very difficult to check that Steps 2, 3, 6 and 7 can also be done within these bounds. Step 5 is a bit trickier. We use the location in the array and the level of each symbol to sort the e-paths. This requires another application of parallel prefix. Now assign to each symbol p_i in the pattern $\log n \times level(p_i)$ processors. Divide the symbols with $level(p_i)$ in the subject into $\log n$ groups of size $n / \log n$. Each set of $level(p_i)$ processors compares e-path(p_i) with the e-path of one symbol in each group in constant time. The $\log n$ results are collected in constant time and the search (if unsuccessful) narrows down to one of the groups of size $n / \log n$. Thus in time $O(\log n / \log \log n)$ all the corresponding symbols have been identified. Thus step 5 requires $n^2 \log n$ processors and we are done.

Complexity on the CREW PRAM: We note that there is an algorithm for parallel prefix on the CREW PRAM which requires n processors and $O(\log n)$ time [8]. Thus we have the following result: There is an algorithm for term matching with strings which requires $O(n^3)$ processors and runs in $O(\log n)$ time. In fact, there is an algorithm for any positive constant $\epsilon, 0 < \epsilon < 1$ which runs in $O((\log n)/\epsilon)$ time and requires $n^{2+\epsilon}$ processors. This can be done by assigning to each symbol in the pattern $n^{1+\epsilon}$ processors for step 5, which dominates the complexity.

4.1 Tree Representation

We use the algorithm devised in [9] for converting trees into strings. On the CREW model this conversion can be done by n processors in $O(\log n)$ time. Hence we have optimal time algorithms for term matching with trees, also. However, we note that a direct optimal time algorithm is also possible but we shall omit it here, for lack of space.

5 Conclusion

In this paper we have presented lower bounds for term matching with several different representations on several models of parallel computation. These lower bounds have been shown to be tight by giving matching bounds for some representations and several models. Some interesting problems remain open. The complexity of term matching on trees on the CRCW model is still $O(\log n)$. Thus the lower bound here is slightly weak. We feel that a matching upper bound on trees would imply a sub-logarithmic algorithm for list ranking. No such algorithm is known. Another important direction is to find non-trivial lower bounds and upper bounds for term matching with interpreted function symbols, e.g. if function symbols are associative or commutative.

References

[1] PAUL BEAME AND JOHN HASTAD, Optimal bounds for decision problems on the CRCW PRAM, In *Proceedings of the ACM Symposium on Theory of Computing*, pages 83–93, 1987.

[2] ASHOK K. CHANDRA, L. STOCKMEYER, AND U. VISHKIN, Constant depth reducibility, *SIAM Journal of Computing*, 13:423–439 (1984).

[3] RICHARD COLE AND UZI VISHKIN, Approximate and exact parallel scheduling with applications to list, tree and graph problems, In *Proceedings of the IEEE Conference on Foundations of Computer Science*, pages 478–491, 1986.

[4] STEPHEN COOK AND CYNTHIA DWORK, Bounds on the time for parallel RAM's to compute simple functions, In *Proceedings of the ACM Symposium on Theory of Computing*, pages 231–233, 1982.

[5] C. DWORK, P. KANELLAKIS, AND J.C. MITCHELL, On the sequential nature of unification, *Journal of Logic Programming*, 1:35–50 (1984).

[6] C. DWORK, P. KANELLAKIS, AND L. STOCKMEYER, Parallel algorithms for term matching, In *Eighth CADE*, Springer-Verlag LNCS vol. 230, 1986.

[7] STEVEN FORTUNE AND JAMES WYLLIE, Parallelism in random access machines, In *Proceedings of the ACM Symposium on Theory of Computing*, pages 114–118, 1978.

[8] C.P. KRUSKAL, L. RUDOLPH, AND M. SNIR, Efficient parallel algorithms for graph problems, In *Proceedings of International Conference on Parallel Processing*, pages 180–185, 1985.

[9] R. RAMESH AND I.V. RAMAKRISHNAN, Optimal speedups for parallel pattern matching in trees, In *Second RTA*, Springer-Verlag LNCS vol. 256, 1987.

[10] R. RAMESH, R.M. VERMA, T. KRISHANPRASAD, AND I.V. RAMAKRISHNAN, Term matching on parallel computers, In *Fourteenth ICALP*, Springer-Verlag LNCS vol. 267, 1987.

[11] Y. SHILOACH AND U. VISHKIN, Finding the maximum, merging and sorting in a parallel computation model, *Journal of Algorithms*, 2:88–102 (1981).

[12] RAKESH M. VERMA, T. KRISHNAPRASAD, AND I. V. RAMAKRISHNAN, An efficient parallel algorithm for Term Matching, In *Sixth FST-TCS*, Springer-Verlag LNCS vol. 241, 1986.

Challenge Equality Problems in Lattice Theory*

William McCune

Mathematics and Computer Science Division
Argonne National Laboratory
Argonne, IL 60439-4844

1. Introduction

This note contains five problems that should be useful for experimentation with equality-oriented automated theorem-proving programs. The problems are formally stated and proved in first-order predicate calculus with equality. The problems are taken from the often-overlooked paper "Semi-automated mathematics" by Guard et al. [Guard69]. That paper focuses on an interactive theorem prover, SAM, which was instrumental in solving an open problem in lattice theory (Theorem L3 below) posed by Bumcroft [Bumcroft65]. That achievement marked an important advance for the field of automated theorem proving.

All of the theorems concern the existence and uniqueness of complements in modular lattices with 0 and 1. According to [Guard69], Theorems L1 and L2 are due to Bumcroft. We include two versions of L1: Theorem L1a states that a particular expression is a complement of a, and Theorem L1b states that there exists a complement of a. In most cases, Theorem L1b will be more more difficult than Theorem L1a for a theorem prover. SAM's Lemma is the result obtained by Guard's group with the aid of their interactive theorem prover. SAM's Lemma led Guard's group directly to the solution of Bumcroft's open question, which we call Theorem L3.

There has been some successful experimentation with SAM's Lemma [McCharen76], but the representation used not entirely satisfactory for the following reason. All of the terms needed for a refutation are named in the input, and no means exists for constructing new terms. In particular, the input contains predicate and constant symbols but does not contain function symbols or an equality symbol.

We use a straightforward equality representation. In addition, we define the predicate COMP(a,b) to assert that a and b are (not necessarily unique) complements, and we define UNICOMP(a,b) to assert that b is a complement of a and the only complement of a. Instead of the standard clause notation, we abbreviate with infix notation for equality and for the meet ($\wedge$) and join ($\vee$) operations. For example, we write

---

*This work was supported by the Applied Mathematical Sciences subprogram of the Office of Energy Research, U.S. Department of Energy, under contract W-31-109-Eng-38.

$(x \wedge z) \neq x \quad | \quad (z \wedge (x \vee y)) = (x \vee (y \wedge z))$

instead of

$\neg EQUAL(meet(x,z),x) \quad | \quad EQUAL(meet(z,join(x,y)),join(x,meet(y,z))).$

In the proofs that follow, the justification [17,12,34] means that the clause was derived by resolving clauses 17, 12, and 34, and the justification [p,2,4,[1,2]] means that the clause was derived by paramodulating the left side of clause 2 into clause 4 at position [1,2]. Also, the proofs use symmetric unification for '=' and 'COMP', and commutative unification for '$\wedge$' and '$\vee$'. Although we include previously proved theorems as lemmas in our proofs, the challenge is to obtain refutations without including lemmas. The proofs were obtained with the aid of an interactive proof-checking program.

2. Modular Lattices

We start with reflexivity of equality, which is required for paramodulation.

    0. $x = x$

The following 8 clauses characterize lattices.

    1. $(x \wedge x) = x$ /* idempotence */
    2. $(x \vee x) = x$
    3. $(x \wedge (x \vee y)) = x$ /* absorption */
    4. $(x \vee (x \wedge y)) = x$
    5. $(x \wedge y) = (y \wedge x)$ /* commutativity */
    6. $(x \vee y) = (y \vee x)$
    7. $((x \wedge y) \wedge z) = (x \wedge (y \wedge z))$ /* associativity */
    8. $((x \vee y) \vee z) = (x \vee (y \vee z))$

Include 0 and 1 in the lattice.

    9. $(x \wedge 0) = 0$
    10. $(x \vee 0) = x$
    11. $(x \wedge 1) = x$
    12. $(x \vee 1) = 1$

Require the lattice to be modular.

    13. $(x \wedge z) \neq x \quad | \quad (z \wedge (x \vee y)) = (x \vee (y \wedge z))$

Define complement and unique-complement.

    14. $\neg COMP(x,y) \quad | \quad (x \wedge y) = 0$
    15. $\neg COMP(x,y) \quad | \quad (x \vee y) = 1$
    16. $(x \wedge y) \neq 0 \quad | \quad (x \vee y) \neq 1 \quad | \quad COMP(x,y)$

    17. $\neg UNICOMP(x,y) \quad | \quad COMP(x,y)$
    18. $\neg UNICOMP(x,y) \quad | \quad \neg COMP(x,z) \quad | \quad z = y$
    19. $UNICOMP(x,y) \quad | \quad \neg COMP(x,y) \quad | \quad COMP(x,f(x,y))$
    20. $UNICOMP(x,y) \quad | \quad \neg COMP(x,y) \quad | \quad f(x,y) \neq y$

2.1. Problem 1

Theorem L1a. *(COMP(x,(u ∨ v)) & COMP(y,(u ∧ v)))* → *COMP(u,(x ∨ (y ∧ v)))*

    <clauses 1-16 from axioms>
21. COMP(r1,(a ∨ b))
22. COMP(r2,(a ∧ b))
23. ¬ COMP(a,(r1 ∨ (r2 ∧ b)))

Proof of Theorem L1a.

24. [p,3,7,[2,2]] ((x ∧ y) ∧ (y ∨ z)) = (x ∧ y)
25. [13,24] ((x ∨ y) ∧ ((z ∧ x) ∨ u)) = ((z ∧ x) ∨ (u ∧ (x ∨ y)))
26. [21,14] (r1 ∧ (a ∨ b)) = 0
27. [22,14] (r2 ∧ (a ∧ b)) = 0
28. [22,15] (r2 ∨ (a ∧ b)) = 1
29. [21,15] (r1 ∨ (a ∨ b)) = 1
30. [p,26,25,[2,2]] ((b ∨ a) ∧ ((x ∧ b) ∨ r1)) = ((x ∧ b) ∨ 0)
31. [p,10,30,[2]] ((b ∨ a) ∧ ((x ∧ b) ∨ r1)) = (x ∧ b)
32. [p,31,7,[2,2]] ((x ∧ (b ∨ a)) ∧ ((y ∧ b) ∨ r1)) = (x ∧ (y ∧ b))
33. [p,3,32,[1,1]] (a ∧ ((x ∧ b) ∨ r1)) = (a ∧ (x ∧ b))
34. [p,7,33,[2]] (a ∧ ((x ∧ b) ∨ r1)) = (x ∧ (b ∧ a))
35. [p,27,34,[2]] (a ∧ ((r2 ∧ b) ∨ r1)) = 0
36. [p,1,7,[2,2]] ((x ∧ y) ∧ y) = (x ∧ y)
37. [36,13] (x ∧ ((y ∧ x) ∨ z)) = ((y ∧ x) ∨ (z ∧ x))
38. [p,28,37,[1,2]] (b ∧ 1) = ((a ∧ b) ∨ (r2 ∧ b))
39. [p,11,38,[1]] ((a ∧ b) ∨ (r2 ∧ b)) = b
40. [p,39,8,[2,2]] ((x ∨ (a ∧ b)) ∨ (r2 ∧ b)) = (x ∨ b)
41. [p,4,40,[1,1]] (a ∨ (r2 ∧ b)) = (a ∨ b)
42. [p,41,8,[1,1]] ((a ∨ b) ∨ x) = (a ∨ ((r2 ∧ b) ∨ x))
43. [p,29,42,[1]] 1 = (a ∨ ((r2 ∧ b) ∨ r1))
46. [16,35,43,23] □

2.2. Problem 2

Theorem L1b. *(COMP(x,(u ∨ v)) & COMP(y,(u ∧ v)))* → *∃w(COMP(u,w))*

    <clauses 1-16 from axioms>
47. COMP(r1,(a ∨ b))
48. COMP(r2,(a ∧ b))
49. ¬ COMP(a,w)

The proof of Theorem L1b is similar to the proof of Theorem L1a.

2.3. Problem 3

Theorem L2. *(COMP(x,(u ∨ v)) & COMP(y,(u ∧ v)) & UNICOMP(u,uc) & UNICOMP(v,vc))* → *COMP((u ∨ v),(uc ∧ vc))*

    <clauses 1-20 from axioms>
50. COMP(r1,(a ∨ b))
51. COMP(r2,(a ∧ b))
52. UNICOMP(a,a2)

53. UNICOMP(b,b2)

54. ¬COMP((a ∨ b),(a2 ∧ b2))

Proof of Theorem L2.

55. ¬COMP(x,(u ∨ v)) | ¬COMP(y,(u ∧ v)) | COMP(u,(x ∨ (y ∧ v))) /* Theorem L1a */

56. ¬COMP(x,(u ∨ v)) | ¬COMP(y,(u ∧ v)) | COMP(v,(x ∨ (y ∧ u))) /* Theorem L1a */

61. [55,50,51] COMP(a,(r1 ∨ (r2 ∧ b)))

62. [56,50,51] COMP(b,(r1 ∨ (r2 ∧ a)))

63. [18,52,61] (r1 ∨ (r2 ∧ b)) = a2

64. [18,53,62] (r1 ∨ (r2 ∧ a)) = b2

65. [p,63,3,[1,2]] (r1 ∧ a2) = r1

66. [p,64,8,[1,1]] (b2 ∨ x) = ((r2 ∧ a) ∨ (r1 ∨ x))

67. [p,2,66,[2,2]] (b2 ∨ r1) = ((r2 ∧ a) ∨ r1)

68. [p,64,67,[2]] (b2 ∨ r1) = b2

69. [65,13] (a2 ∧ (r1 ∨ x)) = (r1 ∨ (x ∧ a2))

70. [p,68,69,[1,2]] (r1 ∨ (b2 ∧ a2)) = (a2 ∧ b2)

71. [p,70,8,[2,2]] ((x ∨ r1) ∨ (b2 ∧ a2)) = (x ∨ (a2 ∧ b2))

72. [15,50] (r1 ∨ (a ∨ b)) = 1

73. [p,72,71,[1,1]] (1 ∨ (b2 ∧ a2)) = ((a ∨ b) ∨ (a2 ∧ b2))

74. [p,12,73,[1]] 1 = ((a ∨ b) ∨ (a2 ∧ b2))

76. [17,14,52] (a ∧ a2) = 0

77. [p,76,7,[2,2]] ((x ∧ a) ∧ a2) = (x ∧ 0)

78. [p,9,77,[2]] ((x ∧ a) ∧ a2) = 0

79. [p,64,69,[1,2]] (a2 ∧ b2) = (r1 ∨ ((r2 ∧ a) ∧ a2))

80. [p,78,79,[2,2]] (a2 ∧ b2) = (r1 ∨ 0)

81. [50,14] (r1 ∧ (a ∨ b)) = 0

82. [p,10,80,[2]] r1 = (a2 ∧ b2)

83. [p,82,81,[1,1]] ((a2 ∧ b2) ∧ (a ∨ b)) = 0

86. [16,83,74,54] □

2.4. Problem 4

SAM's Lemma. *(COMP(x,(u ∨ v)) & COMP(y,(u ∧ v)) → x = ((x ∨ (y ∧ v)) ∧ (x ∨ (y ∧ u)))*

<clauses 1-13 from axioms>

87. COMP(r1,(a ∨ b))

88. COMP(r2,(a ∧ b))

89. r1 ≠ ((r1 ∨ (r2 ∧ b)) ∧ (r1 ∨ (r2 ∧ a)))

Proof of SAM's Lemma.

90. [87,14] (r1 ∧ (a ∨ b)) = 0

91. [88,14] (r2 ∧ (a ∧ b)) = 0

92. [p,3,7,[2,2]] ((x ∧ y) ∧ (y ∨ z)) = (x ∧ y)

93. [92,13] ((x ∨ y) ∧ ((z ∧ x) ∨ u)) = ((z ∧ x) ∨ (u ∧ (x ∨ y)))

94. [p,90,93,[2,2]] ((b ∨ a) ∧ ((x ∧ b) ∨ r1)) = ((x ∧ b) ∨ 0)

95. [p,10,94,[2]] ((b ∨ a) ∧ ((x ∧ b) ∨ r1)) = (x ∧ b)

96. [p,95,7,[2,2]] ((x ∧ (b ∨ a)) ∧ ((y ∧ b) ∨ r1)) = (x ∧ (y ∧ b))

97. [p,3,96,[1,1]] (a ∧ ((x ∧ b) ∨ r1)) = (a ∧ (x ∧ b))

98. [p,7,97,[2]] (a ∧ ((x ∧ b) ∨ r1)) = (x ∧ (b ∧ a))
99. [3,13] ((x ∨ y) ∧ (x ∨ z)) = (x ∨ (z ∧ (x ∨ y)))
100. [p,7,99,[2,2]] ((x ∨ y) ∧ (x ∨ (z ∧ u))) = (x ∨ (z ∧ (u ∧ (x ∨ y))))
101. [p,91,98,[2]] (a ∧ ((r2 ∧ b) ∨ r1)) = 0
102. [p,101,100,[2,2,2]] ((r1 ∨ (r2 ∧ b)) ∧ (r1 ∨ (x ∧ a))) = (r1 ∨ (x ∧ 0))
103. [p,9,102,[2,2]] ((r1 ∨ (r2 ∧ b)) ∧ (r1 ∨ (x ∧ a))) = (r1 ∨ 0)
104. [p,10,103,[2]] ((r1 ∨ (r2 ∧ b)) ∧ (r1 ∨ (x ∧ a))) = r1
105. [104,89] □

2.5. Problem 5

Theorem L3. *(COMP(x,(u ∨ v)) & COMP(y,(u ∧ v)) & UNICOMP(u,uc) & UNICOMP(v,vc)) → UNICOMP((u ∨ v),(uc ∧ vc))*

    <clauses 1-20 from axioms>
106. COMP(r1,(a ∨ b))
107. COMP(r2,(a ∧ b))
108. UNICOMP(a,a2)
109. UNICOMP(b,b2)
110. ¬ UNICOMP((a ∨ b),(a2 ∧ b2))

Proof of Theorem L3.

111. ¬ COMP(x,(u ∨ v)) | ¬ COMP(y,(u ∧ v)) | x = ((x ∨ (y ∧ v)) ∧ (x ∨ (y ∧ u))) /* SAM's Lemma *
112. ¬ COMP(x,(u ∨ v)) | ¬ COMP(y,(u ∧ v)) | COMP(u,(x ∨ (y ∧ v))) /* Theorem L1a */
113. ¬ COMP(x,(u ∨ v)) | ¬ COMP(y,(u ∧ v)) | COMP(v,(x ∨ (y ∧ u))) /* Theorem L1a */
114. ¬ COMP(x,(u ∨ v)) | ¬ COMP(y,(u ∧ v)) | ¬ UNICOMP(u,uc) | ¬ UNICOMP(v,vc) |
                COMP((u ∨ v),(uc ∧ vc)) /* Theorem L2 */
118. [114,106,107,108,109] COMP((a ∨ b),(a2 ∧ b2))
121. [19,110,118] COMP((a ∨ b),g)   /* abbreviation g = f((a ∨ b),(a2 ∧ b2)) */
122. [20,110,118] g ≠ (a2 ∧ b2)
124. [111,121,107] g = ((g ∨ (r2 ∧ b)) ∧ (g ∨ (r2 ∧ a)))
126. [112,121,107] COMP(a,(g ∨ (r2 ∧ b)))
128. [113,121,107] COMP(b,(g ∨ (r2 ∧ a)))
131. [18,108,126] (g ∨ (r2 ∧ b)) = a2
132. [18,109,128] (g ∨ (r2 ∧ a)) = b2
133. [p,131,124,[2,1]] g = (a2 ∧ (g ∨ (r2 ∧ a)))
134. [p,132,133,[2,2]] g = (a2 ∧ b2)
135. [122,134] □

3. Remarks

We close this note with some brief comments on possible attacks on these problems. First, Knuth-Bendix-style completion is not directly applicable because of the presence of the non-unit clauses; however, the related oriented-paramodulation strategies [Peterson83,Hsiang87] may prove helpful. Second, associative/commutative unification is applicable to these problems. Finally, the predicates COMP and UNICOMP are defined strictly for convenience in stating the theorems, and not to enable simpler proofs. It may be helpful to eliminate the corresponding literals at the start of a search for a refutation.

References

[Bumcroft65] Bumcroft, R., *Proc. Glasgow Math. Assoc. 7*, Pt. 1, pp.22-23 (1965).

[Guard69] Guard, J., Oglesby, F., Bennett, J., and Settle, L., "Semi-automated mathematics", *Journal of the ACM* **16,** pp. 49-62 (1969).

[Hsiang87] Hsiang, J., and Rusinowitch, M., "On word problems in equational theories", in *Proceedings of the 14th International Colloquium on Languages, Automata, and Programming*, Springer-Verlag Lecture Notes in Computer Science, Vol. 267, pp. 54-71 (1987).

[McCharen76] McCharen, J., Overbeek, R., and Wos, L., "Complexity and related enhancements for automated theorem-proving programs", *Computers and Mathematics with Applications* **2,** pp. 1-16 (1976).

[Peterson83] Peterson, G. E., "A technique for establishing completeness results in theorem proving with equality", *SIAM Journal of Computing* **12,** pp. 82-100 (1983).

[Robinson69] Robinson, G., and Wos, L., "Paramodulation and theorem-proving in first-order theories with equality", pp. 135-150 in *Machine Intelligence 4*, ed. B. Meltzer and D. Michie, Edinburgh University Press, Edinburgh (1969).

Single Axioms in the Implicational Propositional Calculus

Frank Pfenning

Department of Computer Science
Carnegie Mellon University
Pittsburgh, Pa. 15213-3890
ARPANet: fp@cs.cmu.edu

Abstract

A class of challenge problems derived from a first-order encoding of the implicational propositional calculus is presented.

1. Introduction

In [1] (reprinted as [2]), Łukasiewicz presents the implicational propositional calculus and points out some single formulas which are complete as axioms. The implicational propositional calculus allows a straightfoward encoding in first-order logic. This gives rise to some first-order theorems that seem to very difficult to prove for the human and machine alike. Łukasiewicz himself managed to prove the main theorem, namely that L_1 below is a single axiom.

I tested two of the best automated theorem provers on problems from the set. None were able to show that the Hypothetical Syllogism follows from the single axiom L_1, even when given considerable help with the insertion of lemmas into the initial set of clauses. The table in Section 4 summarizes my results.

2. Implicational Propositional Calculus

2.1. The Tarski-Bernays System

Tarski and Bernays described a system for reasoning in the implicational propositional calculus with three axioms and two inferences rules. The axioms are:

Simplification (S): $p \rightarrow (q \rightarrow p)$

Peirce's Law (P): $((p \rightarrow q) \rightarrow p) \rightarrow p$

Hypothetical Syllogism (H): $(p \to q) \to ((q \to r) \to (p \to r))$

The rules of inference:

Substitution: Substituting arbitrary formulas for propositional variables.

Detachment: From α and $\alpha \to \beta$ infer β for arbitrary formulas α and β.

2.2. Single Axioms

There are single axioms which allow one to derive all of the Tarski-Bernays axioms (and are hence complete). The problem dealt with in [1] is the question what the shortest such axiom would be. The author comes up with an axiom consisting of 13 characters, but also mentions other axioms with more characters, which were discovered earlier. Here are some:

L_1: $((p \to q) \to r) \to ((r \to p) \to (s \to p))$ (the shortest single axiom)

L_4: $((p \to q) \to (r \to s)) \to (t \to ((s \to p) \to (r \to p)))$

L_5: $((p \to q) \to (r \to s)) \to ((s \to p) \to (t \to (r \to p)))$

If we formulate the metatheory in first-order logic, formulas in implicational propositional calculus become terms, $\to$ becomes a binary function symbol, propositional variables become individual variables. We have a single monadic predicate Thm and a single meta-axiom formalizing the rule of *Detachment*:

$$D \equiv \forall p, q. \, \mathsf{Thm}(p) \wedge \mathsf{Thm}(p \to q) \supset \mathsf{Thm}(q)$$

The axiom of *Simplification* then becomes

$$S \equiv \forall p, q \, . \, \mathsf{Thm}(p \to (q \to p))$$

P and H are formalized similarly. In order to show that L_1 is a sufficiently strong single axiom, one has to derive

$$D \wedge \forall p, q, r, s \, \mathsf{Thm}(L_1) \supset S \wedge P \wedge H$$

The sortest known proof that L_1 implies H is given in the paper and consists of 29 applications of *Detachment*. Łukasiewicz notes (in 1947!):

A formalized proof can be checked mechanically but cannot be mechanically discovered. I do not know of any method of finding proofs in the Implicational Propositional Calculus than the method of "trial and error."

3. Problem Summary

I briefly summarize the definitions of the problems, writing the binary function symbol "→" in infix notation.

$$D \equiv \forall p, q \,.\, \mathsf{Thm}(p) \wedge \mathsf{Thm}(p \to q) \supset \mathsf{Thm}(q)$$

$$L_1 \equiv \forall p, q, r, s \,.\, \mathsf{Thm}(((p \to q) \to r) \to ((r \to p) \to (s \to p)))$$
$$L_4 \equiv \forall p, q, r, s, t \,.\, \mathsf{Thm}(((p \to q) \to (r \to s)) \to (t \to ((s \to p) \to (r \to p))))$$
$$L_5 \equiv \forall p, q, r, s, t \,.\, \mathsf{Thm}(((p \to q) \to (r \to s)) \to ((s \to p) \to (t \to (r \to p))))$$

$$I \equiv \forall p \,.\, \mathsf{Thm}(p \to p)$$
$$S \equiv \forall p, q \,.\, \mathsf{Thm}(p \to (q \to p))$$
$$P \equiv \forall p, q \,.\, \mathsf{Thm}(((p \to q) \to p) \to p)$$
$$H \equiv \forall p, q, r \,.\, \mathsf{Thm}((p \to q) \to ((q \to r) \to (p \to r)))$$

D is always used as a clause, and one of the L_i is selected as additional assumption. Then one tries to prove one I, S, P, and H, perhaps using one or more of the others as lemmas.

4. Problem Status

The following is a table indicating the status of various problems that arise as outlined above. These results are not intended to give a measure of power for the theorem provers involved. Rather, they are meant to indicate the order of difficulty of the problems posed. Only the default heuristics were used.

Label	Theorem	Prover A[1]			Prover B[2]		
		Length	Time	Inferences	Length	Time	Inferences
I_1	$L_1 \supset I$	11	70	374	20	28	12,745
S_1	$L_1 \supset S$	11	90	395	14	1	592
P_1	$L_1 \supset P$		failed		30	—[3]	2,352,964
IP_1	$L_1, I \supset P$	14	1585	10,383	18	—[3]	19,359
H_1	$L_1 \supset H$		failed			failed	
IH_1	$L_1, I \supset H$		failed			failed	
IPH_1	$L_1, I, P \supset H$		failed			failed	

The times are in cpu seconds on a Sun 3/260, "Length" is the length of the proof found, "Inferences" is the number of logical inferences done during proof search. "$\supset$"

---

[1]Release 0 of D. Plaisted's C Prolog theorem prover based on the simplified problem reduction format with default heuristics.

[2]Version 1e of M. Stickel's Prolog technology theorem prover in Sun Common Lisp with default heuristics.

[3]Comparable times not available, since run on different systems.

indicates which assumptions were used in addition to the clause D. The subscript to the label indicates which single axiom was used in the experiment (I only used L_1). "Failed" means that the theorem prover ran for several hours without finding a proof.

With respect to the still automatically unproven theorems, note that Łukasiewicz's proof has length 29, its longest formula has 31 characters, and the deepest nesting of implications is 5. Unfortunately we don't know how long he worked on the proof.

References

[1] Jan Łukasiewicz. The shortest axiom of the implicational calculus of propositions. *Proceedings of the Royal Irish Academy*, 52(3):25–33, April 1948.

[2] Jan Łukasiewicz. The shortest axiom of the implicational calculus of propositions. In L. Borowski, editor, *Jan Łukasiewicz, Selected Works*, pages 295–305, North-Holland, 1970.

Challenge Problems Focusing on Equality and Combinatory Logic: Evaluating Automated Theorem-Proving Programs*

Larry Wos   and   William McCune

Mathematics and Computer Science Division
Argonne National Laboratory
Argonne, IL 60439-4844

Abstract

In this paper, we offer a set of problems for evaluating the power of automated theorem-proving programs and the potential of new ideas. Since the problems published in the proceedings of the first CADE conference proved to be so useful, and since researchers are now far more disposed to implementing and testing their ideas, a new set of problems is in order to complement those that have been widely studied. In general, the new problems provide a far greater challenge for an automated theorem-proving program than those in the first set do. Indeed, to our knowledge, five of the six problems we propose for study have never been proved with a theorem-proving program. For each problem, we give a set of statements that can easily be translated into a standard set of clauses. We also state each problem in its mathematical and logical form. In many cases, we provide a proof of the theorem from which a problem is taken so that one can measure a program's progress in its attempt to solve the problem. Two of the theorems we discuss are of especial interest in that they answer previously open questions concerning the constructibility of two types of combinator. We also include a brief description of a new strategy for restricting the application of paramodulation. All of the problems we propose for study emphasize the role of equality. This paper is tutorial in nature.

1. Introduction

To estimate the possible value and power of an automated theorem-proving program or of a new approach, one needs various test problems. One cannot simply make diverse computations in the abstract about CPU cycles, conclusions drawn and discarded, conclusions drawn and retained, and such. Rather, one must attempt to solve problems from various areas—mathematics and logic, for example—with one's program or with the new approach. After all, the value of a discovery such as an inference rule or strategy rests mainly with its effectiveness for problem solving.

To determine how well a given program is doing in its attempt to prove some given theorem or solve some specified problem, one usually requires access to a proof or solution to

*This work was supported by the Applied Mathematical Sciences subprogram of the Office of Energy Research, U.S. Department of Energy, under contract W-31-109-Eng-38.

measure the program's progress. For, without knowing what the answer is, how can one estimate how close the program is to solving the assigned problem? Therefore, to facilitate and encourage the needed experimentation, we offer in this paper various test problems, and for each we include a solution for measuring progress. All of the problems we present emphasize the role of equality.

Each problem focuses on a theorem taken from combinatory logic. In general, the problems we propose for study are far more challenging than those usually used for evaluating a theorem-proving program or a new concept. As evidence of their difficulty, for almost all of them (from what we know) no proof has ever been obtained with a theorem-proving program. These problems are not only hard for a computer program to solve but, in many cases, also hard for a person to solve. One of the theorems (Theorem C3) answers a question that had been open, a question that concerns the constructibility of a particular type of combinator. Theorem C3 is also of interest in that it illustrates the excellent meld between automated theorem proving and combinatory logic, for its proof depends on various properties of unification.

For each problem, we shall first state it as a logician would. To make the presentation self-sufficient, we shall, where necessary, give the needed background. Except in Section 2.1, our discussion of the required concepts will be terse. In that section, we do give a rather lengthy treatment of the problem under discussion. We take this action in part to provide a sample of how one can proceed and to focus on various strategies for restricting paramodulation [RobinsonG69]—especially for those who are new to automated theorem proving—and in part to promote a sharp increase in experimentation in general. In particular, we include three short proofs of a simple theorem (Theorem C1.1) to illustrate the role of different strategies. We conjecture that, with the rapid growth in the interest in automated theorem proving, and with the new breed of researcher who is far more excited about implementing and testing ideas, the field is eager for a set of problems that includes some perhaps beyond the capability of any program now in existence.

To complement the mathematical and logical statement of a problem, we shall give a set of statements—abbreviated clauses—that can be used to submit the problem for attack by a theorem-proving program. If the paradigm on which a particular program is based does not rely on the use of clauses, the mathematical description that we give for each problem will make it possible to map the problem accordingly.

In all cases, we shall supply a mathematical proof, an outline of such a proof, or a proof in abbreviated clause notation—sometimes, more than one of the three. Since almost all of the problems we pose have the property that no proof, as far as we know, has ever been obtained with a theorem-proving program, we include no statistics concerning an attempt to obtain a solution with one of our programs. We would, of course, be very interested in any statistics obtained by a researcher who is successful in solving one of the posed problems—if the solution is obtained with a program. Such statistics provide an important measure of a problem's difficulty and of a program's effectiveness. As an example, the statistics found in the earlier paper [McCharen76]—that focusing on problems and experiments and published in the first CADE conference proceedings—have proved most useful.

In addition to our primary goal of encouraging researchers to test and evaluate programs, we have the secondary goal of causing others to supply various problems for this purpose.

Although an earlier attempt to stimulate such contributions clearly did not succeed, the changes evidenced in the past four years may be of sufficient magnitude that this new attempt will in fact succeed. Consistent with our stated goals, we plan to make available some time in the future a database of test problems, including those presented in this paper and others we use for study. This database will be accessible by electronic mail.

We focus on problems heavily emphasizing equality, in part because of the importance of this relation to so many possible applications of automated theorem proving, and in part because of a view we have concerning the history of automated theorem proving. In particular, were we sipping brandy and having a pleasant conversation with friends, we would suggest that automated theorem proving would have progressed far more rapidly had it not been for the dominant practice of treating equality as just another relation. Indeed, until relatively recently, a large fraction of the discussion, research, and experimentation focusing on problems in which equality naturally plays a vital role (from the viewpoint of mathematics and logic) was in terms of the so-called P-formulation. For example, to avoid the obstacles presented by directly coping with equality, the axiom of left identity in a group was almost always represented as

$P(e,x,x)$

rather than as

$f(e,x) = x$

which is unfortunate. Perhaps this practice was justified by the fact that the field had been in existence for only a few years; on the other hand, perhaps researchers should have been more aggressive. Now, at any rate, we recommend that, when the equality relation naturally dominates the description of a problem domain, the P-formulation be avoided where possible.

With this introduction in hand, let us now turn to a brief discussion of notation, and then to the field of combinatory logic from which we have taken the problems. Combinatory logic, one of the deepest areas of mathematics and logic, offers many problems to test—and perhaps surpass—a theorem-proving program's capacity to solve problems. This field also offers many opportunities to use such a program to make important contributions to mathematics and logic, and challenges one to add to the techniques used in automated theorem proving. With regard to contributions to mathematics and logic, we include problems taken from our successful attack on various previously open questions. To illustrate how one can add to the power of automated theorem-proving programs, we include a brief discussion of a new strategy which was formulated to increase the effectiveness of our programs when used for studying combinators of various types. The object of the new strategy is to sharply restrict the application of paramodulation. Since the strategy did in fact prove very useful for our studies of combinatory logic, it or a variant of it might be of use for studies of other fields of mathematics or logic.

2. Combinatory Logic

Before we discuss the first problem, let us supply the needed background, beginning with notation. In all of the problems we offer, as commented earlier, we heavily emphasize equality and equality-oriented notation. Because we wish to emphasize that the equality relation is treated as a built-in relation, we write = or ≠ between the arguments of a literal, rather than using a predicate such as EQUAL or ¬ EQUAL followed by its two arguments. In other words, we do not precisely present clauses to characterize each problem but, instead, use abbreviated clauses, which we simply call clauses. One can, of course, attack the problems we present within ordinary first-order predicate calculus, which is obviously necessary if one's program lacks the facility for treating equality as built in. Nevertheless, we strongly recommend that, if possible, the problems be considered within the extended first-order predicate calculus where equality does not require axiomatization. If one chooses to follow our recommendation, then one must of course include reflexivity as an axiom if paramodulation is the inference rule to be used. We in fact did use paramodulation heavily in our studies of combinatory logic, the field from which we have taken the problems.

Combinatory logic [Curry58,Curry72,Smullyan85,Barendregt81] is particularly significant for mathematics and logic because it is concerned with the most fundamental aspects of both fields. This field is also of potential interest to automated theorem proving because it challenges the researcher to formulate new strategies to control the reasoning needed to solve problems focusing on combinators and their various properties. In addition, combinatory logic offers one the intriguing opportunity of attempting to answer a number of currently open questions, many of which are amenable to attack with a theorem-proving program playing the role of research assistant. To pique one's curiosity, we shall list some of these open questions.

Combinatory logic can be viewed as an alternative foundation for mathematics—which was Curry's proposal—or viewed as a programming language. On the one hand, the logic offers the same generality and power as set theory in the sense that essentially all of mathematics can be embedded in it. On the other hand, any computable function can be expressed in combinatory logic, and the logic can be used as an alternative to the Turing machine. In fact, combinators have been used as the basis for the design of computers. This logic is concerned with the abstract notion of applying one function to another.

For a more formal definition, we can borrow from Barendregt who defines combinatory logic as a system satisfying the combinators S and K (defined shortly) with S and K as constants, and satisfying reflexivity, symmetry, transitivity, and two equality substitution axioms for the function that exists implicitly for applying one combinator to another. In other words, since the majority of combinatory logic can be studied strictly within the first-order predicate calculus, this logic is clearly within the province of automated theorem proving. Even further, although one can clearly operate within ordinary first-order predicate calculus and rely on inference rules such as hyperresolution and UR-resolution, we recommend that one instead study combinatory logic within the extended calculus and use paramodulation as the inference rule. (Of course, one might prefer to rely on an alternative inference rule for building in equality.) Therefore, to study the entire logic, we need only choose an appropriate function symbol, such as a to stand for "apply", and supply the axiom for reflexivity and those for the combinators S and K.

$x = x$
$a(a(a(S,x),y),z) = a(a(x,z),a(y,z))$
$a(a(K,x),y) = x$

Even though one can study all of combinatory logic in terms of S and K, one can also study the field or subsets of it by choosing other combinators to replace S and K. Indeed, our focus will shift from one set of combinators to another, depending on the type of problem to be studied. For each combinator of interest, we supply an equation that gives the behavior of the combinator. In such an equation, the combinator appears as a constant. Strictly speaking, a combinator is a member of a class of objects that exhibits the behavior given by its equation. For example, were we being more rigorous, we would say that if the combinator E satisfies

$(Ex)y = x$

for all combinators x and y, then E is a K since K satisfies

$(Kx)y = x,$

which we stated earlier in clause form. Therefore, when a problem asks for the construction of a combinator E from a set P of combinators, the object is to find an expression in terms of the elements of P that exhibits the behavior that E does. The solvability of such problems is one of the reasons that the system consisting of S and K alone is studied, for one can always succeed in finding the required expression. Formally, one says that the set consisting of S and K alone is *complete*.

To complete the background—especially for those who are new to automated theorem proving—we point out that, if one follows our recommendation of using paramodulation as the inference rule, one will directly encounter the impressive obstacle of coping with equality-oriented reasoning. Overcoming the difficulties inherent when equality plays a vital role is one of the most important research areas in the field, which is one reason for illustrating (with three proofs of Theorem C1.1) the use of different restriction strategies. The strategies for controlling the application of paramodulation include allowing or preventing paramodulation *from* a variable, *into* a variable, *from* the left side of an equality, *from* the right side of an equality, and *into* terms satisfying some given condition concerning their relative position within a statement. Since the advantages and disadvantages vary widely depending on which combination of strategies is employed, by discussing this aspect in the context of different proofs, we may succeed in increasing the interest in the corresponding research.

2.1. Problem 1

For the first problem in this section, we focus on one of the interesting properties, the *weak fixed point property*, that is sometimes present and sometimes absent for some given set P of combinators. From what we can tell, Raymond Smullyan deserves credit as the one to introduce and then study this property. His book *To Mock a Mockingbird* [Smullyan85] is an excellent source for problems and open questions, and a delight to read. We therefore need the following definition.

Definition. If P is a given set of combinators, then the *weak fixed point property* holds for P if and only if for all combinators x there exists a combinator y such that $y = xy$.

For this paper, because of the complexities that would otherwise be introduced, we can give only the following small hint about why the weak fixed point property and the to-be-defined strong fixed point property are of interest. Gödel's self-referential sentence and Kleene's recursion theorem can be interpreted as applications of fixed point combinators [Barendregt81]. Also, fixed point combinators were known as paradoxical combinators in the early days of combinatory logic, because the Russell Paradox and other paradoxes can be formulated in terms of fixed point combinators.

To study a combinator of the type in which we are interested in here, we need an equation giving the behavior of the combinator. We restrict our attention to combinators that are called *proper*; a combinator is proper if the left side of its equation is left associated and consists of the combinator followed by some nonempty list of distinct variables, and the right side consists of some or all of the variables that occur on the left side. For example, the combinator S is defined with the equation

$$((Sx)y)z = (xz)(yz),$$

which explains why we can, when studying S, use the second clause of the four clauses given earlier, where the function a is given explicitly to show that one combinator is being applied to another.

Theorem C1. The weak fixed point property holds for the set P consisting of the combinators S and K alone, where $((Sx)y)z = (xz)(yz)$ and $(Kx)y = x$.

Problem 1 asks for a proof of Theorem C1. The following clauses, in abbreviated notation, characterize this problem. (In contrast, combinatory logic does not explicitly employ a function symbol such as a and observes the convention that all expressions are left associated unless otherwise indicated.)

$$x = x$$
$$a(a(a(S,x),y),z) = a(a(x,z),a(y,z))$$
$$a(a(K,x),y) = x$$
$$y \neq a(f,y)$$

If one assigns a chosen automated theorem-proving program the task of finding a proof for some specific theorem—in particular, for Theorem C1—with the object of testing and evaluating the program, some means must exist for measuring the program's progress. The most obvious means—and perhaps the only significant one—focuses on what percentage of a proof has been found by the program. One must, therefore, have a proof in hand, or be able to complete a proof by using whatever information the program has found. To meet this requirement for Theorem C1, we shall, as promised earlier, supply our own proof. Before we give that proof, let us focus on some simpler problems to provide additional information that might prove useful for attacking Theorem C1 in various ways—Problem 1 admits a number of distinct solutions—and, even more, might prove useful for formulating general strategies. These simpler problems illustrate some of the interesting aspects of the coupling of strategy and inference rule. Each of the simpler problems focuses on proving Theorem C1.1, but proving it under different restrictions.

Theorem C1.1. The weak fixed point property holds for the set P consisting of the combinators S, B, C, and I, where $((Sx)y)z = (xz)(yz)$, $((Bx)y)z = x(yz)$, $((Cx)y)z = (xz)y$, and $Ix = x$.

As part of the background, one might find it interesting and useful to note that, just as S and K form a complete set of combinators for combinatory logic, S, B, C, and I form a complete set for that part of the logic known as noneliminating. One can find other complete sets of combinators for noneliminating combinatory logic by reading a general text on the subject [Curry58,Curry72,Smullyan85,Barendregt81]. A combinator is noneliminating if all of the variables that appear on the left side of its equation also appear at least once on the right side. In other words, from the set consisting of S, B, C, and I, one can construct a combinator of any desired type from these four combinators provided, of course, that the combinator to be constructed is noneliminating. Naturally, when a given set of combinators is complete for the entire logic or for a large fraction of the logic, one should expect to encounter various hazards when focusing on such a set. Among these hazards are the propensity for requiring the use of long expressions to complete a desired construction, the need to strongly consider permitting paramodulation *from* a variable, and the need to strongly consider permitting paramodulation *into* a variable. Each of the simpler problems focusing on proving Theorem C1.1, which we discuss on the path to giving a proof of Theorem C1, illustrates some of these obstacles.

The following six clauses can be used.

(1) $x = x$
(2) $a(a(a(S,x),y),z) = a(a(x,z),a(y,z))$
(3) $a(a(a(B,x),y),z) = a(x,a(y,z))$
(4) $a(a(a(C,x),y),z) = a(a(x,z),y)$
(5) $a(I,x) = x$
(6) $y \neq a(f,y)$

From this set of clauses, we can quickly and easily give three proofs of Theorem C1.1. Each of the proofs satisfies some given restriction on paramodulation and corresponds to one of the simple problems we promised to examine. We include all three proofs to illustrate the use of different strategies that one might consider using in other studies. Note that, in the first of the three, we use paramodulation in a fashion that is contrary to our usual recommendations for its application—in particular, paramodulation both *from* and *into* variables occurs. As compensation, we do not allow paramodulation *from* the left side of any equality, and only terms in negative clauses are allowed to be *into terms*. In other words, from a technical viewpoint, we place clause (6) only in the set of support and require every paramodulation to be related to what is called in combinatory logic an *expansion*. For the first proof, in fact, the strategy consists of using set of support and restricting paramodulation to expansions into terms confined to the second argument of an inequality.

Proof 1 of Theorem C1.1

(1) $x = x$
(2) $a(a(a(S,x),y),z) = a(a(x,z),a(y,z))$
(3) $a(a(a(B,x),y),z) = a(x,a(y,z))$
(4) $a(a(a(C,x),y),z) = a(a(x,z),y)$
(5) $a(I,x) = x$

(6) $y \neq a(f,y)$

from the second argument of clause (3) into term $a(f,y)$ of clause (6)

(7) $a(y,z) \neq a(a(a(B,f),y),z)$

from the second argument of clause (5) into the second occurrence of the term z in clause (7)

(8) $a(y,z) \neq a(a(a(B,f),y),a(I,z))$

Clause (8) and clause (2) form a unit conflict, which can be seen by letting the variables in clause (2) be x, u, and v in the order in which they occur, and applying the substitution $a(B,f)$ for x, I for u, and $a(a(S,a(B,f)),I)$ for v, y, and z. Therefore, a contradiction is obtained, and the proof is complete.

Proof 1 shows that indeed the weak fixed point property holds for the set P consisting of S, B, C, and I. Note that the combinator C plays no role in this proof.

To use the proof of Theorem C1.1 to obtain the value for the existentially quantified variable y that occurs in the definition of the weak fixed point property, one can adjoin the ANSWER literal—in this case, the literal ANSWER(y)—to the clause corresponding to the denial of the theorem. The ANSWER literal will contain at each point in the proof the current instantiation of the universally quantified variable y that exists because of assuming Theorem C1.1 false. One simply takes the argument of the ANSWER literal when unit conflict is found and, taking into account that we began by assuming the theorem false, replaces the constant f by the variable x. Summarizing, an examination of the unification that establishes unit conflict, the negation of the conclusion of Theorem C1.1, and the path that leads to the unit conflict shows that, for the existentially quantified y that one is seeking, one can choose the square of $((S(Bx))I)$.

As we see in the second proof we give shortly, we can avoid paramodulation *from* and *into* variables if we allow paramodulation *from* the left sides of the various clauses—allow paramodulations related to what are called *reductions* in combinatory logic. The avoidance of paramodulating *from* and *into* variables is often essential for—as is known to those how have experimented with paramodulation or other approaches to building in equality—allowing *from terms* or *into terms* to be variables usually destroys the effectiveness of a theorem-proving program. The reason for such total destruction, for those who may be new to this aspect of the field, is that variables always unify with any chosen expression. This property of never failing to unify would not necessarily be so damaging if a program could separate the needed deductions from the unneeded, but no one has come close as yet to discovering a strategy that produces anything resembling such a separation. Such a discovery would deserve and receive overwhelming acclaim, if it were ever made.

Proof 2 of Theorem C1.1

(1) $x = x$

(2) $a(a(a(S,x),y),z) = a(a(x,z),a(y,z))$

(3) $a(a(a(B,x),y),z) = a(x,a(y,z))$

(4) $a(a(a(C,x),y),z) = a(a(x,z),y)$

(5) $a(I,x) = x$

(6) $y \neq a(f,y)$

from the second argument of clause (3) into term $a(f,y)$ of clause (6)

(7) $a(y,z) \neq a(a(a(B,f),y),z)$

from the second argument of clause (2) into term $a(a(a(B,f),y),z)$ of clause (7)

(8) $a(v,a(u,v)) \neq a(a(a(S,a(B,f)),u),v)$

from the first argument of clause (5) into term $a(u,v)$ of clause (8)

(9) $a(v,v) \neq a(a(a(S,a(B,f)),I),v)$

Clause (9) unit conflicts with clause (1), and the proof is complete.

We can even get a proof that prevents paramodulation from using variables as *from terms* or as *into terms*, and also restricts it to the use of expansions only. However, where the first two proofs fail to use C, the third proof fails to use I.

Proof 3 of Theorem C1.1

(1) $x = x$

(2) $a(a(a(S,x),y),z) = a(a(x,z),a(y,z))$

(3) $a(a(a(B,x),y),z) = a(x,a(y,z))$

(4) $a(a(a(C,x),y),z) = a(a(x,z),y)$

(5) $a(I,x) = x$

(6) $y \neq a(f,y)$

from the second argument of clause (3) into term $a(f,y)$ of clause (6)

(7) $a(y,z) \neq a(a(a(B,f),y),z)$

from the second argument of clause (4) into term $a(a(a(B,f),y),z)$ of clause (7)

(8) $a(z,y) \neq a(a(a(C,a(B,f)),y),z)$

Clause (8) unit conflicts with clause (2)—which can be seen by naming the variables in clause (2) x, u, and v, and by substituting $a(C,a(B,f))$ for x, $a(S,a(C,a(B,f)))$ for u, v, and y, and $a(a(S,a(C,a(B,f))),a(S,a(C,a(B,f))))$ for z—and the proof is complete.

An analysis of this third proof shows that, for the y that must exist for the weak fixed point property to hold for Theorem C1.1, one can choose the cube of $S(C(Bx))$ in contrast to the square of $((S(Bx))I)$ which was found in the first two proofs. Because the set of combinators consisting of S and K alone is complete, we could use either value of y to lead us to a proof of Theorem C1 by expressing certain combinators in terms of S and K. In particular, we could construct two combinators that, respectively, act like B and I, or two that, respectively, act like B and C. One would then use either value of y found by proving Theorem C1.1, but replace the appropriate combinators by the corresponding expressions in terms purely of S and K. However, this indirect path is not what we have in mind when we suggest proving Theorem C1 as the first problem posed in this paper. That indirect path depends on either making a good guess about which other sets of combinators are (sufficiently) complete—for example, the set consisting of S, B, C, and I—or making a good guess about which sets have the weak fixed point property. Because we had found the two given values of *y* in other studies, we were able to pursue the indirect path to proving Theorem C1 quickly and easily with the program ITP. But that is not the objective of Problem 1. Instead, we suggest that the theorem-proving program one is evaluating should attempt to prove Theorem C1 directly—using as axioms that for S, that for K, and that for reflexivity—by simply denying that the weak fixed point property holds for the set P consisting of S and K alone. Our attempt to obtain a computer proof of Theorem C1 along this direct path of inquiry failed, which is one reason why we consider Problem 1 to be an interesting challenge.

Since we are suggesting problems as challenges for various theorem-proving programs rather than for theorem-proving people, we shall complete a proof of Theorem C1 rather than assigning that task to the researcher. We shall use Theorem C1.1 in our proof despite the preceding remarks. We take this action to increase the likelihood of independent and uninfluenced experimentation and, as one might suspect, because of certain pedagogical considerations. We shall employ algebraic notation rather than clause notation for this proof.

A Proof of Theorem C1

From Proof 1 of Theorem C1.1, one can conclude that

(1) $((S(Bx))I)((S(Bx))I) = x(((S(Bx))I)((S(Bx))I))$

is true for all x. One can check this equality by simply applying the equations (given as clauses (2), (3), and (5) in Section 2.1) for S, B, and I to the left side of (1) to obtain the right side. Next, one can prove that

(2) $((SK)K)x = x$

for all x, which translates to the statement that (SK)K is an I, or, equivalently, (SK)K behaves as I does. Then one can show that

(3) $(((((S(KS))K)x)y)z = x(yz)$

for all x, y, and z. Equation (3) says that (S(KS))K is a B, or, equivalently, behaves like B. Because of equations (2) and (3) and the remarks made concerning their meaning, we can in effect substitute into equation (1) for both B and I to obtain

(4) $((S(((S(KS))K)x))((SK)K))((S(((S(KS))K)x))((SK)K)) =$
    $x(((S(((S(KS))K)x))((SK)K))((S(((S(KS))K)x))((SK)K))),$

which holds for all x, and the proof of Theorem C1 is complete.

We can improve on the result contained in the proof of Theorem C1 by presenting a simpler value for the y that must exist satisfying $y = xy$, the equation that defines the weak fixed point property. In particular, if we let y be the square of $(S(S(Kx)))((SK)K)$, we can apply the equations for S and for K and show that this y also satisfies the equation for the weak fixed point property. The simpler y can be found by taking the term Bx in the square of $(S(Bx))I$ and using equation (3) to write B in terms of S and K, then reducing both occurrences of the replaced term with S, and finally using equation (2) to write I in terms of S and K. We therefore have two solutions to Problem 1, and could even obtain a third by focusing on the cube of S(C(Bx)). In the context of Problem 1 and its given solutions, we can immediately pose one of the promised open questions. Does the second solution contain the shortest expression for a y satisfying the weak fixed point property, where y is expressed purely in terms of S, K, and a variable?

Having finished with Problem 1, we can turn in the next section to the second problem we suggest for testing and evaluating theorem-proving programs. However, in contrast to the treatment we have just given Problem 1, we shall be far briefer from here on, confining the discussion in most cases to the statement of the problem and a short proof in logical or mathematical terms. The copious details we have given regarding Theorem C1.1 can be used as a guide for interpreting the clauses obtained when attempting to solve later problems. They are included also to provide an example of how one can map a logical or mathematical proof into clause notation.

2.2. Problem 2

The second problem we pose asks one to use an automated theorem-proving program to find a proof of Theorem C2, which we state after introducing another new concept. For this problem, we need the definition of the *strong fixed point property*.

Definition. If P is a given set of combinators, then the *strong fixed point property* holds for P if and only if there exists a combinator y such that, for all combinators x, $yx = x(yx)$.

Theorem C2. The strong fixed point property holds for the set P consisting of the combinators B and W alone, where $((Bx)y)z = x(yz)$ and $(Wx)y = (xy)y$.

To add to the background for studying problems from combinatory logic, we note that the presence for P of the strong fixed point property implies the presence of the weak fixed point property. The converse is not true, as one can see by considering the combinator L with

$(Lu)v = u(vv),$

noting that the expression $(Lx)(Lx)$ is a y that satisfies the equation for the weak fixed point property which establishes that that property holds for the set P consisting of L alone, and by using Theorem C3 to show that the strong fixed point property does not hold for this set P.

The following clauses can be used in the attempt to have a theorem-proving program solve Problem 2.

(1) $x = x$
(2) $a(a(a(B,x),y),z) = a(x,a(y,z))$
(3) $a(a(W,x),y) = a(a(x,y),y)$
(4) $a(y,f(y)) \neq a(f(y),a(y,f(y)))$

Even though we recommend this set of clauses for studying Problem 2, we now give a proof more in the style that an algebraist might give.

A Proof of Theorem C2

Let $N = ((B((B((B(WW))W))B))B)$. Since, with the following sequence of equalities, we can show that $Nx = x(Nx)$ for all x, we can set y equal to N to complete our proof. To obtain the sequence, we begin with Nx, occasionally abbreviate (W(B(Bx))) to R, apply the reduction corresponding to B or that corresponding to W depending on the leading symbol of the expression under consideration, and substitute from an intermediate result to deduce the final step.

Nx = ((B((B((B(WW))W))B))B)x = ((B((B(WW))W))B)(Bx) =
((B(WW))W)(B(Bx)) = (WW)(W(B(Bx))) =
(W(W(B(Bx))))(W(B(Bx))) = ((W(B(Bx)))(W(B(Bx))))(W(B(Bx))) = (RR)R =
(((B(Bx))R)R)R = ((Bx)(RR))R = x((RR)R) = x(Nx)

Here we have an example of how automated theorem proving differs sharply from mathematics. Specifically, a theorem-proving program has no way to magically offer an expression, such as N, for use in completing a proof. The mathematician, on the other hand, often exhibits the disarming capacity to make such offers; the offers are based on experience, intuition, and who knows what else. This dichotomy between the approach that apparently must be taken by a theorem-proving program and that which is frequently taken by a mathematician is precisely why the two make a powerful team for solving problems and answering open questions. Indeed, the combinator N, which answered a question that was once open, is just such an example of effective teamwork. This combinator was discovered while we were studying B and W with the assistance of various theorem-proving programs designed and implemented by members of our group.

That same study also led us to formulate a new strategy, mentioned earlier, for sharply restricting the application of paramodulation. The strategy restricts paramodulation to considering an *into term* only if its position vector consists of all 1's; we express this restriction by saying that all paramodulation steps must satisfy the 1's rule. The position vector of a term gives the position of the term within a literal. For example, the position vector [2,3,1] says that the corresponding term is the first subterm of the third subterm of the second argument. For a concrete illustration of the use of position vectors, the third occurrence of the constant W in the equality

a(a(B,a(a(B,a(a(B,a(W,W)),W)),B)),B) = N

has the position vector [1,1,2,1,2,2]. (The fixed point combinator N played a vital role in our discovery of what turned out to be an astoundingly large family of combinators.) We find this strategy of restricting *into terms* to be chosen from those whose position vector consists of all 1's to be very effective for our studies of combinatory logic. The strategy, as we discovered some time after its formulation, focuses on a generalization of what is known in combinatory logic as a *head reduction*.

2.3. Problem 3

For the third problem, we focus on Theorem C3, a theorem we proved to answer a previously open question concerning the possible constructibility, from the combinators B and L alone, of a fixed point combinator. For the problem under discussion, we depart rather sharply from our usual practice of focusing on proof by contradiction by suggesting instead that an automated theorem-proving program be used to find a model pertinent to Theorem C3. If the program succeeds in finding such a model, then, in a sense that will become obvious upon reading the statement of Theorem C3, the program will have found a proof of that theorem.

Theorem C3. The strong fixed point property fails to hold for the set P consisting of the combinators B and L alone, where $((Bx)y)z = x(yz)$ and $(Lx)y = x(yy)$. Equivalently, from B and L alone, one cannot construct a Q such that $Qx = x(Qx)$ for all x.

A Proof of Theorem C3

Assume, by way of contradiction, that the strong fixed point property holds for B and L. Then there exists a combinator Q, which is constructed from B and L alone, such that, for an arbitrary combinator f, $Qf = f(Qf)$. (We use the constant f rather than F to be consistent with our notational convention when denying some theorem is true.) By the Church-Rosser property for combinatory logic, there exists a combinator E such that Qf reduces to E and $f(Qf)$ also reduces to E. (The reductions that are used are paramodulations from the left sides of B and L.) Since $f(Qf)$ reduces to E, and since the first occurrence of f cannot be affected by any reduction with B or L, E must be of the form fT for some combinator T. Therefore, Qf reduces to fT for that same T. The combination Qf obviously has the form CD, where C contains no occurrences of f. Let us consider a one-step reduction of CD and show by case analysis that the result C′D′ is such that C′ contains no occurrences of f.

Case 1. The reduction involves C only or D only. Obvious.

Case 2. The reduction is with B and involves both C and D. Then C must unify with (Bx)y, D must unify with z, C′ must be the image of x, and D′ must be the image of yz. Therefore, C′ must be a subterm of C, which implies that C′ contains no occurrences of f.

Case 3. The reduction is with L and involves both C and D. Then C must unify with Lx, D must unify with y, C′ must be the image of x, and D′ must be the image of yy. Therefore, C′ must be a subterm of C, which implies that C′ contains no occurrences of f.

We can conclude that, regardless of the number of reductions we apply starting with CD = Qf, we can never obtain a combinator of the form C*D* with C* containing an occurrence of f. In particular, we can never reduce Qf to fT, and we have arrived at a contradiction. In other words, the strong fixed point property fails to hold for the set P consisting of B and L alone.

The object of Problem 3 is to find a model that satisfies B and L but fails to satisfy the strong fixed point property. Of course, such a model would show that indeed the strong fixed point property does not hold for the set consisting of B and L alone. Problem 3 has added interest since, as far as we know, no one has yet succeeded in finding such a model—in other words, Problem 3 is an open problem. Of course, since we have just proved Theorem C3, a model with the desired properties must exist. Before we had proved Theorem C3, as commented earlier, the question focusing on the constructibility of a fixed point combinator from B and L alone was open.

An alternative to Problem 3 asks for an automated theorem-proving program to find a proof of Theorem C3 directly, starting with its denial and proceeding in the standard fashion in our field. Such an achievement would be of great interest since the proof we give is outside first-order predicate calculus. However, various researchers in the field have discussed the possibility of using an automated theorem-proving program to prove theorems of this type—theorems whose proof depends on properties of unification, and theorems about unification.

2.4. Problem 4

For Problem 4, we focus on one of the systems of combinators that is known to be complete. As commented earlier, the set P consisting of the combinators S and K is one of those systems—given a combinator E and an equation that characterizes its behavior, one can construct from S and K alone a combinator that behaves as E does. For such a construction, one can apply the well-known algorithm found in Smullyan's book. Alternatively—as is standard in our field—one could have a theorem-proving program attempt such a construction by denying that any expression exists satisfying the equation given for the combinator E under study. If the program succeeds with this approach, then—as occurred in the various proofs of Theorem C1.1—the desired construction is obtained by analyzing the unifications upon which the proof rests.

The object of Problem 4 is to construct from S and K alone, by following the standard approach in automated theorem proving rather than by applying the well-known algorithm for such constructions, a combinator that behaves as the combinator U does, where the equation

$$(Ux)y = y((xx)y)$$

gives the behavior of U for all x and all y. The idea is to proceed as we illustrated in Section 2.1 and extract the construction from a proof by contradiction. One can use the following clauses.

(1) $x = x$
(2) $a(a(a(S,x),y),z) = a(a(x,z),a(y,z))$
(3) $a(a(K,x),y) = x$
(4) $a(a(z,f(z)),g(z)) \neq a(g(z),a(a(f(z),f(z)),g(z)))$

Similar to our earlier approach, we shall simply give two answers to Problem 4, rather than giving a proof relying on these four clauses. If one affixes the variables x and y to either of the following expressions, and if one then reduces with S and K, one can see that both expressions do indeed behave like U.

$$((S((S(KS))K))((S(K(S((S((SK)K))((SK)K)))))K))$$

$$((S(K(S((SK)K))))((S((SK)K)X(SK)K)))$$

The first of the two expressions can be found with the algorithm for using S and K for such constructions; the second can be found by noting that the combination LO behaves like U, and then reducing a combinator that behaves like LO, where $(Lx)y = x(yy)$ and $(Ox)y = y(xy)$. Question: Is there a combinator, expressed purely in terms of S and K, containing fewer than 13 symbols that satisfies the equation for U?

2.5. Problem 5

Problem 5 focuses on the combinators S and W.

$$((Sx)y)z = (xz)(yz)$$
$$(Wx)y = (xy)y$$

Problem 5, as with Problem 3, has the object of finding a model. The model one is seeking must satisfy S and W and fail to satisfy the weak fixed point property. The following clauses can be used to search for such a model.

(1) $x = x$

(2) $a(a(a(S,x),y),z) = a(a(x,z),a(y,z))$

(3) $a(a(W,x),y) = a(a(x,y),y)$

(4) $y \neq a(f,y)$

Rather than giving a complete proof of the theorem that corresponds to Problem 5, we are content with the following outline. To see that the set consisting of S and W alone does not satisfy the weak fixed point property, we again rely on the Church-Rosser property for combinatory logic. In particular, if the weak fixed point property does hold, then there must exist an E and a T such that both T and fT reduce to E, where f is an arbitrary combinator. The number of leading f's in T is one less than the number in fT. Each so-called reduction with S or W does not increase the number of leading f's in T. Therefore, even with different reduction paths, no expression can exist such that both T and fT reduce to it, which contradicts the existence of E, and the proof outline is complete.

Both the proof we have just outlined and the result concerning S and W, as far as we know, represent a new result in combinatory logic.

2.6. Problem 6

Problem 6 focuses on the combinators S and K.

$((Sx)y)z = (xz)(yz)$

$(Kx)y = x$

The object of Problem 6 is to prove that the strong fixed point property holds for the set P consisting of S and K alone. An appropriate combinator can be found by obtaining a refutation of the following clauses.

(1) $x = x$

(2) $a(a(a(S,x),y),z) = a(a(x,z),a(y,z))$

(3) $a(a(K,x),y) = x$

(4) $a(y,f(y)) \neq a(f(y),a(y,f(y)))$

We give three fixed point combinators that are in effect solutions to Problem 6. We know of no shorter combinator than the one we list third. For readability, we use abbreviated notation with the following abbreviations.

$I = (SK)K, \quad M = (SI)I, \quad B = (S(KS))K, \quad W = (SS)(SK).$

Here are the three solutions.

$((S(K((SI)I)))((S(KW))B))$

$((S(KM))((SB)(KM)))$

$((S(K(((SS)I)W)))B)$

3. Conclusions

One of the most important activities in automated theorem proving is that of experimenting with various problems taken from mathematics and logic. Experimentation is essentially the only way to measure the power of an automated theorem-proving program or the value of a new idea for increasing that power. In this paper, for such experiments, we focused on problems taken from combinatory logic, a field that is unusually amenable to attack with a theorem-proving program. Other areas from which problems can profitably be taken include ring theory (associative and nonassociative), lattice theory, and the algebra of regular expressions. We also have included various open questions since such questions often promote and provoke experimentation. Our emphasis throughout this paper is on equality. Coping effectively with the equality relation is still one of the major obstacles in the field of automated theorem proving.

References

[Barendregt81] Barendregt, H. P., *The Lambda Calculus: Its Syntax and Semantics*, North-Holland, Amsterdam (1981).

[Curry58] Curry, H. B., and Feys, R., *Combinatory Logic I*, North-Holland, Amsterdam (1958).

[Curry72] Curry, H. B., Hindley, J. R., and Seldin, J. P., *Combinatory Logic II*, North-Holland, Amsterdam (1972).

[McCharen76] McCharen, J., Overbeek, R., and Wos, L., "Problems and experiments for and with automated theorem proving programs", *IEEE Transactions on Computers* **C-25,** pp. 773-782 (1976).

[RobinsonG69] Robinson, G., and Wos, L., "Paramodulation and theorem-proving in first-order theories with equality", pp. 135-150 in *Machine Intelligence 4,* ed. B. Meltzer and D. Michie, Edinburgh University Press, Edinburgh (1969).

[Smullyan85] Smullyan, R., *To Mock a Mockingbird,* Alfred A. Knopf, New York (1985).

Challenge Problems from Nonassociative Rings for Theorem Provers [*]

Rick L. Stevens
Argonne National Laboratory
9700 South Cass Avenue
Argonne IL 60439

Abstract

The Moufang Identities are proving to be a challenging set of problems for automated theorem proving programs. Aside from one program that uses a new technique that has the axioms of nonassociative ring theorem built in, I know of no other program able to prove these identities. In this short paper I include the axioms for nonassociative rings, statements of the five moufang identities, a natural hand proof of one of the left identities and a human guided paramodulation proof of the same identity. I hope that this paper will provide a starting point for others to attack these interesting problems.

1 Introduction

In [5] several problems in nonassociative ring theory were presented. One of the more difficult problems was proving the family of Moufang identities. The Moufang identities are very useful in proving theorems about nonassociative rings[3,4]. So far these identities have resisted a completely automated attack with conventional theorem provers, although the recent Z-Module theorem prover [6,7] can do them easily. Since the Z-Module system is not strictly an equality only based system we present these problems as hard problems for "conventional" equality theorem proving.

2 Introduction to Nonassociative Rings

To get the axioms of nonassociative ring theory, start with the axioms for ring theory and replace the associative law for multiplication with the right and left alternative laws. The right alternative law says that $(xy)y = x(yy)$ for all x and y. The left alternative is $(xx)y = x(xy)$ for all x and y. In my representation "f" is product, "j" is sum and "g" is additive inverse. I also introduce notation for associators and commutators. "a" is the *associator* which is the 3–place function $(x, y, z) = (xy)z - x(yz)$ that is a measure of associativity (and lack of it) in the ring and "c" is the *commutator* $[x, y] = xy - yx$ which measures commutativity (and lack of it).

3 Moufang Identities

Because of the flexible law [3], with a a constant xax can be written unambiguously. In a nonassociative ring the following identities hold

---

[*]This research supported in part by the Applied Mathematical Sciences subprogram of the office of Energy Research, U.S. Department of Energy, under contract W-31-109-Eng-38.

$$x(yzy) = [(xy)z]y \quad \text{right Moufang identity,}$$
$$(yzy)x = y[z(yx)] \quad \text{left Moufang identity,}$$
$$(xy)(zx) = x(yz)x \quad \text{middle Moufang identity.}$$

In addition there are two identities in associators that are equivalent to the right and left Moufang identities, respectively.

$$(x, xy, z) = (x, y, z)x \quad \text{right Moufang identity,}$$
$$(x, yx, z) = x(x, y, z) \quad \text{left Moufang identity.}$$

The clauses for the Moufang identities are:

```
m1.   f(x,f(y,f(z,y))) = f(f(f(x,y),z),y) ;  * right Moufang  ;
m2.   f(f(y,f(z,y)),x) = f(y,f(z,f(y,x))) ;  * left Moufang   ;
m3.   f(f(x,y),f(z,x)) = f(f(x,f(y,z)),x) ;  * middle Moufang ;
m1'.  a(x,f(x,y),z) = f(a(x,y,z),x) ;  * right Moufang in assoc ;
m2'.  a(x,f(y,x),z) = f(x,a(x,y,z)) ;  * left Moufang in assoc ;
```

4   Manual Proof

To give a clue about how to proceed with these problems here is a manual proof of the left identity, the proofs of the others are similar.

Proof of left identity $(xax)y - x(a(xy)) = 0$.

Let's prove the left hand side = 0.

$$(xax)y - x(a(xy))$$

First add $(xa)(xy) - (xa)(xy) = 0$ giving $((xa)x)y - (xa)(xy) + (xa)(xy) - x(a(xy))$ then go to the associator form giving

$$(xa, x, y) + (x, a, xy).$$

Permute elements within the associators using $(x, y, z) = -(x, z, y)$ to get

$$-(x, xa, y) - (x, xy, a).$$

Expand the associators producing

$$-(x(xa))y + x((xa)y) - (x(xy))a + x((xy)a)$$

then factor x from positive terms and apply left alternative law to the negative terms with the result

$$-((xx)a)y - ((xx)y)a + x((xa)y + (xy)a).$$

Add $(xx)(ay) - (xx)(ay) = 0$ and $(xx)(ya) - (xx)(ya) = 0$ then go to the associator form

$$-(xx, a, y) - (xx, y, a) - (xx)(ay) - (xx)(ya) + x((xa)y + (xy)a)$$

since $(xx, a, y) + (xx, y, a) = 0$, we have $-(xx)(ay) - (xx)(ya) + x((xa)y + (xy)a)$

apply the left alternative law and factor x to get

$$x(-x(ay) - x(ya) + (xa)y + (xy)a)$$

regroup the summands to produce two associators

$$x((x, a, y) + (x, y, a))$$

but $(x, a, y) + (x, y, a) = 0$. QED.

This hand proof should give the reader enough detailed understanding of the manipulations typically applied in nonassociative rings for one to follow the machine checked proof.

5 Axioms used in Computer Proof

The axioms given here are slightly different from those in [5]. A number of lemmas (clauses 26-34) were added to attempt to provide the needed basic operations for completing the proof. The clause identifiers also match those in the computer proof.

```
 1. x1 = x1                              * functional reflexive ;
 2. j(x1,x2) = j(x2,x1)                  * Commutativity for addition ;
 3. j(x1,j(x2,x3)) = j(j(x1,x2),x3)      * Associativity for addition ;
 4. j(x1,0) = x1                         * Additive identity ;
 5. j(0,x1) = x1
 6. j(x1,g(x1)) = 0                      * definition of inverse ;
 7. j(g(x1),x1) = 0
 8. g(0) = 0
 9. j(g(x1),j(x1,x2)) = x2
10. j(x1,j(g(x1),x2)) = x2
11. g(j(x1,x2)) = j(g(x2),g(x1))
12. g(g(x1)) = x1
13. f(x1,0) = 0                          * multiplicative zero ;
14. f(0,x1) = 0
15. f(g(x1),g(x2)) = f(x1,x2)
16. f(g(x1),x2) = g(f(x1,x2))
17. f(x1,g(x2)) = g(f(x1,x2))
18. f(x1,j(x2,x3)) = j(f(x1,x2),f(x1,x3))
* Distributivity of addition over multiplication;
19. f(j(x1,x2),x3) = j(f(x1,x3),f(x2,x3))
20. f(x1,j(x2,g(x3))) = j(f(x1,x2),g(f(x1,x3)))
21. f(j(x1,g(x2)),x3) = j(f(x1,x3),g(f(x2,x3)))
22. f(g(x1),j(x2,x3)) = j(g(f(x1,x2)),g(f(x1,x3)))
23. f(j(x1,x2),g(x3)) = j(g(f(x1,x3)),g(f(x2,x3)))
* Distributivity of subtraction;
24. f(f(x1,x2),x2) = f(x1,f(x2,x2))
25. f(f(x1,x1),x2) = f(x1,f(x1,x2))
 *------------ lemmas to get from step 1 to 2 -----------;
26. a(x1,x2,x3) = j(f(f(x1,x2),x3),g(f(x1,f(x2,x3))))
27. f(f(x1,x2),x3) = j(a(x1,x2,x3),f(x1,f(x2,x3)))
28. g(f(x1,f(x2,x3))) = j(g(f(f(x1,x2),x3)),a(x1,x2,x3))
 *---------- lemmas to get from step 2 to 3 --------------;
29. a(f(x1,x2),x3,x4) = g(a(x3,f(x1,x2),x4))
30. a(f(x1,x2),x3,x4) = g(a(x4,x3,f(x1,x2)))
31. a(x1,f(x2,x3),x4) = g(a(f(x2,x3),x1,x4))
32. a(x1,f(x2,x3),x4) = g(a(x1,x4,f(x2,x3)))
33. a(x1,x2,f(x3,x4)) = g(a(f(x3,x4),x2,x1))
34. a(x1,x2,f(x3,x4)) = g(a(x1,f(x3,x4),x2))
```

6 Computer Checked Proof of Moufang identity

The following set of clauses are a proof of the left Moufang identity obtained by guiding ITP [2] manually through each paramodulation [1] step. ITP without such guidance fails to find a proof. This problem is a good example of a proof that is fairly easy for a human but very difficult for traditional equality based systems. In producing this proof I used the manual proof given earlier as a guide. A different type of challenge for automated provers would be to automatically allow the use of such proof outlines to guide it towards finding a proof.

```
{ clauses 1 - 34 are input axioms }
 1.  x1 = x1
 2.  j(x1,x2) = j(x2,x1)
 3.  j(x1,j(x2,x3)) = j(j(x1,x2),x3)
18.  f(x1,j(x2,x3)) = j(f(x1,x2),f(x1,x3))
```

```
20.  f(x1,j(x2,g(x3))) = j(f(x1,x2),g(f(x1,x3)))
25.  f(f(x1,x1),x2) = f(x1,f(x1,x2))
26.  a(x1,x2,x3) = j(f(f(x1,x2),x3),g(f(x1,f(x2,x3))))
27.  f(f(x1,x2),x3) = j(a(x1,x2,x3),f(x1,f(x2,x3)))
28.  g(f(x1,f(x2,x3))) = j(g(f(f(x1,x2),x3)),a(x1,x2,x3))
29.  a(f(x1,x2),x3,x4) = g(a(x3,f(x1,x2),x4))
30.  a(f(x1,x2),x3,x4) = g(a(x4,x3,f(x1,x2)))
32.  a(x1,f(x2,x3),x4) = g(a(x1,x4,f(x2,x3)))
34.  a(x1,x2,f(x3,x4)) = g(a(x1,f(x3,x4),x2))
```

{ clause 35 is the denial of the theorem }
```
35.  j(f(f(f(f(B,C),B),A),g(f(B,f(C,f(B,A))))) != 0
```

{ clauses 37 - 54 are demodulators }
```
37.  j(x1,0) = x1
40.  j(g(x1),x1) = 0
43.  j(x1,j(g(x1),x2)) = x2
44.  g(j(x1,x2)) = j(g(x2),g(x1))
45.  g(g(x1)) = x1
46.  f(x1,0) = 0
54.  j(a(x1,x2,x3),a(x1,x3,x2)) = 0
```

{ clauses 61 - 545 are derived from above }
```
61.  j(j(a(f(B,C),B,A),f(f(B,C),f(B,A))),g(f(B,f(C,f(B,A))))) != 0 From:  35 27
77.  j(a(f(B,C),B,A),j(f(f(B,C),f(B,A)),g(f(B,f(C,f(B,A)))))) != 0 From:  61 3
133. j(a(f(B,C),B,A),a(B,C,f(B,A))) != 0 From:  77 28 43
137. j(g(a(B,f(B,C),A)),a(B,C,f(B,A))) != 0 From:  133 29
143. j(g(a(B,f(B,C),A)),g(a(B,f(B,A),C))) != 0 From:  137 34
148. j(g(a(B,f(B,C),A)),j(f(B,f(f(B,A),C)),g(f(f(B,f(B,A)),C)))) != 0 From:  143 26 44 45
151. j(j(f(B,f(f(B,C),A)),g(f(f(B,f(B,C)),A))),j(f(B,f(f(B,A),C)),g(f(f(B,f(B,A)),C)))) != 0
     From:  148 26 44 45
161. j(j(g(f(f(B,f(B,C)),A)),f(B,f(f(B,C),A))),j(f(B,f(f(B,A),C)),g(f(f(B,f(B,A)),C)))) != 0
     From:  151 2
174. j(j(g(f(f(B,f(B,C)),A)),f(B,f(f(B,C),A))),j(g(f(f(B,f(B,A)),C)),f(B,f(f(B,A),C)))) != 0
     From:  161 2
178. j(j(g(f(f(f(B,B),C),A)),f(B,f(f(B,C),A))),j(g(f(f(B,f(B,A)),C)),f(B,f(f(B,A),C)))) != 0
     From:  174 25
194. j(j(j(g(f(f(f(B,B),C),A)),f(B,f(f(B,C),A))),g(f(f(B,f(B,A)),C))),f(B,f(f(B,A),C))) != 0
     From:  178 3
198. j(j(j(f(B,f(f(B,C),A)),g(f(f(f(B,B),C),A))),g(f(f(B,f(B,A)),C))),f(B,f(f(B,A),C))) != 0
     From:  194 2
208. j(j(j(f(B,f(f(B,C),A)),g(f(f(f(B,B),C),A))),g(f(f(f(B,B),A),C))),f(B,f(f(B,A),C))) != 0
     From:  198 25
221. j(f(B,f(f(B,A),C)),j(j(f(B,f(f(B,C),A)),g(f(f(f(B,B),C),A))),g(f(f(f(B,B),A),C)))) != 0
     From:  208 2
230. j(f(B,f(f(B,A),C)),j(f(B,f(f(B,C),A)),j(g(f(f(f(B,B),C),A)),g(f(f(f(B,B),A),C))))) != 0
     From:  221 3
240. j(j(f(B,f(f(B,A),C)),f(B,f(f(B,C),A))),j(g(f(f(f(B,B),C),A)),g(f(f(f(B,B),A),C)))) != 0
     From:  230 3
244. j(f(B,j(f(f(B,A),C),f(f(B,C),A))),j(g(f(f(f(B,B),C),A)),g(f(f(f(B,B),A),C)))) != 0
     From:  240 18
257. j(f(B,j(f(f(B,A),C),f(f(B,C),A))),j(j(g(f(f(B,B),f(C,A))),g(a(f(B,B),C,A))),
     g(f(f(f(B,B),A),C)))) != 0 From:  244 27 44
273. j(f(B,j(f(f(B,A),C),f(f(B,C),A))),j(j(g(f(f(B,B),f(C,A))),g(a(f(B,B),C,A))),
     j(g(f(f(B,B),f(A,C))),g(a(f(B,B),A,C))))) != 0 From:  257 27 44
283. j(f(B,j(f(f(B,A),C),f(f(B,C),A))),j(j(g(f(B,f(B,f(C,A)))),g(a(f(B,B),C,A))),
     j(g(f(f(B,B),f(A,C))),g(a(f(B,B),A,C))))) != 0 From:  273 25
309. j(f(B,j(f(f(B,A),C),f(f(B,C),A))),j(j(g(f(B,f(B,f(C,A)))),g(a(f(B,B),C,A))),
     j(g(f(f(B,B),f(A,C))),a(C,A,f(B,B))))) != 0 From:  283 30 45
324. j(f(B,j(f(f(B,A),C),f(f(B,C),A))),j(j(g(f(B,f(B,f(C,A)))),a(C,f(B,B),A)),
     j(g(f(f(B,B),f(A,C))),a(C,A,f(B,B))))) != 0 From:  309 29 45
```

```
341.  j(f(B,j(f(f(f(B,A),C),f(f(B,C),A))),j(j(g(f(f(B,B),f(C,A))),a(C,f(B,B),A)),
      j(g(f(f(B,B),f(A,C))),a(C,A,f(B,B)))))) != 0 From:  324 25
362.  j(f(B,j(f(f(f(B,A),C),f(f(B,C),A))),j(j(g(f(f(B,B),f(C,A))),g(a(C,A,f(B,B)))),
      j(g(f(f(B,B),f(A,C))),a(C,A,f(B,B)))))) != 0 From:  341 32
388.  j(f(B,j(f(f(f(B,A),C),f(f(B,C),A))),j(j(g(f(f(B,B),f(C,A))),g(a(C,A,f(B,B)))),
      j(a(C,A,f(B,B)),g(f(f(B,B),f(A,C)))))) != 0 From:  362 2
410.  j(f(B,j(f(f(f(B,A),C),f(f(B,C),A))),j(j(j(g(f(f(B,B),f(C,A))),g(a(C,A,f(B,B)))),
      a(C,A,f(B,B))),g(f(f(B,B),f(A,C))))) != 0 From:  388 3
433.  j(j(f(B,j(f(f(f(B,A),C),f(f(B,C),A))),j(j(g(f(f(B,B),f(C,A))),g(a(C,A,f(B,B)))),
      a(C,A,f(B,B)))),g(f(f(B,B),f(A,C))))) != 0 From:  410 3
449.  j(j(f(B,j(f(f(f(B,A),C),f(f(B,C),A))),g(f(f(B,B),f(C,A)))),g(f(f(B,B),f(A,C)))) != 0
      From:  433 3 40 37
460.  j(j(f(B,j(f(f(f(B,A),C),f(f(B,C),A))),g(f(B,f(B,f(C,A))))),g(f(f(B,B),f(A,C)))) != 0
      From:  449 25
472.  j(f(B,j(j(f(f(f(B,A),C),f(f(B,C),A)),g(f(B,f(C,A)))),g(f(f(B,B),f(A,C)))) != 0
      From:  460 20
482.  j(f(B,j(f(f(f(B,A),C),j(f(f(B,C),A),g(f(B,f(C,A))))),g(f(f(B,B),f(A,C)))) != 0 From:  472 3
494.  j(f(B,j(f(f(f(B,A),C),j(f(f(B,C),A),g(f(B,f(C,A))))),g(f(B,f(B,f(A,C)))) != 0
      From:  482 25
505.  f(B,j(f(f(f(B,A),C),j(f(f(B,C),A),g(f(B,f(C,A))))),g(f(B,f(A,C)))) != 0 From:  494 20
508.  f(B,j(j(f(f(f(B,A),C),a(B,C,A)),g(f(B,f(A,C)))) != 0 From:  505 28 43
520.  f(B,j(f(f(f(B,A),C),j(a(B,C,A),g(f(B,f(A,C)))))) != 0 From:  508 3
525.  f(B,j(j(a(B,C,A),g(f(B,f(A,C)))),f(f(B,A),C))) != 0 From:  520 2
532.  f(B,j(a(B,C,A),j(g(f(B,f(A,C))),f(f(B,A),C)))) != 0 From:  525 3
538.  f(B,j(a(B,C,A),j(f(f(B,A),C),g(f(B,f(A,C)))))) != 0 From:  532 2
544.  0 != 0 From:  538 28 43 54 46
545.  null From:  544 1
***<end of proof>***
```

I would like to hear from others who might have success with these problems and techniques used in overcomming the difficulties.

References

[1] L. Wos and G. A. Robinson *Paramodulation and set of support.* in Proceedings of the IRIA Symposium on Automatic Demonstration, Versailles, France, Springer-Verlag Publ. 1968, 276-310.

[2] E. Lusk and R. Overbeek. *The Automated Reasoning System ITP.* Argonne National Laboratory, ANL-84-27, 1984.

[3] R. D. Schafer. *An Introduction to Nonassociative Algebras.* Academic Press, New York, 1966.

[4] K. A. Zhevlakov, et. al. *Rings that are Nearly Associative.* Academic Press, New York, 1982.

[5] R. L. Stevens, *Some Experiments in Nonassociative Ring Theory with an Automated Theorem Prover,* Journal of Automated Reasoning, Vol 3 No. 2, 1987.

[6] T. C. Wang, *Case Studies of Z-module Reasoning: Proving Benchmark Theorems from Ring Theory,* Journal of Automated Reasoning, Vol 3 No. 4, 1987.

[7] T. C. Wang and R. L. Stevens, *Solving Open Problems in Right Alternative Rings with Z-Module Reasoning,* Submitted to Journal of Automated Reasoning.

An Interactive Enhancement to the
Boyer-Moore Theorem Prover

Matt Kaufmann[1]
Computational Logic, Inc.
1717 W. 6th Street, Suite 290
Austin, TX 78703

This is a system for checking the provability of terms in the Boyer-Moore logic. (That logic is described in [2] and (more recently) updated in [1].) The system runs in Common Lisp, and is loaded on top of the Boyer-Moore Theorem Prover. Thus, the user can give commands at a low level (such as deleting a hypothesis, diving to a subterm of the current term, expanding a function call, or applying a rewrite rule) or at a high level (such as calling the Boyer-Moore Theorem Prover). Commands also exist for displaying useful information (such as printing the current hypotheses and conclusion, displaying the currently applicable rewrite rules, or showing the current abbreviations) and for controlling the progress of the proof (such as undoing a specified number of commands, changing goals, or disabling certain rewrite rules). A notion of *macro commands* lets the user create compound commands, roughly in the spirit of the *tactics* and *tacticals* of LCF [3] and others. An on-line help facility is provided, and a user's manual [4] exists.

As with a variety of proof-checking systems, this system is goal-directed: a proof is completed when the main goal and all subgoals have been proved. Upon completion of an interactive proof, the lemma with its proof may be stored as a Boyer-Moore *event* which can be added to the user's current library of definitions and lemmas. This event can later be replayed in "batch mode". Partial proofs can also be stored.

A number of theorems have been checked with this system, including correctness of a transitive closure program, correctness of a Towers of Hanoi program, the exponent two version of the finite Ramsey Theorem, irrationality of the square root of two, correctness of an algorithm of Gries for finding the size of the largest "true

---
[1]This work was supported in part by ONR Contract N00014-81-K-0634; Department of the Navy Contract N00039-85-K-0085; IBM grant award "ICSCA--Research in Hardware Verification, Various Purposes"; and the Defense Advanced Research Projects Agency, DARPA Orders 6082 and 9151.

square" submatrix of a boolean matrix, and an interpreter equivalence theorem for part of the proof of correctness of an assembler. There is an experimental extension which handles first-order quantification and set theory.

References

[1]   Robert S. Boyer and J Strother Moore.
        The User's Manual for A Computational Logic.
        Technical Report 18, Computational Logic, Inc., Austin, Texas, February,
            1988.

[2]   Robert S. Boyer and J Strother Moore.
        A Computational Logic.
        Academic Press, New York, 1979.

[3]   M. J. Gordon, A. J. Milner, and C. P. Wadsworth.
        Edinburgh LCF.
        Springer-Verlag, New York, 1979.

[4]   Matt Kaufmann.
        *A User's Manual for an Interactive Enhancement to the Boyer-Moore
            Theorem Prover.*
        Technical Report 60, Institute for Computing Science, University of Texas at
            Austin, Austin, Texas, August, 1987.

A Goal Directed Theorem Prover

David A. Plaisted

Department of Computer Science
University of North Carolina
Chapel Hill, North Carolina 27514

The modified problem reduction format of Plaisted [87] has been implemented in C Prolog and compiled in Quintus Prolog. This complete first-order strategy is based on a sequent-style proof system which does not require contrapositives of clauses to be used. This strategy is implemented in a back chaining manner with caching of solutions to avoid repeated work on the same subgoal. The prover also permits forward chaining and term rewriting. Paramodulation is simulated by unification and term rewriting using a special representation for the input. The C Prolog source code is about 35,000 bytes long and has been distributed to probably over 100 sites together with some example files. One of the main advantages of this prover is that it is compact and relatively easy to understand at a high level. This has made it possible to perform numerous experiments and test numerous modifications rapidly. This also makes the prover suitable for instructional use. The prover has been tested on a wide variety of problems and has obtained automatic proofs in set theory, combinatory logic, ternary Boolean algebra, and many other areas. We are now studying the addition of a priority mechanism to the basic depth-first iterative deepening search strategy used by the prover. This has reduced the running time of many problems, often by an order of magnitude or more.

Reference

[1] Plaisted, D., "Non-Horn clause logic programming without contrapositives," 1987, *Journal of Automated Reasoning,* to appear.

m-NEVER System Summary [*]

Bill Pase and Sentot Kromodimoeljo

I.P. Sharp Associates Limited
265 Carling Avenue, Suite 600
Ottawa, Ontario K1S 2E1
Canada
Arpanet bill@ipsa.arpa   sentot@ipsa.arpa

Key words and Phrases. Automatic induction, decision procedures, forward rules, interactive theorem proving, program verification, rewrite rules.

m-NEVER is an interactive theorem prover developed by I.P. Sharp Associates for use in the m-EVES verification system. m-NEVER consists of six components: a simplifier, a rewriter, an invoker, a reducer, user commands, and the required support for input/output and database management. For the most part, the design is not radical, relying on ideas which have been published in the field of Automated Deduction. The design has been primarily influenced by four theorem proving systems with which the authors have had some experience. These four systems are:

- the Bledsoe-Bruell prover,
- the Stanford Pascal Verifier,
- the Boyer-Moore theorem prover, and
- the Affirm theorem prover.

Each of these systems has had a distinct influence on the theorem prover. Most of the ideas for the deduction techniques used for handling equalities and integers were developed from the Stanford system. The Affirm and Boyer-Moore systems inspired the general strategies for the traversal of formulas and the automatic capabilities. Finally, much of the philosophy with respect to incorporating both manual and automatic capabilities was derived from the early work of Bledsoe and Bruell.

m-NEVER operates by applying its capabilities to transform the *current formula*. For most proof steps, the transformed formula is logically equivalent to the

---

[*]The development of m-EVES was jointly funded by the Canadian Department of National Defense under contract number W2207-6-AF08/01-SV and the United States Navy under contract number 05SR.70B101-84-1.

original. For other proof steps, the transformed formula is valid if and only if the original is valid.

The simplifier recognizes propositional tautologies and fallacies. In addition, it can find elementary proofs involving integers, equalities, and instantiation of quantified variables. When a simplification does not result in a proof or refutation, it can result in a formula that is more elementary (according to the prover's measure) than the original.

Axioms can be marked as rewrite or forward rules when they are introduced. These axioms can be used automatically by the prover. The rewriter uses rewrite rules by replacing occurrences of the *pattern* by the *result*, when the *condition* of the rule (if present) can be shown to hold. In addition, the rewriter uses forward rules by adding them as *assumptions* when the forward rules are *triggered*.

The axioms associated with function definitions can be used as rewriting rules. They can be explicitly applied using the INVOKE command, or the automatic part of the prover can use heuristics to choose when to apply them. The DISABLE command can be applied to functions and rules to prevent the automatic application of their axioms.

The prover can use induction to prove a formula. Boyer-Moore heuristics are used to select a plausible induction by examining the definitions of recursively-defined functions appearing in the formula. Alternatively, an induction can be specified using the INDUCT command.

The reduction commands apply various strategies combining simplifying, rewriting, and invoking. They provide the basic level of automatic capabilities of the prover. More sophisticated capabilities are provided by the macro proof steps, which may perform any combination of basic proof steps.

Bledsoe 73 W.W. Bledsoe, P. Bruell. *A man-machine theorem proving system.* 3rd IJCAI, Stanford U., 1973; also Artificial Intelligence 5(1):51-72, 1974.

Boyer 79 R.S. Boyer, J S. Moore. *A Computational Logic.* Academic Press, NY, 1979.

EVES 87 D. Craigen, S. Kromodimoeljo, I. Meisels, A. Neilson, B. Pase, M. Saaltink. *m–EVES: A Tool for Verifying Software.* TR-87-5402-26, I.P. Sharp Associates Limited, August 1987. In proceedings of the "10th International Conference on Software Engineering" (11–15 April 1988), Singapore.

Luckham 79 D.C. Luckham, et al. *Stanford Pascal verifier user manual.* Report STAN-CS-79-731, Stanford U. Computer Science Dept., March 1979.

Pase 87 Bill Pase, Sentot Kromodimoeljo. *m-NEVER User's Manual.* TR-87-5420-13, I.P. Sharp Associates Limited, November 1987.

Thompson 81 D.H. Thompson, R.W. Erickson (eds.). *AFFIRM Reference Manual, USC Information Sciences Institute.* Marina Dey Ray, CA, 1981.

EFS — An Interactive Environment for Formal Systems

Timothy G. Griffin*

Cornell University

The **EFS** is an **E**nvironment for **F**ormal **S**ystems that supports the interactive definition of formal systems and the construction of proofs in systems so defined. The EFS is fully documented in [4].

Implementation. The EFS was implemented with Cornell Synthesizer Generator [8]. The Generator is an implementation tool that takes as input a high-level specification written in the Synthesiser Specification Language (SSL) and produces as output a language-based editing environment. SSL specifications utilize an attribute grammar formalism to define context-sensitive constraints on objects to be edited. Generated editors incrementally update attribute values after each modification of an object to achieve an attribution consistent with the specification. The EFS is among the sample specifications included with the distribution of the Synthesizer Generator, Release 2.0.

Representation of formal systems. The EFS supports two AUTOMATH [3] related formalisms for encoding logical systems: the Edinburgh Logical Framework [6] and the Calculus of Constructions [2]. An EFS user can choose to work in the type system of the Edinburgh Logical Framework or in that of the Calculus of Constructions. These systems employ similar typed λ-calculi with dependent types to encode expressions, rules and proofs, albeit using different approaches.

Proofs. A structure-oriented proof editor supports two methods of proof construction. "Bottom-up" proofs are constructed by building λ-expression that represent proofs. "Top-down" (goal-directed) proof construction proceeds from a goal, represented as a type, to subgoals in the style of Nuprl refinement [1]. A λ-representation of the proof of a goal is automatically synthesized from a refinement proof. Goal-directed proofs are modeled by an attribute grammar having one grammar production for each refinement rule with the relationship between goals and subgoals defined by attribute equations in a manner that extends the scheme introduced by Reps and Alpern [7]. Proof editors implemented using this scheme allow

---

*This work was supported in part by NSF grant no. MCS-83-03327, NFS/ONR grant no. DCR-85-14862, and The British Science and Engineering Research Council. The author's e-mail address is tgg@svax.cs.cornell.edu

for a flexible style of interactive proof construction. Proof trees can pass through inconsistent states and contain incomplete terms. For example, a proof can proceed even when the initial goal is incomplete.

The EFS provides a facility for the interactive definition of new refinement rules that are checked for validity at declaration time.

Notational definitions. The EFS supports a facility for the interactive definition of notational definitions with a declaration style based on that of the Nuprl system but with an implementation based on a formal account of the conventional mathematical practice of introducing and using notational (syntactic) definitions presented in [5].

References

[1] Robert L. Constable, et al. *Implementing Mathematics with the Nuprl Proof Development System.* Prentice-Hall, Englewood Cliffs, New Jersey, 1986.

[2] Thierry Coquand and Gérard Huet. Constuctions: a higher order proof system for mechanizing mathematics. In Bruno Buchberger, editor, *EUROCAL '85,* pages 151–184, Springer-Verlag, 1985.

[3] N. G. de Bruijn. A survey of the project AUTOMATH. In J. P. Seldin and J. R. Hindley, editors, *Essays in Combinatory Logic, Lambda Calculus, and Formalism,* pages 589–606, Academic Press, 1980.

[4] Timothy G. Griffin. *An Environment for Formal Systems.* Technical Report 87-846, Department of Computer Science, Cornell University, 1987. (also LFCS report ECS-LFCS-87-34, Department of Computer Science, University of Edinburgh).

[5] Timothy G. Griffin. Notational definitions — a formal account. In *Proceedings of the Third Symposium on Logic in Computer Science,* July 1988. To appear.

[6] Robert Harper, Furio Honsell, and Gordon Plotkin. A framework for defining logics. In *Proceedings of the Second Symposium on Logic in Computer Science,* 1987.

[7] Thomas W. Reps and Bowen Alpern. Interactive proof checking. In *POPL11,* 1984.

[8] Thomas W. Reps and T. Teitelbaum. *The Synthesizer Generator Reference Manual.* Dept. of Computer Science, Cornell University, Ithaca, NY,14853, 1985. Second Edition, 1987.

Ontic: A Knowledge Representation System for Mathematics

David McAllester
Computer Science Department
Cornell University
Ithaca N.Y. 14853

Ontic is a computer system for verifying mathematical arguments [1]. Starting with the axioms of Zermelo-Fraenkel set theory, including Zorn's lemma as a version of the axiom of choice, the Ontic system has been used to verify a proof of the Stone representation theorem for Boolean lattices. This theorem involves an ultrafilter construction and is similar in complexity to the Tychonoff theorem in topology which states that an arbitrary product of compact spaces is compact. The individual steps in the proof were verified with an automated theorem prover. The Ontic theorem prover automatically accesses a lemma library containing hundreds of mathematical facts; as more facts are added to the system's lemma library the system becomes capable of verifying larger inference steps.

The Ontic theorem prover is based on object-oriented inference. Object-oriented inference is a forward chaining inference process applied to a large lemma library and guided by a set of *focus objects*. The focus objects are terms in the sense of first order predicate calculus. It is well known that unrestricted forward chaining starting with a large lemma library leads to a combinatorial explosion. However, the Ontic theorem prover is guided by the focus objects; the inference process is restricted to statements that are, in a technical sense, about the focus objects. Thus the inference process is "object-oriented".

The Ontic system is able to automatically find and use lemmas in a large lemma library. The library built for the Stone representation theorem contains over 500 definitions and lemmas. Most of these lemmas concern basic properties of sets, pairs, maps, relations, and "structures"; over half the library is dedicated to notions which are prior to the notion of a partial order. Ontic finds relevant definitions and lemmas automatically; there is no mechanism by which the human user can tell the system which lemmas are relevant. The mechanism for finding relevant definitions and lemmas is based on Ontic's type system.

Ontic uses a very expressive "naive" type system; any predicate of one argument is an acceptable type. For example, Ontic allows the user to give classical set-theoretic definitions of the *types* GROUP, LATTICE, and RIEMANNIAN-MANIFOLD. Ontic also allows for a naive form of dependent type, i.e. a type expression which contains a free variable. The expression (LESS-THAN X P), for example, is a type expression whose instances are elements of the partial order P which are less than the element X. Types play three roles in the Ontic system. First, types are used in a wide variety of different syntactic contexts and provide a concise formal notation. Second, types are central to the way Ontic encodes the axioms of set theory; sets are just reified types. Third, types are central to the way Ontic finds relevant information in the

Lemma	Predicate Count Expansion Factor	Word Count Expansion Factor
If arbitrary least upper bounds exist then arbitrary greatest lower bounds also exist.	.9	1.0
Every filter is contained in an ultrafilter.	1.3	1.2
If F is an ultrafilter and $x \vee y \in F$ then $x \in F$ or $y \in F$.	2.1	2.7
Every Boolean algebra is isomorphic to a field of sets.	2.0	1.7

Table 1: Various Measurements of the Expansion Factor

lemma library — Ontic restricts its attention to facts about the types which apply to the given focus objects.

One method of estimating the reasoning power of a formal system is to compare the length of a machine verified proof, e.g. the number of symbols in the proof, with the length of a previously published English proof, e.g the number of words. Because people are generally better than machines at seeing when a statement is valid, machine readable proofs are usually longer than corresponding English proofs. The ratio of the length of the machine readable proof to the length of the original English is called the *expansion factor*. Table 1 gives expansion factor measurements for several of the lemmas proven during the development of the Stone representation theorem. Two expansion factors measurements are given corresponding to two different measures of length.

The machine readable proofs underlying table 1 relied on an extensive lemma library and the expansion factor measurements are thus open to the criticism that parts of the machine readable proof have been hidden in the lemma library. However, once a sufficiently large lemma library has been constructed, it should be possible to prove new theorems without extending the basic lemma library. I believe that the numbers listed in table 1 are accurate in that, with a mature lemma library, new theorems can be verified with small expansion factors even if the expansion factor accounts for the proofs of all lemmas.

References

[1] McAllester, David, Ontic: A Knowledge Representation System for Mathematics, MIT Press, 1988.

Some Tools for an Inference Laboratory (ATINF)

Thierry Boy de la Tour, Ricardo Caferra, Gilles Chaminade
LIFIA-INPG
46, Avenue Félix Viallet
38031 Grenoble Cedex
FRANCE

This software is a component of the ATINF project which is presently under development at LIFIA in Grenoble, France. ATINF is an abbreviation of ATelier d'INFérence, which stands in French for "Inference Laboratory". The aim of this project is to build a set of tools oriented to the cooperation of different inference systems. The following programs are the main available tools.

Formula transformer

It accepts a many-sorted first order formula written in a rather free way (containing ⇔, the arity for ∧ and ∨ is undetermined, prefixed and infixed notations are possible, …) and contains various efficient transformations programmed in Common Lisp (some of them are only available in the non sorted case). Using these transformations, one can put a formula into clausal form, disjunctive form, etc. Some well-known transformations are improved, as for prenex_nf, which tends to minimize the prefix, and thus the number of variables in the matrix.

Miniscoping is also available and may be used in order to decrease the arity of the skolem functions. It has the same interest in another transformation which has some similarities with skolemization: it is a linear transformation into conjunctive (or disjuntive) form in which the number of conjuncts (or disjuncts) is less than in the ordinary transformation. This is an improvement of the algorithms by D. Plaisted [Plaisted 86] and E. Eder [Eder 84].

In the non sorted case, the formula transformer is interfaced by means of a window (running under suntools) in a highly interactive way. The formula to transform is displayed, pretty-printed, in a subwindow, and a part of it, called the selected sub-formula, appears in inverse video. The transformations, only applied on the selected sub-formula, are called by "button pushing" using the mouse, as for the selection itself (with buttons labelled "up", "down" ,…) An interesting feature is the possibility to "lock" some sub-formulas so that they will not be changed in any transformation ("locked" subformulas are considered as atomic ones, and displayed inside square brackets).

But the main interest of this interface is that, despite the great degree of freedom that is given to the user, it can ensure that a particular property is preserved along all these transformations. This property can be the "preservation of models", which means that all the formulas in the sequence are equivalent, and in this case skolemization is not allowed. The property can also be the preservation of (only) satisfiability or validity, which is not as easy to deal with as in the previous case: for example, if the preservation of satisfiability is required, then skolemization would be available, but only on sub-formulas of positive polarity, and dual-skolemization (elimination of ∀'s) is available on sub-formulas of negative polarity, as well as the other validity preserving transformations. There is also the null polarity case (inside ⇔'s), where only model preserving transformations can be applied.

To allow or forbid validity preserving, or satisfiability preserving transformations, the corresponding buttons are displayed or discarded, depending on the property that must be preserved and on the polarity of the selected sub-formula.

Resolution in order sorted logic

The order sorted calculus which has been implemented is similar to the one defined by [Walther 83] or [Schmidt-Schauss 85].

The theorem prover has been designed in order to handle Theory-Resolution [Stickel 85]. The Boyer and Moore's sharing of structure has been modified in order to be able to resolve on an arbitrary number of clauses and on an arbitrary number of literals in each clause.

The input formulas are written in standard notation for order sorted first order logic and are transformed into clausal form by the formula transformer of ATINF.

The user declares:

- the structure of sorts (see for example [Walther 83]).
- the signatures of predicate and function symbols. Function symbols may be polymorphic in the sense of [Schmidt-Schauss 85].

Several well-known unification algorithms have been implemented and have been adapted to order sorted logic. Those algorithms may be used on their own. Furthermore, the post-processing of unsorted substitutions into well-sorted ones, which enables to solve unification problem in a many-sorted equational theory using an unification algorithm for the corresponding unsorted equational theory (see [Schmidt-Schauss 86]), is available. All these algorithms strongly rely on structure sharing.

Apart from selecting an unification algorithm, the user can also interactively choose:

- to use resolution or hyperresolution
- to use different strategies: semantic resolution, ordered linear, set of support, lock resolution,...

Using windows and a mouse, he can also define and use

- heuristic functions
- restrictions on the selection of the literals (e.g. predicate ordering ...).

Higher-order pattern matching

A second order patten matching due to G. Huet and B. Lang has been also implemented and is available. A new polymorphic weak second order algorithm which extracts information from failures is currently under implementation.

References

[Eder 84] **E. Eder**: "An implementation of a theorem prover based on the connection method". Proc. AIMSA'84. Varna, Bulgaria, September 1984. North-Holland, 121-128.

[Plaisted 86] **D. Plaisted**: "A structure preserving clause form translation". Journal of Symbolic Computation (1986) **2**, 293-304.

[Schmidt-Schauss 85] **M. Schmidt-Schauss**: "A many-Sorted calculus with polymorphic functions based on resolution and paramodulation". Interner Bericht. Fachbereich Informatik. Universitat Kaiserslautern.

[Schmidt-Schauss 86] **M. Schmidt-Schauss**: "Unification in many sorted equational theories". Proc. 8th. CADE, Oxford, England, July 1986, 538-552.

[Stickel 85] **M. Stickel**: "Automated deduction by theory resolution". Journal of Automated Reasoning, Vol. 1, N° 4, 1985, 333-376.

[Walther 83] **C. Walther**: "A many sorted calculus based on resolution and paramodulation". Proc. 8th. IJCAI, Karlsruhe, W. Germany, August 1983, 882-891.

QUANTLOG: A SYSTEM FOR APPROXIMATE REASONING IN INCONSISTENT FORMAL SYSTEMS

V.S.Subrahmanian†, and Zerksis D. Umrigar *

The semantics for Quantitative Logic Programs (QLPs) defined in [1] provides a formal framework for mechanical reasoning in the presence of inconsistency. Syntactically, QLPs are similar to ordinary logic programs, except that each atom A is annotated with a truth value (which is a quantitative certainty factor C) between 0.0 and 1.0, with 0.0 representing false and 1.0 representing true. (The annotated atom is represented as $A : C$). A partial order ($\ll$) is defined over the set $\mathcal{T} = [0,1] \cup \{\top\}$ of truth values. Interpretations assign a truth value from $\mathcal{T}$ (intuitively $\top$ stands for "contradictory ") to each member of the Herbrand Base of the QLP. The meaning of a QLP is formally defined in terms of these quantitative interpretations.

Prolog is a logic programming language which has many of the attributes needed for developing expert systems. Unfortunately it does not have any formal way of dealing with uncertainty. Usually, uncertainty factors are added to a Prolog program in an *ad hoc* manner and the resulting semantics often have unintended effects. For example, if a approach similar to that of MYCIN is used, 20 experts being 10% sure of a particular conclusion would lead the system to treat that conclusion as a virtual certainty [4]. By contrast, the semantics of QLPs are rigorously defined and this formal investigation of the semantics (cf. [1,2]) has preceded any attempts to implement a working system.

QUANTLOG integrates together several algorithms from [1] and [3] to allow the user to develop, and utilise programs in Quantitative Logic. In particular, the user may ask existential (and confidence request) queries of QLPs, and also reason about beliefs using the notion of *s-consistency* introduced in [1]. It provides a shell which gives the user access to these algorithms as well as to any user-defined or system Prolog predicates.

Most quantitative logic programming systems are not highly efficient because they use a breadth first search (cf. [6]). This is also the case for theorem proving systems for relevant logics [5] (relevant logics may also be used for mechanical reasoning in inconsistent systems). One of the unique features of Quantlog is that a version of SLD-resolution called

† *School of Computer & Information Science, 313 Link Hall, Syracuse University, Syracuse, NY 13244-1240.*

* *Computer Science Department, SUNY at Binghamton, Binghamton, NY.*

SLDq-resolution is sound and complete for a recursive class of QLPs called *nice* QLPs. What is surprising is that the class of nice QLPs is a *strict* extension of the class of *general* logic programs. QUANTLOG allows the user to use SLDq-resolution [1] to attempt to establish a query like $A : C$ where C is a truth value. Such a query will succeed if there is a ground instance of A which is assigned $C1$ in the least model of the QLP Q and $C \ll C1$. If contradictory information is encountered in the course of the proof, then the system will print a warning message to the user. However, to process queries to non-nice QLPs, a breadth first search technique is used.

QUANTLOG also permits the user to find out the "maximum" confidence possible for some ground atom A by asking a confidence-request query $A :?$. One of the unique features of nice QLPs is that the greatest supported model is semi-computable. Knowing the truth value assigned a variable free atom A in the greatest model of a program allows reasoning about beliefs ([1], Theorem 56). A procedure for computing the greatest supported model of a program is incorporated in the QUANTLOG shell – this procedure is complete for a class of QLPs, and sound for a larger class of QLPs (cf. [3]).

QuantLog is currently implemented as an interpreter written in VM/Prolog. We have programmed several small expert systems in it. We have also used it to solve some "murder mysteries" which involved contradictory information. To our knowledge, this is the first formal system for approximate reasoning in inconsistent formal systems. In the future, we may be able to compile some subset of QLPs directly into Prolog.

References

[1] Subrahmanian,V.S. "On the Semantics of Quantitative Logic Programs", *Proc. 4th IEEE Symp. on Logic Programming*, pps 173-182, San Francisco, Sep. 1987.

[2] Subrahmanian,V.S. "Towards a Theory of Evidential Reasoning in Logic Programming", *Logic Colloquium '87*.

[3] Subrahmanian,V.S. "Query Processing in Quantitative Logic Programming", these proceedings.

[4] B.G. Buchanan and E.H. Shortcliffe, "Rule-Based Expert Systems", Addison-Wesley, Reading, MA., 1985.

[5] Thistlewaite,P.,McRobbie,M.,Meyer,R.K. "The KRIPKE Automated Theorem Proving System", *Proc. 8th CADE*, Lecture Notes in Computer Science, pps 705-706.

[6] Van Emden,M.H. "Quantitative Deduction and its Fixpoint Theory", *J. of Logic Programming*, 4,1, pps 37-53, 1986.

LP: The Larch Prover *

Stephen J. Garland and John V. Guttag †

LP is based primarily on equational term-rewriting, and is a descendant of REVE. LP departs from conventional equational term-rewriting in two important ways: it supports a variety of proof methods beyond normalization and completion; and it allows users to supply non-equational rules of deduction.

Through our work on LP we wish to explore how much can be done efficiently in a framework close to equational logic. The primary uses of LP are to analyze formal specifications written in Larch and to reason about algorithms involving concurrency. In these applications, LP is most often used to debug a specification or a set of invariants. Hence, it is more important to fail quickly and to report when and why a proof breaks down than it is to try all avenues for pushing a proof through to a successful conclusion. For this reason, LP does not employ heuristics to derive subgoals automatically from conjectures to be proved. Instead, it relies largely on forward rather than backward inference, with the user rather than the program being responsible for inventing useful lemmas.

The proof techniques in LP are orthogonal to each other and to the underlying rewriting-based inference engine. In this respect, LP differs from earlier rewriting-based theorem provers (e.g., that of Boyer and Moore) in which the underlying logic and proof tactics are more closely related. We expect the power of LP to grow both by incorporating improved rewriting techniques and (more importantly) by adding more powerful rules of inference.

The basis for proofs in LP is a logical system containing the following types of user-supplied information.

- *Rewrite rules*, which LP uses to reduce terms to normal form and which LP keeps normalized with respect to one another.

- *Equations*, which users can convert into rewrite rules by various *ordering* procedures, some of which guarantee the termination of the resulting set of rules.

- *Assertions* about operators, e.g., that + is associative and commutative. Logi-

*This research was supported in part by the Advanced Research Projects Agency of the Department of Defense, monitored by the Office of Naval Research under contract N00014-83-K-0125, by the National Science Foundation under grant DCR-8411639, and by NYNEX.

†MIT Laboratory for Computer Science, 545 Technology Square, Cambridge, MA 02139

cally, these assertions are merely abbreviations for equations. Operationally, LP uses them in equational term-rewriting to avoid nonterminating rules such as $x + y \rightarrow y + x$.

- *Deduction rules*, which are used to deduce new equations from existing equations and rewrite rules. Logically, these rules are equivalent to $\forall\exists$ axioms such as set extensionality: $\forall x \forall y [\, \forall z [(z \in x) = (z \in y)] => (x = y) \,]$. Operationally, they can be used to deduce equations such as $insert(x, insert(y, s)) = insert(y, insert(x, s))$ from equations such as $(z \in insert(x, insert(y, s))) = (z \in insert(y, insert(x, s)))$.

- *Induction schemas*, which LP uses to generate subgoals to be proved for the basis and induction steps in proofs by induction. LP supports proofs both by traditional induction and by "inductionless" induction.

LP provides both forward and backward rules of inference. Some forward rules are applied automatically each time new information is added to the system; others are invoked explicitly by the user. Backward rules are always invoked explicitly by the user. Among the inference rules in LP are the following.

- *Reduction to normal form:* a forward rule, invoked by users to prove theorems and applied automatically by LP to reduce rewrite rules, equations, and deduction rules in the system whenever a new rewrite rule, assertion, or deduction rule is added.

- *Induction:* a backward rule, for which LP generates subgoals (i.e., lemmas to be proved, sometimes assuming additional axioms as induction hypotheses) from the induction schemas.

- *Proofs by cases and contradiction:* backward rules, for which LP generates subgoals.

- *Critical pairs:* a forward rule, invoked by users. The Peterson-Stickel variation of the Knuth-Bendix completion procedure is the closure of this rule.

- *Instantiation:* a forward rule invoked by users, applicable to rewrite rules, equations, and deduction rules.

- *Deduction:* a forward rule, applied automatically when rewrite rules or equations are added.

LP also provides a wide variety of user amenities. There is extensive on-line help as well facilities for naming objects and sets of objects, logging and replaying input, generating transcripts of sessions, taking checkpoints, getting statistics, and for displaying, adding, and deleting information in a variety of ways.

We are currently (February 1988) in the process of extending LP by adding mechanisms for backward application of deduction rules and for explicit introduction and use of Skolem constants and functions. LP is written in CLU and runs on both DEC VAXes and Sun workstations. It is currently in use at MIT, DEC SRC, and Aarhus (DK).

The KLAUS Automated Deduction System [*]

Mark E. Stickel

Artificial Intelligence Center
SRI International
Menlo Park, California 94025

The KLAUS Automated Deduction System (KADS) was developed as part of the KLAUS project for research on the interactive acquisition and use of knowledge through natural language. KADS is now being used, with further extensions, in other natural-language and reasoning projects at SRI-AIC.

The principal inference operation is nonclausal resolution. The nonclausal representation eliminates redundancy introduced by translating formulas to clause form and improves readability. Special control connectives can be used to restrict use of formulas, e.g., to forward- or backward-chaining uses of implications.

KADS can use either a connection-graph or standard control strategy. Evaluation functions determine the sequence of inference operations. When using a conection graph, KADS resolves on the highest rated link at each step. The resolvent is then evaluated for retention and links to the new formula are evaluated for retention and priority. The alternative standard control strategy, which is essentially the same as that used in Argonne's ITP, schedules formulas rather than links for inferences operations. At each step, all possible user-selected inference operations are performed between the highest rated formula and previous formulas.

KADS supports incorporating theories for more efficient deduction, including by demodulation, associative and commutative unification, many-sorted unification, and theory resolution [3].

Theory-resolution rules, in conjunction with a context mechanism, support reasoning methods for modal logics such as Konolige's B-resolution [2]. For example, if we have the formulas $\mathbf{B}(P) \vee A_1$, $\mathbf{B}(P \supset Q) \vee A_2$, $\neg\mathbf{B}(Q) \vee A_3$, where $\mathbf{B}(P)$ means P is believed, a refutation (in the $\mathbf{B}$ context) of P, $P \supset Q$, $\neg Q$, which proves that it is inconsistent to believe P and $P \supset Q$ and not believe Q, leads to the formation of the theory-resolution rule $\neg\mathbf{B}(P) \vee \neg\mathbf{B}(P \supset Q) \vee \mathbf{B}(Q)$. This rule can be used to derive $A_1 \vee A_2 \vee A_3$ from the original three formulas.

---

[*]This research was supported by the Defense Advanced Research Projects Agency under Contract N00039-84-K-0078 with the Naval Electronic Systems Command.

751

The context mechanism allows association of context names with formulas. Context matching operations specify which formulas are allowed to resolve with, subsume, simplify, etc., which others. The effect is similar to running multiple copies of the theorem prover, but data structures such as term indexes and, in particular, the schedule of pending inference operations, are shared.

KADS is being extended to perform abductive inference with Horn sets of clauses. Literals in the goal can be marked as *assumed* or *assumable* with an assumption cost. Ordinary resolution and factoring operations are employed, with assumption marks and costs being propagated to literals in descendant clauses. A complete derivation is one whose final clause consists entirely of assumed literals. Assuming them would make the goal a consequence of the axioms. Assumption costs are used to compare the desirability of set of assumptions and to order inference operations to favor less costly assumptions. Numerical specifications for propagating the assumption cost of a goal to the antecedents of an implication used in solving that goal can be adapted to specify a range of behavior, including most specific abduction (i.e., in which only pure (unresolvable) literals can be assumptions) and least specific abduction (in which a subset of the initial goals must be assumed).

Our natural-language-interpretation tasks have favored behavior closer to least specific abduction than to most specific abduction that is commonly used in diagnostic tasks [1]. With drastic oversimplification, the import of a declarative sentence is to state the truth of something not previously known by the hearer. This truth is what must be assumed in the process of proving the sentence's logical form. Most specific abduction would unreasonably try to assume its causes rather than its simple truth. It would also make unnecessarily specific taxonomic assumptions, e.g., assume something was mercury when assuming it was simply a liquid would suffice.

KADS includes two other theorem provers constructed using many of the same lower-level functions: an interpreter-based Prolog technology theorem prover and an implementation of the Knuth-Bendix method with associative-commutative completion.

References

[1] Cox, P.T. and T. Pietrzykowski. Causes for events: their computation and applications. *Proc. 8th CADE,* Oxford, 1986, 608–621.

[2] Konolige, K. Resolution and quantified epistemic logics. *Proc. 8th CADE,* Oxford, 1986, 199–208.

[3] Stickel, M.E. Automated deduction by theory resolution. *J. Automated Reasoning 1* (1985), 333–355.

A Prolog Technology Theorem Prover *

Mark E. Stickel

Artificial Intelligence Center
SRI International
Menlo Park, California 94025

A Prolog technology theorem prover (PTTP) is an extension of Prolog that is complete for the full first-order predicate calculus [2]. It differs from Prolog in its use of unification with the occurs check for soundness, the model-elimination reduction rule that is added to Prolog inferences to make the inference system complete [1], and depth-first iterative-deepening search instead of unbounded depth-first search to make the search strategy complete. A Prolog technology theorem prover has been implemented by an extended Prolog-to-LISP compiler that supports these additional features. It is capable of proving theorems in the full first-order predicate calculus at a rate of thousands of inferences per second.

PTTP attains this high performance principally because its inferences are compiled and derived clauses are efficiently represented. PTTP is better able to take advantage of compilation than many other systems because its inference system is an input procedure. The absence of inference operations between derived clauses makes compilation of just the input clauses sufficient.

PTTP extends Prolog's inference system by the addition of the model-elimination reduction rule: If the current goal matches the complement of one of its ancestor goals, then apply the matching substitution and treat the current goal as if it were solved. This inference rule, used as an additional alternative goal-solution method to standard Prolog inference, results in a complete inference system for full first-order predicate calculus. Contrapositives of Prolog clauses must be provided so that any literal of a clause can be resolved on. Deriving indefinite answers requires in addition the inclusion of the negation of the theorem among the axioms.

Prolog-style efficient representation of derived clauses is possible in PTTP because it retains Prolog's depth-bounded search. Only a single derived clause need be represented (on the stack) at a time; alternatives are generated by undoing unifications upon backtracking. For completeness, depth-first iterative deepening, i.e., a sequence of bounded depth-first searches, is performed. This is expected to cost only about

---

*This research is being supported by the National Science Foundation under Grant CCR-8611116.

$\frac{b}{b-1}$ times as many inferences as breadth-first search, where b is the branching factor.

Recent work has tried to improve the effectiveness of the iterative-deepening search by using better estimators of expected remaining cost of a solution than just counting the subgoals. Recognizing that completing a proof using the clause p <- q, r may really require more more than two steps to solve q and r can result in more cutoffs and a diminished search space. In the case of Prolog-like problems consisting entirely of Horn clauses, when the reduction operation is impossible, any predicate defined entirely by nonunit clauses will always require more than a single step to solve. A goal will require in addition at least the number of subgoals in the shortest clause in the goal predicate's definition.

In the case of non-Horn-clause problems, however, even if a predicate is defined entirely by nonunit clauses, a goal with that predicate might still be solved in a single step by a reduction operation. A solution is to include the costs of q or r in the depth-bound computation only if no ancestor goal with complementary predicate ~q or ~r exists. This is a quick check that q or r cannot be removed by reduction. This better estimator is guaranteed to result in no more inferences than the standard one with little extra run-time cost.

Another recent refinement of the search process is to treat some subgoals that cannot lead to infinite deduction sequences (taking into account the model-elimination restriction that a goal not be identical to an ancestor goal) as zero-cost subgoals for the purpose of iterative-deepening search. Some good candidates for zero-cost subgoals are propositional subgoals, taxonomic subgoals (e.g., wolf(X) in animal(X) <- wolf(X)), and subgoals that permute the head goal's arguments (e.g., eq(Y,X) in eq(X,Y) <- eq(Y,X)).

The use of zero-cost subgoals is a heuristic that builds in a bias toward using rules with zero-cost subgoals (since they consume fewer search levels) while it increases the branching factor. Proofs using zero-cost subgoals will be found with a lower depth bound with this refinement than without, often with great savings in the number of inferences performed. However, the increased branching factor can result in a greater number of inferences being performed if zero-cost subgoals appear insufficiently often in the proof.

References

[1] Loveland, D.W. A simplified format for the model elimination procedure. *J. ACM 16* (1969), 349–363.

[2] Stickel, M.E. A Prolog technology theorem prover: implementation by an extended Prolog compiler. To appear in *J. Automated Reasoning.* Earlier, shorter version appeared in *Proc. 8th CADE,* Oxford, 1986, 573–587.

λProlog:
An Extended Logic Programming Language

Amy Felty,[1] *Elsa Gunter,*[1] *John Hannan,*[1]
Dale Miller,[1] *Gopalan Nadathur,*[2] *Andre Scedrov*[3]

[1]Computer and Information Science, University of Pennsylvania
[2]Computer Science, Duke University
[3]Mathematics, University of Pennsylvania

The logic programming language λProlog is an extension of conventional Prolog in several different directions. These extensions provide higher-order functions, λ-terms, a polymorphic typing discipline, modules, and a mechanism for providing secure abstract datatypes. Our original goal in developing λProlog was to understand the essential logical and proof theoretic nature of these extensions. This work has led us to describe a class of formulas called *higher-order hereditary Harrop formulas* which play a role in λProlog that is similar to the role of positive Horn clauses in Prolog. This extended class of formulas permits stronger forms of logical reasoning than can be found in Prolog. For example, it allows universal quantification and implications into the bodies of program clauses as well as some forms of higher-order quantification. Higher-order hereditary Harrop formulas, therefore, significantly extend positive Horn clauses, and, consequently, λProlog significantly enriches Prolog.

Part of our efforts have also been directed at understanding how this enrichment can be used to write programs. In this direction we have been investigating how the mechanisms of λProlog can be used to implement theorem provers, program transformers, and natural language understanding systems.

We have also implemented most features of higher-order hereditary Harrop formulas in a prototype interpreter, called λProlog V2.6. This interpreter comprises roughly 4100 lines of C-Prolog code and has been distributed to about 30 sites in North America and Europe since August 1987. The performance of this interpreter leaves much to be desired, owing, in part, to the underlying implementation language and, to a much larger extent, to the fact that efficiency has not been a major concern in this experimental version. Despite this drawback, the system has been used by us and others for serious experimentation and prototype implementations. Two ongoing Ph.D. theses make use of it as their primary implementation language.

Below is an outline of the major research aspects of this effort to date.

Logic programming foundations The theory of higher-order Horn clauses is outlined in [6] and given in detail in [9]. The higher-order unification process of [2] plays a central role in this analysis. The first-order theory of a language which includes implications in the bodies of program clauses is contained in [4]. Higher-order hereditary Harrop formulas were introduced in [8] to describe program clauses which are both higher-order and contain implications and universal quantification in clause bodies.

Modules for logic programming The extended use of implications and universal quantifiers permits notions of modules and abstract datatypes to be supported. See [4] and [8].

Theorem prover implementation λProlog has proved to be a very natural implementation environment for writing tactic style theorem provers [1].

Program transformation Extending the work of [3], we have experimented with implementing several program transformation algorithms. The relationship between the actual implementation code in λProlog and the semantics of the programs being transformed seems very tight. We hope this will permit us to establish formal correctness proofs for these transformers [7].

Computational linguistics Natural language understanding systems must bring together syntactic and semantic processing. Very often syntactic processing can be identified with first-order operations while semantic processing is often higher-order. λProlog seems to offer a good environment for bringing these two kinds of processing together [5].

References

[1] A. Felty and D. Miller, Specifying Theorem Provers in a Higher-Order Logic Programming Language. Ninth International Conference on Automated Deduction, 23 – 26 May 1988, Argonne Ill.

[2] G. P. Huet, A Unification Algorithm for Typed λ-Calculus. Theoretical Computer Science 1, 1975, 27 – 57.

[3] G. P. Huet and B. Lang, Proving and Applying Program Transformations Expressed with Second-Order Patterns. Acta Informatica 11, (1978), 31 – 55.

[4] D. Miller, A Logical Analysis of Modules for Logic Programming. To appear in the Journal of Logic Programming.

[5] D. Miller and G. Nadathur, Some Uses of Higher-Order Logic in Computational Linguistics. Proceedings of the 24th Annual Meeting of the Association for Computational Linguistics, 1986, 247 – 255.

[6] D. Miller and G. Nadathur, Higher-Order Logic Programming. Proceedings of the Third International Logic Programming Conference, London, June 1986, 448 – 462.

[7] D. Miller and G. Nadathur, A Logic Programming Approach to Manipulating Formulas and Programs. IEEE Symposium on Logic Programming, San Franciso, September 1987.

[8] D. Miller, G. Nadathur, and A. Scedrov, Hereditary Harrop Formulas and Uniform Proofs Systems. Second Annual Symposium on Logic in Computer Science, Cornell University, June 1987, 98 — 105.

[9] G. Nadathur, A Higher-Order Logic as the Basis for Logic Programming. Ph.D. Dissertation, University of Pennsylvania, May 1987.

SYMEVAL: A Theorem Prover Based on the Experimental Logic

Dr. Frank M. Brown
Artificial Intelligence
Research Institute, Inc.
and
University of Kansas

Seung S. Park
Artificial Intelligence
Research Institute, Inc.
and
University of Texas, Austin

I. Introduction

SYMEVAL is a theorem prover which is based on a rather simple, yet effective principle about deduction, namely that deduction is fundamentally a process of replacing expressions by logically equivalent simpler expressions. To facilitate this principle, an experimental logic called *Symmetric Logic* has been developed. The symmetric logic is a collection of rewrite rules regarding propositional and quantificational logic, equality and abstraction logic. The deduction process is implicitly controlled by the way the rewrite rules of the system are constructed. In particular, quantification rules of the symmetric logic are designed so as to reduce the scope of quantifiers in order to eliminate them. [Brown1] gives a detailed description of the symmetric logic, the theorem prover SYMEVAL, the user interface and some example runs. SYMEVAL is implemented in Zetalisp on a Symbolics lisp machine.

II. Applications

Set Theory
We used SYMEVAL to prove various theorems of set theory in [Quine]. The proof of one of those theorems, Weiner-Kurotowski's ordered pair theorem, which is given in [Brown1], is a good example to understand how the symmetric logic handles and eliminates the quantifiers without the instantiation step. It also demonstrates the ability of automatic lemma generation during the course of its proof.

Logic Programming
A symmetric logic specification of a problem can be served as a logic programming statement with SYMEVAL supply the answers and proofs. Unlike PROLOG which uses Horn clauses, there is no restriction about using negation. It also allows both functional and relational notations.

Ontology
Lesniewski's ontology[Luschei] is a set theory closely related to a natural language. This non-standard set theory has a peculier property that {X} = X. We apply ontology to verify a nondeterministic program such as INSERTIONS-OF, PERMUTATION-OF.

Natural Language
Schwind's linguistic theory for natural language specifies how a small subset of English may be translated into the symmetric logic. Definitions given to the system, grammatical relations, specify not only a parsing for English text of a grammatical category but also how to translate it into a meaning structure which is the symmetric logic itself.

Program Verification
SYMEVAL has been applied to verify the eqivalence and correctness of simple
logic programs such as some of the logic programs used in the natural language
system.

Complexity Analysis
The complexity analysis system is a reasoning system to automatically analyze and
determine the complexity of computer programs. This system is capable of
analyzing the complexity of simple recursive LISP functions, such as REVERSE,
APPEND, EQUAL and FRINGE.

REFERENCES

Brown1,F.M. "An Experimental Logic Based on the Fundamental Deduction
Principle" Artificial Intelligence Vol.30, 1986

Brown2,F.M. "Automatic Deduction in Set Theory"
Department of Computer Science TR-86-6, University of Kansas

Brown,F.M. and Liu,P. "A Logic Programming and Verification System
for Recursive Quantificational Logic" Proceedings IJCAI-85, 1985

Liu,P. *A Logic-based Programming System,* Ph.D Thesis.
Department of Computer Science, The University of Texas at Austin
1986

Liu,P. and Chang,R. "A New Structural Induction Scheme for Proving
Properties of Mutually Recursive Concepts" Proceedings of AAAI-87
1987

Luschei,E.C. *The logical systems of Lesniewski,* 1962

Schwind,C.B. Ein Formalismus zur Beschreibung der Syntax und Bedeutung
von Frage-Antwort-Systemen, Ph.D. Thesis, Technische Universitat,
1977

Quine,W.V.O. SET THEORY AND ITS LOGIC, Harvard University Press, 1969

ZPLAN: An Automatic Reasoning System for Situations

Dr. Frank M. Brown
Artificial Intelligence
Research Institute, Inc.
and
University of Kansas

Seung S. Park
Artificial Intelligence
Research Institute, Inc.
and
University of Texas, Austin

Jim Phelps

University of Kansas

I. Introduction

We developed an experimental program, ZPLAN, which is based on the nonmonotonic reasoning formalisms of Brown's modal logic Z[Brown1,2,3]. ZPLAN uses the sequential frame axiom[Brown&Park] and applies it to the reasoning involving belief and action. ZPLAN is a menu-driven system in which most of the operations are selected from the mouse sensitive menus. The main menu is used to select one of the four application subsystems. The subsystems of ZPLAN are:

1. **Knowledgebase Manipulation:** Given a knowledgebase, ASSERT and ERASE operations are performed while keeping the knowledgebase consistent with respect to a given constraint. Each operation can result in non-trivial changes which can be checked either by inspecting the explicit resulting list or by using an inquire operation given in the menu to which the system responds YES, NO or MAY BE.

2. **Forward Planning:** After actions, initial situation, general laws and default laws are defined by the user, the program responds to an action command by predicting its non-obvious results.

3. **Event History Interpretation:** Given a history of an action sequence that has been done, and observation data, the program produces a plausible interpretation on what actually happened.

4. **Belief Revision[Park1]:** Initial beliefs and its hierarchies are given by the user. Belief gathering actions and hierarchy levels are also defined. The program predicts the consistent new beliefs when actions are typed in.

II. An Example Run

Fig.1 is a snap shot of a session of the Forward Planning subsystem. It shows an example session of the Duct example[Ginsberg&Smith] to predict the state of the room after a moving action of TV from its original place to a duct to block it. Note that the room becomes stuffy, because the plant automatically remains on Duct1 after the TV is moved to Duct2 even though this is not stated as a result of the moving action.

III. Implementation

The sequential frame axiom on which current system is based is a simplified version of the normal frame axiom which requires reflective reasoning. The simplification of the modal terms in the sequential frame axiom has been facilitated by attaching a resolution theorem prover to the POS(logical possibility) operator. A general theorem prover for the modal logic Z along with a proof checker is being constructed.

REFERENCES

Brown1,F.M. "A Commonsense Theory of Nonmonotonic Reasoning"
8th International Conference on Automated Deduction, Oxford, 1986
LECTURE NOTES IN COMPUTER SCIENCE 230, Springer-Verlag

Browm2,F.M. "A Comparison of the Commonsense and Fixed Point Theories
of Nonmonotonicity" Proceedings of AAAI-86, 1986

Brown3,F.M. "A Modal Logic for the Representation of Knowledge"
Proceedings of Workshop on Logical Solutions of the Frame Problem
1987

Brown,F.M. and S.Park "Action, Reflective Possibility, and the Frame
Problem" Proceedings of Workshop on Logical Solutions of the
Frame Problem, 1987

Ginsberg,M. and D. Smith "Reasoning about Action I: A Possible Worlds
Approach", Proceedings of Workshop on Logical Solutions of the
Frame Problem, 1987

Park1,S. "Doubting Thomas: Action and Belief Revision",Proceedings of
Workshop on Logical Solutions of the Frame Problem, 1987

Park2,S. *On Formalizing Commonsense Reasoning using the Modal Situation
Logic and Reflective Reasoning*, Ph.D. Thesis, Under Preparation.

```
  Do Action          Inquire            Defaults          P-Law            Refresh
General Laws:
  ((NOT ON X Y) (NOT ON X Z) (CEQ Y Z))
  ((NOT ON X Y) (NOT ON Z Y) (CEQ Z X))
  ((NOT ROUND X) (NOT ON Y X))
  ((NOT DUCT D) (NOT ON X D) (BLOCKED D))
  ((NOT BLOCKED DUCT1) (NOT BLOCKED DUCT2) (STUFFY ROOM))
  ((NOT STUFFY ROOM) (BLOCKED DUCT1))
  ((NOT STUFFY ROOM) (BLOCKED DUCT2))
Default Laws:
(PK (NOT STUFFY ROOM))=>(NOT STUFFY ROOM)
Enter Proposition to be Checked:(BLOCKED DUCT2)
-- No!
Enter Proposition to be Checked:(BLOCKED DUCT1)
-- Yes!
Enter Proposition to be Checked:(STUFFY ROOM)
-- No!
Enter Proposition to be Checked:(ON CUP TABLE)
-- May be.
Enter Action to be Done:(MOVE ?)
   Action     : (MOVE X A B)
   Precondition: ((ON X A))
   Result     : ((ON X B))
Enter Action to be Done:(MOVE TV BTM DUCT2)
Done
Enter Proposition to be Checked:(BLOCKED DUCT2)
-- Yes!
Enter Proposition to be Checked:(STUFFY ROOM)
-- Yes!


(ON TV DUCT2)(DUCT DUCT1)(DUCT DUCT2)(ROUND PLANT)(ON PLANT DUCT1)(ON VCR TOP)(ON CHAIR FLOOR)


Current State

Previous State:
(DUCT DUCT1)(DUCT DUCT2)(ROUND PLANT)(ON PLANT DUCT1)(ON VCR TOP)(ON CHAIR FLOOR)(ON TV BTM)(NOT STUFFY ROOM)
Current Default:
NIL
```

Fig.1 A sample run of Forward Planning subsystem.

The TPS Theorem Proving System

Peter B. Andrews, Sunil Issar, Daniel Nesmith, Frank Pfenning

Carnegie Mellon University, Pittsburgh, Pa. 15213, U.S.A.

TPS is a theorem proving system for first- and higher-order logic. It provides for automatic, semi-automatic, and interactive modes of proof, and contains various facilities useful for research on theorem proving. As its logical language TPS uses typed λ-calculus, in which most theorems of mathematics can be expressed very directly. A new version of TPS, called TPS3, is being developed in Common Lisp. It has grown naturally out of the system TPS1 discussed in [2].

TPS3 is being designed as a general system for proving theorems of higher-order logic. It has two basic components. The first searches for an *expansion proof (ET-proof)* [6], which represents in a nonredundant way the basic combinatorial information required to construct a proof of the theorem in any style. The second contains facilities related to natural deduction proofs. TPS provides a simple metalanguage in which natural deduction systems and tactics for building proofs in them can be defined. It will generate an environment for interactive theorem proving from such a definition. Such an environment is used under the name ETPS (Educational Theorem Proving System) by students in logic courses to construct formal proofs interactively. Facilities based on the ideas in [5], [6], [7], and [8] are being developed for translating back and forth between expansion proofs and natural deduction proofs. In this translation process, natural deduction proofs are constructed using tactics guided by expansion proofs, so no search is necessary.

In the search for an expansion proof, an expansion tree is grown gradually, and a search for an acceptable mating [1] is made at each stage in the development of the tree. In order to preserve information about the matingsearch process when the tree is expanded, a record (called the *failure record*) of incompatible sets of connections (partial matings) is maintained. The failure record represents what has been learned during the search process, and grows continually. It can be enhanced by an analysis of symmetries in the expansion tree, which inevitably occur when variables are duplicated.

Insofar as possible, substitutions are found by Huet's higher-order unification

This work is supported by NSF grant CCR-8702699.

algorithm [4]. The implementation of this algorithm uses lazy β-reduction; it maintains an environment with bindings of variables to terms instead of explicitly applying substitutions. Since not all necessary substitution terms can be generated by unification of formulas in the expansion tree, projective and primitive substitutions (such as those which substitute $[\lambda S_{o\alpha} \ .S \ [P^1_{\alpha(o\alpha)} \ S]]$, $[\lambda S_{o\alpha} \ .P^2_{o(o\alpha)} \ S \ \wedge \ P^3_{o(o\alpha)} \ S]$, and $[\lambda S_{o\alpha} \ .\exists X_\varepsilon \ [P^4_{o\varepsilon(o\alpha)} \ S \ X]]$ for a variable $P_{o(o\alpha)}$) are applied intermittently to expand the expansion tree in an incremental way. They introduce a small amount of new structure, and contain variables for which additional substitutions can be made at a later stage.

TPS runs in CMU Common Lisp on IBM RT's, in Sun Common Lisp on Sun Workstations, and on Dec-20's. An interface to the X window system is being implemented. TPS is interfaced with Scribe and TeX so that output can contain mathematical and logical symbols. Much documentation is available on line and can be produced automatically.

References

1. Peter B. Andrews, *Theorem Proving via General Matings*, Journal of the ACM **28** (1981), 193-214.

2. Peter B. Andrews, Dale A. Miller, Eve Longini Cohen, Frank Pfenning, "Automating Higher-Order Logic," in *Automated Theorem Proving: After 25 Years*, edited by W. W. Bledsoe and D. W. Loveland, Contemporary Mathematics series, vol. 29, American Mathematical Society, 1984, 169-192.

3. Peter B. Andrews, "Connections and Higher-Order Logic," in *8th International Conference on Automated Deduction*, edited by Jorg H. Siekmann, Oxford, England, Lecture Notes in Computer Science 230, Springer-Verlag, 1986, 1-4.

4. Gerard P. Huet, *A Unification Algorithm for Typed λ-Calculus*, Theoretical Computer Science **1** (1975), 27-57.

5. Dale A. Miller. *Proofs in Higher-Order Logic*, Ph.D. Thesis, Carnegie Mellon University, 1983. 81 pp.

6. Dale A. Miller, *A Compact Representation of Proofs*, Studia Logica **46** (1987), 345-368.

7. Frank Pfenning, "Analytic and Non-analytic Proofs," in *7th International Conference on Automated Deduction*, edited by R. E. Shostak, Napa, California, USA, Lecture Notes in Computer Science 170, Springer-Verlag, 1984, 394-413.

8. Frank Pfenning. *Proof Transformations in Higher-Order Logic*, Ph.D. Thesis, Carnegie Mellon University, 1987. 156 pp.

MOLOG: a Modal PROLOG

Pierre Bieber
Luis Fariñas del Cerro
Andreas Herzig

L.S.I,Université Paul Sabatier
31062 Toulouse cedex FRANCE
uucp: mcvax!inria!geocub!farinas

MOLOG is a modal extension of PROLOG. It provides an efficient tool for the mechanization of a large class of non-classical logics.

Modal logics (or more generally non-classical logics) enable us to express concepts of belief, knowledge or assumption thanks to modal operators qualifying classical formulas. Hence MOLOG handles PROLOG clauses qualified by modal operators. Formally we extend the definition of Horn clauses to modal logics.

A clause F is a **Modal Horn Clause** if F is a formula such that if we erase all the modal operators in F the remaining clause is a Horn clause.

We note `knows[X]` the modal operator denoting X's knowledge and `comp[X]` the dual modal operator denoting compatibility with X's knowledge. MOLOG can handle Fact clauses such that:

```
knows[paul] ( come_back(X) <-- comp[paul] it_rains & tired(X)).
```

Paul knows that if someone is tired and it is compatible with Paul's knowledge that it is raining then this person should come back home.

or Goal clauses such that:

```
? comp[paul] come_back(jane).
```

Is it compatible with Paul's knowledge that Jane should come back home?

In order to reason about the facts and goals we just defined, we extend the classsical resolution with **modal resolution rules** that define the operations that can be performed on the modal operators. Examples of modal resolution rules are:

-If from fact `F` and goal `G` I can infer the new goal `NG` then from fact `knows[X]F` and goal `knows[X]G` I can infer `knows[X]NG`,
-If from goal `knows[X]knows[X]G` I can infer `NG` then from goal `knows[X]G` I can infer `NG`

Modal resolution rules depend on the modal operators, or more precisely on the logical system you want to use. To construct a theorem prover for a modal system with MOLOG means to write the set of resolution rules corresponding to this system.

The MOLOG programmer may create his own set of resolution rules using the rule compiler, and he also may use the already existing set of resolution rules belonging to the theorem Prover Library (T.P.L.).

The T. P. L. contains the well known modal systems as T, S4, S5 and the epistemic logic as S4(n) . We also have mechanized applied modal systems that allow us to use the intuitionistic implication, to reason about modules in PROLOG, to handle uncertain information...

The MOLOG interpreter is written in C_PROLOG on VAX780/UNIX 4.3bsd. Recently an abstract machine based on the Warren Abstract Machine has been designed in order to compile some modal systems. This work will lead to a more efficient implementation of MOLOG.

This work is done in collaboration with P.Enjalbert of Caen University, France and M.Penttonen of Kuopio University, Finland.

References

Arthaud, R., Bieber, P., Fariñas del Cerro, L.,Henry, J.,Herzig,A., Automated modal reasonning, proceedings of the Int. Conf. on Information Processing and Management of Uncertainty in Knowledge-Based systems, Paris, July 1986.

Balbiani, Ph., Bieber, P., Bricard, M., Fariñas del Cerro, L., Herzig, A., MOLOG User Manual, May 1987.

Balbiani, Ph., Fariñas del Cerro, L., Herzig, A.,Declarative semantics for modal logic programs. Rapport LSI, November 1987.

Bricard, M., Une machine abstraite pour compiler MOLOG. Rapport LSI, June 1987.

Cialdea M., Une méthode de déduction automatique en logique modale. Thèse Université Paul Sabatier, Toulouse, 1986.

Fariñas del Cerro, L., MOLOG: a system that extends PROLOG with modal logic. The New Generation Computer Journal,4, 1986, pp.35-50.

Fariñas del Cerro, L., Penttonen, M., A note on the complexity of the satisfiability of modal Horn clauses. The Journal of Logic Programming,4, March 1987, pp. 1-10.

PARTHENON: A Parallel Theorem Prover for Non-Horn Clauses

P. E. Allen, S. Bose, E. M. Clarke, S. Michaylov
Department of Computer Science, Carnegie-Mellon University
Pittsburgh, Pa. 15213, Telephone: 412-268-2628

The goal of the PARTHENON project is to develop a general-purpose, parallel, resolution theorem prover that will run on shared memory multiprocessors. The major issues that the project addresses are the selection of an appropriate resolution proof procedure, the partitioning of the search so as to minimize synchronization overhead, and the efficient representation of the search space. In particular, we are attempting to use technology that has been developed for both sequential and parallel Prolog implementations. In this abstract we briefly report on a prototype implementation that currently runs on a sixteen processor Encore Multimax. On some preliminary examples, this program has exhibited near-linear speedup with respect to the number of processors utilized.

Resolution is an inherently parallel procedure, since the branches from an OR-node in a resolution search tree can be explored independently. The naive method for implementing this type of search (*OR-parallelism*) would be to allocate a separate processor for each new branch at a node in the search tree. However, since the number of processors is always limited, this strategy is usually infeasible and a more efficient approach is necessary. For our prototype implementation we deliberately chose a very simple strategy for allocating work to processors in which a global priority queue is used for maintaining resolvents. The resolvents are ordered in the queue according to a heuristic value. Each processor withdraws the best clause for resolution from the queue and inserts the resolvents back into the queue. This procedure is repeated by each processor until some processor derives the empty clause. Interlocks are used for process synchronization instead of general semaphores in order to avoid the expense associated with system calls. Since contention for access to shared memory is light, the time required for an individual lock operation is in the order of tens of microseconds. Because the time required for a processor to determine the set of resolvents is a substantial fraction of a millisecond, the synchronization overhead is not excessive.

The straightforward approach to resolution theorem proving, forming all resolvents possible until the empty clause is found, is infeasible, as the branching factor of the search tree increases exponentially with the depth of the search. Many methods have been proposed which limit the size of the search tree, either by pruning the search tree to remove undesirable paths, or by reducing the branching factor by using other means to retain information about new clauses produced by resolution. We selected the *model elimination* procedure of Loveland [2] since this procedure is an input procedure and therefore avoids the exponential increase in the branching factor of the search. In addition, the model elimination procedure is relatively simple to implement and is complete without the need for factoring. Stickel's Prolog Technology Theorem Prover [5] also employs this approach and has heavily influenced our work.

This research was partially supported by NSF Grant MCS-82-16706.

The Encore Multimax that we are using has 16 32-bit processors and 32 Mbytes of shared memory. It is
itable for medium and coarse grained parallel applications with a synchronization interval of 20 to 2000
structions [4]. The prototype prover is written in C and uses the *C-Threads* package [1], which allows
rallel programming under the MACH operating system [3]. The preliminary results that we have ob-
ned with our prototype implementation are encouraging. The table below shows the speedup that we get
, some examples collected by Stickel [5].

Parthenon on the Encore Multimax *elapsed time in seconds (speedup factor)*				
	Number of Processors			
Theorem	1	5	10	15
wos6	23.459	6.578 (3.57)	2.356 (9.96)	2.085 (11.25)
wos32	3.971	0.768 (5.17)	0.560 (7.09)	0.512 (7.76)
ls65	15.021	3.143 (4.78)	1.689 (8.89)	1.246 (12.05)
ls75	13.335	2.900 (4.60)	1.539 (8.66)	1.201 (11.10)
ls116	23.921	5.051 (4.74)	2.884 (8.29)	2.170 (11.02)
haspartst2	159.720	32.835 (4.86)	17.247 (9.26)	12.187 (13.11)

There is still much to be done before our prototype implementation can be turned into a useful tool.
R-parallel Prolog technology is not yet as mature as the sequential Prolog technology adapted by Stickel
5] in the Prolog Technology Theorem Prover. Nevertheless, it seems likely that the techniques developed
r maintaining bindings in OR-parallel Prolog may be useful for general theorem proving. We plan to
vestigate the Argonne, SRI, Manchester-SRI, Argonne-SRI and Versions-Vector strategies [6]. However,
is important to realize that typical theorems are likely to have quite different characteristics from typical
olog programs, so a strategy that is appropriate for OR-parallel Prolog may not be appropriate for general
eorem proving. In particular, we conjecture that many theorems will have a larger branching factor than
common in Prolog programs and that we will be able to obtain better speedup than OR-parallel Prolog
nplementations.

References

.]   E. C. Cooper. C Threads. 1987.Available as a document at the Department of Computer Science,
arnegie Mellon University, Pittsburgh, Pennsylvania.

²]   D. Loveland. A Simplified Format for the Model Elimination Procedure. In *JACM*, pages 349-363.
ily, 1969.

!]   R. V. Baron, R. F. Rashid, E. Siegel, A.Tevanian, and M. W. Young. MACH-1: A Multiprocessor
riented Operating System and Environment. In *New Computing Environments: Parallel, Vector and
ymbolic*. SIAM, 1986.

!]   *Multimax Technical Summary* Encore Computer Corporation, 1986.

5]   M. E. Stickel. A Prolog Technology Theorem Prover. In *New Generation Computing 2, 4*, pages
71-383. 1984.

5]   D. H. D. Warren. *Or-Parallel Execution Models of Prolog*. Technical Report, Department of
omputer Science, University of Manchester, 1987.

An nH-Prolog Implementation

Bruce T. Smith
University of North Carolina
Chapel Hill, NC 27514

Donald W. Loveland
Duke University
Durham, NC 27706

A pilot implementation of *near-Horn Prolog* (*nH-Prolog*) has been in use ʑ
Duke University since this past summer (1987). Based on the system outlined i
Loveland [1], the system is written in Quintus Prolog and runs on a VAX 8650. Iι
speed results from the procedural reading given to clause sets — in our opinion,
straightforward extension of Prolog's. Because it makes essentially no use of tɧ
underlying Prolog system's `assert` and `retract` (other than when consulting usǝ
programs) the interpreter itself enjoys significant speedup from compilation.

The nH-Prolog paradigm was developed to execute logic programs (clause sets
containing non-definite clauses. A *non-definite program clause* has multiple heaɗ
or zero heads (i.e. a disjunction of negative literals). Negation is treated classicallʏ
Even though the nH-Prolog paradigm is based on a complete proof procedure, tɧ
system is tuned to be most effective when the number of negations plus the numbǝ
of additional heads is small. We do not view this system as a general theorem provǝ
and it is clear that on logic puzzles, for example, several theorem proving designs aɿ
more effective. We have emphasized the programming language aspect, includiɲ
a Prolog-like trace facility and built-in predicates for I/O, arithmetic, etc. It is oᴜ
intention to extend the system to include cut, `assert`, `retract`, negation-as-failuɿ
and, optionally, true unification (with occurs check).

We can best describe the implementation by its differences from standard Proloɠ
An nH-Prolog computation is a type of analysis-by-cases, comprising a sequence ɵ
blocks, each of which is in essence a conventional Prolog derivation. The user's queɼ
is the first goal of the *initial block*. *Restart blocks* are initialized using *ancestor goaɭ
introduced below, which may include variants of the user's query. Within a bloc
nH-Prolog uses an `nH_solve` predicate, similar to the `solve` predicate in typicɑ
Prolog meta-interpreters. Given a goal, `nH_solve` first attempts unification with
single fact (the block's *active head*). If that *cancellation* step fails, it calls a prograɱ
clause. A multiple head clause is called by unifying the current goal with any oɲ
head; the other heads are *deferred*, i.e. become inactive until a later block. Calliɲ
a clause adds the calling goal to the *ancestor list* and retains a copy of the ancestǝ
list with each deferred head. The leftmost deferred head from the previous bloᴄ

This research was partially supported by Army Research Office under Grant DAAG29-84-Ӏ
0072.

ecomes a restart block's active head. The initial goal of a restart block is a chosen ncestor goal from the ancestor list of the block's active head. In "nearly Horn" lause sets, the number of deferred heads will be very small.

The current implementation runs Horn-clause programs approximately three imes slower than the Quintus Prolog interpreter. (We necessarily run Horn-clause rograms for comparison.) Some of its time is spent maintaining the ancestor ists — information already present in the underlying Prolog system's stack. Since ve have no access to that, we must maintain the lists explicitly. This introduces complexity in our "inner loop" (i.e. the predicate nH_solve) even though we ise the ancestor lists only on block restart. For this reason and others we feel hat a future implementation of nH-Prolog could be significantly faster, perhaps pproaching Prolog in speed.

References

1] D. W. Loveland. Near-Horn Prolog. In J. Lassez, editor, *Logic Programming: Proceedings of the Fourth International Conference*, The MIT Press, 1987.

RRL: A Rewrite Rule Laboratory†

Deepak Kapur
Department of Computer Science
State University of New York
Albany, NY, USA

Hantao Zhang
Department of Computer Science
Rensselaer Polytechnic Institute
Troy, NY, USA

The RRL (Rewrite Rule Laboratory) is a theorem proving environment based on equational logic and rewriting techniques. It currently provides facilities for

(i) automatically proving theorems in first–order predicate calculus with equality,

(ii) generating decision procedures for first–order (equational) theories,

(iii) different approaches for proving formulae by induction – methods based on the inductionless–induction approach as well as the explicit induction approach,

(iv) checking the consistency and completeness of equational specifications,

(v) an interpreter for an equational functional language,

(vi) solving equations modulo a (equational) theory (a la logic programming) using narrowing methods.

Over the past year, RRL has been used at GE Corporate Research and Development as a theorem prover for hardware verification. In particular, RRL has been used to automatically prove simple circuits including a number of leaf cells of a CMOS bit–serial VLSI compiler [4]. With the help of the interactive theorem prover in RRL, two bugs were recently detected in the VHDL description of an image processing chip implementing Sobel's edge detection algorithm designed by Research Triangle Institute; this chip consisted of over 1000 gates. After fixing those bugs, the verification of the chip at the gate level was completed using RRL [5].

The input to RRL is a first–order theory specified by a finite set of axioms – a mixture of equations, conditional equations and arbitrary first–order formulae, in which function symbols can be specified to have properties such as the commutativity property, or the associativity and commutativity property. Every axiom or formula is first transformed into an equation, and is considered as specifying a congruence relation. Reasoning is performed in the context of this congruence relation by relating properties of the original set of formulae to the properties of the congruence relation.

The kernel of RRL is the extended Knuth–Bendix completion procedure [3] which first makes terminating rewrite rules from equations and then attempts to generate a complete (canonical) rewriting system for the congruence relation specified by a finite set of formulae. If RRL is successful in generating a complete set of rules, these rules associate a unique normal (canonical) form for each congruence class in the congruence relation. These rules thus serve as a decision procedure for the first–order theory given as the input. For proving a first–order formula, RRL uses the proof–by–contradiction (refutational) method. The set of hypotheses and the negation of the conclusion are given as the input, and RRL attempts to generate a contradiction, which is a system including the rule true → false. There are two key inference operations used in RRL – superposition among rules, a concept similar to resolution, to generate new rules, and rewriting.

Since the last write–up in CADE–8 [1,2], new algorithms and methods for automated reasoning have been developed and implemented in RRL. RRL has two different methods implementing the inductionless–induction (also called the proof by consistency) approach for proving properties by induction. Recently, a method based on the concept of a *cover–set* of a function definition has been developed for mechanizing proofs by explicit

† This project is partially supported by the National Science Foundation Grant nos. DCR–8211621 and CCR–8408461.

induction for equational specifications. Many ideas from Boyer and Moore's approach have been incorporated in the framework of function definitions by equations. This method has been successfully used to prove over a hundred theorems about lists, sequences, and number theory including the unique prime factorization theorem; see [7].

A method for automatically proving theorems in first–order predicate calculus with equality using conditional rewriting has also been implemented. This method combines ordered resolution and oriented paramodulation of Reiter and Hsiang and Rusinowitch into a single inference rule, called *clausal superposition,* on conditional rewrite rules. Rewriting is used to reduce the search space. Preliminary experiments have been highly successful and the method appears to work very well on a number of examples including Schubert's steamroller example as well as examples from clausal representation of set theory; see [6].

A number of improvements have been made in the completion procedure for handling associative–commutative function symbols. These improvements have resulted in proving that an associative ring in which every element x satisfies $x^3 = x$, is commutative, in nearly 2 minutes on a Symbolics 3600 Lisp machine. In our attempts to prove a more general theorem (i.e., for associative rings, if for every element x, $x^n = x$, the ring is commutative), we have made changes to the completion procedure to specifically handle such problems. With these changes, RRL can prove the theorem in special cases; RRL takes 5 seconds for the case when $n = 3$, 70 seconds for $n = 4$, and 270 seconds for $n = 6$.

Apart from being a powerful theorem prover, RRL provides an environment to facilitate research into the term rewriting approach to automated deduction. By running experiments on RRL, we have been able to introduce a number of improvements in the algorithms in RRL. In particular, we have been led to study normalization strategies and complexity of matching and unification algorithms, identify redundant critical–pairs, in completion procedures, and examine variations in the completion procedure to make it faster. Providing support for such activities and experiments is one of the main purposes of building RRL.

RRL is implemented in Franz Lisp and it runs on Vax computers, SUN workstations, as well as Symbolics Lisp machines. The work on RRL has been going on since the Fall 1983 following a workshop on rewrite rule laboratories in Schenectady. Another theorem prover, called GEOMETER, has also been developed for proving theorems in algebraic geometry based on rewriting techniques and Buchberger's Groebner basis algorithm, a completion procedure for polynomial ideals (see a write–up in this proceedings). Eventually, this theorem prover will also be integrated into RRL.

A Partial List of References

1. Kapur, D., Sivakumar, G., and Zhang, H., "RRL: A Rewrite Rule Laboratory," Proc. of *8th International Conf. on Automated Deduction (CADE-8),* Oxford, England, LNCS 230, Springer Verlag, 691–692, July 1986.
2. Kapur, D., and Zhang, H., *RRL: A Rewrite Rule Laboratory - User's Manual.* Unpublished Manuscript, General Electric Corporate Research and Development, Schenectady, NY, May 1987.
3. Knuth, D.E., and Bendix, P.B., "Simple Word Problems in Universal Algebras," *Computational Problems in Abstract Algebras* (ed. Leech), Pergamon Press, 1970, 263–297.
4. Narendran, P., and Stillman, J., "Hardware Verification in the Interactive VHDL Workstation," in: *VLSI Specification, Verification and Synthesis* (eds. G. Birtwistle and P.A. Subrahmanyam), Kluwer Academic Publishers, 217–235, 1988.
5. Narendran, P., and Stillman, J., "Formal Verification of the Sobel Image Processing Chip," to appear in the Proc. of *IEEE Design Automation* Conf., Anaheim, CA, June 1988.
6. Zhang, H., and Kapur, D., "First-Order Theorem Proving using Conditional Rewrite Rules," to appear in the Proc. of *CADE-9,* Argonne, May 1988.
7. Zhang, H., Kapur, D., and Krishnamoorthy, M.S., "A Mechanizable Induction Principle for Equational Specifications," to appear in the Proc. of *CADE-9,* Argonne, May 1988.

GEOMETER: A THEOREM PROVER FOR ALGEBRAIC GEOMETRY

David A. Cyrluk
Corporate Research & Development
General Electric Co.
Schenectady, NY, USA

Richard M. Harris
Dept. of Computer Science
Rensselaer Polytechnic Institute
Troy, NY, USA

Deepak Kapur
Dept. of Computer Science
State University of New York
Albany, NY, USA

GEOMETER is a reasoning system for algebraic geometry. It implements a refutational approach for proving universally quantified formulae in algebraic geometry developed in [4, 5] which is based on Buchberger's Groebner basis algorithm [1]. This work was inspired by another algebraic method and its succesful implementation. The method, based on Ritt's characteristic set, was developed by Wu [6], and later implemented by Wu and Chou [2].

GEOMETER is implemented on a Symbolics Lisp machine. There are three major components in GEOMETER: (i) a graphical interface using window and mouse for specifying geometric constraints and translating these constraints into an algebraic form, (ii) an implementation of an algorithm to draw the picture to make sure that there is at least one configuration which satisfies these constraints, and (iii) an implementation of a Groebner basis algorithm which is used for theorem proving.

Most geometry problems can be specified graphically in GEOMETER. The user can create points, lines, and circles, which GEOMETER draws on the screen. Constraints or geometric relations among the points can be specified using a menu of predefined constraints. GEOMETER redraws the points, lines, and circles, previously created to satisfy the user's constraints. These constraints include:

1. Equi–distant   2. Co–linear   3. Parallel
4. Perpendicular   5. Not–Co–linear   6. Distinct

Every time a new constraint is specified, the picture is updated to satisfy all the constraints specified so far. This is done by finding a representative solution to geometry relations specified so far. An approximate numerical method based on the Newton–Raphson technique for solving equations is used. If it is possible to draw the picture, then there is at least one geometric configuration which satisfies the specified constraints, thus implying that the hypotheses are consistent. This part of GEOMETER was implemented by Clark Cooper.

The user can give names and symbolic coordinates to the points used in a picture. Symbolic coordinates correspond to generic points in the picture which could take any value. These symbolic coordinates are used to express properties of all possible geometric configurations specified by the constraints on generic points.

Once the picture is drawn by GEOMETER, the user can then conjecture that some additional constraints must necessarily hold in the geometric configuration he created. To validate his conjectures the user can invoke the geometry theorem prover. If GEOMETER returns *true* then the user knows his conjectures are valid. A false conjecture may sometimes be due to the user having missed some trivial constraints to rule out degenerate cases, such as two points having to be distinct. To help the user find these "subsidiary conditions," the program allows the user to look at the Groebner basis generated as the output. The Groebner basis provides information about what additional constraints are needed for the conjecture to hold. The user can then enter the additional constraints and reinvoke the theorem prover.

Nearly a hundred theorems, mostly in plane Euclidean geometry, have been proved using the above approach; these include Simson's theorem, Pascal's theorem, Pappus' theorem, Desargues' theorem, nine–point circle theorem, butterfly theorem, and Gauss's theorem. Even human beings find some of these theorems extremely hard to prove, whereas GEOMETER can prove them in a few seconds. Currently, GEOMETER is also being used in image understanding application for the view consistency problem of extracting 3 dimensional features from multiple images [3].

Typically a geometry formula is specified as consisting of (i) a finite set of hypotheses which translate to polynomial equations, (ii) a finite set of non–degenerate (or subsidiary) conditions, such as two points being distinct, a circle being of non–zero radius, etc., which translate to negations of polynomial equations, and (iii) a conjecture in the form of a polynomial equation to be proved. Since GEOMETER uses a refutational approach, the conjecture is negated and it is checked whether the negated conjecture is not consistent with the constraints consisting of hypotheses and non–degenerate conditions. The key idea in the approach is to transform the negation of a polynomial equation $p \neq 0$ into another polynomial equation $(p\ z) - 1 = 0$ with the introduction of a new variable z; $p \neq 0$ is satisfiable if and only if $(p\ z) - 1 = 0$ is satisfiable. In this way, any universally quantified formula can be transformed into an existentially quantified formula that is a conjunction of polynomial equations, with the property that the original universally quantified formula is a theorem if and only if the existentially quantified formula is unsatisfiable.

A Groebner basis algorithm is used for checking whether a set of polynomial equations are unsatisfiable (or, inconsistent or do not have a solution) in an algebraically closed field. This method is thus a complete decision procedure for universally quantified formulae. It is shown in [5] that this approach is also complete for deducing subsidiary conditions. However, this approach takes much longer to deduce subsidiary conditions missing from a geometry statement than to verify the validity of a geometry statement. The approach also has the limitation that inequalities involving ordering relations cannot be handled. As a result, GEOMETER cannot prove geometry theorems expressed using the betweenness relation.

A Groebner basis of a polynomial ideal is a special basis with which the membership problem of the ideal can be decided easily by polynomial reduction (an ideal generated by a finite basis is the collection of all elements expressed as linear combinations of the basis elements). Buchberger proved that such a basis exists for every polynomial ideal over a field, and gave an algorithm for computing such a basis. A Groebner basis algorithm can also be used to determine the solutions of a system of polynomial equations, in particular the consistency of a system of polynomial equations using Hilbert's Nullstellensatz; see [1, 5].

A Partial List of References

1. Buchberger, B., "Groebner Bases: An Algorithmic Method in Polynomial Ideal Theory," in *Multidimensional Systems Theory* (ed. N.K. Bose), Reidel, 184–232, 1985.
2. Chou, S.C., *Proving and Disproving Theorems in Elementary Geometry using Wu's Method*. Ph.D. Thesis, Dept. of Mathematics, University of Texas, Austin, 1985.
3. Cyrluk, D., Kapur, D., Mundy, J., and Nguyen, V., "Formation of Partial 3D Models from 2D Projections – An Application of Algebraic Reasoning," Proc. *DARPA Workshop on Image Understanding*, Los Angeles, Calif., 798–809, February 1987.
4. Kapur, D., "Using Groebner Bases to Reason about Geometry Problems," *J. of Symbolic Computation*, 2, 399–408, 1986.
5. Kapur, D., "A Refutational Approach to Geometry Theorem Proving," to appear in a special issue of the *Artificial Intelligence* Journal on an International Workshop on Geometry, Oxford, England, June 1986. A preliminary version appeared under the title "Geometry Theorem Proving using Hilbert's Nullstellensatz," Proc. *SYMSAC 1986*, Waterloo, Canada, July 1986, 202–208.
6. Wu, W., "Basic Principles of Mechanical Theorem Proving in Elementary Geometries," *J. of System Sciences and Mathematical Sciences*, 4, 3, 207–223, 1984. Also published in *J. of Automated Reasoning*, 2, 3, September 1986, 221–253.

Isabelle:
The next seven hundred theorem provers

Lawrence C Paulson

Computer Laboratory, University of Cambridge
Cambridge CB2 3QG, England

Isabelle [2] is a theorem prover for a large class of logics. The object-logics are formalized within Isabelle's meta-logic, which is intuitionistic higher-order logic with implication, universal quantifiers, and equality.[1] The implication $\phi \Longrightarrow \psi$ means 'ϕ implies ψ', and expresses logical entailment. The quantification $\bigwedge x.\phi$ means 'ϕ is true for all x', and expresses generality in rules and axiom schemes. The equality $a \equiv b$ means 'a equals b', and allows new symbols to be defined as abbreviations.

Isabelle takes many ideas from LCF [1]. Formulae are manipulated through the meta-language Standard ML; proofs can be developed in the backwards direction via tactics and tacticals. But LCF represents the inference rule $\frac{A \quad B}{A\&B}$ by a function that maps the theorems A and B to the theorem $A \& B$, while Isabelle represents this rule by an axiom in the meta-logic:

$$\bigwedge A. \bigwedge B. [\![A]\!] \Longrightarrow ([\![B]\!] \Longrightarrow [\![A \& B]\!])$$

Observe how object-logic formulae are enclosed in brackets: $[\![A]\!]$.

Higher-order logic uses the typed λ-calculus, whose notions of free and bound variables handle quantifiers. So $\forall x.A$ can be represented by $\text{All}(\lambda x.A)$, where All is a new constant and A is a formula containing x. More precisely, $\forall x.F(x)$ can be represented by $\text{All}(F)$, where the variable F denotes a truth-valued function. Isabelle represents the rule $\frac{A}{\forall x.A}$ by the axiom

$$\bigwedge F.(\bigwedge x. [\![F(x)]\!]) \Longrightarrow [\![\text{All}(F)]\!]$$

The introduction rule is subject to the proviso that x is not free in the assumptions. Any use of the axiom involves proving $F(x)$ for arbitrary x, enforcing the proviso [3]. Similar techniques handle existential quantifiers, the Π and Σ operators of Type Theory, the indexed union operator of set theory, and so forth. Isabelle easily handles induction rules and axiom schemes, like set theory's Axiom of Separation.

---

[1] An early version called Isabelle-86 uses a naive calculus of proof trees as its meta-logic.

Proof trees are derived rules, built by putting rules together. This gives forwards and backwards proof at the same time. Backwards proof is matching a goal with the conclusion of a rule; the premises become the subgoals. Forwards proof is matching theorems to the premises of a rule, making a new theorem.

Isabelle uses unification when joining rules. *Higher-order* unification is solving equations in the typed λ-calculus with respect to α, β, and η-conversion. Unifying $f(x)$ with the constant A gives the two unifiers $\{f = \lambda y.A\}$ and $\{f = \lambda y.y, \ x = A\}$. Multiple unifiers are a reflection of ambiguity: the four unifiers of $f(0)$ with $P(0,0)$ reflect the four different ways that $P(0,0)$ can be regarded as depending upon 0. Isabelle uses Huet's unification procedure.

Logics are proliferating at an alarming rate; there are seven theorem provers descended from Edinburgh LCF. With Isabelle, you need only specify the logic's syntax and rules. To go beyond proof checking, you can implement search procedures using built-in tools. Isabelle consists of 4000 lines of Standard ML. On this base stand object-logics such as Martin-Löf's Type Theory, intuitionistic first-order logic, and classical logic together with Zermelo-Fraenkel set theory.

Constructive Type Theory examples include the derivation of a choice principle and simple number theory: proofs of commutative, associative, and distributive laws for the arithmetic operations, culminating with $(m \bmod n) + (m/n) \times n = m$.

For first-order logic, an automatic procedure can prove many theorems involving quantifiers. The set theory examples include properties of union, intersection, and Cartesian products. One example is a proof that the standard definition of ordered pairs works: define $(a, b) \equiv \{\{a\}, \{a, b\}\}$; if $(a, b) = (c, d)$ then $a = c$ and $b = d$. Two interesting properties of indexed intersection include

$$A \neq \emptyset \ \& \ B \neq \emptyset \quad \rightarrow \quad \bigcap(A \cup B) = \left(\bigcap A\right) \cap \left(\bigcap B\right)$$

$$C \neq \emptyset \quad \rightarrow \quad \bigcap_{x \in C}(A(x) \cap B(x)) = \left(\bigcap_{x \in C} A(x)\right) \cap \left(\bigcap_{x \in C} B(x)\right)$$

References

[1] L. C. Paulson, *Logic and Computation: Interactive Proof with Cambridge LCF* (Cambridge University Press, 1987).

[2] L. C. Paulson, Natural deduction as higher-order resolution, *Journal of Logic Programming* **3** (1986), pages 237–258.

[3] L. C. Paulson, Higher-order logic as the basis of generic theorem provers, Report, Computer Lab., University of Cambridge (1988).

The CHIP System :
Constraint Handling In Prolog

M.Dincbas, P.Van Hentenryck, H.Simonis, A.Aggoun & A.Herold
European Computer-Industry Research Centre (ECRC)
Arabellastrasse 17, 8000 Munich 81, F.R.Germany

Many real-life problems like scheduling, layout, hardware verification and diagnosis are combinatorial problems. No general and efficient algorithms exist to solve these difficult problems (i.e. NP-complete), which can also be viewed as search problems in presence of constraints. Current approaches can be divided into two classes: general tools and specialized programs. General tools (like theorem provers) support the declarative statement of problems but are too inefficient. Specialized programs require much programming effort and are hard to maintain.

Logic Programming, as examplified by Prolog, provides a powerful language for a logical (declarative) formulation of combinatorial problems. Its relational form and the logical variables are very adequate to state problems in a declarative way and its non-deterministic computation liberates the user from the tree-search programming. However, until now, logic programming languages like Prolog have been used to solve only "small" combinatorial problems. The reason is the inefficiency of these languages due to their search procedure based on the *generate-and-test* paradigm.

CHIP (Constraint Handling In Prolog) is a new generation logic programming language combining the declarative aspect of Prolog with the efficiency of constraint solving techniques. It is related to the recent work in the Prolog-III and CLP projects. CHIP differs from Prolog by its "active" use of constraints that enables it to avoid the combinatorial explosion. This active use is achieved through a reasoning mechanism on constraints. This mechanism consists of sophisticated constraint solving and consistency checking (forward and lookahead) techniques. They are combined with an efficient *delay* mechanism providing a *demon-driven* computation. Thus the execution order of constraints is dynamically determined by the system. This introduces, besides *resolution*, a new reasoning paradigm based on *constraint propagation*. Since this paradigm is embedded in a programming language, heuristics specific to particular problems can be added when necessary.

The constraint solving mechanism in CHIP has been further specialized to three important computation domains : *boolean expressions, linear arithmetic terms on natural numbers* and (explicit) *finite domains*. In addition to the primitive constraints (like equality, disequality, inequalities and more complex ones) available in CHIP, the user can define his own constraints (which can be any logic program) and the strategy to use them. Finally, some higher-order predicates for optimization purposes (providing logic programming with a kind of depth-first branch & bound technique) enable combinatorial optimization and integer linear programming problems to be solved.

With these extensions, we have solved in CHIP several problems which were unfeasible within standard Prolog with an efficiency comparable to specific programs written in procedural languages. Some of them are "real-life" problems in the areas of Operations Research (e.g., graph coloring, project management, job-shop scheduling, warehouse location, integer linear programming, ...) and Digital Circuit Design (e.g., simulation, symbolic verification, fault diagnosis, test generation, channel routing, ...). CHIP is also very powerful for solving logical-arithmetic puzzles (e.g., cryptarithmetic problems, n-queens, crosswords, mastermind, ...).

CHIP shows that resolution-based systems can be extended by constraint solving techniques and sophisticated search procedures providing a powerful problem solving tool for a large range of application domains.

An interpreter of CHIP written in C is running on VAX, SUN-3, SPS9 and SPS7.

References

[1] M. Dincbas, H. Simonis, and P. Van Hentenryck "Extending Equation Solving and Constraint Handling in Logic Programming" In MCC, editor, *Colloquium on Resolution of Equations in Algebraic Structures (CREAS)*, Texas, May 1987.

[2] H. Simonis and M. Dincbas. "Using an Extended Prolog for Digital Circuit Design" In *IEEE International Workshop on AI Applications to CAD Systems for Electronics*, pages 165–188, Munich, W.Germany, October 1987.

[3] P. Van Hentenryck and M. Dincbas. "Forward Checking in Logic Programming" In M.I.T Press, editor, *Fourth International Conference on Logic Programming*, pages 229–256, Melbourne, Australia, May 1987.

[4] M. Dincbas, H. Simonis, and P. Van Hentenryck. "Solving Large Combinatorial Problems in Logic Programming" *Journal of Logic Programming*, 1988. (To appear).